Space Sciences Series of ISSI

The Space Sciences Series of ISSI books are coherent reports of the findings, discussions, and ideas that result from international scientific workshops regularly held at the International Space Science Institute (ISSI) in Bern, Switzerland. ISSI's main task is to contribute to the achievement of a deeper understanding of the results from space-research missions, adding value to those results through multi-disciplinary research in an atmosphere of international cooperation. The books are reprints of special issues in the Space Science Reviews journal and occasionally of special issues in the Surveys in Geophysics journal.

More information about this series at https://link.springer.com/bookseries/6592

Anna Milillo · Menelaos Sarantos ·
Benjamin Teolis · Go Murakami · Peter Wurz
Editors

Surface-Bounded Exospheres and Interactions in the Inner Solar System

The book is a spin-off from the Topical Collection
"Surface-Bounded Exospheres and Interactions in the Inner Solar
System" of the journal Space Science Reviews

 Springer

Editors

Anna Milillo
Institute of Space Astrophysics and
    Planetology, INAF
Rome, Italy

Menelaos Sarantos
NASA Goddard Space Flight Center
Greenbelt, MD, USA

Benjamin Teolis
Southwest Research Institute (SwRI)
San Antonio, TX, USA

Go Murakami
Japan Aerospace Exploration Agency
Kangawa, Japan

Peter Wurz
Physics Institute
University of Bern
Bern, Switzerland

ISSN 1385-7525   Space Sciences Series of ISSI
ISBN 978-94-024-2291-7

Cover Image: Schematics of the Sun interaction with the two main airless bodies, i.e. Mercury and the Moon. Artwork by Carmelo Magnafico (INAF/IAPS). For the full version of the image see Figure 1 of chapter 9.

This Springer imprint is published by the registered company Springer Nature B.V.
The registered company address is: Van Godewijckstraat 30, 3311 GX Dordrecht, The Netherlands

Paper in this product is recyclable.

Contents

Space Science Reviews (2023) 219:50
https://doi.org/10.1007/s11214-023-00998-4

Editorial to "Surface-Bounded Exospheres and Interactions in the Inner Solar System"

Anna Milillo[1] · Menelaos Sarantos[2] · Go Murakami[3] · Ben D. Teolis[4] · Peter Wurz[5]

Received: 25 August 2023 / Accepted: 25 August 2023 / Published online: 15 September 2023

Studying the evolution of the surfaces and atmospheres of planetary bodies in the solar system is fundamental to our understanding of the present state of the solar system. Exospheres are the interfaces between the planetary body and the open space, so that, studying the exospheric filling and loss processes is the way to expand knowledge of the body's evolution. This endeavour entails finding variation of the rates of the ongoing processes as a function of the space environment, or, in other words, how the planetary space weather affects these bodies. Aside from occasional catastrophic events, such as volcanic eruptions and geysers in a few bodies or occasional impacts of comets and asteroids, surface and atmospheric changes are caused predominantly by the continuous bombardment of the bodies by photons, energetic ions, and micrometeoroids.

While the exospheres are present around any kind of planetary body, they are quite different if we consider the bodies with an atmosphere and those without a collisional gas envelope. In fact, in the former case the exosphere is the upper part of the gas envelope where collisions become less and less frequent with altitude, so that, the boundary, the exobase, is a thick shell only conventionally defined as the surface where Knudsen number, Kn (the ratio of the mean free path over the atmospheric scale height), is equal to unity. On the contrary, in the latter case the exosphere is directly connected to the surface, thus, it is called surface-bounded exosphere, since the surface release processes are also the exospheric filling ones and atoms and molecules collide with the surface far more frequently than collisions with each other. In this case, the exobase is considered the surface itself, but it has quite different characteristics from the exosphere – atmosphere boundary.

Surface-Bounded Exospheres and Interactions in the Inner Solar System
Edited by Anna Milillo, Menelaos Sarantos, Benjamin D. Teolis, Go Murakami, Peter Wurz and Rudolf von Steiger

---

✉ A. Milillo
anna.milillo@inaf.it

1 Institute of Space Astrophysics and Planetology—INAF, Via del Fosso del Cavaliere, 00133, Rome, Italy

2 NASA Goddard Space Flight Center, Greenbelt, MD, USA

3 Japan Aerospace Exploration Agency, Institute of Space and Astronautical Science, Sagamihara, Kanagawa, 252-5210, Japan

4 Southwest Research Institute, San Antonio, TX, USA

5 Physics Institute, University of Bern, Bern, Switzerland

Since the exospheric constituents experience small interaction among them in a non-collisional condition, we can talk about separate exospheres for different species that can overlap to each other. In fact, some surface-release processes and some exospheric processes are active only for specific species. For example, the observed distributions of refractory species in the exospheres are quite different from the volatile ones and the distributions of specific molecules are expected to be even more peculiar. At the same time, the exospheres at different bodies have different characteristics depending on the external environment, such as the presence of a magnetospheric cavity, the intensity of the solar radiation and solar wind, the micrometeoroid population, as well as on surface composition and mineralogy.

The present collection of papers focuses on the large subset of planetary objects (planets, moons, and small bodies) that are not protected by either strong magnetic fields or thick atmospheres in the inner solar system where the solar influence is stronger. The alteration of the solid surface and the production of the surface-bounded exospheres by the impacts of meteoroids and of the time-varying solar wind over the last 4.54 Gy constitute an essential component of space weathering of exposed bodies such as Mercury, Moon, and asteroids. Furthermore, the detailed investigation of this subject is a paramount element in exo-planet studies. In fact, the observations performed with the new generation of very powerful telescopes could allow to obtain the exospheric composition and shape of exo-planet exospheres.

In the last decade, several space missions provided important new findings for many airless bodies. The missions SELENE, Chandrayaan-1, LADEE and LRO provided important results about the solar wind and Moon surface interaction, MESSENGER provided many important findings about the neutral and ionized environment at Mercury. Furthermore, new ground-based imaging techniques offered the possibility of improved exosphere observations. For example, the use of solar telescopes, such as THEMIS, for imaging the sodium exosphere of Mercury, has allowed to have high spatial resolution images for an observation time of several hours, while with classic nocturnal telescopes it was possible to observe only for a few hours at sunrise or sunset.

In the next decade, the ESA-JAXA BepiColombo mission to Mercury (orbit insertion is scheduled for December 2025) and various orbiters and landers to the Moon are in the space programmes of almost all space agencies in the worldwide. Therefore, this collection is the occasion to gather the present state of knowledge on this subject in preparation for the interpretation of the data to be received from the next generation of missions.

The contributions of this collection deal with different themes related to exospheric processes for different species and bodies in a comparative view.

In the Wurz et al. paper (*Particles and photons as drivers: Comparison between Moon and Mercury*) the external environment of Mercury and the Moon are described and the processes of particle release from the surface as driven by external drivers like thermal radiation, photons, electrons, ions, and dust are detailed.

In the Teolis et al. paper (*Surface Exospheric Interactions*) the surface effects related to exospheric generation processes are described in detail. Current understanding, latest developments, and future directions on studies of sticking and accommodation of exospheric species onto the surface, of diffusion in the regolith and ultimate escape of species from the regolith back into space are reviewed with focus on the Moon and Mercury.

In the Janches et al. paper (*Meteoroids as one of the sources for exosphere formation on airless bodies in the inner solar system*) the meteoroids and dust distributions in the inner Solar System are described by placing them in the context of the exosphere of the Moon, Mercury and other airless bodies. Effects onto the exosphere components of meteoroid streams and large impactors are also discussed.

In the Grava et al. paper (*Volatiles and refractories in surface-bounded exospheres in the inner Solar System*) the exospheres of extreme volatile species, like noble gas, and those constituted by refractory elements or refractory bearing molecules are considered in a comparative way. The different properties of these two groups, i.e., how they are bounded to the surface, how they are released, and how they migrate or stick onto the regolith, result in exospheres generated by different main agents and having totally different distributions and variabilities.

The Leblanc et al. paper (*Comparative Na and K Mercury and Moon exospheres*) treats in a comparative way the alkali exospheres at Mercury and Moon. These two main air-less bodies in the inner Solar System have surface bounded exospheres with similarities and important differences when observed in the two alkali main components, Na and K, i.e., the brightest elements that can be observed from the Earth and in space in the visible range. The observations and models are discussed in the frame of the different environments of these bodies with and without a magnetosphere.

The Schorghöfer et al. paper (*Water Group Exospheres and Surface Interactions on the Moon, Mercury, and Ceres*) focuses on the special case of water groups in the exospheres linked to the likely-present water ice in the cold traps of different bodies. In fact, surprisingly cold traps holding water ice are present from the distant dwarf planet of the main belt, like Ceres, to the innermost planet Mercury. Their formation mechanism is still unclear but future observations of water group species in the exosphere could provide important information on their transport.

In the Lammer et al. paper (*The Exosphere as a Boundary: Origin and Evolution of Airless Bodies in the Inner Solar System and Beyond Including Planets with Silicate Atmospheres*) the exospheric processes are considered from an historical point of view in relation to the parent stars, estimating the history of source and loss rates and surface alterations of the Moon and Mercury. Different possible environments as expected in exoplanets are also discussed and new targeted observations, that will shed light in these topics, are suggested.

Finally, the last paper by Milillo et al. (*Future directions for the investigation of surface-bounded exospheres in the inner Solar System*) reconsiders the main points of the whole collection with a special focus on open questions, expected results from the next space missions and recommended directions for modelling, laboratory experiments and future space and ground observations.

This collection intends to summarize the current state of knowledge on surface-bounded exospheres for the next-generation of scientists that will mainly be involved in the exploration of Mercury, Moon and exoplanets.

Declarations

Competing Interests The authors declare that they have no competing interests.

Publisher's Note Springer Nature remains neutral with regard to jurisdictional claims in published maps and institutional affiliations.

Space Science Reviews (2022) 218:10
https://doi.org/10.1007/s11214-022-00875-6

Particles and Photons as Drivers for Particle Release from the Surfaces of the Moon and Mercury

P. Wurz[1] · S. Fatemi[2] · A. Galli[1] · J. Halekas[3] · Y. Harada[4] · N. Jäggi[1] · J. Jasinski[5] ·
H. Lammer[6] · S. Lindsay[7] · M.N. Nishino[8] · T.M. Orlando[9] · J.M. Raines[10] · M. Scherf[6] ·
J. Slavin[10] · A. Vorburger[1] · R. Winslow[11]

Received: 20 April 2021 / Accepted: 16 February 2022 / Published online: 22 March 2022
© The Author(s) 2022

Abstract

The Moon and Mercury are airless bodies, thus they are directly exposed to the ambient plasma (ions and electrons), to photons mostly from the Sun from infrared range all the way to X-rays, and to meteoroid fluxes. Direct exposure to these exogenic sources has important consequences for the formation and evolution of planetary surfaces, including altering their chemical makeup and optical properties, and generating neutral gas exosphere. The formation of a thin atmosphere, more specifically a surface bound exosphere, the relevant physical processes for the particle release, particle loss, and the drivers behind these processes are discussed in this review.

Keywords Mercury · Moon · Exosphere · Release processes · Sputtering ·
Photon-stimulated desorption · Escape

Surface-Bounded Exospheres and Interactions in the Inner Solar System
Edited by Anna Milillo, Menelaos Sarantos, Benjamin D. Teolis, Go Murakami, Peter Wurz and Rudolf von Steiger

✉ P. Wurz
peter.wurz@space.unibe.ch

1   Physics Institute, University of Bern, Bern, Switzerland

2   Department of Physics, Umeå University, Umeå, Sweden

3   Department of Physics and Astronomy, University of Iowa, Iowa City, IA, USA

4   Kyoto University, Oiwake-cho, Sakyo-ku, Kyoto, Japan

5   NASA Jet Propulsion Laboratory, California Institute of Technology, Pasadena, CA, USA

6   Space Research Institute, Austrian Academy of Sciences, Graz, Austria

7   School of Physics and Astronomy, The University of Leicester, Leicester, UK

8   Japan Aerospace Exploration Agency, Sagamihara, Kanagawa, Japan

9   Georgia Institute of Technology, Atlanta, GA, USA

10   Dept. of Climate and Space Sciences and Engineering, University of Michigan, Ann Arbor, USA

11   Institute for the Study of Earth, Oceans, and Space, University of New Hampshire, Durham, NH, USA

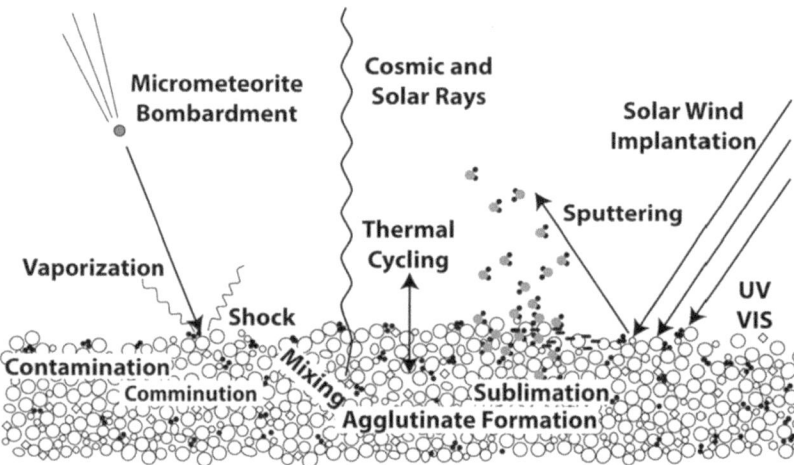

Fig. 1 Summary of the processes acting on the surface of airless planetary bodies, typically covered by regolith of heterogeneous in composition. Figure reproduced from Pieters and Noble (2016), with permission

1 Space Environment

The Moon and Mercury are planetary bodies without a substantial atmosphere, there is only a thin collisionless atmosphere, which is called exosphere. Thus, their surfaces are directly exposed to the ambient plasma (ions and electrons), to energetic particles, to photons mostly from the Sun ranging from the infrared range all the way to X-rays, and to meteoroid fluxes. Direct exposure to these exogenic sources has important consequences for the formation and evolution of planetary surfaces, including altering the chemical makeup of the surface material, formation of the regolith, and significantly modifying optical properties of the surface. These alterations of planetary surfaces are referred to as space weathering in the literature.

Figure 1 shows an overview of the processes acting on the surface of the Moon or Mercury. Actually, the processes illustrated in Fig. 1 apply to the many planetary objects of our solar system that are not protected against these external agents by a sufficient atmosphere.

These external agents are responsible for the formation of a neutral gas exosphere, and the escape of a fraction of particles from this exosphere into space. Since the particles in the exosphere have their origin at the surface of the Moon or Mercury, we speak of a surface bound exosphere. The origin of the exospheric particles, the relevant physical processes for the particle release, the loss of particles from the exosphere to interplanetary space, and the drivers behind these processes are the main topic of this review.

1.1 Space Weathering

The effects of these external agents on the surface of airless planetary bodies are discussed in the literature as space weathering. Space weathering is very important for studies of planetary bodies by remote sensing because it causes major changes in the optical properties of the surfaces over time. Micrometeorites, solar wind plasma and electromagnetic radiation bombard the surfaces of Mercury, moons, and asteroids without atmospheres during billions of years. Therefore, these processes can have important effects on the regolith, which can result in particle implantation, chemical modification of surface material, surface spectral alterations (darkening, reddening and subdued absorption bands), and the distinctive magnetic electron spin resonance caused by single-domain metallic iron particles (e.g., Hapke 2001; Noble et al. 2007; Pieters and Noble 2016).

Space weathering gradually alters unprotected surfaces that are exposed to the harsh space environment to some degree in their chemical composition and physical properties. As illustrated in Fig. 1, there are multiple processes that act simultaneously, at times together, to alter surface materials in different efficiencies. Understanding the causes and the effects, however, is not simple.

Two important parameters that can be found in the space weather-related literature are soil maturity and exposure age. Soil maturity describes the degree to which a given surface material has accumulated space-weathering products (e.g., Morris 1977; Lucey et al. 2000), while the exposure age is a quantitative laboratory measure of how long soil or rock grains have been exposed to space. The latter is ascertained on measurements of accumulated products such as solar wind noble gases or cosmic ray tracks (e.g., Zinner 1980; Berger and Keller 2015). One can group space weathering processes more or less in two broad categories (see Fig. 1) that are related to: i) random impacts by small particles throughout the solar system and ii) irradiation by electromagnetic radiation (e.g., solar X-ray, EUV, flares), and plasma from the Sun (solar wind, Coronal Mass Ejections (CMEs), solar energetic particles (SEPs)), galactic sources (cosmic rays, gamma-ray bursts, etc.), or magnetosphere accelerated ionized particles (magnetic storms, etc.).

Solar X-rays, EUV radiation, solar wind electrons and ions will excite atoms at the surface of airless planetary bodies, which can produce line emission and bremsstrahlung. This makes it possible to infer information on the surface composition from measured X-ray fluorescent spectra. On the Moon, on Mercury, and other solar system bodies, solar-induced X-ray emissions from the surfaces have been used to infer element abundances (e.g., Adler and Trombka 1977; Banerjee and Vadawale 2010; Okada et al. 2009; Starr et al. 2012).

Incident solar wind protons will be implanted in the regolith of an airless body where they can excite and ionize other atoms. Moreover, high-energy particles produce various types of physical and chemical defects and hence cause chemical alteration of the surfaces (e.g. Mura et al. 2009; Tucker et al. 2019). On the Moon it is expected that the diffusion of H atoms is lower when the atoms form metastable bonds with O atoms (Tucker et al. 2019). H atoms that diffuse can also recombine with another H atom, leading to the direct formation of H_2 that is then degassed into the exosphere.

The incident high-energy protons of SEPs ($\sim$MeV) may cause dielectric breakdown of the lunar regolith, in particular in the shadowed regions inside the polar craters (Jordan et al. 2015, 2017; Jordan 2021). The dielectric breakdown may alter the porosity of the subsurface ($\sim$1 mm), facilitating vaporization of volatile elements in the regolith. The same process may also take place at the surface of other airless bodies including Mercury.

1.2 Mercury's Space Environment

The space environment of Mercury is dominated by its interaction with the Sun, which, owing to its close proximity, is the most intense of all the planets in the solar system (Milillo et al. 2020). As the solar wind expands radially outwards throughout the heliosphere, the plasma density and interplanetary magnetic field (IMF) magnitude decrease with distance from the Sun (Russell et al. 1988). This has important consequences on both the plasma conditions at Mercury's orbit as well as the planet's interaction with the solar wind. As Mercury travels through its elliptical orbit between 0.31–0.46 AU distance, it experiences solar wind proton densities that are 5–10 times higher than at the Earth and IMF magnitudes

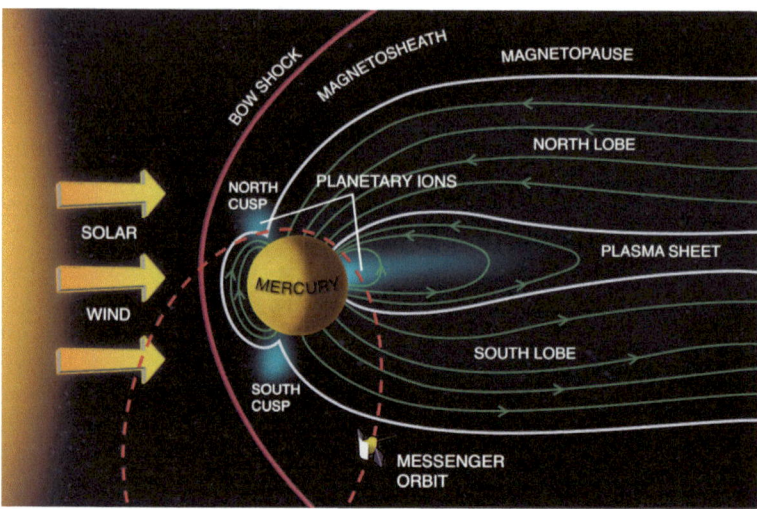

Fig. 2 Mercury's miniature magnetosphere. Figure adapted from Zurbuchen et al. (2011) with permission

that are 3–5 times higher (Masters 2018). Mercury is partially shielded from this interaction by a weak planetary dipole magnetic field of 190 nT-R_M^3, which has an offset northwards by about 490 km (Anderson et al. 2012; Johnson et al. 2012). The magnetic dipole forms a permanent intrinsic magnetosphere similar to that of the Earth, though much smaller in size (Fig. 2). The plasma populations and processes resulting from the interaction between Mercury's magnetic field and the solar wind plasma, leading to the formation of Mercury's magnetosphere, have been reviewed by Seki et al. (2015). The combination of higher solar wind density and IMF strength promotes frequent magnetic reconnection of the IMF with Mercury's planetary magnetic field (DiBraccio 2013; Slavin et al. 2014). Magnetic reconnection is a process that rearranges the topology of magnetic fields, releases magnetic energy and results in the acceleration of plasma; it is the primary mechanism in energizing space plasmas (see book by Gonzalez and Parker 2016). This process brings energy, plasma and magnetic flux into Mercury's magnetospheric system, and produces wide-ranging effects on the surface, exosphere and magnetosphere. The stripping away of the dayside magnetosphere drives Earth-like Dungey cycle plasma convection (Dungey 1963; Slavin et al. 2009) that circulates plasma and magnetic flux into the nightside magnetosphere and drives magnetic reconnection in the magnetotail.

The rate of reconnection at the Earth's magnetopause is very low when the IMF B_Z is positive, but increases rapidly as B_Z becomes strongly negative. The underlying reason for this well-known "half-wave rectifier effect" in the response of Earth's magnetosphere to IMF clock angle is attributed to the relatively high plasma beta, the ratio of kinetic to magnetic pressure, in the magnetosheath. Under these conditions the magnetic field just inside the magnetopause is much larger than that in the magnetosheath and the reconnection rate is reduced relative to situations where the magnetic fields on the inside and outside of the magnetopause are similar (Sonnerup 1974; Koga et al. 2019). The low Alfvenic Mach numbers in the inner heliosphere reduce plasma beta in Mercury's magnetosheath relative to conditions at Earth (Gershman et al. 2013). For this reason, the magnetic fields on the inside and outside of Mercury's magnetopause are similar in contrast to Earth, and the reconnection rates measured by MESSENGER are indeed significantly higher than typically found at Earth (Slavin et al. 2021). As at Earth, magnetopause reconnection rate and related phenomena such as flux transfer events are observed to become more frequent and intense with increasingly negative IMF B_Z at Mercury (DiBraccio 2013; Leyser et al. 2017; Sun et al. 2020). For this reason, the high rates of reconnection at Mercury still drive the injection of large fluxes of solar wind plasma into the magnetospheric cusps and high levels of magnetic flux transfer into the magnetotail even for positive IMF B_Z and more modest IMF clock angles (Slavin et al. 2014; Sun et al. 2020). Overall, however, reconnection at Mercury's magnetopause has been observed to be less sensitive to magnetic shear angle than at Earth (Slavin et al. 2014) primarily due to lower plasma beta values (Gershman et al. 2013; Slavin et al. 2014). High-time resolution magnetic field and plasma measurements from MESSENGER indicate that magnetopause reconnection is dominated by the formation of flux transfer events (FTE)—type flux ropes that channel accelerated solar wind plasma from the reconnection site into the magnetospheric cusps down to the surface (Slavin et al. 2020). FTEs have been observed at Earth before (see Lee and Fu 1985) but occur much more frequent at Mercury. MHD simulations with embedded particle-in-cell computations have shown that FTE-type flux ropes form frequently under the action of highly dynamic reconnection at multiple X-lines at the dayside magnetopauses of planetary magnetospheres even in the presence of only small angular shears between the interplanetary magnetic field and the planetary field (Chen et al. 2017).

Closer to and within the magnetospheric cusps the solar wind plasma takes the form of cusp plasma filaments when channeled downward by the FTE (Slavin et al. 2012; Poh et al. 2016). Multiple FTEs are frequently observed as "showers" with the individual flux rope events separated by only a few seconds and the total number identified while MESSENGER was near the magnetopause reaching 10 to 100 (Slavin et al. 2014). MESSENGER observations have shown that these FTE showers are observed on approximately 50% of the dayside MESSENGER orbits and that the FTE showers contribute up to 85% of the magnetic flux transferred from the dayside to the nightside magnetosphere into the lobes of the magnetotail during Mercury's Dungey cycle (Sun et al. 2020). Examination of MESSENGER FIPS plasma measurements in the vicinity of the northern magnetospheric cusp during FTE showers has revealed the formation of a cusp entry layer with strong enhancements in the downward flux of solar wind protons and Na-group ions (Na through Si) originating from the surface due to sputtering by impacting ions (Sun et al. 2022). The flux of solar wind protons impacting on the surface in and around the cusp is found to increase from order 10^{24} to 10^{25} s^{-1} during FTE showers (Sun et al. 2022).

Magnetic reconnection on the dayside magnetopause allows entry of solar wind plasma which is energized and funneled into the magnetospheric cusps where it travels along magnetic field lines towards the planetary surface. Particles that are injected with sufficiently field-aligned pitch angles (i.e., the angle between the particle velocity direction with respect to the magnetic field) will not be reflected away by the increasing magnetic field strength closer to the planet, but will instead impact on Mercury's surface. Orientations in the $-B_Y$ or $+B_Y$ directions will act to shift the cusp dawnwards or duskwards, respectively (Massetti et al. 2003; Jasinski et al. 2017). Continuous reconnection at $-B_Z$ orientations will act to lower the cusp in latitude, resulting in particle impact on the surface occurring closer to the equator (e.g. Raines et al. 2022). A lower southern extent of the cusp due to reconnection at times of $-B_Z$ would also imply an altogether wider latitudinal extent (e.g., Winslow et al. 2012, 2014, 2017). The IMF magnitude is also important; higher magnitudes will produce higher parallel electric fields, which will accelerate more protons along the magnetic field increasing the fluxes and the energies of the particles impacting on the surface (simulations of reconnection from e.g., Egedal et al. 2012; Li et al. 2017; and Mercury observations e.g.

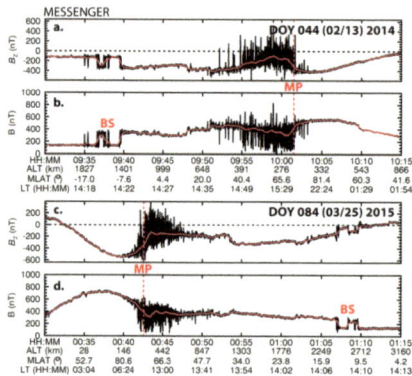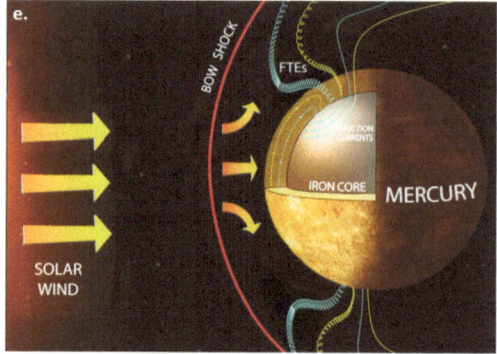

Fig. 3 Total B and the B_z component of the magnetic field measured by MESSENGER during likely ICME intervals in 2014 (panels **a-b**) and 2015 (panels **c-d**). The black trace displays the full rate magnetometer data (20 s^{-1}), while the red trace depicts the smoothed data. The bow shock and magnetopause crossings are labelled "BS: and "MP." Panel **e**: The solar wind interaction with Mercury during these disappearing dayside magnetosphere (DDM) events is illustrated in the right-hand image. The magnetic field lines due to the core dynamo and the induction currents on the core surface are shown in yellow and green, respectively. The incident solar wind and very near bow shock are shown with yellow-orange arrows and red conic section (adapted from Slavin et al. 2019)

Jasinski et al. 2017). By using MAG and FIPS data on MESSENGER it was observed that weaker IMF magnitudes at Mercury are responsible for ion velocity distributions after reconnection that are more likely to be reflected in the cusp fields and therefore these ions do not impact on the surface of Mercury (Jasinski et al. 2017).

During ICME impacts Mercury experiences very high dynamic pressure from the solar wind and strong southward interplanetary magnetic fields (Slavin et al. 2014, 2019; Winslow et al. 2020). Slavin et al. (2019) reported that the usual $B_z > 0$ closed magnetic flux dayside magnetosphere situated between the north and south cusps is replaced by the most intense showers of FTEs observed by MESSENGER (i.e., the large variance in B-total and B_z just sunward of the magnetopause shown in Fig. 3). The magnetopause is observed only at very high latitudes just sunward of Mercury's terminator plane. The average B_z magnetic field (Fig. 3 red trace) becomes positive nowhere indicating, consistent with the very high latitude magnetopause crossings, that MESSENGER did not observe closed dayside magnetic flux even though it reached altitudes as low as ∼300 to 400 km. The magnetic flux removed from the dayside is split between the flux compressed into the crust by the extremely high solar wind pressure and an intensified Dungey circulation of mostly FTE-type flux ropes forming near the magnetopause and the nightside magnetosphere. It is important to note that MESSENGER observed the bow shock to be very close to the Mercury during these DDM events (see Fig. 3e) implying direct solar wind impact and absorption over nearly the entire dayside hemisphere of Mercury (Slavin et al. 2019).

MESSENGER observations of weaker IMCEs and high-speed solar wind interactions with Mercury's magnetosphere have been investigated with global MHD simulations. The results reproduced the measured magnetic field and found that solar wind compression, the generation of induction currents on the surface of Mercury's iron core and reconnection-driven magnetic flux transfer into the tail are all important to understanding these interactions under extreme solar wind forcing (Jia et al. 2015). Similar simulations have been done by Heyner et al. (2016) and follow-up studies were done by Jia et al. (2019). Further global simulations are planned or are underway for these very strong ICMEs reported by Slavin et al. (2019) and Winslow et al. (2020). As suggested by the DDM illustration in Fig. 3, these simulations are expected to provide new windows into the effect of intense solar wind and interplanetary magnetic fields on magnetized planets in close orbits about their stars everywhere. A three-dimensional ten-moment MHD multifluid model which incorporates the nonideal effects including the Hall effect, electron inertia, and tensorial pressures was used for investigating collisionless magnetic reconnection in Mercury's magnetotail and at its magnetopause (Dong et al. 2019).

On the nightside in Mercury's magnetotail, reconnection-driven convection causes plasma impact on the surface there as well (Kallio and Janhunen 2003; Fatemi et al. 2020). Transport of plasma and magnetic flux from the dayside loads the magnetotail, until it is released by reconnection in the tail (Slavin et al. 2021; Imber and Slavin 2017). This process accelerates plasma toward the nightside of the planet, where a fraction of it may impact the surface. Based on models from Earth (Suszcynsky et al. 1993; Shiokawa et al. 1993) the impacting particles, mostly protons and electrons originating from the solar wind, are thought to be deflected by the planetary magnetic field and reach the surface near the open-closed field line boundary in the plasma sheet horns (see also review by Raines et al. 2015). Evidence of this flow braking and subsequent flux pileup has been inferred from analysis of magnetic dipolarization signatures (Dewey et al. 2020). Furthermore, a pattern of X-ray emissions originating from this region were attributed to electrons impacting on the surface (Lindsay et al. 2016), later corroborated by observations of energetic electrons that could be mapped to the same location (Dewey et al. 2017). These bands of impacts on the surface are analogous to the auroral regions at Earth. The latitudinal center and extent of these bands vary considerably, most likely due to space weather conditions and appear to be mostly protons and electrons. In extreme cases, plasma accelerated toward the nightside surface may not be deflected at all and impact broadly on the surface there. The conditions under which this may happen are not fully understood, though there are several measures of the state of the magnetosphere which are part of the puzzle. The magnetic flux content of the magnetotail, which can be observed through the rise and fall in the lobe magnetic field strength, provides some evidence of mis-matched dayside and nightside reconnection rates (Slavin et al. 2021; Imber and Slavin 2017). Several signatures of magnetic reconnection are associated with magnetic dipolarizations (Sun et al. 2016; Dewey et al. 2017), suprathermal protons (Sun et al. 2017, 2018), the substorm current wedge (Poh et al. 2017) and fast plasma flows (Dewey et al. 2018). Regardless of how it is controlled, particle impact on the nightside (see Fig. 15 below) likely contributes to Mercury's exosphere most of the time, possibly significantly under some circumstances.

High solar wind pressures, especially during CMEs, can compress the magnetosphere. Due to Mercury's large internal iron core, variations in the solar wind pressure drive induction currents inside the planet core that act to increase the strength of the dipole field (Slavin et al. 2014). Increases of up to 30% of the magnetic dipole field have been observed during CME events where the dynamic pressure can reach up to ∼90 nPa in comparison to typical values of 10–15 nPa (Jia et al. 2019). This induction effect produces additional magnetic flux to the magnetosphere. In contrast, dayside magnetic reconnection acts to erode the magnetospheric flux. Therefore, the location of the magnetopause and subsequent size of the magnetosphere is the result of the balance between dayside magnetic reconnection and magnetic induction.

Models have been able to shed significant light on the system and reproduce some of the observed behavior. In general, two types of simulation models have been applied to study the global structure of the Hermean magnetosphere: magnetohydrodynamics (MHD), where both ions and electrons are considered as fluid, and hybrid model of plasma, where ions are treated as kinetic macro-particles and electrons are considered as a charge neutralizing fluid. These models have been used to study the Hermean magnetospheric response to the solar wind plasma and IMF configurations, and have found that the structure of the Hermean magnetosphere is highly dynamic and controlled by the upstream solar wind variations (e.g., Kabin et al. 2000; Ip and Kopp 2002; Kallio and Janhunen 2004; Trávníček et al. 2007; Müller et al. 2012; Jia et al. 2015; Fatemi et al. 2018; Exner et al. 2018). In agreement with observations and theoretical investigations, both MHD and hybrid simulations have suggested that the magnetospheric cusps are the main channels where the solar wind plasma impacts on the surface at high latitudes of Mercury while the closed field lines at low latitudes considerably limit the access of plasma to the surface (Kabin et al. 2000; Ip and Kopp 2002; Kallio and Janhunen 2003; Massetti et al. 2007; Benna et al. 2010; Trávníček et al. 2010; Schrijver et al. 2011; Richer et al. 2012; Varela et al. 2015; Hercík et al. 2016; Fatemi et al. 2020). Both of these fundamentally different numerical models have also suggested that the magnetic reconnection, especially during a southward oriented IMF, facilitates the access of plasma to the surface through magnetospheric cusps on the dayside, and to the mid- to high-latitudes on the nightside (e.g., Kallio and Janhunen 2003; Massetti et al. 2017; Mura et al. 2005; Richer et al. 2012; Varela et al. 2015; Hercík et al. 2016; Fatemi et al. 2020).

1.3 Lunar Space Environment

Because of the absence of global shielding by a thick atmosphere or by an intrinsic magnetic field of internal dynamo origin, the bulk of the lunar surface is directly exposed to the ambient space environment. The formation of the lunar exosphere is driven by incoming fluxes of mass and energy from space, including those of meteorites, photons, and charged particles, which will be discussed in Sects. 2–6 below. The bombardment by these drivers may not be homogeneous nor stable in time, with each driver exhibiting different spatial and temporal variabilities. Therefore, the lunar surface is very vulnerable to surface weathering processes, and in fact, the lunar surface is the best place to study the full range of space weathering effects in situ, because of its relatively easy access with spacecraft for detailed observations. The lunar surface might serve as a proxy for space weathering for the many planetary objects that are equally unprotected against space weathering, and which are only observed spectroscopically.

The influx of charged particles to the lunar surface is determined by the plasma surrounding the Moon. The Moon is exposed to plasma environments with very different plasma characteristics as it orbits around the Earth. While the Moon spends nearly three quarters of its orbit in the solar wind, the Moon is also exposed to the terrestrial magnetosheath and magnetotail plasmas. The solar wind (e.g., Marsch 2006) covers much of each lunation and is often thought of as one of the major drivers for the lunar exosphere. The solar wind is tremendously variable in all its properties even at 1 AU, with flow speeds of ~ 250–1000 km s^{-1}, densities of ~ 0.1–200 cm^{-3}, and ion and electron temperatures of ~ 0.1–500 eV (Gosling et al. 1971; Crooker et al. 2000; Wilson et al. 2018). Some of the most extreme solar wind conditions are seen during coronal mass ejections, which typically have high speed and density but low temperature plasma, often surrounded by a hotter "sheath" Lepri and Zurbuchen (2004, 2010). These events can strongly alter the lunar environment, including the near-surface electrostatic characteristics (Farrell et al. 2012). The heavy ion content of the solar wind also varies tremendously during such events (Wurz et al. 2001, 2003), with potential implications for sputtering (Killen et al. 2012).

A notable subset of the upstream region is the terrestrial foreshock, which is the portion of the solar wind magnetically connected to the terrestrial bow shock and thus filled with ions and electrons back-streaming from the shock, which drive a variety of plasma waves (e.g., Eastwood et al. 2005). Downstream of the bow shock exists the magnetosheath (e.g., Lucek et al. 2005), which contains compressed, decelerated, deflected, and heated solar wind plasma. Though slowed, the magnetosheath flow typically remains supersonic at lunar distance since the shock is highly oblique in the flank. Around the full Moon, the Moon is located within the terrestrial magnetotail, in which tenuous plasmas of both solar wind and ionospheric origins are present, with energy spectra that can vary substantially depending on geomagnetic activities and on the Moon's position with respect to the terrestrial plasma sheet and tail lobes.

The incident charged particle fluxes can vary markedly between the dayside and night-side of the Moon. Much of the incoming plasma is absorbed and/or neutralized on the up-stream side of the lunar surface (corresponding to the dayside in the solar wind and magne-tosheath), resulting in the formation of a plasma void and wake structure downstream of the Moon (e.g., Halekas et al. 2015). Because of the release of photoelectrons, the lunar dayside charges up positively, to $< +20$ V, and on the lunar night side in the absence of the photons the surface charges up negatively because the plasma electrons from the tenuous plasma in the wake dominate the charging interaction with the surface (Stubbs et al. 2007). In the wake the surface potential is up -200 V (up to -600 V in the plasma sheet), forming a negative potential structure, thereby decelerating and partly reflecting the electrons directed to the surface (Halekas et al. 2005, 2008a, 2011). In general, ions in a supersonic flow do not have direct access to the near-Moon wake, though a variety of entry mechanisms are discussed (Nishino et al. 2009a, 2009b; Futaana et al. 2010, Dhanya et al. 2013, Halekas et al. 2014a). Thus, the nightside of the Moon is generally subject to much lower (but not completely zero) incident fluxes of ions and electrons compared with those on the dayside when the Moon, which is located in the solar wind and magnetosheath, as has been observed in energetic neutral atom reflection ratios (Vorburger et al. 2016). The situation is more complicated in the terrestrial magnetotail, where both sunward and anti-sunward flows commonly exist (Troshichev et al. 1999; Øieroset et al. 2002).

Additionally, the velocity distributions of impacting ions and electrons can be modified by local shielding effects of the crustal magnetic fields (e.g., Dyal et al. 1974) and by wave-particle interactions (e.g., Harada and Halekas 2016; Nakagawa 2016). As evident from or-bital observations of enhanced fluxes of reflected electrons and ions (Anderson et al. 1975; Lue et al. 2011; Saito et al. 2010, 2012) and decreased fluxes of surface-scattered neutral atoms (Vorburger et al. 2012, 2013) above lunar magnetic anomalies, strongly magnetized areas of the lunar surface are shielded from some fraction of the incident particle fluxes, while proton fluxes may be enhanced in the surrounding regions by the solar wind deflec-tion (Wieser et al. 2010; Futaana et al. 2013). Some observations suggest that the velocity distributions of downward-travelling particles are altered from those of the pristine ambi-ent plasma by interactions with plasma waves in the near-Moon space (Halekas et al. 2012; Harada et al. 2014a, 2014b) and possibly with a shock driven by reflected protons (Halekas et al. 2014b). Details of electron and ion fluxes at the Moon are described in Sects. 5.3 and 6.6, respectively.

2 Thermal Release

2.1 Theoretical Description

Thermal desorption will be responsible for the release volatile species present on the surface into the exosphere. In that case the sublimation rate of the species at the prevailing temperature of the surface will determine the amount of released material, if the reservoir on the surface is not exhausted. Which species is volatile is determined by the surface temperature and the corresponding sublimation rate. For the sublimation flux, Φ_{th}, often the expression:

$$\Phi_{th} = \nu s\,(T)\,C_i \exp\left(\frac{-U}{k_B T}\right) \tag{1}$$

is used (Hunten et al. 1988), where $\nu \approx 10^{13}$ s^{-1} is the assumed vibration frequency of a species at the surface, $s\,(T)$ is the total surface number density, C_i is the fraction of species i on the surface, U is the activation energy, k_B the Boltzmann constant, and T the surface temperature. Equation (1) is derived from the residence time, τ, of an adsorbate atom or molecule on a surface (Bernatowicz and Podosek 1991):

$$\tau = \frac{h}{k_B T} \exp\left(\frac{-U}{k_B T}\right) \tag{2}$$

with h the Planck constant. The vibration frequency of a species at the surface is $\nu = \frac{k_B T}{h}$ and is only a function of temperature in Eq. (2). For Mercury υ is in the range from $2.2 \cdot 10^{12}$ s^{-1} to $1.5 \cdot 10^{13}$ s^{-1}, and for the Moon υ is in the range from $2.2 \cdot 10^{12}$ s^{-1} to $9.4 \cdot 10^{12}$ s^{-1}. Moreover, all the energetics of the release is captured in the surface activation energy U. Unfortunately, Eq. (1) is a severe simplification of sublimation rate, and can be off by orders of magnitude. It is much better to use the measured sublimation data, which are given in the form

$$\ln(p_i) = A_0 + \sum_{j=1}^{n} \frac{A_j}{T^j}, \tag{3}$$

where p_i is the equilibrium vapor pressure of species i, at the temperature T, and the A_j are constants determined experimentally for a substance (e.g. Fray and Schmitt 2009).

Note that the sublimation data are mostly given for pure substances. If the amount of material to evaporate becomes less than a monolayer, the range of binding energies, also referred to as activation energies, to the underlying material has to be considered. If the binding energy changes, the ratio U/T in the exponential term changes (see Eq. (1)), resulting in significant change in the sublimation flux. The underlying surface material ideally is an atomically flat surface, but much more realistically the surface has steps, voids, cracks, defects, and other structures on the surface, which will have different activation energies, likely higher than for the pure substance. These different binding energies, or their range, are not known for any realistic planetary surfaces. Possible ranges for these binding energies (0.6–1.2 eV) have been simulated for solar wind implantation of H, its diffusion, and final release as H_2 from a surface (Farrell et al. 2007).

From the vapor pressure we get of the flux of released volatile species from the surface

$$\Phi_{th} = n_i \frac{1}{4} \sqrt{\frac{8 k_B T}{\pi m_i}} = \frac{p_i}{\sqrt{2\pi k_B T m_i}} \quad \text{with} \quad n_i = \frac{p_i}{k_B T} \tag{4}$$

where n_i is the corresponding number density of the exospheric species i at the surface.

Of course, the reservoir of volatile species on the surface must be able to support this sublimation flux, either by its volume, or by fluxes to the surface, via diffusion from below the surface and the return fluxes from the exosphere. If the sublimation flux cannot be supported, then the actual released volatile flux is source limited to whatever is available at the surface for sublimation. For example, all Ar is in the lunar exosphere on the dayside, but it condenses out on the lunar surface on the lunar nightside (Stern 1999). To maintain a constant Ar density in the lunar or hermean exosphere the Ar lost from the exosphere by escape and ionization has to be replenished by diffusion from the interior (Killen 2002).

Given the large range of surface temperatures on the Moon (Williams et al. 2017), and the even larger range on Mercury (Chase et al. 1976), governed by solar illumination on the dayside and radiation to cold space on the nightside, the thermal release is highly variable with local time. For a rocky body the day-side temperature follows a "1/4" law in a good approximation, with T_{max} the temperature at the sub-solar point and T_{min} the night-side temperature all the way to the terminator. Thus, we can write the local surface temperature as

$$T_0(\phi, \theta) = \begin{cases} T_{min} + (T_{max} - T_{min})(\cos\phi\cos\theta)^{1/4}, & \text{for } 0 < |\theta| < \frac{\pi}{2} \\ T_{min}, & \text{for } \frac{\pi}{2} < |\theta| < \pi \end{cases} \tag{5}$$

with the longitude θ being measured from the planet-Sun axis and the latitude ϕ measured from the planetary equator. The simple "1/4" law, presented in Eq. (5), neglects the thermal inertia of the lithosphere and local albedo and emissivity variations. To determine the effective temperature at the subsolar point, T_{max}, we use the Stefan–Boltzmann law for blackbody radiation to obtain

$$T_{max} \approx T_{Sun} \left[\left(\frac{R_{Sun}}{R_{orb}} \right)^2 \frac{1-\alpha}{\varepsilon} \right]^{1/4} \tag{6}$$

where T_{max} is the effective temperature of the surface, R_{orb} is the distance to the Sun, R_{Sun} is the solar radius, $T_{Sun} \approx 5778$ K is the effective solar surface temperature, α is the bond albedo, with $\alpha_{Mercury} = 0.07$ (Mallama et al. 2002) and $\alpha_{Moon} = 0.14$ (Matthews 2008), and ε is the emissivity, with $\varepsilon_{Mercury} = 0.9$ (Murcray et al. 1970; Saari and Shorthill 1972; Hale and Hapke 2002) and $\varepsilon_{Moon} = 0.9$ (Gaidos et al. 2006).

Typical volatile species released thermally considered for the Moon (Stern 1999) and Mercury (Killen et al. 2007) are H, He, Ar, Ne, H_2, O_2, N_2, H_2O, OH, and CO_2. Some of these volatile species freeze out on the night side of the Moon and Mercury, like H_2O and CO_2 and are released again at dawn. Given the high temperatures at and near the sub-solar point also species like Na or K can be considered volatile.

Diffusion of volatile species to the surface that were trapped in solids, for example noble gases, will contribute to the available inventory of volatiles to be released into the exosphere. Probably H, H_2, He and Ne in the exosphere of the Moon and Mercury originate mostly from the solar wind being implanted into the surface material. Given the typical exposure of the surface to solar wind it will be saturated with solar wind material, which means that the implanted flux of ions matches the released flux of volatiles by diffusion. This was suggested for the Moon already a while ago (e.g., Hinton and Taeusch 1964; Johnson 1971; Hodges 1973, 1980). For example, assuming this scenario in a calculation the obtained H_2 density was 2100 cm^{-3} (Wurz et al. 2012), which was confirmed later by measurements from the LAMP UV spectrograph on the Lunar Reconnaissance Orbiter (Stern et al. 2013). A detailed study of the solar wind proton implantation into the regolith, the diffusion of H inside the regolith, possible chemical reactions of the H to form H_2 and OH, and the diffusion of these volatiles to the surface were presented by Tucker et al. (2019).

We assume a complete thermal accommodation of the sublimating volatile species with the local surface, thus a Maxwell-Boltzmann velocity distribution, $f(v)$, describes the release of sublimated particles.

$$f(v) = 4\pi \left(\frac{m}{2\pi k_B T} \right)^{\frac{3}{2}} v^2 e^{\frac{-mv^2}{2k_B T}} \tag{7}$$

The angular dependence is constant in azimuth angle and has the sine-dependence on the polar angle. For particles falling back to the surface and being re-released (no permanent sticking), the Maxwell-Boltzmann velocity distribution is also valid, even if the particles were initially released by a non-thermal process, because of the fast thermal accommodation at the surface (see Eq. (2)). The reason for the thermal accommodation is that the surface material is a fine-grained soil, regolith, with high porosity and particles sizes of 100 μm and less (Langevin 1997; Cooper et al. 2001), causing the particles to undergo multiple collisions with the highly structured regolith grains, making a thermal accommodation very likely.

One can use Gaussian deviates to sample the Maxwell-Boltzmann distribution for numerical analysis of exospheric particles, at a given temperature, i.e., the local surface temperature or at an elevated temperature. A set of three Gaussian deviates is needed to determine the components of a velocity vector for the thermal particle release at the surface. The Gaussian deviates, denoted G_i, are calculated by using the relation given in (Zelen and Severo 1965; Hodges 1973):

$$G_i = \sqrt{-2\ln(p_i)} \cos(2\pi q_i) \tag{8}$$

where p_i and q_i are uniform deviates for the three spatial directions, ranging between 0 and 1. The variance of each G_i is 1 (for $i = 1, 2, 3$). The initial velocity vector, $\vec{v}_0 = (v_1, v_2, v_3)$, at the start of the particle trajectory on the surface is

$$\vec{v}_0 = \sqrt{\frac{m}{k_B T_0}} \vec{G} + \vec{r} \times \vec{\Omega} \tag{9}$$

with particle mass m, the Boltzmann constant k_B, T_0 the main temperature of the released particle, $\vec{r}$ the release location on the surface, and $\vec{\Omega}$ the rotation vector of the planet. The particle velocity v_0 at the point of origin on the surface is $\sqrt{\vec{v}_0^2}$. The main temperature T_0 is taken either as local surface temperature at the particle release site or as characteristic temperature of the release process as discussed below.

Equation (4) assumes an infinite reservoir of the volatile material at the surface to support their sublimation. For Mercury and the Moon, the reservoir of volatiles on the surface is very limited, at least on the dayside, perhaps even less than a monolayer, thus the released flux from the surface is limited by fluxes of material to the surface. Most of the species released thermally will fall back onto the surface and thus return to the reservoir on the surface. Depending on species, the population of the surface reservoir is different, there are contributions by diffusion of volatiles from the interior (Killen 2002; Wurz et al. 2012), atoms being liberated from the mineral compound (Mura et al. 2009), and infall of volatile material, e.g. from comets (Stern 1999). Also, the sticking of these species on the surface, or the residence time of a volatile on the surface, must be considered. Details of the processing of volatile species between the exosphere, the surface, and the interior are presented by Grava et al. (2021) and will not be discussed further here.

2.2 Thermal Escape

Particles with thermal energies, the Maxwell-Boltzmann velocity distribution (Eq. (7)), rise in altitude against the gravitational force. Since there is no upper limit in velocity in the Maxwell-Boltzmann distribution, there will be a few particles that have initial energies that exceed the escape speed of the planet. The escape speed, v_∞, from the surface (assuming a surface-bound exosphere) is given by

$$v_\infty = \sqrt{\frac{2GM}{R}} \tag{10}$$

where G is the gravitational constant, M the mass of the planet, and R the radius of the planet. The most probable thermal speed is

$$v_0 = \sqrt{\frac{2k_B T}{m}} \tag{11}$$

where T is the temperature of the exospheric gas and m the mass of the species. To calculate the thermal escape (also called Jeans escape) we define the parameter X:

$$X = \left(\frac{v_\infty}{v_0}\right)^2 = \frac{2GM}{R}\frac{m}{2k_B T} = \frac{R}{H} \tag{12}$$

with H the scale height of the exospheric gas. The fraction of gas exceeding the escape speed is given by Lammer and Bauer (2004):

$$f_{esc} = \frac{1}{2\sqrt{\pi}}(1+X)e^{-X} \tag{13}$$

The smaller the value of X the larger the escape fraction f_{esc} will be. The rule of thumb is that for $X < 15$ the species is lost from the exosphere. For example, the escape of H from Titan's exosphere is close to hydrodynamic escape (Hedelt et al. 2010). From Eq. (13) we can calculate the flux of escaping particles for a species i

$$\phi_{esc,i} = \Phi_{th} f_{esc} = \frac{v_0}{2\sqrt{\pi}}n_i(1+X)e^{-X} \tag{14}$$

with n_i the number density of species i at the exobase. In the following we estimate the thermal escape for a few know volatile species on the Moon and Mercury.

Table 1 presents the X parameter (Eq. (12)), the escape fraction f_{esc} (Eq. (13)), the escape flux $\phi_{esc,i}$ (Eq. (14)), and the escaping mass flux for each of several volatile species considered for the exospheres of the Moon and Mercury (Stern 1999; Wurz et al. 2019). There are some significant differences between Mercury and the Moon: on Mercury only the light gases, up to ^{4}He, escape from the exosphere, whereas on the Moon there is significant escape even up to species as heavy as water. Heavier species will remain bound to their object, form permanent gases in the exosphere, if they do not condense at the night side or

Table 1 Thermal escape calculated for volatile species from the Moon and Mercury, with exospheric abundances at the surface from Stern (1999) and Wurz et al. (2019), respectively

Mercury

Species	H	H_2	^3He	^4He	OH	H_2O	Ne	N_2	O_2	Ar	CO_2
mass	1	2	3	4	17	18	20	28	32	40	44
n_i [m^{-3}]	2.29E+07	9.40E+12	2.95E+06	5.90E+09	1.40E+09	2.70E+11	6.00E+09	5.00E+09	1.6E+11	4.40E+10	4.0E+9
v_0 [m/s]	2995.90	2118.42	1729.68	1497.95	726.61	706.14	669.90	566.17	529.60	473.69	451.65
X	2.012	4.025	6.037	8.049	34.209	36.221	40.246	56.344	64.393	80.492	88.541
f_{esc}	1.14E-01	2.53E-02	4.74E-03	8.15E-04	1.38E-14	1.95E-15	3.87E-17	5.48E-24	2.00E-27	2.54E-34	8.91E-38
$\phi_{esc,i}$ [s^{-1}]	7.79E+09	5.04E+14	2.42E+07	7.21E+09	1.41E-02	3.72E-01	1.55E-04	1.55E-11	1.69E-13	5.29E-21	1.61E-25
$\phi_{esc,i}$ [kg/s]	2.08E+05	2.70E+10	1.94E+03	7.70E+05	6.39E-06	1.79E-04	8.31E-08	1.16E-14	1.45E-16	5.66E-24	1.89E-28

Moon

Species	H	H_2	^3He	^4He	CH_4	OH	H_2O	Ne	N_2	Ar	CO_2
mass	1	2	3	4	16	17	18	20	28	40	44
n_i [m^{-3}]	2.40E+07	9.00E+09	3.64E+06	7.27E+09	1.00E+10	1.60E+12	2.70E+11	1.00E+09	8.00E+08	1.00E+11	1.00E+09
v_0 [m/s]	2578.46	1823.24	1488.67	1289.23	644.61	625.37	607.75	576.56	487.28	408.20	388.72
X	0.848	1.697	2.545	3.394	13.575	14.423	15.272	16.969	23.756	33.853	37.331
f_{esc}	2.23E-01	1.39E-01	7.85E-02	4.16E-02	5.23E-06	2.37E-06	1.07E-06	2.17E-07	3.36E-10	1.95E-14	6.63E-16
$\phi_{esc,i}$ [s^{-1}]	1.38E+10	2.29E+12	4.25E+08	3.90E+11	3.37E+07	2.37E+09	1.76E+08	1.25E+05	1.31E+02	7.97E-01	2.58E-04
$\phi_{esc,i}$ [kg/s]	3.69E+05	1.22E+08	3.40E+04	4.17E+07	1.44E+04	1.08E+06	8.45E+04	6.67E+01	9.82E-02	8.52E-04	3.03E-07

in cold traps (like some permanently shadowed polar craters). Of course, all species are also subject to loss from the exosphere via photo-ionization and photo-dissociation.

With regards to the efficiency and type of thermal escape we have to look at the role of the X parameter (see Eq. (12)) in Eq. (13) in detail. Escape from an atmosphere is considered significant for situations where $X \lesssim 15$. Escape increases exponentially for smaller X, up to the point where the escape velocity equals the most probable velocity of the Maxwell-Boltzmann distribution at $X = 3/2$. For $X < 3/2$ the formalism presented above does not apply anymore, and for classical atmospheres escape would transition to hydrodynamic escape (often referred to as blow-off regime), with velocity distributions different from Maxwell-Boltzmann, being a shifted Maxwellian or a modified Maxwellian. The transition to the blow-off regime is not a step function at $X = 3/2$, but it is more gradual and becomes important already for $X < 2$–3 according to a study by Benedikt et al. (2000). According to Volkov et al. (2011) and Erkaev et al. (2015) the thermal escape regime changes to blow-off over a narrow range of the critical escape parameter X_{crit}: the escape is purely hydrodynamic for $X_{crit} \leq 2$–3, and for $X_{crit} \geq 6$ it is purely Jeans escape. Therefore, for Mercury and the Moon we can assume that for $X \lesssim 6$ any outgassed elements are lost immediately to space.

3 Micrometeorite Impact Vaporization

Micrometeorites impacting an unprotected planetary surface cause a range of processes including impact gardening, exospheric generation, surface contamination, and electrostatic effects on surface processes (Szalay et al. 2018). A global, direct consequence of the meteoroid bombardment is the formation of ejecta clouds of solids and gases. Escaping, unbound ejecta becomes a source of planetary or interplanetary meteoroids. We will focus on the contribution to the exosphere in the following.

Micrometeorite impacts on a solid rock or regolith surface result in an impact plume consisting of mostly surface material of broken fragments of minerals or rock, melt, all the way to atoms and molecules. Thus, with each impact, micrometeorites contribute to the production of exospheric densities, including also the low-volatile and refractory species. The details of the loss and source processes of volatiles and refractories in the exosphere are discussed by Grava et al. (2021). The steady flow of micrometeorites causes a steady contribution of particles to the exosphere. At quiet solar times, i.e., without energetic ions from the magnetosphere or the solar wind hitting the surface, it might dominate the particle release process acting over the whole planetary surface, at least for refractory species. During the night it may be the only particle release process acting over the whole planetary surface.

3.1 Theoretical Description

Since the impact of a micrometeorite on the surface creates an impact plume it is natural to model the volatile material of the plume by a thermal velocity distribution. The measured time-averaged temperature in the micrometeorite produced vapor cloud is in the range of 2500–5000 K (Eichhorn 1976, 1978a). Collette et al. (2013) reproduced the time-resolved measurements of dust impacts. Eichhorn (1978a, 1978b) studied the velocities of impact ejecta parameters during hypervelocity particle impacts and found that the velocity of the ejecta increases with increasing impact velocity and decreasing ejection angle, with the ejection angle measured with respect to the plane of the target surface, but the ratio of the maximum ejecta velocity to the primary impact velocity decreases with increasing impact speed.

At Mercury, impact plume temperatures are up to a factor ten higher than Mercury's dayside surface temperature. At the Moon, impact plume temperatures are even higher compared to dayside temperatures. However, the corresponding characteristic energies for micrometeorite impact are still lower than for particles that result from surface sputtering, see below.

For the simulation of trajectories for released particles that have their origin in micrometeorite vaporization a Maxwellian-Boltzmann velocity distribution, as described above (Eq. (7)), is used but with an average temperature of the released material of about 3000–4000 K (Wurz and Lammer 2003; Leblanc and Johnson 2013; Mangano et al. 2007; Gamborino et al. 2019).

The impact of micrometeorites and meteorites will evaporate a certain volume from the lunar or hermean surface, from rocks or from the regolith covered surface, to contribute to the exospheric gas at the impact site. Cintala (1992) calculated impacts for Mercury and the Moon and gave an analytical formula for the release of volatile material. The volume of surface material, V_v, being released into the vapor phase is given by

$$V_v = \frac{m_P}{\varrho_P} \left(c + dv_P + ev_P^2\right) = V_P \left(c + dv_P + ev_P^2\right) \tag{15}$$

where m_P is the projectile's mass, ϱ_P its mass density, V_P its volume, and v_P is the projectile's velocity. The constants c, d, and e are determined for the combination of surface material and projectile composition (Cintala 1992). Note that the released material is highly dependent on the composition and density of both the target and projectile, the heat capacities and enthalpies of melting and vaporization playing a large role. About one to two orders of magnitude more material than that of the impactor is released as vapor phase because of the high impact speed for meteorites at Mercury (Cintala 1992).

Alternatively, the impact of micrometeorites, all the way to large bolides, can be calculated from scaling laws (Holsapple 1993), which are based on a large experimental data set. The scaling laws apply from for impactor speeds from about 3 km/s upwards to much higher velocities (Holsapple and Housten 2020). The scaling laws provide the crater size, the mass of the excavated material, and the vapor mass, among much other data.

Having the volume of material released from the surface as volatiles for an impactor of mass m_P and velocity v_P of from Eq. (15) we can derive the total flux of particles released by micrometeorite impact. Using the formalism from Cintala (1992) we derive:

$$\Phi_{MIV} = \frac{\varrho_{surf}}{\langle \mu \rangle} \frac{1}{\varrho_P} \left(\frac{dm_P}{dAdt}\right) \left(c + dv_P + ev_P^2\right) \tag{16}$$

where ϱ_{surf} is the mass density of the surface material, $\langle \mu \rangle$ is the average atomic weight of the surface material, and $(dm_P/dAdt)$ is the mass flux of micrometeoritic bombardment on the surface. See Table 2 for typical values for the micrometeorite mass flux on Moon and Mercury. Note that the range of the micrometeorite mass fluxes cover a significant range, both for the Moon and Mercury, which has a direct influence on the contribution of this release process to the exosphere (see Eq. (16)).

3.2 Overview of Micrometeorite Fluxes at Mercury and Moon

For larger impacts, one can calculate the temporary contribution to the exosphere for a single impact using the scaling laws (Holsapple 1993), or Eq. (15) (Cintala 1992). Mangano et al. (2007) have shown such calculations where the impact plumes reach their maximum extent after about 1000 s in Mercury's exosphere.

Table 2 Survey of reported micrometeorite fluxes onto the surface of the Moon and Mercury

Object	MIV flux	Size range	Comments	Reference
Mercury	$2.82 \cdot 10^{-16}\,\mathrm{g\,cm^{-2}\,s^{-1}}$	$m < 0.1$ g Size range 5–100 μm	Modelling	Cintala (1992)
Mercury	10.7–23.0 t/day, corres. $(1.66–3.56) \cdot 10^{-16}\,\mathrm{g\,cm^{-2}\,s^{-1}}$	$m < 1$ g	Modelling, aphelion – perihelion	Müller et al. (2002)
Mercury	$2.9 \cdot 10^{-16}\,\mathrm{g\,cm^{-2}\,s^{-1}}$	Size range 10^{-8}–0.1 m		Bruno et al. (2007)
Mercury	$2.382 \cdot 10^{-14}\,\mathrm{g\,cm^{-2}\,s^{-1}}$	$m < 0.1$ g Size range 5–100 μm	Modelling	Borin et al. (2009)
Mercury	$8.982 \cdot 10^{-15}\,\mathrm{g\,cm^{-2}\,s^{-1}}$	$m < 0.1$ g Size range 5–100 μm	Modelling	Borin et al. (2010)
Mercury	12.6 ± 3.5 tons/day, corres. $(1.95 \pm 0.54) \cdot 10^{-16}\,\mathrm{g\,cm^{-2}\,s^{-1}}$	Diameters from 10 μm to 2000 μm	Modelling, averaged over Mercury's orbit	Pokorný et al. (2018)
Moon	$5.12 \cdot 10^{-17}\,\mathrm{g\,cm^{-2}\,s^{-1}}$	$m < 0.1$ g	Modelling	Cintala (1992)
Moon	2.7 t/day, corres. $8.36 \cdot 10^{-17}\,\mathrm{g\,cm^{-2}\,s^{-1}}$	Diameters from 10 nm to about 1 mm	Interplanetary meteorites	Grün et al. (2011)
Moon	1.4 t/day, corres. $4.27 \cdot 10^{-17}\,\mathrm{g\,cm^{-2}\,s^{-1}}$	Diameters from 10 μm to 2000 μm, depending on source populations	Modelling, averaged over lunar orbit	Pokorný et al. (2019)

Since larger impacts are very rare compared to the residence time of atoms and molecules in the exosphere, e.g. the impact frequency of 1-m meteoroids is only 2 events per Earth year on Mercury (Mangano et al. 2007), we are more interested in the steady contribution to the exosphere by the continuous flux of micro-meteorites. A survey of these fluxes is given in Table 2. More detailed information on the micrometeorite fluxes onto the Moon and Mercury is given in the accompanying paper (Janches et al. 2021).

Table summarizes observations of impacts rates on Mercury and the Moon. Each year the Moon is bombarded by about 10^6 kg of interplanetary micrometeoroids of cometary and asteroidal origin. For Mercury, the bombardment is about $(4 – 20) \cdot 10^6$ kg per year, although the meteoroid impact rates at very small distances from the Sun are not very well known (see also Table 2). Most of these projectiles range from 10 nm to about 1 mm in size and they impact the lunar surface at speeds of 10–72 km/s (Marchi et al. 2009). Some meteoroid populations are calculated to have higher energies (Pokorný et al. 2018), which is important since the vaporized volume is proportional to the square of the impact speed (see Eq. (15)). The mean impact velocity on Mercury is about $20\ \mathrm{km\,s^{-1}}$, and on the Moon it is about $14\ \mathrm{km\,s^{-1}}$ (Langevin 1997).

The impactors are delivering their kinetic energy to a point below the surface down to a depth comparable to the size of the impactor (Holsapple 1993). The total yield of excavated material, $Y = M_{ej}/M_{imp}$, from a hypervelocity impact (and the impact speed larger than the sound speed of the sample material) where M_{ej} is the total ejected mass created in an impact and M_{imp} is the impactor mass, can be anywhere from $Y = 10^3$ to 10^6 depending on the sample material properties. Impacts into solid rock result in lower yields and higher ejecta velocities than impacts into unconsolidated sand or powder (Housen et al. 1983; Hartmann 1985; Holsapple 1993).

Pokorný et al. (2018) developed a model that combined four distinctive sources of meteoroids in the solar system: main-belt asteroids, Jupiter-family comets, Halley-type comets, and Oort Cloud comets to characterize the meteoroid environment around Mercury. From this model of meteorite fluxes impinging on the surface they derived the contribution to the planet's exosphere. For Mercury's year they obtained good agreement with previously reported Ca observations in Mercury's exosphere by the MESSENGER spacecraft during the primary and first extended missions (March 2011–March 2013) (Burger et al. 2014).

3.3 Exospheric Escape via Mircometeorite Vaporization

Like the volatile species, also a fraction of the vaporized material from micrometeorite impact vaporization (MIV) will be lost from the exosphere. We can use the same formalism as for thermal escape to estimate the exospheric loss from MIV.

Table 3 presents the X parameter (Eq. (12)), the escape fraction f_{esc} (Eq. (13)), the escape flux $\phi_{esc,i}$ (Eq. (14)), and the escaping mass flux for several species released from the regolith via micro-meteorite impact vaporization. As discussed above, MIV is modelled as thermal release with temperatures of 3500 K. Exospheric abundances at the surface are taken from Wurz et al. (2010) for Mercury, and for low-Ti mare soils composition of the Moon from Wurz et al. (2007). Because of the high temperature of the vapor plume the X parameter is low, resulting in a larger fraction of escaping particles compared to thermal escape of volatiles. Moreover, there is a strong mass dependence, with a preferred loss of light species, and heavier species more likely to fall back onto the respective surfaces. As discussed above (Sect. 2.2), particles with $X_{crit} \leq 6$ will be immediately lost into space. For the Moon, because of the low gravity there, basically all species released by MIV are immediately lost into space, for Mercury only the species up to O are lost immediately, but a large fraction of the heavier elements returns to the surface. Thus, over geological time scales, there are chemical changes of Mercury's regolith resulting from MIV, but for the Moon these changes are much less. In addition, all species are subject to loss from the exosphere via photoionization and photo-dissociation, which has been discussed earlier (Wurz and Lammer 2003; Wurz et al. 2010).

3.4 Key Observations of Micrometeorite Impact Vaporization

Since the release processes are often operating at the same time, it is difficult to find observations of exospheric populations that can be solely attributed to micrometeorite bombardment. The estimates of the meteoritic flux range cover more than two decades (see Table 2) which directly scales the produced species in the exosphere (see Eq. (16)), making MIV possibly a dominating process or unimportant release process for a species, depending which meteorite flux is chosen for the data interpretation. In the following we give a few examples where this was possible or likely.

Kameda et al. (2009) studied the Na exosphere over a Mercury year via ground-based telescopic observations. Results of past observations have revealed that the atmospheric Na density has no or low correlation with the solar flux, sunspot number, heliocentric distance, or solar radiation pressure. Kameda et al. (2009) showed that the variability of Mercury's atmospheric Na density depends strongly on the IPD distribution. Since Mercury's orbit plane is inclined by 7° to the symmetry plane of the interplanetary dust particles (IPD) they found a corresponding temporal variability of the Na density in Mercury's atmosphere, that is, the Na density is low when Mercury is far away from the symmetry plane of IPDs and is high when Mercury is close to the symmetry plane. Actually, the authors could infer the IPD distribution near Mercury orbit from the temporal variability of Na density in Mercury's atmosphere. Exospheric Ca observed by MESSENGER could partly be explained by Mercury moving in and out the dusk disk (Killen and Hahn 2015).

Table 3 Thermal escape calculated for species from released from the Moon (low-Ti mare soils) and Mercury via micrometeorite impact vaporization, with exospheric abundances at the surface from Wurz et al. (2007) and Wurz et al. (2010), respectively

Mercury

Species	C	OH	O	Na	Mg	Al	Si	S	K	Ca	Ti	Fe
mass	12	17	16	23	24	27	28	32	39	40	48	56
n_i [m^{-3}]	3.87E+06	8.84E+04	7.77E+07	2.12E+06	2.50E+07	4.33E+06	2.95E+07	1.07E+06	7.61E+04	3.51E+06	3.07E+05	2.01E+06
v_0 [m/s]	2201.78	1849.86	1906.79	1590.38	1556.89	1467.85	1441.40	1348.31	1221.33	1205.96	1100.89	1019.22
X	3.726	5.278	4.967	7.141	7.451	8.383	8.693	9.935	12.108	12.419	14.902	17.386
f_{esc}	3.21E-02	9.04E-03	1.17E-02	1.82E-03	1.38E-03	6.06E-04	4.59E-04	1.49E-04	2.04E-05	1.53E-05	1.51E-06	1.46E-07
$\phi_{esc,i}$ [s^{-1}]	2.74E+08	1.48E+06	1.74E+09	6.13E+06	5.39E+07	3.85E+06	1.95E+07	2.16E+05	1.90E+03	6.48E+04	5.11E+02	2.99E+02
$\phi_{esc,i}$ [kg/s]	8.78E+04	6.72E+02	7.42E+05	3.77E+03	3.46E+04	2.78E+03	1.46E+04	1.84E+02	1.98E+00	6.92E+01	6.56E-01	4.48E-01

Moon

Species	O	Na	Mg	Al	Si	S	K	Ca	Ti	Cr	Mn	Fe
mass	16	23	24	27	28	32	39	40	48	52	55	56
n_i [m^{-3}]	1.52E+08	7.90E+05	1.54E+07	1.57E+07	5.06E+07	1.98E+05	1.98E+06	1.59E+07	2.52E+06	5.08E+05	3.30E+05	2.47E+07
v_0 [m/s]	1906.79	1590.38	1556.89	1467.85	1441.40	1348.31	1221.33	1205.96	1100.89	1057.70	1028.45	1019.22
X	1.551	2.230	2.327	2.618	2.715	3.103	3.782	3.879	4.654	5.042	5.333	5.430
f_{esc}	1.53E-01	9.80E-02	9.16E-02	7.45E-02	6.94E-02	5.20E-02	3.07E-02	2.85E-02	1.52E-02	1.10E-02	8.63E-03	7.95E-03
$\phi_{esc,i}$ [s^{-1}]	4.42E+10	1.23E+08	2.20E+08	1.72E+09	5.06E+09	1.39E+08	7.43E+08	5.46E+08	4.21E+07	5.92E+06	2.93E+06	2.00E+08
$\phi_{esc,i}$ [kg/s]	1.89E+07	7.57E+04	1.41E+06	1.24E+06	3.79E+06	1.19E+05	7.75E+03	5.83E+05	5.41E+04	8.22E+03	4.30E+03	3.00E+05

Optical spectroscopy measurements of Na in Mercury's exosphere near the subsolar point by MESSENGER Mercury Atmospheric and Surface Composition Spectrometer Ultraviolet and Visible Spectrometer, MASCS/UVVS (Cassidy et al. 2015). These observations have been interpreted Monte Carlo (MC) exosphere model (Wurz and Lammer 2003) to calculate the subsolar Na content of the exosphere for the observation conditions *ab initio*. The observed Na tangential column density profile as a function of altitude could be reproduced by the model using two components (Gamborino et al. 2019): i) below 500 km altitude, the dominant release mechanism of Na is thermal desorption with the local surface temperature of 594 K, and ii) at altitudes above the contribution by MIV prevails up to the observed 4000 km characterized by a temperature of about 3500 K.

However, one has to be careful with the interpretations of Na observations in Mercury's exosphere, because Na is released into the exosphere by all release processes considered (i.e., thermal release, PDS, sputtering, micrometeorite impact vaporization) and furthermore Na is affected by radiation acceleration, which explains a large fraction of the variability of the Na signal observed by ground-based telescopes (Potter et al. 2007). Usually, a mix of the release processes is occurring and the attribution of the Na signal, or a part of the Na signal, to a single release process is often difficult. The observations of Na and K at Mercury and the Moon, and their interpretation, are discussed in detail by Leblanc et al. (2022).

Burger et al. (2012, 2014) explained their Ca observations with the UVVS instrument on MESSENGER being the result of MIV. They observed very high temperatures of the Ca atoms of > 50'000 K, which cannot be the direct result of meteoritic impact release of Ca atoms. The authors argue that the high energy might result from the CaO being released by MIV and dissociated into Ca and O atoms in the exosphere, but they admit that the excess energies of these species upon dissociation are not well understood (Burger et al. 2014).

Merkel et al. (2018) found good correlation between Mg observations in the exosphere with the Mg abundance in the terrain on the underlying surface being the result of MIV. They also observed high temperatures of the Mg atoms in the range between 5000 to 10 000 K, which cannot be the direct result of meteoritic impact release of the Mg atoms, using the same argument of MgO being the initial species released by MIV and the additional energy arises from the breakup into atomic constituents. From the regularity in the year-to-year variations, and no short time variations, they conclude that sputtering is an unimportant contributor at these latitudes, and MIV is the likely source for the observed Mg in the exosphere.

More information on the micrometeorite fluxes, and their temporal variation over the year, onto the surface of the Moon and Mercury is given in the accompanying paper (Janches et al. 2021).

4 Photon Stimulated Desorption

Photon Stimulated Desorption (PSD) has been discussed as a release process mostly for Na and K for the exospheres of Mercury and the Moon. Several processes promote Na and K into these exospheres, and it has been difficult to isolate the PSD release process from the other processes in the observations. Note, from laboratory experiments we know that PSD only releases atoms or molecules adsorbed on the surface, i.e., species which are not chemically bound within a mineral.

4.1 Mechanism of Photon Stimulated Desorption

Photon stimulated desorption (PSD) using photon energies just above the bandgap of the lunar and Mercury regolith surfaces can occur due to direct photon absorption and subsequent stimulated desorption of excited surface states. Typically, photon energies have to be

Fig. 4 The PSD processes can be explained by 1) a single-photon absorption and Frank-Condon transition of the surface species to a repulsive excited state potential $V^*(z)$, 2) motion of the excited species along the potential energy curve, and finally quenching of the excited state and either 3a) return to the ground state $V(z)$ or 3b) desorption. The inset shows the final $E(t_r)$ energy as a function of excited state resonance time, t_r. Figure reproduced from Schaible et al. (2020) with permission

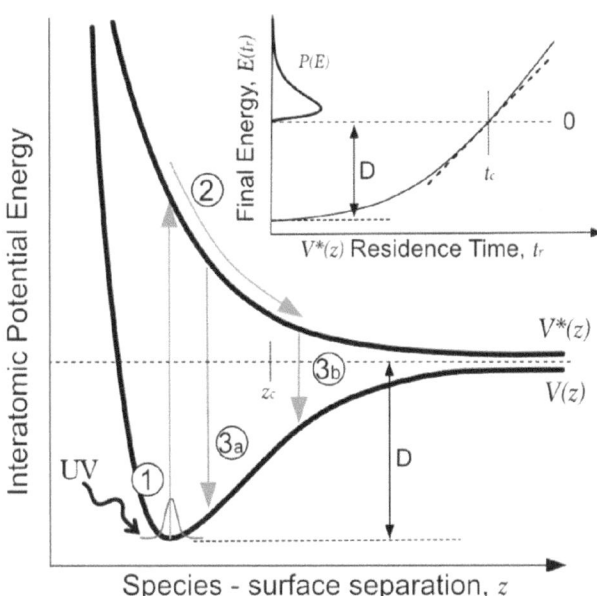

in the range of about 4–10 eV. Photo-excitations invoke an electronic transition, which must be localized at the surface, leading to an anti-bonding state that results in the release of the excited atom (Na, K, ...) or molecule (e.g. H_2O) from the surface, with non-thermal release energies of the desorbed species. For single centered excitons, self-trapping will occur at the surface mainly at defect sites with the ejection force primarily along the surface normal.

A qualitative diagram depicting the PSD process is shown in Fig. 4. PSD of neutral atoms or molecules residing on the surface of a substrate, often referred to as adatoms or molecules, is initiated by 1) a single photon excitation of a Franck-Condon transition from a ground state $V(z)$ to an excited surface state $V^*(z)$. As a result, 2) the excited atom or molecule gains kinetic energy by moving away from the surface along the excited state potential energy surface. The excited state is quenched and depending on the quenching rate, either 3a) the absorbed energy is dissipated and lost to the bulk or 3b) the displacement reaches a critical point z_c, corresponding to an excited state residence time t_c, which supplies sufficient kinetic energy to overcome the surface bond energy D and the atom or molecule desorb. The translation energy distribution of the desorbing atom(s) depend(s) upon the final energy and excited state residence time as shown in the inset of Fig. 4. The energy of the surface species after the excitation-quenching process is given by

$$E(t_r) = \frac{p(t_r)^2}{2m} + V(z(t_r)) \tag{17}$$

where $V(z)$ is the ground state potential energy curve and $p(t_r)$ is the momentum, which is determined by solving the classical equations of motion on the excited state potential $V^*(z)$. Though the potential energy surfaces are not generally known, the translational energy distribution can be approximated classically by evaluating the energy dependence on excited state residence time near t_c with a linear approximation (dotted line in inset of Fig. 4). This approximation holds for strongly quenched systems where the probability of the system remaining in the excited state beyond t_c decays exponentially (Zimmermann and Ho 1994).

4.2 PSD Velocity Distribution

Using the linear approximation for $E(t_r)$ near t_c and assuming the angular distribution of ejected particles is large compared to the detection cone for the velocity distribution, the velocity distributions for PSD of water can be approximated by a flux weighted Maxwellian of the form:

$$P_v(v, T_{trans}) = \left(\frac{m}{\langle E(t_r) \rangle} \right) v^3 \exp \left(\frac{-mv^2}{\langle E(t_r) \rangle} \right) \qquad (18)$$

where v is the velocity, $\langle E(t_r) \rangle$ is the mean translation energy and m is the mass of the desorbing species. Note the temperature is related to the mean translation energy through $\langle E(t_r) \rangle = 2k_B T_{trans}$, where k_B is the Boltzmann constant, and T_{trans} is the apparent translation temperature of the distribution that results from a convolution of the surface temperature and the repulsive potential of the desorbing electronic state (see Fig. 4). Experimental data for H_2O desorption can be fit using this form with a bi-modal distribution consisting of a low temperature thermal component (matching the substrate temperature) and a high temperature 'suprathermal' component due to direct desorption along the normal component (Bennett et al. 2016; DeSimone and Orlando 2014; Zimmermann and Ho 1994; Schaible et al. 2020). Since this distribution function includes the intrinsic physics of the desorption process, this is often used as an experimental fitting procedure in addition to the Weibull distributions discussed below.

One of the few experimental results in the laboratory studying the Na release processes happening on regolith surfaces are the experiments by Yakshinskiy and Madey (2000, 2004), which are relevant for the Moon and Mercury. These authors studied the desorption induced by electronic transitions (DIET) of Na adsorbed on model mineral surfaces and lunar basalt samples, where the laboratory experiments included PSD and Electron-Stimulated Desorption (ESD) as release processes. In particular, they measured velocity distribution functions (VDF) of Na released via ESD from SiO_2 surfaces and found it to be "clearly non-thermal" with respect to the surface temperature, somewhat resembling a Maxwell-Boltzmann distribution at 1200 K, but with a high-energy tail and a positive offset above zero for the lowest velocities. Thus, Eq. (18) derived for PSD measurements of water cannot be used here. A range of distributions, thermal and non-thermal, have been used in the literature to model the particle release by PSD. However, none of these distributions replicates all the features as they are observed in the laboratory experiments: i) non-thermal distribution peaking at energies much higher than the surface temperature, ii) a high energy tail extending to speeds beyond what is possible from non-thermal Maxwell-Boltzmann distributions, iii) an offset in velocity so that there is a minimum release velocity. The details of this have been reviewed by Gamborino et al. (2019).

Gamborino et al. (2019) presented an improved VDF for the PSD process satisfying the mentioned features from the observations. To mathematically best describe the published laboratory measurements (Yakshinskiy and Madey 2000, 2004) and planetary observations (e.g. Cassidy et al. 2015) the sought-after VDF has to have a characteristic energy significantly higher than what corresponds to the surface temperature and that tails towards higher speeds. The second goal of the sought-after VDF is a parametrization that allows for its applications at other surface temperatures than the measured ones in the laboratory, in particular for the surfaces of Mercury and the Moon.

Gamborino et al. (2019) presented an empirical energy distribution function, for PSD at Mercury and the Moon, namely the Weibull distribution, which allows for a wide range of shapes using only two parameters for its definition. The normalized Weibull distribution for the random variable υ is defined as:

$$f(\upsilon; \lambda, \kappa) = \begin{cases} \frac{\kappa}{\lambda}\left(\frac{\upsilon}{\lambda}\right)^{\kappa-1} e^{-(\upsilon/\lambda)^{\kappa}} & : \upsilon \geq 0 \\ 0 & : \upsilon < 0 \end{cases}$$

where κ is the dimensionless shape parameter and $\lambda > 0$ is the scale parameter of the distribution (in m/s). The scale parameter λ is obtained after calculating the mean of the probability distribution function (first central moment):

$$\bar{\upsilon} = \int_{-\infty}^{+\infty} \upsilon f(\upsilon; \lambda, \kappa) \, d\upsilon = \int_{0}^{+\infty} \upsilon \frac{\kappa}{\lambda}\left(\frac{\upsilon}{\lambda}\right)^{\kappa-1} e^{-(\upsilon/\lambda)^{\kappa}} \, d\upsilon = \lambda \Gamma\left(1 + \frac{1}{\kappa}\right) \tag{19}$$

with Γ the Gamma function. The surface, which is the starting point of the desorbed atoms, has a given temperature T_S. This surface temperature will cause an energy broadening of the electronic transition induced by the adsorption of the UV photon (Gamborino et al. 2019). Therefore, the related kinetic energy of the desorbed atom of $\frac{1}{2}m\bar{\upsilon}^2 = \frac{3}{2}k_B T_S$ folds into the Weibull distribution. Since we consider the one-dimensional case, we have

$$\bar{\upsilon} = \sqrt{\frac{3k_B T_S}{m}}$$

with m being the mass of the desorbed atom. Thus, we get for the scale parameter the expression:

$$\lambda = \frac{\bar{\upsilon}}{\Gamma\left(1 + \frac{1}{\kappa}\right)} = \sqrt{\frac{3k_B T_S}{m}} \frac{1}{\Gamma\left(1 + \frac{1}{\kappa}\right)} \tag{20}$$

The normalized Weibull distribution is then:

$$f(\upsilon, \upsilon_0, \kappa) = \kappa \Gamma\left(1 + \frac{1}{\kappa}\right)\left(\frac{m}{3k_B T_S}\right)^{\frac{1}{2}}\left((\upsilon - \upsilon_0)\sqrt{\frac{m}{3k_B T_S}}\Gamma\left(1 + \frac{1}{\kappa}\right)\right)^{\kappa-1} \times$$

$$\times \exp\left(-\left((\upsilon - \upsilon_0)\sqrt{\frac{m}{3k_B T_S}}\Gamma\left(1 + \frac{1}{\kappa}\right)\right)^{\kappa}\right) \tag{21}$$

where υ_0 is the offset speed, and κ is the shape parameter, which is an implicit function that is usually determined by numerical means (see Bhattacharya and Bhattacharjee 2010). υ_0 and κ are the only free parameters for this velocity distribution, λ is derived from Eq. (21) using the actual the surface temperature T_S, with the best fits calculated for all available experimental data sets with a single set of parameters using $\kappa = 1.7$ and $\upsilon_0 = 575$ m/s. Even though the Weibull distribution with the given parameters is currently the best available presentation of the velocity distribution for PSD, it is based on little data from laboratory studies.

4.3 Released Particle Flux by PSD

We can calculate the flux of released atoms from a surface by photon-stimulated desorption of Na atoms Φ_{PSD} for a species i from

$$\Phi_{PSD,i} = f_i N_S \int \phi_{ph}(\lambda) Q_i(\lambda) \, d\lambda \approx \frac{1}{4} f_i N_S \phi_{ph} Q_i \tag{22}$$

where f_i is the fraction of species, N_S is the surface density of the regolith, $\phi_{ph}(\lambda)$ is the incident photon flux, $Q_i(\lambda)$ is the PSD cross section, the factor 1/4 gives the surface-averaged value. The regolith surface density is estimated as $N_S = 7.5 \cdot 10^{14}$ cm^{-2}, and the average photon flux at 1 AU is $\phi_{ph} \approx 2 \cdot 10^{15}$ cm^{-2} s^{-1} for the relevant UV wavelength range, as discussed below.

PSD with solar photons is an important low-energy release process when sufficient UV flux is present and free Na or K is available on the surface. For Mercury and the Moon only Na and K have significant release rates due to PSD. The experimentally determined PSD cross-section for Na is $Q_i = (3 \pm 1) \times 10^{-20}$ cm^2 in the wavelength range of 400–250 nm (Yakshinskiy and Madey 2000), and for K it is $Q_i = 2 \cdot 10^{-20}$ cm^2 (Madey et al. 1998). Note that these cross sections are for Na and K atoms physically bound on a mineral surface, i.e., already freed from the mineral bonds because solar photons lack sufficient energy to release alkaline earth atoms from minerals. Direct PSD of Na or K from minerals has not been observed in laboratory experiments (Yakshinskiy and Madey 2004; Bennett et al. 2013). Thus, neither PSD nor thermal release are primary release mechanisms, meaning that they cannot eject the Na or K atoms that are still bound in the mineral of the rock.

4.4 Exospheric Escape via Photon-Stimulated Desorption

Figure 5, left panel, shows the PSD velocity distribution (Eq. (21)) for Na and K for a surface temperature of 540 K, the case for the sub-solar point Mercury near apocenter. The distribution for the Moon looks similar to the one for Mercury, since the surface temperature only has small effect on the VDF (see Eq. (24)). The VDF for PSD is clearly non-thermal, and it is tailed toward higher velocities. For calculating the escape, we need to estimate the part of the VDF above the escape speed, v_∞, which is easily done numerically. Figure 5, right panel, shows the cumulative function of the PSD velocity distribution for Na and K, which tends to 1 because the function is normalized. The escape fraction is then 1 minus the cumulative function evaluated at the escape speed, v_∞, which is also shown in Fig. 5, right panel. For Mercury $v_\infty = 4250$ m s^{-1} and for the Moon $v_\infty = 2375$ m s^{-1}. This means the escape fraction for Mercury for PSD is below 10^{-3} and for the Moon it is below 10^{-2}. Thus, for a PSD process of Na and K, photoionization is the dominating loss process from the exosphere for both objects.

However, there is radiation pressure acting on the Na and K atoms in the exosphere that accelerate these atoms away from the Sun, resulting in non-Keplerian particle trajectories, thus complicating the situation. Simulations by Schmidt et al. (2012) demonstrate that both photon-stimulated desorption and micrometeoroid impacts can result in an about 20% loss of Mercury's sodium atmosphere, depending on Mercury's orbital phase, and together the two release processes are responsible for the observed comet-like tail as driven by solar radiation pressure.

4.5 Experimental Observations of PSD

In the laboratory, the experiments by Madey et al. (1998) and Yakshinskiy and Madey (1999, 2000, 2004) have been fundamental to the understanding of the PSD release process for Mercury and the Moon. These authors deposited Na or K atoms onto SiO$_2$ surfaces and lunar basalt and measured the PSD cross sections, the temperature dependence, and the velocity distribution of released Na and K atoms. Among other things, they clearly demonstrated that PSD is a non-thermal process, the desorption of an alkali atom is the result of an electronic excitation resulting from the absorption of a UV photon at the surface.

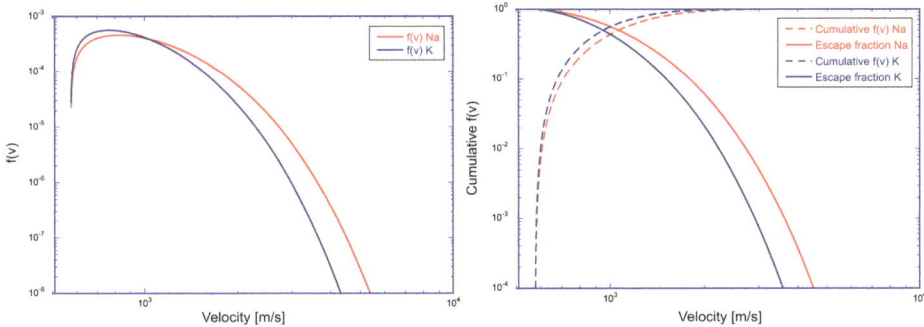

Fig. 5 Left: PSD velocity distribution for Na (red line) and K (blue line). Right: Cumulative velocity distribution for Na (dashed red line) and K (dashed blue line), and escape fraction for Na (red line) and K (blue line)

Bennett et al. (2016) studied PSD of Ca from CaS powder samples, with CaS considered as an analogue material for oldhamite ((Mg,Ca)S), which the authors considered to be a possible component of the Mercury surface, particularly within the hollows identified within craters, and could therefore serve as a source of the observed exospheric calcium of Mercury. They measured cross sections for PSD using 3.4 eV photons for neutral Ca as $Q_i(\text{Ca}^0) = (1.1 \pm 0.7) \cdot 10^{-20}$ cm^2 and for ionized Ca as $Q_i(\text{Ca}^+) = (3.2 \pm 0.9) \cdot 10^{-24}$ cm^2. They also observed non-thermal velocity distributions for Ca, which they fitted with a sum of two Maxwell-Boltzmann distributions, that support PSD as the release process. Therefore, for interpreting Ca observations in the exosphere also PSD must be considered, at least at provinces with a significant oldhamite fraction on the surface.

Schaible et al. (2020) studied PSD of S from powder samples of MgS. They measured cross sections for PSD using 6.42 eV photons for neutral S^0 as $Q_i = 4 \cdot 10^{-22}$ cm^2. Also these authors observed non-thermal velocity distributions for the released S atoms, which they fitted with a sum of two Maxwell-Boltzmann distributions, that support PSD as the release process. Although the MgS (niningerite) used in these experiments is not considered the major sulfur-bearing mineral of Mercury's surface, but is considered similar to sulfur-bearing minerals such as CaS (oldhamite) and FeS (troilite) that were considered for Mercury's surface (Wurz et al. 2010; Nittler and Weider 2019).

The process of photon-stimulated desorption (PSD) has been considered for generating Na and K exospheric species as measured at Mercury and the Moon, when the flux of solar photons is high, e.g. on the dayside. However, since the release processes compete with each other it is often not easy to assign an observation to a single release process. For PSD it is clear that the maximum of the released flux should be located at the sub-solar point. However, since thermal release has also its maximum there, there is a direct competition and since fluxes from evaporation are much higher for Na and K for the temperatures at the sub-solar point of Mercury and the Moon, thermal release wins over PSD there. Since evaporation is an exponential function of the surface temperature (see Sect. 2.1), its importance drops fast for larger solar zenith angles (SZA) and PSD becomes important since it scales only with the cosine of the SZA when going towards higher latitudes. However, the activation energy for thermal release will increase for sub-monolayer coverages of Na and K on the surface, as discussed above. At least for Mercury, higher latitudes are where the solar wind ions have access to the surface because of the structure of its magnetosphere sputtering is contributing to the exospheric particle populations. Thus, the source process of Na and K in the hermean exosphere must be investigated carefully for each observation. The situation is less complicated at the Moon because of the lack of a lunar magnetosphere. For example, Sarantos et al. (2008) have concluded that PSD is responsible for the observed Na in the lunar exosphere when the Moon is in the solar wind.

Although of lesser importance for the exospheres of Mercury and the Moon, there is release of H_2O via PSD, and laboratory work on the PSD of water molecules from lunar surface material has also been reported (DeSimone and Orlando 2014, 2015). PSD of H_2O ($v = 0$) and O ($^3P_{J=2,1,0}$) was measured with resonance-enhanced multiphoton ionization following 157-nm photon irradiation of adsorbed water on a lunar mare basalt or an impact melt breccia. Water removal cross sections and time-of-flight (TOF) distributions were measured at exposures between 0.1 and 10 Langmuir (1 L = $1.33 \cdot 10^{-6}$ mbar s). The average cross section for H_2O ($v = 0$) removal and destruction at 0.1 Langmuir H_2O exposure was measured to be $(7.1 \pm 1.9) \cdot 10^{-19}$ cm^2 and then decreased with increasing coverage. The cross sections were similar for lunar impact melt breccias and mare basalt samples. Additionally, non-resonant ionization was employed to detect photofragments of vibrationally excited H_2O. The OH^+ fragment of H_2O (v^*) and the O(^3P_J) photoproducts increased in intensity during prolonged irradiation as hydroxyl groups accumulated on the surface and then recombined. The formation of excited water molecules via this process simulates the probable water formation during localized meteoroid impact events. For an initial exposure of 5 L H_2O, after reaching maximum signal, the cross sections for H_2O (v^*) and O($^3P_2$) depletion were measured to be $1.2 \cdot 10^{-19}$ cm^2 and $6.7 \cdot 10^{-20}$ cm^2, respectively. These photo-desorption, photo-destruction and photo-formation cross sections are relatively high and indicate that surficial water will not persist in Sun-lit regions of Mercury and the Moon unless a persistent source term exists. More discussion on the formation of H_2O, its distribution and migration over the planet, and its contribution to the exosphere are given in (Schörghofer et al. 2021).

4.6 Photon Fluxes at Mercury and the Moon

The Sun emits a continuous flux of photons (i.e., the solar spectrum), where most energy is contained in the visible (VIS), the infrared (IR), and the ultraviolet (UV) energy range, with relatively small additions in the X-ray and the gamma-ray region. Figure 6 shows the solar spectrum as measured by the SOLar SPECtrometer (SOLSPEC) instrument, part of the SOLAR payload on board the International Space Station (Meftah et al. 2018). The total solar power received at Earth, also referred to as the total solar irradiance (TSI), is on average 1360.96 W m^{-2}.

The total solar power received at Earth, also referred to as the total solar irradiance (TSI), follows closely the Sun's 11-year Schwabe cycle (Fröhlich 2013), as is clearly evident in Fig. 7, with minimum and maximum values of 1357.08 W m^{-2} and 1363.25 W m^{-2}, respectively. Thus, the TSI variation with solar cycle is small compared to the absolute value (<1%). Integrating over the solar photon spectrum, the TSI average value of 1360.96 W m^{-2} corresponds to a total photon flux of $6 \cdot 10^{17}$ cm^{-2} s^{-1}, at solar minimum to a flux of $5.99 \cdot 10^{17}$ cm^{-2} s^{-1}, and at solar maximum to a flux of $6.01 \cdot 10^{17}$ cm^{-2} s^{-1}.

Since the absorbed photon must overcome the binding energy of an atom on the surface there is a minimal necessary photon energy for the release via PSD (see discussion in Sect 4.1). Yakshinskiy and Madey (1999) found that only photons with energies larger than about 4 eV are capable of inducing PSD of Na atoms from surfaces that simulate lunar silicates, as discussed in Sect. 4.1. It is thus only the part of the solar spectrum with $\lambda < 300$ nm that is relevant for PSD. Of all solar photons available only a photon flux of about $\phi_{ph} = 2 \cdot 10^{15}$ cm^{-2} s^{-1} exhibits energies larger than 4 eV (see also the blue arrow in Fig. 6).

Fig. 6 The solar spectrum as measured by SOLSPEC on board the ISS (Meftah et al. 2018). The wavelength range in the UV that contributes to PSD is indicated

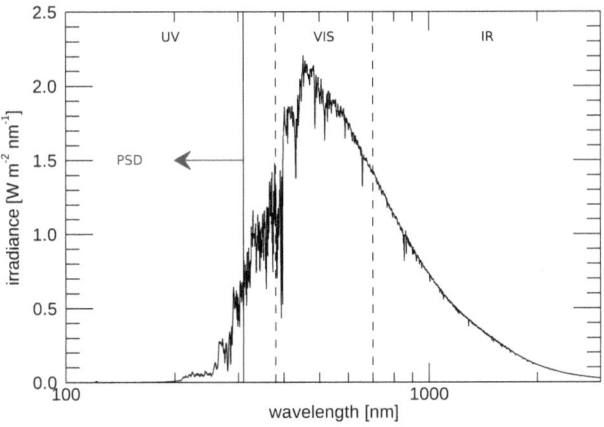

Fig. 7 TSI variation from 1979 until 2017, based on Fröhlich (2013). Data are obtained from PMOD/WRC, Davos, Switzerland, of the updated dataset version 42_65_1709 with new data from the VIRGO Experiment on the cooperative ESA/NASA Mission SoHO

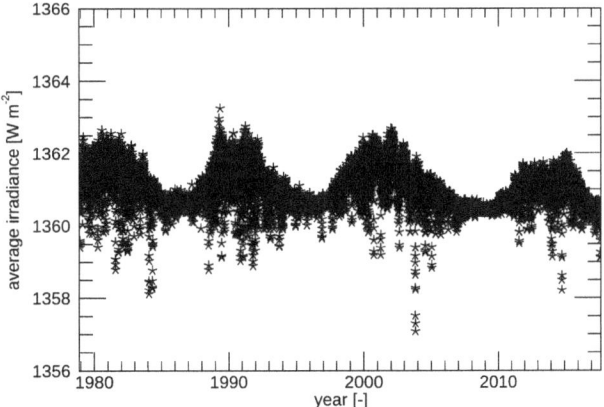

The TSI values measured by SOLSPEC and presented above are valid for a Sun distance of 1 AU (where AU is the astronomical unit, about $1.5 \cdot 10^{11}$ m). As the photon flux expands into space it decreases with $1/R^2$, where R is the distance to the Sun. To determine the flux at the Moon and Mercury, one thus has to factor in these bodies' distances from the Sun.

The Moon's distance to the Sun is close to 1 AU, i.e., the average flux of photons with energies >4 eV at the Moon is close to the $2 \cdot 10^{15}$ cm^{-2} s^{-1} mentioned above. As Earth (and the Moon) orbit the Sun, the Moon's distance to the Sun varies by $\sim 6 \cdot 10^9$ m, introducing a photon flux variation of $\sim 1.5 \cdot 10^{14}$ cm^{-2} s^{-1}. This variation is relatively small (7.5%) compared to the total photon flux of $2 \cdot 10^{15}$ cm^{-2} s^{-1}.

Mercury, being located at a solar distance of $(4.60–6.98) \cdot 10^9$ m, is much closer to the Sun, resulting in Mercury receiving a much higher photon flux than the Moon. On average, the total flux of photons with energies >4 eV at Mercury amounts to $1.34 \cdot 10^{16}$ cm^{-2} s^{-1}, i.e., almost 10 times the photon flux the Moon receives. With Mercury's orbit being quite eccentric, the variation in the photon flux due to the variation in distance between Mercury and the Sun is quite substantial. During one Mercury year, the PSD-relevant solar photon flux (with $E > 4$ eV) varies from $9.2 \cdot 10^{15}$ cm^{-2} s^{-1} at aphelion to $2.12 \cdot 10^{16}$ cm^{-2} s^{-1} at perihelion, i.e., the variation is of the same order as the total photon flux itself. For the Moon, the PSD-relevant solar photon flux (with $E > 4$ eV) varies from $1.93 \cdot 10^{15}$ cm^{-2} s^{-1} at aphelion to $2.08 \cdot 10^{15}$ cm^{-2} s^{-1} at perihelion.

Fig. 8 The figure illustrates various frequency distributions for so-called nano-, micro- and normal solar flares, but also for different types of stellar flares. Figure is adapted from Notsu et al. (2019) with permission

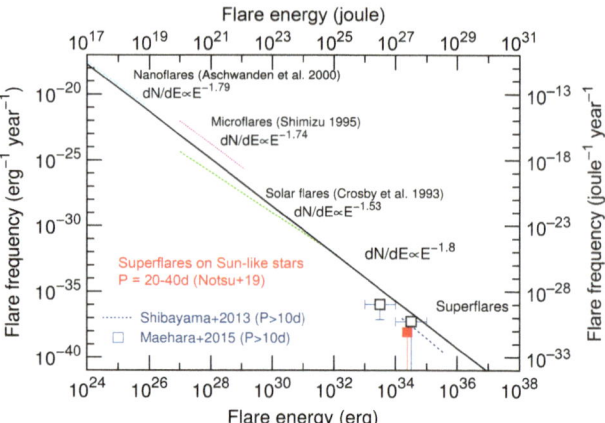

The Sun's photon flux only reaches the sunlit hemisphere of a planetary body and, at the surface, its irradiance scales with the cosine of the zenith angle. The variation in photon flux from the equator to the terminator is thus substantial. On the microscopic scale of individual regolith grains, the variation of the solar irradiance will probably be even larger.

4.7 Fluxes from Solar and Stellar Flares

Flares are energetic events on the Sun and on other stars that are generated when magnetic energy is released through a sudden rearrangement of stressed magnetic field lines on active regions on the star (see e.g. Benz and Güdel 2010; Fletcher et al. 2011; Hudson et al. 2011; Shibata and Magara 2011; Benz 2017 for comprehensive reviews on flares and their underlying physics). When happening, charged particles are accelerated toward the lower solar atmosphere, where they heat the plasma and generate a broad range of intense electromagnetic radiation, ranging from gamma rays to hard and soft X-rays, to UV and optical wavelengths, and up to the radio spectrum. Over their duration of tens of minutes, flares on the Sun were observed to release a total energy of up to 10^{25} J (Emslie et al. 2005, 2012; Moore et al. 2014), and even more on extremely rare occasions (Schrijver et al. 2012) such as the famous Carrington event in 1859 (Carrington 1859).

The occurrence rate of flares decreases with increasing total energy, with flares of the largest energies generally occurring only at solar maximum, and even then less than once a year (e.g., Benz and Güdel 2010). The frequency distribution of flares can be described as a power law function of E_f as $N\left(E_f\right)dE_f \propto E_f^{\alpha}dE_f$, where dE_f is the respective energy range, and α is the number of flares per time within dE_f. For instance, Schrijver et al. (2012) found a value of $\alpha = -2.3 \pm 0.2$, and $N\left(E_f\right)$ and a break in the frequency spectrum for energies above $\sim 10^{25}$ J. Figure 8 illustrates different flare frequencies, including some from stellar flares (see below).

Flares are categorized by observations of the peak brightness in the X-ray range through the NOAA/GOES definition by letters from A ($< 10^{-7}$ W m^{-2}), B (10^{-7}–10^{-6} W m^{-2}), C (10^{-6}–10^{-5} W m^{-2}), M (10^{-5}–10^{-4} W m^{-2}) to X ($> 10^{-4}$ W m^{-2}) followed by numbers from 1 to 9.9, specifying the peak intensity within each range (e.g., Fletcher et al. 2011; Schrijver et al. 2012). They are partitioned into two distinct phases, the impulsive phase, and the gradual phase. The impulsive phase is the initial explosive release of energy mainly characterized by the hard X-ray (HXR; below $\sim$0.1 nm) enhancement, which generally lasts for a few minutes, while the gradual phase peaks shortly after the impulsive phase in soft X-ray (SXR; ~0.1–10 nm) and Hα emission and then gradually declines within up to several hours The impulsive phase is accompanied by strong enhancements in the continuum emission of UV and EUV. Observations of the respective UV continuum of solar flares between 100 and 300 nm—the relevant wavelength range for PSD, however, are surprisingly rare but do exist (e.g., Dominique et al. 2018).

The highest irradiation for flares can be found in the short wavelengths with flux enhancements of over 5 orders of magnitude during flares, while for $\lambda \sim 27$–120 nm it can still be as high as a factor of 100 during the impulsive phase for a few transition region emissions, but only up to a factor of 2 for the gradual phase (e.g., Woods et al. 2004, 2005, 2006). The longer wavelengths generally show lower variations, with the EUV (27–120 nm) and FUV (120–195 nm) having low to very low variations (Woods et al. 2004, 2006). For the X17 flare investigated by Woods et al. (2004), for instance, the EUV and FUV ranges only contribute to the change in total solar irradiance (TSI) by 1.4% and 2.3%, respectively. Woods et al. (2006) further found that for four X-class flares with a total energy of about 10^{25} J each, about 50% of the change in the total TSI is related to flux enhancements above 200 nm in the NUV, optical and infrared (Woods et al. 2006), which was later confirmed by, e.g., Warmuth and Mann (2016). For so-called white light flares that are visible in the optical Kretzschmar (2011) found that wavelengths below 50 nm only contribute between 10% and 20%, while the visible and NUV constitute the bulk of the released flare energy. However, an important part of this energy release might come from $\lambda > 300$ nm, since an increase of the emission from the Balmer continuum at 350 nm by a factor of 2.3 to 5.5 was measured during an X-class flare (Kotrč et al. 2016). The NUV wavelength range is the most relevant for PSD, thus only moderate increases in the PSD releases are expected as a direct result of flares.

Chamberlin et al. (2018) investigated the fluxes from solar flares during the intense September 2017 storm period and compared the radiation from the quiet Sun with the emission from several flares for $\lambda < 190$ nm. They did not find a strong increase for the UV from about 150 nm up to 190 nm. Heinzel and Kleint (2014) were the first to measure the increase in the Balmer continuum during a white-light flare between $\lambda \sim 280$–300 nm and found an enhancement of the observed X1 flare by 100%–200%. The first detection of a solar flare emission in the MUV around 200 nm was finally reported by Dominique et al. (2018), who found an insignificant increase at 190–220 nm of only 0.35% from $6.901 \cdot 10^{-5}$ J s^{-1} cm^{-1} to $6.926 \cdot 10^{-5}$ J s^{-1} cm^{-1}. Similarly, the Lyman α emission only increased by 0.97%. Figure 9 illustrates the flare investigated by Dominique et al. (2018).

Besides the moderate increase in the relevant wavelength range during a flare, there might be another effect related with flares that could increase the photon-stimulated desorption yield. Energetic flares, at least at the Sun, are often occurring together with CMEs (e.g., Schrijver et al. 2012). Since the photon-stimulated desorption yield increases with an increase in the incident particle flux (Mura et al. 2009; Sarantos et al. 2008, 2010), this could have a recognizable effect on PSD, for surface regions of airless bodies (i.e., Mercury, Moon, asteroids, etc.) that are not protected by a magnetic field.

Whether flares at other stars are also often related with CMEs is yet a matter of debate, and they might at least be restricted for magnetically very active stars (e.g. Moschou et al. 2019). Stellar flares, however, can reach energies of up to more than 10^{30} J (e.g. Wu et al. 2015) and were frequently observed on stars observed by the Kepler (Wu et al. 2015; Davenport et al. 2016) and TESS (e.g. Doyle et al. 2020; Günther et al. 2020; Howard et al. 2020) space missions. Flares with such high energies, also called superflares, are not restricted to

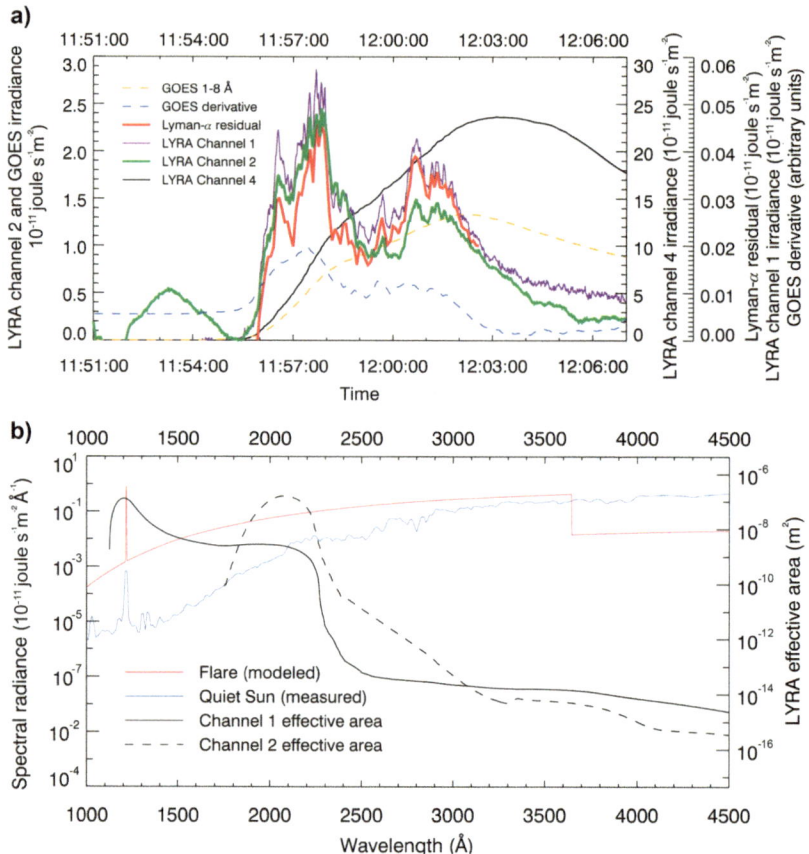

Fig. 9 A typical X-class flare as observed on 6 September 2017. Panel **a**) Solar irradiance during the X9.3 flare with the pre-flare irradiance subtracted, as observed by the GOES satellite (orange), by different channels of the Large-Yield Radiometer (LYRA). LYRA 1 shows 120–123 nm, LYRA 2 shows 190–222 nm, and LYRA 4 shows 0.1–20 nm. Panel **b**) shows the UV wavelength ranges as modeled and observed. Figure is reproduced from Dominique et al. (2018) with permission

magnetically active M-dwarfs (Yang et al. 2017) but can also frequently be found around K- and G-dwarfs (Maehara et al. 2012, 2015; Shibayama et al. 2013, Candelaresi et al. 2014a, 2014b; Namekata et al. 2017; Doyle et al. 2020). However, there seems to be a trend of stars with faster rotation periods ($P_{Rot} < 10$–12 days) exhibiting stronger superflares (e.g. Notsu et al. 2019; Howard et al. 2020), even though some were also found at stars rotating as slowly as the Sun (Nogami et al. 2014; see also Fig. 8). Shibata et al. (2013) further discuss the occurrence rate of superflares at the Sun, and found that superflares with an energy of $\sim 10^{27}$ J could also occur at the Sun about once in 800 years.

Such superflares might also have a pronounced effect on the planetary space weather of airless bodies. Welsh et al. (2006), for instance, investigated stellar flares in the UV (135 nm–275 nm) with NASA's GALEX satellite and found an average UV flare energy of $2.5 \cdot 10^{13}$ J, thereby increasing the UV irradiation of the respective stars by up to 2.7 magnitudes.

5 Electron Stimulated Desorption

5.1 Theoretical Description

Electron Stimulated Desorption (ESD) and PSD are considered to be very similar processes, where desorption of an atom or molecule from the surface is caused by an electronic excitation of the atom or molecule to be desorbed, leading to an anti-bonding state of the atom or molecule, and eventual release of it from the surface (Yakshinskiy and Madey 2000, 2003, 2004, 2005). The desorption cross-sections for ESD are higher than for PSD, for atomic Na it is $Q_i \approx 1 \cdot 10^{-19}$ cm^2 at 11 eV (Yakshinskiy and Madey 2000, 2005), and atomic K it is $Q_i \approx 2 \cdot 10^{-19}$ cm^2 at 11 eV (Yakshinskiy and Madey 2003). However, the ESD desorption cross-sections first increase and then decrease with increasing coverage of the alkali atoms on the surface, with a maximum ESD yield when 0.3 monolayer of the alkali metal is present on the surface. This yield dependence is explained by the formation metallic clusters on the surface at increasing alkali metal coverage. Also, energy thresholds were observed at 3 eV and 4 eV for Na and K, respectively (Yakshinskiy and Madey 2000, 2005).

Since ESD is a charge transfer process leading to electronic excitations similar to PSD, with comparable cross sections and comparable excitation threshold (Yakshinskiy and Madey 2000), the Velocity Distribution Function (VDF) of released Na are quite similar, so that ESD measurements can be substituted for the desorption processes caused by UV photons. The desorbing Na atoms were found to be 'hot' compared to the surface temperature, with suprathermal velocities and non-Maxwellian tails (Yakshinskiy and Madey 2000). The VDF of Electron Stimulated Desorption (ESD) from a lunar basalt sample was found to be similar to that of SiO$_2$, both for offset speed and the peak of the VDF (Yakshinskiy and Madey 2004), and similar to the VDFs measured for PSD. Therefore, we can use the same VDF as used for PSD, Eq. (21).

For the flux of atoms release via ESD for a species i we have to consider

$$\Phi_{ESD,i} = f_i N_S \int \phi_e\left(E_e\right) Q_i\left(E_e\right) dE \qquad (23)$$

where f_i is the fraction of species, N_S is the surface density of the regolith, $\phi_e\left(E_e\right)$ is the incident electron flux as function of electron energy, and $Q_i\left(E_e\right)$ is the ESD cross section. This equation cannot be easily simplified since typically there is a wide electron spectrum in the relevant space plasmas, as discussed below, that must be folded with the energy dependent ESD cross section.

5.2 Experimental Observation of ESD

In the laboratory, ESD is widely known for the release of atoms, molecules, and molecular fragments from a variety of adsorbate/substrate systems, see review by Ramsier and Yates (1991). For Mercury and the Moon again the experiments by Madey et al. (1998) and Yakshinskiy and Madey (1999, 2000, 2003, 2004, 2005) have been fundamental to the understanding of the ESD release process. These authors deposited Na or K atoms onto SiO$_2$ surfaces and lunar basalt and measured the ESD cross sections, the temperature dependence, coverage dependence, and the velocity distribution of released Na and K atoms. Among other things, they clearly demonstrated that also ESD is a non-thermal process, the desorption of an alkali atom is the result of an electronic excitation resulting from the absorption of an energetic electron at the surface. In addition to the release of adsorbed species

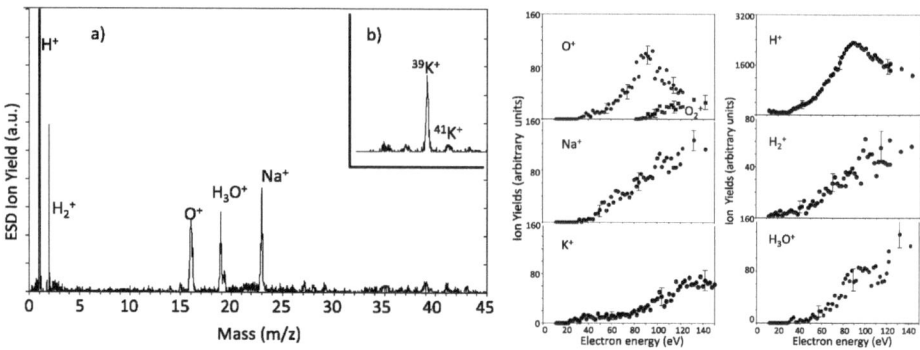

Fig. 10 Left: The 200 eV ESD mass-per-charge (m/z) spectrum of ion yields in arbitrary units (a.u.) obtained during pulsed electron beam irradiation of Na and K bearing glasses. Right: The H^+, H_2^+, H_3O^+, Na^+, K^+, O^+ and O_2^+ yields as a function of incident electron energy. Figure reproduced from McLain et al. (2011) with permission

from the surface by ESD, there are also few reports on the release of neutral atoms from alkali earth surfaces (Wurz et al. 1989, 1991) or ions from silicates (McLain et al. 2011).

ESD has been discussed as a potentially important mechanism for releasing material to the exospheres of the Moon and Mercury. When considered for the Na and K contribution to the lunar exosphere (Madey et al. 2002; Wilson et al. 2006) it was found that contributions ESD cannot be separated from PSD, given the much larger fluxes of solar UV photons compared to solar wind electrons, and the joint presence of relevant populations of electrons and ions in the magnetotail plasma. Similarly, for Mercury's exosphere ESD is considered as one possible source process contributing to the Na exosphere (McClintock et al. 2009), but again in competition with the PSD release process.

The ESD of ionic species from synthetic Na and K silicate glasses have been compared to data from the Fast Imaging Plasma Spectrometer (FIPS) on the MESSENGER spacecraft. From laboratory measurements of ESD ion yields it was concluded that most ions observed by FIPS (Zurbuchen et al. 2008, 2011) can be formed and released by initial single electron scattering events with simple mineral analogs, the from the glassy silicates (McLain et al. 2011). The yields of water group ions are also high, particularly at very low water coverage and the formation and release of O_2^+ (see Fig. 10, middle panel) and Si^+ are very low (McLain et al. 2011); requiring significant surface damage. These ions are less likely to be produced via ESD under the typical flux conditions on Mercury's surface. Thus, any signal at amu/e = 32 in FIPS the data is more likely due to S^+. This is consistent with the proposed stimulated desorption of S^+ from CaS and MgS (Bennett et al. 2016; Schaible et al. 2020).

The ESD ion yields exhibit significant and reversible dependencies as a function of the temperatures present on Mercury. This reversible temperature dependence can be explained by changes in the density and location of defect sites on the surface that become more ESD active, which is also consistent with lattice expansion and increased hole localization at surface defects or defect and vacancy diffusion to the surface. ESD ion yields as function of incident electron energy shown in Fig. 10, right panel, indicate that ion desorption proceeds via ionization of shallow core levels. It is well known that inner shell vacancies (holes) of these shallow core levels decay via Auger electron emission producing two-hole final states that Coulomb explode and eject ions with kinetic energies of several eV. Since the desorption of ions requires a two-hole localized state or an ionized state in the vicinity of an ionized defect or vacancy, increased energy localization and vacancy diffusion will result in increased cross-sections for ion desorption. Under conditions where the impinging electron energy exceeds 15–20 eV, direct ejection of ions or molecules via ESD is highly likely. This may be a higher probability process than neutral desorption followed by photo- or impact ionization and could be a dominant source term for populating exospheres of airless bodies such as the Moon and Mercury with complex ions. Interestingly, the cross sections for ionic desorption from regolith materials also depend on the amount of irradiation damage. A substantial increase in the Na^+ yield is observed with increasing electron dose due the creation of surface defect sites, and these "activated" surface conditions are expected to be present at the surface of the Moon and Mercury.

5.3 Electron Fluxes at Moon and Mercury

5.3.1 Electron Fluxes at the Moon

The Moon is exposed to variable electron fluxes during its orbit as demonstrated in Fig. 11a, which shows the mean electron energy flux at lunar distance as a function of energy and lunar phase compiled from 8 years of ARTEMIS data. The different plasma regions (solar wind, magnetosheath, and magnetotail) are clearly visible in the electron energy spectra. We note that the various magnetotail plasma populations (such as the plasma sheet and the tail lobes) are combined and averaged in Fig. 11a. Additionally, occasional solar transient events temporarily enhance the fluxes of energetic electrons. Most of these transients are smoothed out in the time averaged spectra, but some transients remain identifiable as vertical spikes during the solar wind phase. Figure 11b shows the mean electron energy spectra in units of distribution function for these different plasma environments, providing average electron spectra incident upon the upstream side of the Moon. The incident electron fluxes can be locally modified by multiple processes such as deceleration by the wake potential formed downstream of the plasma flow, deceleration or acceleration by the lunar surface potential, shielding by crustal magnetic fields, and wave-particle interaction. The black arrow and the dashed curve in Fig. 11b show a demonstration of electrostatic deceleration through a -200 V potential, resulting in a 200 eV energy shift to lower energies, conserving the distribution function.

5.3.2 Electron Fluxes at Mercury

The electron environment at Mercury is considerably less well characterized, because only two spacecraft have been able to make in situ measurements to date and electron detection has only been possible at high (>1 keV) electron energies.

During its three flybys of Mercury, Mariner 10's electron spectrometer made observations of several bursts of electrons with unexpectedly high energies. Simpson et al. (1974) reported fluxes of protons with energies of $\sim$550 keV and electrons with energies of $\sim$300 keV, which exceed approximately 10^4 and 10^5 cm^{-2} s^{-1}, respectively, that have been recorded in the magnetosphere of Mercury during the first Mariner 10 flyby. An alternate explanation of these bursts, suggesting that instrument pile-up was likely responsible, was published soon afterwards (Armstrong et al. 1975). This debate was reviewed in detail (Wurz and Blomberg 2001) but for about 30 years the Mariner 10 data were the only observations available of Mercury's plasma environment.

The MESSENGER mission carried an electron and particle spectrometer (EPS) with a lower detection threshold of 25 keV, which detected bursts of electrons with energies up to 100 keV (Ho et al. 2011a). Electrons of lower energies were detected by repurposing data

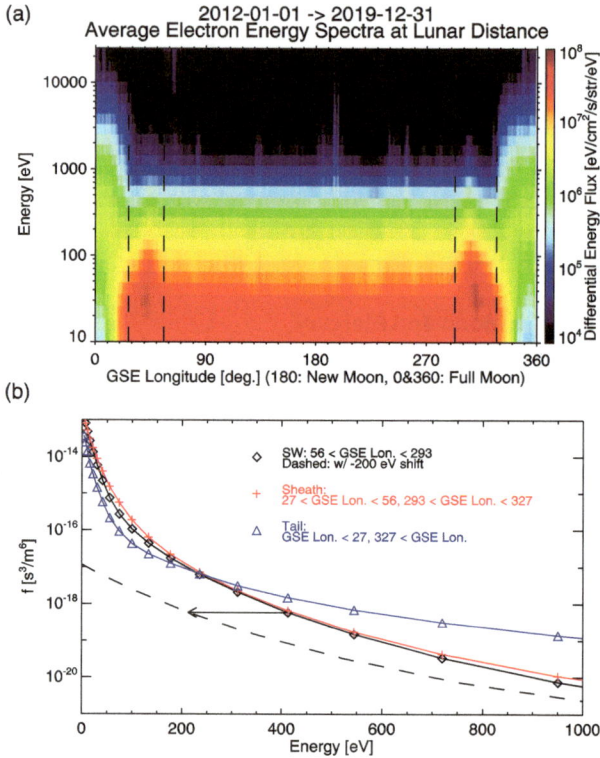

Fig. 11 (**a**) The mean electron energy flux at lunar distance as a function of energy and lunar phase. Spacecraft potential correction was performed in units of distribution function. Possible disturbances of ambient electrons by the Moon and lunar wake are filtered out to first order by using data acquired at $(X^2_{SSE} + Y^2_{SSE} + Z^2_{SSE})^{1/2} > 2R_L$ and $(X^2_{SSE} + Y^2_{SSE} + Z^2_{SSE})^{1/2} > 1.5R_L$. The dashed lines denote the locations of plasma boundaries (bow shock and magnetopause) observed in ion data (Poppe et al. 2018) for reference. (**b**) The mean electron energy spectra for the different plasma environments averaged over the indicated ϕ_{GSE} ranges. A constant background (equivalent to $7 \cdot 10^3$ eV/cm^2/s/str/eV) was subtracted before conversion to distribution function. The dashed curve shows the average solar wind spectrum with -200 eV energy shift, demonstrating electrostatic deceleration by typical wake and nightside surface potentials (e.g., Halekas et al. 2005, 2008a). The ARTEMIS data used to generate this figure are publicly available at http://artemis.ssl.berkeley.edu

from several other instruments, which observed effects driven by electron interactions with matter.

During every one of MESSENGER's three Mercury flybys in January and October 2008 and September 2009, the spacecraft's X-ray Spectrometer (XRS) observed in situ energetic electrons through X-ray fluorescence induced in the instrument's Mg and Al filters, and Cu collimators (Schlemm et al. 2007). These energy spectra are described by a kappa distribution peaking at 0.7–1 keV and with shape factor $\kappa = 7$–8 (Ho et al. 2011b). These electrons were consistently observed throughout MESSENGER's orbital mission, were located in latitudinal groups, and exhibit dawn-dusk asymmetries (Ho et al. 2016).

XRS also observed X-ray fluorescence emitted from the surface of Mercury, stimulated by electrons impacting the surface (Starr et al. 2012). These electrons impact preferentially on the dawnward sector of the nightside of the planet, in aurora-like patterns just equatorward of the open-closed field line boundary at both poles (Lindsay et al. 2016). The energy spectrum of these electrons cannot currently be well characterized, although their energies must exceed 1.9 keV and 4 keV to produce the observed X-ray fluorescence in Si and Ca.

MESSENGER's gamma ray spectrometer (GRS) detected electrons interacting with its anticoincidence shielding (Lawrence et al. 2015), and electrons originating from dipolarization events within the magnetotail (Dewey et al. 2017). These electrons have much higher energies than those detected by XRS either in situ or on the surface, with energies exceeding 100 keV; nevertheless, their spatial distribution matches well with those observed by XRS.

5.3.3 Comparison Between Moon and Mercury

The fundamental differences in electron behaviors and fluxes at the surfaces of Mercury and the Moon stem from the fact that Mercury has an intrinsic global magnetic field, while the Moon does not.

The surfaces of both Mercury and the Moon are accessible to electron impact without an intervening collisional atmosphere, although neither are simply exposed directly to the solar wind at all times in the manner of smaller planetary bodies. At Mercury the locations, energies and fluxes of electron impact are controlled by the structure and processes of the planet's magnetosphere, while at the Moon the dominant driver is the Moon's location within the Sun-Earth-Moon system, and transient variations in solar wind velocity and density. The electron environment at the Moon is also affected by the Earth's magnetosphere; it passes through Earth's current sheet and magnetotail, where the electron energy spectrum increases in energy and hardens. The bulk of the surface is accessible to electron impact, although local crustal magnetic anomalies are very likely to shield small parts of the surface from electron impact.

Acceleration of particles within the Mercury magnetosphere by reconnection and dipolarization events means that electrons impacting at its surface can have significantly higher energies than those typically seen at the Moon, although similar energized electron populations are observed as the Moon passes through Earth's magnetotail.

6 Sputtering by Ion Impact

6.1 The Velocity Distribution of Sputtered Particles

The impact of energetic ions, or neutral atoms, will cause the release of atoms and molecules from the top-most layers of a solid surface, even though the impacting particles will penetrate much deeper. This process is called sputtering, and has been studied in detail in solid state physics for many decades. Recent reviews on sputtering induced by ion bombardment are provided by Sigmund (2012) and Baragiola (2004). For most materials sputtering is the result of the nuclear interaction of the projectile ion with the target material (Betz and Wien 1994), which is the case for the rocks and regolith on Mercury's and the Moon's surface. However, for water also the electronic interaction between the projectile ion with the water ice results in sputtering, with significantly higher sputter yields at higher ion energies (Baragiola et al. 2003).

The sputter yield, Y, is the ratio of the flux of released atoms to the flux of impinging ions. The sputter yield in the nuclear interaction regime has a maximum at an energy around 1 keV/nuc of the impacting ions (Wurz 2012). Towards lower and lower energies, the sputter yield goes to zero because the deposited energy by the impacting ion is not sufficient to overcome the binding energy of atoms at the surface; for energies much higher, the sputter yield also goes to zero because the ions penetrate deeper into the solid without depositing significant energy at and near the surface to cause the release of particles. The energy of 1 keV/nuc corresponds to a velocity of about 440 km/s, which is a typical solar wind velocity. Thus, solar wind plasma, when impinging on a planetary surface, will cause sputtering.

The energy distribution for particles sputtered from a solid, $f(E_e)$, with the energy E_e of the sputtered particle, has been given as Sigmund (1969), Thompson et al. (1968):

$$f(E_e) = 2E_B \frac{E_e}{(E_e + E_B)^3} \quad \text{with} \quad \int_0^\infty f(E_e) \, dE_e = 1 \tag{24}$$

and is known as Sigmund-Thompson energy distribution. Written here in normalized form, where E_B is the binding energy of the sputtered particle, with E_B usually assumed to be the heat of sublimation. Note that the maximum of the energy distribution in Eq. (24) is at $E_{max} = E_B/2$. At higher energies the distribution falls off with E^{-2}, which was observed experimentally (e.g., Thompson et al. 1968; Husinsky et al. 1985). Since the binary collision between the impinging ion and the surface atom is the limiting case for the energetics of sputtering, this limitation has to be considered in the energy distribution (Wurz and Lammer 2003) and results in a cut-off of the energy distribution at higher energies:

$$f(E_e) = \frac{6E_B}{3 - 8\sqrt{E_B/E_{BC}}} \frac{E_e}{(E_e + E_B)^3} \left(1 - \sqrt{\frac{E_e + E_B}{E_{BC}}}\right) \quad \text{with} \quad \int_0^\infty f(E_e) \, dE_e = 1 \tag{25}$$

where the maximum energy, E_{BC}, that can be transferred in a binary collision is given by

$$E_{BC} = E_{in} \frac{4m_{ion}m_{surf}}{(m_{ion} + m_{surf})^2}$$

with m_{ion} the mass of the impacting ion, m_{surf} the mass of the sputtered atom, and E_{in} is the energy of the incident ion. Since in planetary sputtering the sputter agents are mostly H^+ and He^{++} ions, which have low mass compared to the species of a mineral surface, and typical ion energies are keV/nuc, the limit imposed by the maximum transferred energy has to be considered. At low impact energies E_{in} the energy distribution of sputtered atoms will deviate from the E^{-2} dependence, and will peak at lower energies, given by Eq. (25). This deviation of the energy distribution from the E^{-2} for low impact energies has been observed experimentally (e.g., Brizzolara et al. 1988; Goehlich et al. 2000).

The average release velocity, $\langle v \rangle$, is derived from the sputter distribution (Eq. (25)) as

$$\langle v \rangle = \frac{\int vf(v) \, dv}{\int f(v) \, dv} = \frac{1}{2} v_1^2 v_2 \left(\frac{-3v_1^2 + 5v_2^2}{(v_1^2 + v_1^2)^2} + \frac{3\arctan(v_2/v_1)}{v_1 v_2}\right) \tag{26}$$

with the abbreviations

$$v_1 = \sqrt{\frac{2E_B}{m_{surf}}} \quad \text{and} \quad v_2 = \frac{v_{ion}}{m_{ion} + m_{surf}}$$

The limit of the binary collision imposes a high energy cut-off of the energy distribution of the sputtered particles. Given that the solar wind plasma consists mostly of H and He, this high-energy cut-off is significant and has to be considered in the calculations.

The energy distributions of sputtered atoms from monoatomic samples are very well understood, there is good agreement between the theoretical formulation (Eq. (24)) and the experimental results, see for example reviews by Betz and Wien (1994) and Behrisch and Eckstein (2007). For multicomponent samples fewer experimental data are available. For metal alloys, there are small changes the surface binding energies of elements observed, as inferred from the peak of the energy distribution of sputtered species (Behrisch and Eckstein 2007). Also for alkali-halides and earth alkali-halides the energy distributions of sputtered atoms are described by typical binding energies, however, at elevated temperatures an additional thermal component is observed (Betz et al. 1987; Betz and Husinsky 1988). The most dramatic changes of the energy distribution are observed for oxidized surfaces where a broadening of energy distribution has been observed, which can be fitted with Eq. (24) or Eq. (25) very well, if a large E_B is used (Husinsky 1985). The larger E_B for oxides has been attributed to the larger binding energy of atoms in the oxide (Dullni 1984; Kelly 1986).

Not only atoms are sputtered by ion impact, but also polyatomic compounds and clusters. It is found that for sputtered metallic clusters the energy distributions peak at slightly lower energies than for atoms, and their energy spectrum falls off more steeply towards higher energies (Wahl and Wucher 1994; Behrisch and Eckstein 2007). From the sputter release the polyatomic compounds also have substantial internal temperatures, i.e., rotational and vibrational excitation, of several 1000 K (Behrisch and Eckstein 2007), which will result in the unimolecular decay of these polyatomic compounds and thus limit their life time in a planetary exosphere.

Sputtered ions, usually called secondary ions in the literature, are used in surface science for analytics of surfaces a lot, in Secondary Ion Mass Spectrometers (SIMS) for surface chemical analyses (see Benninghofen et al. 1987; van der Heide 2014). In addition to the physical release, i.e., the actual sputtering of the neutral atoms or ions, the sputtered ions can undergo a charge exchange process when leaving the surface, which is energy dependent, thus it modifies the ion sputter yield and the energy distribution. Faster ions survive more likely the charge exchange process, thus the resulting peak of the energy distribution of sputtered ions moves to higher energies. The variations of the secondary ion yield can span five orders of magnitude or more (see Benninghofen et al. 1987; van der Heide 2014). Since the sputter yield of ions is strongly correlated with the oxidation state of the surface (Wurz et al. 1990) the higher binding energy of oxidized species on the surface will also result in an energy distribution peaking at higher energies. The higher energies of sputtered ions are very well known in surface science community for decades (Benninghofen et al. 1987; Chatzitheodoridis and Kiriakidis 2002; van der Heide 2014). The energy distribution of sputtered ions can be written as

$$f(E_e) \propto \frac{E_e}{\left(E_e + E_{B,i}\right)^{3-2x}} \tag{27}$$

with $E_{B,i}$ the releavant binding energy to characterize the energy distribution of the sputtered ions, and x a numerical value less than 0.15 characterizing the screened Coulombic interaction potential (van der Heide 2014). The observed $E_{B,i}$ are often higher than their neutral counterparts (e.g. Pahlke et al. 1982; Tolstogouzova et al. 2002). The exponent in the denominator of Eq. (27) being less than 3 characterizes the wider tails in the energy distribution of sputtered ions compared to sputtered neutrals (see Eq. (24)). The higher energy of sputtered ions has been observed also in studies relevant to planetology (Dukes and Baragiola 2015).

The angular distribution of the sputtered particles is given in general (Wurz and Lammer 2003) as

$$f(\alpha) = \frac{n\Gamma(n/2)}{\sqrt{\pi}\Gamma\left(\frac{1+n}{2}\right)}\cos\alpha^n \text{ with } \int_{-\pi/2}^{+\pi/2} f(\alpha)\,d\alpha = 1 \tag{28}$$

with α the polar angle, and symmetry in the azimuth angle is assumed. This assumption is well justified because exospheric observations are at a much larger distance from the surface than the typical grain size of regolith particles, thus asymmetries in the azimuth distribution, which do exist (Betz and Wien 1994), are averaged out very well. The exponent n, a real number, is a matter of debate in the planetary science literature. In laboratory measurements of polycrystalline samples often values larger than 1 are found (Betz and Wien 1994), e.g. $n = 2$ by Hofer (1991), or a range of n (Ait El Fqih 2010). However, for modelling of sputtering of planetary surfaces, i.e., the porous regolith with its microscopic roughness, $n = 1$ is often used (Cassidy and Johnson 2005), which gives $f(\alpha) = \cos\alpha$.

6.2 The Total Sputter Yield

The total sputter yield, Y_{tot}, is the ratio of sputtered atoms and the incoming ions. The total sputter yield can be calculated analytically (e.g. Behrisch and Eckstein 2007; Lammer et al. 2003; and references therein). The total sputter yield, Y_{tot}, is given by (Sigmund 1969)

$$Y_{tot} \approx \frac{3\alpha}{\pi\sigma_D E_b} \frac{\left(\frac{2m_{SW}}{m_{SW}+m_{surf}} Z_{SW} Z_{surf} e^2\right)^2}{\gamma E_{in}} 2\varepsilon s_n(\varepsilon) \tag{29}$$

where $s_n(\varepsilon)$ is the nuclear elastic stopping cross-section at the reduced energy, ε of the incident ion, E_b is the surface binding energy, σ_D is the average diffusion cross-section, e is the elementary charge, and α is a collision parameter. The reduced energy, ε, is given by

$$\varepsilon = \frac{\gamma E_{in}}{2e^2} \frac{1}{\frac{2m_{SW}}{m_{SW}+m_{surf}} Z_{SW} Z_{surf}} \frac{0.8853 a_0}{Z_{SW}^{0.23} + Z_{surf}^{0.23}} \tag{30}$$

where E_{in} is the energy of the incident particle, m_{SW} is the mass of the incident ion, m_{surf} the mass of the sample atom, and $\gamma = \left(4m_{SW}m_{surf}\right)/\left(m_{SW}+m_{surf}\right)^2$, Z_{SW} and Z_{surf} are the nuclear charges of the incident and sample particles, respectively, and a_0 is the Bohr radius of the hydrogen atom. The nuclear elastic stopping cross-section, $s_n(\varepsilon)$, is given for $\varepsilon < 30$ as

$$s_n(\varepsilon) = \frac{1}{2\varepsilon} \frac{\ln(1 + 1.138\varepsilon)}{1 + 0.0132\varepsilon^{-0.787} + 0.196\varepsilon^{0.5}} \tag{31}$$

and for $\varepsilon > 30$ as

$$s_n(\varepsilon) = \frac{1}{2\varepsilon}\ln(\varepsilon) \tag{32}$$

The analytical expression for the total sputter yield (Eq. (29) through Eq. (32)) contains parameters from the incident ion (E_{in}, m_{SW}, Z_{SW}), which are well known, parameters from the sample atoms (m_{surf}, γ, Z_{surf}), which are also well known, and parameters from the solid that is formed from the sample atoms (E_b, σ_D, α, s_n). The latter parameters are more difficult to obtain, and some are not known at all from experiments. An important parameter of this group is the surface binding energy E_b that inversely affects the sputter yield (Eq. (29)) for which often the sublimation energy of the solid is taken. A problem with this theoretical formulation is that it is only given for single element solids. Compounds, which are the normal samples of interest in planetary science, are not covered by this formulation. Another problem are the values of the surface binding energy for the elements contained in compound samples, like minerals, which are not known. This is also a problem for the numerical methods discussed below. To obtain these binding energies for Na Morrissey et al. (2022) performed molecular dynamics calculations of sputtering of Na bearing minerals to derive these binding energies and obtained a range from 2.6 to 8.4 eV for the studied minerals. Considering only minerals that are part of mineral groups relevant to Mercury, jadeite (pyroxene), albite (feldspar) and nepheline (feldspatoid), reduces binding energies to range from 4.8 to 8.4 eV. These minerals are the Na-endmembers of their respective mineral group and therefore very unlikely to appear on the lunar or hermean surface. As the mineral structure of the less Na-bearing pyroxenes, feldspars, and feldspatoids are similar to their Na-endmember, the binding energies could be applicable for less Na-rich minerals, but the binding energies from molecular dynamics likely express an upper limit to their respective mineral group. Moreover, the calculated surface binding energies are for the perfect crystal, but the surfaces of the regolith grains are severely space weathered, such that the top layer of about 100 nm has amorphous fractions, nano-phase iron, and agglutinates (Pieters and Noble 2016). Thus, more work on these surface binding energies is needed.

Modern computer codes give more accurate results for the sputter yield, especially considering compound materials. For example, the TRIM software, today known as SRIM (Ziegler et al. 2010), which uses the Binary Collision Approximation (BCA) to calculate the interaction of the projectile atoms with the atoms of the sample material. TRIM is often used in simulations for planetary science problems because it is simple to use. SRIM includes a graphical user interface that allows for anyone with knowledge of the system they wish to simulate to run a TRIM simulation. It is a relatively old code, based on earlier versions of TRIM (Biersack and Haggmark 1980), and is very well tested against experiment and other simulation methods (Ziegler et al. 2010), however, is not the most accurate BCA code that is available. While the simulation tools TRIM and SDTrimSP have shown good accuracy for yields from monatomic substrates, simulation inputs for sputtering from compounds requires further research.

Because the sputter yield, Y_{tot} is a fundamental parameter for the quantitative interpretation of measurements in the exosphere we briefly review a few of the established tools that can be used to derive these parameters. The TRIM software is publicly available and has been widely used in a planetary context to estimate sputter yields of mineral surfaces since relevant laboratory experiments are scarce (see Sect. 6.8). The few available laboratory results on planetary analog surfaces do indicate, however, that TRIM overestimates sputter yields for light ions impacting composite materials at solar wind energies because of unknown surface binding energies and because TRIM does not account for composition changes in the surface introduced by sputtering (Schaible et al. 2017; Szabo et al. 2018, 2020a). To correct for these limitations, improved versions of TRIM have been developed in the past, the TRIDYN code (Möller and Eckstein 1984), and the SDTrimSP, SDTrim-2D, and SDTrimSP-3D codes (Mutzke et al. 2009, 2011, 2019). The SDTrimSP code predicts the stopping and range of ions in matter as in SRIM but considers the change in composition of the kinetically sputtered sample. The 2D version further considers surface roughness whereas the 3D one introduces also spatial variability of concentrations (Stadlmayr et al. 2018; von Toussaint et al. 2017). The OKSANA (Shulga 2018) code addresses the same limitations of TRIM as SDTrimSP does. Cupak et al. (2021) found that the governing parameter for description of the sputtering behavior is the mean value of the distribution of surface inclination angles, rather than the commonly used root mean square roughness. This finding is the basis for an analytical treatment of the effect of surface roughness on the sputter yield (Szabo et al. 2022).

6.3 Released Particle Flux by Sputtering

Combining the total sputter yield with the angular and energy distribution, we obtain

$$Y(E_e, \alpha) = Y_{tot} \cdot f(E_e) \cdot f(\alpha) \tag{33}$$

With the sputter yield, Y_{tot}, we can proceed to calculate the sputtered flux, Φ_{sp}, resulting from an ion flux, Φ_{ion}, impinging onto the surface:

$$\Phi_{sp} = \Phi_{ion} Y_{tot}$$

For a species i in the surface material we get its sputter flux as

$$\Phi_{sp,i} = \Phi_{ion} Y_{tot,i} = \Phi_{ion} C_i Y_{rel,i} \tag{34}$$

with C_i the concentration of species i on the surface and $Y_{rel,i}$ the relative sputter yield. The sputtered flux, $\Phi_{sp,i}$, released from the surface results in a density profile in the exosphere, which can be observed remotely with optical telescopes providing line-of-sight column densities, or in situ by mass spectrometers providing local densities of species. Having measurements of densities in the exosphere one can invert Eq. (34) to obtain the concentration of a species on the surface

$$C_i = \frac{1}{\Phi_{ion}} n_i(r) \left(\frac{n_i(0)}{n_i(r)} \right) \langle v_i \rangle \frac{1}{Y_{rel,i}} \tag{35}$$

where Φ_{ion} is the flux of ions onto the surface during the observation (obtained either from a measurement or a model), $n_i(r)$ is the measured density of species i at an altitude r, the ratio $(n_i(0)/n_i(r))$ relates density to the surface using a model, and $\langle v_i \rangle$ is the average velocity from the release process (Eq. (26)). Similarly, one can use the measurement of a column density to derive the concentration of a species on the surface

$$C_i = \frac{1}{\Phi_{ion}} NC_i(r) \left(\frac{n_i(0)}{NC_i(r)} \right) \langle v_i \rangle \frac{1}{Y_{rel,i}} \tag{36}$$

where $NC_i(r)$ is the measured column density of species i at an altitude r, and the ratio $(n_i(0)/NC_i(r))$ relates the column density to the surface density using a model.

Since the sputtered flux reflects the bulk composition of the solid quite well, after an equilibrium has been established (Behrisch and Eckstein 2007), the measurement of species in the exosphere resulting from sputtering can be used to infer the chemical composition of the surface.

In practice this deriving the surface concentrations is difficult because of usually there is an unknown mix of release processes that contribute to a given density of an exospheric species. In addition to the mix of processes, there are uncertainties in the flux of ions, the sputter yield of the compound(s) at the surface, a possible latitude and longitude dependence of processes, and the actual global distribution of the exosphere. Therefore, one has to select observations carefully where one or several release processes can be excluded from contributing significantly to the exospheric signal, and assure that the necessary input parameters for the inversion are available, with the selection being different for different species.

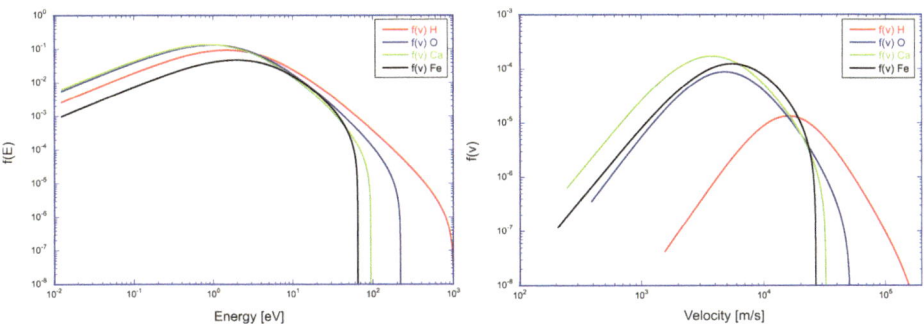

Fig. 12 Left: Sputter energy distribution for H, O, Ca, and Fe from Eq. (25) using a solar wind speed of 440 km s^{-1}. Right: Sputter velocity distribution for H, O, Ca, and Fe

6.4 Exospheric Escape of Sputtered Particles

Figure 12, left panel, shows the energy distribution for sputtered atoms according to Eq. (25) for H, O, Ca, and Fe atoms. Clearly the effect of the high-energy cutoff can be seen, which truncates the energy distribution to lower energies for higher mass atoms. For calculating the escape, it is more convenient to look at the velocity distribution of sputtered atoms, which is shown in Fig. 12, right panel. Because sputtering imparts comparable energies to the different sputtered atoms, they disperse when plotted in velocity space. For the escaping fraction we need to estimate the part of the VDF above the escape speed, which is for Mercury $v_\infty = 4250$ m s^{-1} and for the Moon $v_\infty = 2375$ m s^{-1}.

Figure 12, right panel we immediately see that a considerable fraction of the sputtered atoms is escaping, for Mercury and more so for the Moon. For Mercury, escape fractions range from 31% for Fe to 65% for O, and for the Moon from 90% for Fe to almost 100 % for the light species. To first order there is a mass dependence resulting from the cutoff energy and of course the mass, and to second order there is a species dependence because of the energy distribution is also a function of the binding energy (Eq. (25)). In addition to the direct losses from sputtering, there is also photoionization of the species contributing to the escape.

Note that the energy distributions for the species depend on the binding energy of the atoms on the surface (see Eq. (24) and Eq. (25)), and thus also the calculated escape fractions depend on the binding energy. The binding energies for all species for the range of minerals on the lunar and hermean surface are not known and often binding energies for mono-elemental solids are used or other approximations. Morrissey et al. (2022) performed molecular dynamics calculations of sputtering of Na bearing minerals to derive these binding energies as discussed above. Using these binding energies, they calculated also escape rates for Na, and got similar escape rates as mentioned above. Since they used Eq. (24) rather than Eq. (25) for the energy distributions, their escape rates for Na tend to be somewhat too high.

Comparing the Moon and Mercury, we notice a significant difference: whereas most sputtered atoms escape from the lunar exosphere, a considerable fraction of the sputtered atoms fall back onto Mercury's surface. Since the escape on Mercury is species dependent there will be a change in Mercury's surface composition such that light species are preferentially lost, in particular oxygen, leading to a chemical reduction of the visible surface. Thus, the space weathering is different on the Moon and Mercury.

6.5 Ion Fluxes at Mercury

As described above, plasma regularly impacts Mercury's surface that is primarily by controlled magnetic reconnection and the dynamics of magnetospheric circulation. Early in the MESSENGER mission, Winslow et al. (2014), estimated impact fluxes of protons in the northern and southern magnetospheric cusps as part of proton reflectometry study which estimated the surface magnetic field in those places. Using averages for loss cone and proton properties (density and temperature), they obtained fluxes of $9.8 \cdot 10^3$–$3.9 \cdot 10^8$ cm^{-2} s^{-1} impacting on the surface. Those authors also made a rough estimate of the ion impact flux in the southern cusp, finding about 4 times the northern value, which is a result of the northward shift of the magnetic dipole (Anderson et al. 2012). MESSENGER's highly elliptical orbit with northern hemisphere periapsis, kept it much farther away from the southern cusp than the $\sim$200–1000 km altitudes of northern cusp crossings. This estimate is consistent with one made from small, plasma-filled magnetic structures termed plasma filaments, that were observed passing through the northern cusp (Poh et al. 2016). They estimated the total rate of impacts on the surface from all cusp filaments to be $(2.70 \pm 0.09) \cdot 10^{25}$ s^{-1}. With the entire mission dataset available, Raines et al. (2022) located about 2800 cusp crossings from plasma data enhancements, out of the 4106 orbits containing plasma data. They computed orbit-by-orbit estimates of proton impact flux in the northern cusp ranging from 10^4 cm^{-2} s^{-1} to 10^8 cm^{-2} s^{-1}, with an average of $\sim$1 $\cdot$ 10^7 cm^{-2} s^{-1}. They found that impact fluxes varied on timescales as low as the 10 s plasma measurements, with peaks up to $\sim$1 $\cdot$ 10^9 cm^{-2} s^{-1}. Applying the same technique to the plasma sheet horn, preliminary estimates of proton impact fluxes in the range of 10^4–10^7 cm^{-2} s^{-1} (Raines et al. 2022).

Model estimates of plasma impact fluxes on the surface are at the higher end of the observed range. Kallio and Janhunen (2003) used their hybrid model to simulate proton impact across Mercury's entire surface for a wide range of solar wind conditions. They found ion fluxes impacting on the surface ranging 10^7–10^9 cm^{-2} s^{-1} across the dayside, concentrated in the cusp region in most cases, and about an order of magnitude lower fluxes on the nightside, along the open-closed field line boundary. MHD simulations (e.g., Benna et al. 2010) and test particle calculations in analytical fields (Massetti et al. 2003) have reported similar ion fluxes to the surface. Solar wind conditions mainly changed the area over which the highest fluxes were spread, with strongly southward IMF B_Z conditions opening the entire dayside to direct impact by the solar wind. These conditions may have been observed by MESSENGER during intense CMEs (Slavin et al. 2019; Winslow et al. 2020) but only in less than 10 of the over 4100 orbits. A later hybrid modelling study was able to reproduce the relative insensitivity of dayside reconnection at Mercury to the IMF direction (Fatemi et al. 2020), producing fluxes in the same range for both fully southward and northward IMF configurations. A unique aspect of Mercury's interaction with the solar wind arises from the large ratio of the scale of the planet to the scale of its magnetosphere and the presence of a large-size core composed of highly conducting material. This results in strong feedback between the induction field of the planetary interior and the magnetosphere. Especially under conditions of strong external forcing the global magnetospheric structure is affected changing the extent to which the solar wind directly impacts on the surface, which has been studied in recent MHD simulations (Jia et al. 2015; Dong et al. 2019).

Furthermore, Kallio et al. (2008) studied the impact of multiply charged solar wind Fe^{9+} and O^{7+} ions on Mercury's surface by using a quasi-neutral hybrid model. The results of their simulations showed that these heavy multiply charged ions impacted the surface non-homogenously. The highest flux was near the magnetic cusps, similar to the impacting solar wind protons. However, in contrast to protons, the multiply charged ions did not create high ion impact flux regions near the open-closed magnetic field-line boundary (Kallio et al. 2008), instead there is a dawn–dusk asymmetry and the total ion impact rates increased with respect to the increasing mass per charge ratio for ions. The reason is that the gyroradii for the highly charged ions become relevant with respect to the size of the magnetosphere, with the asymmetry resulting from the gradient and curvature drifts that drive positive ions clockwise when viewed from above the North Pole. This asymmetry indicates that Mercury's magnetosphere acts as a kind of "mass spectrometer" for heavy solar wind ions. Impacting multiply charged heavy ions are energy sources that can result in the generation of ion pairs, electrons and soft X-rays as well as EUV-photons in the upper surface layer. The associated energy release from the neutralization of the multiply charged heavy ions occurring at or in the vicinity of the surface can result in an additional release of ions, electrons, neutrals and photons from the planet's surface, which may contribute to its non-isotropic space-weathering.

6.6 Ion Fluxes at Moon

While the Moon is in the solar wind, it is exposed to the solar wind ions that are composed mainly ($\sim$90–95%) of $\sim$0.5–2 keV protons with minor contributions from multiply charged heavy elements, e.g., He^{2+}, O^{6+}, O^{7+}, Si^{8+}, Fe^{9+}, and others (e.g., Bame et al. 1975; von Steiger et al. 2000; Wurz 2005; Bochsler 2007; Kallio et al. 2008). The solar wind is mainly supersonic and super-Alfvenic and it usually has a non-Maxwellian velocity distribution function (e.g., Marsch et al. 1982; Demars and Schunk 1990; Matteini et al. 2012). The Moon also interacts with the terrestrial bow shock and magnetosheath during its inbound and outbound orbit of the Earth magnetosphere. Plasma composition in those regions is similar to those in the solar wind, but the plasma is highly thermalized (e.g., Halekas et al. 2015). In the magnetotail, the Moon is interacting with earthward and anti-earthward subsonic flow of the solar wind (i.e., H^+ and He^{+2}) and terrestrial ions (mainly O^+, O_2^+, and N_2^+) (e.g., Christon et al. 1994; Seki et al. 1996; Poppe et al. 2016b). The flux of these ions is several orders of magnitude smaller than the solar wind ion flux (e.g., Vaisberg et al. 1996; Poppe et al. 2016b; Artemyev et al. 2017). However, during intense geomagnetic storms and substorms, highly energetic ions have been observed at Moon's distance and beyond (e.g., Zong et al. 1998).

Outside the terrestrial magnetosphere, the Moon may also be exposed to the terrestrial foreshock ions (e.g., Gosling et al. 1978; Greenstadt et al. 1980), foreshock bubbles and cavities (e.g., Sibeck et al. 2002; Turner et al. 2013, 2020; Omidi et al. 2010, 2013; Archer et al. 2015), and hot flow anomalies (HFAs) (e.g., Schwartz et al. 1985; Schwartz 1995; Thomas and Brecht 1988; Eastwood et al. 2008; Zhang et al. 2013; Wang et al. 2013). The foreshock ions, which are the backstreaming particles from the terrestrial bow shock, form three different types of distribution functions: a reflected beam of narrow angular extent, as the kidney-bean shaped distribution, and as a nearly isotropic and diffuse distribution (see review by Eastwood et al. 2005, and references therein). Foreshock bubbles and HFAs have similar characteristics: they have low density plasma with low magnetic field strength in the core, surrounded by high density and thermalized plasma and enhanced field strength. The main difference between them is the HFAs are transient phenomena that move along the bow shock at the intersection between the bow shock and magnetic discontinuity (e.g., Turner et al. 2013).

Nearly 60 years ago, Luna 2 and Explorer X satellites provided the first observation of solar wind ion flux around the Moon in the range of 10^8 to 10^9 $cm^{-2} s^{-1}$ (Gringauz et al. 1961; Bridge et al. 1962). Since then, the solar wind plasma parameters around the Moon have been observed by several spacecraft (e.g., Ogilvie et al. 1996; Lin et al. 1998; Halekas et al. 2008b, 2011; Poppe et al. 2014, 2018). Poppe et al. (2018), have statistically provided the most complete picture of the ion fluxes around the Moon. Thanks to the long-period observations of lunar plasma and electromagnetic environment by the ARTEMIS dual-probe mission, Poppe et al. (2018), calculated mean ion energy flux in near-lunar space by averaging over 5 years of ARTEMIS observations. They separated the data set into periods with and without solar energetic particle (SEP) events. They found that SEP events comprised approximately 15% of the data set, while the remaining 85% of the time is without SEPs. Figure 13 shows the average ion energy spectrogram as a function of lunar Geocentric-Solar-Ecliptic (GSE) longitude, i.e., lunar phase. Figure 13a and Fig. 13b show the low and high energy ion spectrum for non-SEP times, respectively, and Fig. 13 c and Fig. 13d show the low and high energy ion spectrum for SEP times, respectively. The vertical dashed lines in Fig. 13a and Fig. 13b indicate the terrestrial bow shock and magnetopause boundary crossings (Poppe et al. 2018). As a result of the tidal locking of the Moon's rotation with its orbit, the highest cumulative solar wind proton implantation on the lunar surface is located on the lunar farside while the most energetic protons impact on the nearside, and the total ion impact rate was found to be smallest when the Moon is deep in the magnetotail (Kallio et al. 2019).

Figure 14 shows the ions' differential flux at different plasma environments the Moon encounters during its orbit around the Earth (Poppe et al. 2018). The averaged differential flux of all different environments (black curve with diamonds) shows the flux peaks at $\sim 10^4$ cm^{-2} s^{-1} sr^{-1} eV^{-1} near 600 eV. We see the ions cover a broad range of energies, but in general, the most dominant ion fluxes have energies between ~ 0.1–10 keV.

In addition to the different plasma regimes, lunar crustal magnetic fields (also known as crustal magnetic anomalies) add an extra level of complexity to the ion fluxes that impact the lunar surface. Lunar crustal fields are extensively spread over the entire lunar surface with various field intensities, but they are mostly clustered on the southern hemisphere of the lunar far side (Richmond and Hood 2008; Mitchell et al. 2008; Tsunakawa et al. 2010). The fields are mainly non-dipolar and have complex structures and the maximum strength of the fields on the lunar surface is expected to be at least a few hundred nanotesla (Dyal et al. 1974; Hood et al. 2001; Mitchell et al. 2008). Depending on the strength, geographical location, and the plasma properties that interact with the crustal fields, the access of the plasma to the lunar surface can be substantially altered.

In general, the access of the ions and electrons into the lunar nightside surface is considerably blocked by the plasma absorption on the lunar dayside (e.g., Lyon et al. 1967; Whang 1968; Bale 1997; Harada et al. 2010; Fatemi et al. 2012). However, different mechanisms may facilitate the access of the solar wind ions into the low altitude lunar wake, and eventually the lunar surface. Vorburger et al. (2016) observed that the ion impact on the lunar night side reaches from the terminator to up to 30° beyond the terminator. These mechanisms include (1) The gyrating solar wind protons enter the lunar wake perpendicular to the direction of the IMF as a result of ambipolar processes (Nishino et al. 2009a), (2) Scattered protons from the lunar day side are picked-up by the solar wind and enter deep into the wake (Nishino et al. 2009b; Dhanya et al. 2018), (3) The scattered protons at lower deflection angles on the day side are accelerated close to the polar terminator and enter the lunar night side perpendicular to the magnetic field lines (Wang et al. 2010), (4) The solar wind protons intrude into the wake along the magnetic field lines perhaps due to the ambipolar acceleration (Futaana et al. 2010), (5) The high energy solar wind protons from the tail of the proton velocity distribution function can enter deep into the lunar wake even during parallel IMF conditions (Dhanya et al. 2013). Furthermore, a small fraction of solar wind ions

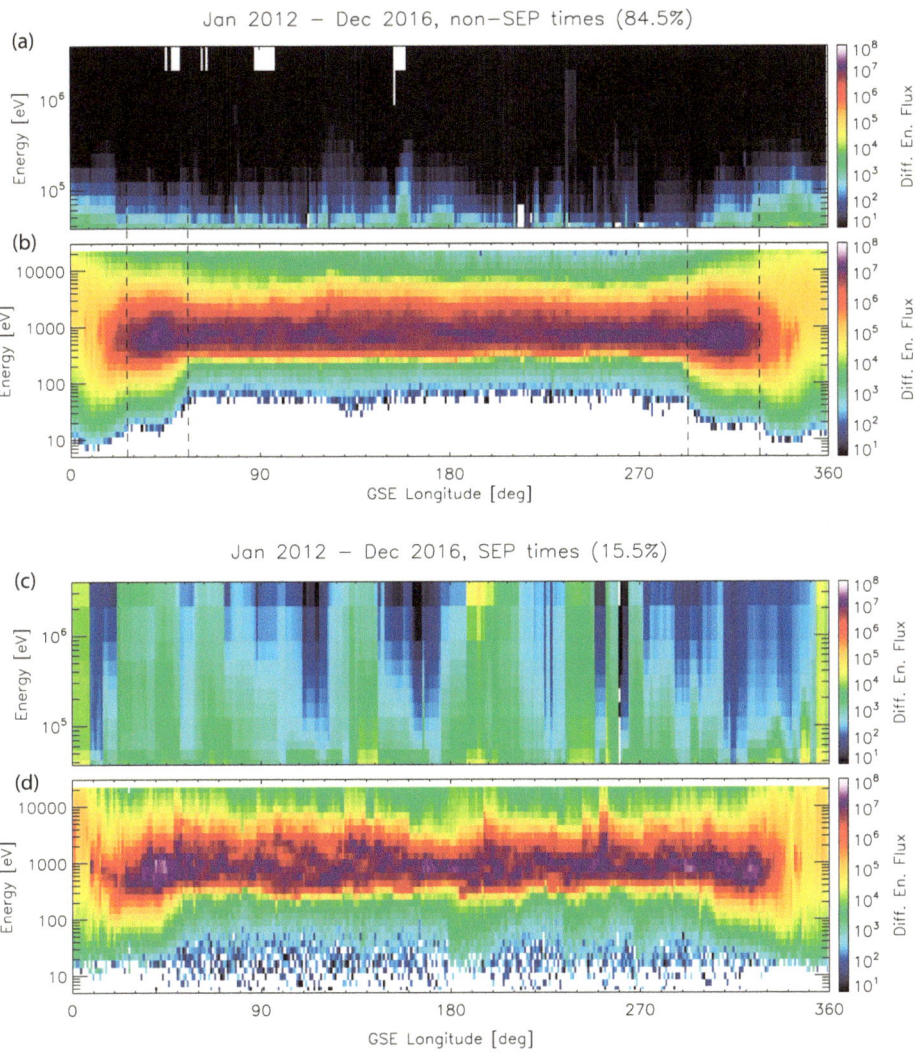

Fig. 13 The mean differential ion energy flux as a function of energy and lunar GSE longitude, where the longitude 0/360° is the "full moon" and 180° is the "new moon". Panels (**a** and **b**) are for non-SEP times and panels (**c** and **d**) SEP times. The dashed lines in panel a and panel b denote observed plasma boundaries (bow shock and magnetopause) as described in the text. Figure reproduced from Poppe et al. (2018) with permission

is reflected at the Earth's bow shock and accelerated to go back to the upstream region along the IMF forming the foreshock (e.g. Eastwood et al. 2005). The typical energy of the reflected ions ranges from several keV to MeV (which is included in the red curve in Fig. 14). When the Moon is located in the foreshock, these high-energy ions backstreaming from the bow shock can directly access the lunar surface (Benson et al. 1975; Nishino et al. 2017). The high-energy ion bombardment on the lunar surface may facilitate sputtering of volatile species there. We still do not exactly know which fraction of these ions impact the nightside lunar surface.

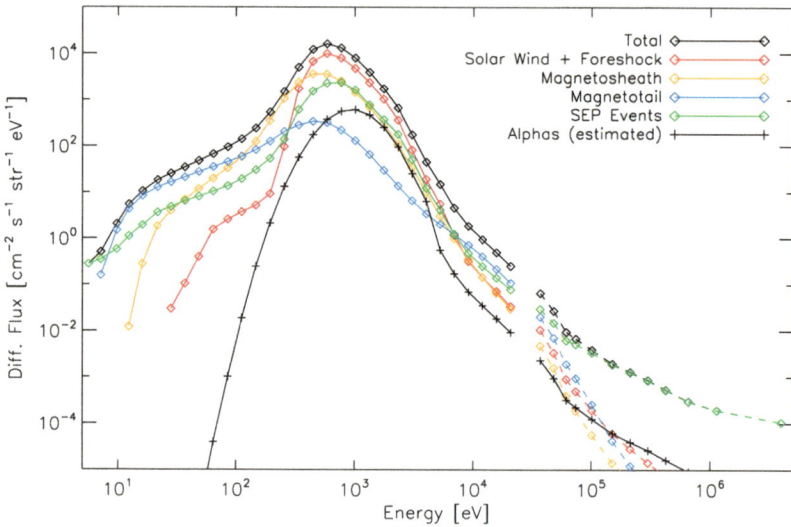

Fig. 14 The mean energy spectrum of ions at the Moon as calculated from the ARTEMIS observations (black diamonds). Colored curves denote the contributions to the mean flux from the solar wind and terrestrial foreshock (red), magnetosheath (orange), magnetotail (blue), and SEP events (green). The fraction of the observed flux due to alpha particles (He^{++}) is estimated as shown by the black crosses. The gap near 30 keV results from the transition of the ESA instrument (1–25,000 eV) to the SST instrument (30 keV–3 MeV). Figure reproduced from Poppe et al. (2018) with permission

6.7 Space Weathering and Sputtering by CME Plasma

Solar wind interacting with exposed surfaces produces energetic exospheric components of sputtered surface minerals (e.g., Lammer and Bauer 1997; Killen and Ip 1999; Wurz and Lammer 2003; Milillo et al. 2005; Killen et al. 2007, 2012; Pfleger et al. 2015), as discussed above. Comparing regular slow solar wind and CME exposure indicates strong enhancements of sputter yields caused by an increased abundance of He and O ions in the CMEs (Kallio et al. 2008). Killen et al. (2012) calculated that CME exposure can enhance the source rates of sputtered elements from the surface such as Ca or Mg from the Moon up to 50 times, depending on the CME plasma parameters, compared to normal solar wind sputtering. Additionally, these researchers found that the released surface minerals have a high probability of escaping the Moon by leaving its Hill sphere either via direct escape or by photoionization.

Due to Mercury's magnetosphere, the usual solar wind plasma is mostly deflected around the planet and does not reach the whole planetary surface on the dayside. During nominal solar wind conditions most of the release of atoms from surface minerals is from the cusp regions where the magnetic field is weakest and particle impact is common (Zurbuchen et al. 2011; Winslow et al. 2012, 2014; Raines et al. 2014; Poh et al. 2016). The release of sputtered atoms from the cusps via sputtering by solar wind ions has been modelled by Mura et al. (2005) in preparation of their direct measurement by the SERENA instrument on BepiColombo (Orsini et al. 2021). As one can see Fig. 15, when a CME collides with Mercury, the magnetosphere gets so compressed that the CME-plasma can reach the surface, releasing surface minerals more efficiently from the planet's surface than during quiet solar wind conditions. Winslow et al. (2017) estimated that during the passage of roughly 30% of CMEs, Mercury's dayside magnetosphere reaches the planet's surface, thereby leaving it

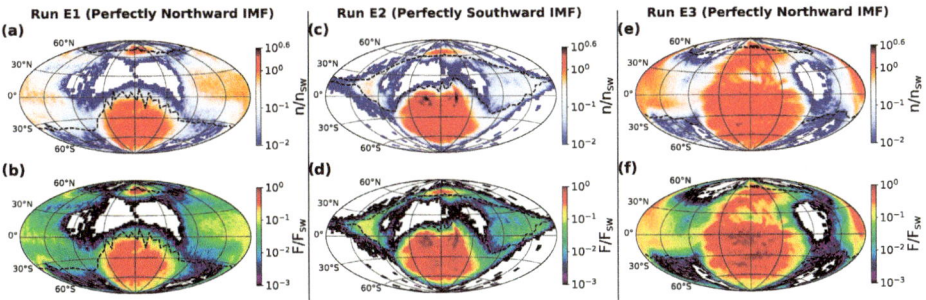

Fig. 15 Contour maps of the solar wind proton impact on the surface of Mercury from hybrid simulations. The density of impacting protons is normalized to the upstream solar wind density (panels **a–c**), and the flux of impacting protons normalized to the upstream solar wind flux (panels **d–f**). The subsolar point is at the center of each panel at latitude 0° and longitude 12 hr. The open-closed magnetic field line boundaries are calculated from magnetic field from the hybrid simulations and are shown by dashed black lines in each panel. For all panels, the IMF magnitude is $B_{IMF} = 18$ nT, solar wind speed is 600 km s^{-1}, and the Mach number is $M_S = 9.2$. For runs E1 and E2 the solar wind density is 70 cm^{-3}, the dynamic pressure is $P_{SW} = 42.2$ nPa, plasma beta $\beta = 2.1$, and Alfvenic Mach number is 12.8. For run E3 the solar wind density is 120 cm^{-3}, the dynamic pressure is $P_{SW} = 72.3$ nPa, plasma beta $\beta = 3.6$, and Alfvenic Mach number is 16.7. Figure reproduced from Fatemi et al. (2020) with permission

temporarily open to bombardment by CME plasma and the interplanetary medium. Slavin et al. (2019) selected four such events in the MESSENGER data and the flux transfer events associated with the passage of a CME. Winslow et al. (2020) provided a thorough analysis of MESSENGER FIPS plasma and the MAG magnetic field data during periods of CME plasma at Mercury, presenting strong evidence for the compression of the magnetopause toward the surface and a simultaneous enhancement in Na$^+$-group ion densities on the whole dayside. Photoionization of sputtered Na is effective (Wurz et al. 2019), allowing to observe this response in sputtering at the foot-point of the cusp. Orsini et al. (2018) compared ground-based Na images taken at the time of the MESSENGER FIPS and MAG data noticing that the Na exospheric emission during CME passage on 20 September 2013 extended over the whole dayside, while under nominal solar wind conditions the distributions mostly had the usual double peaks at mid latitudes.

Figure 16 shows modelled column densities of Mg and Ca (Pfleger et al. 2015). The column densities increase with increased IMF and plasma parameters. One can see that they become significantly larger for stronger particle exposures. In the latter scenario the entire dayside of Mercury experiences intense ion sputtering which results in column densities that are more than an order of magnitude enhanced in comparison with nominal solar wind conditions. Strong ion density enhancement associated with CMEs have been observed for Na-group ions (Winslow et al. 2020), but not for Mg and Ca ions. Such an efficiency enhancement of the sputter yield can be compared with the findings for the Moon by Killen et al. (2012) mentioned above.

Because the Sun was more active in its past, during the first hundreds of million years of their history Mercury and also the Moon experienced frequent CME, SEP and also flare events that may have led to a depletion of moderately volatile elements during their history (Saxena et al. 2019).

These authors reconstructed the possible CME frequency of the young Sun by using the rotation-flare rate relation for Sun-like stars observed by the Kepler satellite (Notsu et al. 2013) for three classes of assumed rotational histories of the young Sun. One should note that the rotational history of the Sun is unknown, but it would control the amount of magnetic

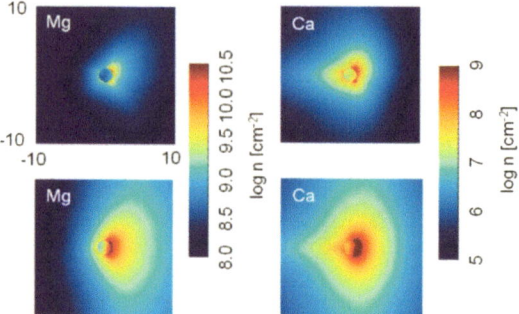

Fig. 16 Modelled column densities for the usual solar wind and CME exposure scenarios described in Fig. 15 obtained by passing the planet along a line parallel to the Sun-planet direction at 10 Mercury radii and projecting the density integrated along the lines of sight onto the noon-midnight plane of Mercury. Each image displays a region of 20×20 Hermean radii. Figure adapted from Pfleger et al. (2015) with permission

Fig. 17 Reconstructed CME collision number with Earth per day for three different young Sun rotational evolutions. CME-rates correspond to X10 and X100 flare energies and frequencies inferred from Kepler-based rotation-flare (Notsu et al. 2013) relationships. Figure reproduced from Saxena et al. (2019) with permission

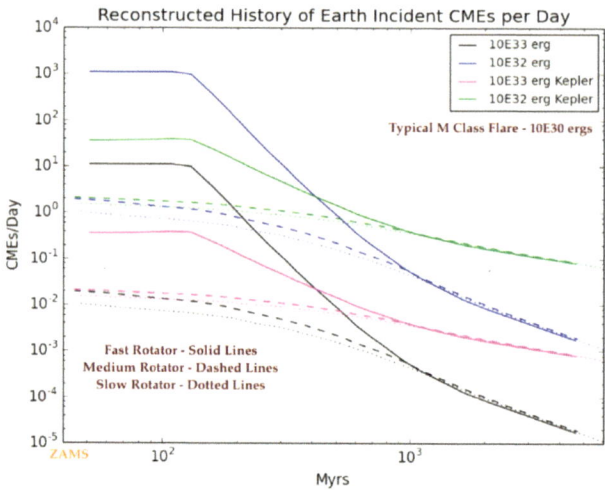

flux emitted and hence the flare and CME activity (Tu et al. 2015; Saxena et al. 2019). Figure 17 shows the reconstructed history of collisions of CMEs with the Earth over time for different evolutionary paths of the Sun (Saxena et al. 2019).

Saxena et al. (2019) concluded that the lunar crust may have stored space weather related evidence from the young Sun due to the expected energetic CMEs and SEPs that interacted with the Moon during the first few hundred million years after its origin. SEPs associated with these CMEs may have induced spallation, the formation of chemical elements from the impact of energetic particles, which could have been recorded in fission tracks, and should be searched for in future samples.

Enhanced spallation effects caused by energetic protons early in the history of the solar system have been discovered in meteoritic grains. High-sensitivity noble gas mass-spectrometric analyses of meteorite grains including solar flare heavy ion tracks show large enrichments of spallation-produced ^{21}Ne and ^{38}Ar when compared to non-irradiated grains from the same meteorite (Caffe et al. 1987; Kööp et al. 2018). These findings can be explained by solar flare irradiation in the early solar system with a proton flux several orders of magnitude higher than contemporary solar flares.

Saxena et al. (2019) suggest that it may be possible to constrain the activity and rotation history of the young Sun by careful analysis of specific samples from different regions of the Moon. We note that similar processes caused by an early period of an even more intense and increased space weather compared to that at the Moon's surface should also have affected Mercury. Orsini et al. (2014) estimated that Sun-induced erosion processes in early times, a combination of ion sputtering, PSD, and enhanced diffusion, caused an erosion of Mercury's top-20 m surface layer, thereby also depleting the moderately volatile element Na.

6.8 Laboratory Experiments for Sputtering of Planetary Surfaces

Laboratory experiments are crucial to verify theoretical predictions for sputtering and to provide models with the required input parameters for yield, angular distribution, energy distribution, and chemical composition of sputtered ejecta (see Sects. 6.1–6.3). First sputter investigations on lunar and lunar analogue material by solar wind ions date back to the early times of lunar exploration (Wehner et al. 1963a, 1963b; Wehner and Kenknight 1967) where lunar solar wind sputter rates between 0.4 and 1.1 Å/year, depending on material, were found. Later, also the ion emission due to solar wind ion impact on lunar analogue material was studied (Elphic et al. 1991), were the range of relative ion yields covered four decades of variation depending on sputtered ion species.

To perform sputtering experiments under laboratory conditions relevant to the surfaces of the Moon and Mercury, a well-defined sputter sample must be created which can then be irradiated. Two main types of samples have been developed by researchers over the past decades: Thin films (micrometers or thinner) that can be deposited on a microbalance and thick samples ($\gg$ micrometers) of loose mineral powder, mineral grains, or pressed pellets. Thin films have the advantage that the sputter yield can be directly measured as the mass loss rate from the microbalance (see Hayderer et al. 1999 for a description of the method) and the chemical and physical properties are easier to characterize. Studies relevant for the Moon and Mercury surface include e.g., Sporn et al. (1997), Küstner et al. (1998, 1999), Tona et al. (2005), Hijazi et al. (2014), Martinez et al. (2017), Hijazi et al. (2017), Stadlmayr et al. (2018), and Szabo et al. (2018). These studies are summarized in Table 4. Thick samples are closer to actual surfaces on the Moon and Mercury, offering the opportunity to directly compare the optical properties of the sample with space observations and to assess the influence of porosity, crystallinity and surface roughness on the sputtering process. Such approaches have been followed by e.g. Loeffler et al. (2009), Meyer et al. (2011), Dukes and Baragiola (2015), Kuhlman et al. (2015), Vyšinka et al. (2016), and Jäggi et al. (2021). The sputter yield from thick samples can only be measured indirectly by catching the ejected particles on a microbalance, in a mass spectrometer or studying them with an atomic force microscope (see e.g. Christoffersen et al. 2012).

The knowledge from sputter experiments relevant for the Moon and Mercury can be summarized in the following very general terms:

- Sputter experiments in laboratory with mineral samples SiO_2, $CaSiO_3$ (wollastonite), and $(Ca,Mg,Fe)_2Si_2O_6$ augite (Schaible et al. 2017; Szabo et al. 2018, 2020a, 2020b) indicate that TRIM overestimates the true sputter yield of solar wind ions sputtering the surfaces of the Moon and Mercury (see Sect. 6.2). The SDTrimSP code reproduces these experimental results better, but some remaining discrepancies indicate that physical processes are still missing in the numerical simulations. This point is illustrated in Fig. 18.

Table 4 Compilation of experimental studies conducted during the last 25 years that are directly relevant to the sputtering of the surface of the Moon and Mercury. This list is non-exhaustive of course

Sample	Sputter agents	Sputter energy	Measurement method(s)	Reference
SiO_2 grains	Ar^+	2 keV	Ion trap m/q and q determination	Vyšinka et al. (2016)
Si films	Mg, Ca, Cr	1 keV	Sputter depth profiling with white light interferometry	Baryshev et al. (2012)
$CaSiO_3$ film	He^{2+}, Ar^{q+}	1 keV/amu	Microbalance	Szabo et al. (2020a)
Fe film	Ar^+	500 eV	Microbalance, AFM to determine surface roughness	Stadlmayr et al. (2018)
$CaSiO_3$ film	Ar^{q+}	1 keV/amu	Microbalance	Szabo et al. (2018)
Glassy film of SiO-undersaturated feldspatoid	Xe^{15+}, Ni^{24+}	225 keV (low-energy example) to 630 MeV (high-energy example)	XY-TOF-SIMS	Martinez et al. (2017)
Anorthite-like glass film	H^+, He^{2+}, Ar^{q+}	0.25 keV/amu and 0.5 keV/amu	Microbalance, measure sputter yield	Hijazi et al. (2014, 2017)
Orthopyroxene	H^+, He^+	1 keV/amu	scanning transmission electron microscope and Electron energy loss spectra to study nanoparticle iron formation	Kuhlman et al. (2015)
Apollo soil samples	He^+	4 keV	Measured sputtered species and their energy distribution incl. Na^+, Mg^+, Al^+, Si^+, Ca^+, Ca^{2+}, Ti^+, Fe^+, and molecular NaO^+, MgO^+, and SiO^+ with mass spectrometer	Dukes and Baragiola (2015)
Na bearing tectosilicate films and Na layers on albite and olivine films	He^+	4 keV	X-ray photoelectron spectroscopy and secondary ion mass spectrometry to measure ionic composition of ejecta	Dukes et al. (2011)
Lunar regolith simulant	H^+, Ar^+, Ar^{6+}, Ar^{9+}	375 eV/amu	Quad. Mass spectrometer, measure sputter yield	Meyer et al. (2011)
Olivine grains	He^+	4 keV	Measure NIR spectra to study nanophase Fe	Loeffler et al. (2009)
SiO_2 and SiH films	I^{q+} ($q = 15$–20)	$E = q \times 3$ keV	TOF SIMS	Tona et al. (2005)
SiO_2 (among other samples)	Highly charged ions ($q > 9$)	Low energies	Various	Schenkel et al. (1999)
Isotropic and pyrolytic graphite	D^+	2 keV	Investigate dependence of Y on angle and surface roughness	Küstner et al. (1998)
SiO_2, LiF	Ar^{q+} ($q \leq 14+$), Xe^{q+}($q \leq 27+$)	$E = (5$–$20) \times q$ keV	Microbalance	Sporn et al. (1997)

- The sputter yield is $Y \sim 0.01$–0.1 atoms/ion for H^+ at solar wind energy, and $Y \sim 0.1$–1.0 atoms/ion for He^+ at solar wind (Schaible et al. 2017; Roth et al. 1979; Wehner and Kenknight 1967; Hijazi et al. 2017).
- Higher charge states of ions existing in solar wind increase the sputter yield due to potential sputtering. For He^{++} ions, this enhancement is roughly 50% compared to kinetic sputtering by He only (for wollastonite (Szabo et al. 2020a), for KREEP (Barghouty et al. 2011), and anorthite samples (Hijazi et al. 2017)). This is particularly important as He is the most abundant among multiply charged solar wind ion. For heavier multiply charged ions there is also a significant enhancement in the sputter yield because of potential sputtering, but the abundance of the heavy ions in the solar wind is too low to make a difference for the total sputter yield. Claims for significant, or even dominant, contributions to the sputter yield by multiply charged ions heavier than He (Shemansky 2003; Killen et al. 2004) are not compatible with the experimental data from the laboratory.
- In various experiments, the sputter yield has been measured and tabulated as a function of energy, species, inclination angle, and charge state of impactor, but the influence of the sample material properties (i.e., mineral composition, surface roughness, porosity, crystallinity, and temperature) on the sputter yield are not well constrained.
- The sputter yield is a strong function of the angle of incidence, see Fig. 18, however only for smooth surfaces. Since the regolith grains have a large roughness on the microscopic scale, actually all angles of incidence will occur, independent on the macroscopic angle of incidence of the ions. Thus, for the calculation of the sputter yield an angle of $45°$ should be chosen, which is representative for rough surfaces (Küster et al. 2000; Wurz et al. 2007).
- The energy distribution of neutral sputtered atoms from monoatomic substrates is well established (Betz and Wien 1994). For multi-component substrates less experimental information is available, see discussion in Sect. 6.1, and review by Behrisch and Eckstein (2007). Typical energies for sputtered atoms are a few eV, and typical energies for sputtered ions are about 10 eV (Benninghofen et al. 1987; Dukes and Baragiola 2015).
- The stoichiometry equivalence of ejecta and irradiated surface is attained in equilibrium because the composition of the sputtered flux has to match the bulk composition of the sample, otherwise the sample would be depleted in a species of disproportionally higher sputter yield (Behrisch and Eckstein 2007). This equilibrium means that the very surface, the top 1–3 atomic layers where the sputtered particles come from, have a different composition than the bulk of the sample. Before equilibrium is reached some species can have a disproportional sputter yield, such as sputtering of oxygen that is augmented by potential sputtering (Szabo et al. 2020a). Also adsorbed layers of material, like Na on minerals, might have a larger sputter yield than if these species would be sputtered from the mineral compound (Dukes et al. 2011).
- Changes in surface properties (porosity, roughness, composition, mineral phases) of analogue samples upon sputtering and the implication for remote sensing of planetary surfaces via reflectance spectroscopy can also be investigated in the laboratory (Jäggi et al. 2021).
- The effects of solar wind electrons (with bulk energy of about 5 eV) on mineral samples have not been investigated in laboratory conditions to our knowledge. However, laboratory studies at higher electron energies (tens to hundreds of eV) resembling electrons in Mercury's magnetosphere showed the release of H^+, H_2^+, O^+, H_3O^+, Na^+, K^+ and O_2^+ from silicate glasses, with the H- and O-bearing species from chemisorbed water on the surface (McLain et al. 2011).

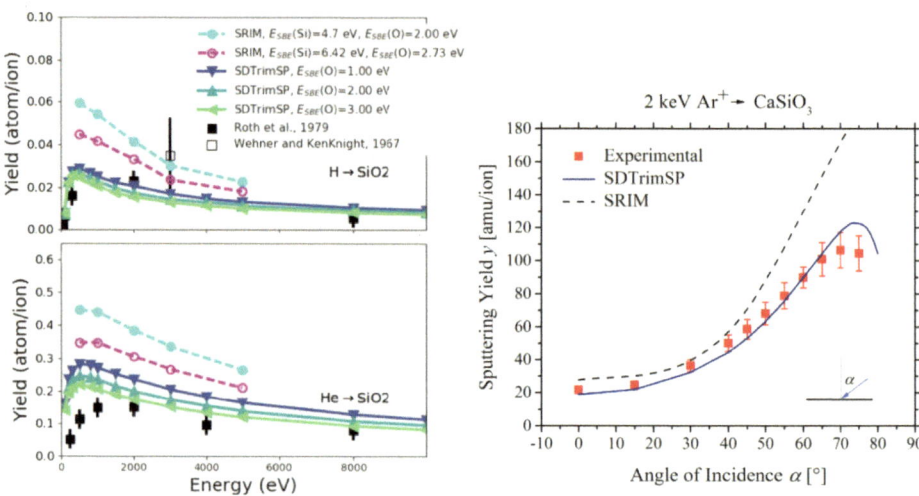

Fig. 18 Measured sputter yields versus numerical predictions. Left panel: Sputter yields of H (top) and He ions (bottom) irradiating a SiO$_2$ sample. Figure reproduced from Schaible et al. (2017) with permission. Right panel: Ar ions irradiating a CaSiO$_3$ sample as a function of ion incidence angle. Figure reproduced from Szabo et al. (2018) with permission

Although the principles of ion sputtering of minerals are known, there remain several open questions related to sputter processes on the surfaces of the Moon and Mercury, which must be examined in future laboratory experiments. These include determination of sputter yields for mineral species relevant for the Moon and Mercury that have not been investigated, in particular for volatile-rich minerals. Moreover, it should be investigated if and how the sputter characteristics of vapor-deposited mineral films (single minerals, compact, chemically pure) usually preferred in laboratory experiments differ from more 'realistic' samples (i.e., lunar regolith, breccias). On the other hand, more systematic assessments are needed on how accurate laboratory sputter yields can be predicted by numerical simulations in general. This is relevant because we will never have laboratory experiments for every possible configuration needed for surface and exosphere models.

7 Chemical Sputtering

We discussed physical sputtering above, where collisions between a primary particle and the atoms of the solid lead to a collision cascade with the release of atoms from the surface in direct consequence. In contrast, in chemical sputtering the surface is changed by the impact of a primary particle in such a way that an atom or molecule from the surface is released, or its bond to the surface is largely reduced so it can be set free by thermal desorption or by PSD. This can be accomplished by changing the chemical composition (the formation of radiolysis products), by changing the mineralogical composition, by the production of crystal defects and voids, and related processes (Haring et al. 1984; Winters and Coburn 1992). In the context of planetary science, in particular for the surfaces of Mercury and the Moon, Potter (1995) gave this example for chemical sputtering for a simple mineral, sodium silicate:

$$2H + Na_2SiO_3 \rightarrow 2Na + SiO_2 + H_2O \tag{37}$$

where the H atoms in Eq. (36) are solar wind protons being implanted in the near-surface volume, which leads to the formation of silicon oxide, water molecules and atomic Na. Note that the solar wind proton comes with about 1 keV of energy, which is much more than the typical binding energies of atoms in a mineral, facilitating rearrangement of atomic bonds and formation of relocations or dislocations of atoms in the mineral structure. Moreover, the proton is a chemical reactive species contributing to the changes in the mineral. However, the simple sodium silicate presented in Eq. (36) is not a likely mineral to be found on the surface of Mercury or the Moon, but Feldspar is (Wurz et al. 2010), and the chemical equation is then:

$$H + NaAlSi_3O_8 \rightarrow Na + AlO + 3SiO_2 + H_2O \tag{38}$$

again providing the atomic Na on the surface. Similarly, for potassium feldspar (orthoclase, $KAl_3Si_3O_8$) the analog chemical equation can be written with atomic K as one of the end products.

Note that the formulation of the process of chemical sputtering presented as a chemical equation in Eq. (36) and Eq. (37) is a severe simplification of the actual physical and chemical processes induced by the solar wind proton in the near surface volume of the mineral. Nevertheless, processes like this may provide the atomic Na (and K) on the surface needed for the PSD and ESD processes. As a side product, water molecules are produced from the implantation of the solar wind protons in sodium silicate, but the efficiency of this process is likely much lower than 1, with alternative pathways like the production of OH (Tucker et al. 2019).

In summary, chemical sputtering can free alkali atoms (perhaps also other species) from their chemical bounds in the mineral, either on the surface or in the near surface region within the penetration range of the particles. Solar wind protons penetrate about 30 nm depth, energetic protons of a few MeV penetrate to about 0.1 mm (which is a typical regolith grain size), and galactic cosmic rays with GeV energies penetrate to about 1 m. The freed alkali atoms, will diffuse to the surface, given the temperatures on the dayside of Mercury and the Moon. Diffusion of Na and K from the interior to the surface was considered already a while ago to explain the Na and K exosphere observations (Cheng et al. 1987; Tyler et al. 1988). However, the diffusion of Na and K in crystalline matter if not enough by orders of magnitude, and only diffusion enhanced by defects, cracks, voids, ... resulting from particle irradiation and micro-meteorite gardening, can explain the observed exospheric densities (Sprague 1990). These alkali metals are the needed supply for the PSD (and ESD) process of releasing Na and K into the exosphere. This concept was successfully used to quantitatively explain the Na observations during Mercury transit in 2004 (Mura et al. 2009).

Alternatively, based on correlations of the ion flux impacting on the surface and the exospheric neutral Na density, Sarantos et al. (2008) concluded for Na observations during the Moon's passage through the magnetotail plasma of the Earth, that the defects created in the surface crystals resulting from the ion bombardment were responsible to enhance other release processes, without invoking the concept of chemical sputtering. As discussed above, these other processes are PSD and thermal desorption. Earlier, Wilson et al. (2006) speculated on a connection between the plasma impact on the lunar surface and the Na in the exosphere observed at some delay.

8 Particle Reflection from Planetary Surfaces

Particle scattering from surfaces has been studied in the laboratory for many years (Niehus et al. 1993). From laboratory experiments it is known that when the ions impact on a surface, a substantial fraction of the ions lose their energies and get absorbed and implanted into the surface. However, depending on the surface composition and roughness as well as the energy of the incident ions, a fraction of them can be backscattered in the form of negative, positive, and/or neutral state particles (McCracken 1975; Niehus et al. 1993; Woodruff 2016). In particular, significant particle reflection is observed for shallow angles of incidence. To first order, the interaction is a binary collision between the projectile ion and the surface atom, moderated by the presence of other atoms on the surface (crystal structure, shadowing, electronic interaction, and other effects). The surfaces used in the laboratory studies were single crystal surfaces, or at least highly polished samples, thus they differ dramatically from the surfaces on the Moon or Mercury made up by regolith, which makes laboratory data often not applicable to the situation on a planetary surface. Thus, we must resort mostly to reports observations from space missions.

Particle reflection from planetary surfaces has been almost exclusively observed for the Moon, because only lunar missions carried the necessary instrumentation for such studies. Particle reflection was observed by the Kaguya, Chandrayaan-1, and IBEX missions, where reflected solar wind ions as well as neutralized solar wind has been reported. In late 2025, the BepiColombo mission will enter Mercury orbit. It also carries instruments for the detection of energetic neutral atoms (Saito et al. 2021; Orsini et al. 2021) so that similar studies will be executed at Mercury. Fortunately, what was learned from the lunar studies on particle reflection applies to other planetary bodies without an atmosphere, for example for Mercury (Lue et al. 2017).

Analysis of observed ENA data from the Chandrayaan-1 mission showed that the directional ENA flux $j_{ENA}(SZA, \theta, \varphi)$ can be described as the product of the solar wind flux j_{SW}, the reflection ratio for perpendicular incidence R_T, and the directional scattering function flux $f_S(SZA, \theta, \varphi)$:

$$j_{ENA}(SZA, \theta, \varphi) = j_{SW} \cdot f_S(SZA, \theta, \varphi) \cdot R_T \qquad (39)$$

where φ is the scattering azimuth angle and θ is the scattering polar. R_T is a property of the surface and has to be found experimentally (Vorburger et al. 2013).

Note that while $j_{ENA}(SZA, \theta, \varphi)$ denotes the ENA flux scattered in one angular direction, $j_{ENA}(SZA) = \int j_{ENA}(SZA, \theta, \varphi) d\theta d\varphi$ gives the total ENA flux scattered in all directions. The angular distribution of the backscattered solar wind protons as ENAs can be described by the product of four separate ad hoc functions (Schaufelberger et al. 2011; Vorburger et al. 2013):

$$f_S(SZA, \theta) = f_0(SZA) \cdot f_1(SZA, \varphi) \cdot f_2(SZA, \varphi) \cdot f_3(SZA, \theta) \qquad (40)$$

where θ is the angle to the surface normal (polar scattering angle) and ranges from 0 to $\pi/2$ angle (i.e., $\theta = 0$ is perpendicular to the surface). φ is the angle between the surface projections of the Sun vector and the vector pointing to the observer (azimuth scattering angle) and ranges from $-\pi$ to $+\pi$. $\varphi = 0$ is the sunward direction, whereas $\varphi = \pm\pi$ is the anti-sunward direction. The solar zenith angle (SZA) is defined as the angle between the surface normal and the vector pointing to the Sun, the latter of which corresponds to the direction from which the solar wind ions impinge onto the surface. While f_0 describes the scattering function's overall amplitude, f_1 through f_3 describe three different features that were seen in the observed data and which are given by:

$$f_1(z_1, \varphi) = z_1 \cdot \cos(2\varphi) + (1 - z_1)$$

$$f_2(z_2, \varphi) = z_2 \cdot \cos(\varphi) + (1 - z_2) \tag{41}$$

$$f_3(z_3, \theta) = \left(1 - \frac{z_3}{\pi/2}\right) \cdot \sin(\theta + z_3) + \frac{z_3}{\pi/2}$$

where the z-terms are functions of the SZA (given in radians), and given by:

$$z_1(SZA) = (0.30 \cdot SZA + 0.03)$$
$$z_2(SZA) = 0.24 \cdot \cos(1.30 - 1.52 \cdot SZA) \tag{42}$$
$$z_3(SZA) = \frac{\pi}{2} - 1.03 \cdot SZA$$

Since, the integral of f_S must be equal to the cosine of the solar zenith angle, and f_0 is only a function of the solar zenith angle and not of the observation angles, thus Eq. (38) leads to

$$f_0(SZA) = \frac{\cos(SZA)}{\iint f_1(SZA, \varphi) \cdot f_2(SZA, \varphi) \cdot f_3(SZA, \theta) \, d\Omega} \tag{43}$$

Inserting Eq. (41) and Eq. (42) in Eq. (43) then gives for f_0:

$$f_0(SZA) = \cos(SZA) / \{0.74 \cdot (SZA - 3.23) \cdot$$
$$\cdot (2 \cdot SZA \cdot \cos(1.03 \cdot SZA) + SZA \cdot \sin(1.03 \cdot SZA) \cdot \pi - 4 \cdot (SZA - 1.53)) \cdot$$
$$\cdot (\cos(1.52 \cdot SZA - 1.30)) - 4.17\} \tag{44}$$

The energy spectrum of backscattered hydrogen ENAs was also determined empirically from observed ENA data from the Chandrayaan-1 mission (Futaana et al. 2012), which showed that the energy spectra of backscattered ENAs are best reproduced by Maxwell-Boltzmann distribution functions:

$$f(E) = n_0 \sqrt{\frac{2E}{m}} \left(\frac{m}{2\pi k_B T}\right)^{3/2} \exp\left(\frac{-E}{k_B T}\right) \tag{45}$$

where $k_B T$ is the characteristic energy of the scattered particles, and n_0 is the number density. The median of the best parameters of 108 data sets resulted in values of $n_0 = 2.98$ cm^{-3} and $k_B T = 93.0$ eV. If the solar wind velocity is known, then the characteristic energy can be calculated from the solar wind velocity (Futaana et al. 2012):

$$k_B T = 0.273 \cdot v_{SW} - 1.99 \tag{46}$$

with the characteristic energy $k_B T$ in units of [eV] and the solar wind speed v_{SW} in units of [km s^{-1}].

Figure 19 shows an energy sprectum for ENAs recorded by the Chandrayaan-1 mission (Futaana et al. 2012). The best fit of the ENA energy spectra is by a Maxwell-Boltzmann distribution (Eq. (45)), which indicates that the back-scattering mechanism is not a single binary collision with a surface atom, because then the ENA energy would be much higher. Most likely the backscattered ENAs are generated via multiple collisions off surfaces of regolith grains, which have a very rough surface on the microscopic scale. Considering the observed average energy spectrum shown in Fig. 19 implies that the ENAs have lost a considerable fraction of their initial energy as solar wind protons. Assuming an energy loss for each collision with a surface atom of typical 10–20% (Niehus et al. 1993), the impinging protons with an energy of 1 keV ($v_{SW} \approx 400$ km s^{-1}) experience 10 to 20 of collisions to end up with a characteristic energy of about 100 eV.

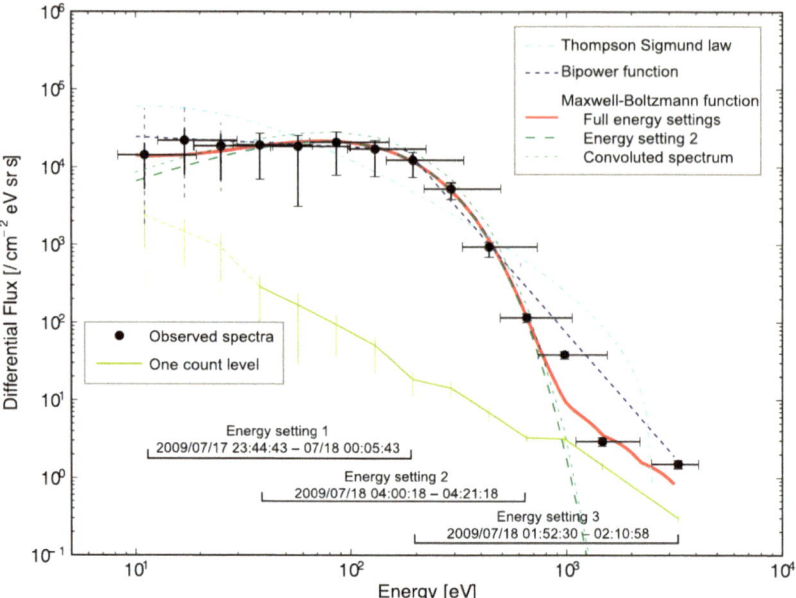

Fig. 19 Differential ENA flux of the observed backscattered ENAs (black circles). Data for three consecutive Chandrayaan-1 orbits around the Moon are averaged. Error bars correspond to the energy resolution (x-axis) and the error in flux (y-axis) mainly due to the uncertainty of the ionization efficiency of the conversion surface of the CENA instrument (Kazama et al. 2009). Particularly, no reliable ionization efficiency is available for low energy channels < 25 eV, and thus, the error bars may be underestimated in the low energy channels < 25 eV (dashed lines). The energy ranges corresponding to the energy settings 1–3 of the CENA instrument are shown. The yellow line gives the one count level based on the accumulation time, the energy resolution, and the ionization efficiency. Three different functions for fitting the data were examined: the Thompson-Sigmund law, Eq. (25) (light blue line), the bi-power law (blue line), and the Maxwell-Boltzmann distribution (red line). In addition, a Maxwell-Boltzmann distribution fitted only to the energy setting 2 data (38–652 eV) is distribution convolved with relatively wide energy also shown by the green dotted line. Figure reproduced from Futaana et al. (2012) with permission

8.1 Observations of Particle Reflection from the Moon

Up until about 2010, it was commonly assumed that almost all (∼99%) plasma ions impinging onto the lunar surface are immediately absorbed there (see e.g., Feldman et al. 2000, Crider and Vondrak 2002). This absorption, and the large obstacle the Moon presents to the plasma flow will cause a plasma vacuum (also known as plasma cavity) behind the Moon (e.g., Lyon et al. 1967). Later, the assumption of total ion absorption was invalidated when the Interstellar Boundary Explorer (IBEX; McComas et al. 2009a) and Chandrayaan-1 (Goswami and Annadurai 2008) observed lunar energetic neutral atoms (ENAs) for the first time (McComas et al. 2009b; Wieser et al. 2009). According to these measurements, a substantial fraction of ∼10–20% of the impinging solar wind protons are neutralized and scattered back as ENAs during their interaction with the surface (Vorburger et al. 2013; Lue et al. 2016). Since then, many more ENA reflection studies conducted by IBEX, Chandrayaan-1, and Chang'E-4 have been published. These studies differ in location of the observing instrument with respect to the lunar surface (and thus size of the observed area) and in energy range, but all three missions measured similar ENA reflection ratios.

Table 5 Compilation of reflection ratios for ENAs of lunar hydrogen

Reference	Mission	Moon distance	Surface area/ Observation	Energy range	ENA albedo
McComas et al. (2009b)	IBEX	$\sim$100 000 km	1/4 of Moon	380 eV–2.5 keV	$\sim$0.1
Wieser et al. (2009)	Chandrayaan-1	100–200 km	few hundred km^2	38–652 eV	0.16–0.20
Futaana et al. (2012)	Chandrayaan-1	100–200 km	few hundred km^2	11 eV–3.3 keV	$0.19^{0.21}_{0.16}$
Rodríguez M. et al. (2012)	Chandrayaan-1	100–200 km	1/4 of Moon	10 eV–2 keV	0.09 ± 0.05
Funsten et al. (2013)	IBEX	$\sim$100 000 km	1/4 of Moon	250 eV–3.6 keV	0.07–0.20
Saul et al. (2013)	IBEX	$\sim$100 000 km	1/4 of Moon	10 eV–2 keV	0.11 ± 0.06
Vorburger et al. (2013)	Chandrayaan-1	100–200 km	few hundred km^2	11 eV–2.2 keV	0.16 ± 0.05
Zhang et al. (2020)	Chang'E-4	48 cm	1.6 m^2	30 eV–10 keV	$0.32^{0.58}_{0.19}$

Table 5 lists all published reflection ratios together with their observation parameters. One notable difference between the individual observations is that Chang'E-4, i.e., the measurements were performed from a platform on the lunar surface, measured substantially higher ENA fluxes below 10% solar wind energy than Chandrayaan-1 and IBEX did (Zhang et al. 2020). The authors propose that this is a result of local surface regolith features (e.g., porosity, grain size, composition, and sputter yield) that only become apparent in the high spatial resolution measurements of Chang'E-4 at the surface. As expected, the ENA reflected flux varies with plasma incidence angle, with the highest values measured at the sub-solar point, and with the lowest values measured at the terminator (Wieser et al. 2009).

Since the solar wind plasma not only consists of protons but also contains alpha particles ($\sim$4%), it can be expected that some of these alpha particles are also reflected back to space from the lunar surface as helium ENAs. Indeed, Chandrayaan-1 measured backscattered He for the first time (Vorburger et al. 2014). Unfortunately, the He signal was not strong enough and the instrument's geometric factor was not known well enough to allow any quantification of the He ENA fluxes.

All ENA observations exhibit a similar ENA energy spectrum: The ENA energy spectrum is almost flat at lower energies, rolls over at a characteristic energy of about 100 eV, and above it exponentially decreases with energy up to the initial solar wind energy, $\sim$0.1–1 keV (Futaana et al. 2012; Rodríguez M. et al. 2012; Allegrini et al. 2013; Funsten et al. 2013). The characteristic ENA energy of 100 eV corresponds to about 10% of the initial solar wind energy. The energy spectrum suggests that the ENAs are mainly produced through backscattering of the solar wind ions and not by the surface sputter processes (e.g., Rodríguez M. et al. 2012). Assuming an average energy loss of 10–20% per surface interaction (Niehus et al. 1993), implies that the particles have undergone 10–20 collisions on the surface before they are scattered back to space. A clear observed linear correlation between the characteristic ENA energy and the solar wind velocity (rather than energy) further hints at the surface interaction process being momentum rather than energy driven.

ENA observations have also indicated that the directional scattering function for hydrogen ENAs depends on the solar wind incident angle to the lunar surface, and it is azimuthally isotropic and mainly sunward at low solar zenith angles (SZAs) but less isotropic and forward-scattering at high SZAs (Schaufelberger et al. 2011). What is contrary to expectations and to laboratory experiments, though, is that ions interacting with the lunar surface are preferably backward-scattered instead of forward-scattered (Schaufelberger et al. 2011). It is assumed that this is a result of the lunar surface regolith not being smooth on an atomic level, but highly porous on a micro-, mini-, and macro-scale instead, resulting in atoms undergoing several collisions before being scattered back to space. This is also supported by the ENAs' energy spectrum, which was discussed above.

The hydrogen ENAs energy spectrum and scattering function are dependent on the plasma temperature (Allegrini et al. 2013; Lue et al. 2016); therefore, their spectrum gets broader in the magnetosheath (Allegrini et al. 2013) and in the terrestrial plasma sheet (Harada et al. 2014a) due to the higher temperature of the ambient plasma compared to that in the solar wind.

ENAs were also observed on the lunar nightside (Vorburger et al. 2016). These nightside ENAs appear as two distinct ring-shaped distributions: the first of which ranges $\sim 6°$ into the nightside and can be related to the solar wind kinetic temperature; and the second of which ranges from the terminator $\sim 30°$ into the nightside, with its maximum located $\sim 12°$ beyond the terminator. The second population is related to ions entering the lunar wake, which are deflected by ambipolar fields at the wake boundary to the lunar night side surface. The nightside ENA populations amount to $\sim 1.5\%$ of the total dayside ENA flux, and exhibit characteristic energies ~ 4 eV lower than the average dayside ENA energy, with an abrupt drop of ~ 10 eV in characteristic energy at the terminator.

As the Moon enters Earth's magnetosphere, the plasma environment changes drastically. The solar wind plasma, which was supersonic upstream, is slowed down, shocked to subsonic velocities, and the plasma is compressed and heated in the process. Consequently, the ions backscattered from the lunar surface as ENAs are also modified to some extent. In Earth's magnetosheath, the H ENA energy spectrum is slightly broader and less peaked because of the increased plasma temperature (Lue et al. 2016), whereas in Earth's plasma sheet the hydrogen ENA reflection ratio is slightly lower than the upstream value (Harada et al. 2014a, 2014b).

A fraction of the backscattered particles is electrically charged. Before the first observations of ENAs from the Moon, the Kaguya spacecraft observed that 0.1–1% of the incident solar wind proton flux was backscattered from the lunar surface as protons (H^+) (Saito et al. 2008; Holmström et al. 2010; Lue et al. 2014). From Chandrayaan-1 and IBEX measurements we know that this is a small component compared to the flux of backscattered H-ENA, with a ratio of H^+/H-ENA < 0.1 (see Table 5). The fraction of the backscattered ions from lunar surface is correlated with the solar wind speed, varies from $\sim 0.01\%$ for low solar wind speed of ~ 250 km/s up to $\sim 1\%$ for high solar wind speed of ~ 550 km/s (Lue et al. 2014). The observed backscattered protons cover a broad range of energies from ~ 10–100% of the incident solar wind energy, with a peak at ~ 70–80% of the solar wind energy, which is higher than the energy peak for hydrogen H-ENAs (Lue et al. 2014). Up until now, the observed ions backscattered from the surface only consist of protons and the backscattering of the heavier solar wind ions (e.g., alpha particles) have not been observed yet (Saito et al. 2008; Lue et al. 2014), perhaps due to the low abundance of the heavier ions in the solar wind results in a backscattering fraction below the observation limits and/or because of the chemistry of their interaction with lunar regolith.

Despite the small fraction of backscattered H^+ ions, this population has important consequences for the lunar plasma environment since these ions flow against the solar wind stream and perturb the solar wind plasma. In turn, the solar wind forces the backscattered

H^+ to cycloid trajectories with curvatures comparable to the Moon itself. Depending on the solar wind and interplanetary magnetic field configurations, these trajectories can bring the ions thousands of kilometers upstream of the Moon (e.g., Holmström et al. 2010; Halekas et al. 2013) or deep into the lunar wake (e.g., Nishino et al. 2009a, 2009b; Dhanya et al. 2016), and may contribute to the H^+ flux onto the nightside surface (see Vorburger et al. 2016).

The backscattered H^+ ions appear to have a similar scattering function to that of the backscattered H-ENAs (Lue et al. 2017, 2018). However, there are clear differences in the energy spectra, where the H^+ ions have a higher mean energy than the H-ENAs (Lue et al. 2014; Lue et al. 2017, 2018). This suggests that the exit speed from the surface affects the charge state fractions of the scattered particles, i.e., the charge state fraction H^+/H-ENA of backscattered particles increases with higher exit speeds from the lunar surface. Since the mean exit speed increases with increasing impact speed (Futaana et al. 2012), Lue et al. (2014) suggested that this exit-speed dependence could explain a positive correlation between the total H^+ scattering rate and the solar wind impact speed, observed in a case study of Chandrayaan-1 data. However, in a larger statistical study of ARTEMIS data, Lue et al. (2018) did not find a similar trend in the total H^+ scattering rate as function of impact speed. Further studies are required to paint a better picture of the relation between the H-ENA and H^+ scattering.

A small fraction of the backscattered solar wind is also expected to be negatively charged, e.g. H^- (Wekhof 1981). H^- photo-detaches into H ENA within a fraction of a second in sunlight at 1 AU, and H^- ions likely do not have a significant impact on the lunar plasma environment. Nevertheless, observations of H^- are required to complete the picture of solar wind scattering from planetary surfaces. Scattered negative ions may be an important component in plasma environments further out in the solar system where the solar photon flux is lower, or in shadowed environments such as the nightside of Mercury. Photo-detachment of negative ions would also contribute to the overall ENA fluxes in a planetary environment.

When incoming ions encounter localized regions of lunar crustal magnetization the interaction between solar wind ions and the lunar surface becomes more complicated (e.g., Kallio et al. 2012; Wieser et al. 2009; Bamford et al. 2012, 2016; Jarvinen et al. 2014; Fatemi et al. 2015; Deca et al. 2015, 2016; Poppe et al. 2016a). As the incoming ions encounter the so-called mini-magnetosphere associated with the crustal magnetization, some ions are decelerated, some are deflected, and some are heated and reflected back to space without interacting with the lunar surface at all. Figure 20 shows a hybrid calculation of the solar wind interaction with a magnetic anomaly on the lunar surface. It demonstrates the deflection of ions at the magnetic anomaly, causing a reduction of ion flux to the surface at the anomaly, and an increase of ion flux surrounding it (an ion "halo"), and a reduction of the ion velocity component towards the surface (Kallio et al. 2012). In the ion "halo" the impact flux and the density of protons are higher than in the undisturbed regions far from the magnetic anomaly (>50 km). The total magnetic field differs from the crustal magnetic field because of the magnetic field associated with the electric currents around the magnetic anomaly. In ENA images this altered interaction process results in the area associated with the crustal magnetization exhibiting less ENA flux while the surrounding area exhibits more ENA flux than the non-magnetized surface nearby (Wieser et al. 2010). Another feature of this modified interaction process is that the mini-magnetosphere exhibits a large electric potential, $+135$ V, because the protons being able to penetrate further into the magnetic field of the mini-magnetosphere than electrons can. This potential was predicted by hybrid modelling (Kallio et al. 2012; Jarvinen et al. 2014) and was observed in Chandrayaan-1 measurements (Futaana et al. 2013).

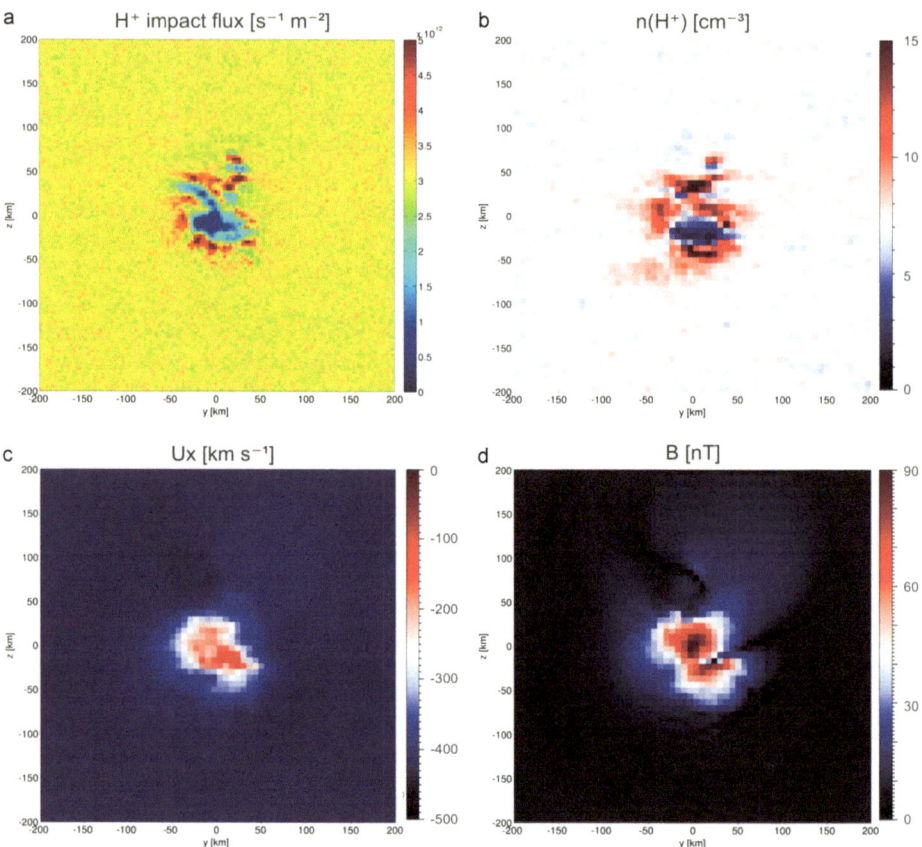

Fig. 20 Properties of the solar wind protons on the lunar surface near a magnetic dipole on the lunar surface. Figure reproduced from Kallio et al. (2012) with permission. (**a**) The impact flux of protons to the surface $[s^{-1} m^{-2}]$, (**b**) the density of protons at the surface $[cm^{-3}]$, (**c**) the vertical velocity of protons $[km\,s^{-1}]$ and (**d**) the total magnetic field [nT]. The magnetic dipole is located 50 km below the surface at $y = z = 0$

From the Chandrayaan-1 observations we know that the shielding efficiency of lunar crustal magnetizations against solar wind ions ranges from 0% to 70%, and strongly depends on the magnetic field strength, the geometric structure of the magnetic anomaly, and on the upstream plasma conditions (Saito et al. 2010; Wieser et al. 2010; Vorburger et al. 2012, 2013; Harada et al. 2014a). Lunar crustal magnetic fields play a crucial role in the incidence and reflection of the solar wind ions on the lunar surface (e.g., Lue et al. 2011; Saito et al. 2012). Observations have shown that on average between 5–10% of the solar wind proton flux is reflected from lunar crustal magnetic fields at an altitude between the surface and the location of the spacecrafts (Lue et al. 2011; Saito et al. 2010, 2012; Poppe et al. 2017; Tsunakawa et al. 2010, 2015). However, depending on the strength and location of the crustal magnetic fields and the energy and angle of the incident solar wind ions, up to 50% reflection over strong crustal fields have been also reported (Lue et al. 2011). Global ENA maps also show a considerable reduction (up to 50%) of backscattered hydrogen ENAs over the magnetized areas on the lunar surface (Vorburger et al. 2013). These ENA observations suggest partial shielding of the lunar surface from the solar wind ions (Saito et al. 2010; Wieser et al. 2010; Vorburger et al. 2013), but the shielding efficiency strongly depends on the strength of the crustal fields and the upstream plasma conditions (Vorburger et al. 2012; Harada et al. 2014a). In addition to the surface shielding, simulations and observations have shown that the solar wind ions are deflected around the magnetized areas and impact the lunar surface (e.g., Kallio et al. 2012; Wieser et al. 2009; Bamford et al. 2012, 2016; Jarvinen et al. 2014; Fatemi et al. 2015; Deca et al. 2015, 2016; Poppe et al. 2016a), resulting in the enhancement of the incident ion flux to the lunar surface in the areas surrounding the magnetic anomaly, and consequently provides a relatively higher backscattering ENA fluxes around some of the strongly magnetized areas compared to the unmagnetized surface (Wieser et al. 2010; Futaana et al. 2013).

The reflected ions from lunar crustal magnetic fields and the backscattered ions from lunar surface interact with the ambient electromagnetic environment and the upstream plasma and generate a broad range of plasma waves from $\sim$0.01–0.1 Hz narrowband ULF (Nakagawa et al. 2012; Halekas et al. 2013) to $\sim$0.1–10 Hz broadband whistlers (Halekas et al. 2008a, 2012; Nakagawa et al. 2011; Tsugawa et al. 2012) and narrowband whistlers from $\sim$1 Hz up to $\sim$100 Hz (e.g., Lin et al. 1998; Halekas et al. 2006, 2008a; Tsugawa et al. 2011, 2012; Harada et al. 2014b). Some of these reflected ions are "picked-up" and move into the lunar wake and they may impact the lunar nightside (e.g., Nishino et al. 2009b; Dhanya et al. 2018). However, the flux of these ions is expected to be very low, thus they have a minor contribution in the surface sputtering on the nightside.

9 Conclusions

A quantitative treatment of the particle release from the surfaces of the Moon and Mercury is a complicated endeavor. Although the theoretical formulation of the release processes is quite mature, a set of external drivers and parameters related to surface physics are needed for a quantitative calculation of the particle fluxes released from the surface into the exosphere, which often is not available in the needed accuracy. Moreover, in the observations of exospheric species, most often there is a mix of release processes active, which makes the comparison to the theoretical calculations complicated. In that respect Na and K in the exosphere are the most complicated species to interpret, since they can be released by all four release processes.

As external drivers one needs the fluxes of photons, of fluxes plasma particles (ions and electrons) and their energies, and fluxes micrometeorites onto these surfaces. Some of these parameters are available from observations by several spacecraft, with more to expected in the future. Often these parameters are not available at the exact time of the exosphere observation, or the exact location of the exosphere observation, requiring extrapolation and adjustments of these parameters. Thus, often modelling the solar wind plasma, its interaction with the magnetosphere, to define the different plasma populations, is needed to derive the necessary plasma parameters for particle release. Moreover, from the plasma data measured at the spacecraft location one must infer the particle flux that actually arrives at the surface, which is more complicated at Mercury than at the Moon because of its magnetosphere, which is quite variable on short time scales. The resulting global magnetic field is affecting the fluxes of charged particles inside the magnetosphere and thus also the plasma fluxes arriving at the surface and their location.

The surface physics on the microscopic scale, even at the atomic scale, governs the actual release of atoms or molecules from the surface. In principle, processes like sputtering, thermal desorption, and PSD have been studied in the surface science community for decades and longer. However, many of the surface physics parameters, like sputter yields, binding energies, roughness, surface charging, ... and the actual mineralogical composition are needed to be known on microscopic spatial scales and thus affect the quantitative prediction of released particle fluxes. Moreover, the systems investigated in these laboratory studies are typically idealized surfaces (e.g. polished sample surfaces) under very controlled laboratory conditions. Thus, their results apply to actual planetary surfaces, which are continuously modified by space weathering, only in a limited way. Bennett et al. (2013) reviewed the processes at planetary surfaces from the surface physics point of view.

Clearly, more laboratory work on the release processes is needed using analogue samples representing the surfaces of the Moon or Mercury, e.g., matching the roughness, the granularity, the mix of minerals. Also, the irradiation by particles or photons should representative of the respective planetary environment. For example, sputtering of regolith-like samples has been investigated by several groups, and promises to be an ongoing activity. If made available, lunar samples from the Apollo missions would be even better for such studies. Release by PSD and ESD is also under investigation by several laboratories. The largest uncertainty is in the exosphere production by micrometeorite impact: on the one hand, the fluxes of micrometeorites onto the surfaces of the Moon and Mercury are not known that well, on the other hand the formation of the impact plasma (its temperature and density), and the release of surface material is the least understood on a quantitative level from the four release processes discussed.

Funding Note Open access funding provided by University of Bern.

References

I. Adler, J.I. Trombka, Orbital chemistry-lunar surface analysis from X-ray and gamma-ray sensing experiments. Phys. Chem. Earth **10**, 17–43 (1977). https://doi.org/10.1016/0079-1946(77)90004-0

M. Ait El Fqih, Angular distribution of sputtered alloy. Experimental and simulated study. Eur. Phys. J. D **56**, 167–172 (2010). https://doi.org/10.1140/epjd/e2009-00272-8

F. Allegrini, M.A. Dayeh, M.I. Desai, H.O. Funsten, S.A. Fuselier, P.H. Janzen, D.J. McComas, E. Möbius, D.B. Reisenfeld, D.F. Rodríguez M., N. Schwadron, P. Wurz, Lunar energetic neutral atom (ENA) spectra measured by the interstellar boundary explorer (IBEX). Planet. Space Sci. **85**, 232–242 (2013). https://doi.org/10.1016/j.pss.2013.06.014

K.A. Anderson, R.P. Lin, R.E. McGuire, J.E. McCoy, Measurement of lunar and planetary magnetic fields by reflection of low energy electrons. Space Sci. Instrum. **1**, 439–470 (1975)

B.J. Anderson, C.L. Johnson, H. Korth, R.M. Winslow, J.E. Borovsky, M.E. Purucker, J.A. Slavin, S.C. Solomon, M.T. Zuber, R.L. McNutt Jr., Low-degree structure in Mercury's planetary magnetic field. J. Geophys. Res. **117**, E00L12 (2012). https://doi.org/10.1029/2012JE004159

M.O. Archer, D.L. Turner, J.P. Eastwood, S.J. Schwartz, T.S. Horbury, Global impacts of a Foreshock Bubble: magnetosheath, magnetopause and ground-based observations. Planet. Space Sci. **106**, 56–66 (2015). https://doi.org/10.1016/j.pss.2014.11.026

T.P. Armstrong, S.M. Krimigis, L.J. Lanzerotti, A reinterpretation of the reported energetic particle fluxes in the vicinity of Mercury. J. Geophys. Res. (1896–1977) **80**, 4015–4017 (1975). https://doi.org/10.1029/JA080i028p04015

A.V. Artemyev, V. Angelopoulos, A. Runov, I.Y. Vasko, Hotion flows in the distant magnetotail: ARTEMIS observations fromlunar orbit to $\sim -200\ R_E$. J. Geophys. Res. Space Phys. **122**, 9898–9909 (2017). https://doi.org/10.1002/2017JA024433

S.D. Bale, Shadowed particle distributions near the Moon. J. Geophys. Res. Space Phys. **102**(A9), 19773–19778 (1997). https://doi.org/10.1029/97JA01676

S.J. Bame, J.R. Asbridge, W.C. Feldman, M.D. Montgomery, P.D. Kearney, Solar wind heavy ion abundances. Sol. Phys. **43**(2), 463–473 (1975). https://doi.org/10.1007/BF00152368

R.A. Bamford, B. Kellett, W.J. Bradford, C. Norberg, A. Thornton, K.J. Gibson, I.A. Crawford, L. Silva, L. Gargaté, R. Bingham, Mini-magnetospheres above the lunar surface and the formation of lunar swirls. Phys. Rev. Lett. **109**(8), 081101 (2012). https://doi.org/10.1103/PhysRevLett.109.081101

R.A. Bamford, E.P. Alves, F. Cruz, B.J. Kellett, R.A. Fonseca, L.O. Silva, R.M.G.M. Trines, J.S. Halekas, G. Kramer, E. Harnett, R.A. Cairns, R. Bingham, 3D PIC simulations of collisionless shocks at lunar magnetic anomalies and their role in forming lunar swirls. Astrophys. J. **830**(2), 146 (2016). https://doi.org/10.3847/0004-637X/830/2/146

D. Banerjee, S. Vadawale, Theoretical modelling of X-ray fluorescence signals for different lunar compositions and dependence on solar activity. Adv. Space Res. **46**, 651–656 (2010). https://doi.org/10.1016/j.asr.2010.04.021

R.A. Baragiola, Sputtering: survey of observations and derived principles. Philos. Trans. - Royal Soc., Math. Phys. Eng. Sci. **362**, 29–53 (2004). https://doi.org/10.1098/rsta.2003.1301

R.A. Baragiola, R.A. Vidal, W. Svendsen, J. Schou, M. Shi, D.A. Bahr, C.L. Atteberrry, Sputtering of water ice. Nucl. Instrum. Methods Phys. Res. B **209**, 294–303 (2003). https://doi.org/10.1016/S0168-583X(02)02052-9

A.F. Barghouty, F.W. Meyer, P.R. Harris, J.H. Adams, Solar-wind protons and heavy ions sputtering of lunar surface materials. Nucl. Instrum. Methods Phys. Res., Sect. B, Beam Interact. Mater. Atoms **269**(11), 1310–1315 (2011). https://doi.org/10.1016/j.nimb.2010.12.033

S.V. Baryshev, A.V. Zinovev, C.E. Tripa, R.A. Erck, I.V. Veryovkin, White light interferometry for quantitative surface characterization in ion sputtering experiments. Appl. Surf. Sci. **258**, 6963 (2012). https://doi.org/10.1016/j.apsusc.2012.03.144

R. Behrisch, W. Eckstein (eds.), *Sputtering by Particle Bombardment, Experiments and Computer Calculations from Threshold to MeV Energies*. Topics in Applied Physics, vol. 110 (Springer, Berlin, 2007). ISBN 978-3-540-44502-9

M.R. Benedikt, M. Scherf, H. Lammer, E. Marcq, P. Odert, M. Leitzinger, N.V. Erkaev, Escape of rock-forming volatile elements and noble gases from planetary embryos. Icarus **347**, 113772 (2000). https://doi.org/10.1016/j.icarus.2020.113772

M. Benna, B.J. Anderson, D.N. Baker, S.A. Boardsen, G. Gloeckler, R.E. Gold, G.C. Ho, R.M. Killen, H. Korth, S.M. Krimigis, M.E. Purucker, R.L. McNutt Jr., J.M. Raines, W.E. McClintock, M. Sarantos, J.A. Slavin, S.C. Solomon, T.H. Zurbuchen, Modeling of the magnetosphere of Mercury at the time of the first MESSENGER flyby. Icarus **209**, 3–10 (2010). https://doi.org/10.1016/j.icarus.2009.11.036

C.J. Bennett, C. Pirim, T.M. Orlando, Space-weathering of solar system bodies: a laboratory perspective. Chem. Rev. **113**(12), 9086–9150 (2013). https://doi.org/10.1021/cr400153k

C.J. Bennett, J.L. McLain, M. Sarantos, R.D. Gann, A. DeSimone, T.M. Orlando, Investigating potential sources of Mercury's exospheric Calcium: photon-stimulated desorption of Calcium Sulfide. J. Geophys. Res., Planets **121**, 137–146 (2016). https://doi.org/10.1002/2015JE004966

A. Benninghofen, F.G. Rüdenauer, H.W. Werner, Secondary ion mass spectrometry–basic concepts, instrumental aspects, applications and trends, Wiley, New York, 1277 pages. Surf. Interface Anal. **10**, 435 (1987). https://doi.org/10.1002/sia.740100811

J. Benson, J.W. Freeman, H.K. Hills, R.R. Vondrak, Bow shock protons in the lunar environment. Earth Moon Planets **14**, 19–25 (1975). https://doi.org/10.1007/BF00562969

A.O. Benz, Flare observations. Living Rev. Sol. Phys. **14**, 2 (2017). https://doi.org/10.1007/s41116-016-0004-3

A.O. Benz, M. Güdel, Physical processes in magnetically driven flares on the Sun, stars, and young stellar objects. Annu. Rev. Astron. Astrophys. **48**(1), 241–287 (2010). https://doi.org/10.1146/annurev-astro-082708-101757

E.L. Berger, L.P. Keller, Solar flare track exposure ages in regolith particles: a calibration for transmission electron microscope measurements. Paper 1543, in *46th Lunar and Planet. Sci. Conf.*, The Woodlands, Texas (2015). https://www.hou.usra.edu/meetings/lpsc2015/pdf/1543.pdf

T.J. Bernatowicz, F.A. Podosek, Argon adsorption and the lunar atmosphere. Proc. Lunar Planet. Sci. **21**, 307–313 (1991).

G. Betz, W. Husinsky, Sputtering of Insulators. Nucl. Instrum. Methods Phys. Res. B **32**, 331–340 (1988)

G. Betz, K. Wien, Energy and angular distributions of sputtered particles. Nucl. Instrum. Methods Phys. Res. B **140**, 1–110 (1994)

G. Betz, E. Wolfrum, P. Wurz, K. Mader, B. Strehl, W. Husinsky, R.F. Haglund, N.H. Tolk, Ground state and excited state atom production by electron and ion bombardment of NaCl and CaF$_2$, in *Desorption Induced by Electronic Transitions, DIET III*, ed. by R.H. Stuhlen, M.L. Knotek (Springer, Berlin, 1987), pp. 278–283

P. Bhattacharya, R. Bhattacharjee, A study on weibull distribution for estimating the parameters. J. Appl. Quant. Methods **5**(2), 234–241 (2010)

J.P. Biersack, L.G. Haggmark, A Monte Carlo computer program for the transport of energetic ions in amorphous targets. Nucl. Instrum. Methods **174**, 257–269 (1980). https://doi.org/10.1016/0029-554X(80)90440-1

P. Bochsler, Minor ions in the solar wind. Astron. Astrophys. Rev. **14**, 1–40 (2007). https://doi.org/10.1007/s00159-006-0002-x

P. Borin, G. Cremonese, F. Marzari, M. Bruno, S. Marchi, Statistical analysis of micrometeoroids flux on Mercury. Astron. Astrophys. **503**, 259–264 (2009). https://doi.org/10.1051/0004-6361/200912080

P. Borin, M. Bruno, G. Cremonese, F. Marzari, Estimate of the neutral atoms' contribution to the Mercury exosphere caused by a new flux of micrometeoroids. Astron. Astrophys. **517**, A89 (2010). https://doi.org/10.1051/0004-6361/201014312

H.S. Bridge, C. Dilworth, A.J. Lazarus, E.F. Lyon, B. Rossi, F. Scherb, Direct observations of the interplanetary plasma. J. Phys. Soc. Jpn. **17**, 553 (1962). http://adsabs.harvard.edu/abs/1962JPSJS..17B.553B

R.A. Brizzolara, C.B. Cooper, T.K. Olson, Energy distribution of neutral atoms sputtered by very low energy heavy ions. Nucl. Instrum. Methods Phys. Res. B **35**, 36–42 (1988)

M. Bruno, G. Cremonese, S. Marchi, Neutral sodium atoms release from the surfaces of the Moon and Mercury induced by meteoroid impacts. Planet. Space Sci. **55**, 1494–1501 (2007). https://doi.org/10.1016/j.pss.2006.10.006

M.H. Burger, R.M. Killen, W.E. McClintock, R.J. Vervack Jr., A.W. Merkel, A.L. Sprague, M. Sarantos, Modeling MESSENGER observations of calcium in Mercury's exosphere. J. Geophys. Res. **117**, E00L11 (2012). https://doi.org/10.1029/2012JE004158

M.H. Burger, R.M. Killen, W.E. McClintock, A.W. Merkel, R.J. Vervack, T.A. Cassidy, M. Sarantos, Seasonal variations in Mercury's dayside calcium exosphere. Icarus **238**, 51–58 (2014). https://doi.org/10.1016/j.icarus.2014.04.049

M.W. Caffe, C.M. Hohenberg, T.D. Swindle, J.N. Goswami, Evidence in meteorites for an active early Sun. Astrophys. J. **313**, L1 (1987). https://doi.org/10.1086/184826

S. Candelaresi, A. Hillier, H. Maehara, A. Brandenburg, K. Shibata, Superflare occurrence and energies on G-, K-, and M-type dwarfs. Astrophys. J. **792**, 67 (2014a). https://doi.org/10.1088/0004-637X/792/1/67

S. Candelaresi, A. Hillier, H. Maehara, A. Brandenburg, K. Shibata, Superflare occurrence and energies on G-, K-, and M-type dwarfs. Astrophys. J. **792**, 67 (2014b). https://doi.org/10.1088/0004-637X/792/1/67

R.C. Carrington, Description of a singular appearance seen in the Sun on September 1. Mon. Not. R. Astron. Soc. **20**, 13–15 (1859). https://doi.org/10.1093/mnras/20.1.13

W. Cassidy, R.E. Johnson, Monte Carlo model of sputtering and other ejection processes within a regolith. Icarus **176**, 499–507 (2005). https://doi.org/10.1016/j.icarus.2005.02.013

T.A. Cassidy, A.W. Merkel, M.H. Burger, M. Sarantos, R.M. Killen, W.E. McClintock, R.J. Vervack, Mercury's seasonal sodium exosphere: MESSENGER orbital observations. Icarus **248**, 547–559 (2015). https://doi.org/10.1016/j.icarus.2014.10.037

P.C. Chamberlin, T.N. Woods, L. Didkovsky, F.G. Eparvier, A.R. Jones, J.L. Machol, J.P. Mason, M. Snow, E.M.B. Thiemann, R.A. Viereck, D.L. Woodraska, Solar ultraviolet irradiance observations of the solar flares during the intense September 2017 storm period. Space Weather **16**, 1470–1487 (2018). https://doi.org/10.1029/2018SW001866

S.C. Chase Jr., E.D. Miner, D. Morrison, G. Münch, G. Neugebauer, Mariner 10 infrared radiometer results: temperatures and thermal properties of the surface of Mercury. Icarus **28**(4), 565–578 (1976). https://doi.org/10.1016/0019-1035(76)90130-5

E. Chatzitheodoridis, G. Kiriakidis, Secondary ion mass spectrometry and its application to thin film characterization, in *Handbook of Thin Film Materials*, ed. by H.S. Nalwa. Characterization and Spectroscopy of Thin Films, vol. 2 (Academic Press, San Diego, 2002)

Y. Chen, G. Toth, P. Cassak, X. Jia, T.I. Gombosi, J.A. Slavin, M.G. Henderson, Global three-dimensional simulation of Earth's dayside reconnection using a two-way coupled magnetohydrodynamics with embedded particle-in-cell model: initial results. J. Geophys. Res. **122**, 10318–10335 (2017). https://doi.org/10.1002/2017JA024186

A.F. Cheng, R.E. Johnson, S.M. Krimigis, L.J. Lanzerotti, Magnetosphere, exosphere and surface of Mercury. Icarus **71**, 430–440 (1987). https://doi.org/10.1016/0019-1035(87)90038-8

R. Christoffersen, Z. Rahman, L.P. Keller, Solar ion sputter deposition in the lunar regolith: experimental simulation using focused-ion beam techniques. LPSC Conference Paper (2012). https://ntrs.nasa.gov/citations/20120003591

S.P. Christon, G. Gloeckler, D.J. Williams, T. Mukai, R.W. McEntire, C. Jacquey, V. Angelopoulos, A.T.Y. Lui, S. Kokubun, D.H. Fairfield, M. Hirahara, T. Yamamoto, Energetic atomic and molecular ions of ionospheric origin observed in distant magnetotail flow-reversal events. Geophys. Res. Lett. **21**(25), 3023–3026 (1994). https://doi.org/10.1029/94GL02095

M.J. Cintala, Impact-induced thermal effects in the lunar and Mercuria regoliths. J. Geophys. Res. **97**, 947–973 (1992). https://doi.org/10.1029/91JE02207

A. Collette, K. Drake, A. Mocker, Z. Sternovsky, T. Munsat, M. Horanyi, Time-resolved temperature measurements in hypervelocity dust impact. Planet. Space Sci. **89**, 58–62 (2013). https://doi.org/10.1016/j.pss.2013.02.007

B. Cooper, A. Potter, R.K. Illen, T. Morgan, Midinfrared spectra of Mercury. J. Geophys. Res. **106**(E12), 32803–32814 (2001). https://doi.org/10.1029/2000JE001377

D.H. Crider, R.R. Vondrak, Hydrogen migration to the lunar poles by solar wind bombardment of the Moon. Adv. Space Res. **30**, 1869–1874 (2002). https://doi.org/10.1016/S0273-1177(02)00493-3

N.U. Crooker, S. Shodhan, J.T. Gosling, J. Simmerer, R.P. Lepping, J.T. Steinberg, S.W. Kahler, Density extremes in the solar wind. Geophys. Res. Lett. **27**, 3769–3772 (2000). https://doi.org/10.1029/2000GL003788

C. Cupak, P.S. Szabo, H. Biber, R. Stadlmayr, C. Grave, M. Fellinger, J. Brötzner, R.A. Wilhelm, W. Müller, A. Mutzke, M.V. Moro, F. Aumayr, Sputter yields of rough surfaces: importance of the mean surface inclination angle from nano- to microscopic rough regimes. Appl. Surf. Sci. **570**, 151204 (2021). https://doi.org/10.1016/j.apsusc.2021.151204

J.R.A. Davenport, D.M. Kipping, D. Sasselov, J.M. Matthews, C. Cameron, MOST observations of our nearest neighbor: flares on Proxima Centauri. Astrophys. J. Lett. **829**, L31 (2016). https://doi.org/10.3847/2041-8205/829/2/L31

J. Deca, A. Divin, B. Lembège, M. Horányi, S. Markidis, G. Lapenta, General mechanism and dynamics of the solar wind interaction with lunar magnetic anomalies from 3-D particle-in-cell simulations. J. Geophys. Res. Space Phys. **120**(8), 6443–6463 (2015). https://doi.org/10.1002/2015JA021070

J. Deca, A. Divin, X. Wang, B. Lembège, M. Horányi, S. Markidis, G. Lapenta, Three-dimensional full-kinetic simulation of the solar wind interaction with a vertical dipolar lunar magnetic anomaly. Geophys. Res. Lett. **43**(9), 4136–4144 (2016). https://doi.org/10.1002/2016GL068535

H.G. Demars, R.W. Schunk, Solar wind proton velocity distributions: comparison of the bi-Maxwellian based 16-moment expansion with observations. Planet. Space Sci. **38**(9), 1091–1103 (1990). https://doi.org/10.1016/0032-0633(90)90018-L

A.J. DeSimone, T.M. Orlando, Photodissociation of water and $O(^3P_J)$ formation on a lunar impact melt breccia. J. Geophys. Res., Planets **119**, 894–904 (2014). https://doi.org/10.1002/2013JE004598

A.J. DeSimone, T.M. Orlando, H_2O and O (^3P_J) photodesorption from amorphous solid water deposited on a lunar mare basalt. Icarus **255**, 44–50 (2015). https://doi.org/10.1016/j.icarus.2014.08.023

R.M. Dewey, J.A. Slavin, J.M. Raines, D.N. Baker, D.J. Lawrence, Energetic electron acceleration and injection during dipolarization events in Mercury's magnetotail. J. Geophys. Res. Space Phys. **122**, 12,170–12,188 (2017). https://doi.org/10.1002/2017JA024617

R.M. Dewey, J.M. Raines, W. Sun, J.A. Slavin, G. Poh, MESSENGER observations of fast plasma flows in Mercury's magnetotail. Geophys. Res. Lett. **45**, 10,110–10,118 (2018). https://doi.org/10.1029/2018GL079056

R.M. Dewey, J.A. Slavin, J.M. Raines, A.R. Azari, W. Sun, MESSENGER observations of flow braking and flux pileup of dipolarizations in Mercury's magnetotail: evidence for current wedge formation. J. Geophys. Res. **125**, e2020JA028112 (2020). https://doi.org/10.1029/2020JA028112

M.B. Dhanya, A. Bhardwaj, Y. Futaana, S. Fatemi, M. Holmström, S. Barabash, M. Wieser, P. Wurz, A. Alok, R.S. Thampi, Proton entry into the near-lunar plasma wake for magnetic field aligned flow. Geophys. Res. Lett. **40**(12), 2913–2917 (2013). https://doi.org/10.1002/grl.50617

M.B. Dhanya, A. Bhardwaj, Y. Futaana, S. Barabash, A. Alok, M. Wieser, M. Holmström, P. Wurz, Characteristics of proton velocity distribution functions in the near-lunar wake from Chandrayaan-1/SWIM observations. Icarus **271**, 120–130 (2016). https://doi.org/10.1016/j.icarus.2016.01.032

M.B. Dhanya, A. Bhardwaj, A. Alok, Y. Futaana, S. Barabash, M. Wieser, M. Holmström, P. Wurz, First observation of transport of solar wind protons scattered from magnetic anomalies into the near lunar wake: observations by SARA/Chandrayaan-1. Geophys. Res. Lett. **45**(17), 8826–8833 (2018). https://doi.org/10.1029/2018GL079330

G.A DiBraccio, J.A. Slavin, S.A. Boardsen, B.J. Anderson, H. Korth, T.H. Zurbuchen, J.M. Raines, D.N. Baker, R.L. McNutt Jr., S.C. Solomon, MESSENGER observations of magnetopause structure and dynamics at Mercury. J. Geophys. Res. **118**997–1008 (2013). https://doi.org/10.1002/jgra.50123

M. Dominique, A.N. Zhukov, P. Heinzel, I.E. Dammasch, L. Wauters, L. Dolla, S. Shestov, M. Kretzschmar, J. Machol, G. Lapenta, W. Schmutz, First detection of solar flare emission in mid-ultraviolet Balmer continuum. Astrophys. J. Lett. **867**, L24 (2018). https://doi.org/10.3847/2041-8213/aaeace

C. Dong, L. Wang, A. Hakim, A. Bhattacharjee, J.A. Slavin, G.A. DiBraccio, K. Germaschewski, Global ten-moment multifluid simulations of the solar wind interaction with Mercury: from the planetary conducting core to the dynamic magnetosphere. Geophys. Res. Lett. **46**, 11588–11596 (2019). https://doi.org/10.1029/2019GL083180

L. Doyle, G. Ramsay, J.G. Doyle, Superflares and variability in solar-type stars with TESS in the Southern hemisphere. Mon. Not. R. Astron. Soc. **494**, 3596–3610 (2020). https://doi.org/10.1093/mnras/staa923

C.A. Dukes, R.A. Baragiola, The lunar surface-exosphere connection: measurement of secondary-ions from Apollo soils. Icarus **255**, 51 (2015). https://doi.org/10.1016/j.icarus.2014.11.032

C.A. Dukes, W.-Y-. Chang, M. Famá, R.A. Baragiola, Laboratory studies on the sputtering contribution to the sodium atmospheres of Mercury and the Moon. Icarus **212**, 463 (2011). https://doi.org/10.1016/j.icarus.2011.01.027

E. Dullni, Velocity distributions of the metal atoms sputtered from oxygen and nitrogen covered Ti- and Al-surfaces. Nucl. Instrum. Methods Phys. Res. B **2**, 610–613 (1984). https://doi.org/10.1016/0168-583X(84)90276-3

J.W. Dungey, Interactions of solar plasma with the geomagnetic field. Planet. Space Sci. **10**, 233–237 (1963). https://doi.org/10.1016/0032-0633(63)90020-5

P. Dyal, C.W. Parkin, W.D. Daily, Magnetism and the interior of the Moon. Rev. Geophys. **12**(4), 568–591 (1974). https://doi.org/10.1029/RG012i004p00568

J.P. Eastwood, E.A. Lucek, C. Mazelle, K. Meziane, Y. Narita, J. Pickett, R.A. Treumann, The foreshock. Space Sci. Rev. **118**(1–4), 41–94 (2005). https://doi.org/10.1007/s11214-005-3824-3

J.P. Eastwood, D.G. Sibeck, V. Angelopoulos, T.D. Phan, S.D. Bale, J.P. McFadden, C.M. Cully, S.B. Mende, D. Larson, S. Frey, C.W. Carlson, K.-H. Glassmeier, H.U. Auster, A. Roux, O. Le, Contel, THEMIS observations of a hot flow anomaly: solar wind, magnetosheath, and ground-based measurements. Geophys. Res. Lett. **35**, L17S03 (2008). https://doi.org/10.1029/2008GL033475

J. Egedal, W. Daughton, A. Le, Large-scale electron acceleration by parallel electric fields during magnetic reconnection. Nat. Phys. **8**, 321–324 (2012). https://doi.org/10.1038/nphys2249

G. Eichhorn, Impact light flash studies: temperature, ejecta, vaporization, in *Interplanetary Dust and Zodiacal Light*, vol. 48 (1976), pp. 243–247. https://doi.org/10.1007/3-540-07615-8_490

G. Eichhorn, Heating and vaporization during hypervelocity particle impact. Planet. Space Sci. **26**, 463–467 (1978a). https://doi.org/10.1016/0032-0633(78)90067-3

G. Eichhorn, Primary velocity dependence of impact ejecta parameters. Planet. Space Sci. **26**, 469–471 (1978b). https://doi.org/10.1016/0032-0633(78)90068-5

R.C. Elphic, H.O. Funsten III, B.L. Barraclough, D.J. McComas, M.T. Paffett, D.T. Vaniman, G. Heiken, Lunar surface composition and solar wind-induced secondary ion mass spectrometry. Geophys. Res. Lett. **18**(11), 2165–2168 (1991). https://doi.org/10.1029/91GL02669

A.G. Emslie, B.R. Dennis, G.D. Holman, H.S. Hudson, Refinements to flare energy estimates: a follow-up to "Energy partition in two solar flare/CME events" by A.G. Emslie et al. J. Geophys. Res. **110**, A11103 (2005). https://doi.org/10.1029/2005JA011305

A.G. Emslie, B.R. Dennis, A.Y. Shih, P.C. Chamberlin, R.A. Mewaldt, C.S. Moore, G.H. Share, A. Vourlidas, B.T. Welsch, Global energetics of thirty-eight large solar eruptive events. Astron. J. **759**, 71 (2012). https://doi.org/10.1088/0004-637X/759/1/71

N.V. Erkaev, H. Lammer, P. Odert, Yu.N. Kulikov, K.G. Kislyakova, Extreme hydrodynamic atmospheric loss near the critical thermal escape regime. Mon. Not. R. Astron. Soc. **448**, 1916–1921 (2015). https://doi.org/10.1093/mnras/stv130

W. Exner, D. Heyner, L. Liuzzo, U. Motschmann, D. Shiota, K. Kusano, T. Shibayama, Coronal mass ejection hits Mercury: A.I.K.E.F. hybrid-code results compared to MESSENGER data. Planet. Space Sci. **153**, 89–99 (2018). https://doi.org/10.1016/j.pss.2017.12.016

W.M. Farrell, T.J. Stubbs, R.R. Vondrak, G.T. Delory, J.S. Halekas, Complex electric fields near the lunar terminator: the near-surface wake and accelerated dust. Geophys. Res. Lett. **34**, L14201 (2007). https://doi.org/10.1029/2007GL029312

W.M. Farrell, J.S. Halekas, R.M. Killen, G.T. Delory, N. Gross, L.V. Bleacher, D. Krauss-Varben, P. Travnicek, D. Hurley, T.J. Stubbs, M.I. Zimmerman, T.L. Jackson, Solar-Storm/Lunar Atmosphere Model (SSLAM): an overview of the effort and description of the driving storm environment. J. Geophys. Res. **117**, E00K04 (2012). https://doi.org/10.1029/2012JE004070

S. Fatemi, M. Holmström, Y. Futaana, The effects of lunar surface plasma absorption and solar wind temperature anisotropies on the solar wind proton velocity space distributions in the low-altitude lunar plasma wake. J. Geophys. Res. Space Phys. **117**, A10 (2012). https://doi.org/10.1029/2011JA017353

S. Fatemi, C. Lue, M. Holmström, A.R. Poppe, M. Wieser, S. Barabash, G.T. Delory, Solar wind plasma interaction with Gerasimovich lunar magnetic anomaly. J. Geophys. Res. Space Phys. **120**(6), 4719–4735 (2015). https://doi.org/10.1002/2015JA021027

S. Fatemi, N. Poirier, M. Holmström, J. Lindkvist, M. Wieser, S. Barabash, A modelling approach to infer the solar wind dynamic pressure from magnetic field observations inside Mercury's magnetosphere. Astron. Astrophys. **A614**, A132 (2018). https://doi.org/10.1051/0004-6361/201832764

S. Fatemi, A.R. Poppe, S. Barabash, Hybrid simulations of solar wind proton precipitation to the surface of Mercury. J. Geophys. Res. **125**, e2019JA027706 (2020). https://doi.org/10.1029/2019JA027706

W.C. Feldman, D.J. Lawrence, R.C. Elphic, B.L. Barraclough, S. Maurice, I. Genetay, A.B. Binder, Polar hydrogen deposits on the Moon. J. Geophys. Res. **105**, 4175–4176 (2000). https://doi.org/10.1029/1999JE001129

L. Fletcher, B.R. Dennis, H.S. Hudson, S. Krucker, K. Phillips, A. Veronig, M. Battaglia, L. Bone, A. Caspi, Q. Chen, P. Gallagher, P.T. Grigis, H. Ji, W. Liu, R.O. Milligan, M. Temmer, An observational overview of solar flares. Space Sci. Rev. **159**, 19 (2011). https://doi.org/10.1007/s11214-010-9701-8

N. Fray, B. Schmitt, Sublimation of ices of astrophysical interest: a bibliographic review. Planet. Space Sci. **57**, 2053–2080 (2009). https://doi.org/10.1016/j.pss.2009.09.011

C. Fröhlich, Total solar irradiance: what have we learned from the last three cycles and the recent minimum? Space Sci. Rev. **176**, 237–252 (2013). https://doi.org/10.1007/s11214-011-9780-1

H.O. Funsten, F. Allegrini, P.A. Bochsler, S.A. Fuselier, M. Gruntman, K. Henderson, P.H. Janzen, R.E. Johnson, B.A. Larsen, D.J. Lawrence, D.J. McComas, E. Möbius, D.B. Reisenfeld, D. Rodríguez, N.A. Schwadron, P. Wurz, Reflection of solar wind hydrogen from the lunar surface. J. Geophys. Res. **118**(2), 292–305 (2013). https://doi.org/10.1002/jgre.20055

Y. Futaana, S. Barabash, M. Wieser, M. Holmström, A. Bhardwaj, M.B. Dhanya, R. Sridharan, P. Wurz, A. Schaufelberger, K. Asamura, Protons in the near-lunar wake observed by the sub-keV atom reflection analyzer on board Chandrayaan-1. J. Geophys. Res. Space Phys. **115**(A10), A10248 (2010). https://doi.org/10.1029/2010JA015264

Y. Futaana, S. Barabash, M. Wieser, M. Holmström, C. Lue, P. Wurz, A. Schaufelberger, A. Bhardwaj, M.B. Dhanya, K. Asamura, Empirical energy spectra of neutralized solar wind protons from the lunar regolith. J. Geophys. Res. **117**, E05005 (2012). https://doi.org/10.1029/2011JE004019

Y. Futaana, S. Barabash, M. Wieser, C. Lue, P. Wurz, A. Vorburger, A. Bhardwaj, K. Asamura, Remote energetic neutral atom imaging of electric potential over a lunar magnetic anomaly. Geophys. Res. Lett. **40**, 262–266 (2013). https://doi.org/10.1002/grl.50135

E. Gaidos, N. Moskovitz, D.M. Williams, Terrestrial exoplanet light curves, in *Proc. International Astronomical Union Colloquium*, vol. 200 (2006), pp. 153–158. https://doi.org/10.1017/S1743921306009239

D. Gamborino, A. Vorburger, P. Wurz, Mercury's sodium exosphere: an ab initio calculation to interpret MESSENGER observations. Ann. Geophys. **37**, 455–470 (2019). https://doi.org/10.5194/angeo-2018-109

D.J. Gershman, J.A. Slavin, J.M. Raines, T.H. Zurbuchen, B.J. Anderson, H. Korth, D.N. Baker, S.C. Solomon, Magnetic flux pileup and plasma depletion in Mercury's subsolar magnetosheath. J. Geophys. Res. **118**, 7181–7199 (2013). https://doi.org/10.1002/2013JA019244

A. Goehlich, D. Gillmann, H.F. Döbele, Angular resolved energy distributions of sputtered atoms at low bombarding energy. Nucl. Instrum. Methods Phys. Res. B **164–165**, 834–839 (2000). https://doi.org/10.1016/S0168-583X(99)01106-4

W. Gonzalez, E. Parker (eds.), *Magnetic Reconnection: Concepts and Applications*. Astrophysics and Space Science Library, vol. 427 (Springer, Cham, 2016). 549 p. ISBN 9783319264301

J.T. Gosling, R.T. Hansen, S.J. Bame, Solar wind speed distributions: 1962–1970. J. Geophys. Res. **76**(7), 1811–1815 (1971). https://doi.org/10.1029/JA076i007p01811

J.T. Gosling, J.R. Asbridge, S.J. Bame, G. Paschmann, N. Sckopke, Observations of two distinct populations of bow shock ions in the upstream solar wind. Geophys. Res. Lett. **5**(11), 957–960 (1978). https://doi.org/10.1029/GL005i011p00957

J.N. Goswami, M. Annadurai, Chandrayaan-1 mission to the Moon. Acta Astronaut. **63**(11), 1215–1220 (2008). https://doi.org/10.1016/j.actaastro.2008.05.013

C. Grava, R.M. Killen, M. Benna, A.A. Berezhnoy, J.S. Halekas, F. Leblanc, M.N. Nishino, C. Plainaki, J.M. Raines, M. Sarantos, B.D. Teolis, O.J. Tucker, R.J. Vervack, A. Vorburger, Extreme volatiles and refractories in surface-bounded exospheres. Space Sci. Rev. **217**, 61 (2021). https://doi.org/10.1007/s11214-021-00833-8

E.W. Greenstadt, C.T. Russell, M. Hoppe, Magnetic field orientation and suprathermal ion streams in the Earth's foreshock. J. Geophys. Res. Space Phys. **85**(A7), 3473–3479 (1980). https://doi.org/10.1029/JA085iA07p03473

K.I. Gringauz, V.V. Bezrukikh, V.D. Ozerov, R.E. Ribchinsky, Some results of experiments in interplanetary space by means of charged particle traps on Soviet space probes. Space Res. **2**, 539–553 (1961)

E. Grün, M. Horanyi, Z. Sternovsky, The lunar dust environment. Planet. Space Sci. **59**, 1672–1680 (2011). https://doi.org/10.1016/j.pss.2011.04.005

M.N. Günther, Z. Zhan, S. Seager, P.B. Rimmer, S. Ranjan, K.G. Stassun, R.J. Oelkers, T. Daylan, E. Newton, M.H. Kristiansen, K. Olah, E. Gillen, S. Rappaport, G.R. Ricker, R.K. Vanderspek, D.W. Latham, J.N. Winn, J.M. Jenkins, A. Glidden, M. Fausnaugh, A.M. Levine, J.A. Dittmann, S.N. Quinn, A. Krishnamurthy, E.B. Ting, Stellar flares from the first TESS data release: exploring a new sample of M dwarfs. Astron. J. **159**, 60 (2020). https://doi.org/10.3847/1538-3881/ab5d3a

A.S. Hale, B. Hapke, A time-dependent model of radiative and conductive thermal energy transport in planetary regoliths with applications to the Moon and Mercury. Icarus **156**, 318–334 (2002). https://doi.org/10.1006/icar.2001.6768

J.S. Halekas, S.D. Bale, D.L. Mitchell, R.P. Lin, Electrons and magnetic fields in the lunar plasma wake. J. Geophys. Res. **110**, A07222 (2005). https://doi.org/10.1029/2004JA010991

J.S. Halekas, D.A. Brain, D.L. Mitchell, R.P. Lin, Whistler waves observed near lunar crustal magnetic sources. Geophys. Res. Lett. **33**, L22104 (2006). https://doi.org/10.1029/2006GL027684

J.S. Halekas, G.T. Delory, R.P. Lin, T.J. Stubbs, W.M. Farrell, Lunar Prospector observations of the electrostatic potential of the lunar surface and its response to incident currents. J. Geophys. Res. **113**, A09102 (2008a). https://doi.org/10.1029/2008JA013194

J.S. Halekas, D.A. Brain, R.P. Lin, D.L. Mitchell, Solar wind interaction with lunar crustal magnetic anomalies. Adv. Space Res. **41**(8), 1319–1324 (2008b). https://doi.org/10.1016/j.asr.2007.04.003

J.S. Halekas, V. Angelopoulos, D.G. Sibeck, K.K. Khurana, C.T. Russell, G.T. Delory, W.M. Farrell, J.P. McFadden, J.W. Bonnell, D. Larson, R.E. Ergun, F. Plaschke, K.H. Glassmeier, First results from ARTEMIS, a new two-spacecraft lunar mission: counter-streaming plasma populations in the lunar wake. Space Sci. Rev. **165**, 93–107 (2011). https://doi.org/10.1007/s11214-010-9738-8

J.S. Halekas, A.R. Poppe, W.M. Farrell, G.T. Delory, V. Angelopoulos, J.P. McFadden, J.W. Bonnell, K.H. Glassmeier, F. Plaschke, A. Roux, R.E. Ergun, Lunar precursor effects in the solar wind and terrestrial magnetosphere. J. Geophys. Res. **117**, A05101 (2012). https://doi.org/10.1029/2011JA017289

J.S. Halekas, A.R. Poppe, J.P. McFadden, K.-H. Glassmeier, The effects of reflected protons on the plasma environment of the moon for parallel interplanetary magnetic fields. Geophys. Res. Lett. **40**(17), 4544–4548 (2013). https://doi.org/10.1002/grl.50892

J.S. Halekas, A.R. Poppe, J.P. McFadden, The effects of solar wind velocity distributions on the refilling of the lunar wake: ARTEMIS observations and comparisons to one-dimensional theory. J. Geophys. Res. Space Phys. **119**, 5133–5149 (2014a). https://doi.org/10.1002/2014JA020083

J.S. Halekas, A.R. Poppe, J.P. McFadden, V. Angelopoulos, K.-H. Glassmeier, D.A. Brain, Evidence for small-scale collisionless shocks at the Moon from ARTEMIS. Geophys. Res. Lett. **41**, 7436–7443 (2014b). https://doi.org/10.1002/2014GL061973

J.S. Halekas, D.A. Brain, M. Holmström, M. Plasma Wake, in *Magnetotails in the Solar System*, ed. by e.A. Keiling, C.M. Jackman, P.A. Delamere (2015), pp. 149–167. https://doi.org/10.1002/9781118842324.ch9

B. Hapke, Space weathering from Mercury to the asteroid belt. J. Geophys. Res. **106**, 10039–10073 (2001). https://doi.org/10.1029/2000JE001338

Y. Harada, J.S. Halekas, Upstream waves and particles at the Moon, in *Low-Frequency Waves in Space Plasmas*, ed. by e.A. Keiling, D.-H. Lee, V. Nakariakov (2016). https://doi.org/10.1002/9781119055006.ch18

Y. Harada, S. Machida, Y. Saito, S. Yokota, K. Asamura, M.N. Nishino, T. Tanaka, H. Tsunakawa, H. Shibuya, F. Takahashi, M. Matsushima, H. Shimizu, Interaction between terrestrial plasma sheet electronsand the lunar surface: SELENE (Kaguya) observations. Geophys. Res. Lett. **37**, L19202 (2010). https://doi.org/10.1029/2010GL044574

Y. Harada, Y. Futaana, S. Barabash, M. Wieser, P. Wurz, A. Bhardwaj, K. Asamura, Y. Saito, S. Yokota, H. Tsunakawa, S. Machida, Backscattered energetic neutral atoms from the Moon in the Earth's plasma sheet observed by Chandarayaan-1/sub-keV atom reflecting analyzer instrument. J. Geophys. Res. **119**(5), 3573–3584 (2014a). https://doi.org/10.1002/2013JA019682

Y. Harada, J.S. Halekas, A.R. Poppe, S. Kurita, J.P. McFadden, Extended lunar precursor regions: electron-wave interaction. J. Geophys. Res. **119**, 9160–9173 (2014b). https://doi.org/10.1002/2014JA020618

R.A. Haring, A.W. Kolfschoten, A.E. De Vries, Chemical sputtering by keV ions. Nucl. Instrum. Methods Phys. Res. B **2**(1–3), 544–549 (1984). https://doi.org/10.1016/0168-583X(84)90263-5

W.K. Hartmann, Impact experiments, 1, ejecta velocity distributions and related results from regolith targets. Icarus **63**, 69–98 (1985). https://doi.org/10.1016/0019-1035(85)90021-1

G. Hayderer, M. Schmid, P. Varga, H. Winter, F. Aumayr, A highly sensitive quartz-crystal microbalance for sputtering investigations in slow ion–surface collisions. Rev. Sci. Instrum. **70**, 3696 (1999). https://doi.org/10.1063/1.1149979

P. Hedelt, Y. Ito, H.U. Keller, R. Reulke, P. Wurz, H. Lammer, H. Rauer, L. Esposito, Titan's atomic hydrogen corona. Icarus **210**, 424–435 (2010). https://doi.org/10.1016/j.icarus.2010.06.012

P. Heinzel, L. Kleint, Hydrogen balmer continuum in solar flares detected by the Interface Region Imaging Spectrograph (IRIS). Astrophys. J. Lett. **794**, L23 (2014). https://doi.org/10.1088/2041-8205/794/2/L23

D. Hercík, P.M. Trávníček, Š. Štverák, P. Hellinger, Properties of Hermean plasma belt: numerical simulations and comparison with MESSENGER data. J. Geophys. Res. **121**, 413–431 (2016). https://doi.org/10.1002/2015JA021938

D. Heyner, C. Nabert, E. Liebert, K.-H. Glassmeier, Concerning reconnection-induction balance at the magnetopause of Mercury. J. Geophys. Res. Space Phys. **121**, 2935–2961 (2016). https://doi.org/10.1002/2015JA021484

H. Hijazi, M.E. Bannister, H.M. Meyer III, C.M. Rouleau, A.F. Barghouty, D.L. Rickman, F.W. Meyer, Anorthite sputtering by H^+ and Ar^{q+} (q=1-9) at solar wind velocities. J. Geophys. Res. Space Phys. **119**, 8006 (2014). https://doi.org/10.1002/2014JA020140

H. Hijazi, M.E. Bannister, H.M. Meyer III, C.M. Rouleau, F.W. Meyer, Kinetic and potential sputtering of an anorthite-like glassy thin film. J. Geophys. Res., Planets **122**, 1597 (2017). https://doi.org/10.1002/2017JE005300

F.L. Hinton, D.R. Taeusch, Variation of the lunar atmosphere with the strength of the solar wind. J. Geophys. Res. **69**(7), 1341–1347 (1964)

G.C. Ho, S.M. Krimigis, R.E. Gold, D.N. Baker, J.A. Slavin, B.J. Anderson, H. Lorth, R.D. Starr, D.J. Lawrence, R.R. McNutt, S.C. Solomon, MESSENGER observations of transient bursts of energetic electrons in Mercury's magnetosphere. Science **333**, 1865–1868 (2011a). https://doi.org/10.1126/science.1211141

G.C. Ho, R.D. Starr, R.E. Gold, S.M. Krimigis, J.A. Slavin, D.N. Baker, B.J. Anderson, R.L. McNutt, L.R. Nittler, S.C. Solomon, Observations of suprathermal electrons in Mercury's magnetosphere during the three MESSENGER flybys. Planet. Space Sci. **59**, 2016–2025 (2011b). https://doi.org/10.1016/j.pss.2011.01.011

G.C. Ho, R.D. Starr, S.M. Krimigis, J.D. Vandegriff, D.N. Baker, R.E. Gold, B.J. Anderson, H. Korth, D. Schriver, R.L. McNutt Jr., S.C. Solomon, MESSENGER observations of suprathermal electrons in Mercury's magnetosphere. Geophys. Res. Lett. **43**, 550–555 (2016). https://doi.org/10.1002/2015GL066850

R.R. Hodges Jr., Differential equation for exospheric lateral transportation and its application to terrestrial hydrogen. J. Geophys. Res. **78**(31), 7340–7346 (1973)

R.R. Hodges Jr., Methods for Monte Carlo simulation of the exospheres of the Moon and Mercury. J. Geophys. Res. **85**(A1), 164–169 (1980)

W.O. Hofer, Angular, energy, and mass distribution of sputtered particles, in *Sputtering by Particle Bombardment III*, ed. by R. Behrisch, K. Wittmaack. Topics in Applied Physics, vol. 64 (Springer, Berlin, 1991). https://doi.org/10.1007/3540534288_16

M. Holmström, M. Wieser, S. Barabash, Y. Futaana, A. Bhardwaj, Dynamics of solar wind protons reflected by the Moon. J. Geophys. Res. **115**, A06206 (2010). https://doi.org/10.1029/2009JA014843

K.A. Holsapple, The scaling of impact processes in planetary sciences. Annu. Rev. Earth Planet. Sci. **21**, 333–373 (1993). https://doi.org/10.1146/annurev.ea.21.050193.002001

K.A. Holsapple, K.R. Housten (2020). http://keith.aa.washington.edu/craterdata/scaling/index.htm

L.L. Hood, A. Zakharian, J. Halekas, D.L. Mitchell, R.P. Lin, M.H. Acuña, A.B. Binder, Initial mapping and interpretation of lunar crustal magnetic anomalies using Lunar Prospector magnetometer data. J. Geophys. Res. **106**(E11), 27825–27839 (2001). https://doi.org/10.1029/2000JE001366

K.R. Housen, R.M. Schmidt, K.A. Holsapple, Crater ejecta scaling laws: fundamental forms based on dimensional analysis. J. Geophys. Res. **88**(B3), 2485–2499 (1983). https://doi.org/10.1029/JB088iB03p02485

W.S. Howard, H. Corbett, N.M. Law, J.K. Ratzloff, N. Galliher, A.L. Glazier, R. Gonzalez, A.V. Soto, O. Fors, D. del Ser, J. Haislip, EvryFlare. III. Temperature evolution and habitability impacts of dozens of superflares observed simultaneously by evryscope and TESS. Astrophys. J. **902**, 115 (2020). https://doi.org/10.3847/1538-4357/abb5b4

H.S. Hudson, T.N. Woods, P.C. Chamberlin, L. Fletcher, G. Del Zanna, L. Didkovsky, N. Labrosse, D. Graham, The EVE Doppler sensitivity and flare observations. Sol. Phys. **273**, 69–80 (2011). https://doi.org/10.1007/s11207-011-9862-y

D.M. Hunten, T.M. Morgan, D.M. Shemansky, The Mercury Atmosphere, in *Mercury (A89-43751 19-91)* (University of Arizona Press, Tucson, 1988), pp. 562–612

W. Husinsky, The application of Doppler shift laser fluorescence spectroscopy for the detection and energy analysis of particles evolving from surfaces. J. Vac. Sci. Technol., B Microelectron. Process. Phenom. **3**, 1546 (1985). https://doi.org/10.1116/1.582983

W. Husinsky, I. Girgis, G. Betz, Doppler shift laser fluorescence spectroscopy of sputtered and evaporated atoms under Ar+ bombardment. J. Vac. Sci. Technol. B **3**, 1543–1545 (1985). https://doi.org/10.1116/1.582982

S.M. Imber, J.A. Slavin, MESSENGER observations of magnetotail loading and unloading: implications for substorms at Mercury. J. Geophys. Res. **122**(11), 402–411 (2017). https://doi.org/10.1002/2017JA024332. p. 412

W.-H. Ip, A. Kopp, MHD simulations of the solar wind interaction with Mercury. J. Geophys. Res. **107**(A11), 1348 (2002). https://doi.org/10.1029/2001JA009171

N. Jäggi, A. Galli, P. Wurz, H. Biber, P.S. Szabo, J. Bröotzner, F. Aumayr, P.M.E. Tollan, K. Mezger, Creation of Lunar and Hermean analogue mineral powder samples for solar wind irradiation experiments and thermal infrared spectra analysis. Icarus **365**, 114492 (2021). https://doi.org/10.1016/j.icarus.2021.114492

D. Janches, A.A. Berezhnoy, A.A. Christou, G. Cremonese, T. Hirai, M. Horányi, J.M. Jasinski, M. Sarantos, Meteoroids as one of the sources for exosphere formation on airless bodies in the inner solar system. Space Sci. Rev. **217**, 50 (2021). https://doi.org/10.1007/s11214-021-00827-6

R. Jarvinen, M. Alho, E. Kallio, P. Wurz, S. Barabash, Y. Futaana, On vertical electric fields at lunar magnetic anomalies. Geophys. Res. Lett. **41**, 2243–2249 (2014). https://doi.org/10.1002/2014GL059788

J.M. Jasinski, J.A. Slavin, J.M. Raines, G.A. DiBraccio, Mercury's solar wind interaction as characterized by magnetospheric plasma mantle observations with MESSENGER. J. Geophys. Res. Space Phys. **122**, 12,153–12,169 (2017). https://doi.org/10.1002/2017JA024594

X. Jia, J.A. Slavin, T.I. Gombosi, L.K.S. Daldorff, G. Toth, B. van der Holst, Global MHD simulations of Mercury's magnetosphere with coupled planetary interior: induction effect of the planetary conducting core on the global interaction. J. Geophys. Res. Space Phys. **120**, 4763–4775 (2015). https://doi.org/10.1002/2015JA021143

X. Jia, J.A. Slavin, G. Poh, G.A. DiBraccio, G. Toth, Y. Chen, J.M. Raines, T.I. Gombosi, MESSENGER observations and global simulations of highly compressed magnetosphere events at Mercury. J. Geophys. Res. **124**, 229–247 (2019). https://doi.org/10.1029/2018JA026166

F.S. Johnson, Lunar atmosphere. Rev. Geophys. Space Phys. **9**(3), 813–823 (1971)

C.L. Johnson, M.E. Purucker, H. Korth, B.J. Anderson, R.M. Winslow, M.M.H. Al Asad, J.A. Slavin, I.I. Alexeev, R.J. Phillips, M.T. Zuber, S.C. Solomon, MESSENGER observations of Mercury's magnetic field structure. J. Geophys. Res. **117**, E00L14 (2012). https://doi.org/10.1029/2012JE004217

A.P. Jordan, Evidence for dielectric breakdown weathering on the Moon. Icarus **358**, 114199 (2021). https://doi.org/10.1016/j.icarus.2020.114199

A.P. Jordan, T.J. Stubbs, J.K. Wilson, N.A. Schwadron, H.E. Spence, Dielectric breakdown weathering of the Moon's polar regolith. J. Geophys. Res., Planets **120**, 210–225 (2015). https://doi.org/10.1002/2014JE004710

A.P. Jordan, T.J. Stubbs, J.K. Wilson, N.A. Schwadron, H.E. Spence, The rate of dielectric breakdown weathering of lunar regolith in permanently shadowed regions. Icarus **283**, 352–358 (2017). https://doi.org/10.1016/j.icarus.2016.08.027

K. Kabin, T.I. Gombosi, D.L. DeZeeuw, K.G. Powell, Interaction of Mercury with the solar wind. Icarus **143**, 379–406 (2000). https://doi.org/10.1006/icar.1999.6252

E. Kallio, P. Janhunen, Solar wind and magnetospheric ion impact on Mercury's surface. Geophys. Res. Lett. **30**(17), 1877 (2003). https://doi.org/10.1029/2003GL017842

E. Kallio, P. Janhunen, The response of the Hermean magnetosphere to the interplanetary magnetic field. Adv. Space Res. **33**, 2176–2181 (2004). https://doi.org/10.1016/S0273-1177(03)00447-2

E. Kallio, P. Wurz, R. Killen, S. McKenna-Lawlor, A. Milillo, A. Mura, S. Massetti, S. Orsini, H. Lammer, P. Janhunen, W-H. Ip, On the impact of multiply charged heavy solar wind ions on the surface of Mercury, the Moon and Ceres. Planet. Space Sci. **56**, 1506–1516 (2008). https://doi.org/10.1016/j.pss.2008.07.018

E. Kallio, R. Järvinen, S. Dyadechkin, P. Wurz, S. Barabash, F. Alvarez, V. Fernandes, Y. Futaana, A.M. Harri, J. Heilimo, C. Lue, J. Mäkelä, N. Porjo, W. Schmidt, T. Silli, Kinetic simulations of finite gyroradius effects in the Lunar plasma environment on global, meso, and microscales. Planet. Space Sci. **74**, 146–155 (2012). https://doi.org/10.1016/j.pss.2012.09.012

E. Kallio, S. Dyadechkin, P. Wurz, M. Khodachenko, Space weathering on the Moon: farside-nearside solar wind precipitation asymmetry. Planet. Space Sci. **166**, 9–22 (2019). https://doi.org/10.1016/j.pss.2018.07.013

S. Kameda, I. Yoshikawa, M. Kagitani, S. Okano, Interplanetary dust distribution and temporal variability of Mercury's atmospheric Na. Geophys. Res. Lett. **36**, L15201 (2009). https://doi.org/10.1029/2009GL039036

Y. Kazama, S. Barabash, M. Wieser, K. Asamura, P. Wurz, in *An LENA Instrument Onboard BepiColombo and Chandrayaan-1*. AIP Conf. Proc. CP1144 (2009), pp. 109–113. https://doi.org/10.1063/1.3169273

R. Kelly, The surface binding energy in slow collisional sputtering. Nucl. Instrum. Methods Phys. Res. B **18**(1–6), 388–398 (1986). https://doi.org/10.1016/S0168-583X(86)80063-5

R.M. Killen, Source and maintenance of the argon atmospheres of Mercury and the Moon. Meteorit. Planet. Sci. **37**, 1223–1231 (2002). https://doi.org/10.1111/j.1945-5100.2002.tb00891.x

R.M. Killen, J.M. Hahn, Impact vaporization as a possible source of Mercury's calcium exosphere. Icarus **250**, 230–237 (2015). https://doi.org/10.1016/j.icarus.2014.11.035

R.M. Killen, W.-H. Ip, The surface-bounded atmospheres of Mercury and the Moon. Rev. Geophys. **37**, 361–406 (1999). https://doi.org/10.1029/1999RG900001

R.M. Killen, M. Sarantos, A.E. Potter, P. Reiff, Source rates and ion recycling rates for Na and K in Mercury's atmosphere. Icarus **171**, 1–19 (2004). https://doi.org/10.1016/j.icarus.2004.04.007

R. Killen, G. Cremonese, H. Lammer, S. Orsini, A.E. Potter, A.L. Sprague, P. Wurz, M. Khodachenko, H.I.M. Lichtenegger, A. Milillo, A. Mura, Processes that promote and deplete the exosphere of Mercury. Space Sci. Rev. **132**, 433–509 (2007). https://doi.org/10.1007/s11214-007-9232-0

R.M. Killen, D.M. Hurley, W.M. Farrell, The effect on the lunar exosphere of a coronal mass ejection passage. J. Geophys. Res. **117**, E00K02 (2012). https://doi.org/10.1029/2011JE004011

D. Koga, W.D. Gonzalez, V.M. Souza, F.R. Cardoso, C. Wang, Z.K. Liu, Dayside magnetopause reconnection: its dependence on solar wind and magnetosheath conditions. J. Geophys. Res. **124**, 8778–8787 (2019). https://doi.org/10.1029/2019JA026889

L. Kööp, P.R. Heck, H. Busemann, A.M. Davis, J. Greer, C. Maden, M.M.M. Meier, R. Wieler, High early solar activity inferred from helium and neon excesses in the oldest meteorite inclusions. Nat. Astron. **2**, 709–713 (2018). https://doi.org/10.1038/s41550-018-0527-8

P. Kotrč, O. Procházka, P. Heinzel, New observations of Balmer continuum flux in solar flares instrument description and first results. Sol. Phys. **291**, 779–789 (2016). https://doi.org/10.1007/s11207-016-0860-y

M. Kretzschmar, The Sun as a star: observations of white-light flares. Astron. Astrophys. **530**, A84 (2011). https://doi.org/10.1051/0004-6361/201015930

K.R. Kuhlman, K. Sridharan, A. Kvit, Simulation of solar wind space weathering in orthopyroxene. Planet. Space Sci. **115**, 110 (2015). https://doi.org/10.1016/j.pss.2015.04.003

M. Küster, W. Eckstein, V. Dose, J. Roth, The influence of surface roughness on the angular dependence of the sputter yield. Nucl. Instrum. Methods B **145**, 320–331 (2000)

M. Küstner, W. Eckstein, V. Dose, J. Roth, The influence of surface roughness on the angular dependence of the sputter yield. Nucl. Instrum. Methods Phys. Res., Sect. B, Beam Interact. Mater. Atoms **145**(3), 320–331 (1998). https://doi.org/10.1016/S0168-583X(98)00399-1

M. Küstner, W. Eckstein, E. Hechtl, J. Roth, Angular dependence of the sputtering yield of rough beryllium surfaces. J. Nucl. Mater. **265**, 22 (1999). https://doi.org/10.1016/S0022-3115(98)00648-5

H. Lammer, S.J. Bauer, Mercury's exosphere: origin of sputtering and implications. Planet. Space Sci. **45**, 73–79 (1997). https://doi.org/10.1016/S0032-0633(96)00097-9

H. Lammer, S. Bauer, *Planetary Aeronomy* (Springer, Berlin, 2004). ISBN-13 9783540214724

H. Lammer, P. Wurz, M.R. Patel, R. Killen, C. Kolb, S. Massetti, S. Orsini, A. Milillo, The variability of Mercury's exosphere by particle and radiation induced surface release processes. Icarus **166**(2), 238–247 (2003). https://doi.org/10.1016/j.icarus.2003.08.012

Y. Langevin, The regolith of Mercury: present knowledge and implications for the Mercury Orbiter mission. Planet. Space Sci. **45**(1), 31–37 (1997). https://doi.org/10.1016/S0032-0633(96)00098-0

D.J. Lawrence, B.J. Anderson, D.N. Baker, W.C. Feldman, G.C. Ho, H. Korth, R.L. McNutt Jr., P.N. Peplowski, S.C. Solomon, R.D. Starr, J.D. Vandegriff, R.M. Winslow, Comprehensive survey of energetic electron events in Mercury's magnetosphere with data from the MESSENGER Gamma-Ray and Neutron Spectrometer. J. Geophys. Res. Space Phys. **120**, 2851–2876 (2015). https://doi.org/10.1002/2014JA020792

F. Leblanc, R.E. Johnson, Mercury's sodium exosphere. Icarus **164**, 261–281 (2013). https://doi.org/10.1016/S0019-1035(03)00147-7

F. Leblanc, C. Schmidt, V. Mangano, A. Mura, G. Cremonese, J.M. Raines, J.M. Jasinski, M. Sarantos, A. Milillo, R.M. Killen, T. Cassidy, R.J. Vervack Jr., S. Kameda, M.T. Capria, M. Horanyi, D. Janches, A. Berezhnoy, A. Christou, T. Hirai, P. Lierle, J. Morgenthaler, Comparative Na and K Mercury and Moon exospheres. Space Sci. Rev. **218**, 2 (2022)

L.C. Lee, Z.F. Fu, A theory of magnetic flux transfer at the Earth's magnetopause. Geophys. Res. Lett. **12**(2), 105–108 (1985). https://doi.org/10.1029/GL012i002p00105

S.T. Lepri, T.H. Zurbuchen, Iron charge state distributions as an indicator of hot ICMEs: possible sources and temporal and spatial variations during solar maximum. J. Geophys. Res. **109**, A01112 (2004). https://doi.org/10.1029/2003JA009954

S.T. Lepri, T.H. Zurbuchen, Direct observational evidence of filament material within interplanetary coronal mass ejections. Astrophys. J. Lett. **723**, L22–L27 (2010). https://doi.org/10.1088/2041-8205/723/1/L22

R.P. Leyser, S.M. Imber, S.E. Milan, J.A. Slavin, The influence of IMF clock angle on dayside flux transfer events at Mercury. Geophys. Res. Lett. **44**, 10,829–10,837 (2017). https://doi.org/10.1002/2017GL074858

X. Li, F. Guo, H. Li, G. Li, Particle acceleration during magnetic reconnection in a low-beta plasma. Astrophys. J. **843**, 1 (2017). https://doi.org/10.3847/1538-4357/aa745e

R.P. Lin, D.L. Mitchell, D.W. Curtis, K.A. Anderson, C.W. Carlson, J. McFadden, M.H. Acuña, L.L. Hood, A. Binder, Lunar surface magnetic fields and their interaction with the solar wind: results from lunar prospector. Science **281**(5382), 1480–1484 (1998). https://doi.org/10.1126/science.281.5382.1480

S.T. Lindsay, M.K. James, E.J. Bunce, S.M. Imbera, H. Korth, A. Martindale, T.K. Yeoman, MESSENGER X-ray observations of magnetosphere–surface interaction on the nightside of Mercury. Planet. Space Sci. **125**, 72–79 (2016). https://doi.org/10.1016/j.pss.2016.03.005

M.J. Loeffler, C.A. Dukes, R.A. Baragiola, Irradiation of olivine by 4 keV He$^+$: simulation of space weathering by the solar wind. J. Geophys. Res. **114**, E03003 (2009). https://doi.org/10.1029/2008JE003249

E.A. Lucek, D. Constantinescu, M.L. Goldstein, J. Pickett, J.L. Pinçon, F. Sahraoui, R.A. Treumann, S.N. Walker, The magnetosheath. Space Sci. Rev. **118**, 95–152 (2005). https://doi.org/10.1007/s11214-005-3825-2

P.G. Lucey, D.T. Blewett, G.J. Taylor, B.R. Hawke, Imaging of lunar surface maturity. J. Geophys. Res. **105**(E8), 20377–20386 (2000). https://doi.org/10.1029/1999JE001110

C. Lue, Y. Futaana, S. Barabash, M. Wieser, M. Holmström, A. Bhardwaj, M.B. Dhanya, P. Wurz, Strong influence of lunar crustal fields on the solar wind flow. Geophys. Res. Lett. **38**, L03202 (2011). https://doi.org/10.1029/2010GL046215

C. Lue, Y. Futaana, S. Barabash, M. Wieser, A. Bhardwaj, P. Wurz, Chandrayaan-1 observations of backscattered solar wind protons from the lunar regolith: dependence on the solar wind speed. J. Geophys. Res., Planets **119**, 968–975 (2014). https://doi.org/10.1002/2013JE004582

C. Lue, Y. Futaana, S. Barabash, Y. Saito, M. Nishino, M. Wieser, K. Asamura, A. Bhardwaj, P. Wurz, Scattering characteristics and imaging of energetic neutral atoms from the Moon in the terrestrial magnetosheath. J. Geophys. Res. Space Phys. **121**(1), 432–445 (2016). https://doi.org/10.1002/2015JA021826

C. Lue, Y. Futaana, S. Barabash, M. Wieser, A. Bhardwaj, P. Wurz, K. Asamura, Solar wind scattering from the surface of Mercury: lessons from the Moon. Icarus **296**, 39–48 (2017). https://doi.org/10.1016/j.icarus.2017.05.019

C. Lue, J.S. Halekas, A.R. Poppe, J.P. McFadden, ARTEMIS observations of solar wind proton scattering off the lunar surface. J. Geophys. Res. Space Phys. **123**, 5289–5299 (2018). https://doi.org/10.1029/2018JA025486

E.F. Lyon, H.S. Bridge, J.H. Binsack, Explorer 35 plasma measurements in the vicinity of the Moon. J. Geophys. Res. **72**(23), 6113–6117 (1967). https://doi.org/10.1029/JZ072i023p06113

T.E. Madey, B.V. Yakshinskiy, V.N. Ageev, R.E. Johnson, Desorption of alkali atoms and ions from oxide surfaces: relevance to origins of Na and K in atmospheres of Mercury and the Moon. J. Geophys. Res. **103**(E3), 5873–5887 (1998). https://doi.org/10.1029/98JE00230

T.E. Madey, R.E. Johnson, T.M. Orlando, Far-out surface science: radiation-induced surface processes in the solar system. Surf. Sci. **500**, 838–858 (2002). https://doi.org/10.1016/S0039-6028(01)01556-4

H. Maehara, T. Shibayama, S. Notsu, Y. Notsu, T. Nagao, S. Kusaba, S. Honda, D. Nogami, K. Shibata, Superflares on solar-type stars. Nature **485**, 478–481 (2012). https://doi.org/10.1038/nature11063

H. Maehara, T. Shibayama, Y. Notsu, S. Notsu, S. Honda, D. Nogami, K. Shibata, Statistical properties of superflares on solar-type stars based on 1-min cadence data. Earth Planets Space **67**, 59 (2015). https://doi.org/10.1186/s40623-015-0217-z

A. Mallama, D. Wang, R.A. Howard, Photometry of Mercury from SOHO/LASCO and Earth: the phase function from 2 to 170°. Icarus **155**(2), 253–264 (2002). https://doi.org/10.1006/icar.2001.6723

V. Mangano, A. Milillo, A. Mura, S. Orsini, E. DeAngelis, A.M. DiLellis, P. Wurz, The contribution of impact-generated vapour to the hermean atmosphere. Planet. Space Sci. **55**(11), 1541–1556 (2007). https://doi.org/10.1016/j.pss.2006.10.008

S. Marchi, S. Mottola, G. Cremonese, M. Massironi, E. Martellato, A new chronology for the Moon and Mercury. Astron. J. **137**, 4936–4948 (2009). https://doi.org/10.1088/0004-6256/137/6/4936

E. Marsch, Kinetic physics of the solar corona and solar wind. Living Rev. Sol. Phys. **3**, 1 (2006). https://doi.org/10.12942/lrsp-2006-1

E. Marsch, K.-H. Mühlhäuser, R. Schwenn, H. Rosenbauer, W. Pilipp, F.M. Neubauer, Solar wind protons: three-dimensional velocity distributions and derived plasma parameters measured between 0.3 and 1 AU. J. Geophys. Res. Space Phys. **87**(A1), 52–72 (1982). https://doi.org/10.1029/JA087iA01p00052

R. Martinez, Th. Langlinay, C.R. Ponciano, E.F. da Silveira, M.E. Palumbo, G. Strazzulla, J.R. Brucato, H. Hijazi, A.N. Agnihotri, P. Boduch, A. Cassimi, A. Domaracka, F. Ropars, H. Rothard, Sputtering of sodium and potassium from nepheline: secondary ion yields and velocity spectra. Nucl. Instrum. Methods Phys. Res. B **406**, 523 (2017). https://doi.org/10.1016/j.nimb.2017.01.042

S. Massetti, S. Orsini, A. Milillo, A. Mura, E. De Angelis, H. Lammer, P. Wurz, Mapping of the cusp plasma precipitation on the surface of Mercury. Icarus **166**, 229–237 (2003). https://doi.org/10.1016/j.icarus.2003.08.005

S. Massetti, S. Orsini, A. Milillo, A. Mura, Modelling Mercury's magnetosphere and plasma entry through the dayside magnetopause. Planet. Space Sci. **55**, 1557–1568 (2007). https://doi.org/10.1016/j.pss.2006.12.008

S. Massetti, V. Mangano, A. Milillo, A. Mura, S. Orsini, C. Plainaki, Short-term observations of double-peaked Na emission from Mercury's exosphere. Geophys. Res. Lett. **44**, 2970–2977 (2017). https://doi.org/10.1002/2017GL073090

A. Masters, A more viscous-like solar wind interaction with all the giant planets. Geophys. Res. Lett. **45**, 7320–7329 (2018). https://doi.org/10.1029/2018GL078416

L. Matteini, P. Hellinger, S. Landi, P.M. Trávníček, M. Velli, Ion kinetics in the solar wind: coupling global expansion to local microphysics. Space Sci. Rev. **172**(1–4), 373–396 (2012). https://doi.org/10.1007/s11214-011-9774-z

G. Matthews, Celestial body irradiance determination from an underfilled satellite radiometer: application to albedo and thermal emission measurements of the Moon using CERES. Appl. Opt. **47**, 4981–4993 (2008). https://doi.org/10.1364/AO.47.004981

W.E. McClintock, R.J. Vervack Jr., E.T. Bradley, R.M. Killen, N. Mouawad, A.L. Sprague, M.H. Burger, S.C. Solomon, N.R. Izenberg, MESSENGER observations of Mercury's exosphere: detection of magnesium and distribution of constituents. Science **324**, 610–613 (2009). https://www.jstor.org/stable/20493836

D.J. McComas, F. Allegrini, P. Bochsler, M. Bzowski, M. Collier, H. Fahr, H. Fichtner, H. Funsten, S. Fuselier, G. Gloeckler, M. Gruntman, V. Izmodenov, P. Knappenberger, M. Lee, S. Livi, D. Mitchell, E. Möbius, T. Moore, S. Pope, D. Reisenfeld, E. Roelof, J. Scherrer, N. Schwadron, R. Tyler, M. Wieser, M. Witte, P. Wurz, G. Zank, IBEX—the interstellar boundary explorer. Space Sci. Rev. **146**, 11–33 (2009a). https://doi.org/10.1007/s11214-009-9499-4

D.J. McComas, F. Allegrini, P. Bochsler, P. Frisch, H.O. Funsten, M. Gruntman, P.H. Janzen, H. Kucharek, E. Möbius, D.B. Reisenfeld, N.A. Schwadron, Lunar backscatter and neutralization of the solar wind: first observations of neutral atoms from the Moon. Geophys. Res. Lett. **36**, L12105 (2009b). https://doi.org/10.1029/2009GL038794

G.M. McCracken, The behaviour of surfaces under ion bombardment. Rep. Prog. Phys. **38**(2), 241 (1975). https://doi.org/10.1088/0034-4885/38/2/002

J.L. McLain, A.L. Sprague, G.A. Grieves, D. Schriver, P. Travinicek, T.M. Orlando, Electron-stimulated desorption of silicates: a potential source for ions in Mercury's space environment. J. Geophys. Res. **116**, E03007 (2011). https://doi.org/10.1029/2010JE003714

M. Meftah, L. Damé, D. Bolsée, A. Hauchecorne, N. Pereira, D. Sluse, G. Cessateur, A. Irbah, J. Bureau, M. Weber, K. Bramstedt, R. Thiéblemont, M. Marchand, F. Lefèvre, A. Sarkissian, S. Bekki, SOLAR-ISS: a new reference spectrum based on SOLAR/SOLSPEC observation. Astron. Astrophys. **611**, A1 (2018). https://doi.org/10.1051/0004-6361/201731316

A.W. Merkel, R.J. Vervack Jr., R.M. Killen, T.A. Cassidy, W.E. McClintock, L.R. Nittler, M.H. Burger, Evidence connecting Mercury's magnesium exosphere to its magnesium-rich surface terrane. Geophys. Res. Lett. **45**, 6790–6797 (2018). https://doi.org/10.1029/2018GL078407

F.W. Meyer, P.R. Harris, C.N. Taylor, H.M. Meyer III, A.F. Barghouty, J.H. Adams, Sputtering of lunar regolith simulant by protons and singly and multicharged Ar ions at solar wind energies. Nucl. Instrum. Methods Phys. Res. B **269**, 1316 (2011). https://doi.org/10.1016/j.nimb.2010.11.091

A. Milillo, P. Wurz, S. Orsini, D. Delcourt, E. Kallio, R.M. Killen, H. Lammer, S. Massetti, A. Mura, S. Barabash, G. Cremonese, I.A. Daglis, E. DeAngelis, A.M. Di Lellis, S. Livi, V. Mangano, K. Torkar, Surface-exosphere-magnetosphere system of Mercury. Space Sci. Rev. **117**, 397–443 (2005). https://doi.org/10.1007/s11214-005-3593-z

A. Milillo, M. Fujimoto, G. Murakami, J. Benkhoff, J. Zender, S. Aizawa, M. Dósa, L. Griton, D. Heyner, G. Ho, S.M. Imber, X. Jia, T. Karlsson, R.M. Killen, M. Laurenza, S.T. Lindsay, S. McKenna-Lawlor, A. Mura, J.M. Raines, D.A. Rothery, N. André, W. Baumjohann, A. Berezhnoy, P.A. Bourdin, E.J. Bunce, F. Califano, J. Deca, S. de la Fuente, C. Dong, C. Grava, S. Fatemi, P. Henri, S.L. Ivanovski, B.V. Jackson, M. James, E. Kallio, Y. Kasaba, E. Kilpua, M. Kobayashi, B. Langlais, F. Leblanc, C. Lhotka, V. Mangano, A. Martindale, S. Massetti, A. Masters, M. Morooka, Y. Narita, J.S. Oliveira, D. Odstrcil, S. Orsini, M.G. Pelizzo, C. Plainaki, F. Plaschke, F. Sahraoui, K. Seki, J.A. Slavin, R. Vainio, P. Wurz, S. Barabash, C.M. Carr, D. Delcourt, K.-H. Glassmeier, M.N. Grande, M. Hirahara, J. Huovelin, O. Korablev, H. Kojima, H. Lichtenegger, S. Livi, A. Matsuoka, R. Moissl, M. Moncuquet, K. Muinonen, E. Quèmerais, Y. Saito, S. Yagitani, I. Yoshikawa, J.-E. Wahlund, Investigating Mercury's environment with the two-spacecraft BepiColombo mission. Space Sci. Rev. **216**, 93 (2020). https://doi.org/10.1007/s11214-020-00712-8

D.L. Mitchell, J.S. Halekas, R.P. Lin, S. Frey, L.L. Hood, M.H. Acuna, A. Binder, Global mapping of lunar crustal magnetic fields by Lunar Prospector. Icarus **194**(2), 401–409 (2008). https://doi.org/10.1016/j.icarus.2007.10.027

W. Möller, W. Eckstein, Tridyn—a TRIM simulation code including dynamic composition changes. Nucl. Instrum. Methods Phys. Res. B **2**(1–3), 814–818 (1984). https://doi.org/10.1016/0168-583X(84)90321-5

C.S. Moore, P.C. Chamberlin, R. Hock, Measurements and modelling of total solar irradiance in X-class solar flares. Astron. J. **787**, 32 (2014). https://doi.org/10.1088/0004-637X/787/1/32

R.V. Morris, Fine-grained metal distribution in grain-size separates of lunar soils: production and evolution of the fine-grained metal. Lunar Planet. Sci. Conf. **8**, 682–684 (1977)

L.S. Morrissey, O.J. Tucker, R.M. Killen, S. Nakhla, D.W. Savin, Solar wind ion sputtering of sodium from silicates using molecular dynamics calculations of surface binding energies. Astrophys. J. Lett. **925**, L6 (2022). https://doi.org/10.3847/2041-8213/ac42d8

S.-P. Moschou, J.J. Drake, O. Cohen, J.D. Alvarado-Gómez, C. Garraffo, F. Fraschetti, The stellar CME–flare relation: what do historic observations reveal? Astrophys. J. **877**, 105 (2019). https://doi.org/10.3847/1538-4357/ab1b37

M. Müller, S.F. Green, N. McBride, D. Koschny, J.C. Zarnecki, M.S. Bentley, Estimation of the dust flux near Mercury. Planet. Space Sci. **50**, 1101–1115 (2002). https://doi.org/10.1016/S0032-0633(02)00048-X

J. Müller, S. Simon, J.-C. Wang, U. Motschmann, D. Heyner, J. Schüle, W.-H. Ip, G. Kleindienst, G.J. Pringle, Origin of Mercury's double magnetopause: 3D hybrid simulation study with A.I.K.E.F. Icarus **218**, 666–687 (2012). https://doi.org/10.1016/j.icarus.2011.12.028

A. Mura, S. Orsini, A. Milillo, M. Delcourt, S. Massetti, E. De Angelis, Dayside H^+ circulation at Mercury and neutral particle emission. Icarus **175**(2), 305–319 (2005). https://doi.org/10.1016/j.icarus.2004.12.010

A. Mura, P. Wurz, H.I.M. Lichtenegger, H. Schleicher, H. Lammer, D. Delcourt, A. Milillo, S. Orsini, S. Massetti, M.L. Khodachenko, The sodium exosphere of Mercury: comparison between observations during Mercury's transit and model results. Icarus **200**, 1–11 (2009). https://doi.org/10.1016/j.icarus.2008.11.014

F.H. Murcray, D.G. Murcray, W.J. Williams, Infrared emissivity of lunar surface features: 1. Balloon-borne observations. J. Geophys. Res. **75**(14), 2662–2669 (1970). https://doi.org/10.1029/JB075i014p02662

A. Mutzke, R. Schneider, I. Bizyukov, SDTrimSP-2D studies of the influence of mutual flux arrangement on erosion and deposition. J. Nucl. Mater. **390–391**, 115–118 (2009). https://doi.org/10.1016/j.jnucmat.2009.01.133

A. Mutzke, R. Schneider, W. Eckstein, R. Dohmen SDTrimSP Version 5.00. IPP-Report (2011). https://pure.mpg.de/rest/items/item_2139848/component/file_2139847/content

A. Mutzke, R. Schneider, W. Eckstein, R. Dohmen, K. Schmid, U.V. Toussaint, G. Badelow *SDTrimSP Version 6.00 (IPP 2019-02)* (Max-Planck-Institut für Plasmaphysik, Garching, 2019). https://doi.org/10.17617/2.3026474

T. Nakagawa, ULF/ELF waves in near-Moon space, in *Low-Frequency Waves in Space Plasmas*, ed. by A. Keiling, D.-H. Lee, V. Nakariakov (2016). https://doi.org/10.1002/9781119055006.ch17

T. Nakagawa, F. Takahashi, H. Tsunakawa, H. Shibuya, H. Shimizu, M. Matsushima, Non-monochromatic whistler waves detected by Kaguya on the dayside surface of the Moon. Earth Planets Space **63**(1), 37–46 (2011). https://doi.org/10.5047/eps.2010.01.005. 3rd Kaguya Science Meeting on Earth, Planets and Space, Tokyo, January 14–15, 2009

T. Nakagawa, A. Nakayama, F. Takahashi, H. Tsunakawa, H. Shibuya, H. Shimizu, M. Matsushima, Large-amplitude monochromatic ULF waves detected by Kaguya at the Moon. J. Geophys. Res. **117**, A04101 (2012). https://doi.org/10.1029/2011JA017249

K. Namekata, T. Sakaue, K. Watanabe, A. Asai, H. Maehara, Y. Notsu, S. Notsu, S. Honda, T.T. Ishii, K. Ikuta, D. Nogami, K. Shibata, Statistical studies of solar white-light flares and comparisons with superflares on solar-type stars. Astrophys. J. **851**, 91 (2017). https://doi.org/10.3847/1538-4357/aa9b34

H. Niehus, W. Heiland, E. Taglauer, Low-energy ion scattering at surfaces. Surf. Sci. Rep. **17**(4–5), 213–303 (1993). https://doi.org/10.1016/0167-5729(93)90024-J

M.N. Nishino, K. Maezawa, M. Fujimoto, Y. Saito, S. Yokota, K. Asamura, T. Tanaka, H. Tsunakawa, M. Matsushima, F. Takahashi, T. Terasawa, H. Shibuya, H. Shimizu, Pairwise energy gain-loss feature of solar wind protons in the near-Moon wake. Geophys. Res. Lett. **36**(12), L12108 (2009a). https://doi.org/10.1029/2009GL039049

M.N. Nishino, M. Fujimoto, K. Maezawa, Y. Saito, S. Yokota, K. Asamura, T. Tanaka, H. Tsunakawa, M. Matsushima, F. Takahashi, T. Terasawa, H. Shibuya, H. Shimizu, Solar-wind proton access deep into the near-Moon wake. Geophys. Res. Lett. **36**(16), L16103 (2009b). https://doi.org/10.1029/2009GL039444

M.N. Nishino, Y. Harada, Y. Saito, H. Tsunakawa, F. Takahashi, S. Yokota, M. Matsushima, H. Shibuya, H. Shimizu, Kaguya observations of the lunar wake in the terrestrial foreshock: surface potential change by bow-shock reflected ions. Icarus **293**, 45–51 (2017). https://doi.org/10.1016/j.icarus.2017.04.005

L.R. Nittler, S.Z. Weider, The surface composition of Mercury. Elements **15**, 33–38 (2019). https://doi.org/10.2138/gselements.15.1.33

S.K. Noble, C.M. Pieters, L.P. Keller, An experimental approach to understanding the optical effects of space weathering. Icarus **192**, 629–642 (2007). https://doi.org/10.1016/j.icarus.2007.07.021

D. Nogami, Y. Notsu, S. Honda, H. Maehara, S. Notsu, T. Shibayama, K. Shibata, Two Sun-like superflare stars rotating as slow as the Sun. Publ. Astron. Soc. Jpn. **66**(2), L4 (2014). https://doi.org/10.1093/pasj/psu012

Y. Notsu, T. Shibayama, H. Maehara, S. Notsu, T. Nagao, S. Honda, T.T. Ishii, D. Nogami, K. Shibata, Superflares on solar-type stars observed with Kepler II. Photometric variability of superflare-generating stars: a signature of stellar rotation and starspots. Astrophys. J. **771**, 127 (2013). https://doi.org/10.1088/0004-637X/771/2/127

Y. Notsu, H. Maehara, S. Honda, S.L. Hawley, J.R.A. Davenport, K. Namekata, S. Notsu, K. Ikuta, D. Nogami, K. Shibata, Do Kepler superflare stars really include slowly rotating Sun-like stars?—Results using APO 3.5 m telescope spectroscopic observations and Gaia-DR2 data. Astrophys. J. **876**, 58 (2019). https://doi.org/10.3847/1538-4357/ab14e6

K.W. Ogilvie, J.T. Steinberg, R.J. Fitzenreiter, C.J. Owen, A.J. Lazarus, W.M. Farrell, R.B. Torbert, Observations of the lunar plasma wake from the WIND spacecraft on December 27, 1994. Geophys. Res. Lett. **23**(10), 1255–1258 (1996). https://doi.org/10.1029/96GL01069

M. Øieroset, T.D. Phan, R.P. Lin, B.U. Sonnerup, Walén and variance analyses of high-speed flows observed by Wind in the midtail plasma sheet: evidence for reconnection. J. Geophys. Res. **105**(A11), 25247–25263 (2002). https://doi.org/10.1029/2000JA900075

T. Okada, K. Shirai, Y. Yamamoto, T. Arai, K. Ogawa, H. Shiraishi, M. Iwasaki, T. Kawamura, H. Morito, M. Grande, M. Kato, X-ray fluorescence spectrometry of Lunar surface by XRS onboard SELENE (Kaguya). Trans. Jpn. Soc. Aeronaut. Space Sci. **7**, 39–42 (2009). https://doi.org/10.2322/tstj.7.Tk_39

N. Omidi, J.P. Eastwood, D.G. Sibeck, Foreshock bubbles and their global magnetospheric impacts. J. Geophys. Res. Space Phys. **115**, A6 (2010). https://doi.org/10.1029/2009JA014828

N. Omidi, D. Sibeck, X. Blanco-Cano, D. Rojas-Castillo, D. Turner, H. Zhang, P. Kajdič, Dynamics of the foreshock compressional boundary and its connection to foreshock cavities. J. Geophys. Res. Space Phys. **118**(2), 823–831 (2013). https://doi.org/10.1002/jgra.50146

S. Orsini, V. Mangano, A. Mura, D. Turrini, S. Massetti, A. Milillo, C. Plainaki, The influence of space environment on the evolution of Mercury. Icarus **239**, 281–290 (2014). https://doi.org/10.1016/j.icarus.2014.05.031

S. Orsini, V. Mangano, A. Milillo, C. Plainaki, A. Mura, J.M. Raines, E. De Angelis, R. Rispoli, F. Lazzarotto, A. Aronica, Mercury sodium exospheric emission as a proxy for solar perturbations transit. Sci. Rep. **8**, 928 (2018). https://doi.org/10.1038/s41598-018-19163-x

S. Orsini, S. Livi, H. Lichtenegger, S. Barabash, A. Milillo, E. De Angelis, M. Phillips, G. Laky, M. Wieser, A. Olivieri, C. Plainaki, G. Ho, R.M. Killen, J.A. Slavin, P. Wurz, J.-J. Berthelier, I. Dandouras, M. Dosa, E. Kallio, S. McKenna-Lawlor, K. Torkar, O. Vaisberg, F. Allegrini, I.A. Daglis, C. Dong, C.P. Escoubet, S. Fatemi, M. Fränz, S. Ivanovski, N. Krupp, H. Lammer, F. Leblanc, V. Mangano, A. Mura, H. Nilsson, J.M. Raines, R. Rispoli, M. Sarantos, H.T. Smith, K. Szego, A. Varsani, A. Aronica, F. Camozzi, A.M. Di Lellis, G. Fremuth, F. Giner, R. Gurnee, J. Hayes, H. Jeszenszky, F. Tominetti, B. Trantham, J. Balaz, W. Baumjohann, D. Brienza, U. Bührke, M.-D. Bush, M. Cantatore, S. Cibella, L. Colasanti, G. Cremonese, L. Cremonesi, M. D'Alessandro, D. Delcourt, M. Delva, M. Desai, M. Famá, M. Ferris, H. Fischer, A. Gaggero, D. Gamborino, P. Garnier, B. Gibson, R. Goldstein, M. Grande, V. Grishin, D. Haggerty, M. Holmström, I. Horvath, K.C. Hsieh, A. Jacques, R.E. Johnson, A. Kazakov, K. Kecskemety, H. Krüger, C. Kürbisch, F. Lazzarotto, F. Leblanc, M. Leichtfried, R. Leoni, A. Loose, D. Maschietti, S. Massetti, F. Mattioli, G. Miller, D. Moissenko, A. Morbidini, R. Noschese, F. Nuccilli, C. Nunez, N. Paschalidis, S. Persyn, D. Piazza, M. Oja, J. Ryno W, Schmidt, J.A. Scheer, A. Shestakov, S.S. Shuvalov, K. Seki, S. Selci, K. Smith, R. Sordini, F. Stenbeck, J. Svensson, L. Szalai, K. Szego, D. Toublanc, C. Urdiales, N. Vertolli, R. Wallner, P. Wahlstroem, P. Wilson, S. Zampieri, SERENA: particle instrument suite for determining the Sun-Mercury interaction from BepiColombo. Space. Sci. Rev. **217**(11) (2021). https://doi.org/10.1007/s11214-020-00787-3

C. Pahlke, H. Düsterhöft, U. Müller-Jahreis, Measurements of the energy distributions of positive secondary ions in the energy range from 0 eV to about 500 eV, in *Secondary Ion Mass Spectrometry SIMS III* (Springer, Berlin, 1982), pp. 124–127

M. Pfleger, H.I.M. Lichtenegger, P. Wurz, H. Lammer, E. Kallio, M. Alho, A. Mura, S. McKenna-Lawlor, J.A. Martín-Fernández, 3D-modeling of Mercury's solar wind sputtered surface-exosphere environment. Planet. Space Sci. **115**, 90–101 (2015). https://doi.org/10.1016/j.pss.2015.04.016

C.M. Pieters, S.K. Noble, Space weathering on airless bodies. J. Geophys. Res. **121**, 1865–1884 (2016). https://doi.org/10.1002/2016JE005128

G. Poh, J.A. Slavin, X. Jia, G.A. DiBraccio, J.M. Raines, S.M. Imber, D.J. Gershman, W.-J. Sun, B.J. Anderson, H. Korth, T.H. Zurbuchen, R.L. McNutt Jr., S.C. Solomon, MESSENGER observations of cusp plasma filaments at Mercury. J. Geophys. Res. Space Phys. **121**, 8260–8285 (2016). https://doi.org/10.1002/2016JA022552

G. Poh, J.A. Slavin, X. Jia, J.M. Raines, S.M. Imber, W.-J. Sun, D.J. Gershman, G.A. DiBraccio, K.J. Genestreti, A.W. Smith, Mercury's cross-tail current sheet: structure, X-line location and stress balance. Geophys. Res. Lett. **44**, 678–686 (2017). https://doi.org/10.1002/2016GL071612

P. Pokorný, M. Sarantos, D. Janches, A comprehensive model of the meteoroid environment around Mercury. Astrophys. J. **863**, 31 (2018). https://doi.org/10.3847/1538-4357/aad051

P. Pokorný, D. Janches, M. Sarantos, J.R. Szalay, M. Horányi, D. Nesvorný, M.J. Kuchner, Meteoroids at the Moon: orbital properties, surface vaporization, and impact ejecta production. J. Geophys. Res. **124**, 752–778 (2019). https://doi.org/10.1029/2018JE005912

A.R. Poppe, S. Fatemi, J.S. Halekas, M. Holmström, G.T. Delory, ARTEMIS observations of extreme diamagnetic fields in the lunar wake. Geophys. Res. Lett. **41**(11), 3766–3773 (2014). https://doi.org/10.1002/2014GL060280

A.R. Poppe, S. Fatemi, I. Garrick-Bethell, D. Hemingway, M. Holmström, Solar wind interaction with the Reiner Gamma crustal magnetic anomaly: connecting source magnetization to surface weathering. Icarus **266**, 261–266 (2016a). https://doi.org/10.1016/j.icarus.2015.11.005

A.R. Poppe, M.O. Fillingim, J.S. Halekas, J. Raeder, V. Angelopoulos, ARTEMIS observations of terrestrial ionospheric molecular ion outflow at the Moon. Geophys. Res. Lett. **43**, 6749–6758 (2016b). https://doi.org/10.1002/2016GL069715

A.R. Poppe, J.S. Halekas, C. Lue, S. Fatemi, ARTEMIS observations of the solar wind proton scattering function from lunar crustal magnetic anomalies. J. Geophys. Res., Planets **122**(4), 771–783 (2017). https://doi.org/10.1002/2017JE005313

A.R. Poppe, W.M. Farrell, J.S. Halekas, Formation timescales of amorphous rims on lunar grains derived from ARTEMIS observations. J. Geophys. Res., Planets **123**, 37–46 (2018). https://doi.org/10.1002/2017JE005426

A.E. Potter, Chemical sputtering could produce sodium vapor and ice on Mercury. Geophys. Res. Lett. **22**(23), 3289–3292 (1995). https://doi.org/10.1029/95GL03181

A.E. Potter, R.M. Killen, T.H. Morgan, Solar radiation acceleration effects on Mercury sodium emission. Icarus **186**, 571–580 (2007). https://doi.org/10.1016/j.icarus.2006.09.025

J.M. Raines, D.J. Gershman, J.A. Slavin, T.H. Zurbuchen, H. Korth, B.J. Anderson, S.C. Solomon, Structure and dynamics of Mercury's magnetospheric cusp: MESSENGER measurements of protons and planetary ions. J. Geophys. Res. Space Phys. **119**, 6587–6602 (2014). https://doi.org/10.1002/2014JA020120

J.M. Raines, G.A. DiBraccio, T.A. Cassidy, D.C. Delcourt, M. Fujimoto, X. Jia, V. Mangano, A. Milillo, M. Sarantos, J.A. Slavin, P. Wurz, Plasma sources in planetary magnetospheres: Mercury. Space Sci. Rev. **192**(1), 1–54 2015. https://doi.org/10.1007/s11214-015-0193-4

J.M. Raines, R.M. Dewey, N.M. Staudacher, P.J. Tracy, C.M. Bert, M. Sarantos, D.J. Gershman, J.M. Jasinski, J.M. Slavin, Proton precipitation in Mercury's northern magnetospheric cusp. J. Geophys. Res. (2022), submitted

J.M. Raines, R.M. Dewey, N.M. Staudacher, P.J. Tracy, C.M. Bert, M. Sarantos, D.J. Gershman, J.M. Jasinski, J.M. Slavin, Proton precipitation in Mercury's northern magnetospheric cusp. J. Geophys. Res. (2022, submitted)

R.D. Ramsier, J.T. Yates, Electron-stimulated desorption: principles and applications. Surf. Sci. Rep. **6–8**, 243–378 (1991). https://doi.org/10.1016/0167-5729(91)90013-N

E. Richer, R. Modolo, G.M. Chanteur, S. Hess, F. Leblanc, A global hybrid model for Mercury's interaction with the solar wind: case study of the dipole representation. J. Geophys. Res. **117**, A10228 (2012). https://doi.org/10.1029/2012JA017898

N.C. Richmond, L.L. Hood, A preliminary global map of the vector lunar crustal magnetic field based on Lunar Prospector magnetometer data. J. Geophys. Res. **113**, E02010 (2008). https://doi.org/10.1029/2007JE002933

D.F. Rodríguez M., L. Saul, P. Wurz, S.A. Fuselier, H.O. Funsten, D.J. McComas, E. Möbius, IBEX-Lo observations of energetic neutral hydrogen atoms originating from the lunar surface. Planet. Space Sci. **60**(1), 297–303 (2012). https://doi.org/10.1016/j.pss.2011.09.009

J. Roth, J. Bohdansky, W. Ottenberger, *Data on Low Energy Light Ion Sputtering* (Max-Planck-Inst. für Plasmaphysik, Garching bei München, 1979). http://hdl.handle.net/11858/00-001M-0000-0027-6AE8-3

C.T. Russell, D.N. Baker, J.A. Slavin, The magnetosphere of Mercury, in *Mercury* (University of Arizona Press, Tucson, 1988), pp. 514–561

J.M. Saari, R.W. Shorthill, The sunlit lunar surface. I. Albedo studies and full Moon temperature distribution. Moon **5**(1–2), 161–178 (1972)

Y. Saito, S. Yokota, T. Tanaka, K. Asamura, M.N. Nishino, M. Fujimoto, H. Tsunakawa, H. Shibuya, M. Matsushima, H. Shimizu, F. Takahashi, T. Mukai, T. Terasawa, Solar wind proton reflection at the lunar surface: low energy ion measurement by MAP-PACE onboard SELENE (KAGUYA). Geophys. Res. Lett. **35**, L24205 (2008). https://doi.org/10.1029/2008GL036077

Y. Saito, S. Yokota, K. Asamura, T. Tanaka, M.N. Nishino, T. Yamamoto, Y. Terakawa, M. Fujimoto, H. Hasegawa, H. Hayakawa, M. Hirahara, M. Hoshino, S. Machida, T. Mukai, T. Nagai, T. Nagatsuma, T. Nakagawa, M. Nakamura, K. Oyama, E. Sagawa, S. Sasaki, K. Seki, I. Shinohara, T. Terasawa, H. Tsunakawa, H. Shibuya, M. Matsushima, H. Shimizu, F. Takahashi, In-flight performance and initial results of plasma energy angle and composition experiment (PACE) on SELENE (Kaguya). Space Sci. Rev. **154**, 265–303 (2010). https://doi.org/10.1007/s11214-010-9647-x

Y. Saito, M.N. Nishino, M. Fujimoto, T. Yamamoto, S. Yokota, H. Tsunakawa, H. Shibuya, M. Matsushima, H. Shimizu, F. Takahashi, Simultaneous obseravation of the electron acceleration and ion deceleration over lunar magnetic anomalies. Earth Planets Space **64**, 83–92 (2012). https://doi.org/10.5047/eps.2011.07.011

Y. Saito, D. Delcourt, M. Hirahara, S. Barabash, N. André, T. Takashima, K. Asamura, S. Yokota, M. Wieser, M.N. Nishino, M. Oka, Y. Futaana, Y. Harada, J.-A. Sauvaud, P. Louarn, B. Lavraud, V. Génot, C. Mazelle, I. Dandouras, C. Jacquey, C. Aoustin, A. Barthe, A. Cadu, A. Fedorov, A.-M. Frezoul, C. Garat, E. Le Comte, Q.-M. Lee, J.-L. Médale, D. Moirin, E. Penou, M. Petiot, G. Peyre, J. Rouzaud, H.-C. Séran, Z. Němeček, J. Safránková, M.F. Marcucci, R. Bruno, G. Consolini, W. Miyake, I. Shinohara, H. Hasegawa, K. Seki, A.J. Coates, F. Leblanc, C. Verdeil, B. Katra, D. Fontaine, J.-M. Illiano, J.-J. Berthelier, J.-D. Techer, M. Fraenz, H. Fischer, N. Krupp, J. Woch, U. Bührke, B. Fiethe, H. Michalik, H.M.T. Yanagimachi, Y. Miyoshi, T. Mitani, M. Shimoyama, Q. Zong, P. Wurz, H. Andersson, S. Karlsson, M. Holmström, Y. Kazama, W.-H. Ip, M. Hoshino, M. Fujimoto, N. Terada, K. Keika the BepiColombo Mio/MPPE Team, Pre-flight calibration and near-Earth commissioning results of the Mercury Plasma Particle Experiment (MPPE) onboard MMO (Mio). Space Sci. Rev. **217**, 70 (2021). https://doi.org/10.1007/s11214-021-00839-2

M. Sarantos, R.M. Killen, A.S. Sharma, J.A. Slavin, Influence of plasma ions on source rates for the lunar exosphere during passage through the Earth's magnetosphere. Geophys. Res. Lett. **35**, L04105 (2008). https://doi.org/10.1029/2007GL032310

M. Sarantos, R.M. Killen, A. Surjalal Sharma, J.A. Slavin, Sources of sodium in the lunar exosphere: modeling using ground-based observations of sodium emission and spacecraft data of the plasma. Icarus **205**, 364–374 (2010). https://doi.org/10.1016/j.icarus.2009.07.039

L. Saul, P. Wurz, A. Vorburger, D.F. Rodríguez M., S.A. Fuselier, D.J. McComas, E. Möbius, S. Barabash, H. Funsten, P. Janzen, Solar wind reflection from the lunar surface: the view from far and near. Planet. Space Sci. **84**, 1–4 (2013). https://doi.org/10.1016/j.pss.2013.02.004

P. Saxena, R.M. Killen, V. Airapetian, N.E. Petro, N.M. Curran, A.M. Mandell, Was the Sun a slow rotator? Sodium and potassium constraints from the lunar regolith. Astrophys. J. Lett. **876**, L16 (2019). https://doi.org/10.3847/2041-8213/ab18fb

M.J. Schaible, C.A. Dukes, A.C. Hutcherson, P. Lee, M.R. Collier, R.E. Johnson, Solar wind sputtering rates of small bodies and ion mass spectrometry detection of secondary ions. J. Geophys. Res., Planets **122**, 1968–1983 (2017). https://doi.org/10.1002/2017JE005359

M.J. Schaible, M. Sarantos, B.A. Anzures, S.W. Parman, T.M. Orlando, Photon-stimulated desorption of MgS as a potential source of sulfur in Mercury's exosphere. J. Geophys. Res., Planets **125**, e2020JE006479 (2020). https://doi.org/10.1029/2020JE006479

A. Schaufelberger, P. Wurz, S. Barabash, M. Wieser, Y. Futaana, M. Holmström, A. Bhardwaj, M.B. Dhanya, R. Sridharan, K. Asamura, Scattering function for energetic neutral hydrogen atoms off the lunar surface. Geophys. Res. Lett. **38**, L22202 (2011). https://doi.org/10.1029/2011GL049362

T. Schenkel, A.V. Hamza, A.V. Barnes, D.H. Schneider, Interaction of slow, very highly charged ions with surfaces. Prog. Surf. Sci. **61**, 23 (1999). https://doi.org/10.1016/S0079-6816(99)00009-X

C.E. Schlemm, R.D. Starr, G.C. Ho, K.E. Bechtold, S.A. Hamilton, J.D. Boldt, W.V. Boynton, W. Bradley, M.E. Fraeman, R.E. Gold, J.O. Goldsten, J.R. Hayes, S.E. Jaskulek, E. Rossano, R.A. Rumpf, E.D. Schaefer, K. Strohben, R.G. Shelton, R.E. Thompson, J.I. Trombka, B.D. Williams, The X-ray spectrometer on the MESSENGER spacecraft, in *The Messenger Mission to Mercury*, ed. by D.L. Domingue, C.T. Russell (Springer, New York, 2007). https://doi.org/10.1007/s11214-007-9248-5

C.A. Schmidt, J. Baumgardner, M. Mendillo, J.K. Wilson, Escape rates and variability constraints for high-energy sodium sources at Mercury. J. Geophys. Res. **117**, A03301 (2012). https://doi.org/10.1029/2011JA017217

N. Schörghofer, M. Benna, A.A. Berezhnoy, B. Greenhagen, B.M. Jones, S. Li, T.M. Orlando, P. Prem, O.J. Tucker, C. Wöhler, Water group exospheres and surface interactions on the Moon, Mercury, and Ceres. Space Sci. Rev. **217**, 74 (2021). https://doi.org/10.1007/s11214-021-00846-3

C.J. Schrijver, P.M. Trávníček, B.J. Anderson, M. Ashour-Abdalla, D. Baker, M. Benna, S.A. Boardsen, R.E. Gold, P. Hellinger, G.C. Ho, H. Korth, S.M. Krimigis, R.L. McNutt Jr., J.M. Raines, R.L. Richards, J.A. Slavin, S.C. Solomon, R.D. Starr, T.H. Zurbuchen, Quasi-trapped ion and electron populations at Mercury. Geophys. Res. Lett. **38**, L23103 (2011). https://doi.org/10.1029/2011GL049629

C.J. Schrijver, J. Beer, U. Baltensperger, E.W. Cliver, M. Güdel, H.S. Hudson, K.G. McCracken, R.A. Osten, T. Peter, D.R. Soderblom, I.G. Usoskin, E.W. Wolff, Estimating the frequency of extremely energetic solar events, based on solar, stellar, lunar, and terrestrial records. J. Geophys. Res. **117**, A08103 (2012). https://doi.org/10.1029/2012JA017706

S.J. Schwartz, Hot flow anomalies near the Earth's bow shock. Adv. Space Res. **15**(8–9), 107–116 (1995). https://doi.org/10.1016/0273-1177(94)00092-F

S.J. Schwartz, C.P. Chaloner, P.J. Christiansen, A.J. Coates, D.S. Hall, A.D. Johnstone, P.M. Gough, A.J. Norris, R.P. Rijnbeek, D.J. Southwood, L.J. Woolliscroft, An active current sheet in the solar wind. Nature **318**(6043), 269–271 (1985). https://doi.org/10.1038/318269a0

K. Seki, M. Hirahara, T. Terasawa, I. Shinohara, T. Mukai, Y. Saito, S. Machida, T. Yamamoto, S. Kokubun, Coexistence of Earth-origin O^+ and solar wind-origin H^+/He^{++} in the distant magnetotail. Geophys. Res. Lett. **23**(9), 985–988 (1996). https://doi.org/10.1029/96GL00768

K. Seki, A. Nagy, C.M. Jackman, F. Crary, D. Fontaine, P. Zarka, P. Wurz, A. Milillo, J.A. Slavin, D.C. Delcourt, M. Wiltberger, R. Ilie, X. Jia, S.A. Ledvina, R.W. Schunk, A review of general processes related to plasma sources and losses for solar system magnetospheres. Space Sci. Rev. **192**(1), 27–89 (2015). https://doi.org/10.1007/s11214-015-0170-y

D.E. Shemansky, The role of solar wind ions in the space environment, in *Proc. 23rd Int. Symp. Rarefied Gas Dynamics*. AIP Conf. Proc., vol. 663 (2003), pp. 687–696. https://doi.org/10.1063/1.1581610

K. Shibata, T. Magara, Solar flares: magnetohydrodynamic processes. Living Rev. Sol. Phys. **8**, 6 (2011). https://doi.org/10.12942/lrsp-2011-6

K. Shibata, H. Isobe, A. Hillier, A.R. Choudhuri, H. Maehara, T.T. Ishii, T. Shibayama, Y. Notsu, S. Notsu, T. Nagao, S. Honda, D. Nogami, Can superflares occur on our Sun? Publ. Astron. Soc. Jpn. **65**(3), 49 (2013). https://doi.org/10.1093/pasj/65.3.49

T. Shibayama, H. Maehara, S. Notsu, Y. Notsu, T. Nagao, S. Honda, T.T. Ishii, D. Nogami, K. Shibata, Superflares on solar-typee stars observed with Kepler. I. Statistical properties of superflares. Astrophys. J. Suppl. Ser. **209**, 5 (2013). https://doi.org/10.1088/0067-0049/209/1/5

K. Shiokawa, T. Ogino, K. Hayashi, D.J. McEwen, Quasi-periodic poleward motions of morningside Sun-aligned arcs: a multievent study. J. Geophys. Res. **102**(A11), 24325–24332 (1993). https://doi.org/10.1029/97JA02383

V.I. Shulga, Note on the artefacts in SRIM simulation of sputtering. Appl. Surf. Sci. **439**, 456–461 (2018). https://doi.org/10.1016/j.apsusc.2018.01.039

D.G. Sibeck, T.-D. Phan, R. Lin, R.P. Lepping, A. Szabo, Wind observations of foreshock cavities: a case study. J. Geophys. Res. Space Phys. **107**(A10), SMP-4 (2002). https://doi.org/10.1029/2001JA007539

P. Sigmund, Theory of sputtering. I. Sputtering yield of amorphous and polycrystalline targets. Phys. Rev. **184**, 383–416 (1969)

P. Sigmund, Recollections of fifty years with sputtering. Thin Solid Films **520**, 6031–6049 (2012). https://doi.org/10.1016/j.tsf.2012.06.003

A. Simpson, J.H. Eraker, J.E. Lamport, P.H. Walpole, Electrons and protons accelerated in Mercury's magnetic field. Science **185**, 160–166 (1974). https://doi.org/10.1126/science.185.4146.160

J.A. Slavin, M.H. Acuña, B.J. Anderson, D.N. Baker, M. Benna, S.A. Boardsen, G. Gloeckler, R.E. Gold, G.C. Ho, H. Korth, S.M. Krimigis, R.L. McNutt Jr., J.M. Raines, M. Sarantos, D. Schriver, S.C. Solomon, P. Trávníček, T.H. Zurbuchen, MESSENGER observations of magnetic reconnection in Mercury's magnetosphere. Science **324**(5927), 606–610 (2009). https://doi.org/10.1126/science.1172011

J.A. Slavin, S.M. Imber, S.A. Boardsen, G.A. DiBraccio, T. Sundberg, M. Sarantos, T. Nieves-Chinchilla, A. Szabo, B.J. Anderson, H. Korth, T.H. Zurbuchen, J.M. Raines, C.L. Johnson, R.M. Winslow, R.M. Killen, R.L. McNutt Jr., S.C. Solomon, MESSENGER observations of a flux-transfer-event shower at Mercury. J. Geophys. Res. **117**, A00M06 (2012). https://doi.org/10.1029/2012JA017926

J.A. Slavin, G.A. DiBraccio, D.J. Gershman, S.M. Imber, G.K. Poh, J.M. Raines, T.H. Zurbuchen, X. Jia, D.N. Baker, K.-H. Glassmeier, S.A. Livi, S.A. Boardsen, T.A. Cassidy, M. Sarantos, T. Sundberg, A. Masters, C.L. Johnson, R.M. Winslow, B.J. Anderson, H. Korth, R.L. McNutt Jr., C. Solomon, MESSENGER observations of Mercury's dayside magnetosphere under extreme solar wind conditions. J. Geophys. Res. **119**, 8087–8116 (2014). https://doi.org/10.1002/2014JA020319

J.A. Slavin, H.R. Middleton, J.M. Raines, X. Jia, J. Zhong, W.-J. Sun, S. Livi, S.M. Imber, G.-K. Poh, M. Akhavan-Tafti, J.M. Jasinski, G.A. DiBraccio, C. Dong, R.M. Dewey, M.L. Mays, MESSENGER observations of disappearing dayside magnetosphere events at Mercury. J. Geophys. Res. Space Phys. **124**, 6613–6635 (2019). https://doi.org/10.1029/2019JA026892

J.A. Slavin, D.N. Baker, D.J. Gerhsman, G.C. Ho, S.M. Imber, S.M. Krimigis, T. Sundberg, Mercury's dynamic magnetosphere, in *Mercury: The View after MESSENGER*, ed. by S.C. Solomon, L.R. Nittler, B.J. Anderson (Cambridge University Press, London, 2020), pp. 461–496

J.A. Slavin, S.M. Imber, J.M. Raines, A Dungey cycle in the life of Mercury's magnetosphere, in *Space Physics and Aeronomy Collection Volume 2: Magnetospheres in the Solar System*, ed. by R. Maggiolo,

N. André, H. Hasegawa, D.T. Welling. Geophysical Monograph, vol. 259 (Am. Geophys. Union/Wiley, Washington/New York, 2021). https://doi.org/10.1002/9781119815624.ch34

B.U. Sonnerup, Magnetopause reconnection rate. J. Geophys. Res. **79**(10), 1546–1549 (1974). https://doi.org/10.1029/JA079i010p01546

M. Sporn, G. Libiseller, T. Neidhart, M. Schmid, F. Aumayr, H.P. Winter, P. Varga, M. Grether, D. Niemann, N. Stolterfoht, Potential sputtering of clean SiO_2 by slow highly charged ions. Phys. Rev. Lett. **79**, 945 (1997). https://doi.org/10.1103/PhysRevLett.79.945

A.L. Sprague, A diffusion source for sodium and potassium in the atmospheres of Mercury and the Moon. Icarus **84**, 93–105 (1990). https://doi.org/10.1016/0019-1035(90)90160-B

R. Stadlmayr, P.S. Szabo, B.M. Berger, C. Cupak, R. Chiba, D. Blöch, D. Mayer, B. Stechauner, M. Sauer, A. Foelske-Schmitz, M. Oberkofler, T. Schwarz-Selinger, A. Mutzke, F. Aumayr, Fluence dependent changes of surface morphology and sputtering yield of iron: comparison of experiments with SDTrimSP-2D Nucl. Instrum. Methods Phys. Res., Sect. B, Beam Interact. Mater. Atoms **430**, 42 (2018). https://doi.org/10.1016/j.nimb.2018.06.004

R.D. Starr, D. Schriver, L.R. Nittler, S.Z. Weider, P.K. Byrne, G.C. Ho, E.A. Rhodes, C.E. Schlemm, S.C. Solomon, P.M. Trávníček, MESSENGER detection of electron-induced X-ray fluorescence from Mercury's surface. J. Geophys. Res. **117**, E00L02 (2012). https://doi.org/10.1029/2012JE004118

S.A. Stern, The lunar atmosphere: history, status, current problems, and context. Rev. Geophys. **37**(4), 453–491 (1999). https://doi.org/10.1029/1999RG900005

S.A. Stern, J.C. Cook, J.-Y. Chaufray, P.D. Feldman, G.R. Gladstone, K.D. Retherford, Lunar atmospheric H_2 detections by the LAMP UV spectrograph on the Lunar Reconnaissance Orbiter. Icarus **226**, 1210–1213 (2013). https://doi.org/10.1016/j.icarus.2013.07.011

T.J. Stubbs, J.S. Halekas, W.M. Farrell, R.R. Vondrak, R. Richard, Lunar surface charging: a global perspective using lunar prospector data, in *Workshop on Dust in Planetary Systems (ESA SP-643)*, ed. by H. Krueger, A. Graps 26–30 September 2005, Kauai, Hawaii, USA (2007), pp. 181–184

W.J. Sun, S.Y. Fu, J.A. Slavin, J.M. Raines, Q.G. Zong, G.K. Poh, T.H. Zurbuchen, Spatial distribution of Mercury's flux ropes and reconnection fronts: MESSENGER observations. J. Geophys. Res. Space Phys. **121**, 7590–7607 (2016). https://doi.org/10.1002/2016JA022787

W.J. Sun, J.M. Raines, S.Y. Fu, J.A. Slavin, Y. Wei, G.K. Poh, Z.Y. Pu, Z.H. Yao, Q.G. Zong, W.X. Wan, MESSENGER observations of the energization and heating of protons in the near-Mercury magnetotail. Geophys. Res. Lett. **44**, 8149–8158 (2017). https://doi.org/10.1002/2017GL074276

W.J. Sun, J.A. Slavin, R.M. Dewey, J.M. Raines, S.Y. Fu, Y. Wei, T. Karlsson, G.K. Poh, X. Jia, D.J. Gershman, Q.G. Zong, W.X. Wan, Q.Q. Shi, Z.Y. Pu, D. Zhao, A comparative study of the proton properties of magnetospheric substorms at Earth and Mercury in the near magnetotail. Geophys. Res. Lett. **45**, 7933–7941 (2018). https://doi.org/10.1029/2018GL079181

W.J. Sun, J.A. Slavin, A.W. Smith, R.M. Dewey, G.K. Poh, X. Jia, J.M. Raines, S. Livi, Y. Saito, D.J. Gershman, G.A. DiBraccio, S.M. Imber, J.P. Guo, S.Y. Fu, Q.G. Zong, J.T. Zhao, Flux transfer event showers at Mercury: dependence on plasma β and magnetic shear and their contribution to the Dungey cycle. Geophys. Res. Lett. **47**, e2020GL089784 (2020). https://doi.org/10.1029/2020GL089784

W. Sun, J.A. Slavin, A. Milillo, S. Orsini, X. Jia1, J.M. Raines, S. Livi, J.M. Jasinski, R.M. Dewey, S. Fu, J. Zhao, Q.-G. Zong, Y. Saito, C. Li, MESSENGER observations of planetary ion enhancements at Mercury's northern magnetospheric cusp during Flux Transfer Event Showers (2022). arXiv:2201.03987

D.M. Suszcynsky, J.T. Gosling, M.F. Thomsen, Ion temperature profiles in the horns of the plasma sheet. J. Geophys. Res. **98**(A1), 257–262 (1993). https://doi.org/10.1029/92JA01733

P.S. Szabo, R. Chiba, H. Biber, R. Stadlmayr, B.M. Berger, D. Mayer, A. Mutzke, M. Doppler, M. Sauer, J. Appenroth, J. Fleig, A. Foelske-Schmitz, H. Hutter, K. Mezger, H. Lammer, A. Galli, P. Wurz F. Aumayr, Solar wind sputtering of wollastonite as a lunar analogue material—comparisons between experiments and simulations. Icarus **314**, 98 (2018). https://doi.org/10.1016/j.icarus.2018.05.028

P.S. Szabo, H. Biber, N. Jäggi, M. Brenner, D. Weichselbaum, A. Niggas, R. Stadlmayr, D. Primetzhofer, A. Nenning, A. Mutzke, M. Sauer, J. Fleig, A. Foelske-Schmitz, K. Mezger, H. Lammer, A. Galli, P. Wurz, F. Aumayr, Dynamic potential sputtering of lunar analog material by solar wind ions. Astrophys. J. **891**, 100 (2020a). https://doi.org/10.3847/1538-4357/ab7008

P.S. Szabo, H. Biber, N. Jäggi, M. Wallner, R. Stadlmayr, M.V. Moro, A. Nenning, A. Mutzke, K. Mezger, H. Lammer, D. Primetzhofer, J. Fleig, A. Galli, P. Wurz, F. Aumayr, Experimental insights into space weathering of phobos: laboratory investigation of sputtering by atomic and molecular planetary ions. J. Geophys. Res., Planets **125**, e2020JE006583 (2020b). https://doi.org/10.1029/2020je006583

P.S. Szabo, C. Cupak, H. Biber, N. Jäggi, A. Galli, P. Wurz, F. Aumayr, A theoretical model for the sputtering of rough surfaces. Surf. Interfaces (2022), submitted

J.R. Szalay, A.R. Poppe, J. Agarwal, D. Britt, I. Belskaya, M. Horányi, T. Nakamura, M. Sachse, F. Spahn, Dust phenomena relating to airless bodies. Space Sci. Rev. **214**, 98 (2018). https://doi.org/10.1007/s11214-018-0527-0

V.A. Thomas, S.H. Brecht, Evolution of diamagnetic cavities in the solar wind. J. Geophys. Res. **93**(A10), 11341–11353 (1988). https://doi.org/10.1029/JA093iA10p11341

M.W. Thompson, B.W. Farmery, P.A. Newson, A mechanical spectrometer for analyzing the energy distribution of sputtered atoms of copper and gold. Philos. Mag. **18**(152), 361–383 (1968). https://doi.org/10.1080/14786436808227357

A. Tolstogouzova, S. Daolio, C. Pagura, C.L. Greenwood, Energy distributions of secondary ions sputtered from aluminium and magnesium by Ne^+, Ar^+ and O_2^+: a comprehensive study. Int. J. Mass Spectrom. **214**, 327–337 (2002)

M. Tona, S. Takahashi, K. Nagata, N. Yoshiyasu, C. Yamada, N. Nakamura, S. Ohtani, Coulomb explosion potential sputtering induced by slow highly charged ion impact. Appl. Phys. Lett. **87**, 224102 (2005). https://doi.org/10.1063/1.2136361

P.M. Trávníček, P. Hellinger, M.G.G.T. Taylor, P. Escoubet, I. Dandouras, E. Lucek, Magnetosheath plasma expansion: hybrid simulations. Geophys. Res. Lett. **34**, L15104 (2007). https://doi.org/10.1029/2007GL029728

P.M. Trávníček, D. Schriver, P. Hellinger, D. Hercík, B.J. Anderson, M. Sarantos, J.A. Slavin, Mercury's magnetosphere-solar wind interaction for northward and southward interplanetary magnetic field: hybrid simulation results. Icarus **209**, 11–22 (2010). https://doi.org/10.1016/j.icarus.2010.01.008

O. Troshichev, S. Kokubun, Y. Kamide, A. Nishida, T. Mukai, T. Yamamoto, Convection in the distant magnetotail under extremely quiet and weakly disturbed conditions. J. Geophys. Res. **104**(A5), 10249–10263 (1999). https://doi.org/10.1029/1998JA900141

Y. Tsugawa, N. Terada, Y. Katoh, T. Ono, H. Tsunakawa, F. Takahashi, H. Shibuya, H. Shimizu, M. Matsushima, Statistical analysis of monochromatic whistler waves near the Moon detected by Kaguya. Ann. Geophys. **29**, 889–893 (2011). https://doi.org/10.5194/angeo-29-889-2011

Y. Tsugawa, Y. Katoh, N. Terada, T. Ono, H. Tsunakawa, F. Takahashi, H. Shibuya, H. Shimizu, M. Matsushima, Y. Saito, S. Yokota, M.N. Nishino, Statistical study of broadband whistler-mode waves detected by Kaguya near the Moon. Geophys. Res. Lett. **39**(16), L16101 (2012). https://doi.org/10.1029/2012GL052818

H. Tsunakawa, H. Shibuya, F. Takahashi, H. Shimizu, M. Matsushima, A. Matsuoka, S. Nakazawa, H. Otake, Y. Iijima, Lunar magnetic field observation and initial global mapping of lunar magnetic anomalies by MAP-LMAG onboard SELENE (Kaguya). Space Sci. Rev. **154**(1–4), 219–251 (2010). https://doi.org/10.1007/s11214-010-9652-0

H. Tsunakawa, F. Takahashi, H. Shimizu, H. Shibuya, M. Matsushima, Surface vector mapping of magnetic anomalies over the Moon using Kaguya and Lunar Prospector observations. J. Geophys. Res., Planets **120**, 1160–1185 (2015). https://doi.org/10.1002/2014JE004785

L. Tu, C.P. Johnstone, M. Güdel, H. Lammer, The extreme ultraviolet and X-ray Sun in time: high-energy evolutionary tracks of a solar-like star. Astron. Astrophys. **577**, L3 (2015). https://doi.org/10.1051/0004-6361/201526146

O.J. Tucker, W.M. Farrell, R.M. Killen, D.M. Hurley, Solar wind implantation into the lunar regolith: Monte Carlo simulations of H retention in a surface with defects and the H_2 exosphere. J. Geophys. Res., Planets **124**, 278–293 (2019). https://doi.org/10.1029/2018JE005805

D.L. Turner, N. Omidi, D.G. Sibeck, V. Angelopoulos, First observations of foreshock bubbles upstream of Earth's bow shock: characteristics and comparisons to HFAs. J. Geophys. Res. Space Phys. **118**, 1552–1570 (2013). https://doi.org/10.1002/jgra.50198

D.L. Turner, T.Z. Liu, L.B. Wilson III, I.J. Cohen, D.G. Gershman, J.F. Fennell, J.B. Blake, B.H. Mauk, N. Omidi, J.L. Burch, Microscopic, multipoint characterization of foreshock bubbles with Magnetospheric Multiscale (MMS). J. Geophys. Res. Space Phys. **125**, e2019JA027707 (2020). https://doi.org/10.1029/2019JA027707

A.L. Tyler, R.W.H. Kozlowski, D.M. Hunten, Observations of sodium in the tenuous lunar atmosphere. Geophys. Res. Lett. **15**(10), 1141–1145 (1988). https://doi.org/10.1029/GL015i010p01141

O.L. Vaisberg, L.A. Avanov, J.L. Burch, J.H. Waite Jr., Measurements of plasma in the magnetospheric tail lobes. Adv. Space Res. **18**(8), 63–67 (1996). https://doi.org/10.1016/0273-1177(95)00998-1

P. van der Heide, Sputtering and ion formation, Chap. 3, in *Secondary Ion Mass Spectrometry: An Introduction to Principles and Practices* (Wiley, New York, 2014)

J. Varela, F. Pantellini, M. Moncuquet, The effect of interplanetary magnetic field orientation on the solar wind flux impacting Mercury's surface. Planet. Space Sci. **119**, 264–269 (2015). https://doi.org/10.1016/j.pss.2015.10.004

A.N. Volkov, R.E. Johnson, O.J. Tucker, J.T. Erwin, Thermally driven atmospheric escape: transition from hydrodynamic to Jeans escape. Astrophys. J. Lett. **729**, L24 (2011). https://doi.org/10.1088/2041-8205/729/2/L24

R. von Steiger, N.A. Schwadron, L.A. Fisk, J. Geiss, G. Gloeckler, S. Hefti, B. Wilken, R.R. Wimmer-Schweingruber, T.H. Zurbuchen, Composition of quasi-stationary solar wind flows from Ulysses/Solar

Wind Ion Composition Spectrometer. J. Geophys. Res. **105**(A12), 27217–27238 (2000). https://doi.org/10.1029/1999JA000358

U. von Toussaint, A. Mutzke, A. Manhard, Sputtering of rough surfaces: a 3D simulation study. Phys. Scr. T **170**, 014056 (2017). https://doi.org/10.1088/1402-4896/aa90be

A. Vorburger, P. Wurz, S. Barabash, M. Wieser, Y. Futaana, M. Holmström, A. Bhardwaj, K. Asamura, Energetic neutral atom imaging of the lunar surface. J. Geophys. Res. **117**, A07208 (2012). https://doi.org/10.1029/2012JA017553

A. Vorburger, P. Wurz, S. Barabash, M. Wieser, Y. Futaana, C. Lue, M. Holmström, A. Bhardwaj, M.B. Dhanya, K. Asamura, Energetic neutral atom imaging of the lunar surface. J. Geophys. Res. **118**, 3937–3945 (2013). https://doi.org/10.1002/jgra.50337

A. Vorburger, P. Wurz, S. Barabash, M. Wieser, Y. Futaana, M. Holmström, A. Bhardwaj, K. Asamura, First direct observation of sputtered lunar oxygen. J. Geophys. Res. **119**(2), 709–722 (2014). https://doi.org/10.1002/2013JA019207

A. Vorburger, P. Wurz, S. Barabash, Y. Futaana, M. Wieser, A. Bhardwaj, M.B. Dhanya, K. Asamura, Transport of solar wind plasma onto the lunar nightside surface. Geophys. Res. Lett. **43**, 10586–10594 (2016). https://doi.org/10.1002/2016GL071094

M. Vyšinka, Z. Němeček, J. Šafránková, J. Pavlů, J. Vaverka, J. Lavková, Sputtering of spherical SiO$_2$ samples. IEEE Trans. Plasma Sci. **44**, 1036 (2016). https://doi.org/10.1109/TPS.2016.2564502

M. Wahl, A. Wucher, VUV photoionization of sputtered neutral silver clusters. Nucl. Instrum. Methods Phys. Res. B **94**, 36–46 (1994). https://doi.org/10.1016/0168-583X(94)95655-3

X.-D. Wang, W. Bian, J.-S. Wang, J.-J. Liu, Y.-L. Zou, H.-B. Zhang, C. Lü, J.-Z. Liu, W. Zuo, Y. Su, W.-B. Wen, M. Wang, Z.-Y. Ouyang, C.-L. Li, Acceleration of scattered solar wind protons at the polar terminator of the Moon: results from Chang'E-1/SWIDs. Geophys. Res. Lett. **37**(7) (2010). https://doi.org/10.1029/2010GL042891

S. Wang, Q. Zong, H. Zhang, Hot flow anomaly formation and evolution: cluster observations. J. Geophys. Res. Space Phys. **118**(7), 4360–4380 (2013). https://doi.org/10.1002/jgra.50424

A. Warmuth, G. Mann, Constraints on energy release in solar flares from RHESSI and GOES X-ray observations II. Energetics and energy partition. Astron. Astrophys. **588**, A116 (2016). https://doi.org/10.1051/0004-6361/201527475

G.K. Wehner, C.E. Kenknight, Investigation of sputtering effects on the Moon's surface, Final report (Technical Report). Minneapolis, MN, Litton Systems Inc., Applied Science Div. (1967). https://ntrs.nasa.gov/citations/19670028248

G.K. Wehner, C. Kenknight, D.L. Rosenberg, Sputtering rates under solar-wind bombardment. Planet. Space Sci. **11**(8), 885–895 (1963a). https://doi.org/10.1016/0032-0633(63)90120-X

G.K. Wehner, C.E. Kenknight, D. Rosenberg, Modification of the lunar surface by the solar-wind bombardment. Planet. Space Sci. **11**(11), 1257–1258 (1963b). https://doi.org/10.1016/0032-0633(63)90229-0

A. Wekhof, Negative ions in the ionospheres of planetary bodies without atmospheres. Moon Planets **24**(1), 45–52 (1981). https://rdcu.be/ciKNl

B.Y. Welsh, J. Wheatley, S.E. Browne, O.H.W. Siegmund, J.G. Doyle, E. O'Shea, A. Antonova, K. Forster, M. Seibert, P. Morrissey, Y. Taroyan, GALEX high time-resolution ultraviolet observations of dMe flare events. Astron. Astrophys. **458**, 921–930 (2006). https://doi.org/10.1051/0004-6361:20065304

Y.C. Whang, Interaction of the magnetized solar wind with the Moon. Phys. Fluids **11**(5), 969–975 (1968). https://doi.org/10.1063/1.1692068

M. Wieser, S. Barabash, Y. Futaana, M. Holmström, A. Bhardwaj, R. Sridharan, M.B. Dhanya, P. Wurz, A. Schaufelberger, K. Asamura, Extremely high reflection of solar wind protons as neutral hydrogen atoms from regolith in space. Planet. Space Sci. **57**, 14–15 (2009). https://doi.org/10.1016/j.pss.2009.09.012

M. Wieser, S. Barabash, Y. Futaana, M. Holmström, A. Bhardwaj, R. Sridharan, M.B. Dhanya, A. Schaufelberger, P. Wurz, K. Asamura, First observation of a mini-magnetosphere above a lunar magnetic anomaly using energetic neutral atoms. Geophys. Res. Lett. **37**, L05103 (2010). https://doi.org/10.1029/2009GL041721

J.-P. Williams, D.A. Paige, B.T. Greenhagen, E. Sefton-Nash, The global surface temperatures of the Moon as measured by the Diviner Lunar Radiometer Experiment. Icarus **283**, 300–325 (2017). https://doi.org/10.1016/j.icarus.2016.08.012

L.B. Wilson III, M.L. Stevens, J.C. Kasper, K.G. Klein, B.A. Maruca, S.D. Bale, T.A. Bowen, M.P. Pulupa, C.S. Salem, The statistical properties of solar wind temperature parameters near 1 AU. Astrophys. J. Suppl. Ser. **236**, 41 (2018). https://doi.org/10.3847/1538-4365/aab71c

J.K. Wilson, M. Mendillo, H.E. Spence, Magnetospheric influence on the Moon's exosphere. J. Geophys. Res. **111**, A07207 (2006). https://doi.org/10.1029/2005JA011364

R.M. Winslow, C.L. Johnson, B.J. Anderson, H. Korth, J.A. Slavin, M.E. Purucker, S.C. Solomon, Observations of Mercury's northern cusp region with MESSENGER's magnetometer. Geophys. Res. Lett. **39**, L08112 (2012). https://doi.org/10.1029/2012GL051472

R.M. Winslow, C.L. Johnson, B.J. Anderson, D.J. Gershman, J.M. Raines, R.J. Lillis, H. Korth, J.A. Slavin, S.C. Solomon, T.H. Zurbuchen, M.T. Zuber, Mercury's surface magnetic field determined from proton-reflection magnetometry. Geophys. Res. Lett. **41**, 4463–4470 (2014). https://doi.org/10.1002/2014GL060258

R.M. Winslow, L. Philpott, C.S. Paty, N. Lugaz, N.A. Schwadron, C.L. Johnson, H. Korth, Statistical study of ICME effects on Mercury's magnetospheric boundaries and northern cusp region from MESSENGER. J. Geophys. Res. Space Phys. **122**, 4960–4975 (2017). https://doi.org/10.1002/2016JA023548

R.M. Winslow, N. Lugaz, L. Philpott, C.J. Farrugia, C.L. Johnson, B.J. Anderson, C.S. Paty, N.A. Schwadron, M. Al Asad, Observations of extreme ICME ram pressure compressing Mercury's dayside magnetosphere to the surface. Astrophys. J. **889**, 184 (2020). https://doi.org/10.3847/1538-4357/ab6170

H.F. Winters, J.W. Coburn, Surface science aspects of etching reactions. Surf. Sci. Rep. **14**, 162–269 (1992). https://doi.org/10.1016/0167-5729(92)90009-Z

D.P. Woodruff, *Modern Techniques of Surface Science* (Cambridge University Press, Cambridge, 2016). ISBN 9781107023109

T.N. Woods, F.G. Eparvier, J. Fontenla, J. Harder, G. Kopp, W.E. McClintock, G. Rottman, B. Smiley, M. Snow, Solar irradiance variability during the October 2003 solar storm period. Geophys. Res. Lett. **31**, L10802 (2004). https://doi.org/10.1029/2004GL019571

T.N. Woods, F.G. Eparvier, S.M. Bailey, P.C. Chamberlin, J. Lean, G.J. Rottman, S.C. Solomon, W.K. Tobiska, D.L. Woodraska, Solar EUV experiment (SEE): mission overview and first results. J. Geophys. Res. **110**, A01312 (2005). https://doi.org/10.1029/2004JA010765

T.N. Woods, G. Kopp, P.C. Chamberlin, Contributions of the solar ultraviolet irradiance to the total solar irradiance during large flares. J. Geophys. Res. **111**, A10S14 (2006). https://doi.org/10.1029/2005JA011507

C.-J. Wu, W.-H. Ip, L.-C. Huang, A study of variability in the frequency distributions of the superflares of G-type stars observed by the Kepler mission. Astrophys. J. **798**, 92 (2015). https://doi.org/10.1088/0004-637X/798/2/92

P. Wurz, Solar Wind Composition, in The Dynamic Sun: Challenges for Theory and Observations. ESA SP-600, 5.2, pp. 1–9 (2005). https://ui.adsabs.harvard.edu/abs/2005ESASP.600E..44W

P. Wurz, Erosion processes affecting interplanetary dust grains, in *Nano Dust in the Solar System: Discoveries and Interpretations*. Astrophysics and Space Science Library, vol. 385 (Springer, Berlin, 2012), pp. 161–178. https://doi.org/10.1007/978-3-642-27543-2_8

P. Wurz, L. Blomberg, Particle populations in Mercury's magnetosphere. Planet. Space Sci. **49**(14–15), 1643–1653 (2001). https://doi.org/10.1016/S0032-0633(01)00102-7

P. Wurz, H. Lammer, Monte-Carlo simulation of Mercury's exosphere. Icarus **164**(1), 1–13 (2003). https://doi.org/10.1016/S0019-1035(03)00123-4

P. Wurz, E. Wolfrum, W. Husinsky, G. Betz, L. Hudson, N.H. Tolk, ESD thresholds for excited atoms desorbed from Alkali-Halides. Radiat. Eff. Defects Solids **109**, 203–212 (1989)

P. Wurz, W. Husinsky, G. Betz, Sputtering of clean and oxidized Cr and Ta metal targets using SNMS and SIMS, in *Proceedings of Symposium on Surface Science*, ed. by J.J. Ehrhardt, C. Launois, B. Mutaftschiev, M.R. Tempère (La Plagne, France, 1990), pp. 181–185

P. Wurz, J. Sarnthein, W. Husinsky, G. Betz, P. Nordlander, Y. Wang, Electron-stimulated desorption of neutral ground-state lithium atoms from LiF due to excitation of surface excitons. Phys. Rev. B **43**, 6729–6732 (1991). https://doi.org/10.1103/PHYSREVB.43.6729

P. Wurz, R.F. Wimmer-Schweingruber, K. Issautier, P. Bochsler, A.B. Galvin, F.M. Ipavich, Composition of magnetic cloud plasmas during 1997 and 1998, in *American Institute Physics on Solar and Galactic Composition* vol. CP-598 (2001), pp.145–151. https://doi.org/10.1063/1.1433993

P. Wurz, R. Wimmer-Schweingruber, P. Bochsler, A. Galvin, J.A. Paquette, F. Ipavich, Composition of magnetic cloud plasmas during 1997 and 1998. AIP Conf. Proc. **679**, 685–690 (2003). https://doi.org/10.1063/1.1618687

P. Wurz, U. Rohner, J.A. Whitby, C. Kolb, H. Lammer, P. Dobnikar, J.A. Martín-Fernández, The lunar exosphere: the sputtering contribution. Icarus **191**, 486–496 (2007). https://doi.org/10.1016/j.icarus.2007.04.034

P. Wurz, J.A. Whitby, U. Rohner, J.A. Martín-Fernández, H. Lammer, C. Kolb, Self-consistent modelling of Mercury's exosphere by sputtering, micro-meteorite impact and photon-stimulated desorption. Planet. Space Sci. **58**, 1599–1616 (2010). https://doi.org/10.1016/j.pss.2010.08.003

P. Wurz, D. Abplanalp, M. Tulej, M. Iakovleva, V.A. Fernandes, A. Chumikov, G. Managadze, Mass spectrometric analysis in planetary science: investigation of the surface and the atmosphere. Sol. Syst. Res. **46**, 408–422 (2012). https://doi.org/10.1134/S003809461206007X

P. Wurz, D. Gamborino, A. Vorburger, J.M. Raines, Heavy ion composition of Mercury's magnetosphere. J. Geophys. Res. **124**, 2603–2612 (2019). https://doi.org/10.1029/2018JA026319

B.V. Yakshinskiy, T.E. Madey, Photon-stimulated desorption as a substantial source of sodium in the lunar atmosphere. Nature **400**, 642–644 (1999). https://doi.org/10.1038/23204

B.V. Yakshinskiy, T.E. Madey, Desorption induced by electronic transitions of Na from SiO_2: relevance to tenuous planetary atmospheres. Surf. Sci. **451**, 160–165 (2000). https://doi.org/10.1016/S0039-6028(00)00022-4

B.V. Yakshinskiy, T.E. Madey, DIET of alkali atoms from mineral surfaces. Surf. Sci. **528**, 54–59 (2003). https://doi.org/10.1016/S0039-6028(02)02610-9

B.V. Yakshinskiy, T.E. Madey, Photon-stimulated desorption of Na from a lunar sample: temperature-dependent effects. Icarus **168**, 53–59 (2004). https://doi.org/10.1016/j.icarus.2003.12.007

B.V. Yakshinskiy, T.E. Madey, Temperature-dependent DIET of alkalis from SiO_2 films: comparison with a lunar sample. Surf. Sci. **593**, 202–209 (2005). https://doi.org/10.1016/j.susc.2005.06.062

H. Yang, J. Liu, Q. Gao, X. Fang, J. Guo, Y. Zhang, Y. Hou, Y. Wang, Z. Cao, The flaring activity of M dwarfs in the Kepler field. Astrophys. J. **849**, 36 (2017). https://doi.org/10.3847/1538-4357/aa8ea2

M. Zelen, N.C. Severo, Probability functions, in *Handbook of Mathematical Functions*, ed. by M. Abramowitz, A. Stegun (Dover, New York, 1965)

H. Zhang, D.G. Sibeck, Q.-G. Zong, N. Omidi, D. Turner, L.B.N. Clausen, Spontaneous hot flow anomalies at quasi-parallel shocks: 1. Observations. J. Geophys. Res. Space Phys. **118**, 3357–3363 (2013). https://doi.org/10.1002/jgra.50376

A. Zhang, M. Wieser, C. Wang, S. Barabash, W. Wang, X. Wang, Y. Zou, L. Li, J. Cao, L. Kalla, L. Dai, J. Svensson, L. Kong, M. Oja, B. Liu, V. Alatalo, Y. Zhang, J. Talonen, Y. Sun, M. Emanuelsson, C. Xue, L. Wang, F. Wang, W. Liu, Emission of energetic neutral atoms measured on the lunar surface by Chang'E-4. Planet. Space Sci. **189**, 104970 (2020). https://doi.org/10.1016/j.pss.2020.104970

J.F. Ziegler, M.D. Ziegler, J.P. Biersack, SRIM – the stopping and range of ions in matter. Nucl. Instrum. Methods Phys. Res., Sect. B, Beam Interact. Mater. Atoms **268**(11–12), 1818–1823 (2010). https://doi.org/10.1016/j.nimb.2010.02.091

F.M. Zimmermann, W. Ho, Velocity distributions of photochemically desorbed molecules. J. Chem. Phys. **100**(10), 7700–7706 (1994). https://doi.org/10.1063/1.466864

E. Zinner, On the constancy of solar particle fluxes form track, thermoluminescence and solar wind measurement in lunar rocks, in *The Ancient Sun: Fossil Record in the Earth, Moon and Meteorites*, ed. by R.O. Pepin, J.A. Eddy, R.B. Merrill (Pergamon Press, New York, 1980), pp. 201–226. http://adsabs.harvard.edu/full/1980asfr.symp..201Z

Q.-G. Zong, B. Wilken, J. Woch, T. Mukai, T. Yamamoto, G.D. Reeves, T. Doke, K. Maezawa, D.J. Williams, S. Kokubun, S. Ullaland, Energetic oxygen ion bursts in the distant magnetotail as a product of intense substorms: three case studies. J. Geophys. Res. Space Phys. **103**(A9), 20339–20363 (1998). https://doi.org/10.1029/97JA01146

T.H. Zurbuchen, J.M. Raines, G. Gloeckler, S.M. Krimigis, J.A. Slavin, P.L. Koehn, R.M. Killen, A.L. Sprague, R.L. McNutt, S.C. Solomon, MESSENGER observations of the composition of Mercury's ionized exosphere and plasma environment. Science **321**, 90–92 (2008). https://doi.org/10.1126/science.1159314

T.H. Zurbuchen, J.M. Raines, J.A. Slavin, D.J. Gershman, J.A. Gilbert, G. Gloeckler, B.J. Anderson, D.N. Baker, H. Korth, S.M. Krimigis, M. Sarantos, D. Schriver, R.L. McNutt Jr., S.C. Solomon, MESSENGER observations of the spatial distribution of planetary ions near Mercury. Science **333**, 1862–1865 (2011). https://doi.org/10.1126/science.1211302

Space Science Reviews (2023) 219:4
https://doi.org/10.1007/s11214-023-00951-5

Surface Exospheric Interactions

Ben Teolis[1] · Menelaos Sarantos[2] · Norbert Schorghofer[3] · Brant Jones[4] ·
Cesare Grava[1] · Alessandro Mura[5] · Parvathy Prem[6] · Ben Greenhagen[6] ·
Maria Teresa Capria[5] · Gabriele Cremonese[7] · Alice Lucchetti[7] · Valentina Galluzzi[5]

Received: 27 January 2022 / Accepted: 7 November 2022 / Published online: 23 January 2023
© The Author(s) 2023

Abstract

Gas-surface interactions at the Moon, Mercury and other massive planetary bodies constitute, alongside production and escape, an essential element of the physics of their gravitationally bound exospheres. From condensation and accumulation of exospheric species onto
the surface in response to diurnal and seasonal changes of surface temperature, to thermal
accommodation, diffusion and ultimate escape of these species from the regolith back into
space, surface-interactions have a drastic impact on exospheric composition, structure and
dynamics. The study of this interaction at planetary bodies combines exospheric modeling
and observations with a consideration of fundamental physics and laboratory experimenta-
tion in surface science. With a growing body of earth-based and spacecraft observational
data, and a renewed focus on lunar missions and exploration, the connection between the
exospheres and surfaces of planetary bodies is an area of active and growing research, with
advances being made on problems such as topographical and epiregolith thermal effects on
volatile cold trapping, among others. In this paper we review current understanding, latest
developments, outstanding issues and future directions on the topic of exosphere-surface
interactions at the Moon, Mercury and elsewhere.

Surface-Bounded Exospheres and Interactions in the Inner Solar System
Edited by Anna Milillo, Menelaos Sarantos, Benjamin D. Teolis, Go Murakami, Peter Wurz and Rudolf
von Steiger

✉ B. Teolis
  ben.teolis@swri.org

✉ A. Mura
  alessandro.mura@inaf.it

1   Southwest Research Institute, San Antonio, TX, USA

2   NASA Goddard Space Flight Center, Greenbelt, MD, USA

3   Planetary Science Institute, Tuscon, AZ, USA

4   Georgia Institute of Technology, Atlanta, GA, USA

5   Instituto Nazionale Di Astrofisica, Rome, Italy

6   Johns Hopkins Applied Physics Laboratory, Laurel, MD, USA

7   INAF-Osservatorio Astronomico di Padova, Padova, Italy

Fig. 1 Schematic diagram of gas-regolith interactions, showing possible paths taken by exospheric gas species upon arrival at a surface regolith, including sticking to regolith particles, surface diffusion, vaporization, sputtering, and inter-grain hops. These processes together determine the retention, abundance, distribution and diffusion of exospheric gas within the surface regolith, and thereby impact the structure and dynamics of a planetary body's surface bounded exosphere

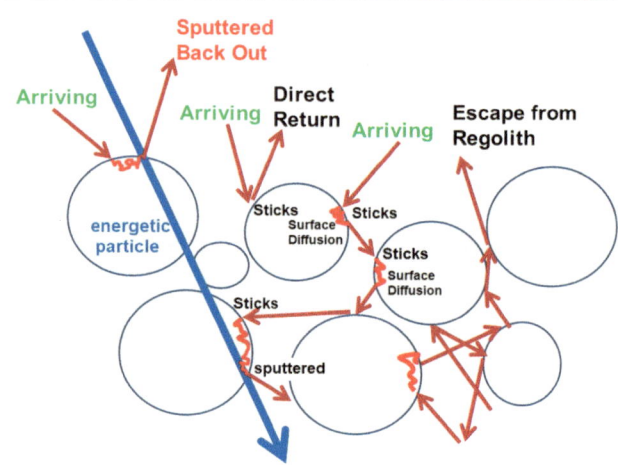

1 Introduction

In the gravitationally bound exospheres of massive airless planetary bodies like the Moon and Mercury, atoms and molecules undergo far more collisions with the planet surface than with each other. For this reason gas-surface atomic/molecular interactions, including (1) thermal accommodation, (2) sticking, (3) volatile diffusion within the regolith, (4) cold trapping, and (5) impact and radiation induced diffusion and desorption, have an outsized influence on the distribution, composition, and time variability of these atmospheres. The properties of an exosphere are therefore determined to a much greater degree by the exosphere's interaction with the surface than with itself.

Since the Apollo era the observations of spacecraft and earth-based telescopes have revealed spatial and temporal variations of exospheric gas and surface frost distributions over many orders of magnitude in response to diurnal and, at some bodies, seasonal changes of surface temperature. Examples include the distribution of argon in the lunar atmosphere and on its surface, first investigated by Apollo and later by the NASA Lunar Atmosphere and Dust Environment Explorer (LADEE) spacecraft, e.g., (Benna et al. 2015; Hodges 2018; Hodges and Mahaffy 2016), and the distribution of Sodium in Mercury's exosphere in relation to this planet's surface 'cold poles' (Cassidy et al. 2016) (Sect. 8). Across the solar system, massive airless planetary bodies tilted to the ecliptic, including Saturn's moons Dione and Rhea (Teolis and Waite 2016), Pluto's moon Charon (Grundy et al. 2016; Teolis et al. 2022), and likely the massive Uranian moons – exhibit extreme seasonal changes in their exospheres on much longer timescales, driven by bi-annual condensation and evaporation of exospheric gases frozen at their poles. Yet remarkably, as we discuss in this chapter, even the Moon's argon exosphere, despite its minor 1.5° angle to the ecliptic, exhibits some seasonal variability likely owing to seasonal condensation and evaporation of frozen argon to and from polar cold traps. In general bodies like the Moon and Mercury appear to represent the limiting case of a weakly seasonal exosphere, in which a day-night adsorption-desorption cycle is the primary effect, while the preponderance of permanent over seasonally shadowed polar cold traps act as ultra-long term exospheric sinks.

Other processes, such as impact and or radiation enhanced diffusion and desorption of semi-volatile species frozen to the surface regolith, and the chemical activation of surface regolith by impacting ions, e.g. from the solar wind, are only just beginning to be explored. Finally, surface microstructure, diffusion of adsorbed gas between regolith grains (Fig. 1), and surface topographical effects, including permanent, seasonal, and diurnal cold trapping, all influence the residence time of exospheric atoms and molecules on the surface material. The total surface interaction/residence time of exospheric species within a granular planetary surface, including the combined effect of surface diffusion and multiple accommodation, adsorption and desorption events between regolith grains (Fig. 1), is a critical parameter controlling the exospheric distribution between the day and night hemispheres, as well as any seasonal variability.

In this chapter we discuss the physics, open questions and current knowledge of three aspects of the surface-exospheric interactions at large airless bodies of the inner solar system: (1) accommodation, sticking and desorption, (2) regolith diffusion, and (3) surface impact/radiation processing and induced diffusion/ejection of adsorbed material.

2 Thermal Accommodation & Sticking

Energy accommodation is a fundamental and ubiquitous process common to surface-bounded exospheres. Accommodation constitutes the first step in the surface interaction of exospheric atomic or molecular species, and determines the local energy and altitude distribution of the exosphere above the surface. In simplest terms, any surface bounded exosphere consists of both a non-accommodated and an accommodated contribution. The non-accommodated *supra-thermal* component represents the exospheric source; i.e. neutral species freshly injected into the exosphere by processes such as impact vaporization, photodesorption, sputtering or radiogenic decay/outgassing from the surface, or charge-exchange at altitude. Some atoms from these high-energy processes have sufficient speeds to escape the planet's gravity, or at least to reach high altitudes before falling back to the surface (e.g., Killen et al. 2018). The *accommodated* component, by contrast, interacts with the planetary surface to acquire an approximate Maxwell-Boltzmann velocity distribution at a temperature close to that of the local surface. The accommodated component has two sources: (1) exospheric neutrals that fall back to the surface under gravity and interact with the regolith, and (2) exospheric source atoms/molecules initially emitted from the surface in an accommodated state; e.g. off-gassing radiolytic or radiogenic species, or sputtered energetic species that impact other regolith grains and lose energy before ever entering the exosphere.

The supra-thermal and accommodated components may segregate into high and low altitude layers, as can be seen in Fig. 2 showing the predicted altitude profile for multiple exospheric species in Europa's exosphere, and for sodium and argon in the Hermean and lunar exospheres, respectively. Compared to the Moon's fully accommodated argon exosphere, energetic release processes at Mercury produce a more extended exospheric sodium altitude profile (Fig. 2) modified to a lesser degree by surface temperature-dependent sticking and accommodation (e.g., Burger et al. 2010; Leblanc and Johnson 2010; Tenishev et al. 2013). At both rocky and icy planetary bodies sufficiently volatile exospheric species may exhibit (1) an extended high-altitude exosphere from freshly sputtered molecules, and (2) a much denser low-altitude thermally accommodated layer (see Fig. 2; Europa's O_2 and H_2O) with molecules lacking the kinetic energy to attain high altitudes in the exosphere. Hence the exosphere at high-altitude gives a more representative sampling of species immediately after ejection from the local surface below, whereas the low-altitude exosphere may be dominated by species which have interacted and accommodated with the surface many times. Surface gas densities may in fact be greatest at night owing to the colder surface there, which cools the exosphere, and reduces the gas thermal speed and exospheric scale height such as to

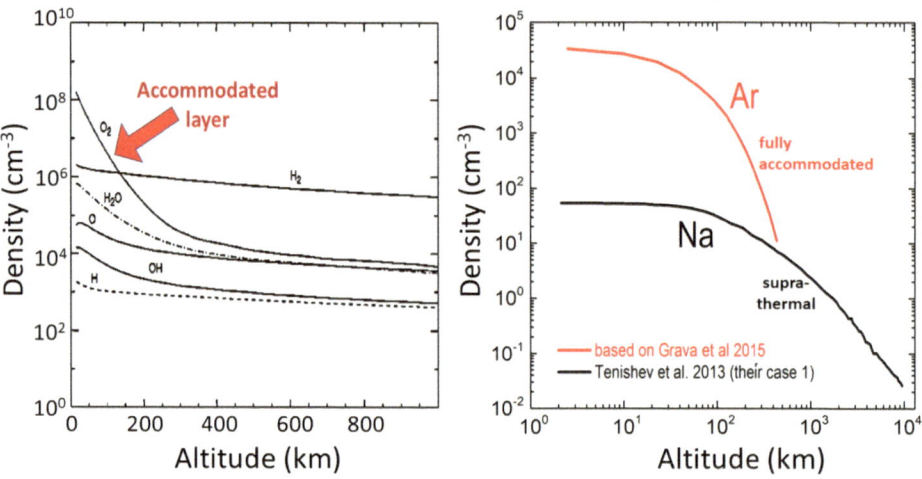

Fig. 2 Left: Estimated average Europan exospheric densities for selected species from the model of Smyth and Marconi (2006). Sufficiently volatile species like O_2 and H_2O exhibit a low altitude thermally accommodated layer, but at high altitudes the dominance of supra-thermal (non-accommodated) molecules newly produced/ejected from the surface results in a shallow slope of density versus altitude. Right: Lunar Ar and Na gas densities at sub-solar noon, from the models used by Grava et al. (2015) and Tenishev et al. (2013) to fit LACE Ar and remote Na observations. Na density taken from Tenishev et al. (2013); their case 1 that assumes full Na sticking to the surface. Ar density versus altitude is extracted from the Grava et al. (2015) model. Lunar Ar is fully thermally accommodated resulting in a steep slope versus altitude. Lunar Na is freshly ejected largely by photon-stimulated desorption. Therefore Na is not accommodated, producing a shallow slope

concentrate the gas there. Examples of this interaction regime are some noble gases at the Moon (Benna et al. 2015). However, the opposite may happen for species such as lunar argon (Sect. 7) that can *stick* to the surface at the given surface temperature; surface gas densities will be *lower* at night if sufficiently cold for exospheric atoms or molecules to freeze to the surface.

The degree to which exospheric species accommodate with the surface on a single collision is quantified by the energy accommodation coefficient, which is the ratio (0 to 1) of energy transferred in a surface collision relative to that required for full accommodation. The accommodation coefficient depends on the mass, kinetic energy, and internal degrees of freedom (e.g. molecular rotations and vibrations) of the impacting species, and the magnitude of the interaction potential or binding energy with the surface (Bonfanti and Martinazzo 2016). Another major consideration is regolith porosity, which enhances the 'effective' accommodation of exospheric species undergoing multiple inter-grain collisions within the regolith before returning to the exosphere (Fig. 1).

The angular dependence of the velocity of a gas thermally accommodated to the planetary surface follows a Knudsen cosine-law (Comsa 1968; Knudsen 1909), which implies that the direction in which an atom or molecule rebounds from a surface is independent of the direction at which it approaches. The Knudsen cosine law can be understood in terms of the dispersing micro-geometry of the solid surface (Feres and Yablonsky 2004), but the same decorrelation is expected from thermal accommodation during contact. Armand (1977) determined the desorption velocity distribution based on lattice vibrations, and arrives at the Maxwell-Boltzmann flux distribution, which obeys the Cosine Law.

Fig. 3 Energy schematic of the adsorption process: (a) back-scattering, (b) chemisorption, (c) physisorption. From Kolasinski (2012)

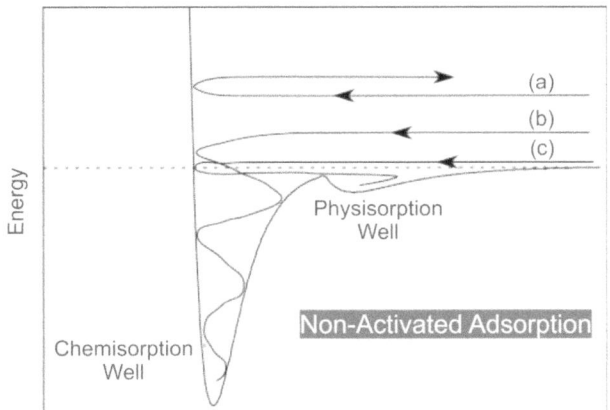

The interaction of gas phase molecules with surfaces can involve vastly complex chemical physics and depends on not only the molecule and the surface, but also the presence of co-adsorbates, surface temperature, incoming molecular energy (translational, rovibrational), and the crystal structure (defects, step edges, crystallographic plane, terraces). There are numerous high-level texts with more in-depth coverage of the material (e.g. Billing 2000; Ibach 2006; Kolasinski 2012; Morrison 2013; Zangwill 1988). Molecular beam experiments are reviewed in e.g. Arumainayagam and Madix (1991). The following is meant as only a general introduction to adsorption.

Energy transfer is a pre-requisite for adsorption. For trapping or sticking to occur, the incoming molecule must lose energy, with the molecule finally transferred into a bound adsorbed state. Adsorption is further divided into two categories, activated and non-activated. An easy way to visualize this is to utilize a Lennard Jones potential as shown in Fig. 3. If an impinging molecule has too much energy, the molecule will reflect as in a typical elastic collision (a). A molecule or atom may lose some energy after reflecting off the potential barrier where it then proceeds down well to a chemisorbed state (b). Alternatively, a small barrier may exist such that a slow molecule will trap in the shallow well resulting in a physisorbed state (c). The height of this barrier is what controls the activated or non-activated adsorption process.

At 'low' kinetic energies of order 0.1 eV or below, typical of most simple exospheric species at or below a few km/s speeds, gaseous species will transiently stick to the surface, with ample 'sticking time' (picoseconds or longer) in the surface interaction potential well to fully accommodate before thermally desorbing. At these low kinetic energies (which are albeit still up to thousands of Kelvins) the gas species are fully adsorbed during their interaction with the surface, with near unity accommodation and sticking coefficients. By contrast, sputtered species ejected with 'high' energies just shy of the planetary escape speed, or species accelerated by radiation pressure, may fall back to the planetary surface with eVs of energy, depending on the planet's gravity (or escape speed) and species mass. At eV energies or above the impact of gaseous atomic species with surfaces may be better approximated as binary collisions between the impactor and a single target atom. Backscattered impactors may lose energy to the surface atoms elastically and inelastically in accordance with the back-scattering angle and relative projectile and target masses. Whereas light species (such as H, H_2 or He) may have relatively low accommodation coefficients at eV energies, heavier impactors with mass similar to the target material atoms may exhibit accommodation coefficients closer to unity.

A caveat is that atomic or other reactive species, such as Na at the Moon or Mercury, may have high eV binding energies owing to chemisorption with the surface. Even at eV kinetic energies these reactive species may transiently stick and accommodate with the surface irrespective of their mass. Another consideration for molecular species at eV energies is the absorption of some of the collision impact energy into internal interatomic motions within the impacting molecule, such that the molecule loses translational kinetic energy in the collision with the surface by becoming internally excited.

2.1 Adsorption Models

Numerous adsorption models have been developed over the years that have resulted in analytical expressions of the adsorption process that include the energy dependence of accommodation. The first theoretical treatment of energy exchange/accommodation came from Baule (1914) and later by Zwanzig (1960), who treated the surface as harmonic oscillators that are excited by collision allowing the incoming molecules to lose energy elastically. Under these conditions the accommodation coefficient $\alpha = \frac{4\mu}{(1+\mu)^2}$, where μ is the ratio of the incoming projectile mass to that of the surface species $\mu = \frac{m_p}{m_s}$. A variation of this follows the hard cube model, where the incoming molecule is accelerated by an attractive potential (V) resulting in a simple modification to the Baule equation $\alpha = \frac{4\mu + V/\varepsilon}{(1+\mu)^2}$, where ε is the collision energy. These models assume that the surface is fixed at 0 K. Taking into consideration the average thermal distribution of surface velocities results in $\alpha = \frac{4\mu(\varepsilon + V - kT/2)/\varepsilon}{(1+\mu)^2}$ (Bonfanti and Martinazzo 2016). Other classical models of energy accommodation as a function of gas and surface temperature were described by Fan and Manson (2010). For a quantum mechanical description, see texts by Billing (2000) and Zangwill (1988). Accommodation coefficient have often also been measured in the laboratory (e.g., Haynes et al. 1992; Persad and Ward 2016).

The well-known Langmuir model (Langmuir 1918), applicable for sub-monolayer coverage, uses a single binding energy and assumes that adsorption and desorption are at equilibrium and reversible processes without an activation barrier. Using these assumptions, the fractional coverage is derived as $\theta = \frac{K_{eq}p}{1 + K_{eq}p}$, where $K_{eq} = \frac{k_{ad}}{k_{des}}$ is the ratio of the rate of adsorption to desorption and p is the partial pressure of the adsorbate. As evident from this equation, the sticking coefficient (S) is not dependent on the incoming incident energy, but simply the available surface coverage $S \sim (1 - \theta)$. The Langmuir model was a big leap forward in thinking about gas-surface interactions, but it deviated from experimental results more often than not (Becker and Hartman 1953; King and Wells 1974; Singh-Boparai et al. 1975). Moreover, the Langmuir model does not take into consideration the energy of the incoming particle.

Ideally, if the adsorption process does not have an activation barrier, the fraction of atoms or molecules that adsorb to the surface will decrease with increasing incoming energy. Here, the sticking probability for a non-activated process will be unity until the incident energy reaches a critical value, at which point it rapidly diminishes to zero. In reality, this apparent step function is smoothed out by thermal effects from the surface. The sticking probability for an ideal surface without an activation barrier can be modelled treating the surface atoms as hard cubes (Sipkens and Daun 2017). The cube has an attractive square well which accelerates the molecule as it approaches the surface to a velocity u_n. As an approximation only the cube's motion in the direction normal to the adsorption surface is considered, with the velocity dictated by the surface temperature T. The normal component of the particle velocity v after collision with the hard cube has a modified Maxwell-Boltzmann distribution,

Fig. 4 From Bowker (2016), showing processes of physisorption, chemisorption and diffusion over occupied (extrinsic) and unoccupied (intrinsic) adsorption sites. P_{desi} and P_{dese} denote the desorption probabilities from intrinsic and extrinsic sites, while P_{ai} and P_{diffe} are the probabilities of adsorption and of surface migration/diffusion to an intrinsic site

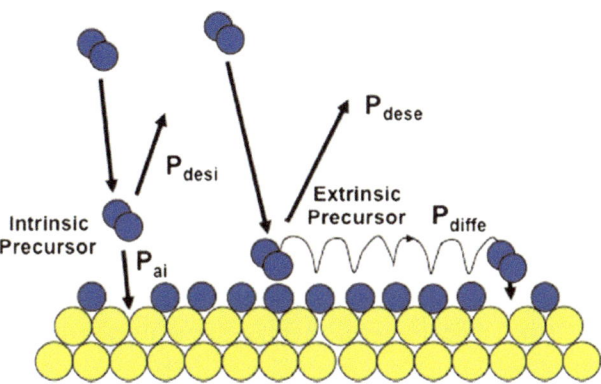

$P(v)dv = (u_n - v)e^{(\frac{-m}{2kT}v^2)}dv$. Sticking occurs if the normal velocity of the molecule after collision falls below a critical value which cannot escape the well. The sticking coefficient is derived from the total fraction of molecules below this critical velocity, v_c, and can be analytically expressed as $S = 1 + \text{erf}(bv_c) + \frac{e^{(-b^2v^2)}}{bu_n\pi^{1/2}}$, where $b^2 = \frac{m}{2kT}$.

In the case of an activated adsorption process, the system will exhibit low sticking coefficients at low energies and low coverages. In this instance, a low energy molecule will reflect off the barrier and not adsorb. Here, the sticking coefficient is practically the inverse of the non-activated case and will be zero until it approaches the barrier height where it should approach unity and then drop off again as it approaches some critical value in energy where it approaches the elastic regime. Quantum tunneling and thermal effects will smooth this step function.

A follow-up to the Langmuir model was established by Kisliuk (1957), whose model allowed for the understanding of atypically high adsorption rates at high surface coverage (Kisliuk 1957). This model is best known as the precursor adsorption model. The theory behind the precursor adsorption model is relatively simple and is outlined in Fig. 4. Here, there are two types of precursors, intrinsic (adsorbate over an empty site allowing chemisorption) and extrinsic (adsorbate physisorbed over a filled site). The intrinsic molecules have a probability of adsorbing, desorbing, or migrating to an empty site, while the extrinsic molecules have a probability of desorbing or migrating. The sum of these probabilities yields an analytical expression of the sticking coefficient, $S = S_0(1 + \frac{\theta}{1-\theta}K)^{-1}$, where K is the precursor parameter and is given by $K = S_0\frac{P'_{des}}{P_{ads}}$ with S_0 being the sticking constant at zero coverage, P'_{des} the desorption probability for the extrinsic molecule, and P_{ads} the probability for intrinsic chemisorption; both of which are products of the general Arrhenius equation with activation energy (E), prefactor f, and the average residence time spent at a particular site (τ) i.e. $P = fe^{(\frac{-E}{kT})}\tau$. Other widely used adsorption models are the Freundlich isotherm (with multiple binding energies), the Brunauer-Emmett-Teller (BET) isotherm (for multilayer adsorption), and the Zeta Adsorption isotherm (Narayanaswamy and Ward 2020).

3 Volatile Diffusion in the Regolith

Gas atoms and molecules can travel in the voids between grains and penetrate centimeters into the ground (Reiss et al. 2021). Furthermore, adsorbed atoms can also travel along a grain and onto other grains following a slower diffusion mechanism called surface diffusion

(Fig. 1). Although slow, surface diffusion can be the dominant mechanism of diffusion for species that bind strongly with the surface such as sodium (Sarantos and Tsavachidis 2020). We consider as an example the case of water vapor in the following discussion, to elucidate the physics of volatile migration and diffusion in a porous regolith.

3.1 Surface Bonding and Vapor Migration

Water molecules and other volatile species can migrate in the porous regolith via a repeated sequence of jumps. At room temperature the diffusion time is approximately the sum of times for each molecular flight, but at low temperature, migrating molecules spend most of their time residing on grain surfaces. In either case, the migration process can be thought of as a random walk, and it is thus described by a diffusion equation. The main uncertainty in quantifying the mass flux lies with the surface-vapor interactions.

At low temperature H_2O can be in crystalline or amorphous form, or adsorbed at monolayer or sub-monolayer thicknesses onto a substrate. All naturally occurring ice and snow on Earth's surface is crystalline with a hexagonal lattice structure. Although many other crystal structures are known for water ice, few are stable at low pressure. Amorphous ice is solid (condensed) H_2O that lacks a crystal structure, and forms at very low temperatures ($<\sim$ 140 K at laboratory time scales; Sack and Baragiola 1993). It takes about a dozen monolayers of H_2O until the properties of macroscopic ice are reached (Cadenhead and Stetter 1974). Lunar grains have a high specific surface area on the order of 500 m^2/kg as indicated by measurements of grain surface areas in lunar soil samples (Heiken et al. 1991). The areal density of H_2O molecules for a monolayer of solid ice is $\vartheta = (\rho/\mu)^{2/3} \approx 10^{19}$ molecules/m^2, where ρ is the density of ice and μ the mass of a water molecule. Multiplied, a monolayer of H_2O corresponds to $\sim 5 \times 10^{21}$ molecules adsorbed per kg of lunar regolith, which equates to about a hundred parts per million by mass. Whereas most of the lunar surface is extremely desiccated, much higher water concentrations have been measured at some locations in the mid-latitudes (Honniball et al. 2021) and lunar polar regions (Colaprete et al. 2010; Feldman et al. 2001).

Vapor pressure, sublimation, and molecular residence times are closely related. The sublimation rate of ice in vacuum F_{ice} is given by the Hertz-Knudsen formula (Persad and Ward 2016; Watson et al. 1961b)

$$F_{ice} = \frac{p_v}{\sqrt{2\pi k_B T \mu}}$$

with p_v the equilibrium vapor pressure, k_B the Boltzmann constant, T temperature, and μ the mass of an H_2O molecule. When the vapor is in equilibrium with ice, the evaporation rate equals the condensation rate, and the above formula is the flux of a dilute ideal gas at the equilibrium vapor pressure. The flux of water molecules from the ice surface is the same with or without incoming water molecules, so this equation describes the sublimation rate into vacuum. A dependence $F = F(\theta, T)$ of the sublimation rate on the surface coverage θ is also typical for sub-monolayer coverage. Conversely, when θ is much more than a monolayer, $F = F_{ice}(T)$, and the equilibrium vapor pressure of a solid is of the form, $p_v \propto \exp(-\Delta H_{subl}/k_B T)$, where ΔH_{subl} is the sublimation enthalpy (the heat of sublimation). The average sticking time τ of an atom or molecule on the substrate surface can also be derived from the inverse of the Polanyi-Wigner equation

$$\tau = \frac{1}{\nu} \exp\left(\frac{Q}{k_B T}\right)$$

Fig. 5 Molecular residence time τ of water molecules on crystalline ice as a function of temperature. Several relevant time scales are indicated

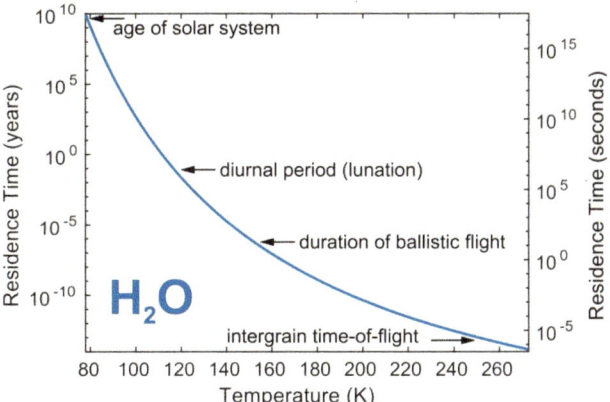

where ν corresponds to the vibrational frequency of the bond between the adsorbate and the substrate surface, typically 10^{12}–10^{14} Hz, and Q is the energy of adsorption (Adamson 1982; Atkins 1986). Not all incident molecules stick to the surface. A fraction $(1 - S)$ of incident molecules are reflected from the surface (Persad and Ward 2016). For ice, the sticking coefficient at 40–120 K temperatures is in the range 1–0.7 (Haynes et al. 1992).

The binding energy of molecules adsorbed to the regolith or on amorphous ice differs from that in crystalline ice (Speedy et al. 1996). The sublimation enthalpy of crystalline ice is 51 kJ/mol, while adsorbed water has a 60–180 kJ/mol range of desorption energies on lunar regolith simulants (Hibbitts et al. 2011; Poston et al. 2013) and lunar samples (Jones et al. 2020; Poston et al. 2015) at sub-monolayer abundances. de Leeuw et al. (2000) report adsorption energies for H_2O on forsterite surfaces of 100–172 kJ/mol. Other atoms such as sodium and potassium bind much more strongly with the silicate surface, with binding energies in excess of 174 kJ/mol (or 1.8 eV) (Yakshinskiy et al. 2000). Irradiation can increase the effective degree of bonding by creating defects and vacancies, while diffusion in powders may also increase the effective binding energy by slowing desorption (Sarantos and Tsavachidis 2020). Finally, surface microstructure prolongs the residence time of adsorbates on the surface by reducing the desorption yields via readsorption on adjacent grains (e.g., Cassidy and Johnson 2005; Sarantos and Tsavachidis 2020).

Figure 5 shows the residence time of water molecules (harmonic mean) as a function of temperature. Below roughly 250 K, the molecules spend most of their time residing on the surface and a comparatively short time in flight (assuming a 1 mm inter-grain flight distance). Additionally, the likely mobility of water adsorbates on surfaces with a distribution of desorption energies further changes the residence time by enabling motion between low and high energy sites. Sarantos and Tsavachidis (2021) have recently suggested that surface diffusion affects the desorption rate in a nonlinear manner, suppressing desorption at lower temperatures yet enhancing desorption at higher temperatures.

Similar conclusions can be drawn for less volatile elements like sodium, for which it can be shown that the residence time of atoms against thermal desorption is ~ 1000 years at maximum lunar surface temperatures, whereas on average several days elapse between desorption events for the more efficient photon-stimulated desorption mechanism (Sarantos and Tsavachidis 2020 and references therein). These timescales are much shorter at Mercury due to the higher surface temperatures. The surface residence times, along with the escape ratio, control the evolution of the atmosphere following episodic brightening events such as meteor showers (Colaprete et al. 2016).

3.2 Models of Subsurface Migration

Mathematically, the migration process can be described at three levels:

- Molecular hops (random walk; discrete formulation)
- Diffusion equation (continuum formulation; partial differential equation)
- Boundary-value problem (for stationary solutions)

The first type of model is a statistical simulation of individual molecules (random walk). Molecular random walk leads to diffusive migration where molecules reside on grain surfaces and then hop a distance ℓ, which is the molecular free inter-grain path within the regolith (Schorghofer and Taylor 2007). The mean grain size in lunar soil samples is typically 45–100 µm (Heiken et al. 1991), which may be taken (Schorghofer and Taylor 2007) as a crude approximation of ℓ.

The migration process can also be described by a continuum equation. The net flux J of molecules is given by differences in sublimation rates between two grain surfaces, and therefore

$$J(t, z) = -\ell \nabla F\big[\theta(t, z), T(t, z)\big]$$

where t and z are time and depth. The gradient in sublimation or desorption rate F is due to gradients in T or θ. Transport can be caused by differences in surface concentrations or by differences in temperature. At constant temperature, only concentration-driven migration occurs. Note that the time average of J, the net flux, is given by the gradient of the time average of F.

For situations that are stationary (time-independent, periodic, or quasi-periodic) the boundary-value formulation leads to insightful results. An example is the determination of the loss rate of ice buried by a layer of regolith of thickness Δz. Mass conservation dictates that after a transient period, the flux J is constant between an ice table and the surface. Thereafter, the value of J is determined by the difference between F on the surface (which is essentially zero, because θ is almost zero) and F_{ice} at the ice-regolith boundary. The loss rate is given by

$$J = \frac{\ell}{\Delta z} F_{\mathrm{ice}}$$

The flux is reduced relative to sublimation from an exposed ice surface by a factor of the order $\ell/\Delta z$. This equation is equivalent to the Knudsen flow through a porous medium. The Knudsen diffusion coefficient is proportional to the mean free inter-grain path ℓ, and the vapor density is related to p_v, and therefore to F_{ice}, via the ideal gas law.

Schorghofer and Taylor (2007) obtained two more solutions to the vapor migration transport equations for constant temperature. For slow continuous water delivery to the surface, there is an optimum temperature for which the amount of adsorbed subsurface water is at a maximum. At low temperature little H_2O accumulates, because the migration is too slow. At temperatures exceeding the optimum the loss from the surface is so fast that few source molecules reside on the surface at any time. Similarly, for an ice cover at constant temperature, there is an optimum temperature which maximizes the subsurface H_2O content. In both cases, the grain coverage is limited to about a monolayer of water, because, at constant temperature, differences in surface concentration are necessary for net transport of additional molecules. For time-varying temperature, subsurface vapor migration can lead to a "pumping effect". Transport of H_2O molecules can result from differences in sublimation rates with depth, even without change of mean temperature with depth. This phenomenon is described in more detail in Sect. 5.

The effect of inward migration on desorption of other adsorbates was studied with a three-dimensional model by Sarantos and Tsavachidis (2020). They used random walk simulations of desorption inside a granular medium consisting of spherical particles of different size. Using kinetic parameters that represent alkali, argon, and water adsorbates, it was demonstrated that surface and Knudsen diffusion, processes competitive to desorption, reduce the desorption rates from a porous medium beyond the rate reduction due to re-adsorption. Sarantos and Tsavachidis (2021) pointed out that thermal desorption from a granular medium is a decelerating process because it initiates Knudsen diffusion which in turn slows down desorption. These authors proposed that thermal desorption of adsorbates from a powder is not a first-order (i.e. constant rate) process, but is instead better approximated by a square or even higher order dependence of the desorption rate on the amount of adsorbate when diffusion is considered. This finding means that models consistently underestimate how long it takes for regolith to outgas volatiles such as CH_4, CO and CO_2 adsorbed weakly on the lunar night side because Knudsen diffusion is not properly accounted for. It was also demonstrated by Sarantos and Tsavachidis (2020) that diffusion leads to a non-linear and non-monotonic relationship between the surface temperature and alkali photodesorption rates from regolith.

It should be noted that bulk diffusion, i.e., diffusion within a solid structure, controls the source of fresh atoms for many of the atmospheric species found around Mercury and the Moon. For example, Killen et al. (2004) have described how diffusion of sodium in the grain rims controls the release rate for photon-stimulated desorption, thermal desorption and ion sputtering, processes that, unlike meteoroid impacts, can extract atoms only from the top nm of grains. Continuous diffusion of argon through the pore space between rocks in the top 25 km of lunar crust has been considered as a source for the lunar argon atmosphere (Killen 2002). And, finally, bulk diffusion followed by re-combinative desorption is responsible for converting the solar wind protons into the lunar H_2 atmosphere (e.g., Tucker et al. 2019).

4 Cold Trapping

At low temperature, sublimation rates into vacuum become negligible. Figure 6 summarizes sublimation rates for a variety of volatiles (Zhang and Paige 2009). For example, the H_2O loss rate at 115 K is around 0.1 m/Gyr, and the ice is said to be "stable" over geologic time periods. This is the phenomenon of cold trapping. Among volatiles that are abundant in comets (e.g. H_2O, CO_2) or volcanic outgassing (CO_2, SO_2, H_2O) the one with the lowest sublimation rate is H_2O (Watson et al. 1961b). It has long been realized that craters near the poles of Mercury and the Moon can be permanently shadowed and volatiles could have accumulated in these eternally cold regions (Thomas 1974; Urey 1952). Recently, cold traps have also been predicted and then observed on Ceres (Hayne and Aharonson 2015; Platz et al. 2016; Schorghofer et al. 2016).

The concept of "stability" (negligible sublimation rate) can be extended from surface cold traps to the subsurface. However, whereas surface cold traps can be supplied with water directly through an exosphere, processes that emplace ice at depth are rarer, e.g., burial by impact ejecta. Maps of subsurface ice stability of the lunar polar regions have been published by Paige et al. (2010) and Schorghofer and Williams (2020).

Another form of trapping for less volatile elements is provided by the surface microstructure. Sarantos and Tsavachidis (2020) demonstrated that at lunar temperatures, conditions

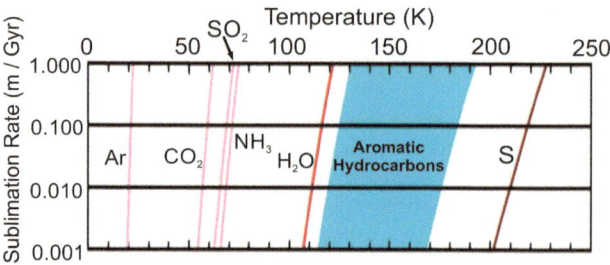

Fig. 6 Summary of sublimation loss rates into vacuum for various species. Adapted from Zhang and Paige (2009)

for which thermal desorption of alkalis has a mean lifetime of thousands of years, about half the adsorbates never participate in photodesorption because they hide in shadows the size of the lunar grains (i.e. the underside of grains). This is equivalent to assuming in models that every atom has a 50% probability of reaction between desorption events. On the other hand, if surface diffusion is fast due to mobility of these adsorbates along the grain surface, all atoms can dislodge from the microshadows and participate in desorption but at reduced rates.

5 Vapor Pumping by Temperature Cycles

Under suitable conditions, H_2O molecules can be "pumped down" into the regolith by periodic (day-night) temperature cycles, leading to an enrichment of H_2O in excess of the surface concentration. The amplitude of temperature oscillations quickly decays with depth, and the strongly nonlinear dependence of molecular residence times on temperature leads to vertical drift processes, see Fig. 7. The concept of an ice pump (i.e., vapor pumping that leads to the accumulation of macroscopic quantities of ice in the subsurface) was first developed in the context of Mars (Mellon and Jakosky 1993), where pumping can occur from a humid atmosphere into a porous subsurface. For the tenuous atmosphere surrounding the Moon, volatile H_2O molecules on the surface are the source of the pump. If the surface concentration is high enough and the temperature amplitude is significant, H_2O will be sequestered (Schorghofer and Aharonson 2014).

The simplest form of an ice pump involves pumping from an ice cover. Crucial for the pumping mechanism is the damping of the temperature amplitude with depth. Since the vapor pressure at the ice is the saturation pressure, which has a convex shaped temperature dependence, larger temperature amplitude implies a larger vapor pressure, thus creating a gradient in vapor pressures that preferentially moves molecules downward. If the concentration is sufficiently high that the nonlinear temperature amplitude effect on the top surface compensates for the reduced saturation pressure of ice at depth, pumping occurs. A weaker form of the lunar ice pump occurs when it does not produce ice but only excess adsorbate.

If the time averaged sublimation rate of adsorbed water on the surface is sufficiently high, it can balance that of pure ice at depth. A pumping differential may be defined as

$$\Delta F = \text{mean}_t \left(F_{\text{ads}}(\text{surface}) \right) - \text{mean}_t \left(F_{\text{ice}}(\text{ice table}) \right)$$

where the "ice table" corresponds to the shallowest depth with ice. If $\Delta F > 0$, then downward pumping exceeds the upward loss. If $\Delta F < 0$, the pumping is too weak and any subsurface ice experiences net loss.

The surface temperature and H_2O surface concentration vary over time. Water molecules can accumulate at night (by delivery from an exosphere), but are lost when the surface warms

Fig. 7 Schematic illustration of a subsurface temperature profile (solid line = instantaneous, dashed lines = minimum and maximum). Any volatile water molecule has a certain probability to hop up or down. A water molecule on the surface (upper dot) has a different mobility than a molecule at depth (lower dot) where temperature and temperature amplitude are different. In the long term, this leads to a net vertical flux of water molecules. When that flux is downward, this acts as an "ice pump". The nonlinear dependence of the sublimation rate on temperature is also illustrated (gray line). From Schorghofer and Aharonson (2014)

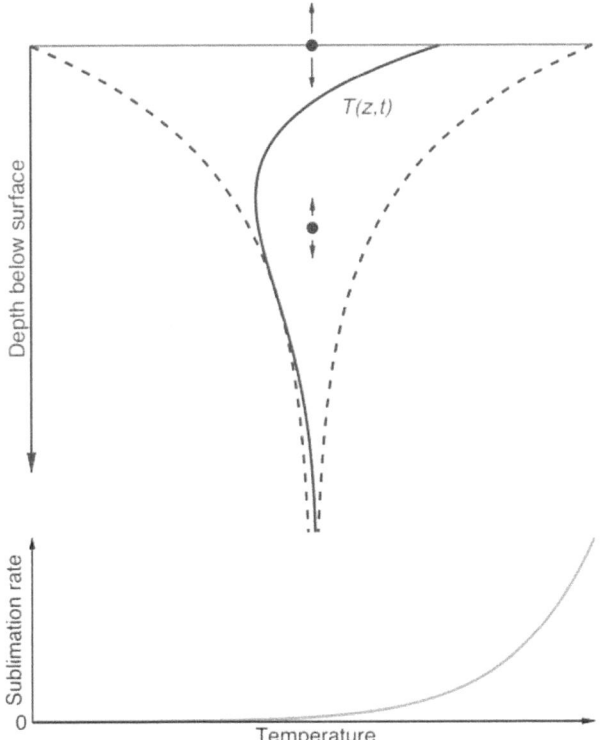

in the morning. Quantifying the strength of an ice pump requires a quantitative model of the population of water molecules on the surface over time. Schorghofer and Aharonson (2014) have carried out model calculations for the surface population of adsorbed water molecules on the Moon. In their model, the temperature varies sinusoidally with time for half a solar day, which mimics daytime, and it is constant for the other half of the solar day, which mimics nighttime. Figure 8 shows the pumping differential as a function of mean and peak temperature. This phase diagram can be divided into three nearly complementary regions. At peak temperatures below about 120 K there is very weak pumping, and this parameter region nearly coincides with the temperature conditions for classical cold trapping. At peak temperatures larger than about 120 K and mean temperatures lower than about 105 K, significant pumping occurs. When the mean temperature is above ≈ 110 K, neither pumping nor cold trapping occurs. Classical cold trapping and strong pumping are nearly complementary. Areas of strong pumping exhibit large variations in surface water concentration over one lunation.

On bodies without an atmosphere, pumping is an inefficient process due to the rapid loss of surface molecules to space relative to their downward diffusion. For every molecular hop from the surface downward there is probabilistically one hop upward and thus one molecule lost. The column-integrated subsurface ice density is smaller than the time-integrated supply of water by at least a factor of $\ell/\Delta z \ll 1$. A rocky surface layer with large pore spaces (large ℓ) favors fast diffusion, whereas dust (small ℓ) is an inefficient medium for pumping. At most a few percent of the H_2O delivered to the surface could have accumulated in the near-surface layer.

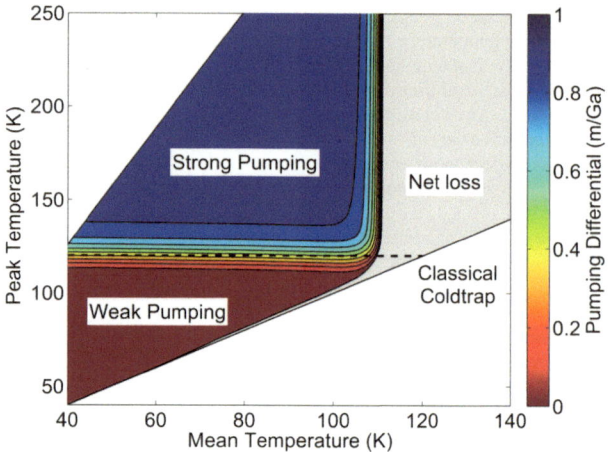

Fig. 8 Pumping differential (defined in the text) color coded, according to model calculations for the Moon that assume a supply rate of 1 m/Ga, a space weathering rate of 1 m/Ga, and prescribed temperature variations. Below the dash line, maximum temperature is lower than 120 K, corresponding to the classical definition of a cold trap. There are three nearly-complementary regions: weak pumping and classical cold trapping (red), strong pumping (blue), and net loss (grey). From Schorghofer and Aharonson (2014)

Maps of the pumping differential have been published by Schorghofer and Aharonson (2014) and Schorghofer and Williams (2020). The total area where substantial pumping occurs (a pumping differential larger than half the supply rate) is estimated to be more than five times the area of surface cold traps. Typically pumping occurs on pole facing slopes in polar areas, but within a few degrees of each pole the equator facing slopes are preferred.

6 Epiregolith Thermal Gradients

The surface layer on airless bodies is heated via solar radiation. To first order the distance from the sun, latitude, and albedo determine the average temperature at depth. In more detail, the thermophysical properties of the surface (e.g. thermal inertia, density, particle size, packing density) and rotational rate determine the depth at which this equilibrium is achieved and the magnitude of the thermal gradients that develop between the surface and subsurface. This 'surface thermal gradient' is time-varying, with the subsurface warmer than the surface at night and cooler than the surface during the day. In addition, for airless body surfaces dominated by fine particulates, the upper few *particles* radiate disproportionally to space relative to deeper particles and create an additional, persistent 'epiregolith thermal gradient' within the uppermost few 100 μm at times of day when the near-surface is sufficiently warm that radiation (rather than conduction) is the dominant heat transfer mechanism (e.g. Hale and Hapke 2002; Henderson and Jakosky 1994, 1997; Logan and Hunt 1970; Logan et al. 1973). The surface and epiregolith thermophysical properties influence the stability of volatile reservoirs and the time-varying nature of exosphere transport.

6.1 Thermophysical Properties: Moon, Mercury & Asteroids

Studies of lunar surface temperatures and physical properties benefit from an unprecedented amount of data compared to other airless bodies, including telescopic, orbital, in situ, and analyses of returned samples. Properties on small scales were determined from laboratory measurements and Apollo drill cores, and showed the lunar regolith to have a highly insulating upper layer and a more conducting lower layer (e.g. Carrier et al. 1991; Jones et al. 1975; Langseth et al. 1976). Analytical and numerical models based on these characteristics predict the lunar surface to be extremely insulating and capable of maintaining temperature

Fig. 9 Recent thermal models of the lunar surface have been revised using orbital observations from the Diviner lunar radiometer and use continuously increasing density and conductivity rather than one or two distinct layers. The new models (dotted line) are a better fit for Diviner nighttime data than a single layer (dashed line) or two-layer (solid line) models. From Vasavada et al. (2012)

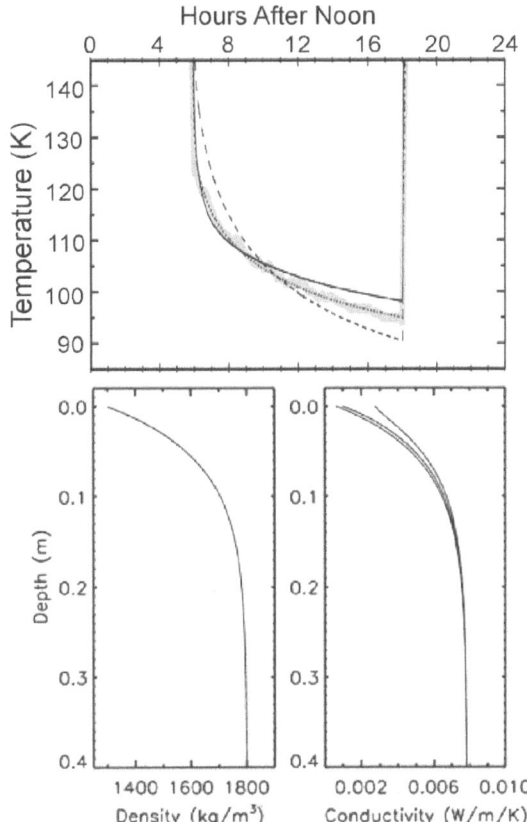

extremes on short distances to the extent that shadowed regions near the lunar poles could harbor water-ice deposits (e.g. Vasavada et al. 1999; Watson et al. 1961a).

High resolution temperature maps are provided by the *Lunar Reconnaissance Orbiter* Diviner Lunar Radiometer and confirmed the extreme nature of the lunar thermal environment (Paige et al. 2010; Vasavada et al. 2012; Williams et al. 2019, 2017). Diviner also provides inferred physical properties of the surface including rock abundance (Bandfield et al. 2011), surface roughness (Bandfield et al. 2015), and regolith thermal inertia (Hayne et al. 2017). Critical to the interpretation of this dataset was a revision to the predominant two-layer lunar thermal model (e.g. Vasavada et al. 1999) to a continuously varying density and conductivity model (Hayne et al. 2017; Vasavada et al. 2012) that more accurately represents the regolith structure over depth scales of order 1 m (Fig. 9).

As described above, the thermal and diffusive properties of the lunar regolith afford significant protection to buried volatile reserves and thus increase their stability on geologic time scales. The same models used to derive surface properties from Diviner temperature observations and lunar topography can be used to predict subsurface temperatures (e.g. Hayne et al. 2017; Paige et al. 2010). In addition, sub-millimeter instruments (e.g. Chang'e 1 and 2 Microwave Radiometer) offer a direct temperature measurement of depths up to ~ 2 m. These data predict an area $\sim 10\times$ larger where ice is stable in the near subsurface than would be stable on the surface (Fig. 10) (Paige et al. 2010). A diurnal or seasonal connection between this potential reservoir and any putative surface reservoir or exosphere has not

Fig. 10 Near surface ice stability is modeled to be possible over vastly larger areas than strict permanently shadowed regions (white areas). The colored regions represent different depths to reach ice stability in the top 1 meter of regolith. From Paige et al. (2010)

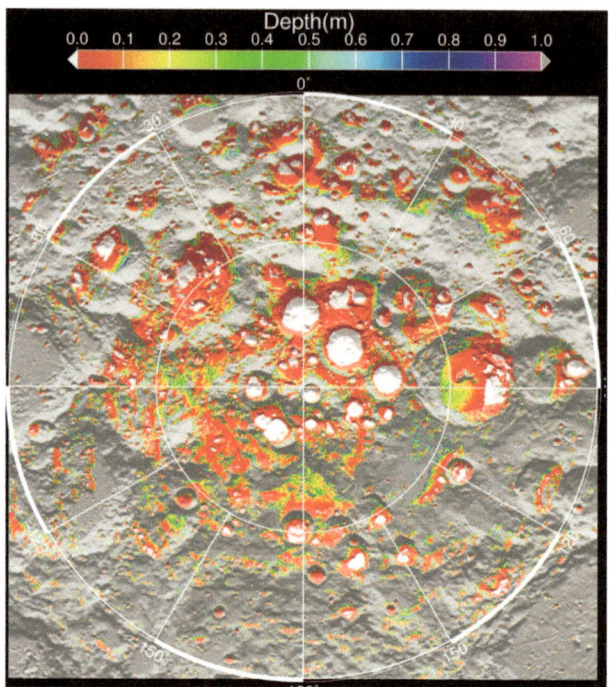

been established. However, the subsurface reservoir can be accessed via impacts, as was demonstrated by the LCROSS mission (e.g. Colaprete et al. 2010; Hayne et al. 2010).

The surface thermophysical properties of Mercury are currently poorly constrained. As such, the estimation of subsurface temperatures is uncertain. Numerical thermal modeling of Mercury was used to investigate temperatures associated with bright and dark regions suspected to be associated with surface and subsurface ice deposits (Neumann et al. 2013; Paige et al. 2013). However, these models assumed modified lunar thermophysical parameters as inputs. The Mercury Radiometer and Thermal infrared Imaging Spectrometer (MERTIS) instrument on BepiColombo will directly measure global surface temperatures at multiple times of day and allow accurate constraints on Mercury surface thermophysical parameters for the first time (Hiesinger and Helbert 2010).

Asteroids can differ significantly from the Moon with regards to regolith formation, age and gardening and thus a wide range of surface thermophysical properties are expected on different types of asteroids. Inferred thermal inertias of asteroids range from < 20 to $700\,\mathrm{Jm}^{-2}\,\mathrm{K}^{-1}\,\mathrm{s}^{-1/2}$ (Hanuš et al. 2018).

6.2 Temperature Gradients Inside the Epiregolith

Early laboratory measurements of returned Apollo soils showed significantly different emission behavior when samples were measured in typical laboratory conditions and under a simulated lunar environment (Logan and Hunt 1970; Logan et al. 1973). Since those first measurements, experiments have grown more sophisticated and spectral effects associated with varying sample composition, albedo, temperature, porosity, and roughness have been investigated (e.g. Donaldson Hanna et al. 2016; Henderson et al. 1996). Several environmental and sample characteristics explain these discrepancies. First, fine particulates under

Fig. 11 Numerical models of the epiregolith consistently show thermal gradients but differ in their predicted amplitude. The magnitude of the thermal gradient also varies as a function of solar incidence angle and/or local time. Top: Noontime equatorial illumination of basalt for different regolith grain sizes from Henderson and Jakosky (1997). Bottom: Noontime and predawn equatorial illumination of a simulated regolith from Hale and Hapke (2002). Here kE_T has units of J/m^2/s/K, with k the solid state thermal conductivity and E_T the thermal radiation extinction coefficient

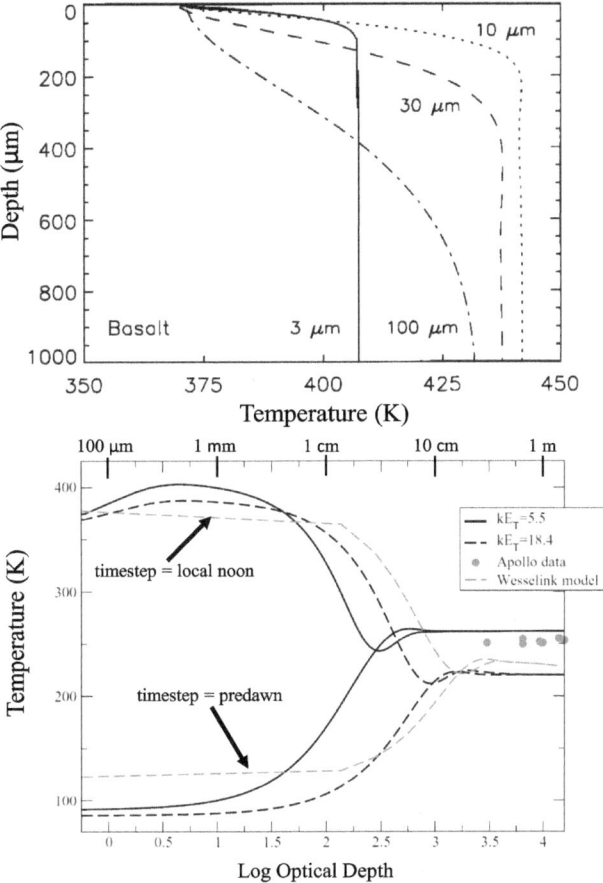

vacuum are extremely insulating. Second, when exposed to a cold shroud the surface particles rapidly cool off. Third, a broadband lamp deposits heat to depths greater than a few particles. And fourth, emission in the thermal infrared is depth-dependent on wavelength. In total, these simulated lunar conditions produce an emission spectrum where some radiation comes from the warmer interior (at the Christiansen feature emissivity maximum) and some radiation comes from the cold exterior (at the Reststrahlen Band absorption bands). Diviner orbital measurements have confirmed the spectral effects of these thermal gradients on the lunar surface (e.g. Donaldson Hanna et al. 2016; Greenhagen et al. 2010).

Efforts to model the epiregolith thermal gradients have demonstrated the underlying physics, including a dependence on scattering properties, particle size, and packing density. There remains some uncertainty as to the actual magnitude of near-surface thermal gradients on the Moon, Mercury and other airless bodies. Using similar but different computational methods, Henderson and Jakosky (1997) predict gradients of 40–50 K per 100 µm under lunar-like conditions, whereas Hale and Hapke (2002) predict gradients closer to $\sim$ 10 K per 100 µm (Fig. 11). Comparing laboratory measurements of quartz under simulated lunar conditions to models, Millán et al. (2011) infer thermal gradients similar to those calculated by Henderson and Jakosky (1997). Epiregolith temperature varies with time of day (Fig. 11, bottom) and latitude. Models indicate that even during the day, the lunar epiregolith is cooler at the surface than at depth, owing to the (wavelength-dependent) optical transmission of thermal infrared radiation through the upper few 100 µm of regolith to space. As a result, temperature rises relatively steeply within the topmost hundreds of microns of regolith before gradually decreasing to approach the diurnal average temperature at ~ 1 m depth (Fig. 11). At cooler, night-time temperatures, radiative cooling is less important and epiregolith temperature is nearly constant with depth.

The ability of Mercury and asteroids to produce epiregolith thermal gradients of a magnitude large enough to significantly affect the exosphere is unknown. Certainly, some asteroids have coarse particle regolith that would inhibit the formation of a thermal gradient in the epiregolith. In addition, even fine-grained bodies further from the sun would experience lower maximum temperatures where the magnitude of the epiregolith thermal gradient is largest and potentially reduce the effects of this phenomenon. Mercury, on the other hand, is expected to have fine-grained regolith, rotates slowly, and is near to the sun. However, the albedo of Mercury is darker than the Moon and this is expected to reduce epiregolith thermal gradient development.

6.3 Epiregolith Effects on Surface Reservoirs

Epiregolith thermal gradients on the Moon, Mercury or asteroids have the potential to be significant drivers of exospheric transport. Remote sensing reports in the UV and NIR often consider surface temperature when discussing the potential for diurnally varying (or invariant) surface hydration (e.g. Hendrix et al. 2019; Li and Milliken 2017). Variations of temperature with depth have been used to estimate the "pumping" of water vapor into the regolith as described above. Additionally, an increase of temperature with depth in the first mm of soil has been shown in simulations to keep sodium adsorbates near the top of the surface, to decrease inward diffusion, and to increase desorption flux (Sarantos and Tsavachidis 2020).

However, more important than a generalized 'surface' temperature, is the *distribution* of temperatures of the *actual grains* from which the light is scattered. Due to epiregolith thermal gradients, these temperatures must be lower than the temperatures predicted by typical thermal models (e.g. Hayne et al. 2017; Vasavada et al. 2012) that do not resolve the epiregolith. Observations of thermal emission at short wavelengths (~ 3 microns) may also be biased towards higher temperatures due to surface anisothermality effects (e.g. Bandfield et al. 2018). As a result, a wide range of different 'thermal corrections' to these types of data have been presented in literature (e.g. Bandfield et al. 2018; Li and Milliken 2017; Wöhler et al. 2017). The significance of this variance on exosphere transport and surface volatile reservoirs is currently unknown and represents an area of ongoing research.

7 Lunar Exosphere

The Moon's ^{40}Ar exosphere, first studied over a half century ago during the Apollo missions as one of the first species detected in the lunar exosphere, is one of the oldest and most spectacular examples of a planetary atmosphere dominated by adsorption and desorption to and from the surface. The effect is best illustrated in Fig. 12, which shows the diurnal profile of exospheric surface ^{40}Ar gas density as a function of time measured during two different lunations by the surface mass spectrometer LACE (Hoffman et al. 1973). The argon gas density decreases from dusk to dawn as a result of the adsorption into the progressively colder nighttime surface (dayside LACE measurements were deemed unreliable). This implies that the dominant regolith interaction is adsorption-desorption (Hodges and Mahaffy 2016). The

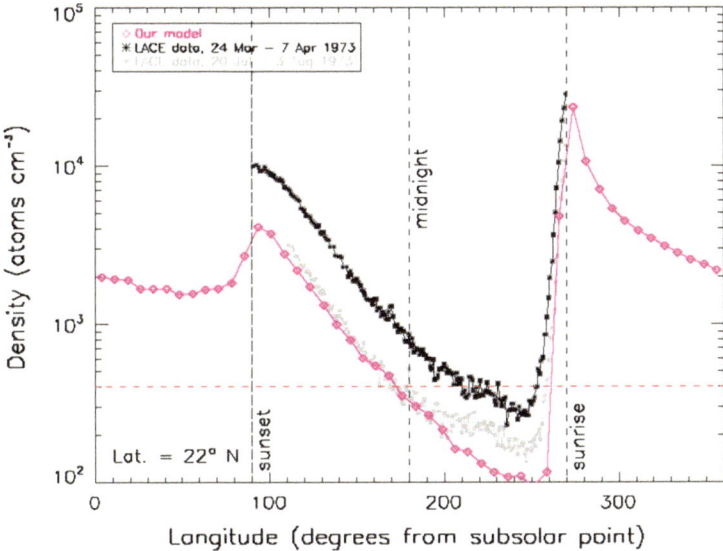

Fig. 12 ^{40}Ar exospheric surface densities over a lunar day, measured by LACE from dusk to dawn over two different lunations. The density declines over night as ^{40}Ar increasingly sticks to the cooling surface, before rapidly rising near dawn as sunrise prompts ^{40}Ar to desorb back into the exosphere. Also shown is an exospheric model fit (Grava et al. 2015) that assumes a pre-existing ^{40}Ar exosphere augmented by a prompt ^{40}Ar release before the 24 Mar–7 Apr lunation. From Grava et al. (2015), data from Hodges (1975)

dawn surge in the exospheric density (Fig. 12) is consistent with desorption of ^{40}Ar as the cold night surface is rapidly heated by the rising sun.

The shape of the density decline during the night is controlled by the strength of the interaction of the argon atoms with the surface. The residence time (time spent by an argon atom on a surface grain) has an exponential dependence on its heat of adsorption Q (Sect. 3.1), which is a still poorly constrained value that determines the shape of the density decline at night. The exospheric simulations performed by Hodges (1980) to explain the argon density decline revealed that the heat of adsorption was 0.26 eV (Grava et al. 2015 derived a similar value of 0.28 eV). This value is much higher than the value derived by adsorption experiments of ^{40}Ar on glass (0.16 eV; DeBoer 1968). The very high value of Q for ^{40}Ar inferred from LACE may be a radiation effect; e.g. trapping of ^{40}Ar atoms in radiation defects. Hodges (1980, 2018) suggested that ^{40}Ar's high heat of adsorption may be a result of the exceptional cleanliness of the lunar soil grains unattainable in laboratory, due to the removal of water molecules from high-energy adsorption sites by prolonged exposure to solar wind or serpentinization (sequestration of water molecules by reaction with olivine).

According to Hodges and Mahaffy (2016) this adsorption of argon on cold lunar surfaces enables ^{40}Ar to be trapped in Permanently Shaded Regions (PSRs) which never receive direct sunlight, or seasonal cold traps shielded from direct sunlight only during winter. Regular seasonal deviations of the exospheric ^{40}Ar abundance from a steady state exospheric model were observed by LADEE/NMS as shown in Fig. 13, consistent with adsorption of argon in seasonal cold traps. The 1.5° tilt of Moon's rotation axis gives rise to "seasons" in the argon atmosphere of the Moon, where argon atoms freeze in the seasonal cold traps near one pole, and then desorb back into the exosphere later around lunar equinoxes. In Fig. 13, the peak exospheric density measured by LADEE was in January 2014, about 1 month after the lunar vernal equinox (in December), when the seasonal cold traps at the north pole begin to release

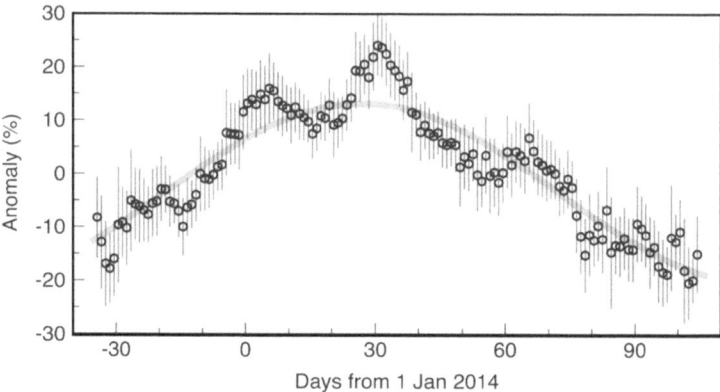

Fig. 13 The deviation of argon exospheric surface density measured from LADEE/NMS (circles with 1-σ error bars) from a steady state model based on thermal desorption (gray line). From Hodges and Mahaffy (2016)

the adsorbed atoms. This delay may be accounted for by variations in transport time through the exosphere and in thermal inertia (Hodges and Mahaffy 2016). Similar behavior may be expected for other condensable species such as H_2O. It's also possible that sub-monolayer deposits of argon exist together with other condensable species on the lunar PSRs perennially, making the measurement of ^{40}Ar valuable for inferring the possible behavior of other condensable species such as water that are more difficult to measure.

A dawn surge (Fig. 12) was also detected in the exospheric methane density (Hodges 2016) by LADEE/NMS, indicating that this species also condenses to the cold lunar night side. However the CH_4 sunrise density maximum is shifted from dawn by almost 1 hour into the day (7:00 am local time). Argon, too, has a delayed sunrise density maximum, although the delay is shorter (at 6:40 am). The delay was interpreted by Kegerreis et al. (2017) to be due to temporary sequestration of argon atoms that migrate deep into the regolith during the night. This argon reservoir is at first unable to re-migrate upwards after sunrise as the sub-surface remains for a time colder than the surface. As a result, the argon atoms are released later, possibly as late as mid-morning. Evidence for this interpretation was obtained by Sarantos and Tsavachidis (2021) with a three-dimensional simulation (Sects. 3.1–3.2) of argon transport inside piles made of spherical grains of porosity ~ 0.55. They found that inward diffusion does not occur at night, but rather as the regolith warms at dawn, with no temperature gradient required. If that is the case, then other condensable species can display the same behavior. Figure 14 shows the exospheric density measured at the lunar surface by LACE for several species: ^{40}Ar and ^{36}Ar, CO_2, and CO or N_2. Of these, only ^{40}Ar exhibits a clear sunrise exospheric density maximum. The lack of a sunrise density maximum at mass 28 is understandable for CO, which is less condensable than ^{40}Ar (Zhang and Paige 2009), and for N_2, which is not expected to condense. However the lack of a sunrise surge for mass 44, if it is CO_2, is less understandable, since this less volatile species may remain adsorbed up to substantially higher temperatures than ^{40}Ar. More measurements from orbit or from the lunar surface are needed to understand this discrepancy and to reveal the true identity of these detections.

Examples of the type of variation expected from gases that only weakly interact with the surface are provided by the detections of lunar He and Ne, which are illustrated in Fig. 15. Instead of the Ar vapor condensation that follows the night time cooling of the surface from dusk to dawn, the surface exospheric gas densities for the more volatile He and Ne atoms are

Fig. 14 Surface gas densities measured near lunar sunrise from LACE mass channels 28 (N_2 or CO_2), 36 (^{36}Ar), 40 (^{40}Ar), and 44 (CO_2). "T" is the time at which the dawn terminator crossed the landing site, while "S" is the actual sunrise time at the landing site. The ^{40}Ar density rises as dawn approaches, as exospheric ^{40}Ar atoms "hop" across the terminator from the dayside and then freeze to the colder night side surface. The lack of a similar pre-dawn rise in CO_2 is surprising, as CO_2 is expected to also stick at the Moon's night side surface temperatures. However the ^{36}Ar signal is heavily contaminated by instrumental HCl, and all four mass channels rise on the dayside owing to outgassing of hydrocarbon contaminants (Hoffman et al. 1973). Therefore, more measurements are needed to verify LACE's CO_2 detection and the contribution of instrumental effects to the data. Figure from Hoffman et al. (1973)

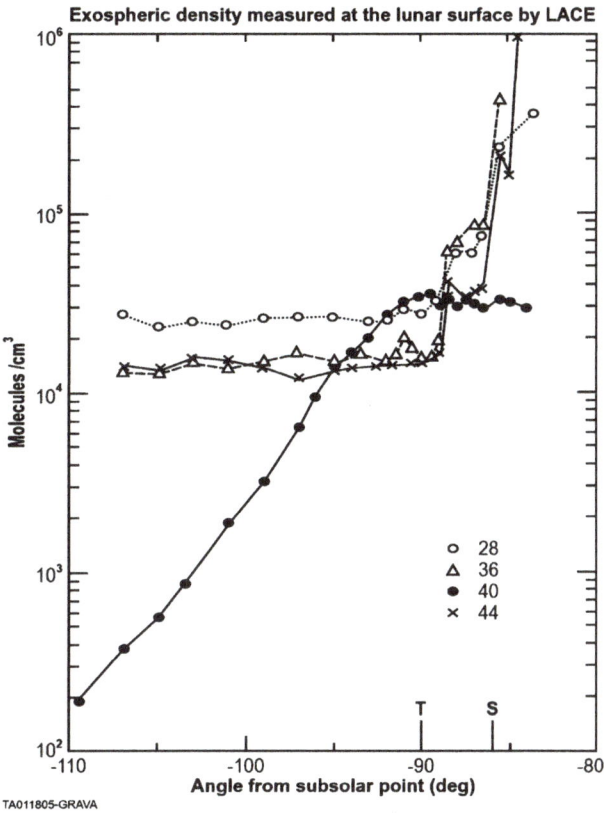

Fig. 15 Surface gas densities of exospheric species that due to weak gas-surface interactions desorb even at the coldest night side temperatures of the Moon and exhibit peak densities at night. From Benna et al. (2015)

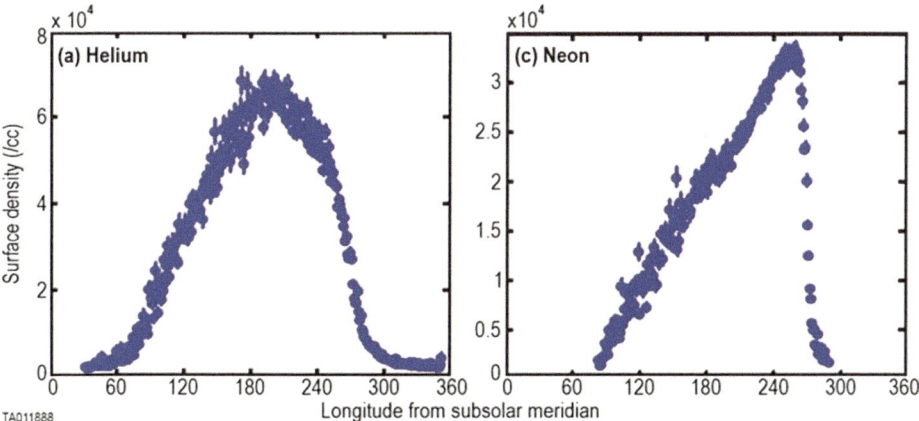

most dense on the night side, showing an anti-correlation to surface temperature T (which decreases continuously during the night due to thermal inertia), and follow approximately a law $n \propto T^{-5/2}$ at flux balance (Hodges and Johnson 1968).

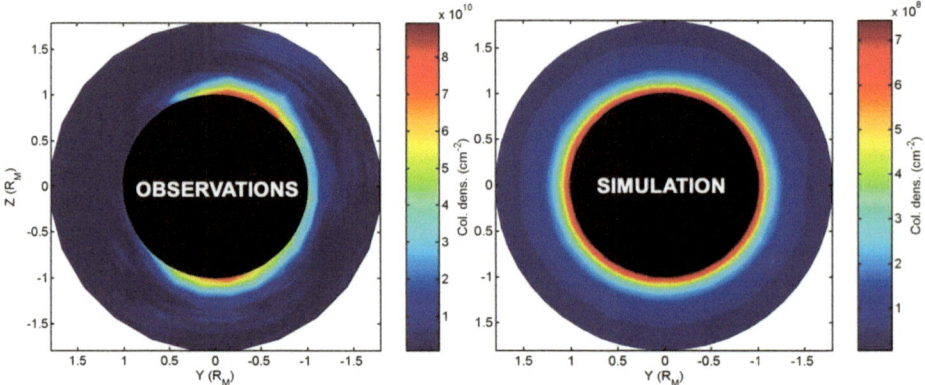

Fig. 16 Left: Observed Na tangential column density during the Mercury transit of May 7, 2003. Data are from Schleicher et al. (2004). North is up, dawn is to the left. Right: Simulated Na tangential column density with a model of successive sodium adsorption-desorption cycles, from Mura et al. (2009)

8 Mercury's Exosphere

Although numerous species have been observed in Mercury's exosphere as reviewed in the companion chapters by Grava et al. and Leblanc et al., sodium is a species that exemplifies circulation as a manifestation of gas-surface interactions discussed in this chapter. Mercury's sodium exosphere has been observed since its discovery in 1985 (Potter and Morgan 1985) and is produced by photon-stimulated desorption, and thermal and impact vaporization (Killen et al. 2007), with a relatively minor contribution from ion sputtering (McGrath et al. 1986). Since this first detection, several further observations from ground telescopes were performed, with Mercury at different True Anomaly angles and phase angle, with the aim of determining the temporal (i.e. function of True Anomaly Angle) and spatial (i.e. function of local time) variability of exospheric sodium. These observations are summarized in the companion paper by Leblanc et al. Almost twenty years after the first detection, two particular observations by Potter et al. (2002) and Schleicher et al. (2004) shed new light on the sodium cycle on Mercury and represented a milestone in the study of the its exosphere.

The observation by Potter et al. (2002) showed a long tail of escaping sodium neutral particles. This observation revealed how the radiation pressure of the Sun increases both the probability of sodium escape and the net sodium transport from the day side to the night side. Since the radiation pressure depends on the Doppler shift between neutral atoms and solar spectrum, the radiation pressure is not constant, but varies over the course of the Mercury year (Smyth and Marconi 1995) – and also varies, with some positive feedback effect, with the acceleration of the particles towards the night side.

Shortly thereafter, Schleicher et al. (2004), published the first sodium observation obtained during the 2003 transit of Mercury in front of the solar disc. Transit observations have the peculiarity of immediately providing information about the distribution of sodium at the terminator, thus revealing any dawn-dusk asymmetries and other latitudinal effects. The observations (Fig. 16, right) actually showed an excess of sodium at dawn, while sodium seemed virtually absent on the dusk side. Furthermore, it was noted that sodium increases in correspondence with the polar regions. Mura et al. (2009) reproduced these observations by means of an exospheric model with a rotating surface (Fig. 16, right), starting from the evidence, already noted by Sprague (1992), that the major processes of release occur on the day side, while the nocturnal surface necessarily fills with sodium, thanks to the exospheric migration and the radiation pressure. The sodium particles that precipitate on the night side are released only as the night side reaches the terminator, i.e. at dawn. This nocturnal filling and dawn release mechanism can explain both the asymmetry and, almost quantitatively, the column density of sodium at dawn observed from ground. Deviations between the observations and the model can easily ascribed to other replenishment mechanisms such as volume diffusion (Killen et al. 2004) or precipitation of micrometeorites.

In the observations of Schleicher et al. (2004), another feature that reveals much about the sodium cycle on Mercury is the presence of two bulges close to the north and south poles. Because of the observation geometry, it is not possible to tell the exact region of emission of these populations but it looks quite obvious that they may be ejected from the cusp regions, i.e. where the solar wind enters the magnetosphere and precipitates onto the surface (e.g. Mangano et al. 2013).

To test this hypothesis, the solar wind conditions at the time of observation were modeled to estimate the flux of the solar wind precipitating onto the cusps, and the resulting augmentation of exospheric sodium over these regions (Sarantos et al. 2008, 2010). Such simulations also provide the correct N/S asymmetry (modulated by the x component of the Interplanetary Magnetic Field), thus increasing the testability of the proposed mechanism. Finally, measurements of sodium's energy distribution from observations of the Doppler shift in its emission spectrum (of neutral sodium in the radial direction), and of the exospheric scale height, provide constraints on the Na exospheric temperature. Such Doppler measurements of the velocity distribution of the Na particles are reproduced only if the ejection source is assumed to have a temperature around 1000 K–1500 K. Hence, the simplest explanation is that the source is photon-stimulated desorption, modulated by proton precipitation through the cusps that induces a more intense diffusion (Sarantos et al. 2008, 2010) and, eventually, release from the surface. In principle, ion sputtering (IS) could work as well, but it would be necessary to review its energy distribution with respect to known laboratory measurements (Wurz and Lammer 2003).

Regarding the annual variability, and the persistence of the dawn-dusk asymmetry at different True Anomaly Angles (TAA) angles, Leblanc and Johnson (2010) have extensively studied the annual trend of Mercury's exosphere and, according to their model, the asymmetry is substantially permanent. Note that dawn, technically, changes position – from 90° east of the sub-solar point, to 90° west – for a short time close to perihelion, because in this period Mercury has a slightly retrograde motion. This work suggests that current models fail to reproduce the seasonal variability as established from existing ground-based observations. More detail about these models is provided in the companion paper by Leblanc et al.

At the time of the 2006 transit of Mercury, new observations were performed by Potter et al. (2013), who did not observe the dawn/dusk asymmetry. This would suggest a strong variability as a function of the TAA. This effect has been studied extensively (Leblanc and Johnson 2010; Mangano et al. 2013; Milillo et al. 2020). However, the main issues with ground-based observations are that they suffer from the inevitable discrepancies between the calibrations of different telescopes, from the phase angle's high variability, and from the difficulty in disentangling the trend with true anomaly. Hence, before the MESSENGER (Mercury Surface, Space Environment, Geochemistry and Ranging) mission, it was thought that the day-night cycle (and the resulting dawn/dusk asymmetry) was a fairly robust paradigm, since the effect that brings the sodium particles onto the night side is basically permanent. The first global observations by MESSENGER (Cassidy et al. 2016, Fig. 2, right) have substantially modified the previous understanding of the sodium day/night cycle, or at least implied the presence of some effects not yet considered. In fact, such observations indicate that at certain true anomalies the sodium atmosphere peaks in the afternoon. A peak in the exospheric sodium column density persists over the cold planetary longitudes, known as "cold poles".

The surface temperature plays an important role in modelling the sodium exosphere. First, thermal desorption produces a thermalized, low altitude population of sodium particles quickly evaporating from the dayside surface. Surface temperature also controls the diffusion rate (Killen et al. 2004) which brings fresh sodium atoms to the uppermost layer of regolith grains, ready for photon-stimulated desorption to release them. In simplified models, the Mercury's surface temperature is assumed to depend only on some power of the solar zenith angle in the dayside (e.g., 1/4), and assumed to be uniform on the night side (Leblanc and Johnson 2003; Vilas et al. 1988; Wurz and Lammer 2003). As such, the sub-solar point and night side temperatures depend only on the distance from the Sun. However, due to the 3:2 spin/orbit resonance, the longitudes at 90° and 270° are at the subsolar point always in correspondence with the aphelion, causing these regions to receive, on a two-yearly cycle, less heat than any other equatorial region.

Cassidy et al. (2016), observing that the sodium exosphere increases above the longitudes of the cold pole of Mercury, deduced that the surface at these longitudes functions as reservoirs/sinks for Na particles. The average temperature being lower there, over long timescales, causes less Na release on average, which may result in the formation of two sodium reservoirs at 90° and 270°. According to these authors, the reservoirs are filled when the anti-sun sodium transport is maximum and when the cold poles are close to the terminator.

This observation requires very complex modelling to be reproduced. In principle, to account for the presence of cold poles, one could just improve the surface temperature models, using for example more sophisticated ones. Unfortunately, this approach does not work well, neither by applying the Mura et al. (2009), nor the Leblanc and Johnson (2010) models. In general, the issue is that when a cold pole region is on the night side, it isn't a "sink" more than any other element at other longitudes (any sodium particle falling onto the night side remains close to where it falls). On the other hand, when the "cold pole region" is on the day side, it cannot emit sodium and act as a sink at the same time. In other words, the models are able to accumulate sodium at the cold poles, but, when Mercury is at TAA = 250° (bottom of Fig. 17, right panel) such a "modelled" cold pole has already crossed through the whole dayside, and presumably has lost most of the surface sodium. However, on average, a model like that by Mura et al. (2009), if fed with an updated temperature map, is able to predict an average excess of sodium on the surface at 90° and 270° (on a 2-year/full solar day average). If we assume that such a cold pole enhancement has worked for ages in the past, causing a permanent excess of sodium at 90° and 270°, then we are able to better reproduce the observations (Fig. 17, right panel). More modelling is needed to fully understand these observations, and detailed steps for improving these simulations are offered in the companion paper by Leblanc et al. (this journal). Since it is difficult to separate annual variability from local time by means of ground-based observations, MESSENGER data will remain the only useful dataset for such studies until BepiColombo data becomes available in 2026.

8.1 Surface Features & Exospheric Replenishment

An important consideration in the analysis of Mercury's surface-exospheric interaction is the presence of geological features with unique surface compositions, which may supply the exosphere with inhomogeneously distributed ejected/sputtered or outgassed source species. These localized 'primary' gas sources act together with 'secondary' surface-exospheric interactions to determine Mercury's exospheric structure and composition. It is therefore essential to consider the spatial variations of the exosphere and their possible relationships

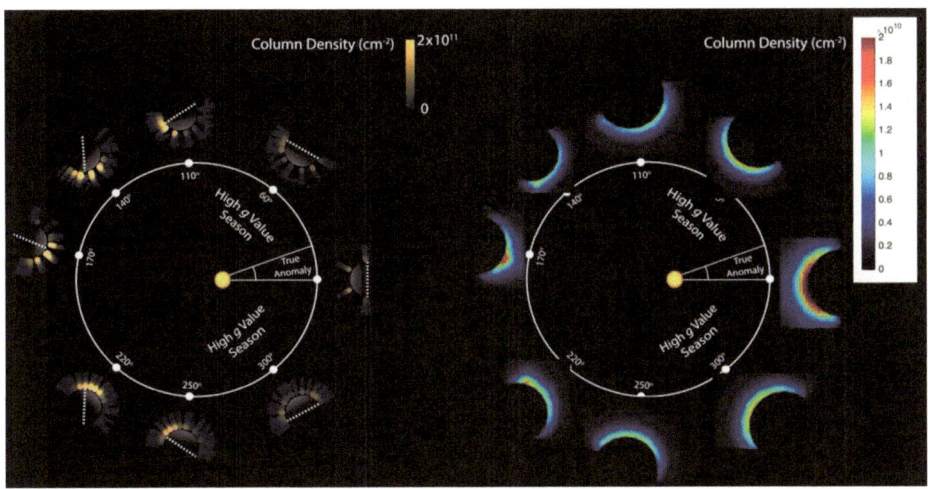

Fig. 17 Left: MESSENGER observations of sodium column density (onto the equatorial plane) at different TAAs and different local times (from Cassidy et al. 2016); right: same, simulated

with the surface composition and its geological features (Wurz et al. 2010). Mercury's surface is characterized by numerous geological features that are evidence for volcanic and tectonic activity in the past (e.g., Byrne 2014; Strom et al. 1975), when the presence of volatiles certainly played a major role in crustal evolution. Widespread effusive volcanic plains cover the planet and several vents of pyroclastic origin are present (Thomas et al. 2014). Thousands of scarps, which are the surface expression of the planet's global contraction (Strom et al. 1975), often form long prominent alignments (e.g., Byrne 2014), some of which likely formed under fluid or gas overpressure (Galluzzi et al. 2019). Such a geological asset certainly caused gas migration and seepage during Mercury's early history, at least up to 1.7 Ga, when the pyroclastic activity, the youngest of these processes, likely ended (Thomas et al. 2014). Today, just a few small fault-scarps are thought to be active (Watters et al. 2016), and albeit too small to be sources of high degassing, they are proof of a planet which is still contracting today. Volatile ascent is known to be particularly effective on a cooling planet, and the surficial layer of Mercury's crust is today a permeable net of fractures caused by fault patterns and impact craters (up to 4.5 km deep) that could easily permit outgassing.

Mercury's exospheric replenishment by surface feature activity likely takes place at several scales today. At the regional scale, Mercury is characterized by large geochemical terranes (Weider et al. 2015) mapped by using the data acquired by the X-Ray Spectrometer (Nittler et al. 2018). The most extensive terrane is the high-Mg region covering a large part of Mercury's surface (Fig. 18a). This same region also appears to be characterized by elevated percentages of Ca (Nittler et al. 2018), although the available map coverage is limited due to the possibility of measuring the abundance of Ca during solar flares only (Weider et al. 2015). MESSENGER measurements highlighted a correlation between the Mg-rich surface and the Mg exosphere (Merkel et al. 2018).

At the local scale, hollows are peculiar features that could contribute to the exosphere. Hollows are shallow irregular and rimless flat-floored depressions with bright interiors and halos, often found on crater walls, rims, floors and central peaks (Fig. 18b) (Blewett et al. 2018; Thomas et al. 2014). These features, from tens of m to tens of km, are scattered all over the surface. Due to their fresh appearance, the hollows are believed to form via a mech-

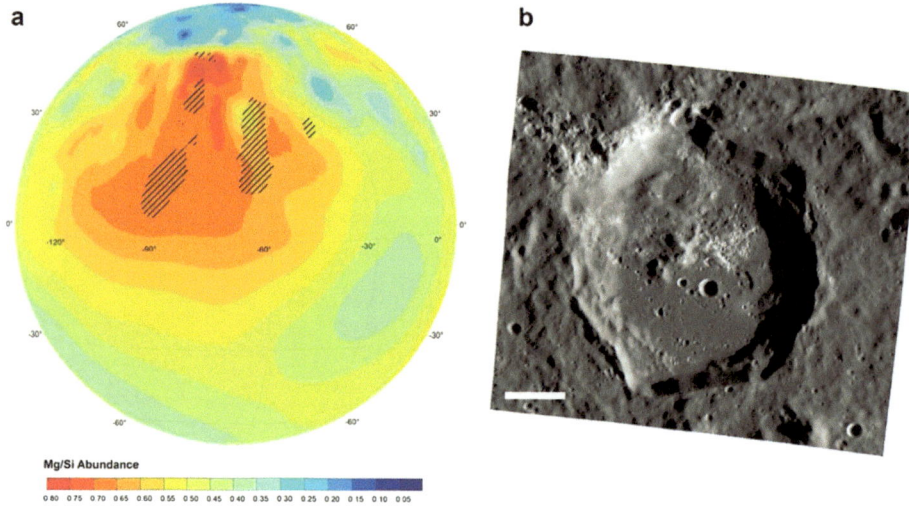

Fig. 18 Mercury surface features that could contribute to exosphere replenishment: (**a**) Mercury in stereographic projection showing the Mg/Si abundance map (data from Nittler et al. 2016) centered on the high-Mg region where the Mg/Si abundance exceeds 60%. Black diagonals indicate areas where the highest Ca/Si abundance was acquired (> 25%, Nittler et al. 2016). (**b**) Canova crater showing fresh hollows within its floor and on its northern rim arc

anism that could involve either depletion of subsurface volatiles (Blewett et al. 2018), such as chlorides and/or sulphurs (Blewett et al. 2018; Lucchetti et al. 2018; Pajola et al. 2021), or when sublimation or destruction of a volatile-bearing phase weakens the host rock as scarp retreats. Several processes have been suggested for the possible release of volatiles for both scenarios (Blewett et al. 2018) such as sublimation, thermal desorption, photon stimulated desorption (PSD) (Schaible et al. 2020), chemical sputtering, micrometeorite impact vaporization and pyroclastic volcanism. To understand if the hollows volatiles could contribute to the exosphere, laboratory measurements on the PSD of CaS powder (an analogue for oldhamite), which could be a predominant component of hollows identified within craters, have been performed. These measurements have been used to model the release of Ca by PSD from the Tyagaraja crater, which is characterized by hollows on its floor. The modeling showed that the localized neutral micro-exosphere produced from Ca PSD can be substantial even if only 1% CaS is assumed in the hollows field located in the crater (Bennett 2016). It is possible that hollows-forming volatile material could contribute to the Hermean exosphere, although the existing measurements do not have sufficient resolution to detect such a correlation.

The observed correlation between the Mg in the exosphere and the surface regions having the highest Mg/Si ratios (Merkel et al. 2018) demonstrates how surface composition data may be important in understanding the exospheric processes and variation. At the Moon the same conclusion was reached by the finding that the potassium exosphere is densest over the potassium-rich KREEP soils (Colaprete et al. 2016; Rosborough et al. 2019). MESSENGER was able to derive information on Mercury's surface composition with a spatial resolution of tens or hundreds of km and a low spectral sampling. Other refractory elements, like Ca, may exhibit a similar correlation, but it is necessary to know the surface composition with better accuracy. Improved measurements will help in distinguishing the contribution of the different source mechanisms, such as meteoroid impact vaporization which is assumed to be relevant for these elements.

9 Future Work

With the vast and complex array of gas-surface interactions expounded upon in this chapter, atmospheric science at the Moon, Mercury, and at other rocky and icy planetary bodies with surface-bounded exospheres is now at a crossroads. With an ever increasing understanding of the energy and material exchange between the atmospheres and surfaces of planetary bodies with surface-bounded exospheres, the field is now positioned for the development of a new generation of exospheric models. The most sophisticated exospheric models of the Moon and Mercury take into consideration solar energetic particle driven regolith grain diffusion as suggested by McGrath et al. (1986), or the surface sticking time of exospheric volatiles onto an assumed smooth surface, and treat cold trapping with a simple stochastic approach (e.g. Grava et al. 2015; Leblanc and Johnson 2010; Mura et al. 2009). Details including local temperature variations, versus location and depth, and the position and time dependence of topographical shadowing are now beginning to be implemented into exospheric models of the Moon (Prem et al. 2018; Schorghofer et al. 2017). Higher-fidelity models at the Moon, Mercury, and elsewhere would also take into account improving knowledge of the effects of porosity on outgassing from regolith (Sarantos and Tsavachidis 2020, 2021). New estimates of the binding energies and ejection energy distributions of sputtered atomic species such as Na from surfaces of different mineral compositions are being furnished by molecular dynamics simulations (Morrissey et al. 2022), which will be highly useful in improving the accuracy of future exospheric models. Epiregolith thermal gradients should also be considered when modeling exosphere transport and surface volatile distributions. Such simulations would have the capability to map the time-evolution of surface 'frosts', in the shadows of mountains or craters, or in PSRs, with the changing solar position, taking into considering both (1) the diffusion of gas into the regolith, including the ice 'pump' phenomenon and epiregolith gradient effects discussed in this chapter, and (2) the removal/addition of volatiles to and from the exosphere.

A detailed understanding of the migration of frozen exospheric volatiles across the surface, and within the regolith, would empower future remote and in situ observers to understand and pinpoint to an unprecedented degree the probable locations and abundances of ices and other volatile repositories on the Moon and Mercury. The answers to these questions is highly dependent on the degree to which these 'ices' are sputtered or otherwise radiation processed; an understanding of which requires knowledge of the distribution versus position and depth of these materials. Exospheric simulations which simultaneously track the detailed surface thermal and frozen volatile distribution, including topographical effects (Paige et al. 2010), require significant computing resources, but ever increasing computing capabilities have for the first time made such models practical and feasible to carry out. The combined power of a new generation of models and experiments to inform and interpret spacecraft remote and in situ observations portends a new phase in the exploration of planetary atmospheres not only at the Moon and Mercury, but at the icy bodies of the outer solar system and beyond.

The effort to understand gas-surface interactions throughout the inner solar system will soon be assisted by a new generation of measurements from solar system exospheres. The renewed emphasis on lunar exploration will provide the international community data from commercial lunar landers equipped with a variety of spectrometers. Constraining the true identity of exospheric detections, such as the mass 44 signal ostensibly attributed to CO_2 by LACE (Sect. 7), will be a high priority for such instruments, and essential to understand the degree to which these species stick to and diffuse on the lunar surface. In situ determination of activation energies for desorption can be expected from the upcoming Volatiles

Investigating Polar Exploration Rover (VIPER) mission. These measurements will eventually be complemented by packages to be deployed by humans during the Artemis missions. BepiColombo's SERENA instrument package (Orsini et al. 2021) will map the density, distribution and composition of Mercury's exospheric gas species through in situ measurements (with Strofio) and remote sensing of charge-exchanged species (with ELENA). Future spacecraft measurement of exospheric Na, Ca, K and S and their spatial relationship with the hollows will contribute to understanding not only of the relationship between the surface and exosphere, but also to the evolution and formation of the hollows. The BepiColombo MERTIS instrument will also provide critical observations to determine the magnitude of epiregolith thermal gradients at Mercury (Hiesinger and Helbert 2010). Additionally, the spacecraft's SIMBIO-SYS remote sensing instrument package (Cremonese et al. 2020) will improve the accuracy of composition data, providing global coverage of Mercury's surface at high (480 m/pixel) spatial resolution. These measurements to be carried out during the coming decade will improve on current understanding from MESSENGER mission data of Mercury's exospheric-surface interactions.

Acknowledgements We are especially grateful to co-author Maria Teresa Capria, who passed away 15 April 2022. Her decades long career in planetary science were an inspiration to the team and the community, and her expertise in planetary surface thermo-physics were essential to this work.

Funding Note Open access funding provided by Istituto Nazionale di Astrofisica within the CRUI-CARE Agreement. BT and CG were supported by LRO/LAMP NASA contract NNG05EC87C. NS was supported by SSERVI cooperative agreement NNH16ZDA001N (TREX). PP was supported by NASA through SSERVI/ICE Five-O (80NSSC20M0027) and LRO/Diviner.

Declarations

Competing Interests The authors declare no competing interests.

References

Adamson AW (1982) Physical Chemistry of Surfaces, 4th edn. Wiley-Interscience, New York
Armand G (1977) Classical theory of desorption rate velocity distribution of desorbed atoms; possibility of a compensation effect. Surf Sci 66(1):321
Arumainayagam CR, Madix RJ (1991) Molecular beam studies of gas-surface collision dynamics. Prog Surf Sci 38(1):1
Atkins PW (1986) Physical Chemistry, 3rd edn. Oxford University Press, Oxford
Bandfield JL, Ghent RR, Vasavada AR, Paige DA, Lawrence SJ, Robinson MS (2011) Lunar surface rock abundance and regolith fines temperatures derived from LRO Diviner radiometer data. J Geophys Res 116:E00H02
Bandfield JL, Hayne PO, Williams JP, Greenhagen BT, Paige DA (2015) Lunar surface roughness derived from LRO Diviner radiometer observations. Icarus 248:357
Bandfield JL, Poston M, Klima R, Edwards C (2018) Widespread distribution of OH/H$_2$O on the lunar surface inferred from spectral data. Nat Geosci 11:173
Baule B (1914) Phenomena in rarefied gases. Ann Phys 44:145

Becker JA, Hartman C (1953) Field emission microscope and flash filament techniques for the study of structure and adsorption on metal surfaces. J Phys Chem 57(2):153–159

Benna M, Mahaffy PR, Halekas JS, Elphic RC, Delory GT (2015) Variability of helium, neon, and argon in the lunar exosphere as observed by the LADEE NMS instrument. Geophys Res Lett 42(10):3723

Bennett CJ (2016) Investigating potential sources of Mercury's exospheric calcium: photon-stimulated desorption of calcium sulfide. J Geophys Res, Planets 121(2):137

Billing GD (2000) Dynamics of Molecule Surface Interaction. Wiley, New York

Blewett DT, Ernst CM, Murchie SL, Vilas F (2018) Mercury's Hollows. Cambridge University Press, Cambridge

Bonfanti M, Martinazzo R (2016) Classical and quantum dynamics at surfaces: basic concepts from simple models. Int J Quant Chem 116(21):1575–1602

Bowker M (2016) The role of precursor states in adsorption, surface reactions and catalysis. Top Catal 59:663

Burger MH, Killen RM, Vervack RJ, Bradley ET, McClintock WE, Sarantos M, Benna M, Mouawad N (2010) Monte Carlo modeling of sodium in Mercury's exosphere during the first two MESSENGER flybys. Icarus 209:63

Byrne PK (2014) Mercury's global contraction much greater than earlier estimates. Nat Geosci 7(4):301

Cadenhead DA, Stetter JR (1974) The interaction of water vapor with a lunar soil, a compacted soil, and a cinder-like rock fragment. In: Fifth Lunar Conference

Carrier WD, Olhoeft GR, Mendell W (1991) Physical properties of the lunar surface. In: Lunar Sourcebook, edited, pp 475–594

Cassidy TA, Johnson RE (2005) Monte Carlo model of sputtering and other ejection processes within a regolith. Icarus 176(2):499–507

Cassidy TA, McClintock WE, Killen RM, Sarantos M, Merkel AW, Vervack RJ Jr, Burger MH (2016) A cold-pole enhancement in Mercury's sodium exosphere. Geophys Res Lett 43(21):11

Colaprete A et al (2010) Detection of water in the LCROSS ejecta plume. Science 330:463

Colaprete A, Sarantos M, Wooden DH, Stubbs TJ, Cook AM, Shirley M (2016) How surface composition and meteoroid impacts mediate sodium and potassium in the lunar exosphere. Science 351(6270):249

Comsa G (1968) Angular distribution of scattered and desorbed atoms from specular surfaces. J Chem Phys 48(7):3235

Cremonese G et al (2020) SIMBIO-SYS: scientific cameras and spectrometer for the BepiColombo mission. Space Sci Rev 216(5):75

de Leeuw NH, Parker SC, Catlow CRA, Price GD (2000) Modelling the effect of water on the surface structure and stability of forsterite. Phys Chem Miner 27:332

DeBoer JH (1968) The dynamical character of adsorption. Soil Sci 76(2):166

Donaldson Hanna KL, Greenhagen BT, Patterson WR III, Pieters CM, Mustard J, Bowles N, Paige DA, Glotch T, Thompson C (2016) Effects of varying environmental conditions on emissivity spectra of bulk lunar soils: application to diviner thermal infrared observations of the Moon. Icarus 283:326

Fan G, Manson JR (2010) Calculations of the energy accommodation coefficient for gas-surface interactions. Chem Phys 370:175

Feldman WC, Maurice S, Lawrence DJ et al (2001) Evidence for water ice near the lunar poles. J Geophys Res, Planets 106(E10):23231

Feres R, Yablonsky G (2004) Knudsen's cosine law and random billiards. Chem Eng Sci 59(7):1541

Galluzzi V et al (2019) Structural analysis of the Victoria quadrangle fault systems on Mercury: timing, geometries, kinematics, and relationship with the high-Mg region. J Geophys Res, Planets 124(10):2543

Grava C, Chaufray JY, Retherford KD, Gladstone GR, Greathouse TK, Hurley DM, Hodges RR, Bayless AJ, Cook JC, Stern SA (2015) Lunar exospheric argon modeling. Icarus 255:135

Greenhagen BT, Lucey PG, Wyatt MB et al (2010) Global silicate mineralogy of the Moon from the Diviner Lunar Radiometer. Science 329:1507

Grundy WM et al (2016) The formation of Charon's red poles from seasonally cold-trapped volatiles. Nature 539(7627):65–68. https://doi.org/10.1038/nature19340

Hale AS, Hapke B (2002) A time-dependent model of radiative and conductive thermal energy transport in planetary regoliths with applications to the Moon and Mercury. Icarus 156(2):318

Hanuš J, Delbo M, Durech J, Ali-Logoa V (2018) Thermophysical modeling of main-belt asteroids from WISE thermal data. Icarus 309:297

Hayne PO, Aharonson O (2015) Thermal stability of ice on Ceres with rough topography. Geophys Res Lett 120:1567

Hayne PO, Greenhagen BT, Foote MC, Siegler MA, Vasavada AR, Paige DA (2010) Diviner Lunar Radiometer observations of the LCROSS impact. Science 330:477

Hayne PO et al (2017) Global regolith thermophysical properties of the Moon from the Diviner Lunar Radiometer Experiment. J Geophys Res, Planets 122:2371

Haynes DR, Tro NJ, George SM (1992) Condensation and evaporation of H_2O on ice surfaces. J Phys Chem 96:8502

Heiken GH, Vaniman DT, French BM (1991) Lunar Sourcebook: A User's Guide to the Moon. Cambridge University Press, Cambridge

Henderson BG, Jakosky BM (1994) Near-surface thermal gradients and their effects on mid-infrared emission spectra of planetary surfaces. J Geophys Res 99(E9):19063

Henderson BG, Jakosky BM (1997) Near-surface thermal gradients and mid-IR emission spectra: a new model including scattering and application to real data. J Geophys Res 102:6567

Henderson BG, Lucey PG, Jakosky BM (1996) New laboratory measurements of mid-IR emission spectra of simulated planetary surfaces. J Geophys Res 101:14969

Hendrix AR, Hurley DM, Farrell WM, Greenhagen BT, Hayne PO, Retherford KD et al (2019) Diurnally migrating lunar water: evidence from ultraviolet data. Geophys Res Lett 46:2417

Hibbitts CA, Grieves GA, Poston MJ, Dyar MD, Alexandrov AB, Johnson MA, Orlando TM (2011) Thermal stability of water and hydroxyl on the surface of the Moon from temperature-programmed desorption measurements of lunar analog materials. Icarus 213(1):64

Hiesinger H, Helbert J (2010) The Mercury Radiometer and Thermal Infrared Spectrometer (MERTIS) for the BepiColombo mission. Planet Space Sci 58:144

Hodges RR (1975) Formation of the lunar atmosphere. Moon 14(1):139

Hodges RR (1980) Lunar cold traps and their influence on argon-40. In: Lunar and Planetary Science Conference

Hodges RR (2016) Methane in the lunar exosphere: implications for solar wind carbon escape. Geophys Res Lett 43(13):6742

Hodges RR (2018) Semiannual oscillation of the lunar exosphere: implications for water and polar ice. Geophys Res Lett 45(15):7409

Hodges RR, Johnson FS (1968) Lateral transport in planetary exospheres. J Geophys Res 73(23):7307

Hodges RR, Mahaffy PR (2016) Synodic and semiannual oscillations of argon-40 in the lunar exosphere. Geophys Res Lett 43(1):22

Hoffman JH, Hodges RR Jr, Johnson FS, Evans DE (1973) Lunar atmospheric composition results from Apollo 17. In: Lunar and Planetary Science Conference

Honniball CI, Lucey PG, Li S, Shenoy S, Orlando TM, Hibbitts CA, Hurley DM, Farrell WM (2021) Molecular water detected on the sunlit Moon by SOFIA. Nat Astron 5(2):121–127. https://doi.org/10.1038/s41550-020-01222-x

Ibach H (2006) Physics of Surfaces and Interfaces. Springer, Berlin

Jones WP, Watkins JR, Calvert TA (1975) Temperatures and thermophysical properties of the lunar outermost layer. Moon 13(4):475

Jones BM, Aleksandrov A, Dyar MD, Hibbitts CA, Orlando TM (2020) Investigation of water interactions with Apollo lunar regolith grains. J Geophys Res, Planets 125(6):e2019JE006147

Kegerreis JA, Eke VR, Massey RJ, Beaumont SK, Elphic RC, Teodoro LF (2017) Evidence for a localized source of the argon in the lunar exosphere. J Geophys Res, Planets 122(10):2163

Killen RM (2002) Source and maintenance of the argon atmospheres of Mercury and the Moon. Meteorit Planet Sci 37(9):1223

Killen RM, Sarantos M, Potter AE, Reiff P (2004) Source rates and ion recycling rates for Na and K in Mercury's atmosphere. Icarus 171(1):1

Killen RM et al (2007) Processes that promote and deplete the exosphere of Mercury. Space Sci Rev 132(2):433–509. https://doi.org/10.1007/s11214-007-9232-0

Killen RM, Burger MH, Farrell WM (2018) Exospheric escape: a parametrical study. Adv Space Res 62(8):2364

King DA, Wells M (1974) Reaction mechanism in chemisorption kinetics: nitrogen on the {100} plane of tungsten. Proc R Soc Lond Ser A, Math Phys Sci 339(1617):245–269

Kisliuk P (1957) The sticking probabilities of gases chemisorbed on the surfaces of solids. J Phys Chem Solids 3(1):95

Knudsen M (1909) Die Gesetze der Molekularstromung und der inneren Reibungsstromung der Gase durch Rohren. Ann Phys 28:75

Kolasinski KW (2012) Surface Science: Foundations of Catalysis and Nanoscience. Wiley, New York

Langmuir I (1918) The adsorption of gases on plane surfaces of glass, mica, and platinum. J Am Chem Soc 40(9):1361–1403. https://doi.org/10.1021/ja02242a004

Langseth MG, Keihm SJ, Peters K (1976) Revised lunar heat-flow values. In: 7th Lunar Sci. Conf.

Leblanc F, Johnson RE (2003) Mercury's sodium exosphere. Icarus 164(2):261

Leblanc F, Johnson RE (2010) Mercury exosphere I. Global circulation model of its sodium component. Icarus 209(2):280

  Springer

Li S, Milliken RE (2017) Water on the surface of the Moon as seen by the Moon Mineralogy Mapper: distribution, abundance, and origins. Sci Adv 3:e1701471

Logan LM, Hunt GR (1970) Emission spectra of particulate silicates under simulated lunar conditions. J Geophys Res 75:6539

Logan LM, Hunt GR, Salisbury JW et al (1973) Compositional implications of Christiansen frequency maximums for infrared remote sensing applications. J Geophys Res 78:4983

Lucchetti A et al (2018) Mercury hollows as remnants of original bedrock materials and devolatilization processes: a spectral clustering and geomorphological analysis. J Geophys Res, Planets 123(9):2365

Mangano V, Massetti S, Milillo A, Mura A, Orsini S, Leblanc F (2013) Dynamical evolution of sodium anisotropies in the exosphere of Mercury. Planet Space Sci 82–83:10

McGrath MA, Johnson RE, Lanzerotti LJ (1986) Sputtering of sodium on the planet Mercury. Nature 323(6090):694–696. https://doi.org/10.1038/323694a0

Mellon MT, Jakosky BM (1993) Geographic variations in the thermal and diffusive stability of ground ice on Mars. J Geophys Res 98(E2):3345

Merkel AW, Vervack RJ, Killen RM, Cassidy TA, McClintock WE, Nittler LR, Burger MH (2018) Evidence connecting Mercury's magnesium exosphere to its magnesium-rich surface terrane. Geophys Res Lett 45(14):6790

Milillo A et al (2020) Exospheric Na distributions along the Mercury orbit with the THEMIS telescope. In: 14th Europlanet Science Congress

Millán L, Thomas I, Bowles N (2011) Lunar regolith thermal gradients and emission spectra: modeling and validation. J Geophys Res 116(E12):E12003

Morrison SR (2013) The Chemical Physics of Surfaces. Springer, Berlin

Morrissey LS, Tucker OJ, Killen RM, Nakhla S, Savin DW (2022) Solar wind ion sputtering of sodium from silicates using molecular dynamics calculations of surface binding energies. Astrophys J Lett 925(1):L6. https://doi.org/10.3847/2041-8213/ac42d8

Mura A, Wurz P, Lichtenegger HIM, Schleicher H, Lammer H, Delcourt D, Milillo A, Orsini S, Massetti S, Khodachenko ML (2009) The sodium exosphere of Mercury: comparison between observations during Mercury's transit and model results. Icarus 200(1):1

Narayanaswamy N, Ward CA (2020) Area occupied by a water molecule adsorbed on silica at 298 K: zeta adsorption isotherm approach. J Phys Chem C 124(17):9269

Neumann GA et al (2013) Bright and dark polar deposits on Mercury: evidence for surface volatiles. Science 339(6117):296

Nittler LR, Frank EA, Weider SZ, Crapster-Pregont E, Vorburger A, Starr RD, Solomon SC (2016) Global major-element maps of Mercury updated from four years of Messenger X-ray observations. In: 47th Lunar and Planetary Science Conference, edited, The Woodlands, Texas, p 1237

Nittler LR, Chabot NL, Grove TL, Peplowski PN (2018) The chemical composition of Mercury. In: Solomon LNS, Anderson B (eds) Mercury: The View After MESSENGER. Cambridge University Press, Cambridge, p 30

Orsini S et al (2021) SERENA: particle instrument suite for determining the Sun-Mercury interaction from BepiColombo. Space Sci Rev 217(1):11. https://doi.org/10.1007/s11214-020-00787-3

Paige DA et al (2010) Diviner Lunar Radiometer observations of cold traps in the Moon's south polar region. Science 330(6003):479

Paige DA, Siegler MA, Harmon JK, Neumann GA, Mazarico EM, Smith DE, Zuber MT, Harju E, Delitsky ML, Solomon SC (2013) Thermal stability of volatiles in the north polar region of Mercury. Science 339:300

Pajola M et al (2021) Lermontov crater on Mercury: geology, morphology and spectral properties of the coexisting hollows and pyroclastic deposits. Planet Space Sci 195:105136

Persad AH, Ward CA (2016) Expressions for the evaporation and condensation coefficients in the Hertz-Knudsen relation. Chem Rev 116(14):7727

Platz T et al (2016) Surface water-ice deposits in the northern shadowed regions of Ceres. Nat Astron 1(1):0007. https://doi.org/10.1038/s41550-016-0007

Poston MJ, Grieves GA, Aleksandrov AB, Hibbitts CA, Darby Dyar M, Orlando TM (2013) Water interactions with micronized lunar surrogates JSC-1A and albite under ultra-high vacuum with application to lunar observations. J Geophys Res, Planets 118(1):105

Poston MJ, Grieves GA, Aleksandrov AB, Hibbitts CA, Dyar MD, Orlando TM (2015) Temperature programmed desorption studies of water interactions with Apollo lunar samples 12001 and 72501. Icarus 255:24

Potter AE, Morgan T (1985) Discovery of sodium in the atmosphere of Mercury. Science 229(4714):651

Potter AE, Killen RM, Morgan TH (2002) The sodium tail of Mercury. Meteorit Planet Sci 37(9):1165

Potter AE, Killen RM, Reardon KP, Bida TA (2013) Observation of neutral sodium above Mercury during the transit of November 8, 2006. Icarus 226(1):172

Prem P, Goldstein DB, Varghese PL, Trafton LM (2018) The influence of surface roughness on volatile transport on the Moon. Icarus 299:31–45. https://doi.org/10.1016/j.icarus.2017.07.010

Reiss P, Warren T, Sefton-Nash E, Trautner R (2021) Dynamics of subsurface migration of water on the Moon. J Geophys Res, Planets 126:e2020JE006742

Rosborough SA et al (2019) High-resolution potassium observations of the lunar exosphere. Geophys Res Lett 46(12):6964

Sack NJ, Baragiola RA (1993) Sublimation of vapor-deposited water ice below 170 K, and its dependence on growth conditions. Phys Rev B, Condens Matter 48(14):9973–9978

Sarantos M, Tsavachidis S (2020) The boundary of alkali surface boundary exospheres of Mercury and the Moon. Geophys Res Lett 47:16

Sarantos M, Tsavachidis S (2021) Lags in desorption of lunar volatiles. Astrophys J Lett 919:L14

Sarantos M, Killen RM, Sharma AS, Slavin JA (2008) Influence of plasma ions on source rates for the lunar exosphere during passage through the Earth's magnetosphere. Geophys Res Lett 35(4):L04105

Sarantos M, Killen RM, Sharma AS, Slavin JA (2010) Sources of sodium in the lunar exosphere: modeling using ground-based observations of sodium emission and spacecraft data of the plasma. Icarus 20(2):36

Schaible MJ, Sarantos M, Anzures BA, Parman SW, Orlando TM (2020) Photon-stimulated desorption of MgS as a potential source of sulfur in Mercury's exosphere. J Geophys Res, Planets 125(8):e2020JE006479. https://doi.org/10.1029/2020JE006479

Schleicher H, Wiedemann G, Wöhl H, Berkefeld T, Soltau D (2004) Detection of neutral sodium above Mercury during the transit on 2003 May 7. Astron Astrophys 425:1119

Schorghofer N, Aharonson O (2014) The lunar thermal ice pump. Astrophys J 788:169

Schorghofer N, Taylor GJ (2007) Subsurface migration of H_2O at lunar cold traps. J Geophys Res 112(E2):E02010

Schorghofer N, Williams J-P (2020) Mapping of ice storage processes on the Moon with time-dependent temperatures. Planet Sci J 1(3):54. https://doi.org/10.3847/psj/abb6ff

Schorghofer N, Mazarico E, Platz T, Preusker F, Schroder SE, Raymond CA, Russell CT (2016) The permanently shadowed regions of dwarf planet Ceres. Geophys Res Lett 43:6783

Schorghofer N, Lucey P, Williams J-P (2017) Theoretical time variability of mobile water on the Moon and its geographic pattern. Icarus 298:111–116. https://doi.org/10.1016/j.icarus.2017.01.029

Singh-Boparai S, Bowker M, King DA (1975) Crystallographic anisotropy in chemisorption: nitrogen on tungsten single crystal planes. Surf Sci 53(1):55–73

Sipkens TA, Daun KJ (2017) Using cube models to understand trends in thermal accommodation coefficients at high surface temperatures. Int J Heat Mass Transf 111:54

Smyth WH, Marconi ML (1995) Theoretical overview and modeling of the sodium and potassium atmospheres of Mercury. Astrophys J 441:839

Smyth WH, Marconi ML (2006) Europa's atmosphere, gas tori, and magnetospheric implications. Icarus 181:510

Speedy RJ, Debenedetti PG, Smith RS, Huang C, Kay BD (1996) The evaporation rate, free energy, and entropy of amorphous water at 150 K. J Chem Phys 105(1):240

Sprague AL (1992) Mercury's atmospheric bright spots and potassium variations a possible cause. J Geophys Res 97(E11):18257

Strom RG, Trask NJ, Guest JE (1975) Tectonism and volcanism on Mercury. J Geophys Res 80(17):2478

Tenishev V, Rubin M, Tucker OJ, Combi MR, Sarantos M (2013) Kinetic modeling of sodium in the lunar exosphere. Icarus 226(2):1538

Teolis BD, Waite JH (2016) Dione and Rhea seasonal exospheres revealed by Cassini CAPS and INMS. Icarus 272:277

Teolis BD, Raut U, Kammer JA, Gimar CJ, Howett CJA, Gladstone GR, Retherford KD (2022) Extreme exospheric dynamics at Charon: implications for the red spot. Geophys Res Lett 49(8):e2021GL097580. https://doi.org/10.1029/2021GL097580

Thomas GE (1974) Mercury: does its atmosphere contain water? Science 183(4130):1197

Thomas RJ, Rothery DA, Conway SJ, Anand M (2014) Hollows on Mercury: materials and mechanisms involved in their formation. Icarus 229:221

Tucker OJ, Farrell WM, Killen RM, Hurley DM (2019) Solar wind implantation into the lunar regolith: Monte Carlo simulations of H retention in a surface with defects and the H_2 exosphere. J Geophys Res 124(2):278

Urey H (1952) The Planets, Their Origin and Development. Yale University Press, New Haven

Vasavada AR, Paige DA, Wood SE (1999) Near-surface temperatures on Mercury and the Moon and the stability of polar ice deposits. Icarus 141(2):179

Vasavada AR, Bandfield JL, Greenhagen BT, Hayne PO, Siegler MA, Williams JP, Paige DA (2012) Lunar equatorial surface temperatures and regolith properties from the Diviner Lunar Radiometer Experiment. J Geophys Res 117:E00H18

Vilas F, Chapman CR, Matthews MS (1988) Mercury. University of Arizona Press, Tucson

Watson K, Murray B, Brown H (1961a) On the possible presence of ice on the Moon. J Geophys Res 66(5):1598

Watson K, Murray BC, Brown H (1961b) The behavior of volatiles on the lunar surface. J Geophys Res 66(9):3033

Watters TR et al (2016) Recent tectonic activity on Mercury revealed by small thrust fault scarps. Nat Geosci 9(10):743

Weider SZ et al (2015) Evidence for geochemical terranes on Mercury: global mapping of major elements with MESSENGER's X-Ray Spectrometer. Earth Planet Sci Lett 416:109

Williams JP, Paige DA, Greenhagen BT, Sefton-Nash E (2017) The global surface temperatures of the Moon as measured by the Diviner Lunar Radiometer Experiment. Icarus 283:300

Williams JP, Greenhagen BT, Paige DA, Schorghofer N, Sefton-Nash E, Hayne PO, Lucey M, Siegler A, Aye KM (2019) Seasonal polar temperatures on the Moon. J Geophys Res, Planets 124(10):2505

Wöhler C, Grumpe A, Berezhnoy AA, Shevchenko VV (2017) Time-of-day-dependent global distribution of lunar surficial water/hydroxyl. Sci Adv 3(9):e1701286. https://doi.org/10.1126/sciadv.1701286

Wurz P, Lammer H (2003) Monte-Carlo simulation of Mercury's exosphere. Icarus 164(1):1

Wurz P et al (2010) Self-consistent modelling of Mercury's exosphere by sputtering, micro-meteorite impact and photon-stimulated desorption. Planet Space Sci 58(12):1599

Yakshinskiy BV, Madey TE, Ageev VN (2000) Thermal desorption of sodium atoms from thin SiO_2 films. Surf Rev Lett 7(01n02):75

Zangwill A (1988) Physics at Surfaces. Cambridge University Press, Cambridge

Zhang JA, Paige DA (2009) Cold-trapped organic compounds at the poles of the Moon and Mercury: implications for origins. Geophys Res Lett 36(16):L16203

Zwanzig RW (1960) Collision of a gas atom with a cold surface. J Chem Phys 32(4):1173–1177

Publisher's Note Springer Nature remains neutral with regard to jurisdictional claims in published maps and institutional affiliations.

Space Science Reviews (2021) 217:50
https://doi.org/10.1007/s11214-021-00827-6

Meteoroids as One of the Sources for Exosphere Formation on Airless Bodies in the Inner Solar System

Diego Janches[1] · Alexey A. Berezhnoy[2,3] · Apostolos A. Christou[4] ·
Gabriele Cremonese[5] · Takayuki Hirai[6] · Mihály Horányi[7,8,9] · Jamie M. Jasinski[10] ·
Menelaos Sarantos[11]

Received: 7 October 2020 / Accepted: 19 March 2021 / Published online: 19 April 2021
© The Author(s) 2021

Abstract

This manuscript represents a review on progress made over the past decade concerning our understanding of meteoroid bombardment on airless solar system bodies as one of the sources of the formation of their exospheres. Specifically, observations at Mercury by MESSENGER and at the Moon by LADEE, together with progress made in dynamical models of the meteoroid environment in the inner solar system, offer new tools to explore in detail the physical phenomena involved in this complex relationship. This progress is timely given the expected results during the next decade that will be provided by new missions such as DESTINY$^+$, BepiColombo, the Artemis program or the Lunar Gateway.

Keywords Meteoroids · Meteoroid streams · Exospheres · Mercury · Moon

1 Introduction

The influence of interplanetary dust on planetary bodies is an ubiquitous phenomenon in the solar system. Planets, moons and asteroids sweep through a cloud of these particles while they move along their orbital path causing a constant bombardment of meteoroids on their surfaces and atmospheres. At planetary atmospheres, meteoroids heat up as they interact with an increasingly denser atmosphere while decelerating. If they are sufficiently large (>30 μm in radius) they will ablate most of their material in the atmospheric aerobraking region, introducing exotic species such as Mg, Fe, and Na into the atmosphere (Carrillo-Sánchez et al. 2020; Crismani et al. 2018, 2017; Grebowsky et al. 2002, 2017; Plane 2003; Berezhnoy and Borovička 2010). For this case, the ablation processes will result in ionization and/or photon production, generating the well known meteor phenomenon (Ceplecha et al. 1998). As these metallic species can be ionized during ablation either by solar ultraviolet photoionization or by charge exchange with existing atmospheric ions, meteoroids therefore affect the structure, chemistry, dynamics, and energetics of planetary ionospheres (Plane et al. 2015, 2018; Plane 2003).

Surface-Bounded Exospheres and Interactions in the Inner Solar System
Edited by Anna Milillo, Menelaos Sarantos, Benjamin D. Teolis, Go Murakami, Peter Wurz and Rudolf von Steiger

Extended author information available on the last page of the article

For the case of airless bodies, meteoroids impact their surfaces directly, producing impact debris and directly shaping the resulting thin exospheres. Several works have shown direct observational evidence of the critical role that meteoroid activity plays in space weathering of airless bodies using observations from NASA missions such as the MErcury Surface, Space ENvironment, GEochemistry, and Ranging (MESSENGER) mission and the Lunar Atmosphere and Dust Environment Explorer (LADEE) (Janches et al. 2018; Pieters and Noble 2016; Pokorný et al. 2019, 2017b, 2018; Szalay et al. 2018). Therefore, knowledge of the meteoroid environment is relevant to planetary science, chemistry of planetary atmospheres, space weathering of airless bodies and even collisional risk assessment to artificial satellites and astronauts (Moorhead et al. 2020).

Recently, Plane et al. (2018) reported a detailed review on the impact and effects of meteoroids in planetary atmospheres. Hence, the principal scientific goal of this review article is to summarize our current knowledge of the role of meteoroids and dust as one of the sources of the hermean and lunar exospheres. A summary of current dynamical models of the most relevant particle populations are presented as well as current observational methods, including both ground-based and satellite observations, and descriptions of laboratory experiments designed to study meteoroids as a source of exospheric formation. Although this review article mostly focuses on the two aforementioned airless bodies, consideration of additional bodies (i.e. asteroids) is also briefly discussed.

For the remainder of this review article, we will adopt the nomenclature used in meteor astronomy that was approved in 2016 by the International Astronomical Union (IAU) Commission F1: Meteors, Meteoroids and Interplanetary Dust. In that respect, a meteoroid is defined as a solid natural object of a size roughly between 30 μm and 1 m moving in, or coming from, interplanetary space. Dust (interplanetary) is finely divided solid matter, with particle sizes generally smaller than meteoroids, moving in, or coming from, interplanetary space. Dust in the solar system is observed e.g. as the zodiacal dust cloud (ZDC), including zodiacal dust bands, and cometary dust trails. In such contexts, the term "dust" also includes smaller meteoroids; i.e., the ZDC and cometary dust trails contain larger particles that can also be called meteoroids. As implied above, small dust particles do not give rise to the meteor phenomenon when they enter planetary atmospheres. A fraction of these particles will be heated to below the melting point and will subsequently sediment to the ground. These unmelted particles can still hold signatures from their atmospheric entry (e.g. vesicles, loss of volatile elements). When collected on the ground, they are called micrometeorites. When collected in the atmosphere, they are called interplanetary dust particles (IDP's). When in interplanetary space, they are simply called dust particles. Particularly relevant to the topic of this review article is that foreign objects on the surfaces of atmosphere-less bodies are not called meteorites (i.e. there is no meteorite without a meteor) and so they are referred to as impact debris.

2 Meteoroid Populations in the Near Earth Environment: Observations and Models

Observing meteors is most likely an activity practiced by humanity since its beginnings, and thus it is natural to start describing the meteoroid environment with what we have learned from Earth-based meteor studies. Meteoroids that produce meteors bright enough to be seen with the naked eye are, in general, larger than the mass range of interest to this chapter. The number of meteors impinging on a solar system body is inversely proportional to their size in a log-log scale (see for example Fig. 25 in Ceplecha et al. 1998). In terms of a

Table 1 In-situ dust measurements in interplanetary space (with updates to Grün et al. (2001))

Spacecraft	Distance range [AU]	Sensitive area [m^2]	Reference
Helios 1/2	0.3 - 1	0.012	Dietzel et al. (1973)
Galileo	0.7 - 5.4	0.1	Grün et al. (1992)
Cassini	0.72 - 10	0.1	Srama et al. (2004)
IKAROS	0.72 - 1.1	0.54	Hirai et al. (2017)
Pioneer 9	0.75 - 0.99	0.0074	Rhee et al. (1974)
Pioneer 8	0.97 - 1.09	0.0094	Berg and Richardson (1969)
Explorer XVI	1	1.6	Hastings (1963)
Explorer XXIII	1	2.1	O'Neal et al. (1965)
Pegasus	1	200	D'Aiutolo et al. (1967)
Hiten	1	0.01	Igenbergs et al. (1991)
HEOS 2	1	0.01	Dietzel et al. (1973)
Nozomi	1 - 1.5	0.014	Sasaki et al. (2007)
Mariner IV	1 - 1.56	0.048	Alexander et al. (1967)
Ulysses	1 - 5.4	0.1	Grün et al. (1992)
Pioneer 10	1 - 18	0.26[a]	Humes (1980)
Pioneer 11	1 - 10	0.56[a]	Humes (1980)
LADEE	1	0.01	Horányi et al. (2014)
New Horizons	1 - 48[b]	0.1	Horányi et al. (2008)

[a]initial area, actual area decreased as the pressurized cells were punctured

[b]as of 6/2020

source capable of producing and maintaining a planetary exosphere, we are interested in a size (or mass) range of particles that are "large" enough to produce sufficient energy to vaporize enough mass from the surface of the meteoroid after colliding with the surface of the airless body and, oftentimes, may trigger a continuous source of collisions. It is widely accepted that this "sweet spot" covers meteoroid masses in the range of 0.1 to 1000 µg with a peak of the flux that varies somewhat depending on the reported work. For example, Fig. 5 in Carrillo-Sánchez et al. (2015) shows that the LDEF mass distribution peaks at 50 µg (around 400 µm in diameter assuming a bulk density of 2.2 g/cm^3 and an impact velocity of 18 km s^{-1}). However, the Zodiacal Cloud Model constrained by the Planck observations Ade et al. (2014) clearly shows that the average diameter range from 0.01 µg (30 µm, JFCs) to 50 µg (400 µm, HTCs).

At Earth, the flux of particles in this mass range have been observed with space-borne detectors such as the Long Duration Exposure Facility (LDEF) (Love and Brownlee 1993; Love and Allton 2006; Miao and Stark 2001; Cremonese et al. 2012), and the Cosmic Dust Experiment onboard the Aeronomy of Ice in the Mesosphere (AIM) mission (Russell et al. 2009; Poppe et al. 2011). In addition, several space missions (see Table 1) carried dedicated instruments to map the spatial and size distributions, and in some cases even the composition of interplanetary dust particles throughout the solar system.

Figure 1 shows the mass influx of cosmic solid material entering the Earth's atmosphere and, by extension bombarding the lunar surface and the surfaces of all asteroids near 1 AU, as a function of mass. Labels in Fig. 1 indicate typical methods detecting interplanetary dust. Undoubtedly, our most detailed knowledge on the origin and morphology of the

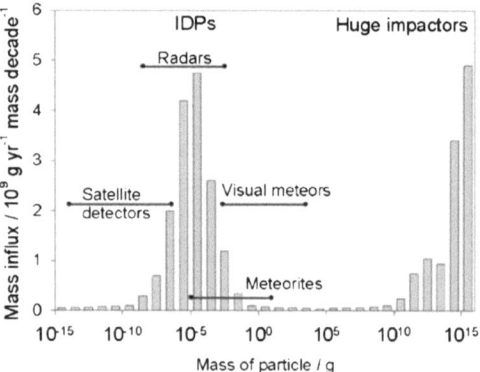

Fig. 1 Mass influx (per decade of mass) plotted against particle mass and the observational methods of interplanetary dust (Flynn et al. 2006; Plane 2012)

near Earth meteoroid environment comes from ground-based observations, specifically using radars and optical measuring techniques (Janches et al. 2017; Pokorný and Brown 2016; Campbell-Brown and Wiegert 2009; Janches et al. 2015; Campbell-Brown 2015; Baggaley 2002; Kero et al. 2019; Koten et al. 2019; Bruzzone et al. 2020). Most of the ZDC particles form the so-called sporadic meteoroid complex (SMC), and unlike meteor showers, their orbits have evolved significantly over time. This evolution is the product of effects such as Poynting– Robertson (PR) drag, radiation pressure, sublimation, mutual collisions, and the dynamical effects of the planets (Nesvorný et al. 2010; Pokorný et al. 2014). Therefore sporadic meteoroids exhibit a broad distribution of physical properties, in particular those of mass, speed, and orbital parameters. Consequently, it is very difficult to associate sporadic meteors with any particular parent body, unlike the case for many meteor showers (Jenniskens 2006; Janches et al. 2020b; Hajduková and Neslusan 2020), and thus they can only be characterized through statistical measurements of the SMC's interaction with the Earth.

The SMC is characterized by six apparent sporadic meteoroid sources, which correspond to different orbital families of interplanetary dust as seen from an Earth-based observing system (Brown and Jones 1995). These are known as: 1) the North and South Apex (NA & SA) sources, composed mainly of dust from Halley Type and Oort Cloud Comets (HTCs & OCCs; Nesvorný et al. 2011b; Pokorný et al. 2014; Sekanina 1976); 2) the Helion (H) and Anti-Helion (AH) sources, composed of dust from Jupiter Family Comets and Main Belt Asteroids (JFCs & MBAs; Hawkins 1956; Nesvorný et al. 2010, 2011a; Weiss and Smith 1960); and 3) the North and South Toroidal (NT & ST) sources, composed of dust from HTCs (Campbell-Brown and Wiegert 2009; Jones and Brown 1993; Pokorný et al. 2014; Janches et al. 2015). Other meteoroid populations such as those produced by Edgeworth-Kuiper Belt objects (EKB) have been shown to be of little significance to the overall flux in the inner Solar System (Poppe 2016).

Figure 2 displays the position of these sporadic meteor radiant distributions as white ellipses on a map in ecliptic coordinates in which the sources are viewed from an Earth-centered frame of reference. Therefore, the radiants are expressed as λ-λ_0, where λ is the heliocentric ecliptic longitude and λ_0 is the true longitude of the Sun, and β, the heliocentric ecliptic latitude. This effectively removes the motion of the Earth relative to the Sun, allowing to display the position of each source fixed in heliocentric ecliptic coordinates throughout the year (e.g. the Earth's apex is always at 270°). The colored image on the map represents a combined composite year of observations from the two main radars which currently provide continuous surveillance of the near Earth meteoroid environment. These are the Canadian Meteor Orbital Radar (CMOR; Brown et al. 2008) and the Southern Argentina

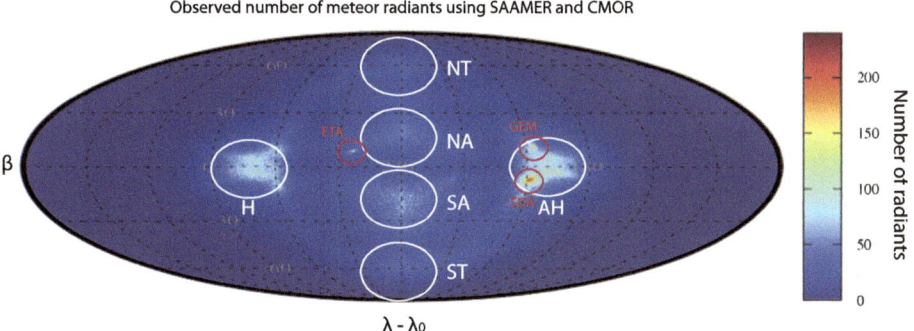

Fig. 2 The position of the sporadic meteor radiant distributions in a map of ecliptic coordinates in which the sources are viewed from an Earth-centered frame of reference. Note that the symmetry in intensity in this figure is artificial as the radars are not cross calibrated. To produce this figure the data were normalized such that SAAMER counts matched CMOR for the Helion/Anti-Helion sources and for the SDA meteor shower. The color bar represents the number of meteors within a box of 0.25×0.25 degree (in $\lambda - \lambda_0$ and β). From Janches et al. (2020)

Agile MEteor Radar - Orbital System (SAAMER-OS; Janches et al. 2015; Bruzzone et al. 2020). Note that the radars are not calibrated with each other and thus the intensity of the sources on the colored image are not absolute. The red circles on the figure identify three meteor showers which, although active only for a limited period of time with respect to the SMC, have activity that is well above the SMC background, even when a full year of observations is combined. These are the η Aquarids (ETA), the Southern δ Aquariids (SDA) and the Geminids (GEM). As can be seen, meteor showers are not only constrained in time but also their radiant distributions are more localized in space, while the sporadic sources result in wider distribution present through out the year.

Because these sources will manifest similarly at other inner Solar System bodies (Pokorný et al. 2018; Janches et al. 2018; Pokorný et al. 2019), it is evident from Fig. 2 that the meteoroid influx on planetary bodies is not isotropic. The anisotropic distribution of meteoroids in arrival direction may produce seasonal, diurnal and planetographic variability of incoming meteoroids (Fentzke and Janches 2008; Janches et al. 2018; Pokorný et al. 2017b; Janches et al. 2020). For a limited number of planetary bodies in the solar system it has indeed been demonstrated that this variability, often referred as Meteoroid Input Function (MIF; Fentzke and Janches 2008; Janches et al. 2020), manifests into the directionality of arrival of meteoroids, providing a specific local time and planetographic dependence (Janches et al. 2006), which measurably influences the composition of planetary atmospheres and space weathering of airless bodies (Marsh et al. 2013; Pokorný et al. 2018; Janches et al. 2018; Pokorný et al. 2019). These studies showed the meteoroid changes from body to body and the relative ratio of SMC populations to strongly depend on the orbital and rotational characteristic of the solar system bodies in question.

A great amount of past effort has been expended to model satellite observations of scattered light and thermal infrared emissions from the ZDC and a detailed list of those results are given by Nesvorný et al. (2010). Until recently, most of the models that described the morphology of the sporadic sources were phenomenological in nature and were driven mostly by specific observations (Grün et al. 1985; Dikarev et al. 2004; Fentzke and Janches 2008; Schult et al. 2017; Moorhead et al. 2020). However, dynamical models are better suited to generalize modeling results at different locations in the solar system, since they take into account the physical properties of interplanetary dust to determine, in principle, the behavior of particles in interplanetary space independently of any observing methodology (Nesvorný et al. 2010). All these models reproduce the directionality of particles, including the spatial, size and velocity distributions required to replicate observations such as those shown in Fig. 2. This level of description is essential for extrapolating the temporal and planetographic variability of the meteoroid influx on planetary surfaces and atmospheres from present measurements, and for assessing collisional risk with satellites (Fentzke and Janches 2008; Janches et al. 2018; Marsh et al. 2013; Thorpe et al. 2019; Moorhead et al. 2020; Pokorný et al. 2020). Predictions from these models can be constrained using a broad spectrum of methodologies such as direct observations of meteors (Pokorný et al. 2014; Janches et al. 2017; Swarnalingam et al. 2019), observations of atmospheric effects related to the meteor influx (on Earth and other planets; Carrillo-Sánchez et al. 2020), results from micrometeorite collection in Antarctica (Carrillo-Sánchez et al. 2016), surface and vaporization effects produced by impact debris (Cintala 1992; Borin et al. 2009, 2016a,b; Pokorný et al. 2018, 2019) and even satellite anomalies (Thorpe et al. 2019).

In particular, in the past decade a set of dynamical models have reproduced well the main characteristics of the four main inner solar system meteoroid sources. These include the model reported by Nesvorný et al. (2010, 2011a) for particles released by JFCs and MBAs, the model reported by Pokorný et al. (2014) for particles released from HTCs, and the model reported by Nesvorný et al. (2011b) for OCCs released meteoroid. Similar dynamical models have been reported by Wiegert et al. (2009) and Poppe (2016). The first one considers a limited number of JFCs to mainly understand CMOR observations, while the latter includes the four population as well as EKB objects and it is mainly constrained with limited statistics from dust detector measurements on board satellites in the outer Solar System.

Dynamical models track the temporal and orbital evolution of particles ejected from sources (JFCs, MBA, HTCs, and OCCs) to sinks (sublimation, impact in a solar system body, collisional destruction, or ejection from the solar system body). They include both gravitational perturbations by planets and relevant non-gravitational effects, such as radiation pressure and PR drag. They also incorporate particle collisions following treatments such as those proposed by Steel and Elford (1986) and/or collisional grooming (Stark and Kuchner 2009). Cloud density is constrained with a size distribution function (SDF) determined with, for example, LDEF (Love and Brownlee 1993) observations assuming a logarithmic shape (at 1 AU well supported by radar observations of meteors, see Janches et al. 2019, and references therein). Some of the models treat the collisional lifetimes and SDF at the source as adjustable parameters, and methods like that proposed by Kessler (1981) are used to calculate the impact probability with planetary bodies. Overall, these models have concluded that a majority ($\sim$85–95%) of the ZDC near 1 AU originates from JFCs (e.g., Levison and Duncan 1997), while MBA, HTC, and OCC meteoroids contribute the remaining 5–15%. Initial constrains using ZDC infrared emissions detected by the InfraRed Astronomical Satellite (IRAS; Nesvorný et al. 2011a); lidar measurements of the global input of neutral Na and Fe measured at Earth (Gardner et al. 2014; Carrillo-Sánchez et al. 2016); and the flux derived from cosmic spherules collected in Antarctica (Taylor et al. 1998) support these contributions, but they are still challenged by other measurements (see Introduction in Janches et al. 2014). Thus using meteoroid related phenomena at airless bodies like Mercury and the Moon, such as dust clouds and exospheric emissions, offer a unique way to further constrain these critical quantities.

Finally, in addition to the parameters mentioned above – number flux, radiant and mass distributions –, the impact velocity of some meteoroid populations may play a significant role in the formation of exospheres. Figure 3 shows modeled velocity distributions at 1 AU for the four dominant meteoroids populations in the inner solar system. In particular, the

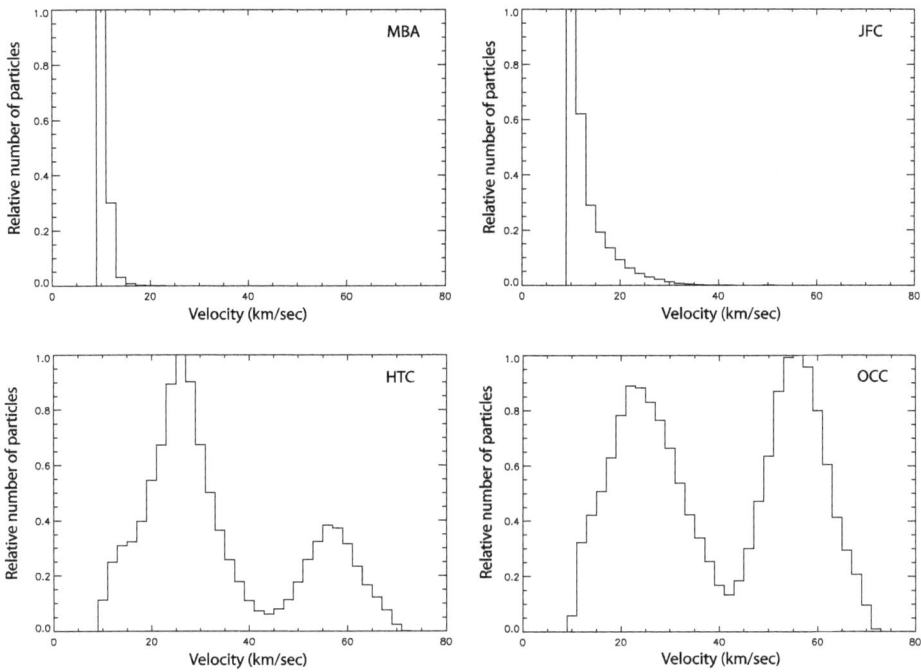

Fig. 3 Velocity distributions at 1 AU from the four main populations of meteoroids

lower velocity peak ($\sim$25 km s^{-1}) of the bi-modal distributions of the long period comets populations (i.e., HTCs and OCCs) corresponds to meteoroids in the toroidal region, while the faster distribution ($\sim$55 km s^{-1}) corresponds to retrograde orbits in the Earth's apex direction. As can be seen in the panels of Fig. 3, these distributions are significantly different at 1 AU and, as it will be shown in the next sections, the physical and orbital differences of Mercury and the Moon are even more extreme.

3 Meteoroid Influx at Mercury

Meteoroid impact vaporization has long been considered to be an important source of metals in Mercury's exosphere. Metal constituents of this exosphere have been observed using ground-based and spaceborne techniques. A detailed summary of observations is presented in Bida and Killen (2017) and Johnson and Hauck (2016). The importance of impact vaporization became clearer when measurements by the UltraViolet and Visible Spectrometer (UVVS) instrument on board the NASA MESSENGER spacecraft permitted the study of the seasonal variations of refractory metals like Mg and Ca (Vervack et al. 2010; Sarantos et al. 2011; Burger et al. 2014; Merkel et al. 2017). Detections of metal emissions of Al and Fe have also been recently reported (Vervack et al. 2016; Bida and Killen 2017) utilizing ground-based telescope and MESSENGER data. Although the relative importance of sources for Mercury's exosphere is detailed in companion papers to this, from data and models, three lines of evidence point to impact vaporization as an important source of these atoms. First, altitude profiles of neutral densities yield source temperatures that appear consistent with meteoroid impacts. Low temperature components are consistent with vaporization of atoms from Mercury's soil, and high temperatures are consistent with molecules that quickly dissociate. Second, the dependence of the inferred source rate for Ca and Mg with Mercury's True Anomaly Angle (TAA) approximately matches predictions of how the impact vapor varies with heliocentric distance (Killen and Hahn 2015; Pokorný et al. 2018). And third, the morphology of Mg and Ca emissions appears to be consistent with impact vaporization because both Ca and Mg exhibit a pronounced dawn–dusk asymmetry (McClintock et al. 2009; Burger et al. 2014; Merkel et al. 2017).

Numerous reports have utilized the numerical modeling of both meteoroid mass and flux distributions to explain their effects on airless bodies, especially at Mercury and the Moon. The main difference between these works is what populations they include. As discussed in Sect. 2, this effectively changes two critical factors to understand exospheric formation: 1) the impact velocity; and 2) the directionality of incoming flux. The first factor was identified by Cintala (1992), while the second one was shown to be crucial at the Moon by LADEE observations (Szalay and Horányi 2015). As more comprehensive data becomes available (e.g., LDEX) more details of these complex interactions is revealed, enabling to discern more accurately the role that each population play in producing the observables.

Asteroidal meteoroids are perhaps the population that has been more widely studied as one of the sources of the hermean exosphere, starting from the work reported by Cintala (1992) and later by Marchi et al. (2005). This is because it was believed that MBA meteoroids dominated the flux at Earth, in comparison to particles from cometary sources (Ceplecha 1992). However, this has been challenged recently by Nesvorný et al. (2010, 2011a) who argued that meteoroids originating from short period comets (i.e., JFCs) dominate the inner solar system in mass flux, number flux, and total cross section in the micrometer to millimeter range. Although this matter is far from settled, it certainly emphasizes the need for considering all populations.

Cintala (1992) considered particles with sizes smaller than 1 cm and a mean impact velocity on Mercury of about 20 km s^{-1} (Fig. 4), while Marchi et al. (2005) treated larger MBA meteoroids ($D > 2$ cm) which arrive with a much broader impact velocity distribution at Mercury, ranging from 4.25 to around 40 km s^{-1} (Fig. 4). The higher velocities are due to the fact that larger particles are only influenced by gravitational forces exerted by mean motion and secular resonances and, thus, in order to reach small heliocentric distances they would need to be ejected on highly eccentric orbits similar to near-Earth asteroids with low perihelia (Pokorný et al. 2018). Nevertheless, the larger particles are likely of little significance as the main source of the permanent exosphere because they are infrequent while the exosphere must be continuously replenished. Small particles originating from MBAs ($D < 2$ cm) were re-visited by Borin et al. (2009, 2016a,b) who showed that the circularization of their orbits by PR drag result in narrow impact velocity distributions (Fig. 4). The analytic velocity distribution of Cintala (1992) is 20.50 km s^{-1} as compared to 16.81 km s^{-1} from Borin et al. (2009, 2016a,b).

More recently, Pokorný et al. (2017b) demonstrated for the first time how the directionality of the meteoroid influx relates to the characteristics of the hermean exosphere. This work pointed out the strong dependence of impact characteristics and fluxes with respect to the planet's TAA. Although such dependence is present in all planets and satellites (Janches et al. 2018, 2020), it is particularly evident at Mercury due to the planet's high eccentricity and orbital inclination (Pokorný et al. 2017b). Furthermore, Pokorný et al. (2018) utilizing the description of MBA-released meteoroids reported by Nesvorný et al. (2010), showed the strong effect that such dependency has on impact velocities. The authors estimated that these particles, as well as JFC-released meteoroids, impact the hermean surface preferentially with lower eccentricities e < 0.2 and small inclinations I < 30°, and much larger impact

Fig. 4 Velocity distribution function of particles, produced in the Main Belt, with radius of 5 μm and 100 μm (Borin et al. 2009)

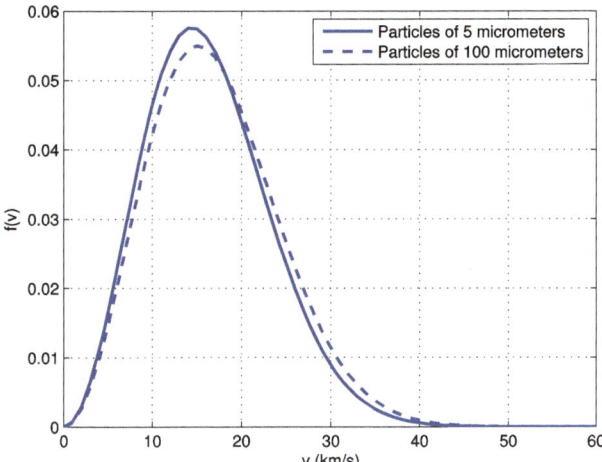

velocities than those estimated by Cintala (1992) and Borin et al. (2009, 2016a,b). Specifically, Pokorný et al. (2018) estimated MBA and JFC meteoroids having impact velocities as large as $V_{imp} < 70$ km s^{-1} at perihelion and $V_{imp} < 50$ km s^{-1} at aphelion.

Regarding JFC released meteoroids, Nesvorný et al. (2010) emphasized the importance of this source in the inner solar system. Both, Borin et al. (2016a,b, 2017) and Pokorný et al. (2017b, 2018) considered for the first time this population as part of the influx at Mercury. Due to the high eccentricity of Mercury and low impact velocities of these meteoroids compared to Mercury's orbital velocity, the impact directions of JFC and MBA meteoroids are expected to experience significant motion in the local time reference frame during Mercury's orbit. This is shown in Fig. 5, where the radiant distributions for JFC meteoroids with diameter D=10 μm are displayed for six TAAs. It can be seen that at perihelion the meteoroid flux is concentrated around (6 hr, 60°) as a result of the nonzero inclination of Mercury's orbit and the orientation of the Hermean velocity vector. Similarly to MBA meteoroids (see Pokorný et al. 2018), there is a shift in the radiant distribution of JFC meteoroids as Mercury moves toward or away from the Sun, caused by the nonzero eccentricity and inclination of Mercury's orbit and a consequent drift of the planet's velocity vector from the ecliptic plane and its perpendicular orientation with regard to the radial vector.

Regarding meteoroids from long period comets (LPC; i.e., HTC and OCC), Pokorný et al. (2017b, 2018) showed for the first time their essential role in the formation of Mercury's exosphere. Dynamically less evolved than MBA or JFC meteoroids and released into highly eccentric orbits, LPC meteoroids can reach orbits intersecting Mercury faster and at high speeds. In fact, Pokorný et al. (2018) showed that these populations, like at Earth, encounter Mercury with a flat eccentricity distribution and a bi-modal distribution of orbital inclinations of prograde and retrograde orbits. In particular, the retrograde portion of the HTC and OCC populations impact Mercury's surface from the apex direction, with velocities as high as 95 km s$^{-1} < V_{imp} < 120$ km s^{-1} at perihelion and 75 km s$^{-1} < V_{imp} < 90$ km s^{-1} at aphelion, and are less influenced by Mercury's orbital motion. The high impact velocity of these particles makes them critical for the morphology of the exosphere at Mercury and, in fact, Pokorný et al. (2017b, 2018) demonstrated a persistent dawn enhancement of the dust/meteoroid environment at Mercury, which should be responsible of the dawn–dusk asymmetry in Mercury's Ca exosphere (Fig. 6).

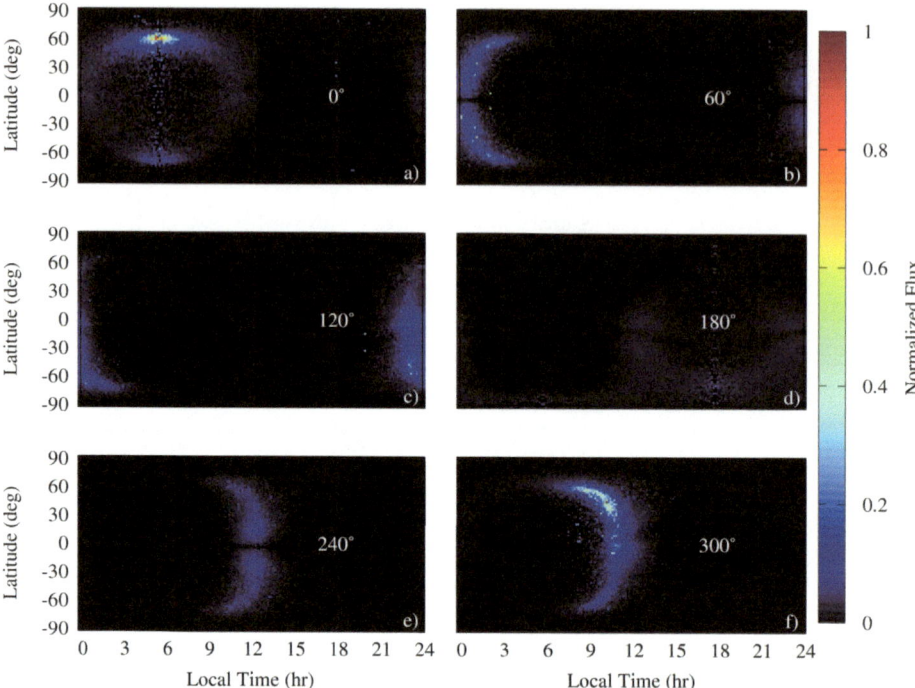

Fig. 5 Normalized radiant distribution of $D=10$ μm JFC meteoroids impacting Mercury's surface for six different TAAs (white number indicated in each panel). The mutual meteoroid collisions are not considered in this case. The x-axis represents the local time on Mercury, and it is fixed with regard to the subsolar point (12 hr). Due to Mercury's eccentricity, the location of the apex (approximately at 6 hr) changes along Mercury's orbit. The latitude is measured from Mercury's orbital plane (not the ecliptic). From Pokorný et al. (2018)

The high impact velocity at Mercury of LPC meteoroids, resulting from retrograde orbits (Levison et al. 2006), makes them the dominant source of physical phenomena regarding the formation and morphology of its exosphere. This is interesting because at the same time they are not considered to be a dominant part of the inner solar system meteoroid budget in terms of the mass flux, number flux, or total meteoroid cross section (Nesvorný et al. 2011b; Pokorný et al. 2014; Carrillo-Sánchez et al. 2016). Specifically, the mass flux of LPC meteoroids at Mercury compared to JFC meteoroids could be as small as $\sim$5% but their impact velocities resulting in values over $100\ \mathrm{km\,s^{-1}}$ makes them the dominant source in terms of the impact vaporization or the impact yield (Pokorný et al. 2017b, 2018).

Besides the orbital and physical characteristics of the meteoroid influx on Mercury and other bodies, a last crucial quantity to consider is the total meteoric mass impinging on the planet, and observations can help constrain better that hotly debated quantity (see Table 1 in Plane 2012). It is important to note that absolute fluxes are heavily dependent on the measurements used to constrain those values (Nesvorný et al. 2010; Carrillo-Sánchez et al. 2016; Janches et al. 2017; Carrillo-Sánchez et al. 2020). In addition, the agreement found with those measurement constraints is highly dependent on the uncertainties of the model parameters such as the assumed collisional lifetimes, particle densities and/or SDF at the source, which can be large (Pokorný et al. 2018, 2019). Naturally, the most convenient

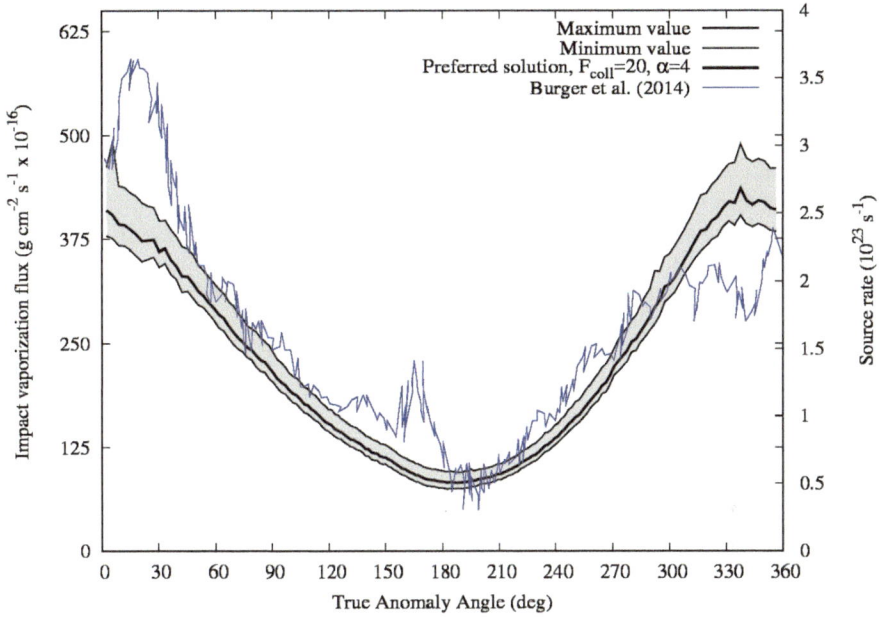

Fig. 6 Seasonal variations of the relative vaporization rate from a dynamical model (solid thick black line is the preferred solution and the confidence interval is marked by the gray area and thin black lines) compared to measurements of exospheric abundance of Ca from Burger et al. (2014) (solid blue lines). From Pokorný et al. (2018)

pivot/anchor point (i.e. something that everything else is relative to) is usually the Earth (and the mass flux on Earth), due to the significant wealth of data for the Earth/Moon region.

Cremonese et al. (2012), for example, considered the measurements of the dimensions of all the hypervelocity impact craters collected on the space-facing end of the gravity-gradient-stabilized LDEF satellite (Love and Brownlee 1993; Miao and Stark 2001), in order to determine the mass flux of extraterrestrial micrometeoroids at Earth in the submillimeter diameter range of 10-500 μm. The first step for translating the crater data on LDEF into a flux estimate is to interpret the crater diameters in terms of projectile size. Love and Brownlee (1993) have calculated the meteoroid mass distribution using a polynomial fit to the crater size-frequency distribution along with the mean depth-diameter ratio as well as a single mean meteoroid impact velocity and angle. It is important to point out that the velocity and impact angle are initially unknown parameters and, therefore, need to be fixed to infer the mass distribution from the crater sizes. In fact, Christiansen (1992) reported laboratory impact experiments of projectiles striking thick aluminum alloy targets at speeds up to 18 km s^{-1}, demonstrating that crater volume under those conditions is nearly proportional to the projectile kinetic energy. Love and Brownlee (1993) chose the average meteoroid speed to be 16.9 km s^{-1}, as found by Erickson (1969) and Kessler (1969) from photographic meteors and supported by crater rate measurements on LDEF. It is important to note that, the latter two assumptions have been revised by several authors (Taylor 1995; Mathews et al. 2001; Miao and Stark 2001). Cremonese et al. (2012) adopted a semiempirical scaling law derived from laboratory experiments that were performed in a range of velocities that were significantly different with respect to the impact speeds of the impacting meteoroids. To improve the analysis of LDEF data the relation between crater diameter and projectile size has been specifically explored with the hydrocode iSALE. The ratio between the depth and diameter of the craters is 0.527, accurately measured by Love and Brownlee (1993) on LDEF, and it has been considered as an important constraint for the iSALE simulations (Cremonese et al. 2012). The resulting average ratio derived by iSALE simulations between depth and diameter is 0.582 for asteroids. This value depends on the precision of the hydrocode simulations, which according to code validation against laboratory experiments (Pierazzo et al. 2008), is 3–4% in radius and 12% in depth.

Using the dynamical model of dust particle orbital evolution described in Borin et al. (2009), the range of impact velocities used in the hydrocode simulations were derived from which Cremonese et al. (2012) calculated the total mass accreted by the Earth of (11.5 ± 1.4) $t.d^{-1}$ assuming cometary dust and $(20.3 \pm 2.7)\,t.d^{-1}$ assuming asteroidal dust, where t represents a metric ton (1000 kg). Borin et al. (2009) extrapolated the curve of the Earth meteoroid flux, obtained by Cremonese et al. (2012), to Mercury assuming the impact velocity distribution shown in Fig. 4. The derived asteroidal mean flux value is $8.20 \pm 0.75 \times 10^{-15}$ g cm^{-2} s^{-1} (Borin et al. 2017).

Pokorný et al. (2018) followed a somewhat different approach by treating the four main populations of meteoroids separately. These authors estimated that Mercury accumulates 7% of MBA meteoroid mass as compared to Earth due to their low-eccentricity orbits, with low relative impact velocities of this population onto both planets. Such low velocities are efficiently attracted by Earth's gravity, while this effect is much smaller for Mercury. In addition, JFC meteoroids, on the other hand, have a broader distribution of eccentricities, which weakens the gravitational focusing and leads to higher mass accretion ($\sim$23%) at Mercury compared to that at Earth. The meteoroids produced by HTCs and OCCs, on the other hand, have significantly higher Mercury-to-Earth mass accretion ratios as compared to JFC meteoroids, $\sim$70% for HTC meteoroids and $\sim$90% for OCC meteoroids. In particular, Pokorný et al. (2018, 2019) explored the effects on the production of the ejecta dust cloud on Mercury and the Moon, respectively, produced by different Size Distribution Functions (SDF) resulting from varying the differential size index, α (i.e. exponent of the SDF) and different collisional lifetimes resulting from using a collisional fudge factor, F_{coll}. The authors defined a preferred solution adopting $\alpha=4$ (or a differential mass index, $s = (\alpha+2)/3 = 2$ in agreement with radar observations; Janches et al. 2019) and $F_{coll}=20$, shown to provide the best agreement with measurements (Pokorný et al. 2019). This solution also used the contribution ratios of the different populations at Earth determined by Carrillo-Sánchez et al. (2016) which were estimated subject to the following assumptions: 1) According to lidar measurements, the global input of neutral Na and Fe measured at Earth (Gardner et al. 2014) is estimated to be $0.3\pm0.1\,t.d^{-1}$ and $2.3\pm1.1\,t.d^{-1}$, respectively; and 2) The flux of cosmic spherules with diameters between 50 μm and 700 μm is estimated to be 4.4 ± 0.8 $t.d^{-1}$ (Taylor et al. 1998). Consequently, Carrillo-Sánchez et al. (2020) determined a total mass input of $27.9\pm8.1\,t.d^{-1}$ for Earth, with JFC meteoroids being the main contributor to this flux ($\sim$70%).

Based on these estimates and assumptions, Pokorný et al. (2018) provided the following values of accreted mass averaged over the entire Hermean orbit: MBA meteoroids $M_{MBA}=0.26\pm0.15\,t.d^{-1}$, JFC meteoroids $M_{JFC}=7.84\pm3.13\,t.d^{-1}$, HTCs $M_{HTC}=1.69\pm0.91$ $t.d^{-1}$ and OCCs $M_{OCC}=2.37\pm1.38\,t.d^{-1}$. This represents a mass influx ratio of JFC/LPC meteoroids of $\sim$2, which is much lower than that at Earth ($\sim$7) according to Carrillo-Sánchez et al. (2016, 2020).

One constraint for these estimated values at Mercury is the impact vaporization flux. The vaporization flux averaged over a hermean year results in $F_{Orbit} = (200 \pm 16) \times 10^{-16}$ g cm^{-1} s^{-1}, with maximum value of $(436\pm57) \times10^{-16}$ g cm^{-2} s^{-1} occurring at

TAA=337°, and a minimum value of $(82\pm12)\times10^{-16}$ g cm^{-2} s^{-1} occurring at TAA=188°. Although, these values represent impact vaporization fluxes of the same order of magnitude as those reported by Borin et al. (2009, 2017), the similarity is misleading because Pokorný et al. (2018) predicted a lower mass flux of meteoroids at Mercury with respect to those modeled by Borin et al. (2009, 2017) (Mercury/Earth ratio less than 1 vs. 35) accompanied by higher-speed distributions. The variations of the impact vaporization flux and the impact directionality also contradict the result of Borin et al. (2009, 2017) because the model reported by Pokorný et al. (2018) predicts a larger perihelion-to-aphelion ratio in the impact vaporization rate (Fig. 7). Regardless of these details, all estimates of impact vaporization rates more than suffice to supply Mercury's exosphere: compared to source rates derived for Ca and Mg from MESSENGER measurements (Sarantos et al. 2011; Burger et al. 2014; Merkel et al. 2017), only ~1% of the estimated total vapor appears to contribute neutrals to Mercury's exosphere.

While future measurements by the Bepi Colombo spacecraft will further constrain our models of the meteoroid environment at Mercury, there are currently two constraints to assess the validity of our estimates of total exospheric context with Mercury's heliocentric distance. Pokorný et al. (2018) showed a strong dawn/dusk asymmetry in both meteoroid impact direction distribution and the impact vaporization pattern on the surface. According to that work, the impact vaporization pattern is expected to undergo some motion during Mercury's orbit, mainly reflecting the precipitation pattern of LPC particles because of their high impact speeds. The total impact vaporization flux integrated over the whole surface follows a similar pattern with TAA to source rates for calcium presented by Burger et al. (2014), with a few enhancements, probably due to meteor shower activities, that will be discussed in Sect. 6.2. As a final note, it is important to note that Ca column density in the exosphere of Mercury obtained by Burger et al. (2014) is not linearly proportional to meteoroid mass flux. The reasons for the deviation of this dependence from a linear function is that Ca delivery rate to the exosphere is a complex function of quenching temperature of condensation of Ca-containing species, photolysis lifetimes of CaO, CaOH, Ca(OH)$_2$ as well as initial temperature and pressure in the impact-produced hot cloud, target-to-impactor mass ratio, typical mass of impacted meteoroids, and the elemental composition of the surface of Mercury and impactors (Berezhnoy 2018).

4 Meteoroid Influx at the Moon

Meteoroids directly reach the surface of airless bodies, becoming impact debris, generating clouds of secondary ejecta particles, and leaving a crater record on the surface. These phenomena provide data to indirectly evaluate the links between measured effects on the lunar surface and meteoroid influx. For example, Grün et al. (1985) used these craters to decipher the meteoroid size distribution impinging on the lunar surface. The majority of the ejecta particles have initial speeds below the escape speed from the Moon (2.4 km s^{-1}) and following ballistic orbits return to the surface, blanketing the lunar crust with a highly pulverized regolith with $\gg$1 m thickness. Micron and sub-micron sized secondary particles that are ejected with speeds up to the escape speed form a highly variable, but permanently present, dust cloud around the Moon. Such tenuous clouds have been observed by the Galileo spacecraft around all lunar-sized Galilean satellites at Jupiter (Krüger et al. 2003). These observations have been the source of progress regarding the lunar dust environment because only limited new observational data at the Earth's natural satellite has been obtained since the Apollo era (Grün et al. 2011). In this manuscript, we will focus on what we have learned

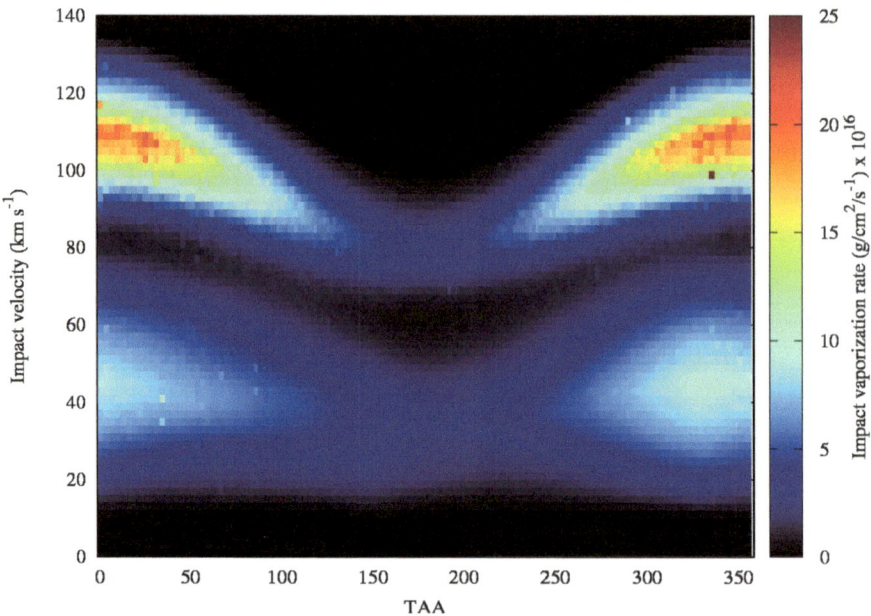

Fig. 7 Total vaporization flux as a function of TAA (x-axis), and the impact velocity (y-axis). The units are g cm^{-2} s^{-1} per 2 km s^{-1} bin. From Pokorný et al. (2018)

during the era following the LADEE mission, which greatly enhanced our view of the high-altitude ($\gg$1 km) lunar dust environment (Elphic et al. 2014). This mission provided key measurements to better constrain our models of the meteoroid environment at 1 AU as well as to define more accurately the connection between meteoroid bombardment and exosphere formation, thus defining a 'before and after' reference point on our knowledge of the meteoroid environment influence on the lunar surface.

LADEE, launched in September 2013, followed a near-equatorial retrograde orbit, with a characteristic orbital speed of 1.6 km s^{-1}. LADEE was designed to make measurements of the dust environment independently from the lunar exosphere and those observations covered a limited latitude range. Even considering these limitations, LADEE is probably the first mission with a synergy in its measuring capabilities such that the connection between meteoroid bombardment and exospheric formation on an airless solar system body could be investigated (see discussion regarding the Geminids meteor shower in Sect. 6.2). LADEE was designed to directly measure the ejecta cloud generated by meteoroid impacts on the lunar surface. This included possible intermittent density enhancements during meteoroid showers, and searching for the putative regions with high densities of dust particles with radii $\ll$1 μm lofted above the terminators (Horányi et al. 2015). The Lunar Dust Experiment (LDEX) on board LADEE was an impact ionization dust detector that measured both the positive and negative charges of the plasma cloud generated when a dust particle strikes its target. The amplitude and shape of the waveforms (signal versus time) recorded from each impact were used to estimate the mass of the dust particles. The instrument had a total sensitive area of 0.01 m^2, gradually decreasing to zero for particles arriving from outside its field-of-view of $\pm$ 68° off from the normal direction (Horányi et al. 2014). LDEX detected a total of approximately 140,000 dust hits (Fig. 8) during about 80 days of cumulative observation time by the end of the mission in April 2014.

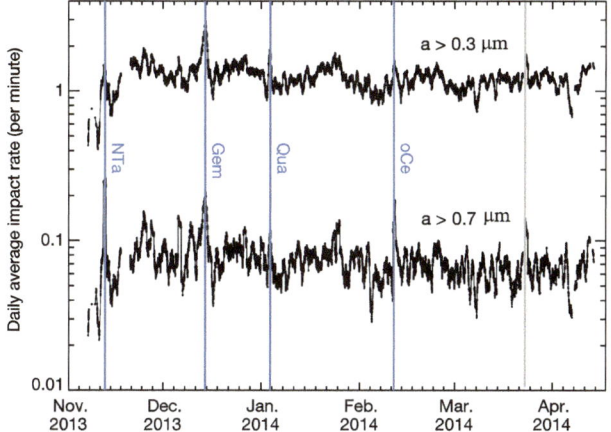

Fig. 8 Impact rates observed by LDEX throughout the duration of the LADEE mission. The panel shows the daily running average of impacts per minute of particles with radii > 0.3 μm and $a > 0.7$ μm. Four of the annual meteoroid showers generated elevated impact rates that lasted several days. The labelled annual meteor showers (blue vertical lines) are: the Northern Taurids (NTA); the Geminids (GEM); the Quadrantids (QUA); and the Omicron Centaurids (oCe). Towards the end of March LDEX data indicated a meteor shower that remained unidentified (vertical grey line) by ground based observers Horányi et al. (2015)

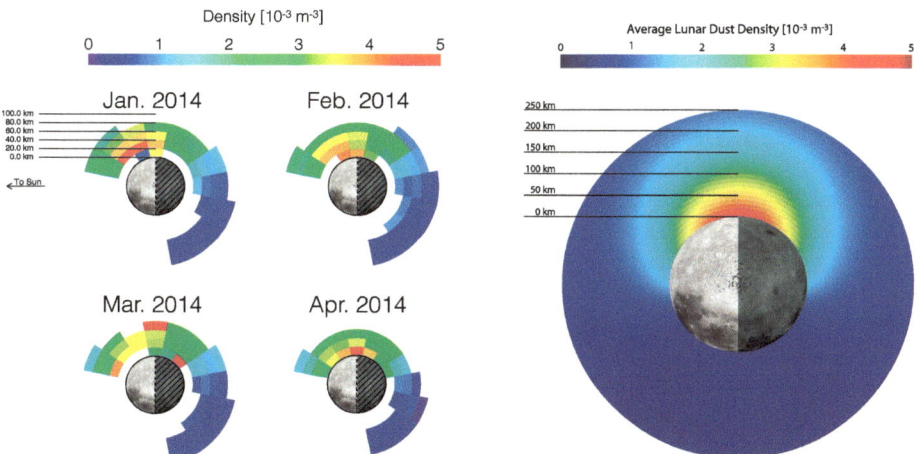

Fig. 9 Left: The average dust ejecta cloud density observed by LDEX for each calendar month for which LADEE was operational in 2014. Each color ring corresponds to the density every 20 km Szalay and Horányi (2015b). Right: The modeled annually averaged lunar dust density distribution for particles with $a \geq 0.3$ μm. These plots are in a reference frame where the Sun is on the left (-x direction) and the apex motion of the Moon about the Sun is towards the top of the page (+y direction) Szalay and Horányi (2016c)

LDEX measurements provided compelling evidence that our understanding of how meteoroids influence the lunar surface must be revisited. First, the measured fluxes showed that the Moon is engulfed in a permanently present, but highly variable dust exosphere (Fig. 9). Specifically, LDEX data showed that the lunar secondary dust ejecta cloud is persistent and asymmetric, and significantly denser at 5–8 hrs of lunar local time, with a peak density tilted somewhat sunward of the dawn terminator. Later, Szalay and Horányi (2015) used an empirical model to show that in order to explain the measurements reported by Horányi et al. (2015), LPC–produced meteoroids (i.e. HTC and OCC) should play a major role in the production of the observed ejecta cloud in the Moon's equatorial plane. The authors argued that the cloud is primarily produced by impacts from a combination of the three known sporadic meteoroid sources (Helion, Anti-Helion, and Apex, See Sect. 2). Furthermore, the cloud density is modulated by both the Moon's orbital motion about the Earth and about the Sun. The tilting of the ejecta cloud toward the Sun was more pronounced earlier in the LADEE mission (November 2013), while the LDEX signal became more centered around the dawn terminator toward the end of the mission (April 2014). From these data features, Szalay and Horányi (2015) inferred a variable relative strength between the Apex, Helion, and Anti-Helion sources to account for the change in the structure of the ejecta cloud throughout the mission. The ejecta mass production rate from the Helion source was found to be approximately twice as strong as the Anti-helion source (Szalay and Horányi 2015), which was a puzzling finding because such asymmetry has never been observed in the distribution of meteors measured by Earth-based radars (Campbell-Brown 2008; Janches et al. 2015) and optical systems (Jenniskens et al. 2016).

Efforts of modeling the influence of meteoroids on the lunar surface parallel those at Mercury and differ again on the meteoroid populations included in the different treatments. For example, Cremonese et al. (2013) reported the production of neutral sodium on the Moon caused by impacts of MBA meteoroids with size between 5–100 μm using the same dynamical model of Mercury and reported by Borin et al. (2009) (see Sect. 3). The authors estimated a production rate of Na atoms of 1.648×10^5 atoms.cm^{-2}.s^{-1}, concluding that the impact process due to meteoroids plays a very important role in the contribution of neutral atoms to the lunar exosphere. Previous reports suggested impact vaporization to be a negligible source to the total column abundance of sodium near noon, with a contribution of $\sim 1\%$, compared to other processes such as photon-stimulated desorption (PSD; Sarantos et al. 2010), with a measurable fraction ($\sim 50\%$) of total column density near the dawn terminator. Similar to the case of Mercury, Borin et al. (2017) updated this model by including cometary particles from short period comets (JFC), but again using a slow velocity distribution (Fig. 4) and lacked sufficient statistics to give distributions of arrival direction.

More recently, Janches et al. (2018) utilized the dynamical models of meteoroids released by JFC, HTC and OCC developed to explain the SMC sources (Nesvorný et al. 2010, 2011a,b; Pokorný et al. 2014) for the purpose of re-interpreting the Szalay and Horányi (2015) results. This effort concluded that $\sim 20\%$ of the asymmetry present in LDEX measurements is due to unaccounted-for biases introduced by the orientation of the LADEE spacecraft orbit with respect to the selenographic latitudes where the ejecta cloud produced by JFC meteoroids is largest. These modeling results also show that the response of the lunar soil to incoming meteoroids should be necessarilly different on the day and night sides of the Moon, in order to fit correctly the ejecta mass production rates measured by LDEX. The work inferred a smaller mass flux ratio between the short and long-period comet meteoroids on the Moon than that reported at Earth (Moon, 1.3:1 vs. Earth, 7:1). This finding was interpreted to indicate that the ejecta rate yield is a steeper function of the velocity of the incident meteoroids than assumed before, since LPC particles have on average higher impact velocities than their short-period comet counterparts.

Later, Pokorný et al. (2019) expanded on this effort and probed the effects of various free parameters intrinsic to the dynamical models used in Janches et al. (2018) including the effect of gravitational focusing that plays a significant role in shaping the lunar and terrestrial meteoroid environment. Although the model was able to reproduce night-side observations

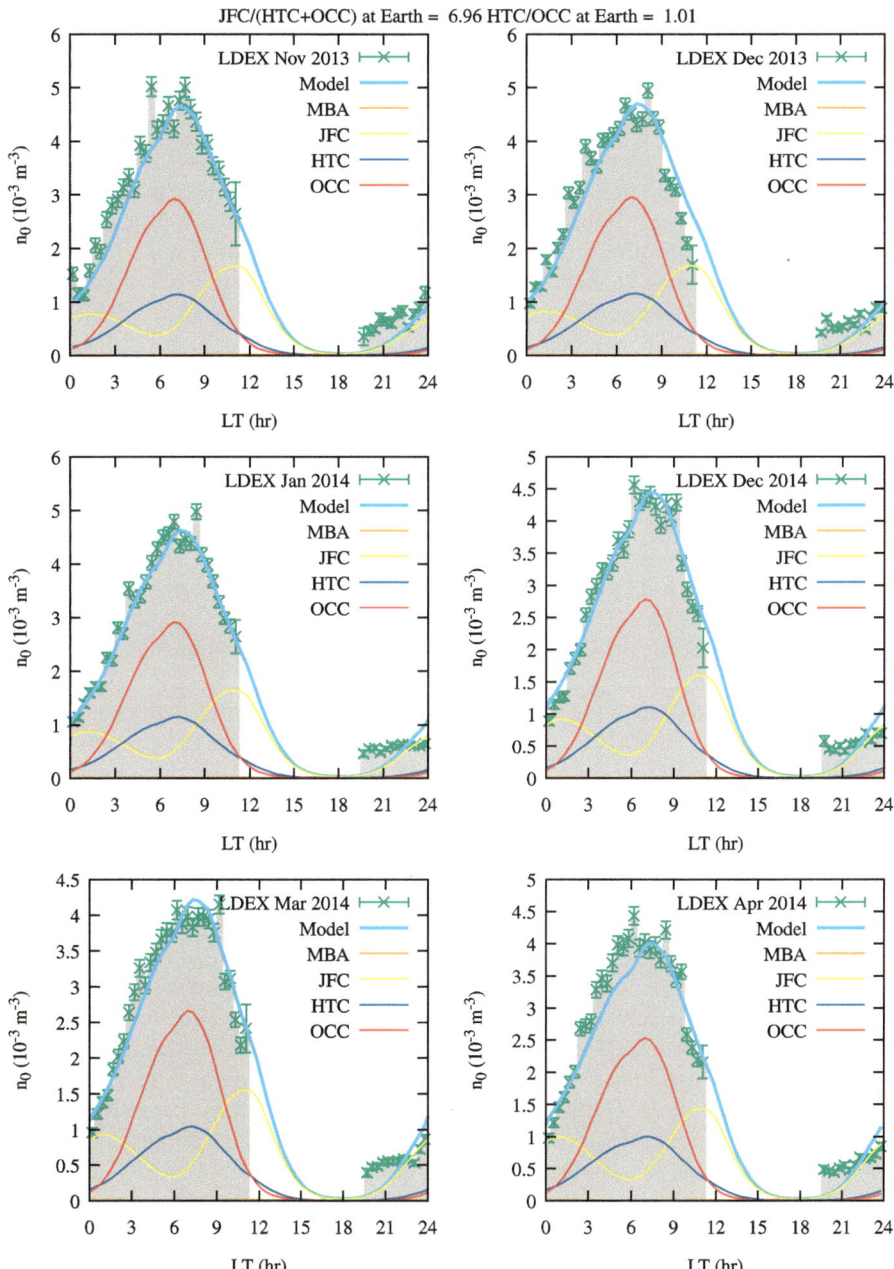

Fig. 10 LDEX data for November 2013–April 2014 (green points with error bars) compared to the model M^+ reported by Pokorný et al. (2019), where the sum of the four populations is represented by the blue solid line. Contributions of individual meteoroid populations are represented by solid lines: MBA (orange), JFC (yellow), HTC (blue), OCC (red). A fit representing a JFC-to-LPC mass ratio at Earth equal to 6.99 was selected, and the HTC/OCC mass influx ratio at Earth equal to 1.04. The free parameters used for this figures are: $F_{coll} = 20$, $\alpha = 4$, $\gamma = 1.23$. This is one out of many representations of the model with a similar goodness of fit ($\chi^2 = 7.99$). The MBA contribution is negligible

of LADEE/LDEX, the authors found that the predicted day-side values were systematically smaller than those measured by LDEX. A linear increase of ejecta mass production rate from 6 AM to noon in lunar local time, convolved with the predicted meteoroid velocities and fluxes, provided significantly better agreement between the model and LDEX observations (Fig. 10) than assuming a similar response on the day and night sides. A different hypothesis was later proposed by Szalay et al. (2020b), who suggested that β-meteoroids (very small meteoroids on hyperbolic trajectories) hitting the Moon's sunward side could explain this asymmetry, since they can impact the Moon at very high speeds ~ 100 km s^{-1} and thus their impact regime may differ from the significantly larger and slower sporadic meteoroids responsible for generating the bulk of the lunar impact ejecta cloud.

In addition to addressing the lunar day-side and night-side asymmetry, Pokorný et al. (2019) adopted the results reported by Carrillo-Sánchez et al. (2016) at Earth to estimate the absolute mass flux of meteoroids onto the Moon, similar to Mercury (Pokorný et al. 2018). The authors found that the total flux of MBA meteoroids cannot be constrained by modeling LDEX observations because they produce a negligible contribution to the total ejecta mass production rate due to their very low velocity. An important finding by the authors was that, in order to stay consistent with Earth-based estimates of the mass flux ratio of short-to-long period comets (Carrillo-Sánchez et al. 2016), the authors needed to revised the functional form of the ejecta mass production rate function, commonly used and suggested by Koschny and Grün (2001), finding that it should be linearly proportional to the meteoroid mass flux. Using constraints from Earth and taking into account the gravitational focusing effects between Earth and the Moon, Pokorný et al. (2019) finally concluded that the total mass accreted at the Moon is approximately $\mathcal{M}_{\mathrm{Moon}} = 1.4$ t day^{-1} assuming 43.3 t day^{-1} at Earth, where the individual contribution of meteoroid populations are: JFCs $\sim$ 72.6%, HTCs $\sim$ 12.8% and OCCs $\sim$ 10.0%. An important note is that these results represent one of many possible fits to the available LDEX measurements and that the solution space to provide a similar or better fit is wide due to the limited selenographic coverage of LADEE.

Both Janches et al. (2018) and Pokorný et al. (2019) showed that JFCs meteoroids are concentrated close to the ecliptic plane, arriving from direction towards and away from the sun (helion/anti-helion sources). HTC and OCC meteoroids impact the Moon mainly towards the apex direction while MBA meteoroids have radiants ranging from all directions and are hence able to populate the anti-apex source. Like at Earth (Fig. 3), the apex source have average impact velocities exceeding 55 km s^{-1}, while the toroidal and helion/anti-helion sources are in general populated by meteoroids a factor of two slower. Due to the smaller gravitational focusing at the Moon, JFC and MBA meteoroids contribute 2.5 and 5 times less in terms of the mass flux to the lunar meteoroid environment, respectively, than at Earth. As a result of the broad latitudinal distribution of cometary impactors, the entire lunar surface can be exposed to impacts with velocities as high as 30 km s^{-1}, where the near ecliptic directions can produce impacts with velocities up to 72 km s^{-1}. This finding was later reiterated by Pokorný et al. (2020) who found that highly inclined orbits can easily access permanently shadowed regions and alter the surface properties via hypervelocity impacts even when the detailed topography of the lunar surface is taken into account.

Finally, Pokorný et al. (2019) showed that the meteoroid mass flux and, consequently, the impact vaporization flux and ejecta mass production rate experience yearly and monthly variations that can be well represented by a sum of two sine functions with periods of one year and 29.5 days (synodic period of the Moon, Fig. 11). Yearly mass flux variations amount to 3.3% of the yearly average mass flux, while monthly variations amount to only 0.2%. These variations are larger for velocity dependent quantities, where yearly variations of the impact vaporization flux account for 6-8%, while monthly variations are around 4-5%.

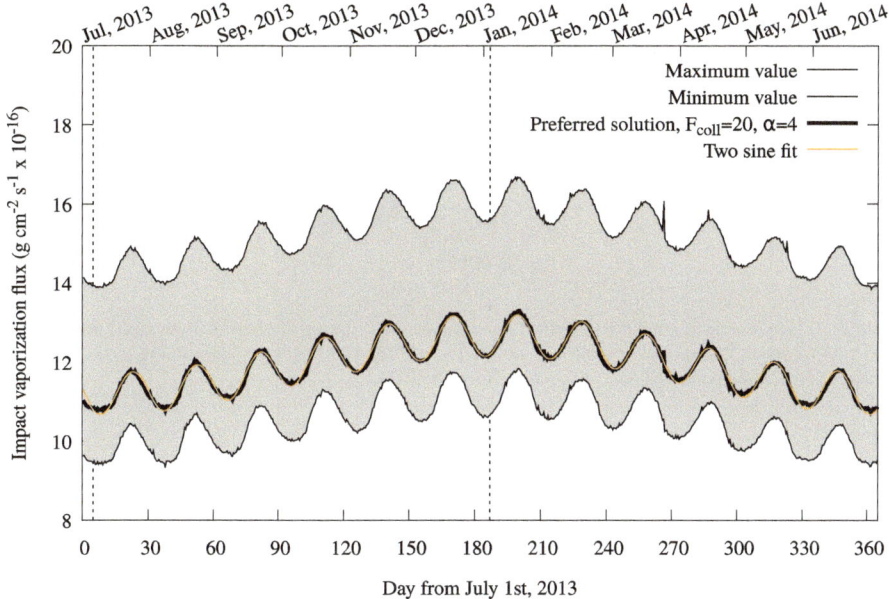

Fig. 11 Variations of the impact vaporization flux rate in $g \cdot cm^{-2} \cdot s^{-1} \times 10^{-16}$. with time from 1 July 2013 to 30 June 2014. The thick black solid line shows a preferred solution, while the confidence interval is shown as the gray area between two black solid curves denoting minimum and maximum variations of the model. The orange thick line is a fit to our preferred solution by a sum of two sines. Two dashed vertical lines represent the time of aphelion (5 July 2013) and perihelion (4 January 2014) passage. From Pokorný et al. (2019)

When the full spectrum of impact velocities is taken into account, the apex/dawn terminator source is dominating both the impact vaporization flux and the ejecta mass production rate for any day of the year. The total vaporization rate was several times higher than estimated by Cremonese et al. (2013) due to the inclusion of LPC particles. This expected total vapor rate is higher than considered in lunar exosphere models (Sarantos et al. 2012b), meaning that the role of impact vaporization in supplying the lunar exosphere with metals may have been previously underestimated, especially for species like Na and K which do not condense.

5 Meteoroid Influx on Other Inner Solar System Airless Bodies

Impact ejecta production occurs on all airless bodies throughout the solar system. Unlike the Moon, which retains a large fraction of its secondary ejecta particles, small asteroids shed most their ejecta and contribute to the interplanetary dust population. These grains carry valuable information about the chemical composition of their parent bodies, which can be measured via in-situ dust detection (Cohen et al. 2019). The LADEE/LDEX measurements (Sect. 4) of the lunar dust cloud can be used to estimate the dust ejecta distribution for any airless body near 1 AU. As shown in Fig. 12, this dust distribution is expected to be highly asymmetric, due to non-isotropic impacting fluxes. Spacecraft flybys near these asteroids would experience many times more dust impacts by transiting the apex side of the body compared to its anti-apex side (Table 2). The ejecta cloud for airless bodies on eccentric orbits is more complex due to their radial velocities modulating the impact speeds of the

Fig. 12 The predicted dust density distribution around asteroids on circular orbits near 1 AU for grains with radii $a \geq 0.3$ μm is shown in the ecliptic frame for selected asteroid sizes. The scale of the dust cloud size is proportional to R^2, with R the radius of the asteroid (Szalay and Horányi 2016b)

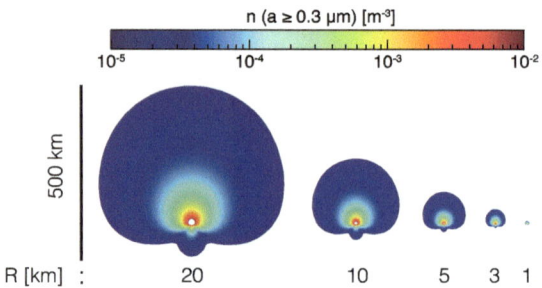

Table 2 Total number of predicted impacts per square meter during spacecraft flybys of an asteroid near 1 AU, with radius R and closest approach distances $b = 15$, 100 km. The first column gives the assumed minimum detectable particle size. (Szalay and Horányi 2016b)

a_{min} [μm]	$R = 1$ km		$R = 10$ km	
	$b = 15$ km	$b = 100$ km	$b = 15$ km	$b = 100$ km
0.1	20	2	2,000	200
0.3	1	<1	90	10
1.0	≪1	≪1	3	<1

bombarding interplanetary dust particles. This is the case for the asteroid 3200 Phaethon (Szalay et al. 2019), the target of the upcoming DESTINY$^+$ JAXA mission (Sarli et al. 2018), which will carry a dust detector capable of chemical composition analysis (Krüger et al. 2019).

Recently, Szalay et al. (2019) used the same dynamical models utilized at the Moon by Janches et al. (2018) and Pokorný et al. (2019, 2020) to predict the morphology of a potential ejecta cloud around 3200 Phaethon, which should be highly asymmetric given the asteroid's high eccentricity. The authors found that at 1 AU, the cloud is canted towards the asteroid's apex direction and its density varies by five orders of magnitude. Furthermore, comparing the peak ejecta density to a body at the same heliocentric distance but with a circular orbit, 3200 Phaethon's peak ejecta cloud density is approximately 30 times higher, largely due to enhanced ejecta production from JFC meteoroids. This calculation implies that eccentric asteroids shed more material than those on near-circular orbits, and are thus more attractive candidates for in-situ dust detection and chemical characterization due to their amplified asymmetric ejecta production.

6 Effects Due to Meteor Showers

Interplanetary space is crisscrossed by streams of meteoroids with typical sizes in the hundreds of microns up to decimeters, which were liberated through the sublimation of cometary ices. Meteoroids are eventually scattered out of the stream due to gravitational perturbations by Jupiter and Saturn, becoming part of the sporadic background (Wiegert et al. 2009; Nesvorný et al. 2010). Initially, however, and for a period which may be as long as 10^4 – 10^5 yr, the material remains confined to the streams, thus retaining a dynamical memory of its birth place. Unlike sporadics, meteoroids within streams move in approximately parallel paths and at nearly the same speed. The Earth intersects numerous streams

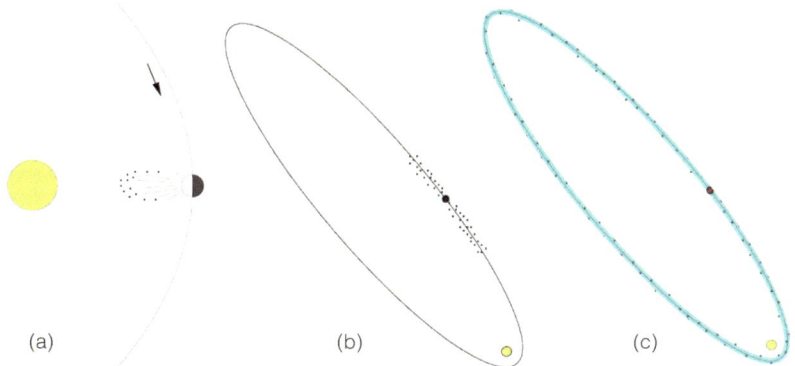

Fig. 13 Creation of a meteoroid stream. Meteoroids are ejected from a parent comet during a perihelion passage (**a**), initially forming azimuthally-confined trails of material along the orbit (**b**). Dust trails from different perihelion passages gradually spread out all around the orbit, forming the stream (**c**)

in its annual trek around the sun, resulting in periods of increased flux, or showers, of meteoroids at the top of the atmosphere (Brown et al. 2010; Pokorný et al. 2017a; Bruzzone et al. 2020). Although the SMC is responsible for providing a constant source of impactors to create and maintain exospheres on airless bodies, events produced by meteoroid streams offer a unique opportunity to study the physical processes involved in this phenomena with a much more constrained input (e.g., velocity distribution, directionality, temporal variability, etc).

Meteoroids newly ejected from the cometary nucleus (Fig. 13, panel a) initially form azimuthally-confined arcs of trails of particles (panel b) that persist over tens of orbital revolutions of the comet until incorporated into the stream by keplerian shear (panel c). Trails are responsible for meteor outbursts: intense, but short-lived, enhancements in the visible meteor rate at Earth, over and above the activity level of the annual shower (Janches et al. 2020b). In many cases, the dynamical evolution of cometary trails is deterministic (Kondrateva and Reznikov 1985; McNaught and Asher 1999; Lyytinen and Van Flandern 2000; Vaubaillon et al. 2005) so meteor outbursts can be reliably forecasted with brute-force numerical simulations of test particles to serve as tracers of the dust evolution.

In contrast to Earth, there is precious little information on meteoroid streams incident on other planetary bodies. In principle, every planet-approaching comet contributes some amount of dust to the local meteoroid environment. In practice, the dynamical mobility and dust properties of the comet will determine the magnitude of its contribution to the flux. The complex relationship between cometary activity and the planet-intersecting component of the dust means that – even for the Earth – it is practically impossible to make reliable quantitative predictions of the flux, except in cases where the dust has been detected previously in the planetary vicinity, for example as a meteor shower. A case in point is the exceptionally close ($\sim$150,000 km) approach of comet C/2013 A1 (Siding Spring) to Mars in October 2014. Despite intense scrutiny of the comet in the year leading up to the encounter, predictions for a meteor shower in Mars's atmosphere ranged from storm-level (10^7 kg of dust mass deposited in the atmosphere; Vaubaillon et al. 2014) to negligible ($<$100 kg deposited; Kelley et al. 2014). Post-encounter estimates inferred from observations of metallic species in the upper atmosphere were closer to the upper end of predictions (Schneider et al. 2015).

6.1 Meteoroid Streams at the Moon

The Moon is a unique natural laboratory for understanding the response of an airless body to passage through a stream and has important lessons to teach us with respect to Mercury. The Moon's proximity to the Earth means that the two bodies share a common meteoroid environment and, because meteor showers are known from terrestrial observations, lunar measurements can be readily linked to showers. Indeed, seismic data from the Apollo Passive Seismic Experiment detector network has shown the lunar impact rate to be highly non-random. Several impact event clusters identified in the seismic data correspond to strong Earth meteor showers (Oberst and Nakamura 1991). One remarkably dense swarm of meteoroids detected by the Apollo network in 1975 (Duennebier et al. 1976) was associated with the so-called Taurid complex, to which we shall return to later.

Flashes caused by kg-class meteoroids impacting the lunar surface were first detected over two decades ago (see Madiedo et al. 2019, for a review on this topic). Early observations, focused on maximising the impact detection rate, were preferentially carried out during meteor showers. These were followed by surveys where observations took place regularly under favourable lunar observing conditions, showing that the occurence of flashes is indeed correlated with showers (Suggs et al. 2014; Liakos et al. 2020). After the impact of a kg-sized meteoroid near the lunar terminator two dust clouds with typical expansion velocities of 0.1 and 3 km s^{-1} were observed (Berezhnoy et al. 2019). The most recent observations of impact flashes detect impactors as small as 1 gr and about 1 cm in size (Avdellidou and Vaubaillon 2019). The derived temperatures from these flashes are consistent with hypervelocity impact experiments (Eichhorn 1975). These flashes not only provide microphysical parameters for exosphere simulations, but also produce stochastic variation in the exosphere (Mangano et al. 2007).

The response of the lunar exosphere to meteoroid bombardment by meteoroid shower activity has been identified in the past. Reports by Verani et al. (1998, 2001), Smith et al. (1999) and Berezhnoy et al. (2014) observed various degrees of exospheric increase of sodium at the Moon during the Leonids, Taurids, Quadrantids, and Perseids. As stated in Sect. 4, LADEE's observations were instrumental at providing detailed measurements of the effects of meteoroids in general and showers in particular on the lunar surface. LDEX detected episodes of enhanced flux of lunar ejecta coinciding with known meteor showers (See Fig. 8 and Szalay and Horányi 2015). The high temporal cadence of LADEE measurements permitted detailed studies of shower effects. For example, Szalay et al. (2018) reported a large enhancement in the lunar impact ejecta cloud while the Moon transited the Geminid meteoroid stream, particularly above the portion of lunar surface normal to the shower mean radiant. The authors found two peaks in the estimated surface density of impact ejecta which coincided with radar observations of shallower mass indices than most of the Geminids, suggesting an enhancement of larger particles. The timing of the main observed peak matched ground-based visual observations of meteors with magnitude of -1 to -3. This finding suggests that LDEX detected ejecta from primary impactors with radii $\sim$2 mm to 2 cm.

LADEE was the first mission to directly observe the link between meteoroid bombardment and exosphere formation (Elphic et al. 2014) because, in addition to the LDEX dust experiment, it also carried an Ultraviolet-Visible Spectrometer (UVS; Colaprete et al. 2014) and a Neutral Mass Spectrometer (NMS; Mahaffy et al. 2014). The comparison between the observations of LDEX and UVS identified a correlation between the meteoroid influx and the Na and K abundances in the lunar exosphere, in particular with shower activity. Specifically, Colaprete et al. (2016) and Szalay et al. (2016) found a strong correlation of exospheric potassium and meteoroid ejecta during the Geminids meteoroid shower, exhibiting a much stronger response than sodium. With the exception of the Geminids, the authors

Fig. 14 Cumulative number of comets as a function of q, the perihelion distance. Comets with $q < 0.05$ AU are not counted. The horizonal segments represent the annual radial excursions of planetary orbits and have been shifted vertically to intersect the curve

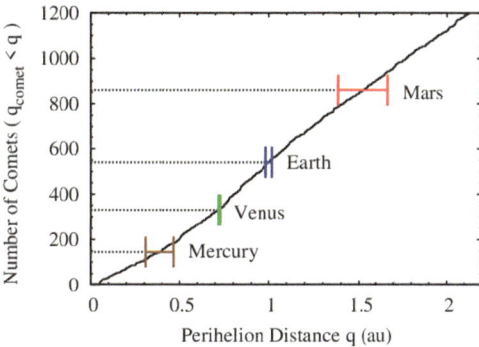

found a weak correlation between the meteoroid influx as measured by LDEX and exospheric density of alkalis as measured by UVS. Similarly with NMS, Benna et al. (2019) reported detections of water vapor released into the lunar exosphere. The timing of 29 water release events agreed with periods when the Moon encountered known meteoroid streams. The authors used these measurements to constrain the hydration state of the lunar soil, arguing that by heating the soil meteoroids release water that is buried below a layer of dry regolith at depths of a few centimeters.

6.2 Meteoroid Streams at Mercury

The meteoroid stream environment at Mercury is effectively unconstrained by direct observation and, at least for the time being, arguments for or against their existence must be based on indirect evidence. Since most known streams are associated with active comets, useful conclusions may be drawn by comparing the number of comets approaching Mercury and the Earth. A necessary condition for two orbits to cross is

$$q_{Comet} < Q_{Planet} \qquad (1)$$

where q_{Comet} is the perihelion distance of the comet and Q_{Planet} the planet's aphelion distance. For a near-circular orbit, Q can be replaced by the semimajor axis a. This is not a sufficient condition because, whether the orbits actually cross depends on the relative geometry between the orbits. Under the working assumption that orbit orientation is – at least within the terrestrial planet region – insensitive to heliocentric distance, examination of the distribution of cometary q should constrain the relative number of streams at Mercury and the Earth.

For this purpose, works that deal with meteor showers in the solar system (e.g., Selsis et al. 2004) utilize the DASTCOM database (https://ssd.jpl.nasa.gov/?sb_elem, retrieved 09 May 2020) available through the JPL online solar system data service (Giorgini et al. 1996). This set contains several clusters of comet fragments sharing the same orbit and, in addition, many sungrazing comets, where q_{Comet} is comparable to the solar radius. Including these clusters would skew the statistics, therefore all but one of the fragment orbits in each cluster were removed. Most sungrazers have $q < 0.05$ AU (Wiegert et al. 2020) so all comets with q below that value are removed.

Figure 14 shows the cumulative distribution of q, $N(q)$, for the DASTCOM entries. The horizontal segments correspond to the radial excursions of the four terrestrial planets. It can be seen in this figure that each of the planetary orbits encompasses the perihelia of 860 (Mars), 540 (Earth) & 330 (Venus) comets while the figure for Mercury is 150. Taken these numbers at face value suggests that Mercury encounters 1/3 to 1/4 of number of streams that would encounter Earth in its annual trek around the Sun. Strictly speaking, this result only applies to comets near the ecliptic plane where the perihelion criterion (Eq. (1)) automatically ensures a close approach. For an isotropic comet population, the distribution should be adjusted for the reduced volume of available heliocentric space enclosed by Mercury's orbit relative to the Earth but also the tendency for stream cross sections to be minimized near perihelion where most stream meteoroids were ejected. Christou (2004) computed Minimum Orbit Intersection Distances (MOIDs) for a sample of 158 multi-apparition comets in DASTCOM. The number of comets with orbital period $P < 20$ yr approaching Mars, Earth and Venus to <0.1 AU was reported to be 31, 12 and 4 for those three planets, in qualitative agreement with Fig. 14. The respective figures for comets with $P > 20$ yr were 3, 3 and 5. Small number statistics notwithstanding, it is reasonable to expect that the number of isotropic comet streams approaching Mercury and the Earth are similar.

Numerical modelling of individual streams is motivated by the need to explain observations (e.g., the Leonid storms; McNaught and Asher 1999). For Mercury, the seasonal modulation of exospheric Ca observed in situ by MESSENGER (Fig. 6 and Burger et al. 2014) motivated modelling of the sporadic background (Killen and Hahn 2015; Pokorný et al. 2018) to investigate meteoroid surface impact vaporization as the source mechanism. As seen in Fig. 6 on this chapter, these models reproduce the overall Ca production rate dependence with TAA but fail to predict a positive feature at TAA = 20–30°. It was suggested by Killen and Hahn (2015) that the cause may be due to an enhanced bombardment of meteoroids by a stream originated from comet 2P/Encke, although this comet crosses Mercury's orbit plane further away from perihelion, at TAA = 45°. Encke has been linked to several strong daytime and nighttime meteor showers at Earth, the so-called Taurid complex (Whipple 1940; Porubčan et al. 2006). Its current orbit is $\sim$0.17 AU from the Earth's but only 0.026 AU from that of Mercury (Selsis et al. 2004). Recently, Stenborg et al. (2018) detected reflected light from the Encke stream near Mercury's orbit in SECCHI/STEREO images.

Christou et al. (2015) investigated the properties of the near-Mercury Encke stream with a two-parameter numerical model. One parameter was age (i.e., the time spent by meteoroids in space since ejection from the nucleus); the other was the particle size, which determines β the strength of solar radiation forces relative to gravity on the meteoroid through the expression

$$\beta = 1150(\rho D)^{-1} \tag{2}$$

where ρ in this context is the bulk density in kg m^{-3} and D is the diameter in μm.

It was found that, despite the proximity of the orbits, Encke-released dust younger than 5 kyr has not undergone sufficient orbital evolution to physically reach Mercury. At the same time, planetary gravitational effects begin to disperse the stream after $\sim$50 kyr. PR drag plays a crucial role in delivering Encke released dust to Mercury by rotating the line of nodes of stream meteoroids away from the comet's (Fig. 15). Indeed, an optimum combination of size and age exists for the orbits to cross Mercury's, at TAA of 350–30°; these are meteoroids ejected 10-20 kyr ago with $\beta=1$–2 10^{-3}, equivalent to particles a few mm in size for $\rho=600$ kg m^{-3} (Eq. (2)). The particles will arrive at Mercury from the antisolar direction and impact on the nightside at 32–37 km s^{-1}. The simulations also showed that the stream intersects Mercury's orbit a second time, at TAA = 135°–165°, with the meteoroids impacting on the dayside and on the outbound leg of their heliocentric trajectories. Christou et al. (2015) reported a minor enhancement in the Ca production rate profile at TAA$\sim$165°

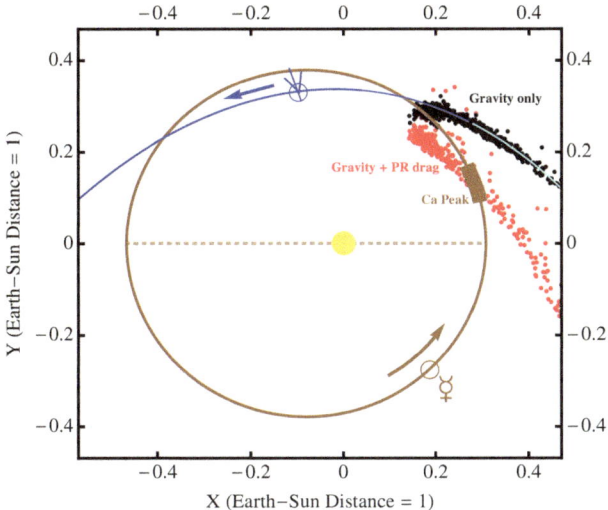

Fig. 15 Locations of descending nodes of 10^4 yr-old Encke particles on Mercury's orbit plane at JD2487500.0. The comet orbit projection on this plane is represented by the blue curve with the part of the orbit lying above the plane shown in a brighter colour. Mercury's orbit is represented by the brown ellipse with a dashed line connecting the orbit apses. The yellow point represents the Sun. The brown rectangle represents the location of the TAA=25° peak in Calcium production rate inferred from MESSENGER observations. Black points correspond to particles subject to planetary gravitational perturbations only; red points are affected by Poynting-Robertson drag with $\beta = 10^{-3}$ in addition to gravity

that may be seen to corroborate this second crossing, yet the evidence is less conclusive than for the peak at TAA = 25°.

7 Laboratory Experiments

Hypervelocity dust impact experiments can be used to establish the efficiency of the production of neutrals and ions by the continual bombardment of the lunar surface by meteoroids. For example, Sugita et al. (1998, 2003) conducted experiments at the NASA Ames Vertical Range using spherical copper projectiles and polycrystalline dolomite targets to record the intensities of emerging atomic lines and molecular bands. The measured emission intensities as function of the speed and mass of the impactor suggested that the impact-vapor contribution to the lunar exosphere involves a complex chain of chemical and physical processes. The first direct laboratory measurement of vapor produced by bombardment with simulated micrometeoroids in the size range of 0.1–1 μm radius and speed range of 1–10 km s^{-1}, used a fast ion gauge (Fig. 16) to quantity the neutrals released per unit projectile mass, N/m (Collette et al. 2014). The results indicated a power-law dependence with the projectile speed v, as $N/m \sim v^{2.4}$ (Fig. 17). At the highest speeds tested, the number of neutral atoms liberated is equivalent to 5% of the atoms in the projectile; complete vaporization is expected at speeds exceeding 20 km s^{-1}. Earlier experiments (Eichhorn 1975) had established the expected temperature of the impact-generated clouds.

At the Moon, the meteoritic source contribution to sustain the dilute exosphere competes with solar wind sputtering (for refractory metals) and photon-stimulated desorption, or PSD (for alkalis). For example, Stern (1999) estimated a total exospheric mass of about

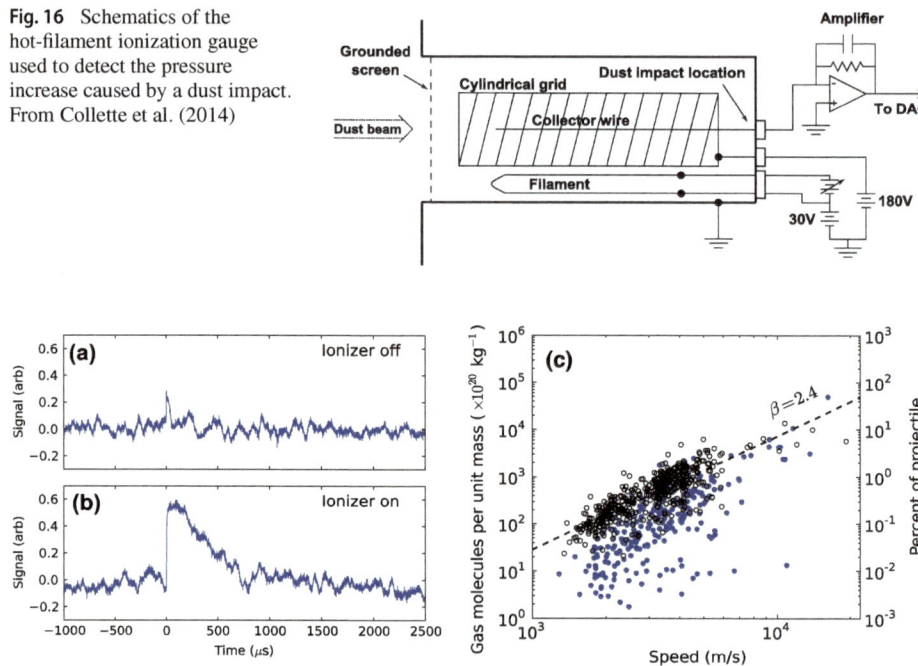

Fig. 16 Schematics of the hot-filament ionization gauge used to detect the pressure increase caused by a dust impact. From Collette et al. (2014)

Fig. 17 The recorded ion gauge (Fig. 16) signals with and without the ionizer turned on for two similar dust particle events: (**a**) $m = 2.3 \times 10^{-12}$ g, $v = 3.2$ km s^{-1}; and (**b**) $m = 2.0 \times 10^{12}$, $v = 3.5$ km s^{-1}. (**c**) The amount of neutral gas molecules recorded (left axis) as function of the dust particle's impact speed with the ionizer on (open black circles) and off (closed blue circles), and the number of neutral atoms as the fraction of the atoms in the projectile (right axis). From Collette et al. (2014)

2×10^7 g, assuming that approximately 5 tons/day (60 g/s) of interplanetary sub-milligram sized meteoroids bombard the lunar surface, about ~2.5 time more mass that recent estimates (Pokorný et al. 2019). The authors argued that the total mass of incoming meteoroids delivered in about 4 days is comparable to the mass of the entire lunar exosphere. The importance of meteoroids increases for less volatile metals such as Mg, Ca, Al, and others, and it is expected to provide about half of the total abundance (Sarantos et al. 2012a).

8 Large Impactors

As described above, much of the scientific research the meteoroid impact effects on an exosphere, and the focus of this manuscript, has been in regards to very small objects diameters smaller than a millimeter, which impact the surface often enough to form and constantly maintain the exosphere (See Sect. 2). As discussed in the earlier sections, the orbital characteristics of meteoroids smaller than this, have evolved dynamically due to various effects. In this section we focus on the rarer impacts made by meteoroids that are much larger in size. As it is the case for meteor showers, when detected, large impactors have a well defined set of physical and dynamical characteristics and thus can help to understand the associated exospheric processes more accurately.

The differential radius distribution, $h(r)$, for impactors in the size range from 1 cm to 100 m is shown in Fig. 18 (Marchi et al. 2005). This shows the number of impacts per

Fig. 18 Differential radius distribution, $h(r)$, which shows the number of impacts for the whole surface per year and size in meters, for Earth and Mercury. From Marchi et al. (2005)

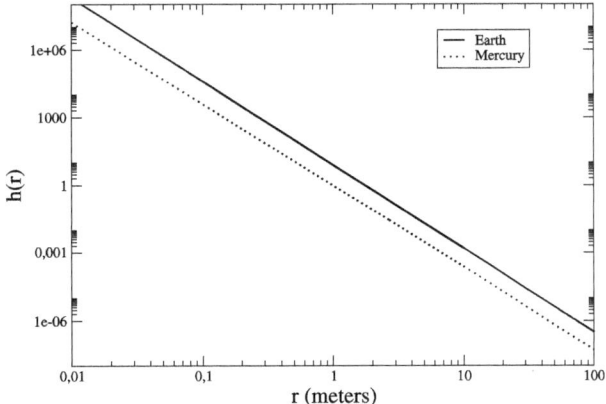

year and size for both Earth and Mercury as a function of radius. It can be seen that the expected rate of impacts decreases with the size of the impactor, being the impact rate of 1 m size meteor of 1 per Earth-year at Mercury, that is, one order-of-magnitude lower than at Earth. These larger meteoroids have a different dynamical evolution and arrive from the asteroid belt due to the 3:1 and ν_6 resonances. The 3:1 resonance lies approximately at 2.5 AU, and objects undergo a change in their eccentricity until they become Earth or Mars orbit-crossing. At this point they can be extracted from the asteroid belt. The ν_6 resonance occurs when the object's longitude of perihelion precession frequency is equal to the sixth secular frequency of the planetary system (for asteroids, this is usually Saturn; Morbidelli and Gladman 1998). This resonance forces the objects to cross the orbits or Mars, Earth or Venus as well as collide with the Sun. These two resonances therefore act to eject objects from the asteroid belt into the inner solar system where they can impact on planetary bodies such as Mercury or the Moon. A study of 59 meter-sized Earth-impactors found that the ν_6 resonance dominates the delivery mechanism with 50% of the probability that the impactors originated from the asteroid belt (Brown et al. 2016). These types of large impacts are rare in comparison to the much smaller scale meteroids (Pokorný et al. 2018, 2019), and are expected to contribute only transient changes to the exospheres of Moon and Mercury.

Figure 19 shows the observations of an impact flash on the Moon on March 17, 2013. The impactor was estimated to have a diameter of 9–15 m, a mass of 16 kg and produced a crater with an estimated rim-to-rim size of 12–20 m (Suggs et al. 2014). The authors analysed a dataset of 126 other observed impact flashes at the Moon in 2006–2011, with impactors ranging in size from 1 cm to 14 cm.

Mangano et al. (2007) modelled the effect of similar impactors at Mercury. They investigated how an impact would generate an enhancement of the local exosphere at altitudes of 400 km and 1500 km, for the most common constituents of Mercury's surface composition. The model showed that the largest density enhancements at 1500 km above the average exospheric density would be for Mg, Si and Al. However, such enhancements are expected to be short-lived, with local enhancements falling back to average densities in timescales of less than an hour.

Such an enhancement was observed by the MESSENGER Fast-Imaging Spectrometer, which measured freshly ionized pick-up ions in the solar wind at high altitudes (Jasinski et al. 2020). The ions were estimated to come from a neutral "plume" of impact vaporized surface material most likely caused by a 1 m sized meteoroid. The neutrals were then photoionized and the ions were picked up by the local solar wind plasma and subsequently

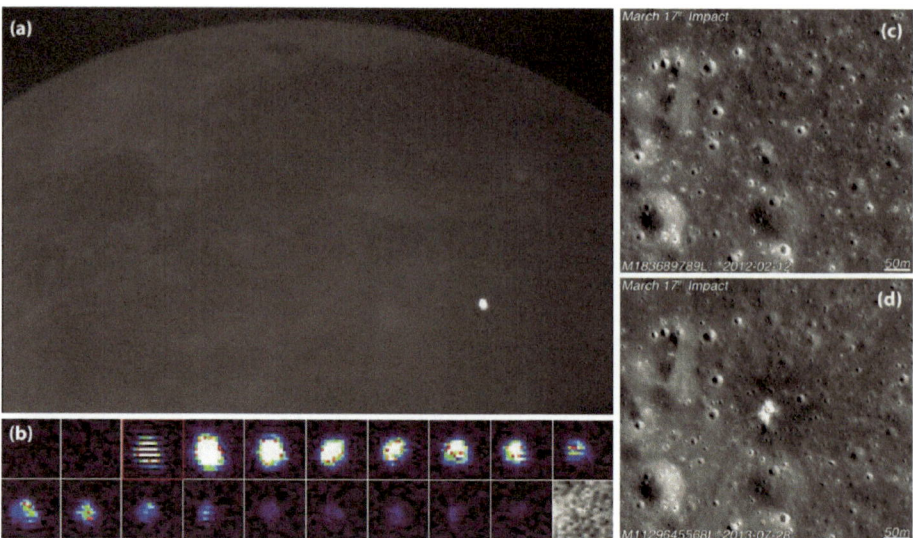

Fig. 19 Images from an impact on the Moon which was observed on March 17, 2013 Suggs et al. (2014). Panel (**a**) the full field of view of the observed flash, (**b**) images taken from the 30 FPS video of the impact flash, (**c**) and (**d**) the before and after images of the impact site observed by the Lunar Reconnaissance Orbiter Camera. From Suggs et al. (2014)

observed spacecraft's Fast-Imaging Plasma Spectrometer. A schematic of the process can be seen in Fig. 20; the Sun is to the left of Mercury, and the magnetospheric boundaries are shown (white curved lines). The orientation of the interplanetary magnetic field and the motional electromagnetic field are shown for the observed event in the top left corner. The pickup ions were measured to be within a mass-per-charge ratio of 21–30 amu/e, which includes Na^+, Mg^+, Al^+, and Si^+. Due to the long and short photoionization lifetimes of Mg and Al, respectively, the composition of the ions from the impactor "plume" was estimated to be primarily of Na^+ and Si^+. BepiColombo, an ESA and JAXA mission with two spacecrafts that will orbit Mercury, is expected to observe more of such events (at the time of writing BepiColombo is in its cruise phase with orbit insertion expected in 2025).

9 Future of the Field

Over the upcoming decade, there are multiple opportunities for spaceborne dust measurements at 1 AU and in the inner Solar System. The $DESTINY^+$ (Demonstration and Experiment of Space Technology for INterplanetary voYage Phaethon fLyby dUSt science; Krüger et al. 2019) and IMAP (Interstellar Mapping and Acceleration Probe; McComas et al. 2018) missions, which will be launched in 2024, carry dust analyzers with the ability to perform chemical composition analysis of interplanetary dust and/or interstellar dust at 1 AU. Besides interplanetary and interstellar dust measurements around 1 AU, $DESTINY^+$ will also observe the ejecta cloud around the active asteroid 3200 Phaethon for the first time allowing us to constrain our current predictions (Szalay et al. 2019). IMAP will focus on the observation of interstellar dust, potentially allowing to link the composition of solar wind ions with the constituents of interstellar dust. The Tanpopo experiment, a sample return mission with a silica aerogel capture medium on the International Space Station, will provide the

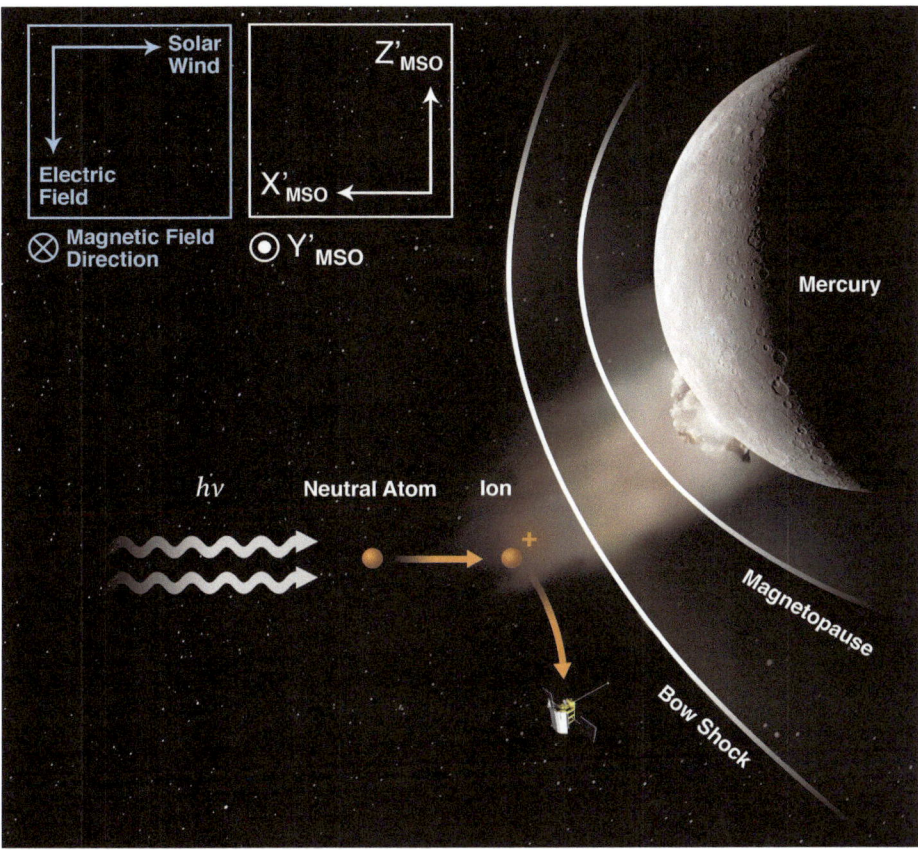

Fig. 20 A schematic from Jasinski et al. (2020) showing the photoionization of neutral particles released from the surface of Mercury due to a large impactor. The newly photoionized particles were observed as pickup ions by the MESSENGER spacecraft in the solar wind upstream of the bow shock

information of chemical composition of dust particles falling onto the Earth (Tabata et al. 2016). For the next space station orbiting around the Moon, the Lunar Orbital Platform-Gateway, a number of international contributions for dust measurements are expected that, taking advantage of the long duration and large infrastructure of this platform, can lead to better understanding of the Earth-Moon meteoroid environment. In contrast to the larger space missions, 6U-class (10 cm × 20 cm × 30 cm) deep-space explorer, EQUULEUS, will detect the first dust impact by a CubeSat during the cruise to a libration orbit around the Earth-Moon Lagrangian point L2 (Funase et al. 2020). In the innermost region of our Solar System, the Parker Solar Probe instrumentation has already reported impacts by very small dust particles which seem to be consistent with the existence of a dust population on hyperbolic trajectories, i.e. β-meteoroids (Mann et al. 2019; Szalay et al. 2020a). Although the mass influx of these dust particles represent, most likely, a minor contribution to the formation of exosphere of airless bodies, it could be responsible for some effects such as the night-day asymmetry observed by LADEE on the lunar dust cloud (Szalay et al. 2020b) and they can impose constraints on the size distribution of mass-dominant meteoroids (Fig. 1). LADEE findings provide a unique opportunity to map the composition of the lunar surface from orbit (Postberg et al. 2011) and identify regions that are rich in volatiles, providing opportunities for future in situ resource utilization (ISRU) (Horányi et al. 2020)

Clearly one of the most exciting opportunities to continue growth in this field is the arrival of the dual-spacecraft ESA-JAXA Bepicolombo mission at Mercury in 2025, which will provide simultaneous observations of meteoroid impact, exospheric neutrals and ions in the Hermean orbit. These observations will set more constraints on models described in this manuscript as well as test the link between Mercury's exosphere and the Encke meteoroid stream (Plainaki et al. 2017). The most direct evidence for temporal dust flux enhancements indicative of a stream would likely come from the Mercury Dust Monitor (Nogami et al. 2010) onboard the Mercury Magnetospheric Orbiter module. This instrument should register ejecta from the surface impact of Encke meteoroids, similarly to LADEE on the Moon (Szalay and Horányi 2016). Concurrent exosphere measurements by the SERENA and PHEBUS (Quémerais et al. 2020) instruments onboard the Mercury Planetary Orbiter (MPO) module and MSASI on the Magnetospheric Orbiter would enable establishing a relationship, if any, between Ca, Na or other exospheric species and meteoroid impacts. An exciting possibility is the direct detection of impact flashes by stream meteoroids, despite the short-lived and upredictable nature of the phenomenon. The observation of the flashes will represent a serendipity project of SIMBIO-SYS (Cremonese et al. 2020), the suite of cameras on board the MPO, when it will observe the dark hemisphere of Mercury for calibration purposes.

Another area which requires further work concerns laboratory experiments. In the last few years significant progress has been made regarding the physical processes that meteoroids undergo upon atmospheric entry (i.e. ablation and ionization; Thomas et al. 2016; DeLuca et al. 2018; Gómez Martín et al. 2017; Bones et al. 2016). However, there are still major unknowns regarding meteoroid impacts on planetary surfaces. As described in Sect. 4, current models utilized experimental results from Koschny and Grün (2001) in order to obtain absolute values of the impact mass production rates M^+ for the lunar surface, which yield a proxy of the gardening rate from meteoroids. Current modeling results suggest that equatorial regions experience three to five times higher ejecta production – and thus gardening – rates than the polar regions. They also predict lunar dust cloud density values higher by four orders of magnitude than those inferred by LDEX (Pokorný et al. 2019). This discrepancy indicates that the ejecta mass production yield of lunar regolith is considerably lower than the experimental data reported by Koschny and Grün (2001) yield which was estimated using very low velocity impacts ($1-12$ km s^{-1}) on ice-rich surfaces. Clearly better designed experiments are needed to advance in this area.

Finally, models are constantly updated as new data almost always provide new paradigms. An important point to raise is that current meteoroid models use the ecliptic plane as a reference plane rather than, for example, the invariable plane of the solar system. For instance, Cambioni and Malhotra (2018) reported that the main asteroid belt is inclined by about a degree relative to the ecliptic. If the sporadic meteoroid complex is symmetric about the invariable plane (or some other plane) rather than the ecliptic, it could significantly change the presented results. It is very likely that the sporadic meteoroid complex is warped or twisted and the plane of symmetry changes between the main belt and the innermost parts of the solar system. For example Nesvorný et al. (2006) treated the dust bands in the main-belt as symmetric about the invariable plane when looking outward. On the other hand, Nesvorný et al. (2010) assumed that the ZDC is symmetric with respect to the ecliptic latitude when trying to reproduce IRAS measurements. Rowan-Robinson and May (2013) also showed that the ecliptic plane works well as the symmetry plane for the ZDC. However, between 0.3 and 1 AU the plane of symmetry seems to be different from the invariable plane and similar to the orbital plane of Venus according to Leinert et al. (1980).

There are several reasons why the assumption of ecliptic symmetry in the dynamical models presented here makes sense. Initially, some of these models were constrained with IRAS measurements, which showed that the Zodiacal cloud is symmetric with respect to the ecliptic (Nesvorný et al. 2010, 2011a,b). This conclusion is also supported by decades of observations using meteor radars which have shown that, as viewed from Earth, the main sporadic sources are symmetric with respect to the ecliptic plane (Fig. 2; Janches et al. 2015; Campbell-Brown and Wiegert 2009). Later, the HTC dynamical model reported by Pokorný et al. (2014) reproduced radar observations assuming ecliptic symmetry. Furthermore, earlier semi-empirical models also concluded that a meteoroid environment that is symmetric with respect to the ecliptic plane reproduced the seasonality and geographical variability of meteor observations using high power and large aperture radars (Janches et al. 2006; Fentzke and Janches 2008; Fentzke et al. 2009; Pifko et al. 2013; Schult et al. 2017). Finally, ecliptic symmetry seems to reproduce, at least to first order, the variabilities observed on the lunar dust cloud (Janches et al. 2018; Pokorný et al. 2019) and the Ca exosphere at Mercury (Pokorný et al. 2018).

It is important to note that most of these works concern meteoroids with cometary origin which have enough energy to produce the reported observed and modeled phenomena, while Cambioni and Malhotra (2018) focused on the MBA population. Meteoroids originating from MBAs are too slow to produce ionization (and thus most of the observed radar meteors), ablate in the atmosphere and produce metallic layers, or even produce a significant contribution to the dust plumes in airless bodies. The Zodiacal Cloud is certainly more complex than current models predict and investigating such effects offers exciting future opportunities.

Acknowledgements DJ and MS were funded through the NASA ISFM and LDAP Programs. A.A.B. was partially supported by Russian Science Foundation, grant no. 20-12-00105. GC contribution is supported by the ASI-INAF agreement 2017-47-H.0. JMJ's contribution to this work was supported by an appointment to the NASA Postdoctoral Program (NPP) Fellowship at the Jet Propulsion Laboratory administered by the Universities Space Research Association through a contract with the National Aeronautics and Space Administration (NASA). Astronomical research at the Armagh Observatory and Planetarium is grant-aided by the Northern Ireland Department for Communities (DfC).

Funding Note Open access funding provided by Istituto Nazionale di Astrofisica within the CRUI-CARE Agreement.

References

P.A.R. Ade, N. Aghanim, C. Armitage-Caplan, M. Arnaud, M. Ashdown, F. Atrio-Barandela, J. Aumont, C. Baccigalupi, J. Banday Astron et al., Planck 2013 results. XIV. Zodiacal emission. Astron. Astrophys. **571**, A14 (2014). https://doi.org/10.1051/0004-6361/201321562. arXiv:1303.5074

W.M. Alexander, O.E. Berg, C.W. McCracken, L. Secretan, J.L. Bohn, Interplanetary dust-particle flux measurements between 1.0 and 1.56 a. u. from Mariner 4 cosmic-dust experiment (abstract). Smithson. Contrib. Astrophys. **11**, 227 (1967)

C. Avdellidou, J. Vaubaillon, Temperatures of lunar impact flashes: mass and size distribution of small impactors hitting the Moon. Mon. Not. R. Astron. Soc. **484**(4), 5212–5222 (2019). https://doi.org/10.1093/mnras/stz355. arXiv:1902.00987

W. Baggaley, Radar observations, in *Meteors in the Earth's Atmosphere*, ed. by E. Murad, I. Williams (Cambridge University Press, Cambridge, 2002), pp. 123–148

M. Benna, D.M. Hurley, T.J. Stubbs, P.R. Mahaffy, R.C. Elphic, Lunar soil hydration constrained by exospheric water liberated by meteoroid impacts. Nat. Geosci. **12**, 333–338 (2019)

A.A. Berezhnoy, Chemistry of impact events on Mercury. Icarus **300**, 210–222 (2018). https://doi.org/10.1016/j.icarus.2017.08.034

A.A. Berezhnoy, J. Borovička, Formation of molecules in bright meteors. Icarus **210**(1), 150–157 (2010). https://doi.org/10.1016/j.icarus.2010.06.036

A.A. Berezhnoy, K.I. Churyumov, V.V. Kleshchenok, E.A. Kozlova, V. Mangano, Y.V. Pakhomov, V.O. Ponomarenko, V.V. Shevchenko, Y.I. Velikodsky, Properties of the lunar exosphere during the Perseid 2009 meteor shower. Planet. Space Sci. **96**, 90–98 (2014). https://doi.org/10.1016/j.pss.2014.03.008. arXiv:1404.2075

A.A. Berezhnoy, Y.I. Velikodsky, E. Zubko, M. Iten, R. Lena, S. Sposetti, A.A. Tereshchenko, S.I. Popel, E.A. Feoktistova, A.P. Golub', Detection of impact-produced dust clouds near the lunar terminator. Planet. Space Sci. **177**, 104689 (2019). https://doi.org/10.1016/j.pss.2019.07.004

O.E. Berg, F.F. Richardson, The Pioneer 8 cosmic dust experiment. Rev. Sci. Instrum. **40**, 1333–1337 (1969). https://doi.org/10.1063/1.1683778

T.A. Bida, R.M. Killen, Observations of the minor species al and fe in Mercury's exosphere. Icarus **289**, 227–238 (2017). https://doi.org/10.1016/j.icarus.2016.10.019. http://www.sciencedirect.com/science/article/pii/S0019103516306704

D.L. Bones, J.C. Gómez-Martín, C.J. Empson, J.D. Carrillo-Sánchez, A.D. James, T.P. Conroy, J.M.C. Plane, A novel instrument to measure differential ablation of meteorite samples and proxies: the Meteoric Ablation Simulator (MASI). Rev. Sci. Instrum. **094**, 504 (2016). https://doi.org/10.1063/1.4962751.

P. Borin, G. Cremonese, F. Marzari, M. Bruno, S. Marchi, Statistical analysis of micrometeoroids flux on Mercury. Astron. Astrophys. **503**(1), 259–264 (2009). https://doi.org/10.1051/0004-6361/200912080

P. Borin, G. Cremonese, F. Marzari, Statistical analysis of the flux of micrometeoroids at Mercury from both cometary and asteroidal components. Astron. Astrophys. **585**, A106 (2016a). https://doi.org/10.1051/0004-6361/201526767

P. Borin, G. Cremonese, F. Marzari, Statistical analysis of the flux of micrometeoroids at Mercury from both cometary and asteroidal components (Corrigendum). Astron. Astrophys. **588**, C3 (2016b). https://doi.org/10.1051/0004-6361/201526767e

P. Borin, G. Cremonese, F. Marzari, A. Lucchetti, Asteroidal and cometary dust flux in the inner solar system. Astron. Astrophys. **605**, A94 (2017). https://doi.org/10.1051/0004-6361/201730617

P. Brown, J. Jones, A determination of the strengths of the sporadic radio-meteor sources. Earth Moon Planets **68**, 223–245 (1995). https://doi.org/10.1007/BF00671512

P. Brown, R.J. Weryk, D.K. Wong, J. Jones, A meteoroid stream survey using the Canadian Meteor Orbit Radar. I. Methodology and radiant catalogue. Icarus **195**, 317–339 (2008). https://doi.org/10.1016/j.icarus.2007.12.002

P. Brown, D.K. Wong, R.J. Weryk, P. Wiegert, A meteoroid stream survey using the Canadian Meteor Orbit Radar. II: Identification of minor showers using a 3D wavelet transform. Icarus **207**, 66–81 (2010). https://doi.org/10.1016/j.icarus.2009.11.015

P. Brown, P. Wiegert, D. Clark, E. Tagliaferri, Orbital and physical characteristics of meter-scale impactors from airburst observations. Icarus **266**, 96 (2016)

J.S. Bruzzone, D. Janches, P. Jenniskens, R. Weryk, J.L. Hormaechea, A comparative study of radar and optical observations of meteor showers using SAAMER-OS and CAMS. Planet. Space Sci. **188**, 104936 (2020). https://doi.org/10.1016/j.pss.2020.104936

M.H. Burger, R.M. Killen, W.E. McClintock, A.W. Merkel, R.J. Vervack, T.A. Cassidy, M. Sarantos, Seasonal variations in Mercury's dayside calcium exosphere. Icarus **238**, 51–58 (2014). https://doi.org/10.1016/j.icarus.2014.04.049

S. Cambioni, R. Malhotra, The mid-plane of the Main Asteroid Belt. Astron. J. **155**(3), 143 (2018). https://doi.org/10.3847/1538-3881/aaab6b. arXiv:1801.08096

M.D. Campbell-Brown, High resolution radiant distribution and orbits of sporadic radar meteoroids. Icarus **196**, 144–163 (2008). https://doi.org/10.1016/j.icarus.2008.02.022

M. Campbell-Brown, A population of small refractory meteoroids in asteroidal orbits. Planet. Space Sci. **118**, 8–13 (2015). https://doi.org/10.1016/j.pss.2015.03.022

M. Campbell-Brown, P. Wiegert, Seasonal variations in the North toroidal sporadic meteor source. Meteorit. Planet. Sci. **44**, 1837–1848 (2009). https://doi.org/10.1111/j.1945-5100.2009.tb01992.x

J.D. Carrillo-Sánchez, J.M.C. Plane, W. Feng, D. Nesvorný, D. Janches, On the size and velocity distribution of cosmic dust particles entering the atmosphere. Geophys. Res. Lett. **42**, 6518–6525 (2015). https://doi.org/10.1002/2015GL065149

J.D. Carrillo-Sánchez, D. Nesvorný, P. Pokorný, D. Janches, J.M.C. Plane, Sources of cosmic dust in the Earth's atmosphere. Geophys. Res. Lett. (2016). https://doi.org/10.1002/2016GL071697

J.D. Carrillo-Sánchez, J.C. Gómez-Martín, D.L. Bones, D. Nesvorný, P. Pokorný, M. Benna, G.J. Flynn, J.M.C. Plane, Cosmic dust fluxes in the atmospheres of Earth, Mars, and Venus. Icarus **335**, 113395 (2020). https://doi.org/10.1016/j.icarus.2019.113395

Z. Ceplecha, Influx of interplanetary bodies onto Earth. Astron. Astrophys. **263**(1–2), 361–366 (1992)

Z. Ceplecha, J. Borovička, W. Elford, D. Revelle, R. Hawkes, V. Porubčan, M. Šimek, Meteor phenomena and bodies. Space Sci. Rev. **84**, 327–471 (1998)

E.L. Christiansen, Performance equations for advanced orbital debris shields, in *Space Programs and Technologies Conference* (1992), pp. 24–27

A.A. Christou, Predicting martian and venusian meteor shower activity. Earth Moon Planets **95**(1–4), 425–431 (2004). https://doi.org/10.1007/s11038-005-9023-0

A.A. Christou, R.M. Killen, M.H. Burger, The meteoroid stream of comet Encke at Mercury: implications for MErcury Surface, Space ENvironment, GEochemistry, and Ranging observations of the exosphere. Geophys. Res. Lett. **42**(18), 7311–7318 (2015). https://doi.org/10.1002/2015GL065361

M.J. Cintala, Impact-induced thermal effects in the lunar and Mercurian regoliths. J. Geophys. Res. **97**, 947–973 (1992). https://doi.org/10.1029/91JE02207

B.A. Cohen, J.R. Szalay, A.S. Rivkin, J.A. Richardson, R.L. Klima, C.M. Ernst, N.L. Chabot, Z. Sternovsky, M. HoráNyi, Using dust shed from asteroids as microsamples to link remote measurements with meteorite classes. Meteorit. Planet. Sci. **54**(9), 2046–2066 (2019). https://doi.org/10.1111/maps.13348. arXiv:1906.00876

A. Colaprete, K. Vargo, M. Shirley, D. Land is, D. Wooden, J. Karcz, B. Hermalyn, A. Cook, An overview of the LADEE Ultraviolet-Visible Spectrometer. Space Sci. Rev. **185**(1–4), 63–91 (2014). https://doi.org/10.1007/s11214-014-0112-0

A. Colaprete, M. Sarantos, D.H. Wooden, T.J. Stubbs, A.M. Cook, M. Shirley, How surface composition and meteoroid impacts mediate sodium and potassium in the lunar exosphere. Science **351**(6270), 249–252 (2016)

A. Collette, Z. Sternovsky, M. Horanyi, Production of neutral gas by micrometeoroid impacts. Icarus **227**, 89–93 (2014). https://doi.org/10.1016/j.icarus.2013.09.009

G. Cremonese, P. Borin, E. Martellato, F. Marzari, M. Bruno, New calibration of the micrometeoroid flux on Earth. Astrophys. J. Lett. **749**(2), L40 (2012). https://doi.org/10.1088/2041-8205/749/2/L40

G. Cremonese, P. Borin, A. Lucchetti, F. Marzari, M. Bruno, Micrometeoroids flux on the Moon. Astron. Astrophys. **551**, A27 (2013). https://doi.org/10.1051/0004-6361/201220541

G. Cremonese, F. Capaccioni, M.T. Capria, A. Doressoundiram, P. Palumbo, M. Vincendon, M. Massironi, S. Debei, M. Zusi, F. Altieri, M. Amoroso, G. Aroldi, M. Baroni, A. Barucci, G. Bellucci, J. Benkhoff, S. Besse, C. Bettanini, M. Blecka, D. Borrelli, J.R. Brucato, C. Carli, V. Carlier, P. Cerroni, A. Cicchetti, L. Colangeli, M. Dami, V. Da Deppo, M.C. De Sanctis S. Erard, F. Esposito, D. Fantinel, L. Ferranti, F. Ferri, I. FicaiÂ Veltroni, G. Filacchione, E. Flamini, G. Forlani, S. Fornasier, O. Forni, M. Fulchignoni, V. Galluzzi, K. Gwinner, W. Ip, L. Jorda, Y. Langevin, L. Lara, F. Leblanc, C. Leyrat, Y. Li, S. Marchi, L. Marinangeli, F. Marzari, E. MazzottaÂ Epifani, M. Mendillo, V. Mennella, R. Mugnuolo, K. Muinonen, G. Naletto, R. Noschese, E. Palomba, R. Paolinetti, D. Perna, G. Piccioni, R. Politi, F. Poulet, R. Ragazzoni, C. Re, M. Rossi, A. Rotundi, G. Salemi, M. Sgavetti, E. Simioni, N. Thomas, L. Tommasi, A. Turella, T. Van Hoolst, L. Wilson, F. Zambon, A. Aboudan, O. Barraud, N. Bott, P. Borin, G. Colombatti, M. ElÂ Yazidi, S. Ferrari, J. Flahaut, L. Giacomini, L. Guzzetta, A. Lucchetti, E. Martellato, M. Pajola, A. Slemer, G. Tognon, D. Turrini, SIMBIO-SYS: scientific cameras and spectrometer for the BepiColombo mission. Space Sci. Rev. **216**(5), 75 (2020). https://doi.org/10.1007/s11214-020-00704-8

M.M.J. Crismani, N.M. Schneider, J.M.C. Plane, J.S. Evans, S.K. Jain, M.S. Chaffin, J.D. Carrillo-Sanchez, J.I. Deighan, R.V. Yelle, A.I.F. Stewart, W. McClintock, J. Clarke, G.M. Holsclaw, A. Stiepen, F. Montmessin, B.M. Jakosky, Detection of a persistent meteoric metal layer in the Martian atmosphere. Nat. Geosci. **10**(6), 401–404 (2017). https://doi.org/10.1038/ngeo2958

M.M.J. Crismani, N.M. Schneider, J.S. Evans, J.M.C. Plane, J.D. Carrillo-Sánchez, S. Jain, J. Deighan, R. Yelle, The impact of comet Siding Spring's meteors on the Martian atmosphere and ionosphere. J. Geophys. Res., Planets **123**(10), 2613–2627 (2018). https://doi.org/10.1029/2018JE005750

C.T. D'Aiutolo, W.H. Kinard, R.J. Naumann, Recent NASA meteoroid penetration results from satellites. Smithson. Contrib. Astrophys. **11**, 239 (1967)

M. DeLuca, T. Munsat, E. Thomas, Z. Sternovsky, The ionization efficiency of aluminum and iron at meteoric velocities. Planet. Space Sci. **156**, 111–116 (2018). https://doi.org/10.1016/j.pss.2017.11.003

H. Dietzel, G. Eichhorn, H. Fechtig, E. Grun, H.J. Hoffmann, J. Kissel, The HEOS 2 and HELIOS micrometeoroid experiments. J. Phys. E, Sci. Instrum. **6**, 209–217 (1973)

V. Dikarev, E. Gruen, J. Baggaley, D. Galligan, M. Landgraf, R. Jehn, The new ESA meteoroid model, in *35th COSPAR Scientific Assembly*, vol. 35, ed. by J.P. Paillé (2004), p. 575

F.K. Duennebier, Y. Nakamura, G.V. Latham, H.J. Dorman, Meteoroid storms detected on the Moon. Science **192**(4243), 1000–1002 (1976). https://doi.org/10.1126/science.192.4243.1000

G. Eichhorn, Measurements of the light flash produced by high velocity particle impact. Planet. Space Sci. **23**, 1519–1525 (1975). https://doi.org/10.1016/0032-0633(75)90005-7

R.C. Elphic, G.T. Delory, B.P. Hine, P.R. Mahaffy, M. Horanyi, A. Colaprete, M. Benna, S.K. Noble, The lunar atmosphere and dust environment explorer mission. Space Sci. Rev. **185**, 3–25 (2014). https://doi.org/10.1007/s11214-014-0113-z

J.E. Erickson, Analysis of the meteoroid flux measured by explorer 16 and lunar orbiter. Astron. J. **74**, 279 (1969). https://doi.org/10.1086/110807

J.T. Fentzke, D. Janches, A semi-empirical model of the contribution from sporadic meteoroid sources on the meteor input function observed at arecibo. J. Geophys. Res. Space Phys. **113**, A03304 (2008). https://doi.org/10.1029/2007JA012531

J.T. Fentzke, D. Janches, J.J. Sparks, Latitudinal and seasonal variability of the micrometeor input function: a study using model predictions, Arecibo, and PFISR observations. J. Atmos. Sol.-Terr. Phys. **71**, 653 (2009)

G.J. Flynn, P. Bleuet, J. Borg, J.P. Bradley, F.E. Brenker, S. Brennan, J. Bridges, D.E. Brownlee, E.S. Bullock, M. Burghammer, B.C. Clark, Z.R. Dai, C.P. Daghlian, Z. Djouadi, S. Fakra, T. Ferroir, C. Floss, I.A. Franchi, Z. Gainsforth, J.P. Gallien, P. Gillet, P.G. Grant, G.A. Graham, S.F. Green, F. Grossemy, P.R. Heck, G.F. Herzog, P. Hoppe, F. Hörz, J. Huth, K. Ignatyev, H.A. Ishii, K. Janssens, D. Joswiak, A.T. Kearsley, H. Khodja, A. Lanzirotti, J. Leitner, L. Lemelle, H. Leroux, K. Luening, G.J. MacPherson, K.K. Marhas, M.A. Marcus, G. Matrajt, T. Nakamura, K. Nakamura-Messenger, T. Nakano, M. Newville, D.A. Papanastassiou, P. Pianetta, W. Rao, C. Riekel, F.J.M. Rietmeijer, D. Rost, C.S. Schwandt, T.H. See, J. Sheffield-Parker, A. Simionovici, I. Sitnitsky, C.J. Snead, F.J. Stadermann, T. Stephan, R.M. Stroud, J. Susini, Y. Suzuki, S.R. Sutton, S. Taylor, N. Teslich, D. Troadec, P. Tsou, A. Tsuchiyama, K. Uesugi, B. Vekemans, E.P. Vicenzi, L. Vincze, J. Westphal Astron, P. Wozniakiewicz, E. Zinner, M.E. Zolensky, Elemental Compositions of Comet 81P/Wild 2 Samples Collected by Stardust. Science **314**, 1731 (2006). https://doi.org/10.1126/science.1136141

R. Funase, S. Ikari, K. Miyoshi, Y. Kawabata, S. Nakajima, S. Nomura, N. Funabiki, A. Ishikawa, K. Kakihara, S. Matsushita, R. Takahashi, K. Yanagida, D. Mori, Y. Murata, T. Shibukawa, R. Suzumoto, M. Fujiwara, K. Tomita, H. Aohama, K. Iiyama, S. Ishiwata, H. Kondo, W. Mikuriya, H. Seki, H. Koizumi, J. Asakawa, K. Nishii, A. Hattori, Y. Saito, K. Kikuchi, Y. Kobayashi, A. Tomiki, W. Torii, T. Ito, S. Campagnola, N. Ozaki, N. Baresi, I. Yoshikawa, K. Yoshioka, M. Kuwabara, R. Hikida, S. Arao, S. Abe, M. Yanagisawa, R. Fuse, Y. Masuda, H. Yano, T. Hirai, K. Arai, R. Jitsukawa, E. Ishioka, H. Nakano, T. Ikenaga, T. Hashimoto, Mission to Earth–Moon Lagrange point by a 6U CubeSat: EQUULEUS. IEEE Aerosp. Electron. Syst. Mag. **35**(3), 30–44 (2020). https://doi.org/10.1109/MAES.2019.2955577. https://ieeexplore.ieee.org/document/9076200?source=authoralert

C.S. Gardner, A.Z. Liu, D.R. Marsh, W. Feng, J.M.C. Plane, Inferring the global cosmic dust influx to the Earth's atmosphere from lidar observations of the vertical flux of mesospheric Na. J. Geophys. Res. Space Phys. **119**(9), 7870–7879 (2014). https://doi.org/10.1002/2014JA020383

J.D. Giorgini, D.K. Yeomans, A.B. Chamberlin, P.W. Chodas, R.A. Jacobson, M.S. Keesey, J.H. Lieske, S.J. Ostro, E.M. Standish, R.N. Wimberly, JPL's on-line solar system data service. Bull. Am. Astron. Soc. **28**, 1158 (1996)

J. Gómez Martín, D. Bones, J. Carrillo-Sánchez, J. Trigo-Rodriguez, B. Fegley, J. Plane, Novel experimental simulations of the atmospheric injection of meteoric metals. Astrophys. J. **836**, 212 (2017)

J. Grebowsky, J. Moser, W. Pesnell, *Meteoric Material – An Important Component of Planetary Atmosphere* (Am. Geophys. Union, Washington, 2002), pp. 235–244

J.M. Grebowsky, M. Benna, J.M.C. Plane, G.A. Collinson, P.R. Mahaffy, B.M. Jakosky, Unique, non-Earthlike, meteoritic ion behavior in upper atmosphere of Mars. Geophys. Res. Lett. **44**(7), 3066–3072 (2017). https://doi.org/10.1002/2017GL072635

E. Grün, H. Zook, H. Fechtig, R.H. Giese, Collisional balance of the meteoric complex. Icarus **62**, 244–272 (1985)

E. Grün, H. Fechtig, M.S. Hanner, J. Kissel, B.A. Lindblad, D. Linkert, D. Maas, G.E. Morfill, H.A. Zook, The Galileo dust detector. Space Sci. Rev. **60**, 317–340 (1992)

E. Grün, M. Baguhl, H. Svedhem, H.A. Zook, In situ measurements of cosmic dust, in *Astrophys. Space Sci. Library* (Springer, Berlin, 2001), p. 295

E. Grün, M. Horanyi, Z. Sternovsky, The lunar dust environment. Planet. Space Sci. **59**(14), 1672–1680 (2011). https://doi.org/10.1016/j.pss.2011.04.005

M. Hajduková, L. Neslusan, The χ-andromedids and January α-ursae majorids: a new and a probable shower associated with Comet C/1992 w1 (Ohshita). Icarus, 113960 (2020). https://doi.org/10.1016/j.icarus.2020.113960. http://www.sciencedirect.com/science/article/pii/S0019103520303316

E.C. Hastings, The explorer XVI micrometeoroid satellite description and preliminary results for the period December 16, 1962 through January 13 1963. Tech. Rep., NASA (1963). https://ntrs.nasa.gov/search.jsp?R=19630002763, tM X-810

G.S. Hawkins, Variation in the occurrence rate of meteors. Astron. J. **61**, 386 (1956)

T. Hirai, H. Yano, M. Fujii, S. Hasegawa, N. Moriyama, C. Okamoto, M. Tanaka, Data screening and reduction in interplanetary dust measurement by ikaros-aladdin. Adv. Space Res. **59**(6), 1450–1459 (2017). https://doi.org/10.1016/j.asr.2016.12.023. http://www.sciencedirect.com/science/article/pii/S027311771630727X

M. Horányi, V. Hoxie, D. James, A. Poppe, C. Bryant, B. Grogan, B. Lamprecht, J. Mack, F. Bagenal, S. Batiste, N. Bunch, T. Chanthawanich, F. Christensen, M. Colgan, T. Dunn, G. Drake, A. Fernandez, T. Finley, G. Holland, A. Jenkins, C. Krauss, E. Krauss, O. Krauss, M. Lankton, C. Mitchell, M. Neeland, T. Reese, K. Rash, G. Tate, C. Vaudrin, J. Westfall, The Student Dust Counter on the New Horizons Mission. Space Sci. Rev. **140**, 387–402 (2008). https://doi.org/10.1007/s11214-007-9250-y

M. Horányi, Z. Sternovsky, M. Lankton, C. Dumont, S. Gagnard, D. Gathright, E. Grün, D. Hansen, D. James, S. Kempf, B. Lamprecht, R. Srama, J.R. Szalay, G. Wright, The Lunar Dust Experiment (LDEX) onboard the Lunar Atmosphere and Dust Environment Explorer (LADEE) mission. Space Sci. Rev. **185**, 93–113 (2014). https://doi.org/10.1007/s11214-014-0118-7

M. Horányi, J.R. Szalay, S. Kempf, J. Schmidt, E. Grün, R. Srama, Z. Sternovsky, A permanent, asymmetric dust cloud around the Moon. Nature **522**, 324–326 (2015). https://doi.org/10.1038/nature14479

M. Horányi, E. Bernardoni, S. Kempf, Z. Sternovsky, J. Szalay, Exploration of resources in lunar polar regions, in *Lunar and Planetary Science Conference* (2020), p. 1465

D.H. Humes, Results of Pioneer 10 and 11 meteoroid experiments - Interplanetary and near-Saturn. J. Geophys. Res. **85**, 5841–5852 (1980). https://doi.org/10.1029/JA085iA11p05841

E. Igenbergs, A. Hüdepohl, K. Uesugi, T. Hayashi, H. Svedhem, H. Iglseder, G. Koller, A. Glasmachers, E. Grün, G. Schwehm, H. Mizutani, T. Yamamoto, A. Fujimura, N. Ishii, H. Araki, K. Yamakoshi, K. Nogami, The Munich Dust Counter — a cosmic dust experiment on board of the MUSES-A mission of Japan (1991), Pp 45–48. https://link.springer.com/chapter/10.1007/978-94-011-3640-2_9

D. Janches, C.J. Heinselman, J.L. Chau, A. Chandran, R. Woodman, Modeling the global micrometeor input function in the upper atmosphere observed by high power and large aperture radars. J. Geophys. Res. Space Phys. **111**, A07317 (2006). https://doi.org/10.1029/2006JA011628

D. Janches, W. Hocking, S. Pifko, J.L. Hormaechea, D.C. Fritts, C. Brunini, R. Michell, M. Samara, Interferometric meteor head echo observations using the Southern Argentina Agile Meteor Radar. J. Geophys. Res. Space Phys. **119**, 2269–2287 (2014). https://doi.org/10.1002/2013JA019241

D. Janches, S. Close, J.L. Hormaechea, N. Swarnalingam, A. Murphy, D. O'Connor, B. Vandepeer, B. Fuller, D.C. Fritts, C. Brunini, The Southern Argentina Agile MEteor Radar Orbital System (SAAMER-OS): an initial sporadic meteoroid orbital survey in the southern sky. Astrophys. J. **809**, 36 (2015). https://doi.org/10.1088/0004-637X/809/1/36

D. Janches, N. Swarnalingam, J.D. Carrillo-Sanchez, J.C. Gomez-Martin, R. Marshall, D. Nesvorný, J.M.C. Plane, W. Feng, P. Pokorný, Radar Detectability Studies of Slow and Small Zodiacal Dust Cloud Particles: III. The Role of Sodium and the Head Echo Size on the Probability of Detection. Astrophys. J. **843**, 1 (2017). https://doi.org/10.3847/1538-4357/aa775c

D. Janches, P. Pokorný, M. Sarantos, J.R. Szalay, M. Horányi, D. Nesvorný, Constraining the ratio of micrometeoroids from short- and long-period comets at 1 AU from LADEE observations of the lunar dust cloud. Geophys. Res. Lett. **45**, 1713–1722 (2018). https://doi.org/10.1002/2017GL076065

D. Janches, C. Brunini, J.L. Hormaechea, A decade of sporadic meteoroid mass distribution indices in the southern hemisphere derived from SAAMER's meteor observations. Astron. J. **157**, 240 (2019). https://doi.org/10.3847/1538-3881/ab1b0f

D. Janches, J.S. Bruzzone, P. Pokorný, J.D. Carrillo-Sanchez, M. Sarantos, A comparative study of the seasonal, temporal, and spatial distribution of meteoroids in the upper atmosphere of Venus, Earth and Mars. Planet. Space Sci. **1**, 59 (2020)

D. Janches, J.S. Bruzzone, R.J. Weryk, J.L. Hormaechea, P. Wiegert, C. Brunini, Observations of an unexpected meteor shower outburst at high ecliptic southern latitude and its potential origin. Astrophys. J. Lett. **895**(1), L25 (2020b). https://doi.org/10.3847/2041-8213/ab9181

J.M. Jasinski, L.H. Regoli L.H. Cassidy, et al., A transient enhancement of Mercury's exosphere at extremely high altitudes inferred from pickup ions. Nat. Commun. **11**, 4350 (2020). https://doi.org/10.1038/s41467-020-18220-2

P. Jenniskens, *Meteor Showers and Their Parent Comets* (2006)

P. Jenniskens, J. Baggaley, I. Crumpton, P. Aldous, P.S. Gural, D. Samuels, J. Albers, R. Soja, A surprise southern hemisphere meteor shower on New-Year's Eve 2015: the Volantids (IAU#758, VOL). J. Int. Meteor Organ. **44**, 35–41 (2016)

C.L. Johnson, S.A. Hauck II, A whole new Mercury: messenger reveals a dynamic planet at the last frontier of the inner solar system. J. Geophys. Res., Planets **121**(11), 2349–2362 (2016). https://doi.org/10.1002/2016JE005150. https://agupubs.onlinelibrary.wiley.com/doi/pdf/10.1002/2016JE005150

J. Jones, P. Brown, Sporadic meteor radiant distribution: orbital survey results. Mon. Not. R. Astron. Soc. **265**, 524–532 (1993)

M.S.P. Kelley, T.L. Farnham, D. Bodewits, P. Tricarico, D. Farnocchia, A study of dust and gas at Mars from Comet C/2013 A1 (Siding Spring). Astrophys. J. Lett. **792**(1), L16 (2014). https://doi.org/10.1088/2041-8205/792/1/L16. arXiv:1408.2792

J. Kero, M.D. Campbell-Brown, G. Stober, J.L. Chau, J.D. Mathews, A. Pellinen-Wannberg, *Radar Observations of Meteors*. (2019), p. 65

D. Kessler, Average relative velocity of sporadic meteoroids in interplanetary space. AIAA J. **7**(12), 2337–2338 (1969). https://doi.org/10.2514/3.5539

D.J. Kessler, Derivation of the collision probability between orbiting objects the lifetimes of Jupiter's outer moons. Icarus **48**, 39–48 (1981). https://doi.org/10.1016/0019-1035(81)90151-2

R.M. Killen, J.M. Hahn, Impact vaporization as a possible source of Mercury's calcium exosphere. Icarus **250**, 230–237 (2015). https://doi.org/10.1016/j.icarus.2014.11.035

E.D. Kondrateva, E.A. Reznikov, Comet Tempel-Tuttle and the Leonid meteor swarm. Astron. Vestn. **19**, 144–151 (1985)

D. Koschny, E. Grün, Impacts into ice-silicate mixtures: ejecta mass and size distributions. Icarus **154**, 402–411 (2001). https://doi.org/10.1006/icar.2001.6708

P. Koten, J. Rendtel, L. Shrbený, P. Gural, J. Borovička, P. Kozak, Meteors and meteor showers as observed by optical techniques, in *Meteoroids: Sources of Meteors on Earth and Beyond*, ed. by G.O. Ryabova, D.J. Asher, M.D. Campbell-Brown (Cambridge University Press, Cambridge, 2019), pp. 90–115. ISBN 9781108426718

H. Krüger, A.V. Krivov, M. Sremčević, E. Grün, Impact-generated dust clouds surrounding the Galilean moons. Icarus **164**(1), 170–187 (2003). https://doi.org/10.1016/S0019-1035(03)00127-1

H. Krüger, P. Strub, R. Srama, M. Kobayashi, T. Arai, H. Kimura, T. Hirai, G. Moragas-Klostermeyer, N. Altobelli, V.J. Sterken, J. Agarwal, M. Sommer, E. Grün, Modelling DESTINY$^+$ interplanetary and interstellar dust measurements en route to the active asteroid (3200) Phaethon. Planet. Space Sci. **172**, 22–42 (2019). https://doi.org/10.1016/j.pss.2019.04.005. arXiv:1904.07384

C. Leinert, M. Hanner, I. Richter, E. Pitz, The plane of symmetry of interplanetary dust in the inner solar system. Astron. Astrophys. **82**, 328–336 (1980)

H.F. Levison, M.J. Duncan, From the Kuiper belt to Jupiter-family comets: the spatial distribution of ecliptic comets. Icarus **127**, 13–32 (1997). https://doi.org/10.1006/icar.1996.5637

H.F. Levison, M.J. Duncan, L. Dones, B.J. Gladman, The scattered disk as a source of Halley-type comets. Icarus **184**, 619–633 (2006). https://doi.org/10.1016/j.icarus.2006.05.008

A. Liakos, A.Z. Bonanos, E.M. Xilouris, D. Koschny, I. Bellas-Velidis, P. Boumis, V. Charmandaris, A. Dapergolas, A. Fytsilis, A. Maroussis, R. Moissl, NELIOTA: methods, statistics, and results for meteoroids impacting the Moon. Astron. Astrophys. **633**, A112 (2020). https://doi.org/10.1051/0004-6361/201936709. arXiv:1911.06101

S.G. Love, J.H. Allton, Micrometeoroid impact crater statistics at the boundary of Earth's gravitational sphere of influence. Icarus **184**, 302–307 (2006). https://doi.org/10.1016/j.icarus.2006.05.023

S. Love, D.E. Brownlee, A direct measurement of the terrestrial mass accretion rate of cosmic dust. Science **262**, 550–553 (1993)

E.J. Lyytinen, T. Van Flandern, Predicting the strength of Leonid outbursts. Earth Moon Planets **82**, 149–166 (2000)

J.M. Madiedo, J.L. Ortiz, M. Yanagisawa, J. Aceituno, F. Aceituno, Impact flashes of meteoroids on the Moon, in *Meteoroids: Sources of Meteors on Earth and Beyond*, ed. by G.O. Ryabova, D.J. Asher, M.D. Campbell-Brown (Cambridge University Press, Cambridge, 2019), pp. 136–158. ISBN 9781108426718

P.R. Mahaffy, R. Richard Hodges, M. Benna, T. King, R. Arvey, M. Barciniak, M. Bendt, D. Carigan, T. Errigo, D.N. Harpold, V. Holmes, C.S. Johnson, J. Kellogg, P. Kimvilakani, M. Lefavor, J. Hengemihle, F. Jaeger, E. Lyness, J. Maurer, D. Nguyen, T.J. Nolan, F. Noreiga, M. Noriega, K. Patel, B. Prats, O. Quinones, E. Raaen, F. Tan, E. Weidner, M. Woronowicz, C. Gundersen, S. Battel, B.P. Block, K. Arnett, R. Miller, C. Cooper, C. Edmonson, The Neutral Mass Spectrometer on the Lunar Atmosphere and Dust Environment Explorer Mission. Space Sci. Rev. **185**(1–4), 27–61 (2014). https://doi.org/10.1007/s11214-014-0043-9

V. Mangano, A. Milillo, A. Mura, S. Orsini, E. De Angelis, P. Di Lellis, A.M. Wurz, The contribution of impulsive meteoritic impact vapourization to the Hermean exosphere. Planet. Space Sci. **55**(11), 1541–1556 (2007). https://doi.org/10.1016/j.pss.2006.10.008

I. Mann, L. Nouzák, J. Vaverka, T. Antonsen, Å. Fredriksen, K. Issautier, D. Malaspina, N. Meyer-Vernet, J. Pavlů, Z. Sternovsky, J. Stude, S. Ye, A. Zaslavsky, Dust observations with antenna measurements and its prospects for observations with parker solar probe and solar orbiter. Ann. Geophys. **37**(6), 1121–1140 (2019). https://doi.org/10.5194/angeo-37-1121-2019. https://www.ann-geophys.net/37/1121/2019/

S. Marchi, A. Morbidelli, G. Cremonese, Flux of meteoroid impacts on Mercury. Astron. Astrophys. **431**(3), 1123–1127 (2005). https://doi.org/10.1051/0004-6361:20041800

D. Marsh, D. Janches, W. Feng, J. Plane, A global model of meteoric sodium. J. Geophys. Res., Atmos. **118**, 11,442–11,452 (2013). https://doi.org/10.1002/jgrd.50870

J.D. Mathews, D. Janches, D. Meisel, Q. Zhou, The micrometeoroid mass flux into the upper atmosphere: Arecibo results and a comparison with prior estimates. Geophys. Res. Lett. **28**(10), 1929–1932 (2001)

W.E. McClintock, R.J. Vervack, E.T. Bradley, R.M. Killen, N. Mouawad, A.L. Sprague, M.H. Burger, S.C. Solomon, N.R. Izenberg, Messenger observations of Mercury's exosphere: detection of magnesium and distribution of constituents. Science **324**(5927), 610–613 (2009). https://doi.org/10.1126/science.1172525. https://science.sciencemag.org/content/324/5927/610.full.pdf

D.J. McComas, E.R. Christian, N.A. Schwadron, N. Fox, J. Westlake, F. Allegrini, D.N. Baker, D. Biesecker, M. Bzowski, G. Clark, C.M.S. Cohen, I. Cohen, M.A. Dayeh, R. Decker, G.A. de Nolfo, M.I. Desai, R.W. Ebert, H.A. Elliott, H. Fahr, P.C. Frisch, H.O. Funsten, S.A. Fuselier, A. Galli, A.B. Galvin, J. Giacalone, M. Gkioulidou, F. Guo, M. Horanyi, P. Isenberg, P. Janzen, L.M. Kistler, K. Korreck, M.A. Kubiak, H. Kucharek, B.A. Larsen, R.A. Leske, N. Lugaz, J. Luhmann, W. Matthaeus, D. Mitchell, E. Moebius, K. Ogasawara, D.B. Reisenfeld, J.D. Richardson, C.T. Russell, J.M. Sokół, H.E. Spence, R. Skoug, Z. Sternovsky, P. Swaczyna, J.R. Szalay, M. Tokumaru, M.E. Wiedenbeck, P. Wurz, G.P. Zank, E.J. Zirnstein, Interstellar Mapping and Acceleration Probe (IMAP): a new NASA mission. Space Sci. Rev. **214**(8), 116 (2018). https://doi.org/10.1007/s11214-018-0550-1

R.H. McNaught, D.J. Asher, Leonid dust trails and meteor storms. J. Int. Meteor Organ. **27**(2), 85–102 (1999)

A.W. Merkel, T.A. Cassidy, R.J. Vervack, W.E. McClintock, M. Sarantos, M.H. Burger, R.M. Killen, Seasonal variations of Mercury's magnesium dayside exosphere from MESSENGER observations. Icarus **281**, 46–54 (2017). https://doi.org/10.1016/j.icarus.2016.08.032

J. Miao, J.P.W. Stark, Direct simulation of meteoroids and space debris flux on LDEF spacecraft surfaces. Planet. Space Sci. **49**, 927–935 (2001). https://doi.org/10.1016/S0032-0633(01)00042-3

A.V. Moorhead, A. Kingery, S. Ehlert, NASA's Meteoroid Engineering Model 3 and its ability to replicate spacecraft impact rates. J. Spacecr. Rockets **57**, 160–176 (2020). https://doi.org/10.2514/1.A34561. arXiv:1909.05947

A. Morbidelli, B. Gladman, Orbital and temporal distributions of meteorites originating in the asteroid belt. Meteorit. Planet. Sci. **33**(5), 999–1016 (1998). https://doi.org/10.1111/j.1945-5100.1998.tb01707.x

D. Nesvorný, D. Vokrouhlický, W.F. Bottke, M. Sykes, Physical properties of asteroid dust bands and their sources. Icarus **181**, 107–144 (2006). https://doi.org/10.1016/j.icarus.2005.10.022

D. Nesvorný, P. Jenniskens, H.F. Levison, W.F. Bottke, D. Vokrouhlický, M. Gounelle, Cometary origin of the zodiacal cloud and carbonaceous micrometeorites. Implications for hot debris disks. Astrophys. J. **713**, 816–836 (2010). https://doi.org/10.1088/0004-637X/713/2/816. arXiv:0909.4322

D. Nesvorný, D. Janches, D. Vokrouhlický, P. Pokorný, W.F. Bottke, P. Jenniskens, Dynamical model for the zodiacal cloud and sporadic meteors. Astrophys. J. **743**, 129 (2011a). https://doi.org/10.1088/0004-637X/743/2/129. arXiv:1109.2983

D. Nesvorný, D. Vokrouhlický, P. Pokorný, D. Janches, Dynamics of dust particles released from Oort Cloud Comets and their contribution to radar meteors. Astrophys. J. **743**, 37 (2011b). https://doi.org/10.1088/0004-637X/743/1/37. arXiv:1109.2981

K. Nogami, M. Fujii, H. Ohashi, T. Miyachi, S. Sasaki, S. Hasegawa, H. Yano, H. Shibata, T. Iwai, S. Minami, S. Takechi, E. Grün, R. Srama, Development of the Mercury Dust Monitor (MDM) onboard the BepiColombo mission. Planet. Space Sci. **58**(1–2), 108–115 (2010). https://doi.org/10.1016/j.pss.2008.08.016

J. Oberst, Y. Nakamura, A search for clustering among the meteoroid impacts detected by the Apollo lunar seismic network. Icarus **91**(2), 315–325 (1991). https://doi.org/10.1016/0019-1035(91)90027-Q

R. O'Neal, L.R. Center, U.S.N. Aeronautics, S. Administration, The Explorer XXIII micrometeoroid satellite description and preliminary results for the period November 6, 1964 through February 15 1965. NASA technical memorandum, National Aeronautics and Space Administration (1965). https://books.google.co.jp/books?id=aVOF141pwvIC

E. Pierazzo, N. Artemieva, E. Asphaug, E.C. Baldwin, J. Cazamias, R. Coker, G.S. Collins, D.A. Crawford, T. Davison, D. Elbeshausen, K.A. Holsapple, K.R. Housen, D.G. Korycansky, K. Wünnemann, Validation of numerical codes for impact and explosion cratering: impacts on strengthless and metal targets. Meteorit. Planet. Sci. **43**(12), 1917–1938 (2008). https://doi.org/10.1111/j.1945-5100.2008.tb00653.x

C.M. Pieters, S.K. Noble, Space weathering on airless bodies. J. Geophys. Res., Planets **121**, 1865–1884 (2016). https://doi.org/10.1002/2016JE005128

S. Pifko, D. Janches, S. Close, J. Sparks, T. Nakamura, D. Nesvorny, The meteoroid input function and predictions of mid-latitude meteor observations by the MU radar. Icarus **223**, 444–459 (2013). https://doi.org/10.1016/j.icarus.2012.12.014

C. Plainaki, A. Mura, A. Milillo, S. Orsini, S. Livi, V. Mangano, S. Massetti, R. Rispoli, E. De Angelis, Investigation of the possible effects of comet Encke's meteoroid stream on the Ca exosphere of Mercury. J. Geophys. Res., Planets **122**(6), 1217–1226 (2017). https://doi.org/10.1002/2017JE005304

J. Plane, Atmospheric chemistry of meteoric metals. Chem. Rev. **103**(12), 4963–4984 (2003)

J.M.C. Plane, Cosmic dust in the Earth's atmosphere. Chem. Soc. Rev. **41**, 6507–6518, 2012 41:6507–6518 (2012). https://doi.org/10.1039/c2cs35132c

J. Plane, W. Feng, E. Dawkins, The mesosphere and metals: chemistry and changes. Chem. Rev. **115**, 4497–4541 (2015). https://doi.org/10.1021/cr500501m

J.M.C. Plane, G.J. Flynn, A. Määttänen, J.E. Moores, A.R. Poppe, J.D. Carrillo-Sánchez, C. Listowski, Impacts of cosmic dust on planetary atmospheres and surfaces. Space Sci. Rev. **214**(1), 23 (2018). https://doi.org/10.1007/s11214-017-0458-1

P. Pokorný, P.G. Brown, A reproducible method to determine the meteoroid mass index. Astron. Astrophys. **592**, A150 (2016). https://doi.org/10.1051/0004-6361/201628134. arXiv:1605.04437

P. Pokorný, D. Vokrouhlický, D. Nesvorný, M. Campbell-Brown, P. Brown, Dynamical model for the toroidal sporadic meteors. Astrophys. J. **789**, 25 (2014). https://doi.org/10.1088/0004-637X/789/1/25

P. Pokorný, D. Janches, P.G. Brown, J.L. Hormaechea, An orbital meteoroid stream survey using the Southern Argentina Agile MEteor Radar (SAAMER) based on a wavelet approach. Icarus **290**, 162–182 (2017a). https://doi.org/10.1016/j.icarus.2017.02.025

P. Pokorný, M. Sarantos, D. Janches, Reconciling the dawn–dusk asymmetry in Mercury's exosphere with the micrometeoroid impact directionality. Astrophys. J. Lett. **842**, L17 (2017b). https://doi.org/10.3847/2041-8213/aa775d. arXiv:1706.01461

P. Pokorný, M. Sarantos, D. Janches, A comprehensive model of the meteoroid environment around Mercury. Astrophys. J. **863**, 31 (2018). https://doi.org/10.3847/1538-4357/aad051. arXiv:1807.02749

P. Pokorný, D. Janches, M. Sarantos, J.R. Szalay, M. Horányi, D. Nesvorný, M.J. Kuchner, Meteoroids at the Moon: orbital properties, surface vaporization, and impact ejecta production. J. Geophys. Res., Planets **124**, 752–778 (2019). https://doi.org/10.1029/2018JE005912

P. Pokorný, M. Sarantos, D. Janches, E. Mazarico, Meteoroid bombardment of lunar poles. Astrophys. J. **894**(2), 114 (2020). https://doi.org/10.3847/1538-4357/ab83ee. arXiv:2003.12640

A.R. Poppe, An improved model for interplanetary dust fluxes in the outer Solar System. Icarus **264**, 369–386 (2016). https://doi.org/10.1016/j.icarus.2015.10.001

A. Poppe, D. James, M. Horányi, Measurements of the terrestrial dust influx variability by the Cosmic Dust Experiment. Planet. Space Sci. **59**, 319–326 (2011). https://doi.org/10.1016/j.pss.2010.12.002

V. Porubčan, L. Kornoš, I.P. Williams, The Taurid complex meteor showers and asteroids. Contrib. Astron. Obs. Skaln. Pleso **36**(2), 103–117 (2006). arXiv:0905.1639

F. Postberg, E. Grün, M. Horanyi, S. Kempf, H. Krüger, J. Schmidt, F. Spahn, R. Srama, Z. Sternovsky, M. Trieloff, Compositional mapping of planetary moons by mass spectrometry of dust ejecta. Planet. Space Sci. **59**(14), 1815–1825 (2011). https://doi.org/10.1016/j.pss.2011.05.001

E. Quémerais, J.Y. Chaufray, D. Koutroumpa, F. Leblanc, A. Reberac, B. Lustrement, C. Montaron, J.F. Mariscal, N. Rouanet, I. Yoshikawa, G. Murakami, K. Yoshioka, O. Korablev, D. Belyaev, M.G. Pelizzo, A. Corso, P. Zuppella, PHEBUS on Bepi-Colombo: post-launch update and instrument performance. Space Sci. Rev. **216**(4), 67 (2020). https://doi.org/10.1007/s11214-020-00695-6

J.W. Rhee, O.E. Berg, F.F. Richardson, Heliocentric distribution of cosmic dust intercepted by Pioneer 8 and 9. Geophys. Res. Lett. **1**(8), 345–346 (1974). https://doi.org/10.1029/GL001i008p00345

M. Rowan-Robinson, B. May, An improved model for the infrared emission from the zodiacal dust cloud: cometary, asteroidal and interstellar dust. Mon. Not. R. Astron. Soc. **429**(4), 2894–2902 (2013). https://doi.org/10.1093/mnras/sts471. arXiv:1212.4759

J.M. Russell, S.M. Bailey, L.L. Gordley, D.W. Rusch, M. Horányi, M.E. Hervig, G.E. Thomas, C.E. Randall, D.E. Siskind, M.H. Stevens, M.E. Summers, M.J. Taylor, C.R. Englert, P.J. Espy, W.E. McClintock, A.W. Merkel, The Aeronomy of Ice in the Mesosphere (AIM) mission: overview and early science results. J. Atmos. Sol.-Terr. Phys. **71**, 289–299 (2009). https://doi.org/10.1016/j.jastp.2008.08.011

M. Sarantos, R.M. Killen, A. Surjalal Sharma, J.A. Slavin, Sources of sodium in the lunar exosphere: modeling using ground-based observations of sodium emission and spacecraft data of the plasma. Icarus **205**, 364–374 (2010). https://doi.org/10.1016/j.icarus.2009.07.039

M. Sarantos, R.M. Killen, W.E. McClintock, E.T. Bradley, R.J. Vervack, M. Benna, J.A. Slavin, Limits to Mercury's magnesium exosphere from MESSENGER second flyby observations. Planet. Space Sci. **59**(15), 1992–2003 (2011)

M. Sarantos, R.E. Hartle, R.M. Killen, Y. Saito, J.A. Slavin, A. Glocer, Flux estimates of ions from the lunar exosphere. Geophys. Res. Lett. **39**, L13101 (2012a). https://doi.org/10.1029/2012GL052001

M. Sarantos, R.M. Killen, D.A. Glenar, M. Benna, T.J. Stubbs, Metallic species, oxygen and silicon in the lunar exosphere: upper limits and prospects for LADEE measurements. J. Geophys. Res. **117**, A03103 (2012b)

B.V. Sarli, M. Horikawa, C.H. Yam, Y. Kawakatsu, T. Yamamoto, DESTINY$^+$ trajectory design to (3200) Phaethon. J. Astronaut. Sci. **65**(1), 82–110 (2018)

S. Sasaki, E. Igenbergs, H. Ohashi, R. Senger, R. Münzenmayer, W. Naumann, E. Grün, K. Nogami, I. Mann, H. Svedhem, Summary of interplanetary and interstellar dust observation by Mars Dust Counter on board NOZOMI. Adv. Space Res. **39**(3), 485–488 (2007). https://doi.org/10.1016/j.asr.2006.11.006

N.M. Schneider, J.I. Deighan, A.I.F. Stewart, W.E. McClintock, S.K. Jain, M.S. Chaffin, A. Stiepen, M. Crismani, J.M.C. Plane, J.D. Carrillo-Sánchez, J.S. Evans, M.H. Stevens, R.V. Yelle, J.T. Clarke, G.M. Holsclaw, F. Montmessin, B.M. Jakosky, MAVEN IUVS observations of the aftermath of the Comet Siding Spring meteor shower on Mars. Geophys. Res. Lett. **42**(12), 4755–4761 (2015). https://doi.org/10.1002/2015GL063863

C. Schult, G. Stober, D. Janches, J.L. Chau, Results of the first continuous meteor head echo survey at polar latitudes. Icarus **297**, 1–13 (2017). https://doi.org/10.1016/j.icarus.2017.06.019

Z. Sekanina, Statistical model of meteor streams. IV—a study of radio streams from the synoptic year. Icarus **27**, 265–321 (1976)

F. Selsis, J. Brillet, M. Rapaport, Meteor showers of cometary origin in the Solar System: revised predictions. Astron. Astrophys. **416**, 783–789 (2004). https://doi.org/10.1051/0004-6361:20031724

S.M. Smith, J.K. Wilson, J. Baumgardner, M. Mendillo, Discovery of the distant lunar sodium tail and its enhancement following the Leonid Meteor Shower of 1998. Geophys. Res. Lett. **26**(12), 1649–1652 (1999). https://doi.org/10.1029/1999GL900314

R. Srama, T.J. Ahrens, N. Altobelli, S. Auer, J.G. Bradley, M. Burton, V.V. Dikarev, T. Economou, H. Fechtig, M. Görlich, M. Grande, A. Graps, E. Grün, O. Havnes, S. Helfert, M. Horányi, E. Igenbergs, E.K. Jessberger, T.V. Johnson, S. Kempf, A.V. Krivov, H. Krüger, A. Mocker-Ahlreep, G. Moragas-Klostermeyer, P. Lamy, M. Landgraf, D. Linkert, G. Linkert, F. Lura, J.A.M. McDonnell, D. Möhlmann, G.E. Morfill, M. Müller, M. Roy, G. Schäfer, G. Schlotzhauer, G.H. Schwehm, F. Spahn, M. Stübig, J. Svestka, V. Tschernjawski, J. Tuzzolino Astron, R. Wäsch, H.A. Zook, The Cassini cosmic dust analyzer. Space Sci. Rev. **114**, 465–518 (2004). https://doi.org/10.1007/s11214-004-1435-z

C.C. Stark, M.J. Kuchner, A new algorithm for self-consistent three-dimensional modeling of collisions in dusty debris disks. Astrophys. J. **707**(1), 543–553 (2009). https://doi.org/10.1088/0004-637X/707/1/543. arXiv:0909.2227

D.I. Steel, W.G. Elford, Collisions in the solar system. III—meteoroid survival times. Mon. Not. R. Astron. Soc. **218**, 185–199 (1986). https://doi.org/10.1093/mnras/218.2.185

G. Stenborg, J.R. Stauffer, R.A. Howard, Evidence for a circumsolar dust ring near Mercury's orbit. Astrophys. J. **868**(1), 74 (2018). https://doi.org/10.3847/1538-4357/aae6cb

S.A. Stern, The lunar atmosphere: history, status, current problems, and context. Rev. Geophys. **37**(4), 453–492 (1999). https://doi.org/10.1029/1999RG900005

R.M. Suggs, D.E. Moser, W.J. Cooke, R.J. Suggs, The flux of kilogram-sized meteoroids from lunar impact monitoring. Icarus **238**, 23 (2014)

S. Sugita, P.H. Schultz, M.A. Adams, Spectroscopic measurements of vapor clouds due to oblique impacts. J. Geophys. Res. **103**(E8), 19,427–19,441 (1998). https://doi.org/10.1029/98JE02026.

S. Sugita, P.H. Schultz, S. Hasegawa, Intensities of atomic lines and molecular bands observed in impact-induced luminescence. J. Geophys. Res., Planets **108**(E12), 5140 (2003). https://doi.org/10.1029/2003JE002156

N. Swarnalingam, D. Janches, J.D. Carrillo-Sanchez, P. Pokorny, J.M.C. Plane, Z. Sternovsky, D. Nesvorny, Modeling the altitude distribution of meteor head echoes observed with HPLA radars: implications for the radar detectability of meteoroid populations. Astron. J. **157**, 179 (2019). https://doi.org/10.3847/1538-3881/ab0ec6

J.R. Szalay, M. Horányi, Annual variation and synodic modulation of the sporadic meteoroid flux to the Moon. Geophys. Res. Lett. **42**, 10 (2015). https://doi.org/10.1002/2015GL066908

J.R. Szalay, M. Horányi, The search for electrostatically lofted grains above the Moon with the Lunar Dust Experiment. Geophys. Res. Lett. **42**(13), 5141–5146 (2015b). https://doi.org/10.1002/2015GL064324

J.R. Szalay, M. Horányi, Detecting meteoroid streams with an in-situ dust detector above an airless body. Icarus **275**, 221–231 (2016). https://doi.org/10.1016/j.icarus.2016.04.024

J.R. Szalay, M. Horányi, The impact ejecta environment of near Earth asteroids. Astrophys. J. Lett. **830**(2), L29 (2016b). https://doi.org/10.3847/2041-8205/830/2/L29. http://stacks.iop.org/2041-8205/830/i=2/a=L29

J.R. Szalay, M. Horányi, Lunar meteoritic gardening rate derived from in situ LADEE/LDEX measurements. Geophys. Res. Lett. **43**(10), 4893–4898 (2016c). https://doi.org/10.1002/2016GL069148

J.R. Szalay, M. Horányi, A. Colaprete, M. Sarantos, Meteoritic influence on sodium and potassium abundance in the lunar exosphere measured by LADEE. Geophys. Res. Lett. **43**(12), 6096–6102 (2016). https://doi.org/10.1002/2016GL069541

J.R. Szalay, P. Pokorný, P. Jenniskens, M. Horányi, Activity of the 2013 Geminid meteoroid stream at the Moon. Mon. Not. R. Astron. Soc. **474**(3), 4225–4231 (2018). https://doi.org/10.1093/mnras/stx3007

J.R. Szalay, P. Pokorný, M. Horányi, D. Janches, M. Sarantos, R. Srama, Impact ejecta environment of an eccentric asteroid: 3200 Phaethon. Planet. Space Sci. **165**, 194–204 (2019). https://doi.org/10.1016/j.pss.2018.11.001

J.R. Szalay, P. Pokorný, S.D. Bale, E.R. Christian, K. Goetz, K. Goodrich, M.E. Hill, M. Kuchner, R. Larsen, D. Malaspina, D.J. McComas, D. Mitchell, B. Page, N. Schwadron, The near-Sun dust environment: initial observations from Parker Solar Probe. Astrophys. J. Suppl. Ser. **246**(2), 27 (2020a). https://doi.org/10.3847/1538-4365/ab50c1. arXiv:1912.02639

J.R. Szalay, P. Pokorný, M. Horányi, Hyperbolic meteoroids impacting the Moon. Astrophys. J. Lett. **890**(1), L11 (2020b). https://doi.org/10.3847/2041-8213/ab7195

M. Tabata, H. Kawai, H. Yano, E. Imai, H. Hashimoto, S. Yokobori A. Yamagishi, Ultralow-density double-layer silica aerogel fabrication for the intact capture of cosmic dust in low-Earth orbits. J. Sol-Gel Sci. Technol. **77**(2), 325–334 (2016). https://doi.org/10.1007/s10971-015-3857-3

A. Taylor, The Harvard radio meteor project meteor velocity distribution reappraised. Icarus **116**, 205–209 (1995)

S. Taylor, J.H. Lever, R.P. Harvey, Accretion rate of cosmic spherules measured at the South Pole. Nature **392**(6679), 899–903 (1998). https://doi.org/10.1038/31894

E. Thomas, M. Horányi, D. Janches, T. Munsat, J. Simolka, Z. Sternovsky, Measurements of the ionization coefficient of simulated iron micrometeoroids. Geophys. Res. Lett. **43**, 3645–3652 (2016). https://doi.org/10.1002/2016GL068854

J.I. Thorpe, J. Slutsky, J.G. Baker, T.B. Littenberg, S. Hourihane, N. Pagane, P. Pokorny, D. Janches, M. Armano, H. Audley, G. Auger, J. Baird, M. Bassan, P. Binetruy, M. Born, D. Bortoluzzi, N. Brandt, M. Caleno, A. Cavalleri, A. Cesarini, A.M. Cruise, K. Danzmann, M. de Deus Silva, R. De Rosa, L. Di Fiore, I. Diepholz, G. Dixon, R. Dolesi, N. Dunbar, L. Ferraioli, V. Ferroni, E.D. Fitzsimons, R. Flatscher, M. Freschi, C. García Marirrodriga, R. Gerndt, L. Gesa, F. Gibert, D. Giardini, R. Giusteri, A. Grado, C. Grimani, J. Grzymisch, I. Harrison, G. Heinzel, M. Hewitson, D. Hollington, D. Hoyland, M. Hueller, H. Inchauspé, O. Jennrich, P. Jetzer, B. Johlander, N. Karnesis, B. Kaune, N. Korsakova, C.J. Killow, J.A. Lobo, I. Lloro, L. Liu, J.P. López-Zaragoza, R. Maarschalkerweerd, D. Mance, V. Martín, L. Martin-Polo, J. Martino, F. Martin-Porqueras, S. Madden, I. Mateos, P.W. McNamara, J. Mendes, L. Mendes, M. Nofrarias, S. Paczkowski, M. Perreur-Lloyd, A. Petiteau, P. Pivato, E. Plagnol, P. Prat, U. Ragnit, J. Ramos-Castro, J. Reiche, D.I. Robertson, H. Rozemeijer, F. Rivas, G. Russano, P. Sarra, A. Schleicher, D. Shaul, C.F. Sopuerta, R. Stanga, T. Sumner, D. Texier, C. Trenkel, M. Tröbs, D. Vetrugno, S. Vitale, G. Wanner, H. Ward, P. Wass, D. Wealthy, W.J. Weber, L. Wissel, A. Wittchen, A. Zambotti, C. Zanoni, T. Ziegler, P. Zweifel, C. Cutler, N. Demmons, C. Dunn, M. Girard, O. Hsu, S. Javidnia, I. Li, P. Maghami, C. Marrese-Reading, J. Mehta, J. O'Donnell, A. Romero-Wolf, J. Ziemer (LISA Pathfinder Collaboration P ST7-DRS Operations Team, Barela), Micrometeoroid events in LISA Pathfinder. Astrophys. J. **883**(1), 53 (2019). https://doi.org/10.3847/1538-4357/ab3649. arXiv:1905.02765

J. Vaubaillon, F. Colas, L. Jorda, A new method to predict meteor showers. II. Application to the Leonids. Astron. Astrophys. **439**(2), 761–770 (2005). https://doi.org/10.1051/0004-6361:20042626

J. Vaubaillon, L. Maquet, R. Soja, Meteor hurricane at Mars on 2014 October 19 from comet C/2013 A1. Mon. Not. R. Astron. Soc. **439**(4), 3294–3299 (2014). https://doi.org/10.1093/mnras/stu160

S. Verani, C. Barbieri, C. Benn, G. Cremonese, Possible detection of meteor stream effects on the lunar sodium atmosphere. Planet. Space Sci. **46**(8), 1003–1006 (1998). https://doi.org/10.1016/S0032-0633(98)00024-5

S. Verani, C. Barbieri, C.R. Benn, G. Cremonese, M. Mendillo, The 1999 Quadrantids and the lunar Na atmosphere. Mon. Not. R. Astron. Soc. **327**(1), 244–248 (2001). https://doi.org/10.1046/j.1365-8711.2001.04748.x. arXiv:astro-ph/0106447

R.J. Vervack, W.E. McClintock, R.M. Killen, A.L. Sprague, B.J. Anderson, M.H. Burger, E.T. Bradley, N. Mouawad, S.C. Solomon, N.R. Izenberg, Mercury's complex exosphere: results from MESSENGER's third flyby. Science **329**(5992), 672 (2010). https://doi.org/10.1126/science.1188572

R.J. Vervack, R.M. Killen, W.E. McClintock, A.W. Merkel, M.H. Burger, T.A. Cassidy, M. Sarantos, New discoveries from MESSENGER and insights into Mercury's exosphere. Geophys. Res. Lett. **43**(22), 11,545–11,551 (2016). https://doi.org/10.1002/2016GL071284

A.A. Weiss, J.W. Smith, A southern hemisphere survey of the radiants of sporadic meteors. Mon. Not. R. Astron. Soc. **121**, 5 (1960)

F.L. Whipple, Photographic meteor studies. III. The Taurid shower. Proc. Am. Philos. Soc. **83**, 711–745 (1940)

P. Wiegert, J. Vaubaillon, M. Campbell-Brown, A dynamical model of the sporadic meteoroid complex. Icarus **201**, 295–310 (2009). https://doi.org/10.1016/j.icarus.2008.12.030

P. Wiegert, P. Brown, P. Pokorný, Q. Ye, C. Craig, K. Lenartowicz, Z. Krzeminski, D. Clark, Supercatastrophic disruption of asteroids in the context of SOHO comet, fireball, and meteor observations. Astron. J. **159**(4), 143 (2020). https://doi.org/10.3847/1538-3881/ab700d

Publisher's Note   Springer Nature remains neutral with regard to jurisdictional claims in published maps and institutional affiliations.

Authors and Affiliations

Diego Janches[1] · Alexey A. Berezhnoy[2,3] · Apostolos A. Christou[4] · Gabriele Cremonese[5] · Takayuki Hirai[6] · Mihály Horányi[7,8,9] · Jamie M. Jasinski[10] · Menelaos Sarantos[11]

✉ G. Cremonese
  gabriele.cremonese@inaf.it

  D. Janches
  diego.janches@nasa.gov

  A.A. Berezhnoy
  a_tolok@mail.ru

  A.A. Christou
  Apostolos.Christou@Armagh.ac.uk

  T. Hirai
  hirai.takayuki@perc.it-chiba.ac.jp

  M. Horányi
  mihaly.horanyi@lasp.colorado.edu

  J.M. Jasinski
  jamie.m.jasinski@jpl.nasa.gov

  M. Sarantos
  menelaos.sarantos-1@nasa.gov

[1] ITM Physics Laboratory, NASA Goddard Space Flight Center, Code 675, 8800 Greenbelt Rd, Greenbelt MD, 20771 USA

[2] Sternberg Astronomical Institute, Moscow State University, Universitetskij pr., 13, Moscow, 119234 Russia

[3] Institute of Physics, Kazan Federal University, Kremlyovskaya Str., 18, Kazan, 420008 Russia

[4] Armagh Observatory and Planetarium, College Hill, Armagh BT61 9DG, UK

[5] INAF-Astronomical Observatory of Padova, Vicolo dell'Osservatorio 5, 35122 Padova, Italy

[6] Planetary Exploration Research Center (PERC), Chiba Institute of Technology, Narashino, Tsudanuma 2-17-1, 275-0016 Chiba, Japan

[7] Department of Physics, University of Colorado Boulder, 392 UCB, Boulder, CO, 80309, USA

[8] Laboratory for Atmospheric and Space Physics, 1234 Innovation Dr., Boulder, CO, 80303, USA

[9] Institute for Modeling Plasma, Atmospheres, and Cosmic Dust, 3400 Marine St., Boulder, CO, 80303, USA

[10] NASA Jet Propulsion Laboratory, California Institute of Technology, Pasadena, CA, USA

[11] Heliophysics Science Division, NASA Goddard Space Flight Center, Code 673, 8800 Greenbelt Rd, Greenbelt MD, 20771 USA

Space Science Reviews (2021) 217:61
https://doi.org/10.1007/s11214-021-00833-8

Volatiles and Refractories in Surface-Bounded Exospheres in the Inner Solar System

Cesare Grava[1] · Rosemary M. Killen[2] · Mehdi Benna[2,3] ·
Alexey A. Berezhnoy[4,5] · Jasper S. Halekas[6] · François Leblanc[7] ·
Masaki N. Nishino[8] · Christina Plainaki[9] · Jim M. Raines[10] ·
Menelaos Sarantos[2] · Benjamin D. Teolis[1] · Orenthal J. Tucker[2] ·
Ronald J. Vervack Jr.[11] · Audrey Vorburger[12]

Received: 22 December 2020 / Accepted: 21 May 2021 / Published online: 16 June 2021
© The Author(s) 2021

Abstract

Volatiles and refractories represent the two end-members in the volatility range of species in any surface-bounded exosphere. Volatiles include elements that do not interact strongly with the surface, such as neon (detected on the Moon) and helium (detected both on the Moon and at Mercury), but also argon, a noble gas (detected on the Moon) that surprisingly adsorbs at the cold lunar nighttime surface. Refractories include species such as calcium, magnesium, iron, and aluminum, all of which have very strong bonds with the lunar surface and thus need energetic processes to be ejected into the exosphere. Here we focus on the properties of species that have been detected in the exospheres of inner Solar System bodies, specifically the Moon and Mercury, and how they provide important information to understand source and loss processes of these exospheres, as well as their dependence on variations in external drivers.

Keywords Moon · Mercury · Exosphere · Refractories · Volatiles · Solar wind · Magnetosphere · Neutrals · Ions

1 Introduction

Volatiles and refractories are subject to different loss and source processes, and each provides different insights on the behavior of the exospheres of such species. Calcium and magnesium, for example, are predominantly ejected via micrometeoroid impact vaporization (probably in molecular compounds) and (to a lesser extent) sputtering; therefore, they are species of interest to study the exospheric response to micrometeoroid flux (Janches et al. 2021). On the other hand, helium is an element of predominantly solar wind origin that has been detected at both Mercury and the Moon. As such, it offers the opportunity to study the response to the same external driver (solar wind flux) of two very different exospheres: one

Surface-Bounded Exospheres and Interactions in the Inner Solar System
Edited by Anna Milillo, Menelaos Sarantos, Benjamin D. Teolis, Go Murakami, Peter Wurz and Rudolf von Steiger

Extended author information available on the last page of the article

(Mercury's) embedded in its own magnetosphere; the other (the Moon's) directly exposed to solar wind bombardment except for $\sim 1/6$ of its orbit when the solar wind is effectively shielded by the Earth's magnetotail. In this regard, it is fortunate that the two most prominent surface-bounded exospheres in the inner Solar System for which we have measurements are so different, as they highlight the relative importance of different source and loss processes. We discuss volatiles and refractories in Sects. 2 and 3, respectively. Section 4 discusses the "missing" species, i.e. those for which a detection has been expected in these exospheres but so far have not been achieved. Section 5 briefly discusses ions and Energetic Neutral Atoms, as they also play an important role in determining the loss rate and composition of a surface-bounded exosphere. Section 6 recaps the overall discussion. Future considerations for needed laboratory measurements, modeling improvements, and further observations are summarized in Sect. 7. Species with different volatility, such as the alkalis Na and K and OH/H_2O, are discussed in Leblanc et al. (2021) and Schörghofer et al. (2021), respectively.

2 Volatiles

This section discusses the species with the highest volatility (and hence mobility), including the two most prominent noble gases, helium (Sect. 2.1) and argon (Sect. 2.2). These are the species for which a solid database of observations exists (for helium at both Mercury and the Moon), and they represent endogenic species (^{40}Ar much more than ^{4}He). Argon, in particular, is important in studying how surface-bounded exospheres are shaped by temporary cold trapping. Section 2.3 closes with a discussion of other volatiles, most of which give insights into how the exosphere reacts to the variations in the solar wind.

2.1 Helium

Helium (^{4}He) has been detected on both the Moon and Mercury. In both cases, the dominant source of exospheric helium is implantation of solar wind alpha particles (He^{++}) on the surface and their subsequent release into the exosphere as neutrals.

On the Moon, helium was one of the first exospheric species discovered by the Lunar Atmosphere Composition Experiment (LACE) mass spectrometer deployed during the Apollo 17 mission (Hoffman et al. 1973). The measurements, taken during nine lunations at nighttime (during the day, LACE counts were overwhelmed by outgassing from the instrument itself), showed an increase of exospheric surface density from dusk up to ~ 2 AM local time (peak of $\sim 3 \times 10^4$ cm^{-3}), followed by a decrease towards dawn (see Fig. 1).

This profile was predicted by Hodges and Johnson (1968) and explained as a result of helium atoms not adsorbing even at the cold lunar nighttime surface. As a result, the exospheric density, n, is inversely proportional to the surface temperature T: $n \sim T^{-5/2}$ (Hodges and Johnson 1968). Correlation between the helium exospheric density measured by LACE and the geomagnetic index (a proxy for solar activity) revealed that alpha particles from the solar wind are the main source of lunar ^{4}He (Hodges and Hoffman 1974). These particles continuously bombard the lunar surface unimpeded by a magnetosphere except for when the Moon is inside the Earth's magnetotail (during ~ 2 days around full moon), become neutralized, and finally are released as neutrals into the exosphere. LACE observations were adequately described by an exospheric model in which helium atoms are in thermal equilibrium with the lunar surface and where gravitational escape is the dominant loss process, with photoionization being a secondary but non-negligible loss process (e.g. Hodges 1973).

On Mercury, helium was detected by the UltraViolet Spectrometer (UVS) aboard Mariner 10 (Broadfoot et al. 1974) through observation of the 58.4 nm resonant scattering emission

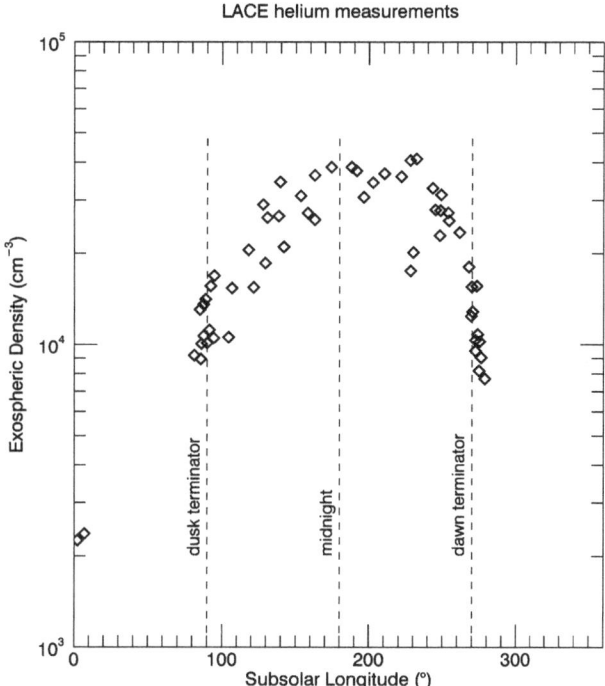

Fig. 1 Exospheric number densities for ^{4}He measured at the lunar surface by the LACE mass spectrometer (Apollo 17) during nine lunations in 1972 and 1973. Subsolar longitudes are angles from the subsolar point. The two points at noon represent sporadic checks when the instrument was briefly turned on at noon. Adapted from Hoffman et al. (1973)

line (HeI). The vertical column density above the subsolar point was 7×10^{11} cm^{-2} for a derived subsolar exospheric surface density of 4.5×10^3 cm^{-3}. The altitude profile observed above the subsolar point could be explained by a relatively simple exospheric model that assumes complete saturation of Mercury's surface with helium and a full thermal accommodation with the surface. However, observations taken close to the terminator could not be explained by the same model (Broadfoot et al. 1976).

The fact that at the Moon helium could be reasonably explained by a full thermal accommodation with the surface, whereas at Mercury this appeared not to be the case, was interpreted to originate from the poor knowledge of the gas-surface interaction. The exchange of energy between exospheric atoms and an airless body's surface is described by the accommodation coefficient α (e.g. Hunten et al. 1988):

$$\alpha = \frac{E_{out} - E_{in}}{E_T - E_{in}}$$

where E_{out} is the energy of the atom or molecule after the collision, E_{in} is its energy prior to the collision, and E_T is the energy of the atom in thermal equilibrium with the surface. When $\alpha = 1.0$, $E_{out} = E_T$ and the atom leaves the surface with an energy corresponding to thermal equilibrium with the surface. In this case, the surface temperature is what controls the energy of the atoms, and therefore the structure (and escape) of the exosphere. Larger hop length on hotter surfaces implies that non-adsorbable species will accumulate in the nightside exosphere. Conversely, with $\alpha < 1.0$ the exosphere is less dependent on the surface temperature. Early modelers of the lunar exospheres (Hartle and Thomas 1974; Hodges 1975) used $\alpha = 1.0$ on the assumption that the lunar surface is saturated with helium, an assumption based on results from the Apollo 11 Solar Wind Composition experiment (Bühler et al. 1969), which measured the solar wind flux impacting the Moon. This experiment revealed that this flux was high enough to establish saturation within just tens of thousands of years (Banks et al. 1970). When the Mariner 10 observations were published, Hartle et al. (1975) proposed that the mismatch between model and observations at terminator could be caused by not knowing the surface temperature close to the terminator with sufficient accuracy, perhaps owing to shadows cast by nearby reliefs (micro-shadows cast by grains, or macro-shadows cast by ridges and crater rims): if $\alpha = 1.0$ and the surface temperature (and thus E_T) is not known accurately, then E_{out} is poorly constrained. This would also explain why the altitude profiles above the subsolar point, where the temperature was better constrained, were better explained by the model. However, Shemansky and Broadfoot (1977) and Smith et al. (1978) noted that the atom-surface interaction involves single phonon collisions rather than multiple ones, and that α depends on the Debye characteristic temperature of the surface lattice. Therefore, they postulated that full thermal accommodation was not justified. As such, helium is an important species for studying the gas-surface interaction in exospheres of airless bodies.

Helium is lost primarily via thermal escape. Simulations of the lunar exospheric helium by Hodges (1977a, 1978) that included solar radiation pressure and the gravitational attraction of the Sun and the Earth (besides that of the Moon) supported the existence of a vast helium corona around the Moon. This corona may extend to tens of lunar radii and is populated by satellite helium atoms whose periapsis is higher than the highest peak on the Moon; hence, they spend their entire lifetime in orbit until they are photoionized (after $\sim$6 months). Some of these atoms may even reach the Earth's exosphere, suggesting the possibility of the existence of a "shared exosphere" between the Moon and the Earth.

Up to 10% of the lunar helium measured by LACE is not accounted for by the solar wind (Hodges 1975). Hodges (1977b) proposed that this is endogenic lunar helium, coming from the radioactive decay of thorium and uranium within the lunar mantle and crust (Kockarts 1973) and finding its way to the exosphere via cracks or fissures (Killen 2002), the same way ^{40}Ar does (see Sect. 2.2). The outgassing rate of endogenic ^{4}He would then constrain the amount of radioactive elements in the lunar crust. The challenge is how to distinguish it from the dominant background, i.e., the solar-wind-derived helium. This intriguing topic has been addressed by spacecraft that detected helium in recent years. The Lyman Alpha Mapping Project (LAMP; Gladstone et al. 2010a) far-ultraviolet (FUV) imaging spectrograph onboard the Lunar Reconnaissance Orbiter (LRO; Chin et al. 2007) made the first spectroscopic detection of helium, by observing the HeI emission line at 58.4 nm (Stern et al. 2012). The retrieved surface densities (obtained around dusk local time) were somewhat lower than those from LACE. Subsequent observations confirmed the 4.5-day decay constant (Feldman et al. 2012). In particular, the helium density was observed to decrease as soon as the Moon entered the Earth's magnetotail, and was thus shielded from the solar wind bombardment. Helium was measured in situ again by the Neutral Mass Spectrometer (NMS; Mahaffy et al. 2014) onboard the Lunar Atmosphere and Dust Environment Explorer (LADEE; Elphic et al. 2014). During LADEE's 7-month mission, NMS measured helium atom densities at a few tens of km altitude around the equator (Benna et al. 2015; see also Fig. 7). At the same time, the twin spacecraft ARTEMIS (Acceleration, Reconnection, Turbulence and Electrodynamics of the Moon's Interaction with the Sun; Angelopoulos 2011) was measuring the flux of solar wind alpha particles around the lunar environment. Therefore, Benna et al. (2015) could make a direct comparison between the direct source (solar wind alpha particles) and the resulting neutrals (helium atoms, measured by NMS), and found a positive correlation between the two. They also derived a value for the helium source rate that is not accounted for by the solar wind alpha particles and interpreted it to be the endogenic population mentioned by Hodges (1977b): $(1.5\text{-}2.0) \times 10^6$ cm^{-2} s^{-1}, or

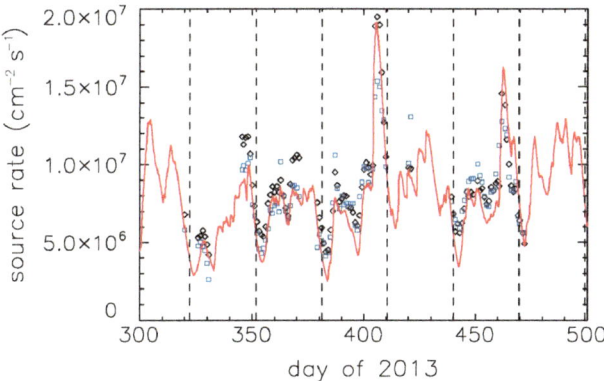

Fig. 2 Three different datasets (neutral helium measured in situ by LADEE/NMS: black diamonds; neutral helium measured remotely by LRO/LAMP: blue squares; solar wind alpha particles measured in situ by ARTEMIS/ESA: red line) show strongly correlated source rates between solar wind alpha particles and lunar exospheric helium. Vertical lines indicate times of full moon, when the geomagnetic tail effectively shields the Moon from the solar wind. Reproduced from Hurley et al. (2016)

about 15-20% of the solar wind alpha particles influx, slightly higher than Hodges' estimate. Benna et al. (2015) also found a 4.5-day escape time constant for lunar exospheric helium, confirming that thermal escape is the major loss process for this exospheric species.

Later, the same two datasets were compared by Hurley et al. (2016) with LAMP surface densities derived from the HeI emission line. The three datasets, which offered three different "views" of the lunar helium (in situ measurements of neutral atoms and solar wind alpha particles, and remote sensing measurements of neutral atoms), agreed well with each other (see Fig. 2). The derived endogenic source rate, however, was considerably higher than previous estimates and consistent with the one derived by Grava et al. (2016) using targeted LRO off-nadir observations with LAMP: 35-40% of the solar wind. Clearly more observations are needed to constrain this important source rate.

Recently, LAMP carried out a more extensive atmospheric campaign to map the lunar helium over several latitudes, longitudes, and local times, comparing the column densities with ARTEMIS solar wind alpha particles. The result of this multi-year long campaign, with more than 170 orbits, points to an endogenic source rate of $1.49 \pm 0.08 \times 10^6$ cm^{-2} s^{-1}, or about 19% of the solar wind (Grava et al. 2021), in agreement with the LADEE/NMS measurements and slightly higher than the estimates of Hodges (1977b) based on the amount of thorium and uranium within the crust estimated by Taylor and Jakeš (1974) and on the assumption that the outgassing rate is the same as that for ^{40}Ar (6% of the total production). The discrepancy might mean that this assumption is wrong (helium is more volatile so its outgassing rate might be higher) or that the outgassing of helium is sporadic, like that of ^{40}Ar. Grava et al. (2021) also found that the same dataset can be adequately reproduced by an exospheric model that assumes full thermal accommodation ($\alpha = 1.0$).

Finally, the mass spectrometer CHACE (CHandra's Altitudinal Composition Explorer; Sridharan et al. 2010) onboard the Moon Impact Probe (MIP) of the Chandrayaan-1 spacecraft (Goswami and Annadurai 2009) attempted the first measurement from a spacecraft of the lunar helium dayside exosphere, but was able to place only an upper limit of 800 cm^{-3} (Das et al. 2017). This low value arises from the combination of several factors: the observations were on the dayside (where the surface density is lowest), obtained during the magnetotail passage of the Moon (when the solar wind — the main source of helium — is deflected by Earth's magnetosphere and thus has no access to the lunar surface), and close in time to the minimum solar wind flux of cycle 24.

An isotope of helium of great interest is ^{3}He, a potential clean energy source. Being scarce in the Earth's atmosphere and mantle yet abundant on the Moon, where it is delivered by the solar wind, it has gained attention particularly in recent times thanks to the renewed interest in lunar exploration. Thus far the only measurements are those from the surface. The ^{3}He content in returned lunar samples correlates well with TiO_2 content and maturity index Is/FeO (Jordan 1989). Taking into account the estimated solar wind flux on the Moon, the correlation coefficient between the measured ^{3}He content and the TiO_2 content, the solar wind flux, and the maturity parameter in the nine Apollo soil samples studied is 0.944 (Johnson et al. 1999). A similar value, 0.938, was found in 25 Apollo soils by Fa and Jin (2007). These authors estimated the ^{3}He content on the surface of the Moon as $C(^3He) = 0.56 * S(TiO_2) * (F/OMAT) + 1.62$, where $C(^3He)$ is in ppb, $S(TiO_2)$ is the TiO_2 content in wt%, F is the normalized solar flux, and $OMAT$ is the maturity index taken from Lucey et al. (2000).

A physically plausible model of the observed correlation between ^{3}He content, TiO_2 content, solar wind flux, and soil maturity in returned lunar samples was developed by Shkuratov et al. (1999). In the returned lunar samples, ^{3}He and ^{4}He are stable at least at room temperature, meaning that these isotopes are strongly bounded in the regolith and have a high activation energy of diffusion in the soil. ^{3}He and ^{4}He are mainly delivered to the regolith by the solar wind, so that the content of these isotopes on the surface of the Moon should be correlated with the solar wind flux. The ^{3}He atoms implanted into the regolith by the solar wind are captured in traps located in vacancies of the crystal grid. This means that the ^{3}He content in the soil increases with increasing concentration of such traps. The degree of damage of the crystal lattice (soil maturity) increases with exposure to the solar wind bombardment, and thus with increasing age of the samples. The concentration of ^{3}He traps depends on the soil maturity and on the volume fraction of minerals with a high content of vacancies (Scherzer 1983). Experimental works show that irradiation of ilmenite ($FeTiO_3$, the main carrier of Ti on the surface of the Moon) by solar wind particles leads to the appearance of radiation-induced defects in the lattice, which are able to trap solar wind ions (Scherzer 1983). Ilmenite is considered to be the most effective He trapper among main lunar minerals because it has a high concentration of vacancies. Incidentally, OH/H_2O content on the surface of the Moon is also correlated with TiO_2 content (Wöhler et al. 2017), providing additional evidence that the $FeTiO_3$ content is the main factor controlling the behavior of many volatiles on the surface of the Moon.

Maps of the ^{3}He content on the lunar surface were calculated using the strong correlations between the ^{3}He content and normalized solar wind flux at the point of collection of lunar samples, the TiO_2 content, and optical maturity in returned lunar samples. Maps from different authors are similar (Johnson et al. 1999; Fa and Jin 2007; Kim et al. 2019). In general, the ^{3}He content is higher in the maria than in the highlands. The ^{3}He content in low-Ti maria such as Mare Frigoris, Mare Imbrium, and Mare Serenitatis is also low, consistent with the TiO_2-^{3}He relationship mentioned earlier. A moderately high ^{3}He content of 10–15 ppb is predicted in Oceanus Procellarum, the Apollo basin, Mare Orientale, Mare Fecunditatis, Mare Crisium, Mare Moscoviense, and Mare Marginis (Kim et al. 2019). The highest ^{3}He concentrations of up to about 24 ppb are predicted for Ti-rich parts of Oceanus Procellarum, Mare Fecundidatis, Mare Tranquillitatis, Mare Crisium, Mare Marginis, and Mare Moscoviense (Kim et al. 2019). Hence, the expected ^{3}He content on the Moon is highest in the western maria. One could therefore expect an enhancement in exospheric helium there. However, no such enhancement could be detected by either LRO/LAMP (Grava

Fig. 3 Peaks in exospheric source rate of ^{40}Ar measured by LACE (histogram) occurred soon after moonquakes recorded by the Apollo seismometers (red triangles). The black line is the argon exospheric loss rate. Adapted from Hodges (1977b)

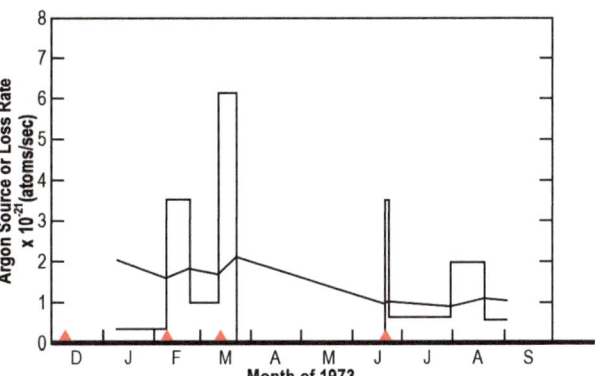

et al. 2021) or LADEE/NMS (Benna et al. 2015). It is noteworthy that LADEE/NMS did detect an enhancement in argon, another endogenic element, in the same region (western maria). A "helium bulge" would be difficult to detect from a single spacecraft, owing to the randomness of the outgassing location and owing to the large scale height and hop length of helium atoms. LAMP is not able to distinguish between ^{3}He and ^{4}He, and LADEE/NMS did not detect ^{3}He. However, a mass spectrometer such as LEMS (Benna et al. 2020), deployed at the lunar surface, would be able to distinguish between the two helium isotopes.

2.2 Argon

Argon (^{40}Ar), like helium was discovered by LACE during the Apollo 17 mission. As opposed to the most common isotope, ^{36}Ar, which comes from the solar wind, ^{40}Ar is an endogenic species, a byproduct of the radiogenic decay of ^{40}K within the lunar crust, which is released into the exosphere following diffusion, melting by impacts, or grinding of rocks (Killen 2002). In fact, spikes in lunar argon-40 density measured by LACE occurred soon after high-frequency teleseismic events, or shallow moonquakes, recorded by the Apollo seismometers (Nakamura 1977; Hodges 1977b; see also Fig. 3).

Shallow moonquakes, which probably occur a few tens of km below the surface (Hodges 1981; Killen 2002), may perturb the upper crust allowing the pockets of gas trapped in voids to diffuse out into the exosphere.

The diurnal profile of ^{40}Ar resembles that of a species that condenses at the cold nighttime surface and is then released at dawn (Fig. 4).

This kind of behavior was not expected from a noble gas. The exospheric model that best reproduced LACE observations required a heat of adsorption Q for ^{40}Ar on the lunar surface of ∼6500 cal mol^{-1}, much higher than the value derived by adsorption experiments of ^{40}Ar on glass (∼3800 cal mol^{-1}; Clausing 1930). The heat of adsorption factors into the equation for the residence time of argon-40 atoms in a grain:

$$t_{res} = \frac{C}{T^2} \exp\left(\frac{4.19 \cdot Q}{RT}\right)$$

where Q is the heat of adsorption, C is a constant (expressed in s K^{-2}), R is the gas constant, T is the surface temperature (in K), and 4.19 is the conversion factor between calories and Joules. Hodges (1980) attributed this very high value of Q for argon-40 (compared to laboratory measurements) to the high cleanliness of soil grains, which have been exposed for

Fig. 4 The diurnal profiles obtained four months apart (four lunations) by LACE in 1973. Measurements were made from dusk (90° subsolar longitude) to dawn (270° subsolar longitude). Adapted from Hodges (1975)

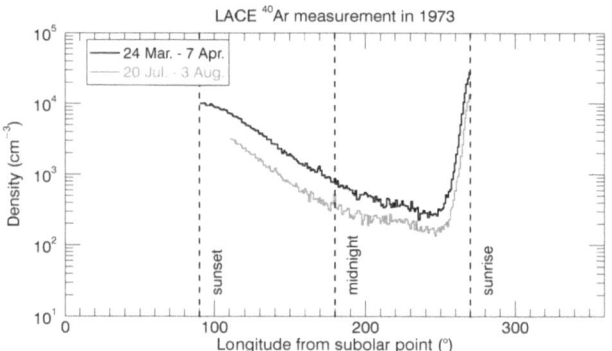

billions of years to the solar wind. Because it sticks efficiently to the cold lunar surface, ^{40}Ar can be trapped in Permanently Shaded Regions (PSRs), areas at the lunar poles that never receive direct sunlight. The facts that argon is an endogenic gas, sticks at the surface, and can be deposited in PSRs where it can reside undisturbed for billions of years (Watson et al. 1961a, 1961b; Arnold 1979) make it a valuable species for studying the behavior of other molecules (most notably, water) that are difficult to measure (^{40}Ar has been detected even at tens of km of altitude by LADEE/NMS). Grava et al. (2015) estimated that, during LACE measurements (∼9 months), 1,900 kg of ^{40}Ar were deposited in PSRs poleward of 85° N/S, corresponding to 30% of the surface-ejected quantity, and that permanent cold trapping is a sink process for the exospheric ^{40}Ar comparable in magnitude to photoionization and charge exchange with solar protons. Roughly four decades later, ^{40}Ar was detected again in the lunar exosphere by LADEE/NMS, which confirmed the exospheric surface density but also revealed a bulge in exospheric density above Oceanus Procellarum (Benna et al. 2015; see also Fig. 5). This area (KREEP terrane) is rich in ^{40}K, as measured by Lunar Prospector (Jolliff et al. 2000), and thus it is postulated that an enhanced diffusion of radiogenic gases occurs there. Two independent and concurrent simulations gave contradictory results, however. Hodges and Mahaffy (2016) found that the argon-40 bulge can be explained by a lower activation energy in that region and a very high activation energy (∼24,000 cal mol^{-1}) everywhere else. On the other hand, Kegerreis et al. (2017) found that the bulge can be explained by an enhanced outgassing rate in that region (the western maria). Modeling LADEE/NMS data, they found that, in general, ^{40}Ar has higher exospheric densities above maria, compared to highlands. This second explanation agrees with the hypothesis that circular fault systems around impact basins (with which the western maria are replete) are the regions where deep moonquakes are more likely to occur (Runcorn 1974).

Not all the argon atoms are readily desorbed at dawn. Some of them are temporarily sequestered at depth (where they arrived after diffusing downwards during the lunar night) and are released much later (mid-day). This mechanism, proposed by Kegerreis et al. (2017), could explain the slight time delay from dawn of the peak ^{40}Ar exospheric density recorded by LACE and LADEE without requiring the high activation energy all over the lunar surface proposed by Hodges and Mahaffy (2016). Interestingly, a similar mechanism (the "thermal pump") has been proposed for other species – most notably water – at the Moon (Schörghofer and Taylor 2007; Schörghofer and Aharonson 2014), Mercury (Vasavada et al. 1999), and Mars (Mellon and Jakosky 1993). It is therefore reasonable to expect that other species can behave the same way. Finally, the adsorbing behavior of ^{40}Ar is such that it makes possible the creation of seasons. Data from LADEE/NMS were in-

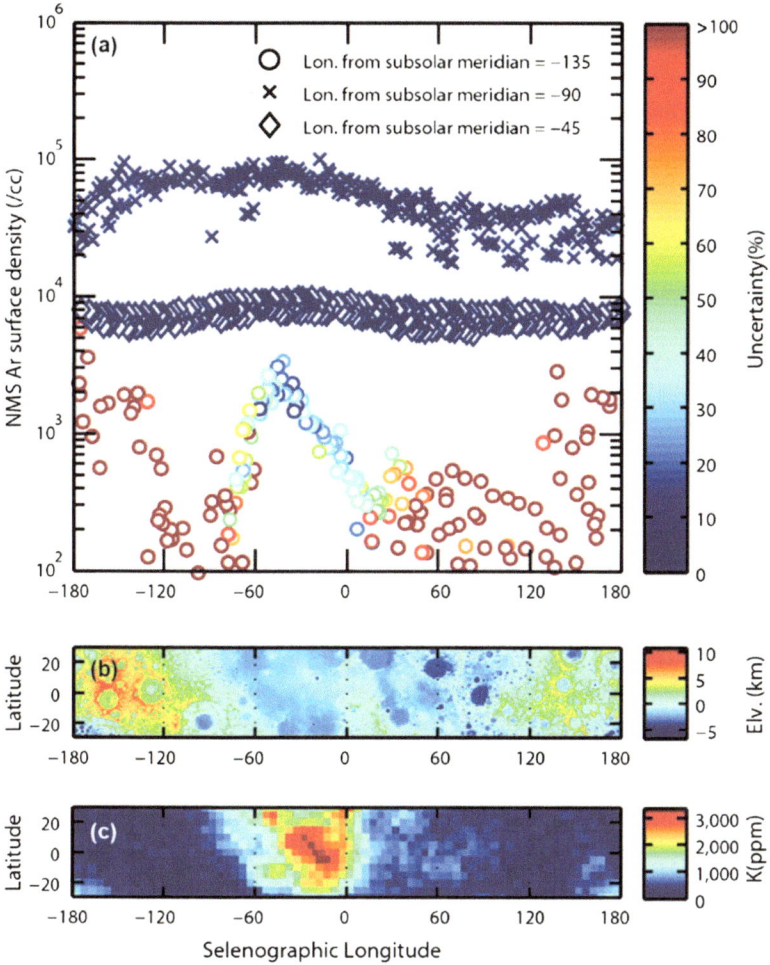

Fig. 5 Exospheric densities of ^{40}Ar measured at dawn (circles in top panel) are greatest above the western maria (middle panel), which are rich in KREEP elements, particularly ^{40}K (bottom panel), which is the radioactive parent of ^{40}Ar. Reproduced from Benna et al. (2015)

terpreted to be the result of seasonal migration of argon from one winter pole to the other (Hodges and Mahaffy 2016; Teolis et al. 2021).

Argon was also detected by CHACE on its route to crash landing into a lunar south polar crater. Thampi et al. (2015) showed densities measured from 100 km altitude at 20° N latitude ($\sim$5,000 cm^{-3}) to $\sim$10 km altitude at the south pole (8,000 cm^{-3}). This was the first detection in the polar regions (Fig. 6).

Argon has not been detected at Mercury. The Mariner 10 UVS placed only an upper limit of 6.6×10^6 cm^{-3} (Shemansky 1988), from the difficult-to-observe emission doublet at 104.8 and 106.7 nm. The MErcury Surface, Space ENvironment, GEochemistry, and Ranging (MESSENGER) spacecraft (Solomon et al. 2007) did not carry a neutral mass spectrometer, and the bandpass of the primary exospheric instrument, the Mercury Atmospheric and Surface Composition Spectrometer (MASCS; McClintock and Lankton 2007) UV spectrograph did not include the wavelength of the ^{40}Ar emission lines. In fact, the only

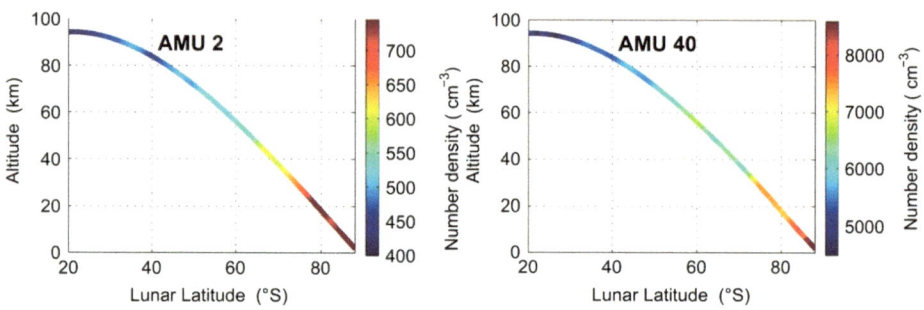

Fig. 6 Number densities of H_2 and ^{40}Ar measured by CHACE onboard Chandrayaan-1 from the ∼100 km altitude above the subsolar point to the surface close to the poles. Reproduced from Thampi et al. (2015)

way neutral argon-40 has been detected in exospheres so far is in situ mass spectrometry (LACE and LADEE/NMS at the Moon). Although Flynn (1998) claimed a detection of the argon doublet at the Moon from the ORPHEUS-SPAS II satellite, this detection was soon dismissed by Parker et al. (1998). Part of this spectroscopic non-detection can be explained by the low intrinsic brightness of the FUV ^{40}Ar emission lines: Parker et al. (1998) found that these lines are optically thick (based on the densities retrieved by LACE), hence tens of times fainter than the HeI emission line observed by LAMP. Stern et al. (2012) note that these lines appear within detection capability of LAMP, but so far it has only placed an upper limit for ^{40}Ar of 2.3×10^4 cm^{-3} (Cook et al. 2013). At Mercury, Killen (2002) estimated the column abundance of ^{40}Ar of $5 \times 10^8 - 2 \times 10^9$ cm^{-2} based on diffusion from anorthite in the top 25 km and a photoionization lifetime of 3.5 days at perihelion and 8 days at aphelion. This estimate of column density would make ^{40}Ar one of the most abundant species in Mercury's exosphere, but it is considerably lower than the estimated upper limit on argon column abundance of $5 \times 10^{12} - 6 \times 10^{13}$ cm^{-2} from the UV spectrometer onboard Mariner 10 (Broadfoot et al. 1976).

At the Moon, LACE detected the less abundant isotope ^{36}Ar, which is of solar wind origin. LACE showed a sunrise peak similar to ^{40}Ar in time but 10 times lower in density: 3×10^3 cm^{-3} (Hoffman et al. 1973). This value of 10 for the ratio $^{40}Ar/^{36}Ar$ in the lunar exosphere is in contrast with the near equality of the two isotopes in returned soil samples (Table 3 in Yaniv and Heymann 1972). Therefore, the soil is not saturated with ^{36}Ar, which means that the solar wind flux of ^{36}Ar is permanently trapped. Excess of so-called "parentless" ^{40}Ar in returned lunar samples, compared to expectations from solar wind composition and in situ decay of ^{40}K, was suggested by Heymann and Yaniv (1970) to be of exospheric origin. This hypothesis was confirmed by Manka and Michel (1970), whose simulations showed that about 10% of the exospheric argon ions ($^{40}Ar^+$) are driven back towards the Moon instead of being entrained in the interplanetary magnetic field. These ions are then implanted into the lunar soil. Because these ions impact the lunar surface with energy of ∼1 keV, much lower than that of solar wind ^{36}Ar ions (∼36 keV), they are not implanted as deeply as $^{36}Ar^+$. Manka and Michel (1970) note that for this reason the $^{40}Ar/^{36}Ar$ ratio should vary with location: higher in surfaces parallel to the ecliptic plane (where mostly of these $^{40}Ar^+$ ions impact); lower in surfaces facing the solar wind (which is rich in ^{36}Ar). The ratio $^{40}Ar/^{36}Ar$ therefore offers the opportunity to study the amount of time a rock has been exposed to the surface and which orientation it had.

2.3 Other Volatiles

Compared to the noble gases discussed above (argon and helium), far fewer observations exist of other volatile species. LACE made tentative detections of neon and methane, but those detections could barely be sifted out from contaminants. Recently, Killen et al. (2019) took advantage of the restoration of LACE neon data on NASA's PDS archive and were able to model its behavior (Sect. 2.3.1). Methane was detected by LADEE/NMS, and Hodges (2016) showed that it can help understand the recycling of solar wind carbon at the Moon (Sect. 2.3.2). Hydrogen was detected at both Mercury and the Moon, but in different forms (molecular at the Moon, atomic at Mercury – see Sect. 2.3.3). Radon and polonium, two more species indicative of radioactivity in the interior of the Moon, were detected by the Apollo orbiters and by Lunar Prospector (Sect. 2.3.4). For several other species, LRO/LAMP provided more stringent upper limits for their lunar exospheric surface densities, most of them several orders of magnitude lower than previous estimates (Cook et al. 2013).

2.3.1 Neon

Neon (^{20}Ne) was predicted to be the most abundant gas of solar wind origin in the lunar exosphere (Hinton and Taeusch 1964). Indeed, it was one of the first species indirectly detected in the lunar exosphere – as an ion – by the series of Suprathermal Ion Detector Experiment (SIDE) detectors deployed during the Apollo 12, 15, and 16 missions (Benson et al. 1975; Freeman and Benson 1977). Subsequently, it was detected in neutral form by LACE (Hoffman et al. 1973). These instruments reported surface densities of $\sim 10^5$ cm^{-3}, confirming ^{20}Ne as one of the most abundant species in the lunar exosphere. However, the ^{20}Ne signature observed by LACE was attributed subsequently to $H_2^{18}O$ (Hodges et al. 1973), so these measurements were not considered further. Later, neon was measured by CHACE, the quadrupole mass spectrometer onboard Chandrayaan-1. The geometry of this spacecraft, en route to its impact point near the lunar South Pole, allowed it to measure neon in the dayside and over different ranges of latitudes. The number density reported varied from $\sim 2,000$ cm^{-3} at the equator at 100 km altitude to $\sim 10,000$ cm^{-3} at the poles close to the surface (Das et al. 2016). Subsequently (although results were published earlier), the LADEE/NMS also detected neon (Benna et al. 2015). During its 7-month long mission timeline, NMS reported neon densities slightly lower than those of helium, with peak density at dawn (2.0-3.5×10^4 cm^{-3}; see Fig. 7).

The NMS diurnal profile show a steady increase in ^{20}Ne exospheric density from dusk to dawn, a sign of its non-condensable nature, but the exospheric density peak was recorded $10°$ (~ 1 hour in local time) before dawn, instead of ~ 2 AM local time in the case of helium. The difference in the two diurnal profiles is the result of the different scale height of the two species, and therefore of their different spatial extent. There is an inconsistency between ^{20}Ne exospheric densities reported in the literature. The NMS surface densities (inferred from orbit) were an order of magnitude greater than the upper limits obtained remotely by LAMP from the emission line at 63.0 nm (4.4×10^3 cm^{-3}; Cook et al. 2013), but lower than those reported in situ by LACE (1.1×10^5 cm^{-3}; Hodges et al. 1974). Recently, LACE Ne data were restored, validated, and re-analyzed by Killen et al. (2019), which corrected the ^{20}Ne measurement, considered to be contaminated by fluorine, using the ^{22}Ne mass bin, supposed to be uncontaminated, and the known isotopic ratios of neon. This re-analysis reported much lower surface densities than those from Hodges et al. (1974): (1.5-$4.5) \times 10^3$ cm^{-3}. One possible explanation of the discrepancy is that the value for Ne reported by Benna et al. (2015) was measured during a Coronal Mass Ejection (CME) passage (7-27

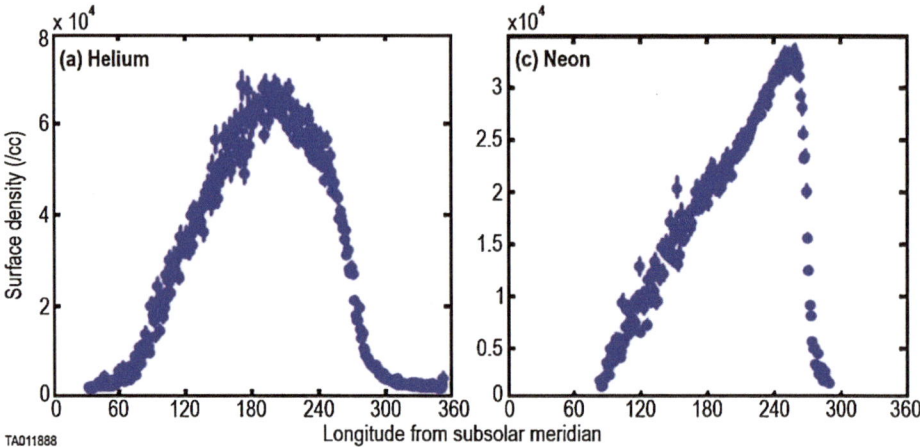

Fig. 7 Surface densities for ^{4}He (left) and ^{20}Ne (right) inferred from LADEE/NMS measurements at altitude. These panels show the different behavior of these two species, mainly attributed to their different scale height. Adapted from Benna et al. (2015)

February 2014), which entails an enhancement in solar wind flux compared to the nominal conditions. If the lifetime of neon is the predicted 100 days for photoionization (Huebner and Mukherjee 2015), the exospheric density would be determined by the averaged solar wind influx during the previous three months. Simulations of the neon density using the photoionization lifetime of 100 days (and nominal solar wind conditions) reproduce LACE measurements, but are twice those from LADEE, taken during a CME. In order to reproduce the estimated surface density of Ne at the morning terminator of $(2.0\text{-}4.5) \times 10^3$ cm^{-3} by LAMP and $(1.5\text{-}4.5) \times 10^3$ cm^{-3} from the re-analyzed LACE data, a lifetime of 4.5 days is required (Killen et al. 2019). Furthermore, the reanalyzed LACE data indicate that the global diurnal distribution of Ne can vary over a lunar day, which is also consistent with a shorter lifetime than 100 days. The discrepancy between the data sets and the lifetimes is unresolved and requires further measurements.

At Mercury, Mariner 10 provided an upper limit for neon of 3×10^{13} cm^{-3} (Broadfoot et al. 1974), from the 73.6 nm emission line. Because MESSENGER/MASCS did not have the capability of measuring the 73.6 nm line of Ne, there is currently no reliable measurement of Ne at Mercury.

2.3.2 Methane and Other Carbon-Bearing Species

Methane (CH_4) has been detected in the lunar exosphere by LADEE/NMS. Hodges (2016) reported observations taken close to the dawn terminator, where exospheric densities peak at a value of 400-450 cm^{-3} at 12 km altitude (see Fig. 8).

The diurnal profile reveals that CH_4, like ^{40}Ar, also adsorbs temporarily at the cold night-time surface. However, the high activation energy (higher than that of argon) means that there is a delay of $\sim$1 hour in morning release ($\sim$7 AM, instead of $\sim$6:30 AM for ^{40}Ar).

Analysis of LADEE/NMS data (Hodges 2016) revealed that methane plays a role in the recycling of solar wind carbon nuclei impacting the lunar surface (as was suggested 40 years earlier by Hodges 1976), which then are lost from the exosphere owing to the low photoionization lifetime of CH_4 (1 day). The delivery of solar wind C to the Moon is substantial: 8 tons/year (Hodges 1976). Because C abundance in returned samples (100

Fig. 8 Methane number density measured by LADEE/NMS (colored lines) referenced to a common altitude of 12 km, around dawn. Black lines are exospheric simulations of methane. This figure shows the pronounced sunrise bulge in exospheric density, indicative of a species that condenses on the cold nighttime surface. Reproduced from Hodges (2016)

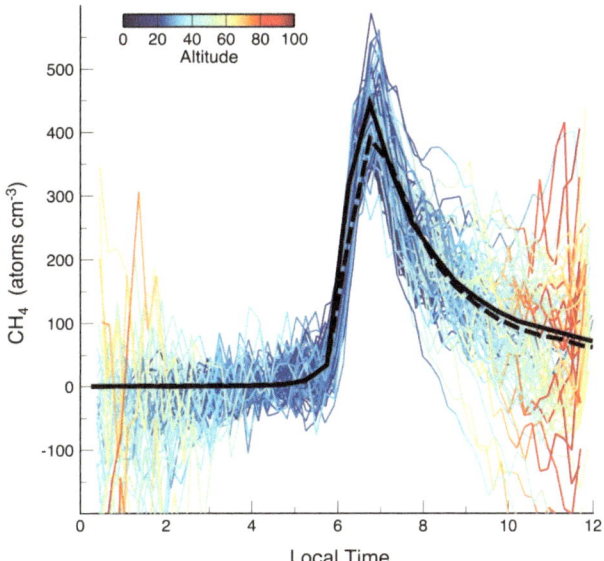

ppm, mostly in CH_4, CO, and CO_2) is less than the saturation level from the solar wind influx (200 ppm; Bibring et al. 1974) and is uniform over the maximum depth probed (250 cm), and because the reworking depth of the regolith owing to micrometeoroid gardening is just 10 cm in 10^9 years (Gault et al. 1974; Costello et al. 2018), it was proposed that the carbon influx must be balanced by a substantial exospheric loss in molecular compounds, especially on the dayside (from the analogy with helium). The most probable candidates are CH_4, CO, and CO_2. These three species were not detected during the nighttime by LACE, most likely because of adsorption at the surface and low exospheric density (LACE minimum threshold was $\sim$100 cm^{-3}; Hoffman et al. 1973). But around dawn LACE recorded peak concentrations at mass bins 28 (CO, but also possibly N_2) and 44 (CO_2) of 10^2-10^3 cm^{-3} close to dawn, with molecules coming from the hot dayside and traveling back towards the night (Hoffman and Hodges 1975; see also Fig. 9).

Hodges (2016) estimated the methane escape rate to be 1.5-4.5 $\times$ 10^{21} s^{-1}, equivalent to 25-76% of the global carbon influx. This can be compared with solar carbon escape of 3.4 $\times$ 10^{21} s^{-1}, obtained separately by analyzing Apollo samples. This led Hodges (2016) to propose that "a significant fraction of C that enters the exosphere as methane escapes as CO". In fact, exothermic reactions between solar wind C and the lunar soil would lead to the creation of CO, whose lifetime against photoionization is nine times that of CH_4 and thus would constitute an even more substantial exosphere than methane itself. LADEE/NMS, which is about four orders of magnitude more sensitive to ions than neutrals, did not detect CO, but it detected CO$^+$ (Halekas et al. 2015). The detection of CH_4 and carbon ions (C$^+$ and CO$^+$), briefly discussed in Sect. 5.1, highlights the existence of a carbon cycle at the Moon.

Other species have been tentatively detected by LACE, as shown in Fig. 9. Mass 28 could be either N_2 or CO. Neither of those adsorbs at equatorial cold nighttime surface temperatures, so no pre-dawn enhancement is expected. But CO_2 (mass 44) does absorb at those temperatures, so it is surprising not to see the pre-dawn enhancement at mass 44 which is seen in ^{40}Ar, another condensable species. From this lack of pre-dawn enhancement, Hoffman et al. (1973) estimated the dawn exospheric density of CO_2 to be 3 $\times$ 10^3 cm^{-3}.

Fig. 9 LACE exospheric density at the surface from four masses. Masses 40 and 36 are interpreted to be argon. Mass 28 could be N_2 or CO. Mass 44 could be CO_2. This species is expected to adsorb at the lunar surface, so the lack of such a bulge at dawn is surprising. T indicates terminator; S indicates sunrise (delayed from the terminator by ~8 hours because of the mountains to the West of the Taurus-Littrow valley). Adapted from Hoffman et al. (1973)

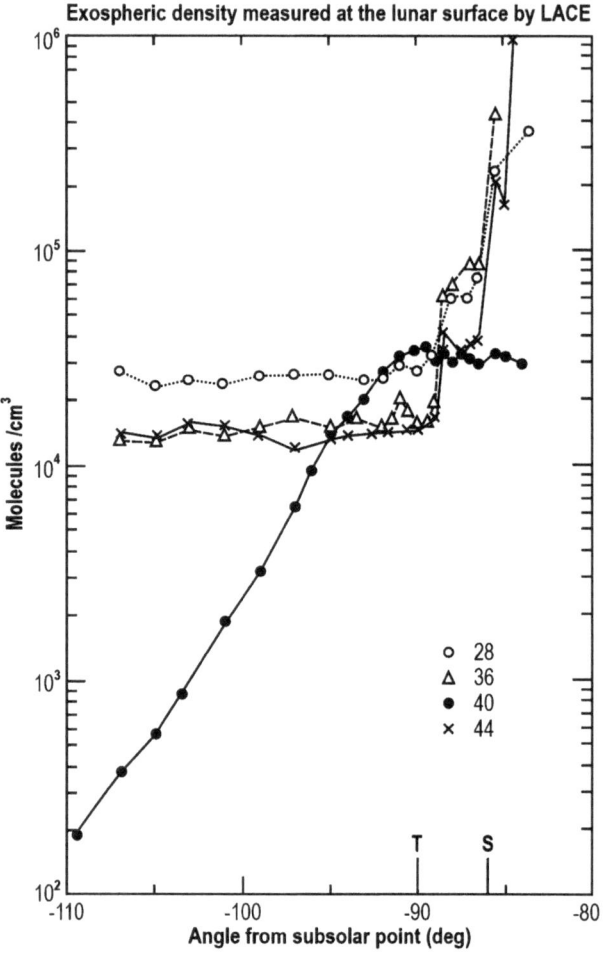

2.3.3 Hydrogen

Given that ~96% of the solar wind is composed mainly by protons, it was assumed that the Moon had a substantial dayside exosphere of hydrogen (at least 3×10^3 cm^{-3}, according to Hartle and Thomas 1974). It was therefore surprising that the Apollo 17 UVS spectrometer onboard the command module did not detect any hydrogen: Fastie et al. (1973) placed an upper limit for H (from the Lyman-alpha emission line at 121.6 nm) of 10 cm^{-3}, and for H_2 (from the Lyman and Werner bands in the FUV) of 1.2×10^4 cm^{-3}. Feldman and Morrison (1991) later revisited the UVS upper limit on H_2 to be 9×10^3 cm^{-3}. It was then speculated by Hodges (1973) that the reaction of solar wind protons with the lunar surface led to the formation of H_2.

Molecular hydrogen is released into the exospheres of the Moon and Mercury by a process referred to as recombinative desorption (e.g. Starukhina 2006), which involves the diffusion to the surface of either bound H atoms released by chemical sputtering (Johnson and Baragiola 1991; Crider and Vondrak 2002), or freshly implanted H atoms (see Fig. 10).

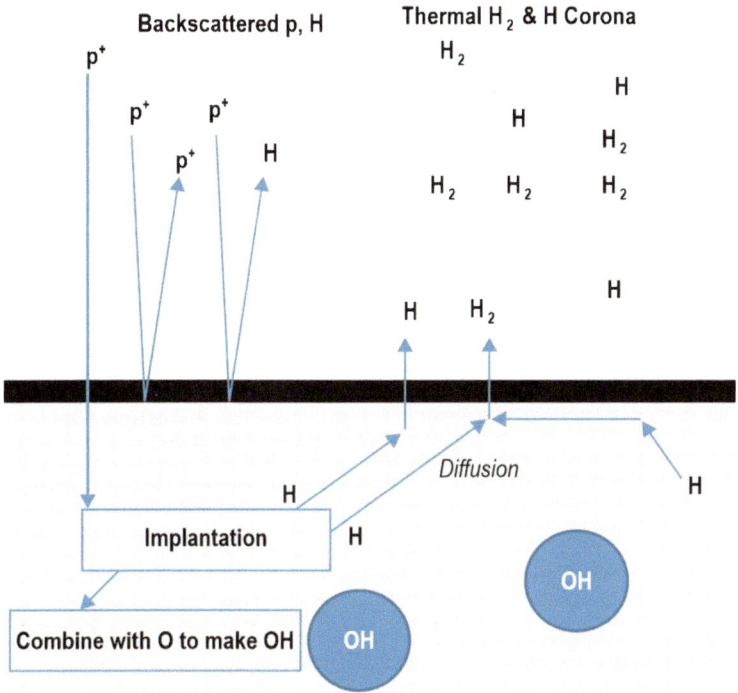

Fig. 10 Mechanism (recombinative desorption) for the creation of H and H₂ exospheres at the Moon or Mercury from solar wind protons and previously implanted H atoms. The diffusion rate depends on the temperature, whereas the implantation rate depends on the solar zenith angle. Reproduced from Tucker et al. (2019)

The global content of H_2 is balanced by the source of incoming solar wind protons, diffusion and formation of H_2 in the surface, and the lifetime of H_2 against thermal (Jeans) escape. The lifetime of H_2 against photoionization ($\sim 10^7$ s) is several orders of magnitude larger than thermal escape (Johnson 1971; Hodges 1974). Because H_2 has a short thermal escape lifetime (hundreds of seconds for subsolar temperatures) compared to the orbital time of the Moon and Mercury, its global distribution is expected to vary directly with changes in the incident proton flux.

The discovery of widespread distribution of H_2O/OH water on the lunar dayside by different instruments – Chandrayaan-1/M³ (Pieters et al. 2009), EPOXI/Deep Impact (Sunshine et al. 2009), Cassini/VIMS (Clark 2009), LRO/LAMP (Hendrix et al. 2019), and the SOFIA airborne telescope (Honniball et al. 2020) – has intensified the debate about the importance of the solar wind in the formation of lunar water (Schörghofer et al. 2021) through reactions between solar wind protons and oxygen (of which the lunar surface is replete). The Lunar Crater Observation and Sensing Satellite (LCROSS) experiment provided additional insight. Molecular hydrogen was detected among the species in the plume following the impact of the LRO Centaur rocket stage in the Permanently Shaded Region (PSR) of Cabeus crater. It was determined that the detected H_2 was not the result of photodissociation of water, but was promptly formed by the impact via combination of two H atoms (Gladstone et al. 2010b; Hurley et al. 2012a). The discovery of energetic neutral hydrogen atoms and solar wind protons backscattered from the lunar surface (see Sect. 5) led Hodges (2011) to postulate that the majority of solar wind protons (98.5%) escapes the Moon as energetic neutral H, a negligible fraction (0.5%) is released as neutral H, and the remaining 1% is simply backscattered as ions. This work discarded the hypothesis that molecular hydrogen was an important constituent of the lunar exosphere. However, H_2 was finally detected by LRO/LAMP on the Moon for the first time (Stern et al. 2013), from the Lyman and Werner bands. It took almost 4 years of twilight observations to build enough signal-to-noise: the spacecraft must be illuminated but the instrument must look at the dark lunar nightside to reduce the background; this geometry only occurs for a few minutes each orbit, near the poles and the terminator, except for when the spacecraft is orbiting along the terminator, but this geometry only occurs for a few days twice a year. The LAMP-derived global H_2 surface density was 1200 $\pm$ 400 cm^{-3} (Stern et al. 2013). Modeling of LAMP observations by Hurley et al. (2017) showed that solar wind chemical sputtering is the dominant source of lunar exospheric H_2, over micrometeoroid impacts and direct physical sputtering. Molecular hydrogen was also detected by the CHACE mass spectrometer onboard Chandrayaan-1, which provided the first detection of H_2 on the dayside. The density was observed to vary in latitude, from $\sim$400 cm^{-3} at $\sim$100 km above the equator to $\sim$800 cm^{-3} at polar latitudes close to the surface (Thampi et al. 2015; see also Fig. 6). The lower densities probably reflect the fact that CHACE observations were carried out when the Moon was inside the geomagnetic tail, which shields the Moon from the solar wind. The LAMP observations showed a dawn/dusk asymmetry in surface density: 1,000 $\pm$ 500 cm^{-3} at dusk and 1,400 $\pm$ 500 cm^{-3} at dawn (Stern et al. 2013). This asymmetry was reproduced by the model of Tucker et al. (2019) which showed that the exospheric concentration of H_2 is increasingly limited by H atom surface diffusion within the subsurface for activation energies $> \sim$0.52 eV. They showed that the variations, over a lunar day, of the rates of diffusion, which depends on temperature, and implantation, which depends on solar zenith angle, combine to give a slight increase of H_2 near dawn compared to dusk. Moreover, using the averaged data of the solar wind flux incident on the surface in and out of the magnetotail, Tucker et al. (2021) showed that the H_2 exospheric density decreases by an order of magnitude when in the magnetotail, a finding consistent with CHACE observations.

Considering the release of H_2 from Mercury to be similar to the Moon, exospheric models have been used to estimate the global surface concentration (Killen and Ip 1999) and altitude profiles of density (Wurz and Lammer 2003). All models agree that H_2 should be one of the most abundant species in Mercury's exosphere, with surface densities on the order of 10^7 cm^{-3}. However, at the time of writing there are no published observational data of H_2 in Mercury's exosphere. Atomic hydrogen (H) has been detected at Mercury by Mariner 10's UVS (Broadfoot et al. 1976) and MESSENGER/MASCS, thanks to the bright Lyman-alpha emission line (121.6 nm; McClintock et al. 2008). Mariner 10 observations revealed two populations, one "hot" at 420 K and one "cold" at 110 K. Work is in progress to model these two populations discovered by Mariner 10 and integrate them with MESSENGER observations, which show a morning enhancement in H above the dayside compared to the afternoon, as well as little emission from H on the nightside (Hurley et al. 2018). It is important to keep in mind that these Lyman-alpha observations are difficult to analyze owing to the substantial background, from both interplanetary hydrogen atoms resonantly scattering solar photons and from dayside scattering of solar H Lyman alpha photons.

2.3.4 Radon and Polonium

Detections of alpha particles resulting from the decay of radon (^{222}Rn) and its radioactive product polonium (^{210}Po) were made by the alpha particle mass spectrometers onboard the

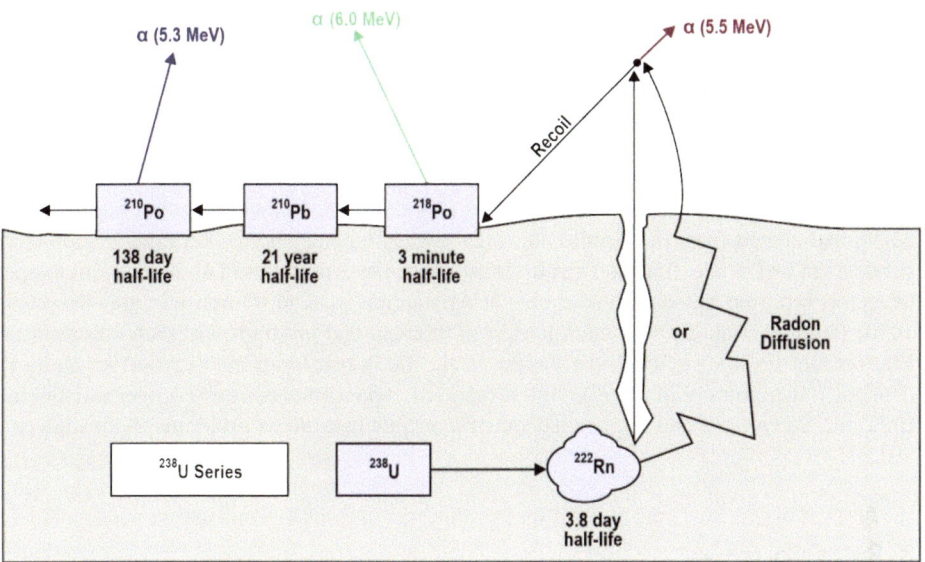

Fig. 11 Scheme of radon decay, with alpha particle energies pertaining to each product. The short half-life of radon makes it a useful species to constrain regions of active outgassing. Adapted from Lawson et al. (2005)

Apollo 15 and 16 command module orbiters (Gorenstein and Bjorkholm 1973; Bjorkholm et al. 1973). Because radon is short-lived (half-life of 3.8 days), it represents another evidence that the Moon is actively outgassing radiogenic elements from its interior. Radon comes ultimately from the radioactive decay of ^{238}U, and ^{210}Po is one of its radiogenic daughters (see Fig. 11).

Because ^{210}Po derives from ^{222}Rn through the intermediate long-term decay of ^{210}Pb, the two species constrain degassing over two different time scales: detection of alpha particles from radon indicates that the outgassing must have happened in the past few days, whereas detection of alpha particles from polonium indicates an outgassing that occurred decades earlier. Friesen and Adams (1976) showed that radon atoms don't migrate directly from grains, where they are formed, to the void, but are carried by other radiogenic elements, for example ^{4}He and ^{40}Ar, during outgassing events. Such events may arise from tidal triggering of fault systems around maria (Runcorn 1977). Also, radon's behavior after it is vented into the lunar exosphere mimics that of other condensable species, with ballistic random hops between one encounter with the surface and the next. The hop length is proportional to the temperature of the surface, so colder surface temperatures results in higher exospheric densities. If radon is vented into the cold nighttime surface, where the temperature is below its freezing point (211 K), it can be adsorbed until dawn, when it is promptly released similar to ^{40}Ar (Heymann and Yaniv 1971; Lambert et al. 1977).

Enhancements of alpha particles from radon were detected above the edges of lunar maria (Gorenstein and Bjorkholm 1973), whereas enhancements of alpha particles from polonium were reported by the Apollo 16 alpha particle spectrometer near Grimaldi crater and the edge of Mare Fecunditatis (Bjorkholm et al. 1973). In a subsequent reanalysis of both spectrometers, Gorenstein et al. (1974) found enhancements of ^{210}Po over edges of all observed maria except Serenitatis.

Other measurements of alpha particles were made by the Alpha Particle Spectrometer (APS) onboard Lunar Prospector (LP). When LP visited the Moon three decades after the

Apollo measurements, it did not detect enhancements of ^{210}Po alpha particles above some regions where detections were made by the Apollo orbiters, such as the Grimaldi crater (Lawson et al. 2005). LP/APS detected enhancements of ^{210}Po alpha particles only above a few maria edges, in contrast with Apollo 15 and 16. One of the few regions that provided an enhancement of polonium in LP/APS data was the Mare Serenitatis, which in contrast was one of the few maria edges without a radon enhancement in the Apollo alpha particle spectrometer data (Gorenstein et al. 1974). This could mean that the radon release mechanism had abated from the Apollo era to LP measurements and/or that other regions have become (more) active (Lawson et al. 2005). Both the Apollo and LP alpha particle spectrometers reported radon release events at Aristarchus plateau (Gorenstein and Bjorkholm 1973; Lawson et al. 2005), which is rich in thorium and uranium. The Selenological and Engineering Explorer (SELENE; Sasaki et al. 2003) spacecraft also carried an alpha-ray detector (Nishimura et al. 2006), which reported enhancements in ^{210}Po over Aristarchus, Imbrium, Serenitatis, and Moscovience maria despite instrument problems (Kinoshita et al. 2012).

3 Refractories

Because of their much stronger bonds with the surface, refractory species are released into the exosphere by more energetic processes than the volatiles discussed earlier. Such processes include micrometeoroid impact vaporization (which peaks near dawn) and sputtering from solar wind and planetary ions. The escape processes for these species are also different. Whereas for light gases such as hydrogen and helium the gravitational (Jeans) escape dominates, photoionization and, to a lesser extent, charge exchange with solar wind ions (mostly protons) and electron impact ionization, are important loss mechanisms for refractories, even though a significant fraction of refractory species ejected by ion sputtering and impact vaporization has sufficient speed to directly escape. As for the volatiles, we concentrate here mostly on species that have been detected – all at Mercury (McClintock et al. 2018; Killen et al. 2018).

3.1 Calcium

Calcium was first discovered in Mercury's exosphere above the polar regions, through high-resolution observations from the Keck telescope of the emission line at 422.7 nm (Bida et al. 2000). MESSENGER/MASCS also observed the Ca emission line at 422.7 nm (McClintock et al. 2008). It was immediately recognized that the calcium in Mercury's exosphere exhibited very high energies, with a scale height consistent with a temperature $>20,000$ K (Killen et al. 2005). Burger et al. (2012), using Monte Carlo simulations of the MASCS data, determined the Ca distribution was consistent with thermal temperatures of as much as 70,000 K (6 eV). Such high energies are necessary to loft the calcium to the high altitudes at which it is observed before it becomes ionized. This conclusion results from the very short photoionization lifetime of the calcium atoms, less than one hour at Mercury's heliocentric distances (Huebner et al. 1992). Killen (2016) suggested that the large scale height of calcium must result from non-thermal processes. Specifically, that calcium is ejected from Mercury's surface by impact vaporization in molecular form and subsequently dissociated by an energetic process such as photodissociation or electron-impact dissociation. The molecular compounds most likely involved are $Ca(OH)_2$, $CaOH$, and/or CaO (Killen et al. 2005; Berezhnoy and

Klumov 2008; Berezhnoy 2018). Using simple photolysis models, Berezhnoy (2013) estimated that the additional energy imparted to Ca-bearing products is 0.6 eV, <0.04 eV, and <0.6 eV for photolysis of CaO, CaOH, and $Ca(OH)_2$, respectively. The photolysis steps are:

1. $Ca(OH)_2 + \gamma = CaOH + OH$
2. $CaOH + \gamma = CaO + H$ or $CaOH + \gamma = Ca + OH$
3. $CaO + \gamma = Ca + O$

Therefore, it seems that even formation of Ca atoms via three steps of photolysis of $Ca(OH)_2$, CaOH, and CaO is unable to produce Ca atoms hotter than about 1.2 eV (the sum of the three imparted energies). This is significantly lower than the 6 eV obtained by Burger et al. (2012). Another possible precursor molecule is CaS. Pfleger et al. (2015) have considered another process to generate energetic calcium: sputtering by solar wind ions precipitating at high latitudes through the magnetic cusps. They found that the Ca exospheric density produced by ion sputtering during nominal solar wind conditions can reach values of 1 cm^{-3}, not insignificant when compared to the $1\text{-}4 \text{ cm}^{-3}$ estimated by Burger et al. (2014). The density can reach even higher values than that if extreme solar events (like coronal mass ejections or high-speed streams) increase the area available to solar wind precipitating ions. Although considered to be a secondary process compared to impact vaporization and subsequent photodissociation, ion sputtering, which at Mercury predominantly occurs at high latitudes, can contribute to the calcium exosphere detected above Mercury's poles by ground-based observations.

The MESSENGER observations confirmed that Mercury's calcium exosphere is centered on the dawn hemisphere and extends anti-sunward of the terminator, consistent with impact vaporization, which peaks at dawn (Pokorný et al. 2018) and indicating that the energization process is probably not photodissociation (Burger et al. 2012). Seasonal variations of the calcium exosphere were modeled by Burger et al. (2014) and subsequently used to determine that the calcium exosphere can be explained by an impact vaporization source centered at dawn. An excess of calcium near TAA = 20° was detected seasonally in the MESSENGER data and is likely due to the intersection of Mercury's orbit with that of the comet 2P/Encke (Killen and Hahn 2015; TAA = True Anomaly Angle is Mercury's angle, along its orbit, from perihelion). Further modeling of the comet 2P/Encke dust torus and its evolution under forces such as Poynting-Robertson drag confirmed the correlation between the position of the calcium excess and the comet Encke dust orbit relative to Mercury's (Christou et al. 2015). Considering different exosphere generation and loss mechanisms, Plainaki et al. (2017) performed simulations of the Ca and CaO neutral environment using the 3-D Monte Carlo exospheric model of Mura et al. (2009). They found that the simulated morphology of the Ca exosphere is consistent with the available MESSENGER observations. According to Plainaki et al. (2017), the generation of a seasonal asymmetric CaO exosphere is expected, with the maximum surface release being on the dawnside-nightside hemisphere, near the equator, because there is where the comet stream particles preferentially impact the planet's surface according to the model by Christou et al. (2015). In addition, an exospheric energetic Ca component, derived from the dissociative ionization and neutralization of CaO, is expected above the same region. The spatial distribution of the thermal Ca exosphere generated by photoionization of the CaO molecules in sunlight is expected to be asymmetric, exhibiting local maxima near the dawn region. Burger et al. (2014) found noticeable differences between the seasonal behavior of calcium and sodium. The Ca exosphere presents a fairly stable year-to-year seasonal dependence, with emission (density) peaks always occurring at dawn near the equator (see Fig. 12).

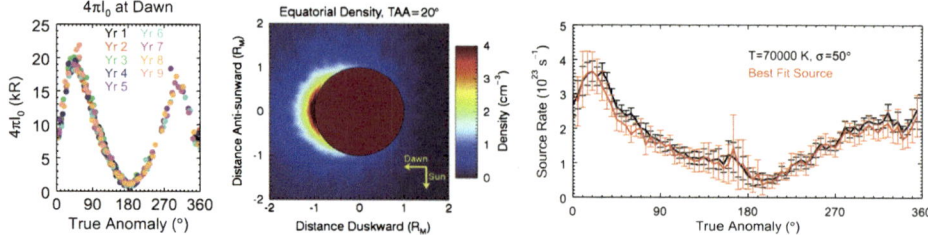

Fig. 12 (Left) Intensity at the surface over Mercury dawn determined from exponential fits to MESSENGER/MASCS limb profiles. Different Mercury years are indicated by different colors. (Center) Ca density in Mercury's equatorial plane at Mercury true anomaly = 20° based on the simple dawn-centered model of Burger et al. (2014) ($T = 70{,}000$ K, $\sigma = 50°$, source rate $= 3.7 \times 10^{23}$ s^{-1}). (Right) Comparison of the source rate determined at all true anomalies using the simple model shown in the center panel to the best-fit source rate at each true anomaly. The simple model works remarkably well. Adapted from Burger et al. (2014)

Thus far, no detection of exospheric calcium has been made at the Moon. The upper limit of the Ca column density in the lunar exosphere is estimated as 9.2×10^7 cm^{-2} (Flynn and Stern 1996). It is possible to estimate the theoretical content of atoms of calcium (or other elements) in the exosphere using a stoichiometric model. A stricter upper limit of Ca column density, 5×10^7 cm^{-2}, was obtained by Berezhnoy et al. (2014) with observations from the Zeiss telescope in Kabardino-Balkaria, Russia, and the Ca depletion factor relative to Na was estimated as >100. This limit is less than that expected from contributions by both impact vaporization and sputtering models (Sarantos et al. 2012). These observations can be explained by condensation of Ca-containing species in impact-produced clouds upon collisions between meteoroids and the Moon (Berezhnoy 2013).

3.2 Magnesium

Magnesium (Mg) was discovered in Mercury's exosphere from the emission line at 285.2 nm during MESSENGER's second flyby (McClintock et al. 2009). Mg was found at high distances from the planet and high altitudes. Sarantos et al. (2011), analyzing the MASCS flyby data, found that the Mg exosphere is consistent with two populations: a hot component ($T > 20{,}000$ K) and a colder component ($T < 5{,}000$ K). MESSENGER orbital data analyzed by Merkel et al. (2017) showed that there is an enhancement in the exospheric Mg in the morning (6–9 AM local time) near perihelion, that the bulk temperature is $\sim$6,000 K, at times as low as $\sim$3,700 K or as high as $\sim$10,400 K, and that the production rate is strongest in the morning on the inbound leg of the orbit, i.e. TAA $> 180°$. Although Merkel et al. found occasional temperatures $>10{,}000$ K, consistent with the hotter component observed during the flybys (Sarantos et al. 2011), no observations from the orbital phase confirmed the colder component, although the lower end of the Merkel et al. temperatures ($\sim$3700 K) is close to the upper end of the Sarantos et al. colder component ($\sim$5,000 K).

In a follow-up paper, Merkel et al. (2018) showed that the Mg column density is greatest over the Mg-rich terrain as measured by MESSENGER's X-Ray spectrometer (XRS; Schlemm et al. 2007). Merkel et al. (2018) concluded that the main Mg source process is impact vaporization. However, the temperature as inferred from the scale height is almost twice that expected from impact vaporization. Figure 13 summarizes the Merkel et al. (2018) findings. Namely, the Mg source rate is higher for those years when the Mg-rich terrain is exposed at dawn at perihelion, compared to those years when the antipodal terrain is exposed at dawn at perihelion (because of the 3:2 spin-orbit resonance, a given longitude is

Fig. 13 Summary of MESSENGER/MASCS observations of Mg over two Mercury years. Top: MASCS observations (circles, color coded by Mercury year) over a Mg/Si elemental weight ratio composite map derived from MESSENGER/XRS measurements (Weider et al. 2015). Middle: temperature fit (using the model of Chamberlain 1963) to MASCS observations. It shows how the temperature from the emission lines (4,000–8,000 K) is independent on the year. Bottom: the retrieved production rate of Mg. It shows how observations in red (years when the Mg-rich terrain is exposed at dawn at perihelion) are consistent with a higher production rate than observations in blue (years when the terrain antipodal to the Mg-rich terrain is exposed at dawn at the perihelion). Adapted from Merkel et al. (2018)

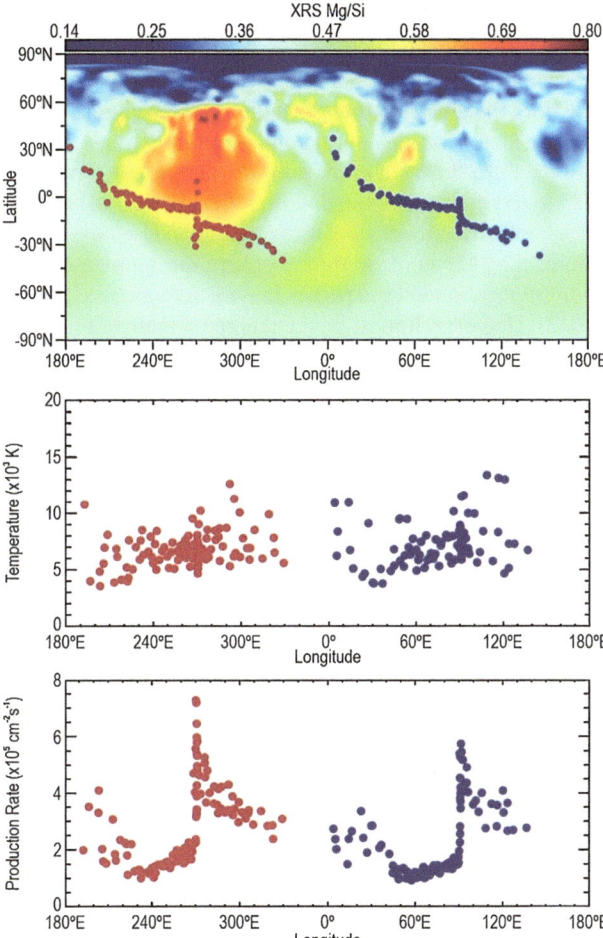

exposed at a given local time every other year; Domingue et al. 2007). This is the first time that a direct link between the composition of Mercury's surface and that of the exosphere has been established.

As with calcium, it is clear that at times an energetic process like ion sputtering or dissociation of a molecular precursor is responsible for ejection of Mg into the exosphere, but at other times impact vaporization dominates. Although the spatial distribution of Mg is not consistent with an ion-sputtering source, a portion of the atomic Mg could be from dissociation of a precursor molecule, similar to Ca. Quenching theory predicts that meteoroid bombardment is an effective source of MgO, Mg, and MgOH in the exosphere of Mercury (Berezhnoy 2018). The energy of Mg atoms produced via photolysis of MgO and MgOH is estimated as 0.4 eV and < 0.6 eV, respectively (Berezhnoy 2013). Agreement between observed and theoretical column density of Mg atoms from photolysis and impact vaporization (2×10^9 cm^{-2}; Merkel et al. 2018) suggests that meteoroid bombardment is the main source of Mg atoms in Mercury's exosphere (Berezhnoy 2018).

There has been no detection of Mg in the lunar exosphere. The upper limit of the intensity of the MgI 285.2 nm emission line in the lunar exosphere was estimated as 53 Rayleighs, corresponding to an exospheric surface density of Mg of 6,000 cm^{-3}, whereas the theoretical value from stoichiometric models is estimated as 476 R (Stern et al. 1997). LRO/LAMP placed an even stricter upper limit for the Mg surface density of 3.4 cm^{-3} near the terminator from the emission line at 182.8 nm (Cook et al. 2013). This value is slightly higher than that predicted by considering only sputtering as a source of Mg atoms in the lunar exosphere (1.0–1.5 cm^{-3}; Wurz et al. 2007), whereas the expected near-surface density from impact vaporization was estimated to be 5 cm^{-3} (Sarantos et al. 2012). The difference between the stoichiometric model and observations can also be explained by less effective delivery of Mg atoms than Na atoms to the exosphere during meteoroid bombardment owing to condensation of Mg-containing species in collisions between meteoroids and the Moon (Berezhnoy 2013). However, it must be recognized that there is a substantial stoichiometric discrepancy between e.g. Na and O in Mercury's exosphere. This discrepancy calls into question whether or not this is a viable assumption to estimate densities for certain species.

3.3 Other Refractories (Al, Fe, Mn)

A handful of other refractory species have been detected at Mercury by ground-based or MESSENGER observations. Aluminum (Al) and iron (Fe) were discovered using the Keck telescope (Bida and Killen 2011), and subsequently manganese (Mn) was discovered by MESSENGER/MASCS (Vervack et al. 2016). Whereas the Keck observations only detected a single line of Al, MESSENGER definitively confirmed the presence of the weaker ground-based detection by observing both lines of Al near 394-396 nm (Vervack et al. 2016). However, MESSENGER did not confirm the detection of Fe despite searches for several Fe lines. Al and Mn were only sporadically observed by MESSENGER, but there was a correlation between the TAA of the Encke-related peak in Ca and the TAA at which MESSENGER observed Al and Mn that suggests these two weakly emitting species may also be related to the comet Encke dust trail (Vervack et al. 2016). If this is the case, we might expect that the release of these species is dominated by meteoroid impact vaporization as with Ca, and that there might be an association, in part, with a molecular origin. Bida and Killen (2017) showed that Fe in Mercury's exosphere increases with altitude, which is evidence for a molecular origin of the neutral atomic species, similar to Ca. On the other hand, in the ground-based observations, Al shows a more normal exponential decrease (Bida and Killen 2017), consistent with a hot exosphere (6,000–8,000 K) like that of Mg but not as extreme as that of Ca. Given that impact vaporization is expected to produce a plume at ∼3,500 K (e.g. Berezhnoy and Klumov 2008), some additional process is necessary to result in a >6,000 K exosphere. In contrast, the MESSENGER observations showed that Al may exhibit a flat to increasing profile with altitude, similar in structure to that found by Bida and Killen (2017) for Fe and thus suggesting a molecular species may be involved. MESSENGER observations of Mn show a completely different altitude distribution from that observed for Al and Ca^{+} (see Fig. 14). Given that the Al and Mn were observed at TAA roughly consistent with the comet Encke dust trail crossing, this different altitude structure may suggest a cometary origin for Mn, or at least a very different process for releasing Mn from Mercury's surface (Vervack et al. 2016). However, both the ground-based and MESSENGER datasets probed the pre-dawn region of the exosphere where the effects of the planet's shadow must be taken into account for the proper interpretation of any observations. Therefore, models need to be constructed to investigate the true profiles for all of these species.

Considering meteoroid bombardment as a source of Fe, Al, and Mn atoms in Mercury's exosphere, the main Fe-, Al-, and Mn-containing species delivered to the exosphere via impacts are Fe, FeO, AlOH, AlO, Al(OH)$_2$, and Mn (Berezhnoy 2018). The theoretical

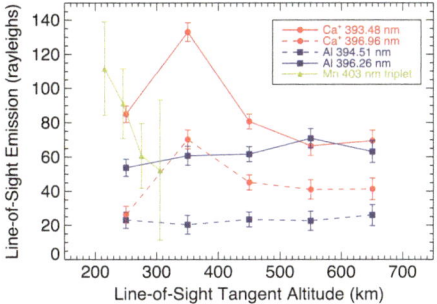

Fig. 14 Line-of-sight tangent altitude profiles of Mn, Al, and Ca detected by MESSENGER/MASCS (spacecraft motion during the measurement of these profiles means they are not strictly radial profiles). The peculiar altitude profile of Mn, different from that of Ca^+ or Al even though observed with similar geometry, when coupled with the timing in Mercury's true anomaly angle, suggests that the Mn may be of cometary origin owing to a possible association with the comet 2P/Encke dust trail. Reproduced from Vervack et al. (2016)

column density of impact-produced Fe atoms, 1.2×10^9 cm^{-2}, agrees well with the observed column density (8.2×10^8 cm^{-2}; Bida and Killen 2017). However, photolysis of FeO leads to production of Fe atoms with energy of about 0.3 eV (Chestakov et al. 2005). This is significantly lower than the typical energy of Fe atoms observed in Mercury's exosphere (~ 1 eV; Bida and Killen 2017). This difference in energy of Fe atoms can be explained if Fe atoms are delivered to Mercury's exosphere mainly by several steps of photolysis of impact-produced FeOH and Fe(OH)$_2$ molecules and its photolysis products. The theoretical column density of photolysis-generated Al atoms, about 10^6 cm^{-2} (Berezhnoy 2018), is significantly lower than the observed value, 7.7×10^7 cm^{-2} (Vervack et al. 2016). Such a low theoretical column density of Al atoms is explained by the effective condensation of Al-containing species during the expansion of impact-produced vapor. The theoretical column density of impact-produced Mn atoms during quiet times is about half the observed value, about 3×10^7 cm^{-2} (Berezhnoy 2018). This difference can be explained by an increased flux of impactors during the MESSENGER observations owing to timing of the observations and Mercury's crossing of the comet Encke dust trail. It is expected that the initial temperature of impact-produced Mn atoms is about 3,000 K because Mn is produced mainly in the form of atoms during impact events (Berezhnoy 2018). However, the temperature of Mn atoms in Mercury's exosphere has not yet been measured.

4 Missing Species

There are several species that are expected to be present in the exospheres of the Moon and Mercury, some in quantities that should have been detected by the past or current instruments, but were not. On the Moon, these include for example nitrogen (N$_2$), carbon dioxide (CO$_2$), magnesium,, and calcium. The last two of these, plus mercury (Hg) and carbon monoxide (CO) were detected by LAMP in the LCROSS impact plume, as species permanently trapped within the Permanently Shadowed Region (PSR) of Cabeus crater and released by the impact (Gladstone et al. 2010b). For some of the other species, LRO/LAMP provided more stringent upper limits for the lunar exosphere, most of them several orders of magnitude lower than previous estimates (Cook et al. 2013).

Lithium (Li) is the third most abundant alkali element in the Solar System after Na and K. The average content of Na, K, and Li in norites in returned lunar samples is equal to 3,000,

1,500, and 12.3 ppm, respectively (Lodders and Fegley 1998). Lithium has a high emission rate (g-factor) for the 670.8 nm emission lines of 16 photons atom^{-1} s^{-1} at 1 AU (Sullivan and Hunten 1964; the g-factor g is the number of solar photons resonantly scattered by each argon atom each second, and in optically thin exospheres it relates the observed intensity I with the column density N with the formula $I = g \cdot N$). This emission rate is higher than that of either the Na 589.0 nm or K 769.9 nm resonance lines, and thus it should favor the search for Li in the exospheres of the Moon and Mercury. However, Li has not been detected so far at either Mercury or the Moon. Several factors decrease the content of exospheric Li atoms. Its photoionization lifetime for quiet Sun, 5100 s, is much shorter than that of sodium (Na), 1.4×10^5 s, and potassium (K), 3.7×10^4 s (Huebner and Mukherjee 2015). Lithium is a light element, and as such it has a faster escape rate from the exosphere (especially at the Moon) in comparison with heavier Na and K atoms.

Spectroscopic searches for Li emission lines at 670.88 nm in the exosphere of Mercury were performed by Sprague et al. (1996) and by Doressoundiram et al. (2009), who reported upper limits for the zenith column density of Li atoms of 8.4×10^7 cm^{-2} and 4×10^7 cm^{-2}, respectively. This column density can be compared to typical Na zenith column densities, 1.5×10^{11} cm^{-2} (Potter and Morgan 1985) to give an upper limit for the Li/Na ratio on the order of 10^{-4}. The Li content on the surface of Mercury is still unknown, so theoretical estimates of Li content in Mercury's exosphere are absent. On the Moon, the upper limit of zenith column density of Li atoms in the exosphere is 1.1×10^6 cm^{-2}, from Flynn and Stern (1996). These authors also reported upper limits of intensities of resonance lines of other alkali atoms (230 Rayleighs for Rb at 780.0 nm and 520 Rayleighs for Cs at 852.1 nm), without converting them to zenith column densities owing to the lack of reliable g-factors (the unit Rayleigh is defined as: $1 \text{ R} = 10^6/4\pi$ photons cm^{-2} s^{-1} sr^{-1}; Hunten et al. 1956). The observations of Flynn and Stern (1996) were performed 20$''$ above the subsolar point near quarter Moon at the most suitable conditions to search for photon-desorbed exospheric atoms. The theoretical intensity of the Li emission lines at 670.8 nm in that region is estimated at 46 R, using a Li-Na stoichiometric model. The assumptions of this model are that the temperature of Na and Li atoms is the same (1,000 K) and that the physical parameters of Na and Li atoms in the exosphere and on the surface of the Moon (sticking coefficients, thermal evaporation rates, accommodation coefficients, diffusion coefficients) are the same. Differences in photoionization rates of Na and Li are also taken into account. However, the observed upper limit of the intensity of the Li 670.8 nm emission lines is only 17 R (Flynn and Stern 1996). Thus, one can tentatively conclude that the behavior of Li in the exosphere of the Moon is different from that of Na. An upper limit of Li zenith column density above the north pole of the Moon during the activity of the 2009 Perseid meteor shower is estimated as 4.9×10^6 cm^{-2} (Berezhnoy et al. 2014). The depletion factor of Li in the lunar exosphere in comparison with Na is found to be >1.6.

The behavior of Li during collisions of meteoroids with the surface of the Moon has been studied theoretically through quenching theory of the chemical composition of impact-produced vapor clouds. Impacts of meteoroids lead to delivery of LiOH, Li, LiO, and LiCl to the exosphere of the Moon (Berezhnoy 2013). LiOH is the main Li-containing impact-produced compound at temperatures of quenching of chemical reactions $<3,700$ K, typical for collisions of meteoroids exceeding 3 cm in radius. Photolysis lifetimes of LiO and LiCl at 1 AU for quiet Sun are equal to 28 and 225 s, respectively, whereas typical velocities of Li atoms produced upon LiO and LiCl photolysis are calculated as 2.6 and 3.8 km/s, respectively (Valiev et al. 2020). The LiOH photolysis lifetime at 1 AU for quiet Sun is estimated as 900 s, and the typical energy of Li atoms produced upon LiOH photolysis is estimated as 1.8 eV (Berezhnoy 2013). Therefore, photolysis lifetimes of the main

Li-containing impact-produced species are shorter than or comparable to typical ballistic flight times of these species ($\sim 10^3$ s). This leads to effective photolysis of impact-produced Li-containing species during the first ballistic flight and therefore to enhancement of hot photolysis-generated Li atoms in the exospheres of the Moon and Mercury during periods of active meteoroid bombardment. Such hot Li atoms could be detected during future observations of Li in the lunar exosphere.

Sulfur (S) is also expected to be present in Mercury's exosphere, especially above the hollows and the Mg-rich areas, but it was not seen in the MESSENGER/MASCS spectra, most likely owing to its small g-factor. The sulfur surface abundance was published for some regions (Weider et al. 2015) and appears to be correlated with regions where Mg and Ca are also enhanced. Moreover, S is enhanced over its average abundance by up to a factor of 5 in the Mg-rich region ($30°$–$60°$ N, $240°$–$300°$ E). In fact, it is speculated that the "light blue" regions surrounding the hollows are sulfur-containing volatiles (Nittler et al. 2011). Hollows are rare in the Caloris Basin (Thomas et al. 2014), where the surface concentration of S is also low (Weider et al. 2015). Theoretical estimates of the S column density in Mercury's exosphere (6×10^7 cm^{-2} from Wurz et al. 2010; 10^9 cm^{-2} from Berezhnoy 2018; 2×10^{10} cm^{-2} from Morgan and Killen 1997; and 2×10^{13} cm^{-2} from Sprague et al. 1995) are inconsistent. Recent laboratory experiments suggest that photon-stimulated desorption of S from MgS, a proxy for the global form of S on Mercury's surface, may provide a global, additional source of S at low altitudes of Mercury's exosphere (Schaible et al. 2020).

Doressoundiram et al. (2009) reported upper limits for the Mercury's exosphere of silicon (Si) of 5×10^{10} cm^{-2}) from the European Southern Observatory – New Technology Telescope in La Silla, Chile. An upper limit of Si from the Moon from Flynn and Stern (1996) appears to have been obtained using an excited line (390.6 nm) that is not expected to be populated (Sarantos et al. 2012).

Oxygen (O) represents a quandary. The published Mariner 10 results provide a generous upper limit for the O column density (emission line at 130.4 nm) of $\sim 10^{11}$ cm^{-2} (Broadfoot et al. 1974), on par with that of sodium. However, no oxygen emission at the 130.4 nm line (or the forbidden line at 135.6 nm) was detected with MESSENGER/MASCS, despite its higher sensitivity compared to the Mariner 10 UVS (Vervack et al. 2016). Column densities reported by Mariner 10 would have been detected by MASCS without difficulty. Vervack et al. (2016) proposed three explanations: the oxygen exosphere was significantly more abundant in 1974 than today; the Mariner 10 "detections" were only upper limits; or the Mariner 10 observations were somehow in error. On the Moon, oxygen has long eluded detection, both from mass spectrometers and from spectrographs. Hodges et al. (1974) noted that the absence of O and O_2 in the lunar exosphere from the LACE mass spectrometer is understandable, if we consider that the Moon is less than fully oxidized, even though O is one of the major constituents of the lunar surface. LACE upper limits for molecular oxygen (O_2) in the lunar exosphere were 100 cm^{-3} (Hoffman and Hodges 1975), which is roughly the sensitivity threshold of LACE (Hoffman et al. 1973). Oxygen has been detected on the Moon (Vorburger et al. 2014), but only as energetic sputtered species (see Sect. 5.2). The derived exospheric surface density (11 cm^{-3} at the subsolar point) is consistent with the LRO/LAMP upper limits (Cook et al. 2013) and predictions based on solar wind sputtering (Wurz et al. 2007).

A number of metallic constituents of the lunar exosphere were expected to be identified by the LADEE mission according to pre-flight calculations (Sarantos et al. 2012). Preliminary detections of Ti, Mg, and Al in the lunar exosphere were reported by Colaprete et al. (2016a) from the LADEE Ultraviolet/Visible Spectrometer (UVS; Colaprete et al. 2014). Line strengths of Ti and Mg decrease shortly after full moon, indicative of a dependence on solar wind. Line strengths of Al show a correlation with Geminids meteoroid stream, indicative of a meteoroid impact vaporization source. However, no density or column abundances have been derived to date from LADEE/UVS. The upper limit for exospheric surface density of Al from LRO/LAMP, 1.1 cm^{-3} (Cook et al. 2013), is close to the range predicted by considering sputtering as the main source of Al atoms in the lunar exosphere: 0.5–1.5 cm^{-3} (Wurz et al. 2007), but is lower than the density expected from impact vaporization (Sarantos et al. 2012). The efficiency of delivery of Al and Fe atoms to the lunar exosphere during meteoroid bombardment is not as high as that for alkali elements Li, Na, and K owing to condensation of Al- and Fe-containing species during expansion of impact-produced cloud and formation of slowly photolyzed Al-containing species in the impact vapor (Berezhnoy 2013).

5 Ions and ENAs

Ions and Energetic Neutral Atoms (ENAs) are important to infer loss rates, interaction between the surface and the solar wind, and even properties of the neutral exospheres. We briefly summarize here the discoveries made on the Moon and Mercury. A more thorough analysis is reported in Wurz et al. (2021).

5.1 Ions

Ions offer the opportunity to study the primary loss process of exospheric neutrals (with the exception of H and He, which escape predominantly with the Jeans mechanism), i.e. photo-ionization, electron-impact excitation, and charge-exchange with the solar wind ions (mainly protons). As Hartle and Killen (2006) have pointed out, with proper modeling tools it is possible to backtrace the ion to its origin at the surface, provided that the solar wind velocity and the interplanetary magnetic field are known. This is the technique used for example to infer exospheric properties from measurements of the lunar ionosphere (e.g. Poppe et al. 2013).

Ions of lunar origin have been measured on the surface by the SIDE detectors (Sect. 2.3.1), in lunar orbit by instruments onboard SELENE, Chang'E-1, LADEE, and ARTEMIS (e.g. Yokota et al. 2014; Saito et al. 2010a; Wang et al. 2011; Halekas et al. 2011, 2012, 2013, 2016; Poppe et al. 2012, 2016), and at more distant locations by instruments on board the WIND and AMPTE spacecraft (Mall et al. 1998; Hilchenbach et al. 1993). Detections or inferred detections to date include H_2^+, He^+, C^+, O^+, Ne^+, Na^+, Al^+, $CO^+/Si^+/N_2^+$, K^+, Ar^+/Ca^+, and Fe^+. The relative abundance of even the most common ion species remains in doubt, in part owing to the different observation geometries, but also to ambiguity regarding the source of the ions.

Ions around the Moon come both from ionization of exospheric neutrals and directly from the surface (Yokota et al. 2009; Tanaka et al. 2009). The interactions of solar photons, solar wind ions, and interplanetary dust with the regolith can all lead to emission of both ions and neutral particles (Elphic et al. 1991; Madey et al. 1998). SELENE, Chandrayaan-1, and ARTEMIS detected low-energy protons reflected from the lunar surface. These measurements showed that between 0.1% and 1.0% of the incoming solar wind protons are backscattered (Saito et al. 2008; Lue et al. 2014, 2018). H_2^+ was detected by LADEE/NMS (Halekas et al. 2015) and Solar Wind Ion Detectors (SWID) onboard Chang'E-1 (Wang et al. 2011). Recent analyses of SELENE data also reveal C^+, apparently derived from the lunar surface (Yokota et al. 2020) and hinting at the importance of a carbon cycle at the Moon

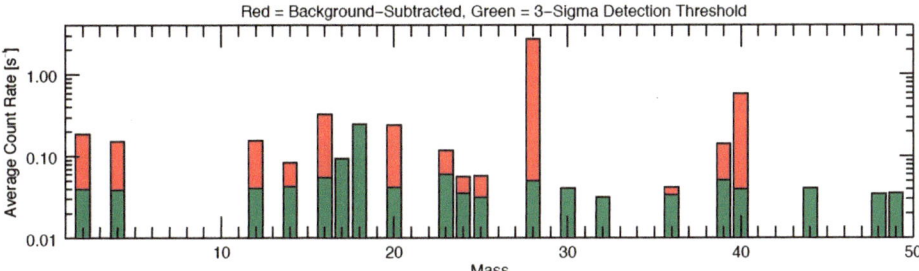

Fig. 15 Mass spectrum of lunar ions detected by LADEE/NMS. Candidates for the substantial peak at m/q = 28 are N_2^+, Si^+, and CO^+, with the latter one being the most plausible given the lower photo-ionization yields of the other two. Adapted from Halekas et al. (2015) with the addition of O^+ signal at mass 16

(see also Sect. 2.3.2). LADEE/NMS, which observed low-energy ions produced locally in the exosphere, found the highest fluxes (in order) for $CO^+/Si^+/N_2^+$, Ar^+/Ca^+, O^+, and Ne^+ (Halekas et al. 2015; see Fig. 15).

The Ar^+ and Ne^+ signals appear consistent with neutral composition data that reveal high abundances of these noble gases (Benna et al. 2015). However, the peak at 28 amu remains puzzling, with CO^+ the most plausible species (as noted in Sect. 2.3.2, neutral CO is difficult to measure owing to the instrumental background of LADEE/NMS). Neutral CO has not been identified in the lunar exosphere or in lunar polar deposits (where it could be released by micrometeoroid impacts or solar wind ion sputtering), but it is a byproduct of exothermic reactions involving solar wind C and the surface (Hodges 2016), and, as mentioned in Sect. 2.3.2, could represent a more substantial exosphere than CH_4 (which peaks at a few hundreds of cm^{-3}). Moreover, since CO can photodissociate to form O^+ and C^+, its presence may help explain the surprising detections of those two ions (also observed by other lunar missions), otherwise difficult to reconcile with spectroscopic limits of their neutral counterparts (Cook et al. 2013 and Sect. 4). SELENE detected O^+ ions with energy 1-10 keV only when the Moon was in Earth's plasma sheet: Terada et al. (2017) concluded that these are terrestrial oxygen ions transported to the Moon by Earth's wind, reminiscent of the "shared" Earth-Moon neutral exosphere mentioned in Sect. 2.1.

At Mercury, like on the Moon, ions of planetary origin come primarily from photoionization of exospheric neutrals and directly through surface processes (Killen et al. 2007). Most observations of Mercury planetary ions come from MESSENGER's Fast Imaging Plasma Spectrometer (FIPS), part of the Energetic Particle and Plasma Spectrometer (EPPS; Andrews et al. 2007). MESSENGER reported He^+, O^+ and Na^+ on essentially every one of the >4,100 orbits as well as in the initial flybys (Zurbuchen et al. 2008, 2011; see Fig. 16).

Two of these ions, O^+ and Na^+, are reported as part of mass per charge (m/q) groups, the O^+ group (m/q 16-20) and the Na^+ group (m/q 21-30), owing to the low resolution of the FIPS instrument. These ions are concentrated in several regions of Mercury's magnetosphere, primarily the cusps and central plasma sheet (Raines et al. 2013). In the central plasma sheet, their density has been estimated at 0.1–1.0 cm^{-3} (Gershman et al. 2014), which is only about 10% of the H^+ number density but up to 50% of the mass density there. Cusp densities have not been published but appear to be at least as high. One of the most surprising results from the first planetary ion measurements was the high energy of planetary ions in the northern magnetospheric cusp, with ions of energy >1 keV being regularly observed (Raines et al. 2014). That study also reported the first indications of ions upwelling in the cusp, possibly owing to solar wind sputtering there. MESSENGER ob-

Fig. 16 Mass spectrum of ions detected at Mercury by FIPS during MESSENGER's first flyby (January 2008). Multiply charged ions (such as O^{++}, Si^{++}, and Mg^{++}) are observed mostly below m/q $\sim$ 12, even though Fe^{++} is observed at m/q = 28. Dashed curves are Gaussian fits to the major peaks, and the solid blue curve is their sum. Adapted from Zurbuchen et al. (2008). Reprinted with permission from AAAS

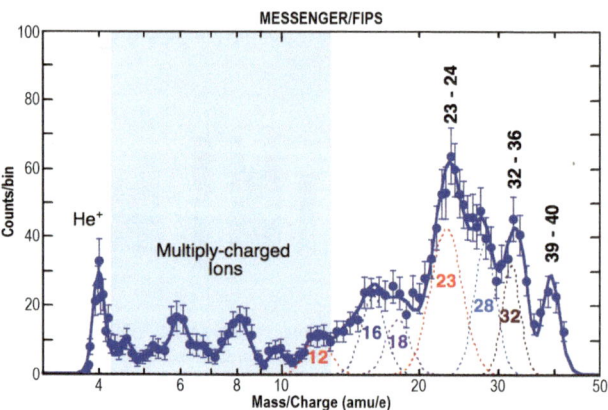

served planetary ions throughout the magnetosphere as well as in the magnetosheath and beyond the bow shock, though lower in numbers than the cusp or plasma sheet. Thermal ions were not observed directly ($\sim$1 eV) as the lower energy bound of the MESSENGER instrument was about 50 eV (Andrews et al. 2007). Calcium ions (Ca^+) have been detected by MESSENGER/MASCS (Vervack et al. 2010, 2016) through emission in the 393.5 and 397.0 nm lines but not with FIPS because of its low mass resolution and possible overlap with other ions such as K^+. MASCS observed Ca^+ emission in two instances. The first was during MESSENGER's third flyby, when emission was observed in the region tailward of the near-planet reconnection line (x-line; Vervack et al. 2010). This implies that a convection mechanism in the magnetosphere may be at play. The similarity between Ca and Ca^+ line of sight column densities for this observation was a surprise, because the two species have very different velocities (Ca^+ 100 s of km/s; Ca: few km/s). The second instance was during the same observations in which MASCS detected Al and Mn (see Fig. 14), suggesting that there might be a connection to the enhanced neutral Ca abundances MESSENGER observed during the interaction of Mercury with comet Encke dust.

Despite Mercury's planetary magnetic field, solar wind ions and electrons can still impinge on its surface, precipitating through the cusps, causing ion sputtering and electron-stimulated desorption (ESD). The behavior and effect of the solar wind precipitation has been modeled extensively (Kallio and Janhunen 2003a, 2003b; Massetti et al. 2007; Benna et al. 2010). It has been difficult to make a definite link between precipitation and exospheric production in observations, due at least in part to the dynamic nature of Mercury's magnetosphere (Milillo et al. 2005), but several studies have provided indications of this connection. Orsini et al. (2018) showed that episodic enhancements in ground-based observations were associated with a passing Coronal Mass Ejection (CME). Jasinski et al. (2020) showed that short-term enhancements in Na-group ions outside Mercury's bow shock could be most logically explained by an episodic and local enhancement in the Na exosphere. Raines et al. (2017) attributed a large but delayed increase in He^+ to a several-day enhancement in the He exosphere, which in turn resulted from the impact of a CME particularly enriched in He^{2+}, contrary to what was reported at the Moon, where the exospheric helium density measured by LADEE/NMS increased promptly with the passage of a CME (Hurley et al. 2016; see Fig. 2, where the passage of the CME is visible in the peak near day 400). Prior to MESSENGER, it was thought that the extreme solar wind environment at Mercury could lead to the stripping away of the entire dayside magnetosphere, causing direct bombardment by the solar wind across the full dayside surface (e.g. Slavin et al. 2007) like on

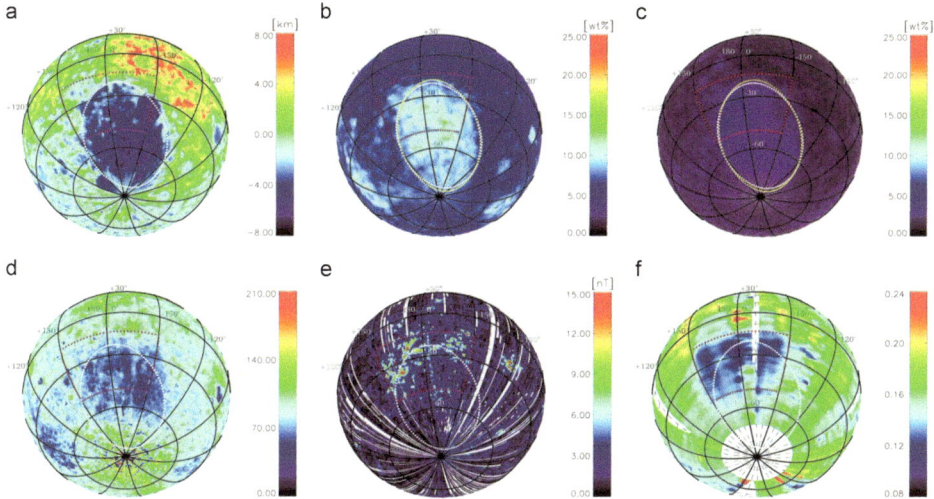

Fig. 17 This composite image illustrates how ENA reflection (map in panel f) is predominantly correlated with lunar magnetic anomalies at the surface (see magnetic field at 30 km altitude from Lunar Prospector in panel e), rather than with topography (Clementine laser altimeter data in panel a), surface composition (Lunar Prospector gamma-ray spectrometer measurements of Fe and Th in panels b and c, respectively), or albedo (Clementine spectral reflectance mosaic at 750 nm in panel d). ENAs are therefore a useful tool for studying the exosphere-surface interaction, particularly on magnetic anomalies. Adapted from Vorburger et al. (2015)

the Moon. Following MESSENGER, it became clear that this was a much rarer condition, as Mercury's planetary field would react via magnetic induction to counteract the effects (Slavin et al. 2014; Jia et al. 2019). However, a small number of "disappearing dayside magnetosphere" events were observed (Slavin et al. 2019; Winslow et al. 2020), where the closed field region of Mercury's dayside magnetosphere was reduced below the altitude of the MESSENGER spacecraft (275–400 km). During these events, substantial portions of Mercury's dayside surface may have been subjected to bombardment by solar wind plasma from Mercury's magnetosheath. Sun et al. (2020), reported an analogous event on Mercury's nightside, where the central plasma sheet may have been forced down to the nightside surface, from its normal position at hundreds km away.

5.2 ENAs

Energetic Neutral Atoms (ENAs) are another useful tool to study the structure of the exosphere and its relationship with the surface. ENAs are solar wind ions that are backscattered as neutrals from the lunar surface with about 10% of the original particles' energy. Traveling at about 140 km/s (~100 eV) and being neutrals, ENAs travel in straight trajectories. Therefore, it is possible to trace detected ENAs back to their place of origin (much more easily than for ions) and to build a map of locations where ENAs are reflected (e.g. Vorburger et al. 2015; Lue et al. 2016; see Fig. 17).

With the ENAs mainly originating in the solar wind (see below for an exception example), most ENAs consist of hydrogen atoms. Interstellar Boundary Explorer (IBEX; McComas et al. 2009a) made the first detection of energetic neutral hydrogen at the Moon (McComas et al. 2009b). IBEX measurements were consistent with 10% of solar wind protons being converted in energetic neutral hydrogen atoms and reflected back with a broad energy range, in any case lower than the solar wind's ~1 keV. The Chandrayaan-1 Energetic

Neutral Atom (CENA), part of the Sub-keV Atom Reflecting Analyzer (SARA; Barabash et al. 2009) onboard Chandrayaan-1, also detected energetic neutral hydrogen atoms, and the inferred fraction of solar wind protons reflected as such was higher: $\sim$20% (Wieser et al. 2009; Futaana et al. 2012).

Helium is the second most abundant element in the solar wind ($\sim$3.8%), surpassed only by hydrogen ($\sim$96%)—see e.g. Table 1 in Von Steiger et al. (2000). It is therefore expected that the total backscattered lunar ENA flux also includes reflected He particles. Indeed, in 2014, CENA measured for the first time alpha particles backscattered from the lunar surface as helium ENAs (Vorburger et al. 2014). The characteristic energy of the helium ENAs is roughly four times the characteristic energy of the hydrogen ENAs, agreeing with particle reflection theory. The measured helium to hydrogen ratio in the CENA mass spectra equaled 0.37×10^{-3}. Unfortunately, CENA's geometric factor (detection efficiency) for helium has not been accurately determined, making it difficult to convert the measured He/H ratio into the He/H ratio actually present in the reflected ENAs. Based on experience with ENA instrumentation, though, the Vorburger et al. (2014) estimate that the actual He content is 10 times higher than determined, implying a He backscatter ratio of 1.4% (compared to the H reflection ratio of 16%).

Chandrayaan-1/CENA also measured lunar surface sputtered oxygen ENAs for the first time (Vorburger et al. 2014). These oxygen atoms do not originate in the solar wind, but are ejected from the topmost surface layer as the surface is irradiated with solar wind ions. Having characteristic energies of a few eV (compared to backscattered particles, which have characteristic energies of $\sim$100 eV; Wurz et al. 2021), these particles are on the lower end of the energy range covered by ENA detectors. Nevertheless, a clear, persistent oxygen signal was observed in the CENA mass spectra, amounting to $\sim$20-40% of the backscattered hydrogen ENA flux. Inferred surface and column densities were on the order of $\sim 10^7$ cm^{-3} and $\sim 10^{13}$ cm^{-2}, respectively. The Advanced Small Analyzer for Neutrals (Wieser et al. 2020a) onboard the Yutu-2 rover of the Chang'E-4 mission also detected ENAs of mass larger than 4 amu at the lunar surface, highly variable in abundance and confined to energies below 100 eV (Wieser et al. 2020b). Whereas this is most probably also sputtered oxygen, the authors note that better statistics and more observations are needed for further characterization.

6 Summary

We have discussed here species that represent the extrema of volatility (mobility) in surface-bounded exospheres in the inner Solar System. Each type of species adds a piece to the puzzle of the complex interaction between the airless bodies and the external environment (solar wind, meteoroids, and solar photons). Both the volatiles He and H$_2$ shed light on the solar wind's role in refilling the lunar exosphere, but each offers its own unique perspective on the exosphere production: H$_2$ addresses the important aspect of what fraction of the lunar water is of solar wind origin; whereas He is a useful species to understand the still poorly known gas-surface interaction. Argon (^{40}Ar), radon, and to a lesser extent helium, offer the tantalizing opportunity to quantify the amount of radiogenic elements of internal origin actively outgassing at present. On the other end of the range of mobility, refractories inform us of the importance of energetic processes (micrometeoroid impact vaporization and ion sputtering) in refilling the exosphere. Other papers that complement the topics discussed here are those on micrometeoroid impact vaporization (Janches et al. 2021), on particles and photons as drivers of exospheres (Wurz et al. 2021), and on surface-exosphere interaction

(Teolis et al. 2021). Table 1 contains the list of species detected so far at the Moon or at Mercury.

We did not discuss asteroids. Although the presence of comae (and even of collisional atmospheres close to the nuclei) is well established for comets (and beyond the scope of this paper), for asteroids, which potentially represent the largest family of surface-bounded exospheres, the observations are still inconclusive. Morgan and Killen (1998) predicted that detection of coronae of two important species, Na and OH, around asteroids would be extremely challenging but not impossible from a spacecraft. There are active asteroids, also called "main-belt comets", which spew dust grains when they are close to perihelion (see review by Jewitt 2012), and recently the Origins, Spectral Interpretation, Resource Identification, Security, Regolith Explorer (OSIRIS-Rex; Lauretta et al. 2017) has even detected cm-sized rocks being flung from asteroid Bennu (Lauretta et al. 2019), but there is a dearth of measurements regarding their exospheres. One exception is Ceres, for which there are detections of exospheric water-group species, including emission lines of hydroxyl (OH) at 309 nm (A'Hearn and Feldman 1992) and water (H_2O) at 556.936 GHz (Küppers et al. 2014). However, the cases of non-detection of water-group species are just as numerous (Rousselot et al. 2011, 2019; Roth et al. 2016; Roth 2018). Right now there is no clear explanation for the origin of Ceres' transient exosphere: sublimation rates from the known distribution of surface ice patches is two orders of magnitude lower than the water production rate derived from the observations (Landis et al. 2019). Villarreal et al. (2017) showed a correlation with solar energetic particle events, but ion sputtering is not effective enough (Küppers 2019).

The Rosetta mission, while en route to comet 67P/Churyumov-Gerasimenko, performed flybys of two asteroids: Steins ($\sim$6 km size) and Lutetia ($\sim$100 km size). Predictions of the exosphere of Steins and Lutetia were made by Schläppi et al. (2008) based on solar wind sputtering and impact vaporization, respectively. They predicted that a sputter-derived exosphere dominates over an impact vaporization-derived exosphere and that magnesium would be the dominant exospheric species after oxygen. They predicted that these detections would be challenging for the ion mass spectrometer ROSINA (Rosetta Orbiter Spectrometer for Ion and Neutral Analysis; Balsiger et al. 2007), but not impossible, at least at Lutetia. But ROSINA did not detect signs of their putative exospheres (Jäckel et al. 2010). Spacecraft outgassing, even after years of interplanetary travel, turned out to be a source of background gas contaminating the tenuous exospheric signal (Schläppi et al. 2010). ROSINA placed an upper limit for water of $\sim$3.5 $\times$ 10^3 cm^{-3} from a closest approach distance of $\sim$3,000 km (Altwegg et al. 2012). Rosetta's UV spectrograph Alice (Stern et al. 2007) also did not detect an exosphere around Lutetia (Stern et al. 2011).

7 Future Steps

Despite the abundant progress made so far in the field of tenuous atmospheres, both observational and theoretical, several uncertainties still hamper our understanding of the inner Solar System exospheres and their interaction with the external drivers and the surface. For example, the column abundances that have been published so far using different observational techniques vary by orders of magnitude, and further observational, modelling, and laboratory advancements are needed. Here we briefly illustrate each of them.

7.1 Remote and in Situ Measurements

The exospheres of Mercury and the Moon are notoriously difficult to study from the ground owing to several reasons: the extremely bright background from sunlight scattered from the

Table 1 List of confirmed detections of neutral volatiles and refractories at the Moon and Mercury. We report here either surface number density (from in situ mass spectrometry) or column density (from spectroscopic observations). Values of densities in italic correspond to extrapolation at the surface from remote sensing or in-orbit mass spectrometer measurements, and they involve the convolution with an exospheric model

Species	Mercury	Moon	Reference
H	$8\ cm^{-3}$ at subsolar point (thermal) $80\ cm^{-3}$ at subsolar point (non-thermal)	–	Broadfoot et al. (1976)
H_2	–	$1.2 \times 10^3\ cm^{-3}$	Stern et al. (2013)
He	$4.5 \times 10^3\ cm^{-3}$ at the subsolar point*	$(5\text{-}30) \times 10^3\ cm^{-3}$ at dawn[+]	*Broadfoot et al. (1976) [+]Hoffman et al. (1973), Benna et al. (2015)
CH_4	–	$450\ cm^{-3}$ at dawn	Hodges (2016)
Ne	–	$(3\text{-}110) \times 10^3\ cm^{-3}$ at dawn	Hodges et al. (1974) Benna et al. (2015) Killen et al. (2019)
Al	$(1.9\text{-}7.7) \times 10^7\ cm^{-2}$	–	Bida and Killen (2017) Vervack et al. (2016)
^{36}Ar	–	$4 \times 10^3\ cm^{-3}$ at dawn	Hoffman et al. (1973)
^{40}Ar	–	$(2\text{-}10) \times 10^4\ cm^{-3}$ at dawn	Hoffman et al. (1973) Benna et al. (2015)
Mn	$4.9 \times 10^7\ cm^{-2}$	–	Vervack et al. (2016)
Fe	$8.2 \times 10^8\ cm^{-2}$	–	Bida and Killen (2017)

surface and, in the case of Mercury, also the proximity to the Sun, which makes it visible for one hour at most during twilight. Nonetheless, ground-based observations have been for most of the time the only way to discover important exospheric species and to study how exospheres vary both in space and time due to variations in the external drivers. For example, observations of exospheric sodium (and to a lesser extent potassium) from ground-based telescopes have proven an essential tool to study source and loss process in the exosphere of both Mercury and the Moon (Potter and Morgan 1988, 1997; Tyler et al. 1988). Unfortunately, several of the species discussed here (Li, Mg, Ne, Ar) cannot be easily observed – or cannot be observed at all – from the ground. But for the few refractories that have been detected from the ground (Ca, Fe, Al), there is the need to perform additional observations to better understand their source and loss processes. For example, observations with adequate temporal coverage (observations over several consecutive nights) of calcium are needed to better constrain the dependence on the external drivers (micrometeoroid flux, solar energetic particle events, etc.). More precise line width measurements provide a more accurate temperature measurement of such gases.

At Mercury, the BepiColombo mission (Benkhoff et al. 2010), composed of two orbiters, the Mercury Planetary Orbiter (MPO) and the Mercury Magnetospheric Orbiter (MMO, also known as Mio), will provide a much anticipated comprehensive in situ study of its exosphere. In particular, the SERENA (Search for Exospheric Refilling and Emitted Natural Abundances) suite of instruments (Orsini et al. 2010, 2021) onboard MPO will make in situ measurements of neutrals and ions in Mercury's environment. This suite of instruments will provide much needed constraints on the high-energy processes (micrometeoroid impact vaporization, ion sputtering) that refill Mercury's exosphere. Spectra obtained by the PHEBUS (Probing of Hermean Exosphere By Ultraviolet Spectroscopy; Chassefière et al.

2010; Quémerais et al. 2020) ultraviolet spectrograph onboard MPO will be useful to supplement the mass spectrometer measurements for species (such as argon) ejected with low energy, and thus unable to reach the periapsis of 400 km of BepiColombo/MPO. PHEBUS large bandpass (55 to 315 nm) will allow it to detect several important species, like He, H, Mg. Complementing SERENA and PHEBUS, the Mercury Plasma Particle Experiment (Saito et al. 2010b) onboard Mio will directly measure charged particles in the exosphere and magnetosphere to quantitatively investigate generation mechanisms of the exosphere of each element. Thanks to the low-altitude orbit of MPO, regional and/or local time dependence of generation, escape, and circulation of heavy elements at the planet will be examined, together with observations from the Mercury Imaging X-ray Spectrometer (Fraser et al. 2010; Bunce et al. 2020) and the Mercury Gamma and Neutron Spectrometer (Mitrofanov et al. 2010), both onboard MPO. These instruments will measure the elemental surface composition of Si, Al, Fe, Mg, Ca, S, Ti, Cr, Mn, Na, K, P, Ni, U, Th, Cl, O, H and possibly C (Milillo et al. 2020). Finally, the Mercury Dust Monitor (Kobayashi et al. 2020) onboard MMO will provide measurements of dust impacts to the planet's surface, much needed in order to constrain source processes of refractories. The synergy, unprecedented in Mercury exploration, of so many instruments in deriving important properties of the planet's surface, exosphere, and magnetosphere will benefit future exospheric models (Milillo et al. 2020).

Regarding the Moon, great benefits would be achieved from orbiters, which would uncover temporal and spatial dependencies of exospheric abundances. This is especially critical in these times of renewed interest in lunar exploration. An assessment of the lunar exospheric composition is needed before it becomes forever changed: in a tenuous surface-bound exosphere like that on the Moon, every landing adds significant amounts of exogenic gases (Prem et al. 2020). As an example, each Apollo mission briefly doubled the mass of the lunar atmosphere (Vondrak 1974, 1992). Mass spectrometers can detect gases whose emission lines are too weak to be promptly detected by a spectrograph. For example, measuring the diurnal Ne abundance could resolve the discrepancy about its lifetime (Sect. 2.3.1). A measurement of ^{40}Ar, coupled with the measurement of the ionized component (^{40}Ar^{+}), providing the loss rate for this element (photo-ionization and electron impact ionization being the major loss processes), would constrain the abundance of ^{40}K within the crust and thus have important implications for the formation of the Moon (as well as that of Mercury). As LRO, SELENE, and LADEE have demonstrated, ultraviolet and visible spectrographs, especially onboard orbiters, also have proven useful to detect species over disparate locations and local times, uncovering temporal and spatial evolution of tenuous exospheres. There are plans to carry mass spectrometers on the lunar surface again, almost five decades since Apollo 17. Thanks to NASA's Commercial Lunar Payload Services program, mass spectrometers (such as LEMS; Benna et al. 2020) will be deployed at the lunar surface. A network of mass spectrometers at different locations on the lunar surface will measure more gases than is possible from orbit, and will monitor their local time dependence (and thus their interaction with the lunar surface).

Regarding ions, published mass composition measurements made around the Moon display little consistency as to the ion species present or the relative abundance of different ions in the lunar exosphere. In part, this results from the wide range of ion mass composition measurement techniques utilized at the Moon, and the very different observational geometries employed by the various missions. In addition, there have been very few studies of the long term variability of ion composition around the Moon, which would provide a window on both the variability of the neutral exosphere and that of the ionization and transport mechanisms. Therefore, there is real value in performing ion composition measurements over a long duration, from a consistent observational platform. This science topic may be addressed at least in part by NASA's HERMES (Heliophysics Environmental and Radiation Measurement Experiment Suite) and ESA's ERSA (European Radiation Sensors Array) suites of plasma instruments planned to fly on the Lunar Gateway.

7.2 Laboratory Measurements

As mentioned earlier, one of the biggest unknowns in the understanding of airless bodies' exospheres is their interaction with the surface. To this regard, more laboratory experiments on gas-surface interaction are needed, for example studies on thermal desorption rates of argon and other adsorbers (e.g. Bernatowicz and Podosek 1991; Dohnálek et al. 2002; Patrick et al. 2015), necessary, for example, to refine the residence time of atoms on regolith grains. Also needed are experiments that refine the yields, cross sections, and threshold energy for photon-stimulated desorption (e.g. Schaible et al. 2020) and electron-stimulated desorption (e.g. McLain et al. 2011).

The dissociation cross sections of possible precursor molecules of Ca and Mg need to be measured or theoretically derived. These precursor molecules include CaO, $CaOH$, $Ca(OH)_2$, CaS, MgO, $MgOH$, and MgS. These cross sections are particularly useful in understanding, for example, Mercury's calcium exosphere and its extremely hot temperature (thousands of K). The energies distributions of the resultant atomic species should be derived. Rough estimates of photolysis lifetimes, as well as energy and velocity distributions of photolysis-generated metal atoms, need to be carried out using correlations between molecular properties of well-studied atmospheric species (e.g. Berezhnoy 2010). Complex modern *ab initio* models of photolysis currently have been applied only to diatomic molecules containing alkali metals (Valiev et al. 2020). Such models should be further developed for application to photolysis of polyatomic species including Ca, Mg, Al, and Fe. Moreover, a refinement of photoionization cross sections should be made for several atomic species, especially Ca, Ne, and Ar.

Finally, the renewed interest in the lunar exploration (like the NASA program Artemis) represents a compelling opportunity to bring back samples from previously unexplored regions of the Moon. For example, the ability to quantify the ^{3}He and ^{4}He content in new lunar samples would allow us to improve our knowledge regarding correlation between the ^{3}He and ^{4}He content and properties of the lunar regolith. This would lead to better constraints of ^{3}He and ^{4}He content on the surface of the Moon on a global scale.

7.3 Simulations

Monte Carlo simulations of the lunar and Mercury's exospheres are usually the best at reproducing the dependence of the exospheres from several parameters (solar radiation pressure, different source processes at the surface, ionization and charge-exchange on the dayside).

Recent works have illustrated the need for more accurate simulations of the surface-exosphere interaction in airless bodies. For example, Sarantos and Tsavachidis (2020) showed that the mobility of alkalis (Na and K) on the surface of regolith grains on both the Moon and Mercury reduce the overall desorption of these species from these grains. On the Moon, this effect might explain why the sodium exosphere reacts more slowly to the changes of the micrometeoroid flux compared to potassium, because the latter, being more massive than the former, has an overall reduced surface mobility, and therefore a higher chance to be photodesorbed. At Mercury this surface-diffusion dependence of photodesorption rate might explain the peak post-noon in the sodium exosphere at aphelion, which is not explained by models that assume that alkali atoms do not move on the grains. Such approach should be applied to other species, notably to another alkali element, lithium. Models should also include the temporary sequestration of adsorbed atoms in the subsurface. As discussed in Sect. 2.2, Kegerreis et al. (2017) showed that this process, by which argon atoms migrate downwards during the night and are released during the day later than dawn, can explain the half-an-hour delay in the sunrise exospheric density bulge measured by both LACE and LADEE.

Simulations of the lunar and Mercury's exospheres should be run using the most-up-to-date information on the surface composition, including its surface variation (as was done e.g. in Colaprete et al. 2016b), using new information on the influx of micrometeorites and cometary material (Pokorný et al. 2018) and including their spatial and temporal variability (Pokorný et al. 2019).

Topography plays an important role in the transport of volatiles in airless bodies (e.g. Hodges 2011; Prem et al. 2018), and as such it should be included in exospheric models. It also affects the surface temperature. For example, it has been shown that the roughness of the lunar surface, casting both micro- and macro-shadows, affects the diurnal temperature profile, especially at the terminators, such that it deviates from a simple function of latitude and local time (Hurley et al. 2015). Therefore, one should include more accurate temperature maps, such as those from LRO's Diviner radiometer (Williams et al. 2017). At Mercury, these maps will eventually be produced by the MERTIS radiometer onboard BepiColombo/MPO (Hiesinger et al. 2010, 2020) but there are already robust models for its surface temperature, validated by lunar parameters (Bauch et al. 2020).

Space weathering should also be included in exospheric modeling, to study how long ice frosts can reside in PSRs without being disturbed. For example, micrometeoroid bombardment, while being one of the source processes of surface-bounded exospheres, can also act as a loss process, especially for frost deposits in a PSR. This process also affects the lateral and vertical distribution of cold-trapped volatile deposits, as studies on micrometeoroid bombardment on water ice have shown (e.g. Crider and Vondrak 2003; Hurley et al. 2012b). Moreover, photo-destruction of adsorbed atoms or molecules by cosmic rays and Lyman-alpha photons from interplanetary hydrogen resonantly scattering sunlight should also be included (e.g. Morgan and Shemansky 1991).

New simulations should be done for ion sputtering loss rates using updated models of the ejecta angular and velocity distributions from experiments (like the SDTRimpSP code; Eckstein et al. 2007) and measurements of the solar wind and Mercury's magnetosphere that will be provided by BepiColombo.

For Mercury, a reanalysis should be made of interplanetary dust and its spatial and temporal variability due to Mercury's orbital parameters. This is especially true for the origin of refractories in Mercury's exosphere. The existence and importance of nano-dust should be considered. Studies of the equilibrium condensation of dust particles were previously performed using limited thermochemical databases including mainly metal oxides. Adding silicate and non-silicate minerals to such thermochemical databases would allow us to study equilibrium condensation of species containing refractory elements during impact events in greater detail. Laboratory and theoretical studies of kinetics of formation of dust particles during impact events are also required for estimates of quenching parameters of condensation in impact-produced clouds.

Acknowledgements Grava was supported by LRO, funded by NASA through contract NNG05EC87C. Berezhnoy was supported by RFBR grant No. 18-03-00726. Halekas was supported by SSERVI and NASA grant 80NSSC20K0311. Raines was supported by the NASA Discovery Data Analysis Program, grants NNX15AL01G and 80NSSC19K0204. Vervack was supported by NASA grant 80NSSC18K0857 (subcontract to JH-APL).

Funding Note Open Access funding provided by Universität Bern.

References

M.F. A'Hearn, P.D. Feldman, Water vaporization on Ceres. Icarus **98**(1), 54–60 (1992)

K. Altwegg, H. Balsiger, U. Calmonte, M. Hässig, L. Hofer, A. Jäckel et al., In situ mass spectrometry during the Lutetia flyby. Planet. Space Sci. **66**(1), 173–178 (2012)

G.B. Andrews et al., The energetic particle and plasma spectrometer instrument on the MESSENGER spacecraft. Space Sci. Rev. **131**, 523–556 (2007). https://doi.org/10.1007/s11214-007-9272-5

V. Angelopoulos, The ARTEMIS mission. Space Sci. Rev. **165**(1–4), 3–25 (2011)

J.R. Arnold, Ice in the lunar polar regions. J. Geophys. Res., Solid Earth (1978–2012) **84**(B10), 5659–5668 (1979)

H. Balsiger, K. Altwegg, P. Bochsler, P. Eberhardt, J. Fischer, S. Graf et al., Rosina–Rosetta orbiter spectrometer for ion and neutral analysis. Space Sci. Rev. **128**(1–4), 745–801 (2007)

P.M. Banks, H.E. Johnson, W.I. Axford, The atmosphere of Mercury. Comments Astrophys. Space Phys. **2**, 214 (1970)

S. Barabash, A. Bhardwaj, M. Wieser, R. Sridharan, T. Kurian, S. Varier, M.B. Dhanya, Investigation of the solar wind–Moon interaction onboard Chandrayaan-1 mission with the SARA experiment. Current Science **96**, 526–532 (2009)

K.E. Bauch, H. Hiesinger, B.T. Greenhagen, J. Helbert, Estimation of surface temperatures on Mercury in preparation of the MERTIS experiment onboard BepiColombo. Icarus **354**, 114083 (2020)

J. Benkhoff, J. van Casteren, H. Hayakawa, M. Fujimoto, H. Laakso, M. Novara et al., BepiColombo—comprehensive exploration of Mercury: mission overview and science goals. Planet. Space Sci. **58**(1–2), 2–20 (2010)

M. Benna, B.J. Anderson, D.N. Baker, S.A. Boardsen, G. Gloeckler, R.E. Gold et al., Modeling of the magnetosphere of Mercury at the time of the first MESSENGER flyby. Icarus **209**(1), 3–10 (2010)

M. Benna, P.R. Mahaffy, J.S. Halekas, R.C. Elphic, G.T. Delory, Variability of helium, neon, and argon in the lunar exosphere as observed by the LADEE NMS instrument. Geophys. Res. Lett. **42**(10), 3723–3729 (2015)

M. Benna, M. Sarantos, N.C. Schmerr, C.A. Malespin, S. Bailey, The lunar environment monitoring station (LEMS), in *Lunar Surface Science Workshop, LPICo* vol. 2241, (2020), p. 5022

J. Benson, J.W. Freeman, H.K. Hills, The lunar terminator ionosphere, in *Lunar and Planetary Science Conference Proceedings*, vol. 6 (1975), pp. 3013–3021

A.A. Berezhnoy, Meteoroid bombardment as a source of the lunar exosphere. Adv. Space Res. **45**(1), 70–76 (2010)

A.A. Berezhnoy, Chemistry of impact events on the Moon. Icarus **226**, 205–211 (2013)

A.A. Berezhnoy, Chemistry of impact events on Mercury. Icarus **300**, 200–212 (2018)

A.A. Berezhnoy, B.A. Klumov, Impacts as sources of the exosphere on Mercury. Icarus **195**, 511–522 (2008)

A.A. Berezhnoy, K.I. Churyumov, V.V. Kleshchenok, E.A. Kozlova, V. Mangano, Y.V. Pakhomov et al., Properties of the lunar exosphere during the Perseid 2009 meteor shower. Planet. Space Sci. **96**, 90–98 (2014)

T.J. Bernatowicz, F.A. Podosek, Argon adsorption and the lunar atmosphere, in *Lunar and Planetary Science Conference Proceedings*, vol. 21 (1991), pp. 307–313

J.P. Bibring, A.L. Burlingame, J. Chaumont, Y. Langevin, M. Maurette, P.C. Wszolek, Simulation of lunar carbon chemistry. I-Solar wind contribution, in *Lunar and Planetary Science Conference Proceedings*, vol. 5 (1974), pp. 1747–1762

T.A. Bida, R.M. Killen, Observations of Al, Fe and Ca+ in Mercury's exosphere, in *EPSC-DPS Joint Meeting Abstracts and Program*, vol. 6 (2011), pp. 2–7

T.A. Bida, R.M. Killen, Observations of the minor species Al and Fe in Mercury's exosphere. Icarus **289**, 227–238 (2017). https://doi.org/10.1016/j.icarus.2016.10.019

T. Bida, R.M. Killen, T.H. Morgan, Discovery of Ca in the atmosphere of Mercury. Nature **404**, 159–161 (2000)

P. Bjorkholm, L. Golub, P. Gorenstein, Detection of a nonuniform distribution of polonium-210 on the Moon with the Apollo 16 alpha particle spectrometer. Science **180**(4089), 957–959 (1973)

A.L. Broadfoot, S. Kumar, M.J.S. Belton, M.B. McElroy, Mercury's atmosphere from Mariner 10: preliminary results. Science **185**(4146), 166–169 (1974)

A.L. Broadfoot, D.E. Shemansky, S. Kumar, Mariner 10: Mercury atmosphere. Geophys. Res. Lett. **3**(10), 577–580 (1976)

F. Bühler, P. Eberhardt, J. Geiss, J. Meister, P. Signer, Apollo 11 solar wind composition experiment: first results. Science **166**(3912), 1502–1503 (1969)

E.J. Bunce, A. Martindale, S. Lindsay, K. Muinonen, D.A. Rothery, J. Pearson et al., The BepiColombo Mercury imaging X-ray spectrometer: science goals, instrument performance and operations. Space Sci. Rev. **216**(8), 1–38 (2020)

M.H. Burger, R.M. Killen, W.E. McClintock, R.J. Vervack Jr., A.W. Merkel, A.L. Sprague, M. Sarantos, Modeling MESSENGER observations of calcium in Mercury's exosphere. J. Geophys. Res. **117**, 0L11B (2012). https://doi.org/10.1029/2012JE004158

M.H. Burger, R.M. Killen, W.E. McClintock, A.W. Merkel, R.J. Vervack, T.A. Cassidy, M. Sarantos, Seasonal variability in Mercury's dayside calcium exosphere. Icarus **238**, 51–58 (2014). https://doi.org/10.1016/j.icarus.2014.04.049

J.W. Chamberlain, Planetary coronae and atmospheric evaporation. Planet. Space Sci. **11**(8), 901–960 (1963)

E. Chassefière, J.L. Maria, J.P. Goutail, E. Quémerais, F. Leblanc, S. Okano et al., PHEBUS: a double ultraviolet spectrometer to observe Mercury's exosphere. Planet. Space Sci. **58**(1), 201–223 (2010)

D.A. Chestakov, D.H. Parker, A.V. Baklanov, Iron monoxide photodissociation. J. Chem. Phys. **122**, 084302 (2005)

G. Chin, S. Brylow, M. Foote, J. Garvin, J. Kasper, J. Keller et al., Lunar reconnaissance orbiter overview: the instrument suite and mission. Space Sci. Rev. **129**(4), 391–419 (2007)

A.A. Christou, R.M. Killen, M.H. Burger, The meteoroid stream of comet Encke at Mercury: implications for MErcury Surface, Space ENvironment,GEochemistry, and Ranging observations of the exosphere. Geophys. Res. Lett. **42**(18), 7311–7318 (2015). https://doi.org/10.1002/2015GL065361

R.N. Clark, Detection of adsorbed water and hydroxyl on the Moon. Science **326**, 562 (2009)

P. Clausing, Über die Adsorptionszeit und ihre Messung durch Strömungsversuche. Ann. Phys. **399**(5), 521–568 (1930)

A. Colaprete, K. Vargo, M. Shirley, D. Landis, D. Wooden, J. Karcz et al., An overview of the LADEE ultraviolet-visible spectrometer. Space Sci. Rev. **185**(1–4), 63–91 (2014)

A. Colaprete, D. Wooden, A. Cook, M. Shirley, M. Sarantos, Observations of titanium, aluminum and magnesium in the Lunar exosphere by LADEE UVS, in *47th Lunar and Planetary Science Conf.* (2016a). Abstract 2635

A. Colaprete, M. Sarantos, D.H. Wooden, T.J. Stubbs, A.M. Cook, M. Shirley, How surface composition and meteoroid impacts mediate sodium and potassium in the lunar exosphere. Science **351**(6270), 249–252 (2016b)

J.C. Cook, S. Alan Stern, P.D. Feldman, G. Randall Gladstone, K.D. Retherford, C.C.C. Tsang, New upper limits on numerous atmospheric species in the native lunar atmosphere. Icarus **225**(1), 681–687 (2013)

E.S. Costello, R.R. Ghent, P.G. Lucey, The mixing of lunar regolith: vital updates to a canonical model. Icarus **314**, 327–344 (2018)

D.H. Crider, R.R. Vondrak, Hydrogen migration to the lunar poles by solar wind bombardment of the Moon. Adv. Space Res. **30**(8), 1869–1874 (2002)

D.H. Crider, R.R. Vondrak, Space weathering of ice layers in lunar cold traps. Adv. Space Res. **31**(11), 2293–2298 (2003)

T.P. Das, S.V. Thampi, A. Bhardwaj, S.M. Ahmed, R. Sridharan, Observation of Neon at mid and high latitudes in the sunlit lunar exosphere: results from CHACE aboard MIP/Chandrayaan-1. Icarus **272**, 206–211 (2016)

T.P. Das, S.V. Thampi, M.B. Dhanya, A. Bhardwaj, S.M. Ahmed, R. Sridharan, Upper limit of helium-4 in the sunlit lunar exosphere during magnetotail passage under low solar wind condition: result from CHACE aboard MIP in Chandrayaan-1. Icarus **297**, 189–194 (2017)

Z. Dohnálek, R.S. Smith, B.D. Kay, Adsorption dynamics and desorption kinetics of argon and methane on MgO (100). J. Phys. Chem. B **106**(33), 8360–8366 (2002)

D.L. Domingue, P.L. Koehn, R.M. Killen, A.L. Sprague, M. Sarantos, A.F. Cheng et al., Mercury's atmosphere: a surface-bounded exosphere, in *The Messenger Mission to Mercury* (Springer, New York, 2007), pp. 161–186

A. Doressoundiram, F. Leblanc, C. Foellmi, S. Erard, Metallic species in Mercury's exosphere: EMMI/New technology telescope observations. Astron. J. **137**, 3859–3863 (2009)

W. Eckstein, R. Dohmen, A. Mutzke, R. Schneider, SDTrimSP: A Monte-Carlo Code for Calculating Collision Phenomena in Randomized Targets. Technical Report IPP 12/3, Max-Planck-Institut für Plasmaphysik, München (2007)

R.C. Elphic, H.O. Funsten, B.L. Barraclough, D.J. McComas, M.T. Paffett, D.T. Vaniman, G. Heiken, Lunar surface composition and solar wind-induced secondary ion mass spectrometry. Geophys. Res. Lett. **18**(11), 2165–2168 (1991)

R.C. Elphic, G.T. Delory, B.P. Hine, P.R. Mahaffy, M. Horanyi, A. Colaprete et al., The lunar atmosphere and dust environment explorer mission. Space Sci. Rev. **185**(1–4), 3–25 (2014)

W. Fa, Y.Q. Jin, Quantitative estimation of helium-3 spatial distribution in the lunar regolith layer. Icarus **190**(1), 15–23 (2007)

W.G. Fastie, P.D. Feldman, R.C. Henry, H.W. Moos, C.A. Barth, G.E. Thomas, T.M. Donahue, A search for far-ultraviolet emissions from the lunar atmosphere. Science **182**(4113), 710–711 (1973)

P.D. Feldman, D. Morrison, The Apollo 17 ultraviolet spectrometer: lunar atmosphere measurements revisited. Geophys. Res. Lett. **18**(11), 2105–2108 (1991)

P.D. Feldman, D.M. Hurley, K.D. Retherford, G.R. Gladstone, S.A. Stern, W. Pryor, J.Wm. Parker, D.E. Kaufmann, M.W. Davis, M.H. Versteeg, Temporal variability of lunar exospheric helium during January 2012 from LRO/LAMP. Icarus **221**(2), 854–858 (2012)

B. Flynn, ORFEUS II far-ultraviolet observations of the lunar atmosphere. Astrophys. J. Lett. **500**(1), L71 (1998)

B.C. Flynn, S.A. Stern, A spectroscopic survey of metallic species abundances in the lunar atmosphere. Icarus **124**, 530–536 (1996)

G.W. Fraser, J.D. Carpenter, D.A. Rothery, J.F. Pearson, A. Martindale, J. Huovelin et al., The Mercury imaging X-ray spectrometer (MIXS) on bepicolombo. Planet. Space Sci. **58**(1–2), 79–95 (2010)

J.W. Freeman, J.L. Benson, A search for gaseous emissions from the Moon. Phys. Earth Planet. Inter. **14**(3), 276–281 (1977)

L.J. Friesen, J.A. Adams, Low pressure radon diffusion: a laboratory study and its implications for lunar venting. Geochim. Cosmochim. Acta **40**(4), 375–380 (1976)

Y. Futaana, S. Barabash, M. Wieser, M. Holmström, C. Lue, P. Wurz, A. Schaufelberger, A. Bhardwaj, M.B. Dhanya, K. Asamura, Empirical energy spectra of neutralized solar wind protons from the lunar regolith. J. Geophys. Res., Planets **117**, E5 (2012)

D.E. Gault, F. Hörz, D.E. Brownlee, J.B. Hartung, Mixing of the lunar regolith, in *Lunar and Planetary Science Conference Proceedings*, vol. 5 (1974), pp. 2365–2386

D.J. Gershman, J.A. Slavin, J.M. Raines, T.H. Zurbuchen, B.J. Anderson, H. Korth, D.N. Baker, S.C. Solomon, Ion kinetic properties in Mercury's pre-midnight plasma sheet. Geophys. Res. Lett. **41** (2014). https://doi.org/10.1002/2014GL060468

G.R. Gladstone, S.A. Stern, K.D. Retherford, R.K. Black, D.C. Slater, M.W. Davis et al., LAMP: the Lyman alpha mapping project on NASA's Lunar Reconnaissance Orbiter mission. Space Sci. Rev. **150**(1–4), 161–181 (2010a)

G.R. Gladstone, D.M. Hurley, K.D. Retherford, P.D. Feldman, W.R. Pryor, J.-Y. Chaufray, M.H. Versteeg, T.K. Greathouse, A.J. Steffl, H. Throop, J.Wm Parker, D.E. Kaufmann, A.F. Egan, M.W. Davis, D.C. Slater, J. Mukherjee, P.F. Miles, A.R. Hendrix, A. Colaprete, S.A. Stern, LRO-LAMP observations of the LCROSS impact plume. Science **330**(6003), 472–476 (2010b)

P. Gorenstein, P. Bjorkholm, Detection of radon emanation from the crater Aristarchus by the Apollo 15 alpha particle spectrometer. Science **179**(4075), 792–794 (1973)

P. Gorenstein, L. Golub, P. Bjorkholm, Detection of radon emission at the edges of lunar Maria with the Apollo alpha-particle spectrometer. Science **183**(4123), 411–413 (1974)

J.N. Goswami, M. Annadurai, Chandrayaan-1: India's first planetary science mission to the Moon. Curr. Sci. **96**(4), 486–491 (2009)

C. Grava, J.-Y. Chaufray, K.D. Retherford, G.R. Gladstone, T.K. Greathouse, D.M. Hurley, R.R. Hodges, A.J. Bayless, J.C. Cook, S.A. Stern, Lunar exospheric argon modeling. Icarus **255**, 135–147 (2015)

C. Grava, K.D. Retherford, D.M. Hurley, P.D. Feldman, G.R. Gladstone, T.K. Greathouse et al., Lunar exospheric helium observations of LRO/LAMP coordinated with ARTEMIS. Icarus **273**, 36–44 (2016)

C. Grava et al., LRO-LAMP observations of the lunar helium exosphere: constraints on thermal accommodation and outgassing rate. Mon. Not. R. Astron. Soc. **501**(3), 4438–4451 (2021). https://doi.org/10.1093/mnras/staa3884

J.S. Halekas, V. Angelopoulos, D.G. Sibeck, K.K. Khurana, C.T. Russell, G.T. Delory et al., First results from ARTEMIS, a new two-spacecraft lunar mission: counter-streaming plasma populations in the lunar wake. Space Sci. Rev. **165**(1–4), 93–107 (2011)

J.S. Halekas, A.R. Poppe, G.T. Delory, M. Sarantos, W.M. Farrell, V. Angelopoulos, J.P. McFadden, Lunar pickup ions observed by ARTEMIS: spatial and temporal distribution and constraints on species and source locations. J. Geophys. Res., Planets **117**, E06006 (2012)

J.S. Halekas, A.R. Poppe, G.T. Delory, M. Sarantos, J.P. McFadden, Using ARTEMIS pickup ion observations to place constraints on the lunar atmosphere. J. Geophys. Res., Planets **118**(1), 81–88 (2013)

J.S. Halekas, M. Benna, P.R. Mahaffy, R.C. Elphic, A.R. Poppe, G.T. Delory, Detections of lunar exospheric ions by the LADEE neutral mass spectrometer. Geophys. Res. Lett. **42**(13), 5162–5169 (2015)

J.S. Halekas, A.R. Poppe, W.M. Farrell, J.P. McFadden, Structure and composition of the distant lunar exosphere: constraints from ARTEMIS observations of ion acceleration in time-varying fields. J. Geophys. Res., Planets **121**(6), 1102–1115 (2016)

R.E. Hartle, R. Killen, Measuring pickup ions to characterize the surfaces and exospheres of planetary bodies: applications to the Moon. Geophys. Res. Lett. **33**(5), L05201 (2006)

R.E. Hartle, G.E. Thomas, Neutral and ion exosphere models for lunar hydrogen and helium. J. Geophys. Res. **79**(10), 1519–1526 (1974)

R.E. Hartle, S.A. Curtis, G.E. Thomas, Mercury's helium exosphere. J. Geophys. Res. **80**(25), 3689–3692 (1975)

A.R. Hendrix, D.M. Hurley, W.M. Farrell, B.T. Greenhagen, P.O. Hayne, K.D. Retherford et al., Diurnally migrating lunar water: evidence from ultraviolet data. Geophys. Res. Lett. **46**(5), 2417–2424 (2019)

D. Heymann, A. Yaniv, Ar40 anomaly in lunar samples from Apollo 11. Geochim. Cosmochim. Acta, Suppl. **1**, 1261 (1970)

D. Heymann, A. Yaniv, Distribution of radon-222 on the surface of the moon. Nat. Phys. Sci. **233**(37), 37–39 (1971)

H. Hiesinger, J. Helbert, M.C.I. Team, The Mercury radiometer and thermal infrared spectrometer (MERTIS) for the BepiColombo mission. Planet. Space Sci. **58**(1–2), 144–165 (2010)

H. Hiesinger, J. Helbert, G. Alemanno, K.E. Bauch, M. D'Amore, A. Maturilli et al., Studying the composition and mineralogy of the hermean surface with the Mercury radiometer and thermal infrared spectrometer (MERTIS) for the BepiColombo mission: an update. Space Sci. Rev. **216**(6), 1–37 (2020)

M. Hilchenbach, D. Hovestadt, B. Klecker, E. Möbius, Observation of energetic lunar pick-up ions near Earth. Adv. Space Res. **13**(10), 321–324 (1993)

F.L. Hinton, D.R. Taeusch, Variation of the lunar atmosphere with the strength of the solar wind. J. Geophys. Res. **69**(7), 1341–1347 (1964)

R.R. Hodges, Helium and hydrogen in the lunar atmosphere. J. Geophys. Res. **78**(34), 8055–8064 (1973)

R.R. Hodges, Model atmospheres for Mercury based on a lunar analogy. J. Geophys. Res. **79**(19), 2881–2885 (1974)

R.R. Hodges, Formation of the lunar atmosphere. Moon **14**(1), 139–157 (1975)

R.R. Hodges Jr., The escape of solar-wind carbon from the Moon, in *Lunar and Planetary Science Conference Proceedings*, vol. 7 (1976), pp. 493–500

R.R. Hodges Jr., Formation of the lunar helium corona and atmosphere, in *Lunar and Planetary Science Conference Proceedings*, vol. 8 (1977a), pp. 537–549

R.R. Hodges, Release of radiogenic gases from the Moon. Phys. Earth Planet. Inter. **14**(3), 282–288 (1977b)

R.R. Hodges, Gravitational and radiative effects on the escape of helium from the Moon, in *Lunar and Planetary Science Conference Proceedings*, vol. 9 (1978), pp. 1749–1764

R.R. Hodges, Lunar cold traps and their influence on argon-40, in *Lunar and Planetary Science Conference Proceedings*, vol. 11 (1980), pp. 2463–2477

R.R. Hodges, Migration of volatiles on the lunar surface, in *Lunar and Planetary Institute Science Conference Abstracts*, vol. 12 (1981), pp. 451–453

R.R. Hodges, Resolution of the lunar hydrogen enigma. Geophys. Res. Lett. **38**(6), L06201 (2011)

R.R. Hodges, Methane in the lunar exosphere: implications for solar wind carbon escape. Geophys. Res. Lett. **43**(13), 6742–6748 (2016)

R.R. Hodges, J.H. Hoffman, Measurements of solar wind helium in the lunar atmosphere. Geophys. Res. Lett. **1**(2), 69–71 (1974)

R.R. Hodges, F.S. Johnson, Lateral transport in planetary exospheres. J. Geophys. Res. **73**(23), 7307–7317 (1968)

R.R. Hodges, P.R. Mahaffy, Synodic and semiannual oscillations of argon-40 in the lunar exosphere. Geophys. Res. Lett. **43**(1), 22–27 (2016)

R.R. Hodges Jr., J.H. Hoffman, F.S. Johnson, D.E. Evans, Composition and dynamics of lunar atmosphere, in *Lunar and Planetary Science Conference Proceedings*, vol. 4 (1973), p. 2855

R.R. Hodges Jr., J.H. Hoffman, F.S. Johnson, The lunar atmosphere. Icarus **21**(4), 415–426 (1974)

J.H. Hoffman, R.R. Hodges Jr., Molecular gas species in the lunar atmosphere. Moon **14**(1), 159–167 (1975)

J.H. Hoffman, R.R. Hodges Jr., F.S. Johnson, D.E. Evans, Lunar atmospheric composition results from Apollo 17, in *Lunar and Planetary Science Conference Proceedings*, vol. 4 (1973), p. 2865

C.I. Honniball, P.G. Lucey, S. Li, S. Shenoy, T.M. Orlando, C.A. Hibbitts et al., Molecular water detected on the sunlit Moon by SOFIA. Nat. Astron. **5**, 121–127 (2020)

W.F. Huebner, J. Mukherjee, Photoionization and photodissociation rates in solar and blackbody radiation fields. Planet. Space Sci. **106**, 11–45 (2015). https://doi.org/10.1016/j.pss.2014.11.022

W.F. Huebner, J.J. Keady, S.P. Lyon, Solar photo rates for planetary atmospheres and atmospheric pollutants. Astrophys. Space Sci. **195**, 1–124 (1992)

D.M. Hunten, F.E. Roach, J.W. Chamberlain, A photometric unit for the airglow and aurora. J. Atmos. Terr. Phys. **8**(6), 345–346 (1956)

D.M. Hunten, T.H. Morgan, D.E. Shemansky, The Mercury atmosphere, in *Mercury* (1988), pp. 562–612

D.M. Hurley et al., Modeling of the vapor release from the LCROSS impact: 2. Observations from LAMP. J. Geophys. Res. **117**, E00H07 (2012a). https://doi.org/10.1029/2011JE003841

D.M. Hurley, D.J. Lawrence, D.B.J. Bussey, R.R. Vondrak, R.C. Elphic, G.R. Gladstone, Two-dimensional distribution of volatiles in the lunar regolith from space weathering simulations. Geophys. Res. Lett. **39**(9), L09203 (2012b)

D.M. Hurley, M. Sarantos, C. Grava, J.P. Williams, K.D. Retherford, M. Siegler et al., An analytic function of lunar surface temperature for exospheric modeling. Icarus **255**, 159–163 (2015)

D.M. Hurley, J.C. Cook, M. Benna, J.S. Halekas, P.D. Feldman, K.D. Retherford et al., Understanding temporal and spatial variability of the lunar helium atmosphere using simultaneous observations from LRO, LADEE, and ARTEMIS. Icarus **273**, 45–52 (2016)

D.M. Hurley, J.C. Cook, K.D. Retherford, T. Greathouse, G.R. Gladstone, K. Mandt et al., Contributions of solar wind and micrometeoroids to molecular hydrogen in the lunar exosphere. Icarus **283**, 31–37 (2017)

D.M. Hurley, R.J. Vervack, W. Pryor, R.M. Killen, Observations and modeling of hydrogen in Mercury's exosphere. LPI Contrib. **2083**, 1723 (2018)

A. Jäckel, K. Altwegg, H. Balsiger, B. Schläppi, B. Fiethe, T. Gombosi et al., ROSINA measurements and interpretations during (2867) Steins and (21) Lutetia flyby, in *EPSC* (2010), p. 400

D. Janches, A. Christou, A.A. Berezhnoy, G. Cremonese, T. Hirai, M. Horany, J.M. Jasinski, M. Sarantos, Meteoroids as one of the sources for exosphere formation on airless bodies in the inner solar system. Space Sci. Rev. **217**, 50 (2021). https://doi.org/10.1007/s11214-021-00827-6

J.M. Jasinski, L.H. Regoli, T.A. Cassidy, R.M. Dewey, J.M. Raines, J.A. Slavin et al., A transient enhancement of Mercury's exosphere at extremely high altitudes inferred from pickup ions. Nat. Commun. **11**(1), 1–9 (2020)

D. Jewitt, The active asteroids. Astron. J. **143**(3), 66 (2012)

X. Jia, J.A. Slavin, G. Poh, G.A. DiBraccio, G. Toth, Y. Chen et al., MESSENGER observations and global simulations of highly compressed magnetosphere events at Mercury. J. Geophys. Res. Space Phys. **124**(1), 229–247 (2019)

F.S. Johnson, Lunar atmosphere. Rev. Geophys. **9**(3), 813–823 (1971)

R.E. Johnson, R. Baragiola, Lunar surface: sputtering and secondary ion mass spectrometry. Geophys. Res. Lett. **18**, 2169–2172 (1991)

J.R. Johnson, T.D. Swindle, P.G. Lucey, Estimated solar wind implanted helium-3 distribution on the Moon. Geophys. Res. Lett. **26**, 385–388 (1999)

B.L. Jolliff, J.J. Gillis, L.A. Haskin, R.L. Korotev, M.A. Wieczorek, Major lunar crustal terranes: surface expressions and crust-mantle origins. J. Geophys. Res., Planets **105**(E2), 4197–4216 (2000)

J.L. Jordan, Prediction of the He distribution at the lunar surface, in *Annual Invitational Symposium on Space Mining and Manufacturing* (UA/NASA Space Engineering Research Center, Univ. of Arizona, Tucson, 1989). pp. VII-38–VII-50

E. Kallio, P. Janhunen, Modelling the solar wind interaction with Mercury by a quasi-neutral hybrid model. Ann. Geophys. **21**(11), 2133–2145 (2003a). Copernicus GmbH

E. Kallio, P. Janhunen, Solar wind and magnetospheric ion impact on Mercury's surface. Geophys. Res. Lett. **30**(17) (2003b)

J.A. Kegerreis, V.R. Eke, R.J. Massey, S.K. Beaumont, R.C. Elphic, L.F. Teodoro, Evidence for a localized source of the argon in the lunar exosphere. J. Geophys. Res., Planets **122**(10), 2163–2181 (2017)

R.M. Killen, Source and maintenance of the argon atmospheres of Mercury and the Moon. Meteorit. Planet. Sci. **37**(9), 1223–1231 (2002)

R.M. Killen, Pathways for energization of Ca and Mg in Mercury's exosphere. Icarus **268**, 32–36 (2016). https://doi.org/10.1016/j.icarus.2015.12.035

R.M. Killen, J.M. Hahn, Impact vaporization as a possible source of Mercury's calcium exosphere. Icarus **250**, 230–237 (2015). https://doi.org/10.1016/j.icarus.2014.11.035

R.M. Killen, W.H. Ip, The surface-bounded atmospheres of Mercury and the Moon. Rev. Geophys. **37**(3), 361–406 (1999)

R.M. Killen, T. Bida, T.H. Morgan, The calcium exosphere of Mercury. Icarus **173**(2), 300–311 (2005)

R. Killen, G. Cremonese, H. Lammer, S. Orsini, A.E. Potter, A.L. Sprague et al., Processes that promote and deplete the exosphere of Mercury. Space Sci. Rev. **132**(2–4), 433–509 (2007)

R.M. Killen, M.H. Burger, R.J. Vervack Jr., T.A. Cassidy, Understanding Mercury's exosphere: models derived from MESSENGER observations, in *Mercury: The View After MESSENGER*, vol. 21, ed. by S.C. Solomon, L.R. Nittler, B.J. Anderson (Cambridge University Press, Cambridge, 2018). Chap. 15

R.M. Killen, D.R. Williams, I. Park, O.J. Tucker, S.J. Kim, The lunar neon exosphere seen in LACE data. Icarus **329**, 246–250 (2019)

K.J. Kim, C. Wöhler, A.A. Berezhnoy, M. Bhatt, A. Grumpe, Prospective ^{3}He-rich landing sites on the Moon. Planet. Space Sci. **177**, 104686 (2019)

K. Kinoshita, K. Yoshida, T. Takashima, J. Nishimura, T. Mitani, S. Okuno et al., Results from Alpha-Ray Detector (ARD) on board SELENE, in *COSPAR*, vol. 39 (2012), p. 929

M. Kobayashi, H. Shibata, K.I. Nogami, M. Fujii, S. Hasegawa, M. Hirabayashi et al., Mercury Dust Monitor (MDM) onboard the Mio orbiter of the BepiColombo mission. Space Sci. Rev. **216**(8), 1–18 (2020)

G. Kockarts, Helium in the terrestrial atmosphere. Space Sci. Rev. **14**(6), 723–757 (1973)

M. Küppers, The mystery of Ceres' activity. J. Geophys. Res., Planets **124**(2), 205–208 (2019)

M. Küppers, L. O'rourke, D. Bockelée-Morvan, V. Zakharov, S. Lee, P. von Allmen et al., Localized sources of water vapour on the dwarf planet (1) Ceres. Nature **505**(7484), 525–527 (2014)

G. Lambert, J.C. Le Roulley, P. Bristeau, Accumulation and circulation of gaseous radon between lunar fines. Philos. Trans. R. Soc. Lond. Ser. A, Math. Phys. Sci. **285**(1327), 331–336 (1977)

M.E. Landis, S. Byrne, J.P. Combe, S. Marchi, J. Castillo-Rogez, H.G. Sizemore et al., Water vapor contribution to Ceres' exosphere from observed surface ice and postulated ice-exposing impacts. J. Geophys. Res., Planets **124**(1), 61–75 (2019)

D.S. Lauretta, S.S. Balram-Knutson, E. Beshore, W.V. Boynton, C.D. d'Aubigny, D.N. DellaGiustina et al., OSIRIS-REx: sample return from asteroid (101955) Bennu. Space Sci. Rev. **212**(1–2), 925–984 (2017)

D.S. Lauretta, C.W. Hergenrother, S.R. Chesley, J.M. Leonard, J.Y. Pelgrift, C.D. Adam et al., Episodes of particle ejection from the surface of the active asteroid (101955) Bennu. Science **366**(6470), eaay3544 (2019)

S.L. Lawson, W.C. Feldman, D.J. Lawrence, K.R. Moore, R.C. Elphic, R.D. Belian, S. Maurice, Recent outgassing from the lunar surface: the Lunar Prospector Alpha Particle Spectrometer. J. Geophys. Res., Planets **110**(E9), E09009 (2005)

F. Leblanc, C. Schmidt, V. Mangano, A. Mura, G. Cremonese, J. Raines, J.M. Jasinski, M. Sarrantos, R. Winslow, S. Fatemi, R. Killen, A. Milillo, T. Cassidy, R. Vervack, D. Kuruppuaratchi, S. Kameda, M.T. Capria, M. Horanyi, D. Janches, A. Berezhnoy, A. Christou, T. Hirai, P. Lierle, J. Morgenthaler, Comparative Na and K (Mercury, Moon and asteroid). Space Sci. Rev. (2021), this journal

K. Lodders, B. Fegley, *The Planetary Scientist Companion* (Oxford University Press, London, 1998), 371 pp.

P.G. Lucey, D.T. Blewett, G.J. Taylor, B.R. Hawke, Imaging of lunar surface maturity. J. Geophys. Res. **105**(E8), 20337–20386 (2000)

C. Lue, Y. Futaana, S. Barabash, M. Wieser, A. Bhardwaj, P. Wurz, Chandrayaan-1 observations of backscattered solar wind protons from the lunar regolith: dependence on the solar wind speed. J. Geophys. Res., Planets **119**(5), 968–975 (2014)

C. Lue, Y. Futaana, S. Barabash, Y. Saito, M. Nishino, M. Wieser et al., Scattering characteristics and imaging of energetic neutral atoms from the Moon in the terrestrial magnetosheath. J. Geophys. Res. Space Phys. **121**(1), 432–445 (2016)

C. Lue, J.S. Halekas, A.R. Poppe, J.P. McFadden, ARTEMIS observations of solar wind proton scattering off the lunar surface. J. Geophys. Res. Space Phys. **123**(7), 5289–5299 (2018)

T.E. Madey, B.V. Yakshinskiy, V.N. Ageev, R.E. Johnson, Desorption of alkali atoms and ions from oxide surfaces: relevance to origins of Na and K in atmospheres of Mercury and the Moon. J. Geophys. Res., Planets **103**(E3), 5873–5887 (1998)

P.R. Mahaffy, R.R. Hodges, M. Benna, T. King, R. Arvey, M. Barciniak et al., The neutral mass spectrometer on the lunar atmosphere and dust environment explorer mission. Space Sci. Rev. **185**(1–4), 27–61 (2014)

U. Mall, E. Kirsch, K. Cierpka, B. Wilken, F. Neubauer et al., Direct observation of lunar pick-up ions near the Moon. Geophys. Res. Lett. **25**(20), 3799–3802 (1998)

R.H. Manka, F.C. Michel, Lunar atmosphere as a source of argon-40 and other lunar surface elements. Science **169**(3942), 278–280 (1970)

S. Massetti, S. Orsini, A. Milillo, A. Mura, Modelling Mercury's magnetosphere and plasma entry through the dayside magnetopause. Planet. Space Sci. **55**(11), 1557–1568 (2007)

W.E. McClintock, M.R. Lankton, The Mercury atmospheric and surface composition spectrometer for the MESSENGER mission. Space Sci. Rev. **131**(1–4), 481–521 (2007)

W.E. McClintock, R.J. Vervack Jr., E. Todd Bradley, R.M. Killen, A.L. Sprague, N.R. Izenberg, Mercury's exosphere: observations MESSENGER's first Mercury flyby. Science **321**, 92–94 (2008)

W.E. McClintock, R.J. Vervack Jr., E. Todd Bradley, R.M. Killen, N. Mouawad, A.L. Sprague, M.H. Burger, S.C. Solomon, N.R. Izenberg, Mercury's exosphere during MESSENGER's second flyby: detection of magnesium and distinct distributions of neutral species. Science **324**, 610–613 (2009)

W.E. McClintock, T.A. Cassidy, A.W. Merkel, R.M. Killen, M.H. Burger, R.J. Vervack Jr., Observations of Mercury's exosphere: composition and structure, in *Mercury: The View After MESSENGER*, vol. 21, ed. by S.C. Solomon, L.R. Nittler, B.J. Anderson (Cambridge University Press, Cambridge, 2018). Chap. 14

D.J. McComas, F. Allegrini, P. Bochsler, M. Bzowski, M. Collier, H. Fahr et al., IBEX—interstellar boundary explorer. Space Sci. Rev. **146**(1), 11–33 (2009a)

D.J. McComas, F. Allegrini, P. Bochsler, P. Frisch, H.O. Funsten, M. Gruntman et al., Lunar backscatter and neutralization of the solar wind: first observations of neutral atoms from the Moon. Geophys. Res. Lett. **36**(12), L12104 (2009b)

J.L. McLain, A.L. Sprague, G.A. Grieves, D. Schriver, P. Travinicek, T.M. Orlando, Electron-stimulated desorption of silicates: a potential source for ions in Mercury's space environment. J. Geophys. Res., Planets **116**(E3), E03007 (2011)

M.T. Mellon, B.M. Jakosky, Geographic variations in the thermal and diffusive stability of ground ice on Mars. J. Geophys. Res., Planets **98**(E2), 3345–3364 (1993)

A.W. Merkel, T.A. Cassidy, R.J. Vervack Jr., W.E. McClintock, M. Sarantos, M.H. Burger, R.M. Killen, Seasonal variations of Mercury's magnesium dayside exosphere from MESSENGER observations. Icarus **281**, 46–54 (2017). https://doi.org/10.1016/j.icarus.2016.08.032

A.W. Merkel, R.J. Vervack Jr., T.A. Cassidy, R.M. Killen, W.E. McClintock, L.R. Nittler, M.H. Burger, Evidence connecting Mercury's Mg exosphere to its Magnesium-rich Surface Terrane. Geophys. Res. Lett. **45**(14) (2018). https://doi.org/10.1029/2018GL078407

A. Milillo, P. Wurz, S. Orsini, D. Delcourt, E. Kallio, R.M. Killen et al., Surface-exosphere-magnetosphere system of Mercury. Space Sci. Rev. **117**(3–4), 397–443 (2005)

A. Milillo, M. Fujimoto, G. Murakami, J. Benkhoff, J. Zender, S. Aizawa et al., Investigating Mercury's environment with the two-spacecraft BepiColombo mission. Space Sci. Rev. **216**(5), 1–78 (2020)

I.G. Mitrofanov, A.S. Kozyrev, A. Konovalov, M.L. Litvak, A.A. Malakhov, M.I. Mokrousov et al., The Mercury Gamma and Neutron Spectrometer (MGNS) on board the planetary orbiter of the BepiColombo mission. Planet. Space Sci. **58**(1–2), 116–124 (2010)

T.H. Morgan, R.M. Killen, A non-stoichiometric model of the composition of the atmospheres of Mercury and the Moon. Planet. Space Sci. **45**, 81–94 (1997)

T.H. Morgan, R.M. Killen, Production mechanisms for faint but possibly detectable coronae about asteroids. Planet. Space Sci. **46**(8), 843–850 (1998)

T.H. Morgan, D.E. Shemansky, Limits to the lunar atmosphere. J. Geophys. Res. Space Phys. **96**(A2), 1351–1367 (1991)

A. Mura, P. Wurz, H.I.M. Lichtenegger, H. Schleicher, H. Lammer, D. Delcourt, A. Milillo, S. Orsini, S. Massetti, M.L. Khodachenko, The sodium exosphere of Mercury: comparison between observations during Mercury's transit and model results. Icarus **200**(1), 1–11 (2009). https://doi.org/10.1016/j.icarus.2008.11.014

Y. Nakamura, HFT events: shallow moonquakes? Phys. Earth Planet. Inter. **14**(3), 217–223 (1977)

J. Nishimura, T. Kashiwagi, T. Takashima, S. Okuno, K. Yoshida, K. Mori et al., Radon alpha-ray detector on-board lunar mission SELENE. Adv. Space Res. **37**(1), 34–37 (2006)

L. Nittler, R.D. Starr, S.Z. Weider, T.J. McCoy, W.V. Boynton, D.S. Ebel, C.M. Ernst et al., The major-element composition of Mercury's surface from MESSENGER X-ray spectrometry. Science **333**, 1847–1850 (2011)

S. Orsini, S. Livi, K. Torkar, S. Barabash, A. Milillo, P. Wurz et al., SERENA: a suite of four instruments (ELENA, STROFIO, PICAM and MIPA) on board BepiColombo-MPO for particle detection in the Hermean environment. Planet. Space Sci. **58**(1–2), 166–181 (2010)

S. Orsini, V. Mangano, A. Milillo, C. Plainaki, A. Mura, J.M. Raines et al., Mercury sodium exospheric emission as a proxy for solar perturbations transit. Sci. Rep. **8**(1), 928 (2018)

S. Orsini, S.A. Livi, H. Lichtenegger, S. Barabash, A. Milillo, E. De Angelis et al., SERENA: particle instrument suite for determining the Sun-Mercury interaction from BepiColombo. Space Sci. Rev. **217**(1), 1–107 (2021)

J.W. Parker, S.A. Stern, G.R. Gladstone, J.M. Shull, The spectroscopic detectability of argon in the lunar atmosphere. Astrophys. J. Lett. **509**(1), L61 (1998)

E.L. Patrick, K.E. Mandt, S.M. Escobedo, G.S. Winters, J.N. Mitchell, B.D. Teolis, A qualitative study of the retention and release of volatile gases in JSC-1A lunar soil simulant at room temperature under ultrahigh vacuum (UHV) conditions. Icarus **255**, 30–43 (2015)

M. Pfleger, H.I.M. Lichtenegger, P. Wurz, H. Lammer, E. Kallio, M. Alho et al., 3D-modeling of Mercury's solar wind sputtered surface-exosphere environment. Planet. Space Sci. **115**, 90–101 (2015)

C.M. Pieters, J.N. Goswami, R.N. Clark, M. Annadurai, J. Boardman, B. Buratti, J.-P. Combe, M.D. Dyar, R. Green, J.W. Head, C. Hibbitts, M. Hicks, P. Isaacson, R. Klima, G. Kramer, S. Kumar, E. Livo, S. Lundeen, S. Malaret, T. McCord, J. Mustard, J. Nettles, N. Petro, C. Runyon, M. Staid, J. Sunshine, L.A. Taylor, S. Tompkins, P. Varanasi, Character and spatial distribution of OH/H2O on the surface of the Moon seen by M3 on Chandrayaan-1. Science **326**(5952), 568–572 (2009)

C. Plainaki, A. Mura, A. Milillo, S. Orsini, S. Livi, V. Mangano et al., Investigation of the possible effects of comet Encke's meteoroid stream on the Ca exosphere of Mercury. J. Geophys. Res., Planets **122**(6), 1217–1226 (2017)

P. Pokorný, M. Sarantos, D. Janches, A comprehensive model of the meteoroid environment around Mercury. Astrophys. J. **863**(1), 31 (2018)

P. Pokorný, D. Janches, M. Sarantos, J.R. Szalay, M. Horányi, D. Nesvorný, M.J. Kuchner, Meteoroids at the Moon: orbital properties, surface vaporization, and impact ejecta production. J. Geophys. Res., Planets **124**(3), 752–778 (2019)

A.R. Poppe, R. Samad, J.S. Halekas, M. Sarantos, G.T. Delory, W.M. Farrell et al., ARTEMIS observations of lunar pick-up ions in the terrestrial magnetotail lobes. Geophys. Res. Lett. **39**(17), L17104 (2012)

A.R. Poppe, J.S. Halekas, R. Samad, M. Sarantos, G.T. Delory, Model-based constraints on the lunar exosphere derived from ARTEMIS pickup ion observations in the terrestrial magnetotail. J. Geophys. Res., Planets **118**(5), 1135–1147 (2013)

A.R. Poppe, J.S. Halekas, J.R. Szalay, M. Horányi, Z. Levin, S. Kempf, LADEE/LDEX observations of lunar pickup ion distribution and variability. Geophys. Res. Lett. **43**(7), 3069–3077 (2016)

A.E. Potter, T.H. Morgan, Discovery of sodium in the atmosphere of Mercury. Science **229**, 651–653 (1985)

A.E. Potter, T.H. Morgan, Discovery of sodium and potassium vapor in the atmosphere of the Moon. Science **241**(4866), 675–680 (1988)

A.E. Potter, T.H. Morgan, Sodium and potassium atmospheres of Mercury. Planet. Space Sci. **45**(1), 95–100 (1997)

P. Prem, D.B. Goldstein, P.L. Varghese, L.M. Trafton, The influence of surface roughness on volatile transport on the Moon. Icarus **299**, 31–45 (2018)

P. Prem, D.M. Hurley, D.B. Goldstein, P.L. Varghese, The evolution of a spacecraft-generated lunar exosphere. J. Geophys. Res. **125** (2020). https://doi.org/10.1029/2020JE006464

E. Quémerais, J.Y. Chaufray, D. Koutroumpa, F. Leblanc, A. Reberac, B. Lustrement et al., PHEBUS on Bepi-Colombo: post-launch update and instrument performance. Space Sci. Rev. **216**(4), 67 (2020)

J.M. Raines, D.J. Gershman, T.H. Zurbuchen, M. Sarantos, J.A. Slavin, J.A. Gilbert et al., Distribution and compositional variations of plasma ions in Mercury's space environment: the first three Mercury years of MESSENGER observations. J. Geophys. Res. Space Phys. **118**(4), 1604–1619 (2013)

J.M. Raines, D.J. Gershman, J.A. Slavin, T.H. Zurbuchen, H. Korth, B.J. Anderson, S.C. Solomon, Structure and dynamics of Mercury's magnetospheric cusp: MESSENGER measurements of protons and planetary ions. J. Geophys. Res. Space Phys. **119**(8), 6587–6602 (2014)

J.M. Raines, K.L. Wallace, M. Sarantos, J.M. Jasinski, P. Tracy, R.M. Dewey et al., First in-situ observations of exospheric response to CME impact at Mercury, in *AGUFM, 2017* (2017), SM43E-02

L. Roth, Constraints on water vapor and sulfur dioxide at Ceres: exploiting the sensitivity of the Hubble Space Telescope. Icarus **305**, 149–159 (2018)

L. Roth, N. Ivchenko, K.D. Retherford, N.J. Cunningham, P.D. Feldman, J. Saur et al., Constraints on an exosphere at Ceres from Hubble Space Telescope observations. Geophys. Res. Lett. **43**(6), 2465–2472 (2016)

P. Rousselot, E. Jehin, J. Manfroid, O. Mousis, C. Dumas, B. Carry et al., A search for water vaporization on Ceres. Astron. J. **142**(4), 125 (2011)

P. Rousselot, C. Opitom, E. Jehin, D. Hutsemékers, J. Manfroid, M.N. Villarreal et al., Search for water outgassing of (1) Ceres near perihelion. Astron. Astrophys. **628**, A22 (2019)

S.K. Runcorn, On the origin of mascons and moonquakes, in *Lunar and Planetary Science Conference Proceedings*, vol. 5 (1974), pp. 3115–3126

S.K. Runcorn, Physical processes involved in recent activity within the Moon. Phys. Earth Planet. Inter. **14**(3), 330–332 (1977)

Y. Saito, S. Yokota, T. Tanaka et al., Solar wind proton reflection at the lunar surface: low energy ion measurements by MAP-PACE onboard SELENE (KAGUYA). Geophys. Res. Lett. **35**(24), L24205 (2008)

Y. Saito, S. Yokota, K. Asamura, T. Tanaka, M.N. Nishino, T. Yamamoto et al., In-flight performance and initial results of plasma energy angle and composition experiment (PACE) on SELENE (Kaguya). Space Sci. Rev. **154**(1–4), 265–303 (2010a)

Y. Saito, J.A. Sauvaud, M. Hirahara, S. Barabash, D. Delcourt, T. Takashima et al., Scientific objectives and instrumentation of Mercury Plasma Particle Experiment (MPPE) onboard MMO. Planet. Space Sci. **58**(1–2), 182–200 (2010b)

M. Sarantos, S. Tsavachidis, The boundary of alkali surface boundary exospheres of Mercury and the Moon. Geophys. Res. Lett. **47**(16), e2020GL088930 (2020)

M. Sarantos, R.M. Killen, W.E. McClintock, E.T. Bradley, R.J. Vervack Jr., M. Benna, J.A. Slavin, Limits to Mercury's magnesium exosphere from MESSENGER second flyby observations. Planet. Space Sci. **59**(15), 1992–2003 (2011)

M. Sarantos, R.M. Killen, A. Glenar, M. Benna, T.J. Stubbs, Metallic species, oxygen and silicon in the lunar exosphere: Upper limits and prospects for LADEE measurements. J. Geophys. Res. **117**(A3) (2012)

S. Sasaki, Y. Iijima, K. Tanaka, M. Kato, M. Hashimoto, H. Mizutani, Y. Takizawa, The SELENE mission: goals and status. Adv. Space Res. **31**(11), 2335–2340 (2003)

J. Schaible, M. Sarantos, B.A. Anzures, S.W. Parman, T.M. Orlando, Photon-stimulated desorption of MgS as a potential source of sulfur in Mercury's exosphere. J. Geophys. Res., Planets **125**(8), e2020JE006479 (2020)

B.M.U. Scherzer, Development of surface topography due to gas ion implantation, in *Sputtering by Particle Bombardment. II. Sputtering of Alloys and Compounds, Electron and Neutron Sputtering, Surface Topography*, ed. by R. Behrisch (Springer, Heidelberg, 1983), pp. 271–355

B. Schläppi, K. Altwegg, P. Wurz, Asteroid exosphere: a simulation for the ROSETTA flyby targets (2867) Steins and (21) Lutetia. Icarus **195**, 674–685 (2008)

B. Schläppi, K. Altwegg, H. Balsiger, M. Hässig, A. Jäckel, P. Wurz et al., Influence of spacecraft outgassing on the exploration of tenuous atmospheres with in situ mass spectrometry. J. Geophys. Res. **115**(A12) (2010)

C.E. Schlemm, R.D. Starr, G.C. Ho, K.E. Bechtold, S.A. Hamilton, J.D. Boldt et al., The X-ray spectrometer on the MESSENGER spacecraft, in *The Messenger Mission to Mercury* (Springer, New York, 2007), pp. 393–415

N. Schörghofer, O. Aharonson, The lunar thermal ice pump. Astrophys. J. **788**(2), 169 (2014)

N. Schörghofer, G.J. Taylor, Subsurface migration of H2O at lunar cold traps. J. Geophys. Res., Planets (1991–2012) **112**(E2), E02010 (2007)

N. Schörghofer, M. Benna, A.A. Berezhnoy, B. Greenhagen, B.M. Jones, S. Li, M. Orlando Th, P. Prem, O.J. Tucker, C. Wöhler, Water group exospheres and surface interactions on the Moon, Mercury, and Ceres. Space Sci. Rev. (2021), this journal

D.E. Shemansky, Revised atmospheric species abundances at Mercury: the debacle of bad g values. Mercury Messenger **2**, 1 (1988). Lunar and Planet. Inst.

D.E. Shemansky, A.L. Broadfoot, Interaction of the surfaces of the Moon and Mercury with their exospheric atmospheres. Rev. Geophys. **15**(4), 491–499 (1977)

Yu.G. Shkuratov, L.V. Starukhina, V.G. Kaidash, N.V. Bondarenko, 3He distribution over the lunar visible hemisphere. Sol. Syst. Res. **33**, 409–420 (1999)

J.A. Slavin, S.M. Krimigis, M.H. Acuña, B.J. Anderson, D.N. Baker, P.L. Koehn et al., MESSENGER: exploring Mercury's magnetosphere, in *The Messenger Mission to Mercury* (Springer, New York, 2007), pp. 133–160

J.A. Slavin, G.A. DiBraccio, D.J. Gershman, S.M. Imber, G.K. Poh, J.M. Raines et al., MESSENGER observations of Mercury's dayside magnetosphere under extreme solar wind conditions. J. Geophys. Res. Space Phys. **119**(10), 8087–8116 (2014)

J.A. Slavin, H.R. Middleton, J.M. Raines, X. Jia, J. Zhong, W.J. Sun et al., MESSENGER observations of disappearing dayside magnetosphere events at Mercury. J. Geophys. Res. Space Phys. **124**(8), 6613–6635 (2019)

G.R. Smith, D.E. Shemansky, A.L. Broadfoot, L. Wallace, Monte Carlo modeling of exospheric bodies: Mercury. J. Geophys. Res. Space Phys. **83**(A8), 3783–3790 (1978)

S.C. Solomon, R.L. McNutt, R.E. Gold, D.L. Domingue, MESSENGER mission overview. Space Sci. Rev. **131**(1–4), 3–39 (2007)

A.L. Sprague, D.M. Hunten, K. Lodders, Sulfur at Mercury, elemental at the poles and sulfides in the regolith. Icarus **118**(1), 211–215 (1995)

A.L. Sprague, D.M. Hunten, F.A. Grosse, Upper limit for lithium in Mercury's atmosphere. Icarus **123**, 345–349 (1996)

R. Sridharan, S.M. Ahmed, T.P. Das, P. Sreelatha, P. Padeepkumar, N. Naik, G. Supriya, The sunlit lunar atmosphere: a comprehensive study by CHACE on the Moon Impact Probe of Chandrayaan-1. Planet. Space Sci. **58**, 1567–1577 (2010)

L.V. Starukhina, Polar regions of the moon as a potential repository of solar-wind-implanted gases. Adv. Space Res. **37**, 50–58 (2006)

S.A. Stern, J.W. Parker, Th.H. Morgan, B.C. Flynn, D.M. Hunten, A. Sprague, M. Mendillo, M.C. Festou, NOTE: an HST search for magnesium in the lunar atmosphere. Icarus **127**, 523–526 (1997)

S.A. Stern, D.C. Slater, J. Scherrer, J. Stone, M. Versteeg, M.F. A'hearn et al., Alice: the Rosetta ultraviolet imaging spectrograph. Space Sci. Rev. **128**(1–4), 507–527 (2007)

S.A. Stern, J.Wm. Parker, P.D. Feldman, H.A. Weaver, A. Steffl, M.F. A'Hearn, L. Feaga, E. Birath, A. Graps, J.-L. Bertaux, D.C. Slater, N. Cunningham, M. Versteeg, J.R. Scherrer, Ultraviolet discoveries at Asteroid (21) Lutetia by the ROSETTA ALICE ultraviolet spectrograph. Astron. J. **141**, 199 (2011)

S.A. Stern et al., Lunar atmospheric helium detections by the LAMP UV spectrograph on the Lunar Reconnaissance Orbiter. Geophys. Res. Lett. **39**(12), L12202 (2012)

S.A. Stern, J.C. Cook, J.-Y. Chaufray, P.D. Feldman, G.R. Gladstone, K.D. Retherford, Lunar atmospheric H$_2$ detections by the LAMP UV spectrograph on the Lunar Reconnaissance Orbiter. Icarus **226**, 1210–1213 (2013)

H.M. Sullivan, D.M. Hunten, Lithium, sodium, and potassium in the twilight airglow. Can. J. Phys. **42**, 937–956 (1964).

W.J. Sun, J.A. Slavin, R.M. Dewey, Y. Chen, G.A. DiBraccio, J.M. Raines et al., MESSENGER observations of Mercury's nightside magnetosphere under extreme solar wind conditions: reconnection-generated structures and steady convection. J. Geophys. Res. Space Phys. **125**(3), e2019JA027490 (2020)

J.M. Sunshine, T.L. Farnham, L.M. Feaga, O. Groussin, F. Merlin, R.E. Milliken, M.F. A'Hearn, Temporal and spatial variability of lunar hydration as observed by the deep impact spacecraft. Science **326**, 565 (2009)

T. Tanaka, Y. Saito, S. Yokota, K. Asamura, M.N. Nishino, H. Tsunakawa et al., First in situ observation of the Moon-originating ions in the Earth's magnetosphere by MAP-PACE on SELENE (KAGUYA). Geophys. Res. Lett. **36**(22), L22106 (2009)

S.R. Taylor, P. Jakeš, The geochemical evolution of the Moon, in *Lunar and Planetary Science Conference Proceedings*, vol. 5 (1974), pp. 1287–1305

B.D. Teolis, N. Schörghofer, C. Grava, M. Sarantos, M.T. Capria, B.T. Greenhagen, T.M. Orlando, Surface exospheric interaction. Space Sci. Rev. (2021), this journal

K. Terada, S. Yokota, Y. Saito, N. Kitamura, K. Asamura, M.N. Nishino, Biogenic oxygen from Earth transported to the Moon by a wind of magnetospheric ions. Nat. Astron. **1**(2), 1–5 (2017)

S.V. Thampi, R. Sridharan, T.P. Das, S.M. Ahmed, J.A. Kamalakar, A. Bhardwaj, The spatial distribution of molecular hydrogen in the lunar atmosphere—new results. Planet. Space Sci. **106**, 142–147 (2015)

R.J. Thomas, D.A. Rothery, S.J. Conway, M. Anand, Hollows on Mercury: materials and mechanisms involved in their formation. Icarus **229**, 221–235 (2014)

O.J. Tucker, W.M. Farrell, R.M. Killen, D.M. Hurley, Solar wind implantation into the lunar regolith: Monte Carlo simulations of H retention in a surface with defects and the H2 exosphere. J. Geophys. Res., Planets **124**(2), 278–293 (2019)

O.J. Tucker, W.M. Farrell, A.R. Poppe, On the effect of magnetospheric shielding on the lunar hydrogen cycle. J. Geophys. Res., Planets **126**(2), e2020JE006552 (2021)

A.L. Tyler, R.W. Kozlowski, D.M. Hunten, Observations of sodium in the tenuous lunar atmosphere. Geophys. Res. Lett. **15**(10), 1141–1144 (1988)

R.R. Valiev, A.A. Berezhnoy, I.D. Gritsenko, B.S. Merzlikin, V.N. Cherepanov, T. Kurten, C. Wöhler, Photolysis of diatomic molecules as a source of atoms in planetary exospheres. Astron. Astrophys. **633**, A39 (2020)

A.R. Vasavada, D.A. Paige, S.E. Wood, Near-surface temperatures on Mercury and the Moon and the stability of polar ice deposits. Icarus **141**(2), 179–193 (1999)

R.J. Vervack Jr., W.E. McClintock, R.M. Killen, A.L. Sprague, B.J. Anderson, M.H. Burger, E. Todd Bradley, N. Mouawad, S.C. Solomon, N.R. Izenberg, Mercury's complex exosphere: results from MESSENGER's third flyby. Science **329**, 672–675 (2010)

R.J. Vervack Jr., R.M. Killen, W.E. McClintock, A.W. Merkel, M.H. Burger, T.A. Cassidy, M. Sarantos, T.A. Cassidy, New discoveries from MESSENGER and insights into Mercury's exosphere. Geophys. Res. Lett. **43**, 11,545–11,551 (2016). https://doi.org/10.1002/2016GL071284

M.N. Villarreal, C.T. Russell, J.G. Luhmann, W.T. Thompson, T.H. Prettyman, M.F. A'Hearn et al., The dependence of the Cerean exosphere on solar energetic particle events. Astrophys. J. Lett. **838**(1), L8 (2017)

R. Von Steiger, N.A. Schwadron, L.A. Fisk, J. Geiss, G. Gloeckler, S. Hefti et al., Composition of quasistationary solar wind flows from Ulysses/Solar Wind Ion Composition Spectrometer. J. Geophys. Res. Space Phys. **105**(A12), 27217–27238 (2000)

R.R. Vondrak, Creation of an artificial lunar atmosphere. Nature **248**(5450), 657–659 (1974)

R.R. Vondrak, Lunar base activities and the lunar environment, in *NASA Conference Publication* (NASA, Washington, 1992), p. 337

A. Vorburger, P. Wurz, S. Barabash, M. Wieser, Y. Futaana, M. Holmström, A. Bhardwaj, K. Asamura, First direct observation of sputtered lunar oxygen. J. Geophys. Res. **119**(2), 709–722 (2014)

A. Vorburger, P. Wurz, S. Barabash, M. Wieser, Y. Futaana, A. Bhardwaj, K. Asamura, Imaging the South Pole–Aitken basin in backscattered neutral hydrogen atoms. Planet. Space Sci. **115**, 57–63 (2015)

X.D. Wang, Q.G. Zong, J.S. Wang, J. Cui, H. Rème, I. Dandouras et al., Detection of m/q= 2 pickup ions in the plasma environment of the Moon: the trace of exospheric H2+. Geophys. Res. Lett. **38**(14), L14204 (2011)

K. Watson, B. Murray, H. Brown, On the possible presence of ice on the Moon. J. Geophys. Res. **66**(5), 1598–1600 (1961a)

K. Watson, B.C. Murray, H. Brown, The behavior of volatiles on the lunar surface. J. Geophys. Res. **66**(9), 3033–3045 (1961b)

S.Z. Weider, L.R. Nittler, R.D. Starr, E.J. Crapster-Pregont, P.N. Peplowski, B.W. Denevi, J.W. Head, P.K. Byrne, S.A. Hauck II., D.S. Ebel, S.C. Solomon, Evidence for geochemical terranes on Mercury: global mapping of major elements with MESSENGER's X-ray spectrometer. Earth Planet. Sci. Lett. **416**, 109–120 (2015). https://doi.org/10.1016/j.epsl.2015.01.023

M. Wieser, S. Barabash, Y. Futaana et al., Extremely high reflection of solar wind protons as neutral hydrogen atoms from regolith in space. Planet. Space Sci. **57**, 2132–2134 (2009)

M. Wieser, S. Barabash, X.D. Wang, A. Grigoriev, A. Zhang, C. Wang, W. Wang, The Advanced Small Analyzer for Neutrals (ASAN) on the Chang'E-4 Rover Yutu-2. Space Sci. Rev. **216**(4), 1–28 (2020a)

M. Wieser, S. Barabash, X.D. Wang, A. Zhang, C. Wang, W. Wang, Solar wind interaction with the lunar surface: observation of energetic neutral atoms on the lunar surface by the Advanced Small Analyzer for Neutrals (ASAN) instrument on the Yutu-2 rover of Chang'E-4, in *EGU General Assembly Conference Abstracts* (2020b), p. 9199

J.P. Williams, D.A. Paige, B.T. Greenhagen, E. Sefton-Nash, The global surface temperatures of the Moon as measured by the Diviner Lunar Radiometer Experiment. Icarus **283**, 300–325 (2017)

R.M. Winslow, N. Lugaz, L. Philpott, C.J. Farrugia, C.L. Johnson, B.J. Anderson et al., Observations of extreme ICME ram pressure compressing Mercury's dayside magnetosphere to the surface. Astrophys. J. **889**(2), 184 (2020)

C. Wöhler, A. Grumpe, A.A. Berezhnoy, V.V. Shevchenko, Time-of-day–dependent global distribution of lunar surficial water/hydroxyl. Sci. Adv. **3**(9), e1701286 (2017)

P. Wurz, H. Lammer, Monte-Carlo simulation of Mercury's exosphere. Icarus **164**(1), 1–13 (2003)

P. Wurz, U. Rohner, J.A. Whitby, C. Kolb, H. Lammer, P. Dobnikar, J.A. Martín-Fernández, The lunar exosphere: the sputtering contribution. Icarus **191**, 486–496 (2007)

P. Wurz, J.A. Whitby, U. Rohner, J.A. Martín-Fernández, H. Lammer, C. Kolb, Self-consistent modelling of Mercury's exosphere by sputtering, micrometeorite impact and photon-stimulated desorption. Planet. Space Sci. **58**, 1599–1616 (2010)

P. Wurz, A. Vorburger, T. Orlando, A. Galli, N. Jäggi, D. Gamborino, M. Horányi, J.M. Raines, S. Fatemi, Y. Harada, M. Scherf, H. Lammer, S. Lindsay, M. Nishino, Particles and photons as drivers. Space Sci. Rev. (2021), this journal

A. Yaniv, D. Heymann, Atmospheric Ar40 in lunar fines, in *Lunar and Planetary Science Conference Proceedings*, vol. 3 (1972), p. 1967

S. Yokota, Y. Saito, K. Asamura, T. Tanaka, M.N. Nishino, H. Tsunakawa et al., First direct detection of ions originating from the Moon by MAP-PACE IMA onboard SELENE (KAGUYA). Geophys. Res. Lett. **36**(11), L11201 (2009)

S. Yokota, T. Tanaka, Y. Saito, K. Asamura, M.N. Nishino, M. Fujimoto et al., Structure of the ionized lunar sodium and potassium exosphere: dawn-dusk asymmetry. J. Geophys. Res., Planets **119**(4), 798–809 (2014)

S. Yokota, K. Terada, Y. Saito, D. Kato, K. Asamura, M.N. Nishino et al., KAGUYA observation of global emissions of indigenous carbon ions from the Moon. Sci. Adv. **6**(19), eaba1050 (2020)

T.H. Zurbuchen, J.M. Raines, G. Gloeckler, S.M. Krimigis, J.A. Slavin, P.L. Koehn et al., MESSENGER observations of the composition of Mercury's ionized exosphere and plasma environment. Science **321**(5885), 90–92 (2008)

T.H. Zurbuchen, J.M. Raines, J.A. Slavin, D.J. Gershman, J.A. Gilbert, G. Gloeckler et al., MESSENGER observations of the spatial distribution of planetary ions near Mercury. Science **333**(6051), 1862–1865 (2011)

Publisher's Note Springer Nature remains neutral with regard to jurisdictional claims in published maps and institutional affiliations.

Authors and Affiliations

Cesare Grava[1] · Rosemary M. Killen[2] · Mehdi Benna[2,3] ·
Alexey A. Berezhnoy[4,5] · Jasper S. Halekas[6] · François Leblanc[7] ·

Masaki N. Nishino[8] (iD) · **Christina Plainaki**[9] (iD) · **Jim M. Raines**[10] (iD) ·
Menelaos Sarantos[2] (iD) · **Benjamin D. Teolis**[1] (iD) · **Orenthal J. Tucker**[2] (iD) ·
Ronald J. Vervack Jr.[11] (iD) · **Audrey Vorburger**[12] (iD)

✉ A. Vorburger

1   Southwest Research Institute, San Antonio, TX, USA

2   NASA Goddard Space Flight Center, Greenbelt, MD, USA

3   University of Maryland Baltimore County, Baltimore, MD, USA

4   Sternberg Astronomical Institute, Moscow State University, Moscow, Russia

5   Institute of Physics, Kazan Federal University, Kazan, Russia

6   Department of Physics and Astronomy, University of Iowa, Iowa City, IA, USA

7   LATMOS/CNRS, Sorbonne Université, UVSQ, IPSL, Paris, France

8   Institute of Space and Astronautical Science, Japan Aerospace Exploration Agency, Sagamihara, Kanagawa, Japan

9   Italian Space Agency, Rome, Italy

10   Department of Climate and Space Sciences and Engineering, University of Michigan, Ann Arbor, MI, USA

11   Johns Hopkins Applied Physics Laboratory, Laurel, MD, USA

12   Physikalisches Institut, University of Bern, Bern, Switzerland

Space Science Reviews (2022) 218:2
https://doi.org/10.1007/s11214-022-00871-w

Comparative Na and K Mercury and Moon Exospheres

F. Leblanc[1] · C. Schmidt[2] · V. Mangano[3] · A. Mura[3] · G. Cremonese[4] · J.M. Raines[5] ·
J.M. Jasinski[6] · M. Sarantos[7] · A. Milillo[3] · R.M. Killen[7] · S. Massetti[3] · T. Cassidy[8] ·
R.J. Vervack Jr.[9] · S. Kameda[10] · M.T. Capria[3] · M. Horanyi[8] · D. Janches[7] ·
A. Berezhnoy[11,12] · A. Christou[13] · T. Hirai[14] · P. Lierle[2] · J. Morgenthaler[15]

Received: 16 March 2021 / Accepted: 7 January 2022 / Published online: 24 January 2022
© The Author(s) 2022

Abstract
Sodium and, in a lesser way, potassium atomic components of surface-bounded exospheres
are among the brightest elements that can be observed from the Earth in our Solar System.
Both species have been intensively observed around Mercury, the Moon and the Galilean
Moons. During the last decade, new observations have been obtained thanks to space missions carrying remote and in situ instrumentation that provide a completely original view of
these species in the exospheres of Mercury and the Moon. They challenged our understanding and modelling of these exospheres and opened new directions of research by suggesting
the need to better take into account the relationship between the surface-exosphere and the
magnetosphere. In this paper, we first review the large set of observations of Mercury and
the Moon Sodium and Potassium exospheres. In the second part, we list what it tells us on
the sources and sinks of these exospheres focusing in particular on the role of their magnetospheres of these objects and then discuss, in a third section, how these observations help
us to understand and identify the key drivers of these exospheres.

Keywords Surface-bounded exosphere · Mercury and the Moon · Sodium and potassium
atomic species · Surface-exosphere-magnetosphere interaction · Ground based observations

1 Observations

1.1 Seasonal/Diurnal Variation of the Lunar Exosphere

Sodium and potassium are the only two species within the lunar exosphere that can be readily measured remotely from the Earth. Na and K within the lunar exosphere produce visible
wavelength emissions that are optically thin. Solar irradiance is invariant at the wavelengths
of their bright D line transitions, so it is straightforward to determine the column density using the incident solar flux and the observation geometry. Scattered moonlight is so intense,
however, that the weak exosphere signal can only be distinguished in off-disk measurements
that sample the tangent column. The atmospheric scale height and tangent column together

---

Surface-Bounded Exospheres and Interactions in the Inner Solar System
Edited by Anna Milillo, Menelaos Sarantos, Benjamin D. Teolis, Go Murakami, Peter Wurz and Rudolf
von Steiger

Extended author information available on the last page of the article

 Springer

Reprinted from the journal

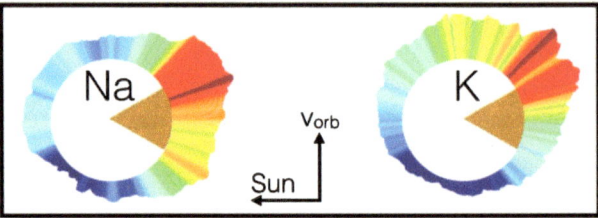

Fig. 1 Alkali column densities in the lunar exosphere as a function of orbital phase. Data are scaled to their dynamic range in radius and color, with blue to red showing increasing abundances. Orange wedges show nominal magnetotail passages, spanning phases of Full Moon $\pm 30°$. From Szalay et al. (2016)

yields an estimate for the zenith column, being about 8×10^8 atoms/cm^2 for Na at the subsolar point (Potter and Morgan 1988a,b). Based on two years of coronagraphic observations, Killen et al. (2021) obtained equatorial column abundances between 2×10^8 and 2×10^{10} atoms/cm^2, strongly dependent on local time of day and position inside or outside of Earth's magnetosphere. Ground-based measurements have estimated Na/K ratios from 4.4 to 5.7 (Potter and Morgan 1988a, 1988b; Hunten and Sprague 1997). More recently, the Lunar Atmosphere and Dust Environment Explorer (LADEE) determined a higher Na/K ratio—if ground-based scale heights are assumed—placing the exospheric Na/K ratio near the stoichiometric value of 7–9, as measured in lunar rocks returned by Apollo (Lodders and Fegley 1998; Colaprete et al. 2016).

Following discovery of the Moon's Na exosphere by Potter and Morgan (1988a), decades of debate have ensued about the relative roles of each source process that sustains it. Sprague et al. (1998) conducted a comprehensive study of sodium's atmospheric scale height. Although a collisionless surface-bound exosphere is not intrinsically thermal, scale heights are nonetheless a useful energy metric, and their measurements corresponded to characteristic energies of 985–1470 K. Concurrently, Potter and Morgan (1998) observations estimated 1280 K, in good agreement. Laboratory experiments by Yakshinskiy and Madey (1999) quickly pointed out that this is the characteristic energy for Na photo-desorption and that the mechanism is amply efficient to explain the observations.

Several measurements suggest that the lunar alkali exosphere cannot be explained by photo-desorption alone. Figure 1 shows the relative orbital variability in the LADEE UVS-derived column densities, averaged over four lunations. These columns are sampled near the subsolar point at tangent altitudes of 40 km. Na and K exhibit a similar and significant structure in Fig. 1 and such structure is unexpected from a photo-desorbed exosphere considering the viewing geometry. Szalay et al. (2016) showed that the potassium behaviour correlates with the local mineral inhomogeneity beneath the tangent point, as the local abundance tracks the K concentration in the lunar regolith measured by Lunar Prospector and Chang'E-2 (Prettyman et al. 2006; Zhu et al. 2013).

Alternative interpretations exist for the orbital behaviour that LADEE measured in the exosphere. Orange regions in Fig. 1 denote lunar orbital phases that are nominally in the Earth's magnetotail. From this perspective, one might infer that the Earth's magnetotail passage primes the lunar surface by shielding it from the solar wind. Hence, Colaprete et al. (2016) theorized that alkali enhancements may reflect drivers in the Moon's plasma environment. Their idea was supported by pre-LADEE models (Sarantos et al. 2010). The mechanism is for ions and electrons to catalyze solid state diffusion, regulating the supply available for release by photo-desorption after the Moon's path through the Earth magnetotail (Fig. 1). These two possibilities exemplify an inherent challenge to understanding the

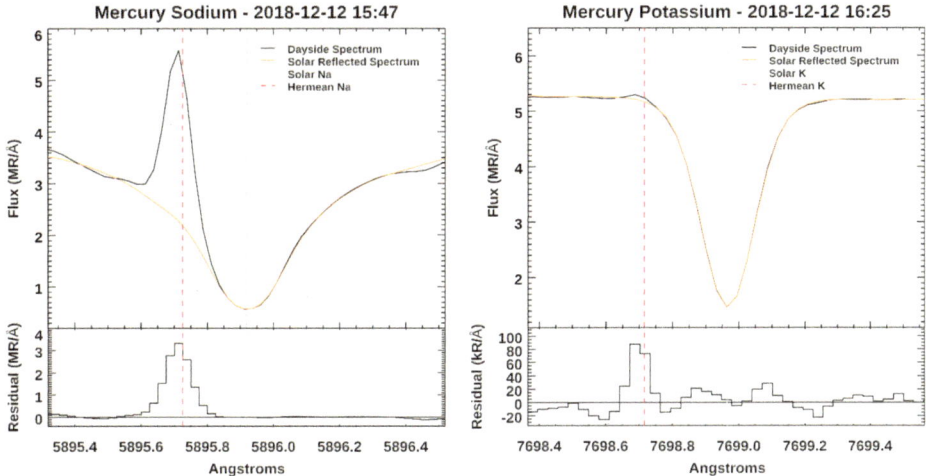

Fig. 2 Na and K emissions over Mercury's dayside at a true anomaly angle of 76°. Because the solar K absorption feature is narrower than that for Na, K scattering occurs in the continuum whereas Na remains partially within the solar absorption well. Emissions are slightly blue-shifted from rest, characteristic of gases leaving the surface (Courtesy, P. Lierle)

lunar exosphere: since our Moon is tidally-locked, it is difficult to know whether observed properties of the exosphere during a lunation are a consequence of coupling from "above" (*i.e.*, the lunar space environment) or from "below" (*i.e.* the lunar surface and gas-surface interactions).

1.2 Seasonal Variation of Mercury's Exosphere

Whereas viable sodium and potassium measurements at the Moon are restricted to off-disk, emissions at Mercury are sufficiently bright for ground-based spectrographs to distinguish the signature above the planet's dayside (Fig. 2). Mercury's 3:2 spin-orbit resonance can effectively break the degeneracy described above for the Moon. Merkel et al. (2018) leveraged this fact in their study of exospheric Mg using MESSENGER. They showed enhancements in the Mg exosphere only appearing during alternating Mercury years, definitive evidence that these enhancements were coupled "from below" to Mg rich terrain, as opposed to the space environment.

To understand the seasonal behaviour of Mercury's alkali exosphere it is first necessary to understand how these atoms interact with sunlight. The planet's eccentric orbit gives heliocentric radial velocities of up to ± 10 km/s. This Doppler shifts the solar spectrum, changing the photon flux that excites the strong D line transitions in Mercury's alkali gases. The Doppler shift is greatest at true anomaly angles of 90° and 270°, where photon excitation occurs not in deep wells of the solar alkali features, but in the strong solar continuum (Fig. 2). In addition to the Doppler shift, the heliocentric radial distance modulates the g-values, hence the maximum g-value is closer to 60° than to 90° (e.g. Killen et al. 2009). This effect strongly modulates the emission brightness over the Mercury year, independent of the actual column of gas that is being observed. Figure 3 nicely demonstrates this seasonal effect for Na. Although poorly characterized owing to the difficulty in making the observations, an even stronger modulation is expected in the potassium brightness because the solar K absorption lines are narrower (Fig. 2, cf. Smyth and Marconi 1995a). Killen et al. (2009)

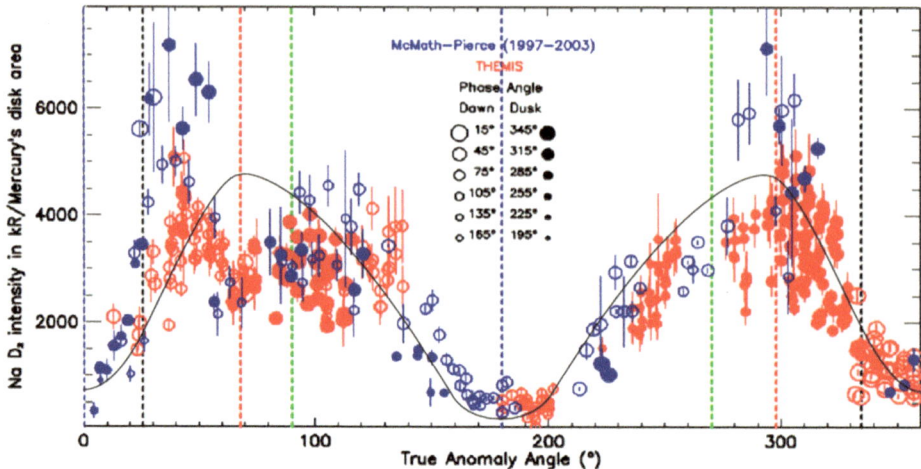

Fig. 3 Disk-averaged Na brightness from ground-based measurements. Adapted from Leblanc and Johnson (2010). Resonance scattering rates are seasonally modulated as the black solid line. Maximal heliocentric velocities occur at the dotted green lines. Maximal radiation pressure occurs at the dotted red lines. Black dotted lines bound the region near perihelion where rotation becomes slightly retrograde. Blue dotted lines correspond to the minimal heliocentric velocities

calculated g-values as a function of true anomaly angle for those emission lines expected to be observable by the MESSENGER MASCS spectrometer, including the 404.5 nm K line and the visible Na lines.

Consider the intensity at aphelion in Fig. 3 (true anomaly 180°). This season shows the faintest alkali emissions because resonance scattering occurs in the solar absorption core with Mercury farthest from the Sun. Note that these aphelion measurements are brighter than theory predicts from scattering rates alone, in some cases by a factor of two or three. This region represents a persistent seasonal enhancement in Mercury's Na column that was not fully realized until after MESSENGER. Cassidy et al. (2015, 2016) analyzed 10 Mercury years of sodium equatorial limb scans with UVVS. They showed Na column peaks at aphelion (Fig. 21a), which is surprising because solar-driven sources of the exosphere are weakest here. They termed this feature the "cold-pole enhancement" because it extends above two geographic longitudes that have the coldest annual surface temperatures, points that alternately face the Sun at aphelion owing to Mercury's 3:2 spin-orbit resonance.

Two interpretations could plausibly explain the cold-pole sodium enhancement. First, as a volatile, the sodium supply in the top-most soil could be sensitive to the maximum annual surface temperature. Na may have simply "baked-out" of the regolith grains, exhausting supplies in all but the coldest regions: high latitudes and cold-pole longitudes where surface temperatures peak ∼130 K below their hot-pole counterparts. Though longitudinal variation in the Na soil concentration remains unknown, variations in abundance within the top few cm of chemically-analogous potassium supports this perspective (Peplowski et al. 2012). A second interpretation involves the bouncing and sticking of exospheric atoms over the surface. Cold-pole longitudes are also located at the terminators during perihelion. Solar driven support of the exosphere peaks at perihelion and could send Na atoms bouncing across the ∼700 K dayside until they stick to the first cold surface they encounter behind the terminator. Surrounding perihelion, Mercury revolves nearly as fast as it rotates, so the progression of local time nearly stands still and the solar sidereal motion even becomes slightly retrograde. The terminators remain at nearly fixed longitudes in seasons between the black dotted lines in Fig. 3, about 15% of Mercury's year. Cold-trapping here could locally enhance the Na reservoir within the topmost regolith, because alkalis are known to stick to ~100 K surfaces (Yakshinskiy and Madey 2005). Over geological timescales, preferential deposition at cold poles may even explain the longitudinal asymmetries in soil concentration that Peplowski et al. reported. Either or both scenarios could be causal to the cold-pole longitudinal enhancements that UVVS observed in the Na exosphere and models have not yet determined which of these two influences dominates.

1.3 Lunar Linewidths & Altitude Profiles

Altitude profiles and linewidths are the two available means to estimate exospheric gas temperatures (or if the exosphere is intrinsically non-thermal, to place useful constrains on its velocity distribution function). As opposed to barometric scale heights, exospheric altitude profiles are traditionally fit using Chamberlain theory (*cf.* Chamberlain and Hunten 1989). A fit for temperature in this way can consider integrated densities that include contributions from all gas particles in ballistic, satellite and escaping orbits. Temperature retrievals based on alkali linewidths must account for hyperfine line structure in order to correctly interpret Doppler broadening, particularly if the gas is cold. Non-thermal treatments have also been made, but generally as forward models and not from analytical theory (Chaufray and Leblanc 2013). If the exosphere has multiple sources, e.g., sputtering and desorption at the Moon, the problem of temperature retrieval quickly becomes intractable. Disentangling a multiple-component thermal exosphere from a non-thermal energy distribution is a formidable undertaking, and even with high quality measurements, this is typically a problem with a non-unique solution. A best approach is of course to obtain concurrent measurements of both altitude and spectral line profiles, but observations of the lunar exosphere have yet not achieved this benchmark.

Reported exospheric temperatures inferred from altitude profiles vary extensively, even within individual publications. Sprague et al. (2012) reported 950 K to 20,000 K near the surface, depending on location and phase. Killen et al. (2019, 2021) reported 2250 K to 6750 K using a coronagraph that sampled higher altitudes above about 450 km. Their observations and others have reported the exosphere is most extended at high latitudes (e.g., Mendillo et al. 1993). Thus, not only is the exosphere not in thermal equilibrium with the surface, but it seems that in the gas-surface interactions superthermal atoms do not thermally accommodate towards local surface temperature. If sources of the exosphere originate predominately at the sub-solar point, however, larger scale heights at high latitudes could merely reflect transport, as only the most energetic atoms can reach the poles. Moreover, radiation pressure acceleration causes the scale height to be shorter on the dayside regardless of the temperature of the source.

Stern and Flynn (1995) have reported observations of a cold exospheric component near the lunar surface temperatures. This was achieved by observing a column just behind the terminator, ~ where the gas is sunlit, but the surface is in shadow. They proposed a two-component exosphere: a cold population in thermal equilibrium with the surface and a spatially extended superthermal population. Potter and Morgan (1988a) also proposed a two component exosphere, with a "cold" component at 543 K above the subsolar limb. Subsequent studies have not yet confirmed this very cold component, but it can be broadly summarized that the scale heights appear more extended farther from the surface, as is expected for a very extended exosphere that is partially escaping.

Na linewidths reported by Potter and Morgan (1988a) are in good agreement with their atmospheric scale height temperatures of ~540K. Although observations have not shown so

Fig. 4 Linewidth derived Na gas velocities at the Moon compared to simulations at various temperatures to assess geometric effects. From Kuruppuaratchi et al. (2018)

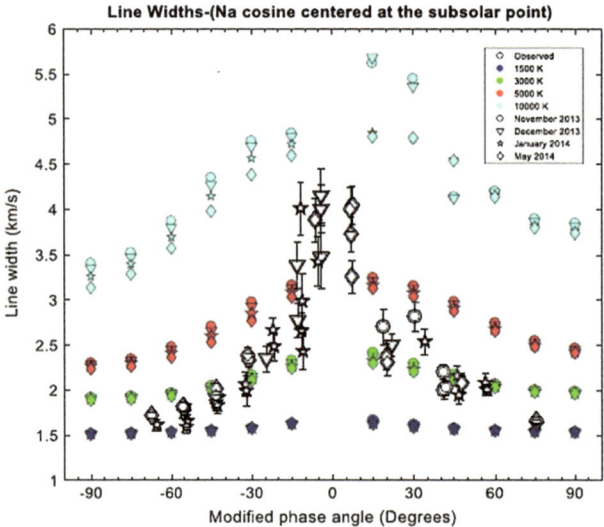

explicitly, the linewidth, like the scale height, probably increases with altitude, because more energetic particles reach higher apex altitudes. Kuruppuaratchi et al. (2018) and Rosborough et al. (2019) have surveyed Na and K linewidths, respectively, using Fabry-Perot techniques. In a 3 arcminute field at tangent altitudes of several hundred km, their Na line profiles corresponded to 1700 K to 9000 K whereas K profiles corresponded to colder temperatures of 980 K–1920 K. Both species show an increase in temperature and a decrease in brightness between Quarter phases and Full Moon. Again, this behavior can be at least partly attributed to viewing geometry and transport; the exosphere's concentration is highest at the subsolar point (Killen et al. 2019; Potter and Morgan 1998) and only the most energetic atoms will reach the limb. Still, kinetic models that account for this partitioning show that a single temperature cannot explain the observations (Fig. 4). Therefore, linewidth variations may reflect the true interplay between multiple sources: relatively cold photo-desorbed sources are diminished near Full Moon, causing the intensity to drop, while the relative contribution from hot sputtered sources becomes more evident. Of course, we are again confronted with the possibility of a single non-thermal source as a potentially viable alternate explanation.

1.4 Mercury Linewidths & Altitude Profiles

Unlike the Moon, ground-based altitude profiles of Mercury's exosphere are not generally possible because the scale height is of smaller angular size than the limitations imposed by atmospheric seeing. Killen et al. (1999) found the Na to be several hundred K above the surface temperature using the Na D_2 linewidth, which were confirmed upon MESSENGER's arrival (Cassidy et al. 2015). The MESSENGER UVVS wavelength range unfortunately did not include the potassium D lines. Furthermore, owing to operational constraints, the majority of the altitude profiles obtained by MESSENGER UVVS were confined to low latitudes, and sampling of profiles at the higher dayside latitudes where the magnetosphere's cusps channel ions to the planet's surface was limited. Leblanc et al. (2008, 2009) reported changes in linewidth alongside dynamic brightening in these regions, however, which together suggest significant contributions from a higher energy plasma sputtering population. MESSENGER did confirm a trace high-energy component above 1000 km altitudes in the

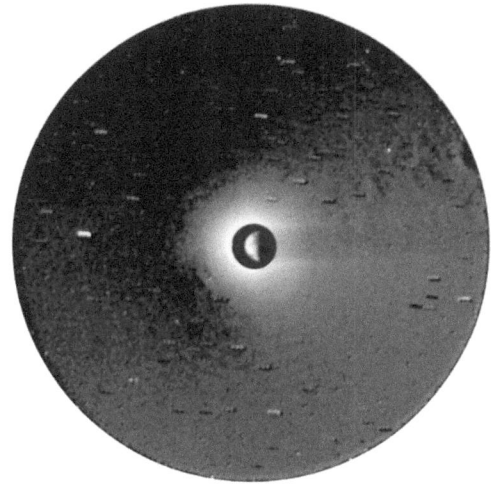

Fig. 5 Image of the lunar sodium exosphere over a 7° field of view. Fifteen on-band/off-band filter pairs of one minute each are co-added, hence the trailing in the stellar residuals. A neutral density mask allows the Moon to be imaged at the same time as its atmosphere. The effect of the Moon's shadow in blocking resonance fluorescence can be seen in the sodium exosphere. From Baumgardner and Mendillo (2009)

low-altitude limb profiles, and Chamberlain models place temperatures of this component near 20,000 K. This is significantly hotter than is expected from meteorite vaporization, indicating a sputtered component or the photodissociation of vaporized molecules imparting extra energy to the released atoms (Burger et al. 2014) or/and a strong influence of the radiation pressure and particle transport on the apparent scale height. A review of these findings is presented in Chapter's 14 and 15 of the book *Mercury: The View after MESSENGER* (eds. Solomon, Nittler, Anderson).

1.5 Lunar Na Tail

Whereas solar photons are incident from the sunward direction, photons interacting with alkalis are scattered nearly isotropically (we say 'nearly' because a small phase function occurs in the D_2 line, *cf.* Chamberlain 2011). Solar photons carry momentum, and so, on average, there is a net momentum in the anti-sunward direction transferred to the atoms. This effect is termed radiation pressure and it contributes significantly to particles overcoming gravity and achieving atmospheric escape. This effect shapes escaping alkalis into a comet-like "tail" pointing directly anti-sunward.

Measurements of the lunar tail are mostly indirect. Figure 5 shows the sole published image of the sodium lunar exo-tail; it has never been detected in potassium. Scattered moonlight from the bright surface competes with the faint tail and were it not for the projection of the Moon's shadow, it would be difficult to discern. A surprising discovery was made by all-sky cameras that the lunar exo-tail can be measured much more readily using the Earth as a lens. The massive comet-like tail has been detected from its backscattered sodium emissions in a direction nearly opposite to New Moon (see Fig. 6). In this geometry, anti-sunward streaming atoms encounter the Earth's gravity field. The trajectories of the Na atoms are bent as they pass by the Earth, and so the broad diffuse tail is focused into a narrow column where the trajectories cross. This forms a spot near the anti-lunar direction for about four days surrounding New Moon, while Earth passes through the lunar exo-tail.

The Na spot's brightness varies measurably with the Earth-Moon distance and its shape depends on the parallax of the observer's topocentric location. Line et al. (2012) observed a mean gas velocity of 12.5 km/s in the spot with measurable emissions of up to 30 km/s originating from down-tail distances of 1.5 million km. Baumgardner et al. (2021) used an

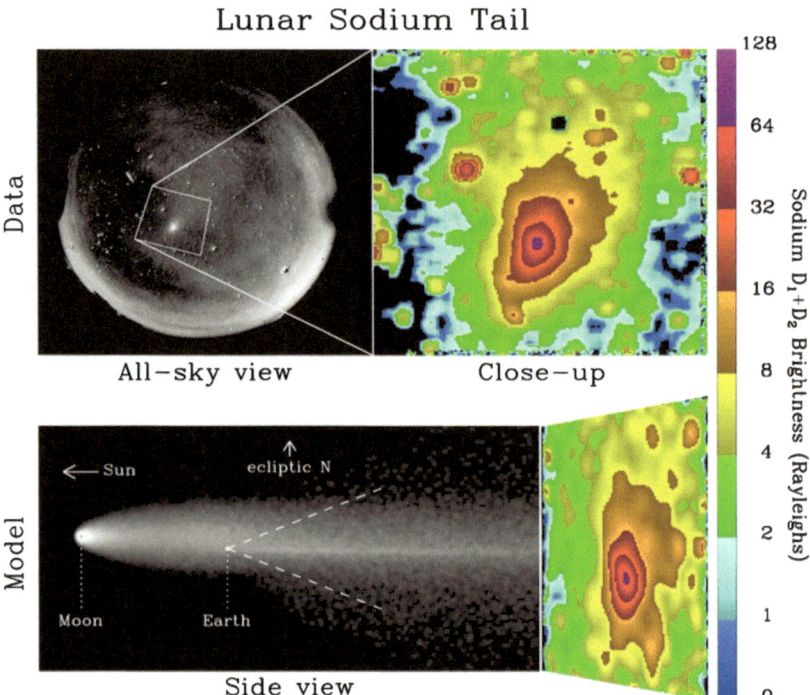

Fig. 6 Observations and modelling of the Moon's sodium tail "spot". Top-left: all-sky image with Na feature identified, with expanded view on top-right. Bottom-left: simulation results (to scale) showing the cloud of cloud of Na atoms passing Earth, gravitationally focused into the tail spot. (Smith et al. 1999; Wilson et al. 1999)

archive of all-sky imaging data to study the spot's apparitions 2006-2019. Figure 7 shows the duration and brightness of the apparition. Brightness in the lower panel effectively represents a cross-section of the lunar tail at a down-tail distance of 60 R_{Earth}. This brightness does not peak at the time of New Moon but 5–6 hours later, with an asymmetrical light curve about this peak. A latitudinal asymmetry is also seen, wherein the spot is brighter when New Moon phases north of the ecliptic. Modelling is needed to clarify if these reflect true axial asymmetries in the exo-tail or merely geometric effects. It is unlikely that an analogous spot of potassium is bright enough to be measured given that the D_2 line is obscured by telluric O_2 absorption and K has both a shorter photo-ionization lifetime and lower escape rate than Na.

Models of the Moon spot's brightness demonstrate that a flux of $\sim 2 \times 10^{22}$ Na atoms/s, escape at New Moon phase (Wilson et al. 1999), which is perhaps 3–15% of the global surface supply (Smyth and Marconi 1995b). This is a minimum, however, of escape rates that are modulated during each lunation cycle. Radiation pressure that propels atmospheric escape has positive or negative feedback, depending on the sign of the lunar heliocentric velocity. Atoms accelerated anti-sunward encounter increasing sunlight as the Doppler shift moves up the solar well (heliocentric radial velocity away from the Sun, as occurs during 1st Quarter), which then further increases the acceleration from solar radiation pressure. Negative feedback occurs at 3rd Quarter, where atoms are Doppler shifted into the solar absorption well, stagnating their acceleration. This effect modulates the fraction of atmosphere stripped away during each phase of the lunation (e.g., Smyth and Marconi 1995b).

Fig. 7 Brightness of the focused lunar tail's spot from 2006-2019 using an all-sky imager in El Leoncito Argentina. Colored dots highlight data within two days of a meteor shower's peak, accounting for the Moon-Earth transit time. The pink plus sign marks the initial discovery of the feature, following the 1998 Leonids shower (Baumgardner et al. 2021)

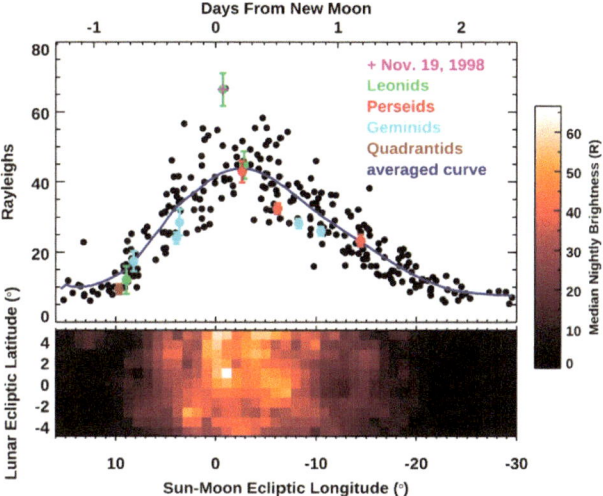

At least in part, it can likely explain the 20% decrease in sodium surface abundance between 1st and 3rd Quarter that was reported from measurements by the SELENE (Kaguya) orbiter (Kagitani et al. 2010). By the time atoms have Doppler shifted into the continuum due to their increasing heliocentric radial velocity when moving away from the Moon, they also scatter up to 20 times more solar photons. This may explain the line-of-sight Doppler shifts that Kuruppuaratchi et al. (2018) reported between 1st and 3rd Quarter, again at least in part. Relating this back to Fig. 4 and the similar potassium finding (Rosborough et al. 2019), it can be deduced that line of sight near New Moon span the full column down the exo-tail. During this configuration, the confluence of surface-ejected gas toward the observer and gas escaping away from the observer naturally produce a broadened line. Still, kinetic models in Fig. 4 (Sarantos et al. 2010; Sarantos and Tsavachidis 2020) demonstrate that this geometric effect is only a partial explanation of the observed linewidth enhancement and consequently the Moon's exosphere is hotter at New Moon phases. Killen et al. (2021) concluded that the largest measured scale heights are at dawn and dusk, and are correlated with local solar time. Observations of the lunar exosphere from Earth are limited to dawn and dusk terminators at both New Moon and Full Moon.

1.6 Mercury's Alkali Tail

Mercury's comet-like tail has been measured beyond 3 million km from the planet. Along with the sodium nebula of Io, it is among the largest structures in our solar system. Seasonal modulation of alkali escape is much stronger at Mercury than at the Moon, owing to the greater range of distances and velocities relative to the Sun experienced over the Mercury orbit. Na radiation pressure peaks near a true anomaly angle of 64°. This is indeed where the largest escape rates have been reported, about 1.3×10^{24} Na atoms/s (Schmidt et al. 2010). This is ~20% of the surface ejection rates modelled by Schmidt et al. (2012) who noted that a dropout from the expected brightness is evident at this true anomaly in Fig. 3. The sodium tail during Mercury's inward bound orbit is far weaker and was not detected until MESSENGER's arrival, emphasizing the importance of negative feedback in radiation acceleration when Mercury's motion is inbound (Potter et al. 2007).

The sodium tail is a valuable observable for understanding Mercury's exosphere overall. It shows a distinct width, as in Fig. 8, that depends on the gas velocity distribution function.

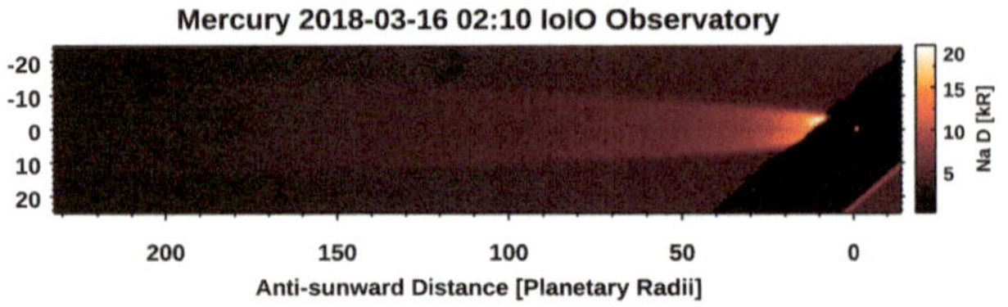

Fig. 8 Image of Mercury's Na tail at 35° true anomaly angle. The disk is visible through a high density filter that serves as a coronagraph. (Courtesy, J. Morgenthaler)

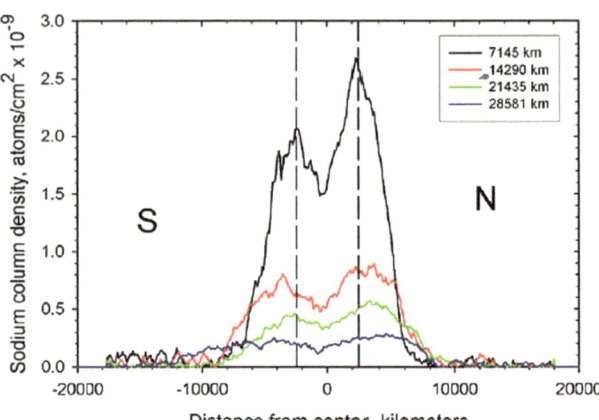

Fig. 9 Cross-tail profiles of Mercury's Na at 110° true anomaly angle. Note the dearth of emission in shadow and the enhancement in the northern lobe, both of which are less pronounced with distance (Potter and Killen 2008)

This width is narrow, consistent with low energy photo-desorption, despite the fact that meteoroid vaporization and sputtering have a much higher fractional escape (Schmidt et al. 2012). Although the tail typically exhibits slow and regular changes in the escape flux, it also records a history up to 15 hours in duration, capturing any high-energy transient events like sputtering during a coronal mass ejection passage, or a vaporization from a large meteor impact.

Despite photons—not plasma—being the source mechanism that supplies the escaping sodium tail, the influence of Mercury's offset magnetosphere is imprinted in the tail's structure. Cross-sections of the tail show an enhancement in the northern lobe (Potter and Killen 2008). Near Mercury, the north/south ratio in the tail is typically 1.2, but varies dynamically. Figure 9 shows an example of this asymmetry, which smooths out to a unity ratio beyond 10 R_M. The N/S asymmetry, combined with the tail's width, is consistent with low-energy atoms originating near the footprint of the southern cusp (Schmidt 2013). The southern surface has a ∼4x larger region for open field lines to channel plasma precipitation, relative to the north (Winslow et al. 2012). Gases escaping from this region drift northward as a natural consequence of the balance between radiation pressure and gravitational forcing as atoms traverse the planet's shadow. Only low-energy atoms that barely reach escape trajectories exhibit this "sloshing" behaviour.

Because atmospheric escape partitions the gas near the escape velocity, the tail is where high-energy sources in Mercury's exosphere should appear most prominently. Yet, the escaping tail show no signs of ion sputtering, imploring the question "*if low-energy desorption is the dominant process and high-energy ion sputtering is absent then why does the magnetosphere affect the exosphere?*" Again, this is thought to be evidence of indirect plasma effects. Low-energy desorption in the cusp regions could be sensitive to plasma bombardment because of gardening. Desorption is a surface process that would be depleted unless fresh alkali atoms are supplied from depth. Ion bombardment can create defects in the micro-structure of regolith grains and thereby catalyse solid-state diffusion. Modellers have independently concluded that this multi-step process is consistent with the available data, both at Mercury (Burger et al. 2010; Schmidt 2013) and at the Moon (Sarantos et al. 2010). However, this process remains hypothetical being not supported by detailed models of sputtering physics. Moreover, the low-energy solar wind particles won't penetrate deeply into the regolith grains limiting the region impacted by enhanced diffusion.

Potassium radiation pressure peaks near $48°$ true anomaly (Smyth and Marconi 1995b). In this portion of the planet's orbit, detections of the faint potassium tail have been reported at conference proceedings. Schmidt et al. (2017) observed a $\sim$95 Na/K ratio at 5 radii downtail. Potassium escape remains poorly characterized and its gas velocity in Mercury's exosphere is unknown. MESSENGER was unable to observe emissions at far red wavelengths. BepiColombo might be able to do so thanks to ViHI/SIMBIO-SYS, an observation that will need, however, specific pointing of MPO that may be difficult to obtain. Future ground-based observations can improve upon this, using instrumentation with high spectral resolution and long-slit capabilities.

1.7 Plasma Drivers in the Lunar Exosphere

Earth's magnetotail provides a convenient experiment to test the influence of plasma bombardment on the lunar exosphere. This sheltered plasma environment is populated at roughly 1/500th the solar wind density. Observations of the exosphere's passage through the magnetotail have differing results, however. Mendillo et al. (1999) showed the exospheric Na abundance was nearly unchanged between Quarter Moon, when the Moon is fully exposed to the solar wind and phases in the magnetosheath, when its surface is shielded from the solar wind. This lack of any response suggested solar wind ion sputtering is not an important driver. Potter et al. (2000) found from low altitude measurements (100–400 km) that the Na density decreased during passage through the magnetotail. Kaguya's measurements showed similar behavior consistent with low altitude measurements, but again no evidence for a magnetotail influence was observed (Kagitani et al. 2010). As noted earlier, decreasing radiation pressure and scattering rates may partly explain these findings. Potter et al. argued that the decrease could be attributed to ion bombardment catalyzing solid-state diffusion, thereby regulating the Na supply available for desorption in the topmost regolith layer. A model by Sarantos et al. (2010) synthesizing multiple data sets agreed with this interpretation. However, the LADEE UVS results in Fig. 1 show quite the opposite effect: alkalis increasing throughout the orange phase in the magnetotail. This led Colaprete et al. (2016) to conclude that the topmost reservoir of alkali atoms accumulates in the absence of solar wind plasma, rather than in its presence. Killen et al. (2019) show, using high altitude measurements ($>$400 km), that the highest column abundance observed in a five-year long study was correlated with ion impact to the lunar surface measured by the Artemis spacecraft. Since observations from Earth are taken off the lunar limb, Killen et al. (2019) also showed that such observations probe local time variations. Viewing geometry and altitude range seems the most likely explanation for discrepancies between the results of this natural experiment, particularly when comparing *in situ* and remote observations taken at different altitude ranges.

Earth's plasma sheet contains a distinct plasma population within the magnetosphere. It is both denser and more energetic than surrounding magnetotail lobes. Crossings of the Earth's plasma sheet have been reported to significantly enhance the exosphere (Wilson et al. 2006). This phenomenon needs to be confirmed by new studies; the plasma sheet at 60 Earth radii is thin and dynamic so it is challenging to predict when crossings occur. The fractional

Fig. 10 Observations of Na during two lunar eclipses in 1993 and 1996. A coronagraph mask remains necessary due to refracted light though Earth's atmosphere. Left was taken when the Moon was near the magnetosheath boundary, two days after plasmasheet crossing. Three crossings occurred in such time preceding the image at right, with the last being ∼4 hours earlier (Mendillo et al. 1999)

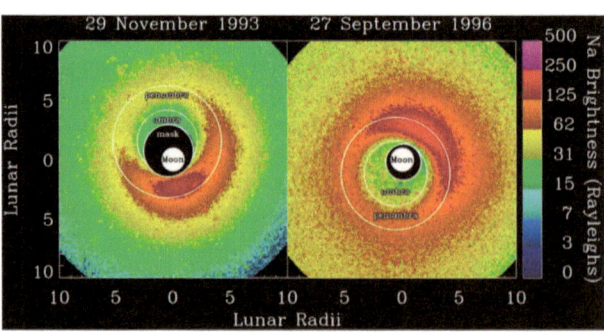

Fig. 11 Enhancements in Mercury's Na exosphere near the magnetic cusp regions (Mangano et al. 2013)

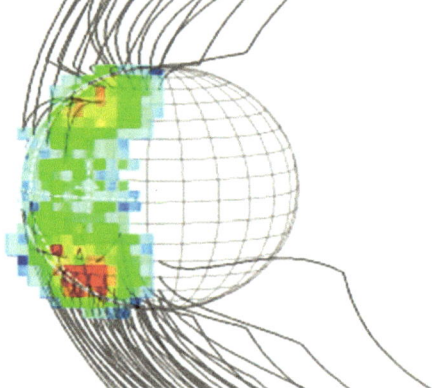

time that our Moon spends in Earth's plasma sheet varies on 18.6-year cycles because of the lunar orbital precession and there is also a weak dependence on lunation number (Hapgood 2007). Using a set of coronagraph observations during lunar eclipses, Wilson et al. (2006) discerned strong enhancements following recent plasma sheet crossings of the Moon (Fig. 10). With only five eclipses surveyed over seven years, statistics are far from robust though. Hapgood (2007) notes that the Wilson et al. study was ideally suited in time for the likelihood of plasma sheet crossings. Correlations between in-situ plasma measurements and ground-based monitoring of the exosphere may be the best way to ascertain the influence of the plasma sheet population.

1.8 Plasma Drivers in the Mercury Exosphere

Mercury's magnetosphere offers more visibility to the effects of plasma drivers. Enhanced emissions associated with plasma precipitation in the magnetosphere's cusps were first reported by Potter and Morgan (1990). MESSENGER later determined the planet's field is, to good approximation, a dipole offset by 484 km northward from the planet's center (Anderson et al. 2011). Consequently, Mercury's weaker field in the southern hemisphere exposes the surface to greater plasma bombardment. This discovery of an offset magnetosphere has helped researchers understand the magnetosphere-exosphere connection. The finding explained multiple reports of persistent asymmetries in the exosphere exhibiting southern maxima as seen in Fig. 11 (e.g., Baumgardner et al. 2008; Leblanc et al. 2008; Mangano et al. 2009). Southern cusp precipitation is thought to occur over a surface area four times greater relative to the north, and the cusp latitudes are centered approximately at 64°S and

Fig. 12 Contours of Na column density, showing near symmetry between north and south, with peaks equatorward of the cusps near 50–60 latitude (Schmidt et al. 2020)

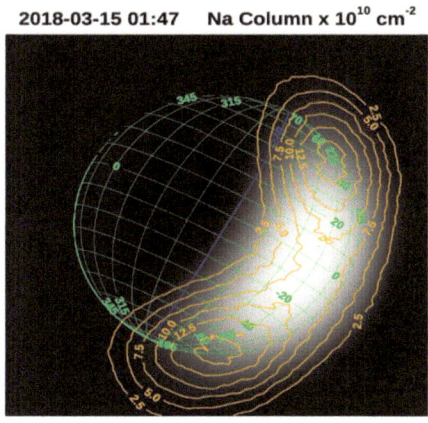

2018-03-15 01:47   Na Column x 10^{10} cm^{-2}

71–75°N (Winslow et al. 2012). However, Na enhancements generally appear equatorward of the magnetic cusps and the north-south ratio is often closer to unity (Fig. 12), or even enhanced in the northern hemisphere (Mangano et al. 2015), an observation which could be partly due to the atmospheric seeing (Killen 2020). It remains uncertain how planetary ions, which are more energetic and precipitate with a different pattern, can influence the exosphere as opposed to solar wind ions. A deeper discussion of exospheric plasma sources follows in Sect. 3 of this chapter. In any case, it is clearly evident that a coupled system between the magnetosphere and exosphere exists and this in turn establishes the capability to observe Mercury as a responsive and dynamic space weather system using the low-cost means of remote sensing.

Several observers have reported short term changes in the exosphere's structure (Potter et al. 1999; Leblanc et al. 2009; Mangano et al. 2013; Massetti et al. 2017). Exospheric structure, in particular the north/south asymmetry, appears sensitive to the interplanetary magnetic field's clock angle, and probably also dynamic pressure (Mangano et al. 2015). Sodium and potassium emissions are therefore a powerful tool in understanding the planet's response to space weather, but much more than a snapshot in time requires daytime observations. Solar telescopes have proven the optimal tool. In principle temporal changes could be on the order of $\sim$10 min, the ballistic flight time of the atoms, although few robust measurements on these timescales have been possible. Hourly changes are certainly evident. Winslow et al. (2015) catalogued coronal mass ejections (CMEs) at Mercury during the MESSENGER mission and Orsini et al. (2018) showed the exosphere rapidly reconfigures during a CME passage. The frequency of CMEs at Mercury is thought to be fairly high, and occasional extreme events are known where strong dynamic pressure pushes the magnetopause below the dayside surface, exposing it to the shocked solar wind (Slavin et al. 2019). BepiColombo will perform continuous observation from the orbit around Mercury using the Mercury sodium atmosphere spectral imager (MSASI) (Yoshikawa et al. 2007).

1.9 Meteoritoid Influence on the Lunar Exosphere

Several observers have reported that the lunar exosphere responds dynamically to strong meteoroid showers including the Leonids (Hunten et al. 1998; Smith et al. 1999), Perseids (Berezhnoy et al. 2014), Quadrantids (Verani et al. 2001) and Geminids (Colaprete et al. 2016). As an example, Na D_1 and D_2 lines in the lunar exosphere at distances of 90, 270, and 455 km from the surface were observed during maximum of the Perseid 2009 meteor shower

(Berezhnoy et al. 2014). A rapid increase of the intensities of Na D1 and D2 resonance lines in the lunar exosphere near the north pole on August 12, 2009, 23:54 UT–August 13, 2009, 1:13 UT) was detected corresponding to an increase in the Na zenith column density of 40%. These observations were explained by numerous impacts of relatively small Perseid meteoroids with masses smaller than 1 kg and total mass flux of impacted meteoroids on the Moon of about 27 kg/hour. The zenith column density of Na atoms on August 13/14, 2009, about $(9.0 \pm 0.2) \times 10^8$ cm^{-2}, was almost the same within error of measurements as the zenith column density of Na atoms on August 12 at 23:13-23:43 UT, $(8.2 \pm 0.5) \times 10^8$ cm^{-2}, before the sudden increase of the Na zenith column density, suggesting a duration of the signature of this episode of meteoroid bombardment in the Na exosphere of less than one day. The photoionization lifetime of Na at the Moon is 2 days, which suggests that the impact-derived sodium exosphere is escaping kinetically.

Each of these streams impacts the lunar surface from a different radiant, so stream impact locations and velocities differ year to year. Both are important: the local alkali soil concentration probably determines its ejecta into the exosphere (Szalay et al. 2016) and energy deposited in the fireball is proportional to the square of the stream velocity. It is possible that higher energy impacts yield not only more ejecta, but ejecta at higher energies. Killen et al. (2010) assumed a 1000 K Na gas temperature to determine the mass of gas released from the low velocity, which is substantially colder than most vaporization studies predict. Temperatures associated with vaporization remain model-dependent. Higher temperatures have higher escape fractions into the anti-sunward tail. Given that several studies show exospheric enhancements during strong meteoroid showers, it is surprising that these events rarely produce measurable column enhancements in the tail spot (see coloured points in Fig. 7). The absence could indicate that vapour produced by showers is relatively cold or that meteoroid showers do not augment the overall bulk exosphere except on rare occasions. Baumgardner et al. (2021) found significant correlations between annual variations in the sporadic influx and the Na tail spot brightness, however, signifying that the sporadic meteor influx is more influential than transient meteoroid showers in driving atmospheric escape.

The LDEX dust detector and UVS on LADEE provided a good proxy to study correlations between meteoroid dust and gas ejecta, and the BepiColombo MDM dust monitor will bring such capabilities to Mercury. LADEE showed potassium was more responsive to meteoroid influx than sodium, as seen in Fig. 13. This might be attributed in part to transport effects: hot Na ejecta could escape the Moon, whereas heavier K remains bound. Berezhnoy et al. (2014) estimated Na temperatures of 3100 K during the Perseids shower. At this temperature, a significant portion of the Na velocity distribution is above the lunar escape speed. Laboratory measurements indicate K ejecta re-impacting the Moon would adsorb to the cold surface whereas Na would bounce (Yakshinskiy and Madey 2005). It is surprising therefore that K residence times in the exosphere are estimated to be several days (Szalay et al. 2016). This may imply re-ejection of adsorbed material, whereas bouncing removes particles from the exosphere via rapid migration to cold traps on the night side.

1.10 Meteoroid Influence on Mercury's Exosphere

A strong meteoroid influence on Mercury's exosphere has been suggested for Ca (Burger et al. 2014), but the connection to alkalis is less evident. Comet Encke's stream has been attributed to Ca enhancements near 30° true anomaly (Killen and Hahn 2015). Like the Moon, impacts preferentially strike Mercury's dawn-side, and although Ca shows persistent enhancement at morning, other exospheric constituents show a more varied behaviour, indicating a mix of sources. Impactors strike Mercury with much higher velocity compared to

Fig. 13 Na and K column densities near noon during the entire LADEE mission (Colaprete et al. 2016). The lunation behaviour in Fig. 1 is evident, as well as long-term trends following the curve in blue. Three meteoroid streams (Leo, Gem, and Qua) are marked with the blue dashed lines. Red arrows mark the New Moon phase

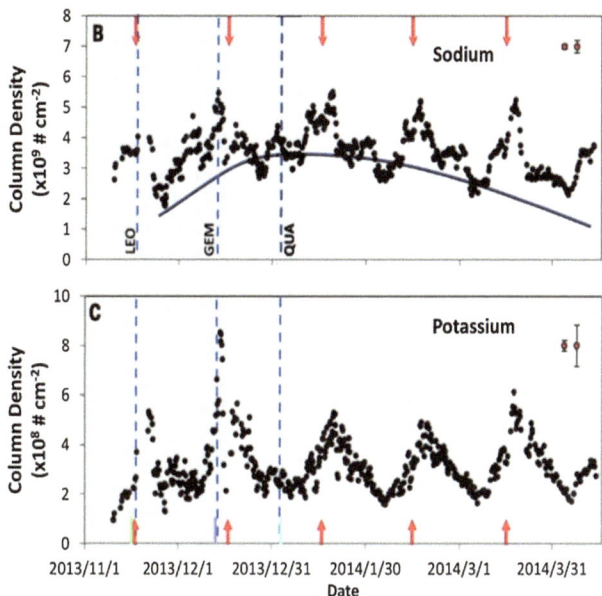

the Moon. Models predict vaporization is dominated by impacting meteoroids >100 km/s, with a factor of five seasonal variation peaking just before perihelion (Pokorny et al. 2018). Detailed overview of this topic can be found in review (Grava et al. 2021). Christou et al. (2015) modelling of the Encke dust stream complicated the story: they found that the Encke dust, depending on its age, can hit nearly anywhere on Mercury's surface except for the dawn hemisphere. Kameda et al. (2009) showed the potential correlation between Mercury sodium density and the interplanetary dust distribution (IPD), as suggested by Killen and Hahn (2015).

1.11 Na & K in Other Surface-Bound Exospheres

Europa is perhaps the best studied icy body with a surface-bound alkali exosphere. The Na/K ratio there is approximately 25, falling between that of the Moon and Mercury (Brown 2001). It has long been debated what fraction of these alkalis originate from the adjacent moon Io, which sprays neutral alkali jets and seeds the magnetosphere's plasma. Simulations have determined that loss rates in Europa's exosphere well exceed realistic implantation rates from Io and its plasma torus, suggesting Europa's alkalis are a native product of its sub-surface ocean salinity (Leblanc et al. 2002). Surface spectroscopy shows signatures of irradiated NaCl (Trumbo et al. 2019) and helps strengthen this argument, because transport from external sources is doubtful in a molecular form. Na emission is too faint to be readily measured at Enceladus, but has been detected with *in situ* mass spectroscopy (Postberg et al. 2009; Schneider et al. 2009). Na and K exospheres have yet to be detected at asteroids, despite a dedicated search at C-type Ryugu by the Hayabusa2 sample and return mission.

2 Sources of Na/K: The Role of Magnetosphere

2.1 Release Processes at Mercury and the Moon

The release of atoms from the surface of Mercury and the Moon essentially occurs at the expense of the energy supplied by the precipitating plasma, neutral atoms, meteoroids, the flow of photons and by the thermal energy of the surface itself. The rate of release of neutral and, in a lesser extent, ion particles and the velocity distribution depend, to first approximation on the energy transferred.

At Mercury, the first energetic process is the release by precipitating solid grains, which is called micro-meteoroid impact vaporization (MMIV) for impactors with the size of a few tens of microns, and meteoroid impact vaporization (MIV) for larger dimensions. The smallest particles precipitate more constantly and uniformly on the surface of Mercury, with an average speed of a few tens of km/s (Cintala 1992; Pokorny et al. 2017, 2018). As the size of impactors increases, events become rarer (and with more variable velocity). Very large meteoroids impact sporadically, but with a higher average velocity (Marchi et al. 2005). The contribution of these larger meteoroids to the Hermean exosphere is, globally, negligible, and their impact is expected to produce strong but localized temporary increases in exospheric density, enriched by species from deeper layers of the surface (Mangano et al. 2007). Regardless of the size of the impactor, refractory species contained in the surface are highly volatile during high-speed impacts owing to the high temperatures and pressures that are produced (Gerasimov et al. 1998). Therefore, atoms released by MIV or MMIV will be almost stoichiometrically representative of the surface composition. Regardless of the size of the impactor, the initial ejection will also be high-temperature vapor ($\sim$5,000 K), followed by the "liquid and vapor" at a slightly lower temperature (2,500 K) (Killen et al. 2007). Hence, the velocity distribution of ejecta can be approximated by a Maxwellian function, with such a temperature parameter (please see Sect. 1.9 for further discussions of this parametrization). Details on the produced species and molecules at different stages of the involved vaporization process are described later in Sect. 2.1.1.

A second energetic release process is caused by the precipitation of plasma on Mercury's surface and is called Ion Sputtering (IS) or electron sputtering (ES) for electron with energy larger than MeV. This causes the release of particles via momentum transfer and, similar to MMIV and MIV, reproduces more or less the local surface composition. At least in the case of MeV electrons and light solar wind ions, sputtering is a two-step process. First, the projectile impacts the surface target and is either implanted into the substrate or backscattered. This impact can also lead to sputtered neutrals from the target due to either direct (from the initial impact) or secondary (from the collision cascade) sputtering. The ejected atom is neutral in most cases, but it could be ionized. The projectile can transmit only a fraction of its energy to the ejected particle:

$$T_m = E_1 \frac{4m_1 m_2}{(m_1 + m_2)^2}$$

where Tm is the maximum transmitted energy from the incident particle (mass m_1 and energy E_1) to the ejecta (mass m_2 and energy T_m). The distribution function (f_S) of the ejection energy can be empirically reproduced by the following function:

$$f_S(E_e, T_m) = c_n \frac{E_e}{(E_e + E_b)^3} \left[1 - \left(\frac{E_e + E_b}{T_m} \right)^{\frac{1}{2}} \right]$$

where E_b is the surface binding energy of the ejected atomic species, E_e is the energy of the emitted particles, and c_n is a normalization constant (Sigmund 1981; Sieveka and Johnson 1984; Johnson and Baragiola 1991; Wurz et al. 2021; Morrissey et al. 2021; see also Eckstein and Preuss 2003). The angular distribution of the ejected particles is difficult to define for space weathered surfaces. Heavy and energetic ions penetrate deeper than lighter and less energetic particle and have multiple scattering impacts (Mura et al. 2005). The resulting angular distribution of ejecta is also dependent on the porosity of the surface, and it may be approximated with a cosine law with an exponent larger than one (Cassidy and Johnson 2005). Angular distributions of sputtered atoms have been measured for H^+ and He^+ onto polycrystalline tungsten and nickel by Bay et al. (1980). For light ions, the angular distribution is related to the ion impact direction, and may exhibit a maximum close to the backscattering angle, with a complex dependence that is usually discarded in Mercury exospheric studies (e.g. Bay et al. 1980; Behrisch and Eckstein 2007), because the plasma precipitation angular distribution is quite wide (Mura et al. 2005; Massetti et al. 2003).

A less energetic release process is Photon-stimulated desorption (PSD), sometimes referred to as photon sputtering, which corresponds to the desorption of surface elements as a result of electron excitation of a surface atom by a photon or by electron of moderate energy (Madey et al. 1998). McGrath et al. (1986) first proposed that PSD was the major source of Na in the Hermean exosphere because the Na distribution peaks at the sub-solar point (Killen et al. 1990). It is nowadays assumed that it is, indeed, one of the major sources, but not for all exospheric species. In fact, this is not a stochiometric process, as it is supposed to release only more volatile species such as Sodium or Potassium. Laboratory measurements by Madey et al. (1998) have shown that this process efficiently releases alkalis from regolith surfaces. The same authors (Yakshinskiy and Madey 1999, 2004) also found that UV photons with energies greater than 5 eV cause desorption of "hot" Na atoms, whereas energies ≤ 4 eV caused little or no Na desorption. Yakshinskiy and Madey (2004) estimated that the cross section of the PSD at photon energies of ≈ 5 eV is about 10^{-20} cm^2, which is about seven times larger than that used by other authors. Killen et al. (2001) claimed that the PSD yield at Mercury is diffusion limited and also reduced by porosity. Yakshinskiy and Madey (2004) also found from their experiments that desorbed Na atoms are super-thermal with a velocity peak in the PSD distribution of about 900 m/s. Cassidy and Johnson (2005) estimated that desorption from a regolith is reduced by about a factor of three compared to that on a flat surface. PSD is induced by electronic excitations rather than by thermal processes or momentum transfer, but the surface temperature is responsible for diffusion of material to the surface (Wurz and Lammer 2003). In fact, because the process is highly efficient, the volatile composition is quickly depleted in the uppermost surface layers, and some refilling mechanism from the innermost layers of the regolith grains is required (McGrath et al. 1986; Mura et al. 2009). The same authors suggested that enhanced diffusion or chemical sputtering is able to produce this refilling and is needed to explain observations of the dawn-dusk asymmetries (Schleicher et al. 2004). The velocity distribution of PSD particles can be either assumed Maxwellian with a temperature of 900–1200 K, or with a Weibull distribution (Johnson et al. 2002):

$$f(E) = \beta (1 + \beta) \frac{E U^{\beta}}{(E + U)^{2+\beta}}$$

where β is the shape parameter (0.7 for Na) and U is the characteristic energy, which is of the order of 0.05 eV, for Na (Wurz et al. 2021). The maximum ejection energy should be lower than the photon energy so that a cut-off function at about 10 eV should be included. Mura

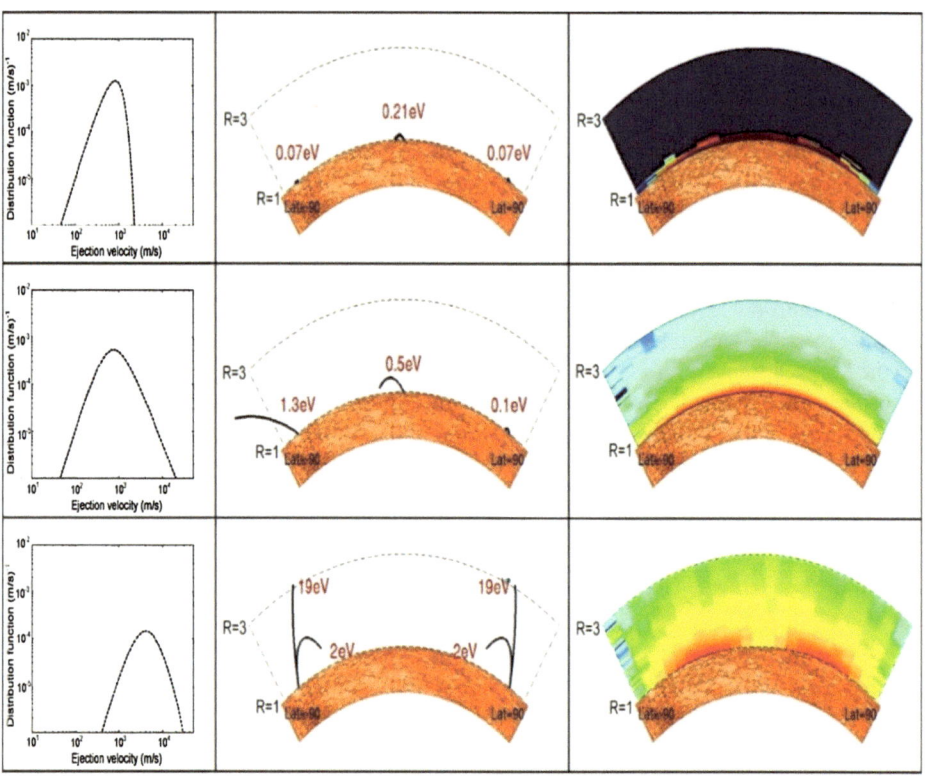

Fig. 14 Energy distribution (left), example of trajectory (center) and exospheric density simulations (right) for three different release mechanism: TD (top), PSD (middle) and IS (bottom)

et al. (2009) found that this second approximation fits much better with Doppler velocity distribution measurements by Schleicher et al. (2004).

Finally, thermal energy of the surface is able to desorb most volatile elements (Thermal desorption, TD). This is the least energetic process and basically results in a population of low-altitude neutrals that are in thermal equilibrium with the surface, and that bounce or stick to the grains in the regolith. In Fig. 14 we show examples of the energy distributions, trajectories and exospheric densities for different sources. However, temperature programmed desorption results of Poston et al. (2015) show high temperature of water desorption, 550 K, for Apollo lunar sample 72501, indicating activation energies approaching 1.5 eV for water. Activation energies for other species on simulated lunar and hermean soils should be examined, particularly to determine the activation energies under the harsh conditions at Mercury.

2.1.1 Meteoroid Bombardment as a Source of Na and K Atoms in the Exospheres of the Moon and Mercury

Chemical processes in hot clouds produced after impacts of meteoroids with the Moon and Mercury have been studied through quenching theory (Berezhnoy and Klumov 2008; Berezhnoy 2013, 2018). Namely, the chemical composition of an impact-produced cloud is in equilibrium soon after an impact when the temperature and pressure in the cloud are

Fig. 15 Equilibrium chemical composition of Na-, K-, Li-containing species during adiabatic cooling of an impact-produced cloud. Initial temperature is 10000 K, initial pressure is 10000 bar, the ratio of specific heats γ is 1.2. This figure is taken from Berezhnoy (2013)

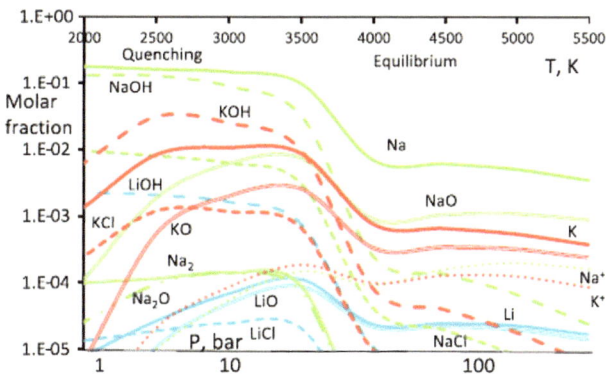

high and chemical reactions occur quickly in comparison with the typical timescale of hydrodynamic processes. During the cloud's expansion, the temperature and pressure decrease and, at the moment of quenching, the timescales of chemical reactions and hydrodynamic processes are comparable. It is assumed that the chemical composition of the cloud remains unchanged after quenching. Such a simple model has some limitations, as discussed in detail by Berezhnoy and Klumov (2008); however, it can be used as a first step for modelling of very complex processes during impact events.

Timescales of the main chemical reactions with participation of Na- and K-containing species are comparable with hydrodynamic timescales during impacts of 1 mm–3 cm meteoroids at about 2500–3500 K (Berezhnoy 2013). Thermodynamic calculations show that atomic Na and K atoms are the main Na-, K- containing species in clouds formed during collisions of meteoroids with the Moon and Mercury at a temperature range between 3500 and 5500 K. The KOH equilibrium content at 2000–3000 K is higher than that of atomic K whereas atomic Na and NaOH contents are comparable at 2500–3000 K (Berezhnoy 2013). NaO and KO molecules are the second or third most abundant Na-, K-containing species at the range of temperatures considered (see Fig. 15). Another important feature of the behavior of alkali metals during impact events is that condensation of atomic Na and K as well as oxides and hydroxides of Na and K does not occur during expansion of impact-produced clouds (Berezhnoy 2013, 2018).

When estimating the probability of photolysis and release of photolysis-generated atoms of alkali metals to the exosphere, the photolysis lifetimes of molecules should be compared with their ballistic flight times. Based on the results of ab-initio calculations of the dependence of photolysis cross sections on wavelength, NaO, KO, NaCl, and KCl photolysis lifetimes at 1 A.U. for quiet Sun conditions were estimated as 5, 14, 42, and 52 s (Valiev et al. 2020). Based on experimental data, NaOH and NaO photolysis lifetimes are estimated as 11 and 42 s (Self and Plane 2002). Ballistic flight times at 3000 K (typical temperature of impact-produced species) of considered species are about 1000 s and much longer than photolysis lifetimes of these species. It means that the probability of photolysis of the main impact-produced Na-, K- containing species during their first ballistic flight in the exospheres of the Moon and Mercury should be close to unity. Therefore, the delivery efficiency of Na and K into the exospheres of the Moon and Mercury should be large contrary to the one of refractory elements (Grava et al. 2021).

2.1.2 Velocity Distribution Function of Na and K Atoms Produced During Photolysis of NaO, KO, NaOH

It is generally accepted that meteoroid bombardment is responsible for production of quite hot Na and K atoms with temperatures of about 2500 to 5000 K in planetary exospheres (see Sect. 2.1). Let us note that a Maxwellian distribution of velocities is valid for the case of atoms produced directly in impact-produced clouds because such atoms were thermalized during numerous collisions before the start of the collisionless regime.

Considering the typical photolysis lifetimes of Na, K-containing species (about 10 s) and the expansion velocities of the impact-produced clouds (about 2 km/s), the photolysis should occur when the size of the impact-produced clouds reaches about 20 km. For the typical radius of impactors (about 0.02 cm), the number density of the impact-produced species decreases down to about 1 cm^{-3} when the photolysis of the molecules occurs. Therefore, the velocity distribution of Na and K atoms produced during photolysis of parent molecules is not Maxwellian because the photolysis occurs during the collisionless regime of the expansion of the impact-produced cloud. Experimental studies of the velocity distribution of Na and K atoms produced during photolysis of Na-, K-containing species are limited. Photolysis cross sections of NaO and NaOH at 200 and 300 K were measured by Self and Plane (2002). Based on these experimental data and taking into account solar flux from Huebner et al. (1992), the velocity distribution of Na atoms produced during NaO and NaOH photolysis were obtained by Berezhnoy (2010). Namely, the velocity of Na atoms produced during NaO photolysis has a maximum at 2200 ± 200 m/s whereas the velocity of Na atoms produced during NaOH photolysis has a broader maximum at 1800 ± 400 m/s. Based on ab-initio calculations of the photolysis process, it was found that peaks of the velocity distribution of photolysis-generated Na and K atoms occur at 1200 ± 200, 1700 ± 200, 1200 ± 200, and 850 ± 100 m/s for photolysis of NaO, NaCl, KO, and KCl, respectively (Valiev et al. 2020). These estimates were performed by assuming that the velocity of parent molecules before photolysis equal to 0. Taking into account the initial velocity of parent molecules can lead to significant increase of velocities of photolysis-generated atoms (Pezzella et al. 2021). So a significant fraction of photolysis-generated Na atoms has velocities exceeding the escape velocity from the Moon, 2380 m/s. For this reason, photolysis-generated Na atoms may be quite abundant in Na lunar tail especially during the maxima of the strong meteoroid showers.

Detection of impact-produced and photolysis-generated Na and K atoms is a difficult task. Such atoms can be seen in the lunar exosphere only during the first ballistic flight time, about 1000 s, because such atoms quickly lose their energy during inelastic collisions with the surface. For this reason, observations of Na and K atoms in the lunar exosphere should be performed during the maxima of the strong meteoroid showers with the highest possible temporal resolution (at least less than 1000 s) and spatial resolution as it was already done during observations of Na plume formed after the LCROSS impact (Killen et al. 2010).

2.2 The Role of the Magnetosphere as a Source of the Exosphere (Focus on Na)

The interaction of energetic charged particles in Mercury's environment with the surface can induce particle release. Evidence of correlation between the dayside regions of plasma precipitation and exospheric Na intensification have been reported from several ground-based observations (Potter and Morgan 1990; Killen et al. 2001; Leblanc et al. 2008; Mangano et al. 2015; Orsini et al. 2018), with temporal variability as small as a few minutes (Leblanc et al. 2009; Massetti et al. 2017).

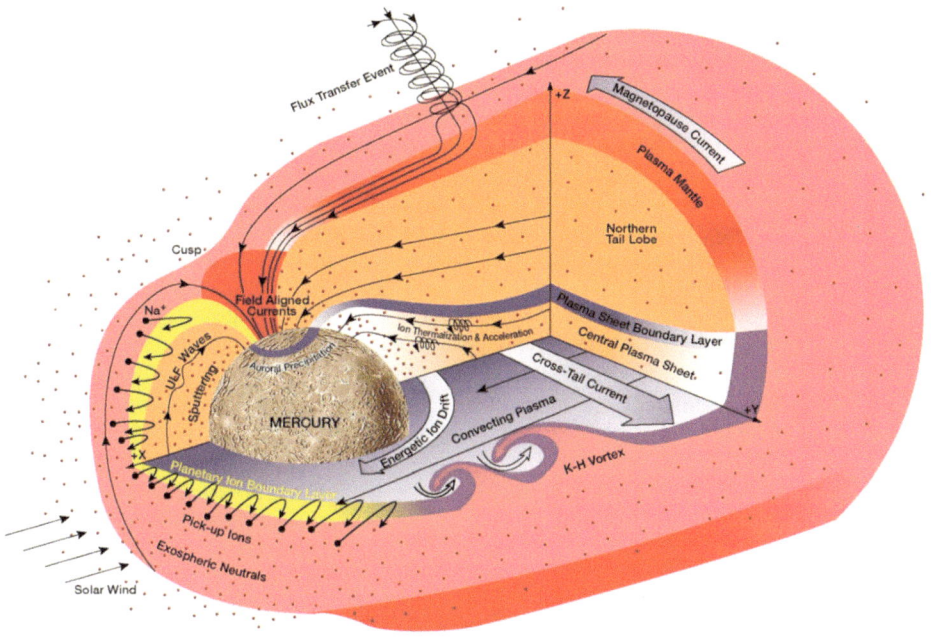

Fig. 16 Schematic of Mercury's magnetosphere (Adapted from Slavin et al. 2009b)

2.2.1 Solar Wind and Magnetosphere Coupling

The interaction of the solar wind with the dipolar magnetic field of Mercury creates a magnetosphere, structurally analogous to that of the Earth's but smaller in size, even with respect to the planetary radius R_M (Fig. 16; see Milillo et al. 2020 and references therein). Recent data from the MAG instrument onboard MESSENGER show that the dipole moment is about 190 nT R_M^3, nearly anti-aligned with the rotation axis (that is, pointing southward), and displaced by 484 km northward with respect the equator (Anderson et al. 2011; Johnson et al. 2012). Computation of the internal magnetic field is complicated by the fact that both magnetic reconnection and magnetic induction affect the amount of magnetic flux above the surface of Mercury (Jia et al. 2019). Consequently, interplanetary transients (as CMEs) can be associated with a variation of the dipolar magnetic moment up to about 25% (Slavin et al. 2019). Under average solar wind conditions, the magnetopause—the current sheet separating Mercury's internal magnetic field from the external Interplanetary Magnetic Field (IMF) dragged by the solar wind—has an average stand-off distance of about 1.4–1.5 R_M in the upstream direction. The magnetotail has a diameter of about 5 R_M at 3 R_M downstream of Mercury (Slavin et al. 2009a; Winslow et al. 2013).

Owing to its heliocentric distance, Mercury's space environment is characterized by a high solar wind density (N_{SW} between 34 to 83 cm^{-3}) and strong IMF ($B \sim$ 15–30 nT), with typical values at 1 AU of $\sim$5–6 cm^{-3} and $\sim$5 nT, respectively. The Parker spiral IMF angle is about 20° at Mercury (that is, half the average value at 1 AU of 45°) with the radial IMF component usually dominating. The solar wind dynamic pressure at Mercury is on average $\sim$10 times larger than at 1 AU. As the solar wind impacts on the planetary magnetosphere it is decelerated and heated, and the IMF moving with the plasma piles up within the magnetosheath region which is between the upstream shock (bow shock) and the magnetopause. The Hermean shock is much weaker than Earth's, as the solar wind has lower plasma β (ratio of the plasma pressure (nk_BT) to the magnetic pressure ($B^2/2\mu_0$)) and Mach number than at Earth. The solar wind Alfvénic Mach number ($M_A = B/(\mu_0\rho)^{1/2}$, where μ_0 is the permeability of vacuum, and ρ is the solar wind density) at Mercury ranges between about 2 and 5 (Slavin and Holzer 1981; Gershman et al. 2013), and it is much lower than at the other magnetized planets, because M_A is typically $\sim$7–10 at 1 AU and it increases with heliocentric distance. Low M_A solar wind interaction drives a low plasma β condition in the Hermean magnetosheath, promoting the development of a strong plasma depletion layer (PDL) just upstream the magnetopause, during nearly any IMF orientation (Gershman et al. 2013). Such a PDL allows magnetic reconnection to frequently occur between magnetic field of similar magnitude for any non-zero shear-angles, and with a higher rate than at the Earth (DiBraccio et al. 2013), driving a series of fast flux transfer events (FTEs) at the magnetopause (Slavin et al. 2012a; Imber and Slavin 2017; Zhong et al. 2020b), as well as flux ropes and plasmoid in the magnetotail (Slavin et al. 2012b; DiBraccio et al. 2015a; Zhong et al. 2020a).

The Dungey convection cycle in the Hermean magnetosphere takes places within only a few minutes (Slavin et al. 2010), which is about 1/60 of the duration observed at Earth where the cycle lasts several hours. The pileup of newly opened magnetic field lines and consequent magnetic merging events in Mercury's magnetotail (loading-unloading cycle) is comparable in duration to the convection cycle as estimated from the reconnection rate at the dayside magnetopause (Slavin et al. 2009b; DiBraccio et al. 2013).

On the basis of the solar wind dynamic pressure distribution at the orbit of Mercury, and including the effect of low latitude erosion on the dayside magnetopause caused by the high reconnection rate (Slavin and Holzer 1979), the solar wind would be expected to impact directly onto the surface frequently. On the other hand, increases of the solar wind pressure were shown to drive induction currents in Mercury's core (Hood and Schubert 1979; Suess and Goldstein 1979) that have the effect to intensify the global dipolar magnetic field of the planet (intrinsic field + induction field). On the dayside, the increased magnetic field acts against the compression caused by the dynamical pressure variation, and also increases the magnetic flux that is available to reconnect with the IMF, thus possibly mitigating— at least in part—the effect of the magnetopause erosion. By analyzing the dependence of the magnetopause compressibility as a function of the distance of Mercury from the Sun, Zhong et al. (2015) found that at the perihelion, where solar wind pressure is higher and Mach number, M_A, is lower, the more effective magnetic reconnection counterbalances the effect of the induction, whereas at aphelion the lower solar wind pressure and higher M_A cause magnetic induction to prevail over reconnection effects.

During extreme solar wind events, such as CMEs, high dynamic pressure was observed to push the magnetopause close to the surface, especially in the southern hemisphere where the intrinsic dipolar field is weaker, thus allowing the direct impact of the solar wind onto the surface. A very low-altitude bow shock was observed when such high solar wind dynamic pressure events ($P_{SW} \sim 140 \div 290$ nPa) are also associated with intense southward magnetic fields ($B_Z \sim -100 \div -400$ nT). The high reconnection-driven erosion of the dayside magnetosphere traced by frequent FTEs indicates that the solar wind should be able to impact a wide portion of the sunlit hemisphere (Slavin et al. 2019). During this class of extreme events, termed "disappearing dayside magnetosphere" (DDM), no dayside magnetopause or magnetosphere signatures were found in MESSENGER data, only magnetosheath plasma was observed down to altitude of about 300 km from the surface of Mercury (four DDM events were identified from the four years of MESSENGER data set; Slavin et al. 2019).

During normal interplanetary conditions, as predicted by numerical modelling studies (e.g.: Sarantos et al. 2001; Kallio and Janhunen 2003, Massetti et al. 2003, 2007), the solar wind plasma can impact on the surface after the magnetic reconnection on the dayside magnetopause injects the plasma into the magnetospheric cusps. Protons in Mercury's (northern) cusp were observed to flow toward the surface (Raines et al. 2014). These protons with energies around 1 keV likely moved into the cusp from the magnetosheath or the dayside reconnection. In the latter case, the protons should have experienced an acceleration along the magnetic field (Cowley and Owen 1989). The pitch-angle distribution shows the existence of a loss cone ($\alpha \geq 40°$), that strongly indicates a fraction of these protons are actually precipitating onto the surface at the footprint of the cusp (Raines et al. 2014). Furthermore, higher magnetic field magnitudes in the IMF are more likely to inject protons with field-aligned pitch-angles, making them more likely to precipitate onto the surface of Mercury rather than mirror in the cusp and travel downtail into the plasma mantle (Jasinski et al. 2017). This occurs because higher magnetic fields magnitudes in the IMF will produce more intense parallel electric fields at the reconnection site, accelerating the particles in the field aligned direction (Egedal et al. 2012; Li et al. 2017).

A clear north-south asymmetry on the night side of Mercury was found by mapping the plasma pressure to invariant latitude (Korth et al. 2014). Because of the northward shifted planetary magnetic dipole and correspondingly a wider southern cusp, an increased particle loss through precipitation in the southern hemisphere is expected. From the analysis of many cusp transits Winslow et al. (2014) found best fit solutions of 121° and 43° pitch angle for the loss cone of the northern and southern cusp, respectively. These observations support the idea that space weathering is actually occurring in the cusp areas.

The planetary ions released from the surface or resulting from the exosphere photoionization, circulate and are accelerated in the magnetosphere. They can be convected back onto the surface mainly on the night side at middle latitudes (Raines et al. 2013, 2014; Wurz et al. 2019). Therefore, these charged particles of solar wind or planetary origin can impact onto the surface over a wide range of local times releasing atoms and molecules from the surface.

Not only positive ions circulate and impact onto the surface of Mercury; in fact, the signature of electron precipitation has been observed through the detection of X-ray emission from Mercury's nightside surface, located mainly between 0 and 6 h local time (Lindsay et al. 2016). Although presently unobserved at Mercury, electron stimulated desorption (ESD, see Sect. 2.1) is likely to occur during these electron precipitation events generally associated with reconnection events in the magnetotail (Starr et al. 2012). This precipitation could provide another possible driver for Na exospheric refilling via the process of ESD.

2.2.2 Magnetosphere–Exosphere–Surface Coupling at Mercury

Despite evidence of the effect of plasma precipitation on the exospheric Na distributions at Mercury (see Sect. 1.8), the details of the mechanism responsible for the surface release of these atoms has still not been identified. In fact, although ion sputtering is the first process that has been advocated, the expected release rate from this process is about two orders of magnitude less than what is needed to explain the observed quantity of Na atoms in Mercury's exosphere. The Na yield, number of released atoms for single impacting ion, for a 1 keV- proton sputtering onto a regolith is between $Y = 0.01$ and 0.1 (Lammer et al. 2003; Johnson and Baragiola 1991). The surface concentration of Na is about $C = 3\%$ in the hotter (low latitudes) regions, and up to 6% in northern smooth plains (Peplowski et al. 2014). The expected Na release due to ion sputtering is $F(\mathrm{Na}_{sp}) = F_{ion} \times Y \times C$, where F_{ion} is the impacting ion flux. If we consider a typical solar wind density of 60 cm^{-3} and a velocity of 400 km/s at Mercury's orbit, the flux of solar wind protons at the magnetopause is 2.5×10^{13} m^{-2} s^{-1}, but it can reach higher values in specific conditions. Only 10% of this flux reaches the surface (Massetti et al. 2003), precipitating through the magnetic field lines in the cusps, so that, the impacting ion flux is $F_{ion} = (2\text{–}20) \times 10^{12}$ m^{-2} s^{-1} on a surface which area below the cusp is of the order of $A = 10^{13}$ m^2. Therefore, $F(\text{Na}_{sp}) = (2\text{–}20) \times 10^{12} \times 0.1 \times (0.03\text{–}0.06) = (8\text{–}200) \times 10^9$ m^{-2} s^{-1} leading to a release rate of $f(\text{Na}_{Sp}) = F(\text{Na}_{Sp}) \times A = (8\text{–}200) \times 10^{22}$ s^{-1}.

The expected induced density at the surface is $n_{\text{Na}} = F(\text{Na}_{Sp})/v_{sp} = (3\text{–}70) \times 10^6$ m^{-3} where $v_{sp} = 3 \times 10^3$ m/s which is the typical velocity of the sputtered Na atoms ejected with a bulk energy around 2 eV (Mura et al. 2007), i.e. about 2 to 4 orders of magnitude less than the typical observed Na equal to 10^{10} m^{-3} (Cassidy et al. 2015).

But ion bombardment can contribute to the formation of the Na exosphere in a different way than direct sputtering. Indeed, Wilson et al. (2006) and Sarantos et al. (2008, 2010) presented an analysis of the Na observations during the Moon's passage in the Earth magnetotail. When the surface is exposed to magnetospheric ions and electrons of the plasma sheet, a significant increase of the Na ejected flux by a factor between 20 and 60 is observed with respect to the period before the path through the magnetospheric plasma sheet. These authors explained these observations by the precipitation-induced diffusion of volatiles inside the regolith up to the upper surface leading to an increase of the reservoir of volatiles available for ejection into the exosphere through other processes. A similar effect was suggested to occur at Mercury (McGrath et al. 1986). Mura et al. (2009) proposed a different mechanism operating at Mercury, being the combined effect of first; ion sputtering onto the surface that would produce a chemical alteration of the surface, and then, second, the reduced binding energy of surface minerals that could allow the Na atoms to be released more efficiently by solar UV photons via PSD. The two mechanisms together may produce the increase of Na release in regions typical of IS, but with the high efficiency and lower energy distribution typical of PSD. This combined mechanism would be active only for volatile species (and not for refractories), which could explain why MESSENGER did not detect a mid-latitude enhancement of Mg and Ca.

Nevertheless, the suggestion that the observed Na exospheric distribution is strongly correlated with the plasma precipitation seem not to fully agree with statistical observations of Sodium distribution. In fact, the northward magnetospheric dipole shift should cause a wider southern region exposed to solar wind impacts. On the contrary, statistical studies based on ground based observations (with limited imaging capabilities owing to the atmospheric seeing) suggested that the southern peaks seen in Na are not more frequent or more extended than the northern ones (Potter et al. 1999; Mangano et al. 2015; Milillo et al. 2021). Milillo et al. (2021) found that the asymmetries in the latitudinal peaks also seem to be related to Mercury's orbital phase, more specifically, the northern peak is more frequent during the outbound leg (roughly coinciding with positions above the ecliptic plane) whereas the southern peak in the inbound leg (coincident with positions below the ecliptic plane). This analysis seems to indicate that the solar effects on the Na exosphere are possibly influenced by another effect due to Mercury's motion or heliocentric position (radiation pressure or distance from the interplanetary dust disk).

A statistical analysis of the Na exospheric distributions obtained by the THEMIS solar telescope by Mangano et al. (2015) concluded that there is not a clear relation between the IMF orientation and the profiles of the Na exosphere, even if double-peak patterns are observed more frequently during positive IMF B_X and negative IMF B_Z components, whereas positive IMF B_Z values are more frequently associated with single-peak equatorial Na emission. This is consistent with the high reconnection rate observed at Mercury not being strongly driven by the IMF orientation (DiBraccio et al. 2013; Slavin et al. 2012a, 2012b).

Springer

Short term Na variabilities accounting for about 10–20% of the total disk intensities have been identified from ground-based observations with time scales of hours down (Killen et al. 1999; Leblanc et al. 2009; Mangano et al. 2013; Massetti et al. 2017). In spite of the rapid changes in the precipitating proton flux related to the fast magnetospheric activity (time scales of 10 s of seconds) and magnetic reconnection processes (DiBraccio et al. 2013; Slavin et al. 2012a,b), the exosphere should show a much smoother response, because of the time delay of the exosphere transport. In fact, the ballistic time scale is about 10 minutes after MIV (Mangano et al. 2007) and the exosphere is expected to recover after a major impulsive ion precipitation event in some hours (Mangano et al. 2013; Mura 2012).

Orsini et al. (2018) reported the unique event of simultaneous in-situ observations of MESSENGER magnetic field (MAG) and ions (FIPS) with the ground-based observations (THEMIS) of the Na exosphere on September 20, 2012 during two strong iCMEs arrivals at Mercury (separated approximately eight hours from each other). Their analysis showed that during solar wind nominal conditions (IMF intensities $\sim$25 nT) the double peak configuration was regularly observed with a Na intensity that seems to be proportional to the ion fluxes observed inside the cusps. During the iCME passage the magnetic field intensity reached values >50 nT and the ion fluxes were so intense that they saturated the FIPS detector. During these events the Na exosphere assumed a more uniform distribution over the whole dayside, and the double peaks were no longer visible. This abrupt change seemed to be an evidence of the effect of plasma precipitation over a wide subsolar region of the Mercury surface, related to the increase of solar wind pressure able to push the magnetopause close to the Mercury surface, as in the cases described above (Slavin et al. 2019). According to this result, the major driver of the Na exosphere configuration may not be the IMF orientation, but the plasma β (the ratio between thermal and magnetic pressures). A possible additional effect occurring during an iCME arrival, is that the iCME plasma is generally rich in heavy and highly charged ions (von Steiger et al. 2000). The yield of multi-charged heavy ion sputtering is higher than the yield of the average solar wind proton (Aumayr and Winter 2003); this fact may increase the surface release and, as a consequence, also the neutral atom, ion and electron density near the surface. The exosphere observed by Orsini et al. (2018) peaked at the subsolar point—this was thought to be unusual, given that Mercury's Na is usually reported to peak at mid to high latitudes. However, the survey of ground-based (THEMIS) data by Milillo et al. (2021) put the observation of the iCME passage in its seasonal context. They found that for that time of year (true anomaly of 128°), exosphere emission peaks frequently near the equator.

2.3 The Lunar Exospheric and Plasma Environment

Na and K exospheres at the Moon were identified from Earth-based telescopes in late 1980 s (Potter and Morgan 1988a,b). In a manner similar to that at Mercury, the two species, though minor, assume the important role of tracers of the exospheric morphology and its dynamics. In particular, the Na component was identified in two components (thermal and suprathermal) with a long and variable tail in the antisunward direction, observed through its gravitational interaction with the Earth, especially at New Moon phase, when it surrounds the Earth and is gravitationally 'refocused' on the night-side (see Sect. 1.5, and references therein). Major lunar exospheric species also include Ar, He and Ne; H is a minor species.

The Neutral population at the Moon has been measured more recently by in-situ missions (Bhardwaj et al. 2005, 2010; Barabash et al. 2009; McComas et al. 2009). In particular, the spacecrafts Chandrayaan-1 and IBEX identified ENAs (Energetic Neutral Atoms) as derived from neutralization of solar wind protons after interaction with the surface. Though the energy spectrum distribution (whether Maxwellian or a power law) (Futaana et al. 2012; Allegrini et al. 2013), and also the dependence or not on solar wind and IMF characteristics are still debated, the intensity of such high energy neutral atoms is clearly related to the solar zenith angle (decreasing as SZA increases) (Wieser et al. 2009).

The Moon primarily differs from Mercury in the fact that it does not possess a global intrinsic magnetic field. Nevertheless, it possesses a series of mini-magnetospheres associated with the numerous magnetic anomalies spread over the surface (Coleman et al. 1972; Hood et al. 2001). A detailed map of magnetic anomalies over the Moon was produced from Kaguya and Lunar Prospector and can be found on Tsunakawa et al. (2015). Magnetic anomalies produce a shielding of specific surface regions, acting on all ionized particles present in the close lunar environment; as a consequence, they cause the double effect of deflecting impinging particles over the core region of the anomaly and of intensifying the flux impinging on the surrounding region (in an annular shape) (Vorburger et al. 2013; Saul et al. 2013). In fact, typical annular regions of higher ENA albedo have been identified almost perfectly superposed over the mini-magnetosphere map. As a consequence, the lunar surface also experiences an inhomogeneous space weathering effect, as well as a differential charging of the different regions of each mini-magnetosphere (Poppe et al. 2016).

The situation is different when the Moon crosses the Earth's magnetotail: during these periods, the plasma in the Earth's magnetosheath has higher velocity and the impacts produce higher ENA flux (and albedo) (Allegrini et al. 2013).

The lunar dayside ion population appears to be composed of three different families of particles: 1. Solar wind protons backscattered from the surface (and not neutralized as ENAs); 2. Protons scattered from the anomalies, and 3. Endogenic ions (Bhardwaj et al. 2015 and references therein). The first family accounts for up to the 1% of backscattered solar wind; the second family shows a scattering efficiency of scattering up to 10% and produces a broadening of the energy spectrum; collisionless shocks are also observed. The third family is composed of heavier ions of lunar origin such as He^+, C^+, O^+, Na^+ and K^+. Low energy measurements by Kaguya showed an evident dawn-dusk asymmetry in Na^+ ions (dusk side being lower by 50%), probably related to higher emission from the surface on the dawn side. Dawn side plasmas were attributed to micrometeoroid impact by Poppe et al. (2021). Pickup ions (probably He^{2+}) are observed when the Moon is in the Earth's magnetotail lobes and also upstream. Pickup ions at dawn with a peak at 16 amu mass have been identified too (probably originating from Earth's OH, CH_4 and O_2).

Nightside protons were identified at an altitude of 100–200 km (in the wake, originally thought not to contain plasma) with trajectories both perpendicular and parallel to the IMF (Halekas et al. 2015). Electron fluxes are also observed, with enhancements in the wake region (Halekas et al. 2011). An overview of the interactions between the solar wind, plasma and the Moon to produce ions and neutrals is shown in Fig. 17.

To conclude, the processes acting on the Moon are similar to those observed at Mercury, with the main difference being that the Moon regularly passes into the Earth's magnetosphere, which produces additional production through the interaction between the surface, the exosphere and the different ion (and electron) populations. Better understanding of the processes involved and of their relative roles, together with the influence of the surface potential and mini-magnetospheres of the magnetic anomalies, can give a global overview of the Lunar environment.

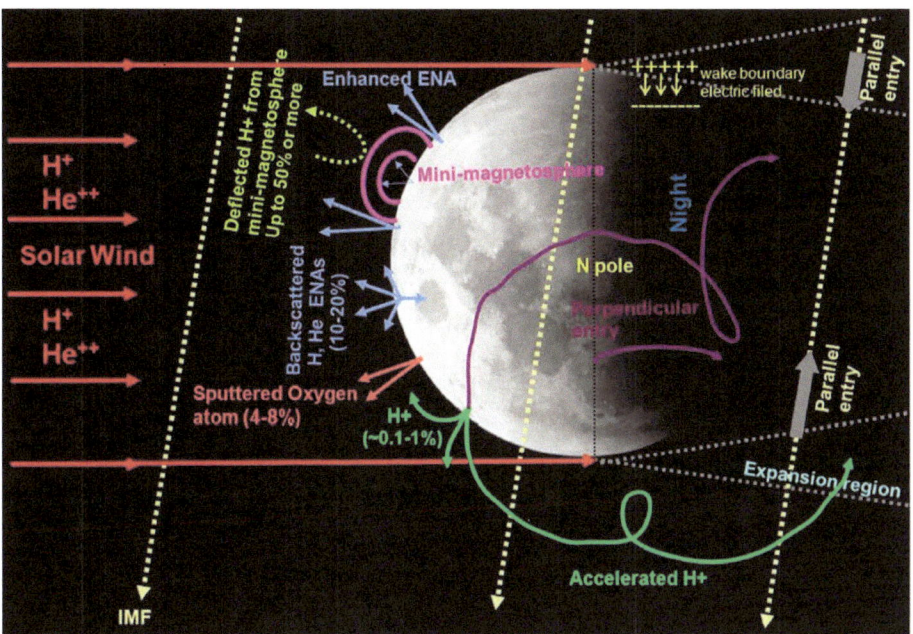

Fig. 17 Overview of the processes known to occur around the Moon and on its surface, owing to interaction with the solar wind (from Bhardwaj et al. 2015)

3 Exospheric Circulation: The Na Example at Mercury

3.1 The Key Observation of the Dawn/Dusk Asymmetry at Mercury and Its Orbital Variability

Sprague (1992) analyzed several set of observations (Sprague et al. 1990; Potter and Morgan 1986, 1990; Killen et al. 1990) and found that Mercury's Na exosphere was spatially organized with systematic brighter emissions when observing the dawn side of Mercury than when looking to the dusk side. These authors explained it as a process of sodium and potassium ions from the dayside being accelerated to energies in excess of 1 keV by electric fields in the plasma sheet, striking the nightside Mercurian surface and being implanted to depths up to it few hundred Angstroms. This recycling of the dayside Na via ion implantation on the nightside could therefore induce a preferential release of the Na atoms in the morning when the surface moved to the dayside. These conclusions were criticized by Killen and Morgan (1993) who argued that the observations did not clearly show a dawn/dusk asymmetry nor could the ion recycling model really explain this asymmetry. These authors also showed that grain/regolith diffusion from depth to the surface is not intense enough to explain the needed supply rate of the exosphere.

Sprague et al. (1997) presented new set of observations, confirming the existence of a dawn/dusk asymmetry. This work was followed by Hunten and Sprague (1997), who suggested that the neutral Na exosphere could follow a cycle similar to that of the lunar Ar. However, Killen and Ip (1999) argued that the two species and environments were significantly different and that a Na cycle like that of Ar should not be expected.

Schleicher et al. (2004) published the first reconstructed image of the sodium exosphere as observed during the May 2003 transit of Mercury. This observation (reproduced

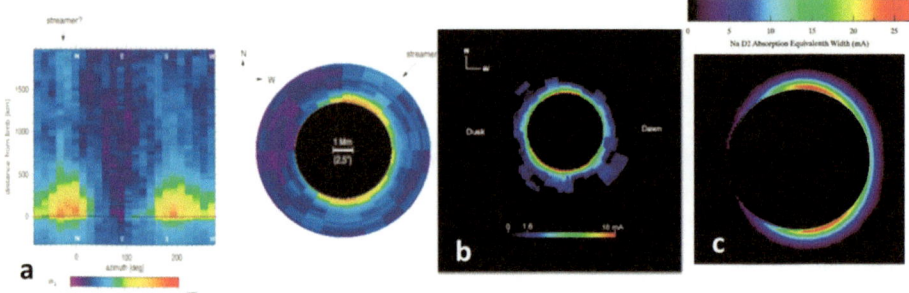

Fig. 18 Reconstructed image of the sodium exosphere as observed during three Mercury transits. Equivalent width above the limb observed during the (**a**) 7 May 2003 transit (Schleicher et al. 2004) at a TAA=149°, (**b**) the 8 November 2006 transit (Potter et al. 2013) at a TAA=328°, and (**c**) the 9 May 2016 transit (Schmidt et al. 2020) at a TAA=149°

in Fig. 18, panel a) clearly showed a strong dawn/dusk asymmetry with a much stronger absorption by the sodium exosphere on the dawn side than on the dusk. During the following transit in 2006, Potter et al. (2013) also observed the sodium exosphere in absorption, producing an image similar to the one published by Schleicher et al. (2004). Contrary to Schleicher et al. (2004), these authors did not observe a dawn/dusk asymmetry (Fig. 18, panel b). A third transit has been observed on 9 September 2016 by Schmidt et al. (2020) at an orbital position very close to that during Schleicher et al. (2004) observation and again displaying a clear dawn/dusk asymmetry. A striking feature of the 2016 transit is that the spatial distribution of the absorption equivalent width is similar to the one observed thirteen years before, consistent with the sodium exosphere being highly repeatable from one year to another as demonstrated by MESSENGER observations (Cassidy et al. 2015).

Given this year-to-year consistency, why didn't Potter et al. (2013) observe a similar dawn/dusk asymmetry? A key difference from the two other transit observations is the orbital position of Mercury. The three transit observations correspond to two different orbital positions associated to two different speeds of rotation of Mercury's surface with respect to the Sun, as explained by Potter et al. (2006, 2013). Indeed, the rate of rotation of the surface *in solar local time* drastically changes along Mercury's orbit, with a much faster prograde rotation at aphelion near TAA=180° (case of Figs. 18 a and c) compared to a much slower retrograde rotation at perihelion near TAA=0° (Fig. 18 b case). Potter et al. (2006, 2013) and Leblanc and Johnson (2003) thus explained this asymmetry by the rotation of Mercury's nightside surface into the dayside. MESSENGER observations showed a definite TAA correlation in the dawn/dusk ratio of Na in the exosphere. This explains the Potter et al. (2013) observation of the 2006 transit at a TAA of 323° (Fig. 18b). The ratio of Na column abundance at 6 AM to that at 18:00, dusk, measured during the outbound leg of the MESSENGER orbit (TAA<180°) shows a strong dawn enhancement, whereas the inbound portion of the orbit (TAA>180°) shows no such dawn/dusk asymmetry.

A preferential bombardment by meteoroids of the trailing hemisphere with respect to the leading could also produce an asymmetry, as actually observed for the Ca exosphere (Burger et al. 2014; Killen and Hahn 2015; Christou et al. 2015). This would produce a permanent dawn/dusk asymmetry, changing along Mercury's orbit. Furthermore, it would produce an equatorial maximum as for Ca and not high-latitude maxima as displayed in Fig. 18 a and c. Finally, the Na released by meteoroid vaporization should be ejected with significantly more energy (with a temperature larger than 3000 K) than as suggested by the altitude profile near the equator observed during these transits (around 1600 K according to Potter et al. 2013). Therefore, Meteoroid Impact Vaporization, MIV, is unlikely to be responsible for any dawn/dusk asymmetry in the Na exosphere.

Another potential explanation is that the release of a significant portion of the exospheric sodium atoms from the surface is controlled by the Sun. In order to induce a dawn/dusk asymmetry this way, at least during a portion of Mercury's orbit, it would be necessary that the interaction between the surface and the Sun ejects a solar local time-dependent amount of sodium atoms into the exosphere with the release of more sodium in the exosphere in the morning than in the evening to yield a dawn peak. The mechanisms of ejection, such as photo-stimulated desorption or thermal desorption, would become less efficient with increasing local time; however this can only be explained by a change in the surface, the solar radiation being the same on both sides of the dayside and the surface temperature as well. This brings us to the second possible explanation which is a change in the upper surface from which the sodium atoms are released. Such a change could be either a variation in the quantity of sodium atom available for ejection (as suggested by Hunten and Sprague 1997) or in the way these atoms are adsorbed in the regolith (a typical signature of space weathering). Both explanations are equivalent, inducing a gradual impoverishment of the surface in sodium atoms available for ejection into the exosphere with increasing solar local time. Moreover, because MESSENGER and transit observations suggest that the sodium exosphere is steady from one year to another, it also implies that a Na-depleted evening surface would need to path back to a Na-enriched surface before rotating back to the dayside, after 88 Earth days in the nightside, to once again produce the observed dawn/dusk asymmetry.

The question of the existence of a TAA dependent dawn/dusk asymmetry is therefore that it provides us an important clue about the fate of the sodium exospheric atoms. As explained before, the best explanation today is that the asymmetry is produced by a permanent but TAA-dependent day to night cycle of the sodium exosphere, involving a layer of the upper surface and release from this layer into the exosphere.

The situation is more complicated than this, however, as shown in Fig. 18 where the dawn/dusk asymmetry is observed to change along Mercury's orbit. Several decades of ground-based observations have led to large set of data allowing us to investigate this variability. Potter et al. (2006) analyzed several years of observations of Mercury's sodium exosphere and extracted the ratio between the limb dayside emission and the terminator, either dawn or dusk, emission from the 94 useful days of observation. These authors further corrected these ratios by dividing them by a theoretical value for a uniformly spatially distributed exosphere, taking into account the phase angle of the observation. As shown in Fig. 19 panel b, the variations of these two ratios during the Mercury year display clear variability and are not always close to one as would be expected for a uniform exosphere. These variations are in good agreement with Leblanc and Johnson (2010), who simulated the dawn-to-dusk ratio (Fig. 19, panel a) and found values close to one around perihelion (below 30°) and which gradually increased up to values significantly larger than one up to 100 to 140° TAA before then being close to one the rest of the orbit. However, they are not in agreement during the second half of Mercury's orbit between 180 to 330° with the model predicting significantly larger dawn-to-dusk asymmetry (panel a) as compared to the observations (panel b; see also panel c of Fig. 21).

Leblanc and Johnson (2010) predicted an almost permanent strong dawn-to-dusk asymmetry because of the cycling of the sodium atoms from dayside to nightside and their accumulation in the nightside surface, leading to a persistent peak of sodium release in the morning, except during the phase of retrograde increase of the local time around perihelion. The intensity of this cycling should be partly influenced by the solar radiation pressure, as

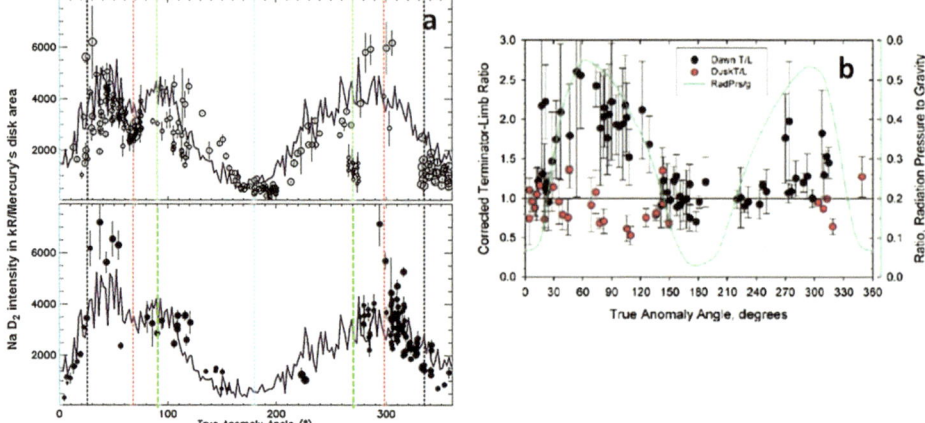

Fig. 19 Panel a: Dawn side (upper) and dusk side (lower) emission intensity in kilo-Rayleigh (kR) per Mercury's disk area along an annual cycle (true anomaly angle) as simulated by Exospheric Global Model—EGM (averaged on half a day period and for a phase angle of 90°). Circles: ground based observations. The size of the circles follows the sine of the phase angle. Increasing size of the circle corresponds to increasing observed proportion of Mercury's exosphere. Vertical blue lines: perihelion and aphelion of Mercury. Vertical dark lines: reversal of the apparent motion of the Sun at Mercury. Vertical red lines: peak of solar radiation pressure and of solar photon scattering efficiency. Vertical green lines: peak of heliocentric radial velocity (Leblanc and Johnson (2010)). Panel b: Variation of the corrected terminator-to-limb ratio with true anomaly angle, black dots: dawn-to-limb ratio, red dots: dusk-to-limb ratio. Green line: ratio between the solar radiation pressure and the gravity surface intensity (Potter et al. 2006)

explained by Potter et al. (2007), because between the inbound and outbound orbital legs, the net effect of this force changes, in an opposite way, the intensity of the solar continuum seen by the sodium atoms (as modelled in Leblanc and Johnson 2010). Mura et al. (2009) added the surface rotation to their Monte Carlo exospheric model and found that the above mentioned mechanism of nightside refilling and dawn release could almost perfectly fit the first available observations at transit (Schleicher et al. 2004) (Fig. 20). The Doppler measurements made during transit and the observed enhancements of sodium column density above the poles also constrained the release mechanism to PSD, partially boosted by solar wind precipitation. It is worth noting, however, that to properly reproduce the transit observation, Mura et al. needed to simulate the evolution of Mercury for more than one Mercury orbit (see Fig. 20), and found that the dawn enhancement is a stationary state, contrary to what is observed along the orbit.

However, this picture of the sodium cycle has been partially contradicted by the long-term observations of the sodium exosphere by MESSENGER (Cassidy et al. 2016). As shown in Fig. 21 panels a, c and d, MESSENGER MASCS instrument reported a persistent peak of the exospheric sodium column density associated with the planetary cold longitudes, see Fig. 21, panel d. The cold longitudes (90° and 270° East longitudes in panel d) are the two planetary longitudes at the dawn and dusk terminators when Mercury is at perihelion. They are called cold longitudes because they received less solar flux than the other planetary longitudes when averaged over one Mercury year (Soter and Ulrichs 1967). Cassidy et al. (2016) explained these observations by a combination of cold surface and solar radiation pressure during a large portion of Mercury's orbit around the perihelion leading to a maximum of surface absorption of exospheric sodium atoms at these specific longitudes. The

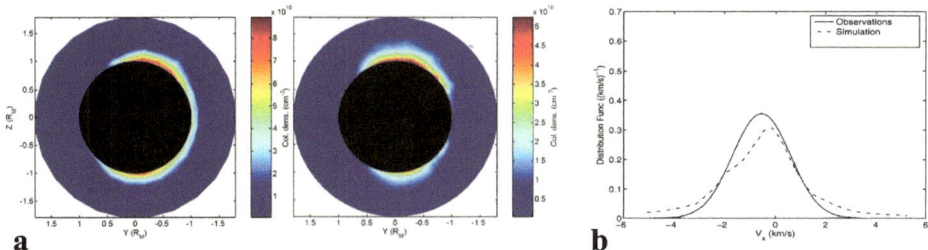

Fig. 20 Panel a: Comparison between transit observations (Schleicher et al. 2004, left) and modelling (Mura et al. 2009, right). The solar wind conditions at the time of the observation were modelled to obtain the proton precipitation onto the cusps, which in turn produces the polar enhancements observed via the mechanism proposed by Sarantos et al. (2008). The Doppler measurements (panel b, solid line) of Na particles' velocity distribution can be properly fitted if the ejection source has a temperature above 1000 K (dashed line)

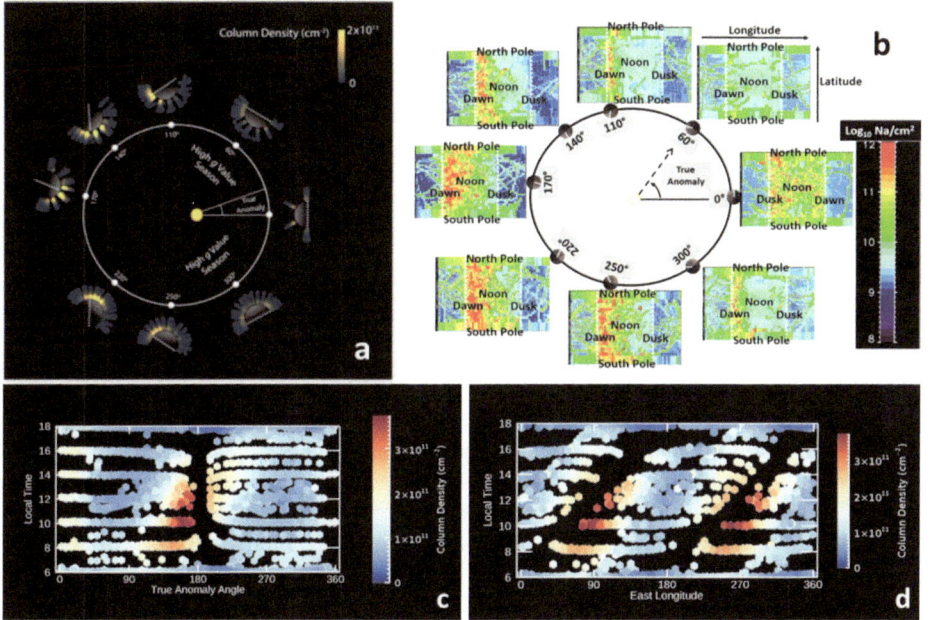

Fig. 21 **a**: Observed sodium limb column density (with line-of-sight tangent points below 30° in latitude) projected onto Mercury's equatorial plane over the course of one Mercury year. The white dashed lines indicate the position of the cold longitudes as Mercury rotates (Cassidy et al. 2016). **b**: EGM-simulated vertical column density along one Mercury year. Each insert represents the vertical column density with respect to longitude (horizontal axis) and latitude (vertical axis) as simulated at the same true anomaly angle (TAA) as in panel a. The two vertical dashed lines in each insert are for the terminators. The centre of the insert corresponds to the subsolar point. **c**: Observed sodium limb column density with respect to True anomaly angle and local time at 300 km altitude. **d**: Observed sodium limb column density at 300 km altitude with respect to East longitude and local time

building of a large surface reservoir around perihelion would then lead to a release of the sodium atoms all along the rest of Mercury's year.

Both adsorption by the cold surface and solar radiation pressure are included in the EGM model (Leblanc and Johnson 2010) and the one developed by Mura et al. (2009). As shown in Fig. 21, panel b, in the case of an EGM-simulated spatial distribution of the vertical column density, the cycle of Na exospheric atoms occurs throughout the Mercury year, so there is always a cycle of Na atoms from the exosphere onto the nightside surface. This in turn leads to the preferential enrichment of the surface pre-dawn regions in sodium atoms and to the existence of a release of Na atoms from the surface into the exosphere at dawn over the majority of Mercury's orbit. As shown in panel b of Fig. 21, the peak of the simulated vertical column density seems to follow the evolution displayed in panel a during the first half of Mercury's orbit, from a true anomaly angle of 0° up to 180°, but continues to appear at dawn during the second part of the orbit, in contradiction with the observations that display no dawn enhancement at TAA=220° and 250°. What could affect the cycle as depicted in panel b during this portion of the orbit remains an open question. Several potential mechanisms that could influence this cycle are discussed in the following sections.

3.2 Surface Abundance as a Signature or a Driver of the Na Cycle?

The surface of Mercury, because of the absence of an atmosphere, has evolved mainly under external forces during the last few billion years. It is saturated with small impact craters and covered by regolith, which is defined as "the entire unconsolidated or secondarily recemented cover that overlies more coherent bedrock, that has been formed by weathering, erosion, transport, and/or deposition of the older material" (Eggleton 2001). In many aspects, the study of the surface of Mercury inherits concepts from that of the Moon, even if the regolith of Mercury is probably more mature than the lunar regolith, with smaller grain sizes and larger proportions of glassy particles (Langevin 1997). This is because some differences exist in the size distribution of micrometeoroids/meteoroids (with relatively more micrometeoroids at Mercury, which also have higher impact energies; Janches et al. 2021), in the solar wind flux, solar cosmic rays' flux and surface temperature.

Mercury's surface is continuously modified by loss processes and by replenishment of new material. Every agent precipitating on the surface transmits energy to bound particles and may produce their release: plasma (IS, ES); photons (PSD), and micrometeoroids (MIV; see Sect. 2.1). Different studies (i.e. Wurz and Lammer 2003; Leblanc and Johnson 2003; Mura et al. 2007) have shown that the particles extracted by the surface just by thermal energy (TD) do not have sufficient energy to escape from the planet; hence, this process may be assumed to be negligible in terms of net surface erosion, although it causes some redistribution of particles over the surface, and it is necessary to take it into account as well. The material on the surface may be characterized in two groups: volatiles (those species with a relatively small equilibrium condensation temperature) and refractories (all others). Compared to refractories, which are primarily sensitive only to very energetic processes such as IS and MIV, volatiles may be extracted by all processes and are more exposed to variations in surface concentration, sometimes in a rapid way. Understanding such short-term variabilities is a key factor in understanding the long-term history of the surface and, hence, of Mercury's planetary evolution. Among volatiles, sodium is the most natural candidate for studying these phenomena, because in addition to being relatively abundant in the exosphere of Mercury, it is also observable from the ground with relative ease.

We therefore assume that volatiles such as Na are extracted by all the surface processes that extract and modify the surface composition of Mercury, and although they mostly reprecipitate, partially they are lost to space. In particular, Na neutral particles in the exosphere are subject to the radiation pressure of the Sun and this increases the escape probability. To add some complexity, the radiation pressure is not constant but varies over the Mercurian year (Fig. 19b). A long tail of escaping sodium particles can be observed (Potter et al. 2002;

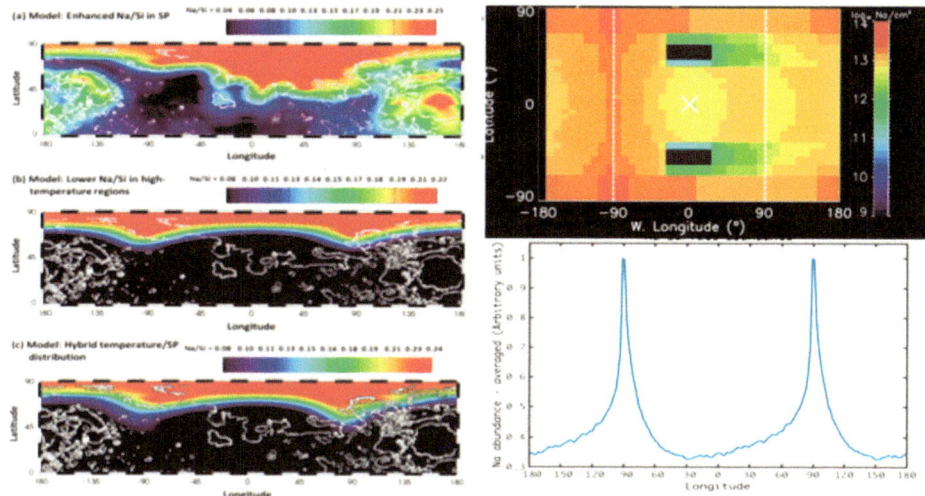

Fig. 22 Left: Example of Na/Si abundance ratio in the Northern hemisphere of Mercury's surface for three different scenario of forward modelling of the observations performed by the Gamma Ray Spectrometer of MESSENGER (see Peplowski et al. 2014, for details). Top-right: Simulated sodium surface density (in $\log_{10}$ of Na/cm^2) in Mercury's surface as modelled by EGM; bottom-right: longitudinal variability of surface Na abundance, as simulated by using the same model as in Mura et al. 2009, and averaged over all latitudes. The two maxima at 90° and 270° are the effect of the updated temperature map with "cold poles" at these longitudes

Fig. 8), and knowing the order of magnitude of the speed (by Doppler measurements—see Schleicher et al. 2004 —or by numerical simulations), it is possible to approximately quantify the amount of sodium that is lost instantly. Some sodium is also lost from the exosphere by Jeans escape, or by photoionization and escape of ions, governed by magnetospheric processes and solar wind pickup. This represents a net loss of sodium, which is, in a small fraction, replenished by that contained in the precipitating micro-meteoroids (Genge et al. 2008).

On a geological timescale, impacts are the dominant mechanism in terms of their effect on the surfaces of the airless bodies of the solar system. The frequency, geographic distribution and type of impact control the evolution of the finer-grained superficial heterogeneous regolith. In addition, larger impacts have an influence on the orography of the terrain. The long-term implications of this net are discussed in Orsini et al. (2013), whereas annual variabilities have been discussed in several studies, as more observational data have become available (e.g. Leblanc and Johnson 2010; Milillo et al. 2021). In any case, to achieve some sort of short-term stability on a yearly cycle, a mechanism that refills the uppermost surface layer is needed. Impacts remix the material in the first layers of the surface in a process called "impact gardening". We may consider that as a "blending" or "tipping" because it repeatedly and stochastically reverses the depth distribution of materials that, otherwise, would be a distinct stratigraphy with depth. Killen et al. (2004) calculated the outward diffusion rates from inside the regolith grains and compared them with the gardening scale times, which are the typical "annealing" timescale for the surface to lose memory of the previous state. The diffusion scale times—to completely unload the grain—must be longer, otherwise the grain empties completely. This implies a threshold for the net outward flux, which is also calculated in Killen et al. (2004), and it is mostly a function of grain size and diffusion coefficient. However, the large uncertainties on these two latter quantities makes

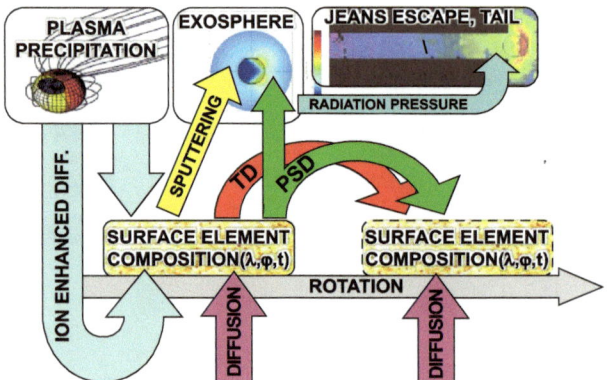

Fig. 23 Scheme of Sodium circulation between Mercury' surface and exosphere

the evaluation quite difficult. Particle precipitation, and the effects of plasma impacts, also adds some complexity to the physics. Sarantos et al. (2008) using plasma and exospheric Na measurements at the Moon, speculated that the plasma bombardment may increase the diffusion rate (enhanced diffusion) by supplying energy to the deepest layers of the grains. At Mercury, that would result in an excess of Na release where solar wind precipitates (day-side magnetospheric cusps), leading to distinct bulges of Na excess in the Northern and Southern dayside exosphere as observed (Schleicher et al. 2004; Mangano et al. 2015). It was also suggested by the observations performed by the Gamma-Ray Spectrometer on-board MESSENGER, that the enrichment in Na of the high latitude surface with respect to the equatorial surface (Fig. 22b) could be a signature of the relation between upper surface composition and exospheric circulation as suggested by the exospheric models (Fig. 22a). On the other hand, the high-latitude enhancement in Na follows the north volcanic plains unit which is a geologic feature. Better sampling of the southern hemisphere by the Bepi-Colombo mission will hopefully provide new insight into this circulation versus geology aspect of the Na exosphere.

A schematic illustration of a model of surface accumulation and release of Na is shown in Fig. 23. Solar wind (or magnetospheric) plasma precipitates onto a surface element and causes immediate ion sputtering or enhanced diffusion of sodium, from deep in the regolith layer (up to fraction of μm). If the element is in the dayside, PSD and TD release Na from the uppermost surface layer, but because these processes are highly efficient at Mercury, the net flux is basically limited by the amount of free Na in that element, which is provided by diffusion and ion-enhanced diffusion. Particles released by TD have low energy and quickly return to a (maybe different) surface element; particles release by PSD have longer ballistic lifetimes, and they may either fall back farther away or be lost to space, depending on the combination of multiple effects (radiation pressure, release energy, etc.). Thus, we expect the free Na content in the surface to be a function of the local time (see as an example Fig. 22b), because surface elements that are brought to the nightside by planetary rotation ceases to release Na particles, so that their Na content can be replenished. Models following Fig. 22 (Mura et al. 2009; Leblanc and Johnson 2010) are in fact able to explain the yearly variability or the excess of Na column density observed close to the dawn terminator during transit measurements (Schleicher et al. 2004), for which the local-time surface Na abundance profile should always peak close to dawn. However, MESSENGER observations show that the picture is more complicated than this. Cassidy et al. (2015) observed that the sodium exosphere is enhanced above Mercury's cold-pole longitudes (Fig. 21a). This stems from the fact that, owing to the 3:2 spin/orbit resonance, the longitudes at 90° and 270° are

Fig. 24 Simulated Sodium column density at different TAAs and projected onto the equatorial plane of Mercury, to be compared with MESSENGER observations in Fig. 21a (see text for details)

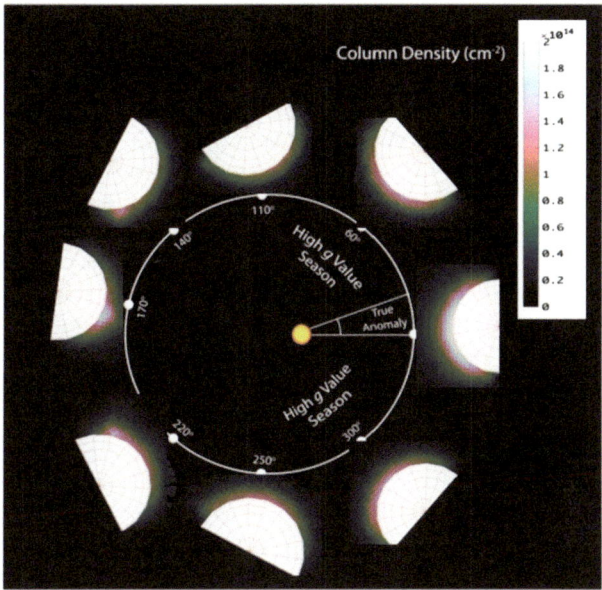

always at the subsolar point at aphelion. On long timescales, this leads to a smaller average temperature close to these longitudes, and smaller temperature means less Na release on average, which will result in the formation of two sodium reservoirs at 90° and 270°. According to Cassidy et al. (2016) the reservoirs are filled during periods of anti-sunward sodium transport, when the cold poles are near the terminator.

To account for the presence of cold poles, one approach could be to adapt the surface temperature models. In most work before the MESSENGER observations, surface temperature was modelled with simple local-time functions with no longitudinal variability. However, although it is straightforward that such reservoirs would work for the surface abundance, it is still debated why the accumulation of Na particles at the cold pole at 270° would, at the same time, result in an enhancement of exospheric sodium over the same region, which seems counterintuitive. For example, we may assume that surface composition maps such as those in Fig. 22 (right), with maxima at 90° and 270°, are valid for all TAAs, and run a typical exospheric model. The result will be similar to Fig. 24, with almost correct asymmetries. The main problem with this approach is that the actual instantaneous surface map is not the average one; in other words, the available models are able to explain the MESSENGER observation in Cassidy et al. (2016) only if we assume that the cold-pole enhancement has worked for ages in the past, causing an excess of sodium at 90° and 270° that is now eroded by PSD release. Because observations from ground are not able to disentangle the yearly variability from the local-time ones, it is likely that a big improvement of our knowledge of the surface variability of sodium will come when BepiColombo orbital data are available.

3.3 What Else Could Control the Na Cycle?

3.3.1 The Role of the Meteoroid Bombardment

The meteoroids arriving on Mercury may have many different sources, from the Main Belt Asteroids (MBA) to comets (Janches et al. 2021). Within the comet populations, these may further break down into Jupiter-family (JFC, short-period) and long-period (LPC) comets, with the LPCs being divided into Halley-type comets (HTC) and Oort Cloud comets (OCC). Small particles originating from MBAs ($D < 2$ cm) were revisited by Borin et al. (2009, 2016a, 2016b) who showed that the circularization of their orbits by Poynting-Robertson drag results in narrow impact velocity distributions. The analytic velocity distribution of Cintala (1992) is 20.50 km s^{-1} as compared to 16.81 km s^{-1} from Borin et al. (2009, 2016a, 2016b). Large MBA meteoroids ($D > 2$ cm) have been treated by Marchi et al. (2005) and provide a much broader impact velocity distribution at Mercury, ranging from 20 to 80 km s^{-1}. These higher velocities are due to the fact that larger particles are influenced only by gravitational forces. The impact of large meteoroids are rare events, with low probability that one might occur during the BepiColombo mission (Mangano et al. 2007) and are briefly described towards the end of this subsection.

More recent models focus on demonstrating that the directionality of the meteoroid influx plays a major role in the characteristics of the Hermean exosphere (Pokorny et al. 2017). This adds a strong dependence of impact characteristics and fluxes with respect to the planet's TAA, which is true for all planets and satellites (Janches et al. 2018, 2020) but is particularly extreme at Mercury owing to the planet's high eccentricity and orbital inclination (Pokorny et al. 2017). Such dependency also influences the impact velocity distributions (Pokorny et al. 2018), which can vary from $V_{imp} < 70$ km s^{-1} at perihelion and $V_{imp} < 50$ km s^{-1} at aphelion for meteoroids originating from MBAs and JFCs. Nesvorny et al. (2010) argued that JFC-released meteoroids represent the majority of the flux incoming into Earth's atmosphere, both in number and mass, emphasizing their importance in the inner solar system. Both Borin et al. (2016a, 2016b, 2017) and Pokorny et al. (2017, 2018) considered for the first time this population as part of the influx at Mercury. Owing to the high eccentricity of Mercury and low impact velocities of these meteoroids compared to Mercury's orbital velocity, the impact directions of JFC and MBA meteoroids are expected to experience significant motion in the local reference time frame during Mercury's orbit.

Similar to that of MBA meteoroids (see Pokorny et al. 2018), there is a shift in the radiant distribution of JFC meteoroids as Mercury moves toward or away from the Sun, caused by the nonzero eccentricity and inclination of Mercury's orbit and a consequent drift of the planet's velocity vector from the ecliptic plane and its perpendicular orientation with regard to the radial vector.

The influence of meteoroids originating from long-period comets, such as HTCs or OCCs, was only treated recently by Pokorny et al. (2017, 2018). These play an essential role in the formation of Mercury's exosphere, even though they represent a small fraction of the meteoroid input budget. This is because they are dynamically less evolved than MBA or JFC meteoroids and are released into highly eccentric orbits intersecting Mercury faster and at high speeds, thus making them the dominant source of physical phenomena regarding the formation and morphology of Mercury's exosphere. Specifically, the mass flux of LPC meteoroids at Mercury compared to JFC meteoroids could be as small as 5% but their impact velocities reach values over 100 km s^{-1} and make them the dominant source in terms of the impact vaporization or the impact yield (Pokorny et al. 2017, 2018).

Recent estimates on the mass accretion of meteoroids (Pokorny et al. 2018), averaged on the Mercury orbit, provided the following values: MBA meteoroids 0.26±0.15 tons per day, JFC meteoroids 7.84±3.13 tons per day, HTC meteoroids 1.69±0.91 tons per day, and OCC meteoroids 2.37±1.38 tons per day (Janches et al. 2021).

The vaporization flux averaged over a Hermean year then results in $F_{orbit} = (200 \pm 16) \times 10^{16}$ g cm^2 s^{-1}, with maximum value of $(436 \pm 57) \times 10^{16}$ g cm^2 s^{-1} occurring at TAA=337°, and a minimum value of $(82 \pm 12) \times 10^{16}$ g cm^2 s^{-1} occurring at TAA=188°.

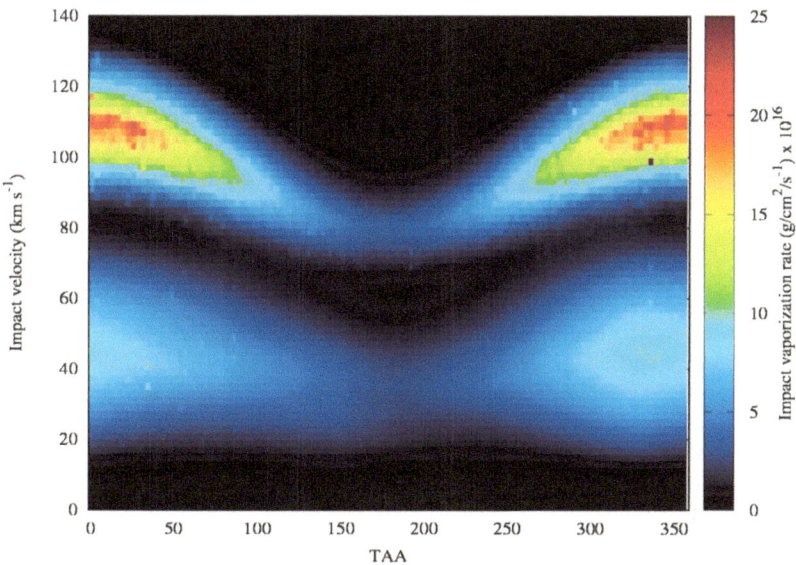

Fig. 25 Total vaporization flux as a function of TAA and the impact velocity. The units are $g\,cm^{-2}\,s^{-1}$ per $2\,km\,s^{-1}$ bin (Pokorny et al. 2018)

Pokorny et al. (2018) state that only $\sim$1% of the estimated total vapor appears to contribute neutrals to Mercury's exosphere. Moreover, the model of Pokorny et al. (2018) predicts an important perihelion-to-aphelion ratio in the impact vaporization rate (Fig. 25).

The meteoroids discussed so far are usually very small in size ($<$cm) and their vaporization at the surface of Mercury provide a mostly continuous source to the exosphere of Mercury. Large impactors do not contribute significantly to the long-term state of the exosphere but instead can provide transient enhancements. A 1 m sized meteoroid is expected to impact the Earth at a rate of 10 per year, and this estimate is an order of magnitude lower for Mercury (Marchi et al. 2005). Mangano et al. (2007) modelled the transient effects to the exosphere of a 1 m and 10 cm sized meteoroid surface impact. It should be noted that the modelled exospheric densities from the impact reported by Mangano et al. (2007) would most likely be an order of magnitude larger for sodium, when adjusted for the surface sodium abundance that has since been updated from MESSENGER observations (McCoy et al. 2019). Mangano et al. (2007) found that an impact by such meteoroids would be noticeable to the exosphere for Al, Mg, Si and Ca, and more likely to be noticeable on the nightside for Na and K owing to the higher background exospheric levels on the dayside. Recently, an observation by MESSENGER of planetary ions that had been detected in the solar wind from an inferred large meteoroid impact was reported by Jasinski et al. (2020). The ions were estimated to be mostly Na^+ and Si^+, just photoionized and picked-up by the solar wind. The neutral densities of the plume at high altitudes of $\sim$5000 km were inferred from the measured ion fluxes to be $<10^2$ cm^{-3}. Such exospheric neutral enhancements are expected to be transient in nature, and only expected to last for $\sim$20 minutes (Mangano et al. 2007).

Meteoroid showers, caused by fresh meteoroids ejected from comet nuclei into heliocentric orbit, represent a significant, but short-lived, enhancement of the flux incident on a planet (Janches et al. 2020). During a shower, the meteoroid velocity distribution and direction remain fixed. Christou et al. (2015) (see also Killen and Hahn 2015) showed that the stream belonging to short-period comet 2P/Encke crosses Mercury's orbit twice, once at TAA$\gtrsim 0°$ with the meteoroids arriving on the planetary nightside and a second time at TAA$\lesssim 180°$ on the dayside (but much farther from Mercury and therefore probably of lesser importance), with a speed of ~ 35 km s^{-1} in both cases. The authors estimated the mass influx on Mercury to be 10.5 g s^{-1} or ~ 0.9 tons per day, which compares favourably with other sources. However, owing to a shower's spatially restricted nature, this influx is only sustained for a short period of time, typically no more than a few tens of degrees of TAA; therefore, the resulting fluence over the entire orbit is likely less significant than that from other sources.

LADEE was the first mission to directly observe the link between meteoroid bombardment and exosphere formation (Elphic et al. 2014) because, in addition to the LDEX dust experiment, it also carried an Ultraviolet-Visible Spectrometer (UVS; Colaprete et al. 2014) and a Neutral Mass Spectrometer (NMS; Mahaffy et al. 2014). The comparison between the observations of LDEX and UVS identified a correlation between the meteoroid influx and the Na and K abundances in the lunar exosphere, in particular with shower activity. Specifically, Colaprete et al. (2016) and Szalay et al. (2016) found a strong correlation of exospheric potassium and meteoroid ejecta during the Geminids meteoroid shower, exhibiting a much stronger response than sodium (Fig. 3). With the exception of the Geminids, the authors found a weak correlation between the meteoroid influx as measured by LDEX and exospheric density of alkalis as measured by UVS. To fully understand why these two species respond differently to meteoritic bombardment, requires to better identify the differences of their reservoirs, sources and sinks on the lunar surface. Similarly, with NMS, Benna et al. (2019) reported detections of water vapor released into the lunar exosphere. The timing of 29 water release events agreed with periods when the Moon encountered known meteoroid streams. The authors used these measurements to constrain the hydration state of the lunar soil, arguing that by heating the soil meteoroids release water that is buried below a layer of dry regolith at depths of a few centimeters.

3.3.2 The Role of the Surface Interaction with the Exosphere

Our inability to explain the observed Na cycle suggests that current models of the surface-exosphere interaction may be incomplete. Some progress towards a better specification of the boundary conditions for models was contained in the recent work of Sarantos and Tsavachidis (2020). To understand how the microstructure of soil affects the exospheric reservoir for alkalis, a simulation of a porous soil (1 mm in depth) was produced in that work using spherical grains sampled with a distribution of grain sizes from Apollo Lunar Soil Sample 72141, a typical mature soil (keeping in mind that Mercury's regolith is probably different from Moon's regolith, as an example, in composition). The competition between desorption and thermal diffusion was studied with a Monte Carlo code, in which the trajectories of 16,000 test particles were tracked in such beds for up to 90 days. Repeated desorption and re-adsorption events were simulated using parameters obtained from thin-film experiments until adsorbates escaped to vacuum or time expired. Diffusion both through the spaces between grains as well as diffusion along the grain surface (surface diffusion) was included in this simulation. The temperature of the soil was adjusted from 100 K, typical of the nightside, up to 590 K, the noon temperature at Mercury's aphelion.

The soil structure was found to change the residence time of atoms on the surface, and the evolution of the surface reservoir, in two ways not previously appreciated: the effect of 1) microshadows and 2) soil thermal gradients. When adsorbed alkalis cannot thermally desorb, as it occurs at lunar temperatures for the usually assumed binding energies, atoms

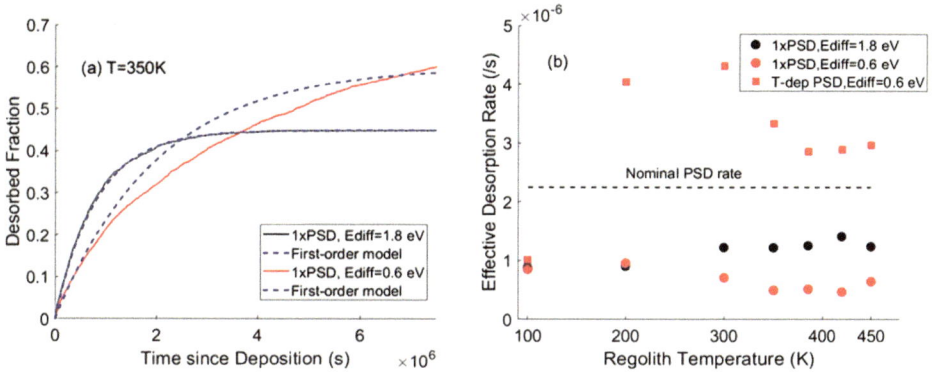

Fig. 26 Simulated desorption of alkali adsorbates from a porous soil. (**a**) Desorbed versus initially adsorbed particles as a function of time for soil temperature $T = 350$ K. Adsorbates that are not on illuminated grains are trapped unless high mobility is considered; (**b**) Effective photodesorption rate for constant and temperature-dependent yield and different mobilities. Mobility in these simulations is defined by a barrier for surface diffusion (Ediff). Figure from Sarantos and Tsavachidis (2020)

that are not in illuminated areas of grains are trapped unless adsorbate mobility is posited (Fig. 26a). This is to say that if adsorbates are assumed to be "frozen" on a grain between the arrivals of UV photons, about half the adsorbates comprising the exosphere reservoir (i.e. equivalent to the porosity of the soil) are trapped in microscopic shadows (the underside of grains) and do not contribute to desorption. Only at different True Anomaly angles, when the soil becomes warmer, would the remaining particles contribute to desorption when thermal desorption or increased mobility along the grain boundary dislodged atoms from shadows. This is different than the assumption in global models of the Na reservoir (e.g., Leblanc and Johnson 2003, 2010; Mura et al. 2009) that all particles returning to the surface are susceptible to UV photons. The simulation suggests that if thermal desorption is suppressed, e.g., when the assumed pre exponential factor is reduced or when the binding energy is increased owing to surface weathering, there is a possibility for increased lags in surface outgassing because half the adsorbates are unavailable for desorption. If the lag is considerable, we would err in our present models of the Na reservoir. A second finding was that the monotonic increase in temperature expected within the first 1-2 cm of the soil helped reduce losses of adsorbates to the subsurface by "biasing" the random walk. This gradient actually increased the ability of adsorbates to enrich the reservoir. In summary, the temperature at which the surface turns from retentive to emissive is a function of the assumed parameters for thermal desorption, the adsorbate mobility, and the soil thermal gradients. More work is required to quantify at which temperature the porous surface starts evaporating when these complications are considered.

In this closer look at the microphysics of soil, diffusion and re-adsorption reduced desorption rates and affected their temperature dependence. The photodesorption rate from an ensemble of grains had a different dependence on temperature than the rate from a single grain as it was found to combine the dependence of photodesorption yield, measured by thin-film experiments for sodium, the sticking coefficient, which is temperature-dependent for sodium, and the diffusion rate. In fact, for high diffusivity of adsorbates the photodesorption rate could be a non-monotonic function of temperature (Fig. 26b). Second, re-adsorption to adjacent grains combined with diffusion was shown to slow thermal desorption at Hermean temperatures as if the effective binding energy from the soil is higher, ~2 eV/atom from the sphere packing as opposed to 1.85 eV/atom used to describe desorption from a single grain.

Fig. 27 Two possible sources for Na$^+$ ions in the cusp: (left) Na$^+$ ions are generated in the cusp, both by solar wind impact and photoionization, and are accelerated by processes there. (Right) Neutral Na atoms are ionized near the magnetopause, picked up in the magnetosheath flow and swept into the cusp. Adapted from Raines et al. (2014)

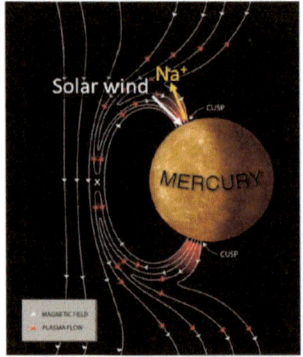
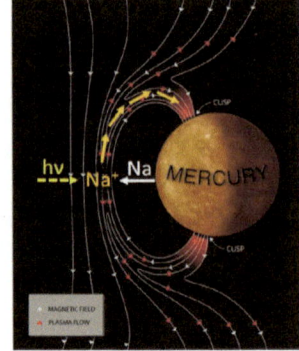

It is clear from these results that 1) parameters from thin-film experiments must be modified when adopted in exosphere models in more ways than the usual assumption of a reduction of the rate for porosity (i.e. the porosity reduction is temperature-dependent), and 2) thermal desorption rates too are subject to this reduction owing to soil porosity. These are additional improvements that we must strive to incorporate in models of the surface reservoir.

3.3.3 Ion Circulation

The picture of planetary ion circulation at Mercury has been assembled from observations, modelling, and intuition gained through decades of research at Earth. The first measurements of these ions were made by MESSENGER's Fast Imaging Plasma Spectrometer (FIPS, Andrews et al. 2007), as the ion instrument on Mariner 10 failed to deploy. These observations quickly revealed that Mercury's magnetosphere is dominated by solar wind plasma that is concentrated mostly in the magnetospheric cusps and central plasma sheet (CPS), much like at Earth (Zurbuchen et al. 2008, 2011; Raines et al. 2011). FIPS observed Na$^+$ ions on every orbit, reported as part of a group of ions in the mass per charge range 21–30 amu/e, including Mg$^+$ and Si$^+$ (Raines et al. 2013). No detection of K$^+$ has been reported from FIPS in the orbital mission, as identification of counts in that mass per charge range was found to be complicated by an instrument background signature. Though most modelling studies have focused on Na$^+$, it is likely that heavier ions, such as K$^+$, behave in a similar manner dynamically, though their source(s) may be different. A wealth of ion measurements was returned from the instrument. Na$^+$ ions at Mercury mostly originate from the exosphere, where photoionization of the neutral exospheric sodium mass-loads the plasma in Mercury's magnetosphere with Na$^+$. Owing to the variation of density and scale height of Mercury's exosphere with the planet's eccentric orbit around the Sun, the Na$^+$ content at Mercury's magnetosphere is also found to largely follow the same trend as the exosphere (Jasinski et al. 2021). Calculations using neutral observations from MESSENGER, estimate that the dayside Na$^+$ production rate from photoionization is $\sim 10^{24}$ ions s^{-1} (Jasinski et al. 2021). With their ~ 1 eV initial energy, Na$^+$ photoions born in Mercury's dayside magnetosphere remain largely trapped on closed planetary magnetic field lines. Their gyroradius is small enough, of order 10 s of km, that they are tightly locked to the dipole-like magnetic field, bouncing between the magnetic mirrors formed in the increasing field near the surface. Those that are very close to the magnetopause may be able to gyrate into that boundary layer, though it would be difficult for them to cross into the magnetosheath as the magnetopause is ~ 100 km thick (DiBraccio et al. 2013). In some cases, they may acquire enough energy to gyrate out

Springer

Fig. 28 Model Na$^+$ trajectories: (Left) projections in the noon-midnight meridian plane (top) and in the equatorial plane (bottom), (right) kinetic energy versus time. The test ions are launched from 65° latitude with 0.1 eV energy on the noon meridian. Three distinct cross-electric magnetospheric potentials are considered: 1 kV, 10 kV, and 30 kV (coded in grey, blue, and red, respectively). Adapted from Delcourt (2013)

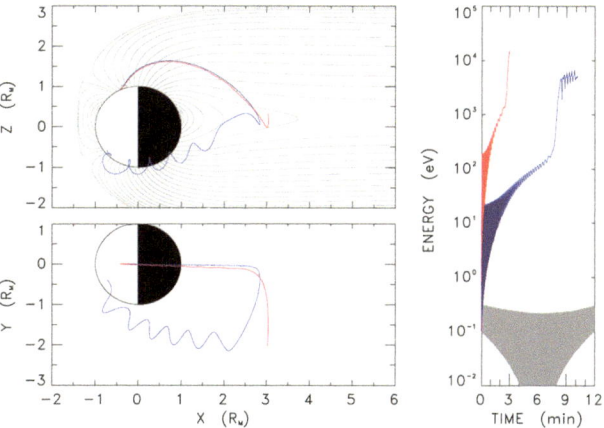

of the dayside. Ions that cross into the magnetosheath, will be picked up by the magnetosheath plasma flow and rapidly energized up to hundreds of eV or even up to several keV (Fig. 27, right). Most of these ions would be lost, carried downstream in the magnetosheath on the flanks of Mercury's magnetotail. A small fraction may be injected into the magnetosphere through the cusps. Ions entering the magnetospheric cusp will either precipitate onto the surface or magnetically mirror and continue travelling away from the planet, forming the plasma mantle in the magnetotail (DiBraccio et al. 2015a; Jasinski et al. 2017). One of the surprises of MESSENGER's first ion measurements was the observation of keV Na$^+$ ions in Mercury's northern cusp (Zurbuchen et al. 2011; Raines et al. 2014). The energization of these ions is not yet well understood, but it is likely to include pickup into the magnetosheath flow prior to cusp entry or energization by reconnection by parallel electric fields. Na$^+$ photoions originating in the cusp may gain energy owing to the curvature of their E x B drift paths (Delcourt et al. 2002), resulting from the combined effect of electrostatic (E) and magnetostatic (B) forces on individual ions. This behaviour was shown convincingly by test particle simulations, tracing trajectories in an analytical model of Mercury's magnetospheric magnetic field by numerical integration of an adiabatic, guiding-center (Delcourt et al. 2002; Delcourt 2013) or full equation of motion (Delcourt et al. 2003). Na$^+$ ions were found to gain energy logarithmically in great arcing trajectories through the northern magnetospheric lobe and into the central plasma sheet (CPS, Fig. 28). Modeled ions were energized from initial energies of 0.1 eV up to several hundred eV through their <10 min. travel time to the CPS (Delcourt et al. 2012). The much smaller size of Mercury's magnetosphere relative to that of the Earth results in substantially increased curvature and thus stronger energization (Delcourt et al. 2002) than in Earth's cleft (high-altitude cusp) ion fountain (Horwitz 1984; Lockwood et al. 1985). Observational evidence of acceleration local to the cusp is very limited. A small portion of Na$^+$-group ions observed by MESSENGER in the northern cusp were found to be moving away from the surface (Fig. 28, left), anti-parallel to the magnetic field, which is mostly radial there (Raines et al. 2014). These ions had energies in the 200–300 eV range, higher than test-particle model predictions but possibly consistent with wave-particle energization mechanisms that may occur in the cusp (see Raines et al. 2014, and references therein).

Na$^+$ ions arrive preferentially on the pre-midnight side of the CPS, a feature shown both in observations (Raines et al. 2013; Jasinski et al. 2017) and test-particle modelling (Delcourt et al. 2003; Delcourt 2013). Once in the CPS, Na$^+$ ions are likely accelerated further, from several keV to up over 10 keV depending on the cross-polar cap potential (Delcourt et al. 2003), which is determined largely by the rate of magnetic reconnection at the dayside magnetopause. Motion of Na^+ ions there is complicated and varies periodically with distance from the planet, owing to the magnetic field geometry and their large and varying gyroradii, of order 0.6–1.3 R_M (2440 km) for 1–5 keV energies. They can undergo quasi-adiabatic, Speiser-type orbits (Speiser 1965), where they gain substantial energy from the dawn-to-dusk electric field as they oscillate around the magnetic field reversal in the central current sheet. At other distances, Na^+ ions behave non-adiabatically, their large gyroradii causing them to be partially demagnetized and drifting substantially across the magnetic field. Gershman et al. (2014) estimated the temperature of Na^+-group ions in the CPS to be predominately in the 5–40 MK range, with a small number of values going higher, up to $\sim$55 MK (4.7 keV). O^+-group ions are found at about the same temperatures, indicating a heating mechanism that is not mass-proportional, consistent with energization in the potential drop of an electric field. Although it is not clear if planetary ions strongly affect the dynamics of the magnetotail, Na^+-group ions were found to make up to 15% of the plasma thermal pressure and 50% of the mass density (Gershman et al. 2014), so clearly that potential exists.

Loss of Na^+ ions from the magnetotail occurs through several processes. Ions travelling downtail and forming the plasma mantle (having previously mirrored in the cusp) will either be lost downtail or will drift equatorward into the plasma sheet (Jasinski et al. 2017). Na^+, owing to its higher mass, will more likely drift into the plasma sheet in comparison to protons—although owing to the observed velocity distributions, most of the ions observed in the plasma mantle are expected to be lost downtail (Jasinski et al. 2017). Reconnection in the CPS creates fast plasma flows that carry heavy ions toward the nightside surface. These flows have been estimated at $\sim$200–300 km/s on average, using a statistical reconstruction from FIPS measurements (Dewey et al. 2018). This plasma may impact broadly across the nightside surface, or be slowed and deflected around the dipolar magnetic field to impact at mid-latitudes near the open-closed field line boundary (like the auroral region at Earth). Studies of reconnection events have shown evidence of the flow breaking that would be associated with this plasma deflection (Poh et al. 2017; Dewey et al. 2020). Energetic electrons ($\sim$100–200 keV) have been shown to be associated with magnetotail reconnection signatures, such as magnetic field dipolarizations (Dewey et al. 2018). When mapped to the surface in a dipolar field, these electrons arrive at magnetic latitudes 10–40°, just below the estimated location of the open-closed field boundary. X-ray fluorescence measurements show substantial enhancements around this region, and are thus attributed to the impact of these energetic electrons upon the surface (Lindsay et al. 2016). Ions that do not precipitate are likely carried around the dusk side of the planet, where their keV-energies result in large gyroradii ($\sim$100–300 km) that likely cause them to be lost either to impact on the surface or crossing into the magnetosheath (Raines et al. 2014). When not carried planetward by magnetotail reconnection, Na^+ ions are likely lost to the tail magnetosheath, either being carried out directly by their large gyroradii or by Kelvin-Helmholtz (K-H) vortices on the magnetopause boundary. Magnetotail plasma is also lost owing to magnetic reconnection and plasmoid formation (DiBraccio et al. 2015a), though Na+ has not yet been observed in these structures. K-H vortices result from the large difference in flow speed across the magnetopause, like the fluid instability of the same name. They have been observed in over 145 orbits, about 6% of those examined, with 93% occurring on the dusk side (Liljeblad et al. 2014). In several large amplitude cases, Na^+-group ions have been found in the K-H vortices themselves (Sundberg et al. 2012). Observations of magnetic field fluctuations, associated with K-H activity, near the Na^+ gyrofrequency indicates that these ions might have a role in determining the size of the vortices (Gershman et al. 2015). Test-particle simulations of

Na^+ ions on MHD-generated K-H vortices showed that the Na^+ ions can be energized by the electric field within the vortices in cases where the IMF is southward (Aizawa et al. 2020a, 2020b). This effect is unique for heavy ions (like Na^+) because of their large gyroradii; protons in the study were not energized in this way. However, when compared to observations of Na^+-group ions from FIPS, the results indicated that these ions might actually *lose* energy in K-H vortices under some conditions, highlighting the complex behavior of heavy ions in Mercury's magnetotail.

4 Future Steps

Future steps to progress in our understanding of surface-bounded exospheres has been described in the final section of the Chapter "Volatiles and refractories in surface-bounded exospheres in the inner Solar System" by Grava et al. (2021). In the following, we described what is specific to the Na/K environments of these objects, in particular in terms of new directions of developments and efforts in the fields of ground based observations, modeling and laboratory measurements.

4.1 Na & K Ground Based Observations

Mercury and the Moon are ideal targets for small aperture telescopes with specialized instrumentation. Numerous studies agree that both bodies possess exospheres that are exceptionally dynamic. Monitoring of their Na and K emissions can therefore provide valuable insights to time-dependent drivers like solar wind and meteoroid influx. An exosphere's response to events like an ICME, a meteor shower, or even individual meteoroid impacts can quantify a wealth of information like yields, temperatures and timescales for exosphere-surface interactions. Such parameters are challenging to constrain from snapshots of an exosphere with multiple sources and losses. Observations of dynamic phenomena in our solar system enables cause-and-effect relationships to be established and characterized, but monitoring with any regularity can become impractical where space-based observation or large aperture telescopes are needed, e.g. for studies of gas giant aurorae. Dedicated monitoring of these exospheres with small and medium-size telescopes at optical wavelengths is low cost means to leverage significant science return, particularly during missions when *in situ* measurements can be compared to better inform our interpretation of both datasets.

4.2 Modeling Development

The spatial organization, composition and temporal evolution of surface-bounded exospheres are highly dependent of their interaction with the surface. The upper layer of the surface of these objects, the regolith, being permanently space weathered by the Sun, the exospheres are an almost direct signature of this surface ageing process. But, the permanent recycling of these exospheres into the surface tends also to modify the upper layer of the regolith. Therefore, to develop a precise description of the interaction between exosphere and regolith is essential to address the formation and evolution of both. Its importance has been highlighted in particular when discussing the modelling of the orbital evolution of the Na and K exospheres at Mercury which does not agree with the observations by MESSENGER discussed in Sect. 3. The discrepancies between the data and models suggests that some additional physical parameters need to be taken into account or updated. For example, a better understanding of the surface-exosphere interaction, especially on a space-weathered surfaces, is necessary. Data from Bepi-Colombo on interplanetary dust flux at Mercury, ion flux and its temporal and spatial variability, and more information on the surface composition in the southern hemisphere will be a definite boon. Calculations and the help of new laboratory experiments regarding parameters such as photon-stimulated desorption and electron-stimulated desorption as well as space weathering are desired.

The complexity of this task is increased by the need to take into account all the environmental parameters that could influence this interaction. As an example, this interaction depends on the existence or not of an intrinsic magnetosphere, on the composition and structure of the upper layer of the surface in direct interaction with these atmospheres, on the solar local time, topography, on the distance to the Sun, on the environment of these objects (embedded or not in the magnetosphere of a larger planetary objects, like our Moon within the Earth's magnetosphere), on the meteoroid environment and its variability and on the variability of the solar conditions either due to the solar cycle or to solar energetic events or due to the orbit of these objects. But more challenging is to take into account the regolith structure and composition. Indeed, the upper surface is not a simple rocky surface but rather a complex assembly of grains whose size, density and microstructure will depend on the depth, solar illumination and residence time in the upper layer as a first order (see Sect. 3.3.2).

Therefore, to meet the present challenge of reconciling models of the exosphere of these objects with the observations, we must integrate in these models: 1) the description of the fate of the exosphere during its interaction with the surface; 2) its dependency on the location at the surface but also the particularities of the planetary environment, 3) to describe the regolith taking into account all the mechanisms that could weather it (solar illumination, radiation, temperature profile with depth, structure and size of the grains, impact gardening...); and 4) to describe how the surface controls the exosphere.

Such type of models integrating both the description of the exosphere and of the fate of its atomic and molecular contents through the regolith remains to be developed for Mercury and the Moon and represents the next step in our capability to describe these environments.

4.3 Towards Unprecedented New Achievements Thanks to BepiColombo

ESA/JAXA BepiColombo mission, with its two spacecraft and a much larger set of instruments, is much better equipped and conceived to address the questions related to the relations between our Sun and Mercury's system than was MESSENGER. BepiColombo mission, with its unprecedented capability to distinguish spatial and temporal variabilities, will be able to track the various signatures of the solar activities in Mercury's magnetosphere, exosphere, ionosphere and surface. As a matter of facts and illustrated in this paper, the Na and K exospheres are among the easiest observable signatures of these relations from both an orbiter and/or ground based telescopes.

We can list two main advantages for BepiColombo mission with respect to what has been done so far by previous space missions at Mercury. First, two spacecraft able to observe simultaneously the far and close environments of Mercury, second a set of instruments with a much better coverage of the different regions of Mercury thanks also to MIO and MPO orbital coverages and also able to observe the Na and K exospheres with various approaches (see Milillo et al. 2020 for a description of BepiColombo instruments):

- BepiColombo will be able to track the solar environment, thanks to SIXS for the radiative environment but more importantly for the exosphere, thanks to MDM for the dust environment, and because the use of two spacecraft will occasionally allow for the simultaneous measurement of conditions in the upstream solar wind and inside the magnetosphere.

- With two spacecraft inside Mercury's magnetosphere, BepiColombo will be able to follow the propagation of solar perturbations through the magnetosphere up to the surface and exosphere/ionosphere, characterizing the magnetospheric response to solar events.
- BepiColombo two spacecraft have an unprecedented set of instruments dedicated to the observation of Mercury's exosphere, from in situ instruments with MPO/SERENA/STROFIO able to in situ measure the Na and K exospheres to UV spectrometer with MPO/PHEBUS, potentially able to measure the Na exosphere, up to the visible imager MIO/MSASI specifically dedicated to the Na exospheric component.
- Tracking Mercury's exospheric variability will be also possible by observing its signatures in its ionosphere and vice-versa, thanks to a much larger set of plasma dedicated instruments as MPO/SERENA/PICAM and MPO/SERENA/MIPA and to MIO/MPPE, all able to measure the Na^+ and K^+ ions as previously observed by MESSENGER/FIPS.
- Thanks to MESSENGER observations, we also know that there are direct observable signatures of Mercury's surface composition in its exosphere. Clearly the many instruments dedicated to the observations of the surface, SIMBIO-SYS, MIXS, BELA, MERTIS and MGN on board MPO, will provide a completely new view on the surface composition and will be essential to complete our view of Mercury's exosphere. Actually, Mercury's surface volatile abundance is also known to be potentially related to Mercury's exospheric circulation.
- At the end, a description of the exosphere won't be completed without considering the relations between Mercury's magnetosphere and its internal structure, in particular without considering the induced currents in Mercury's core.

Few years after the completely new view provided by MESSENGER on Mercury's exosphere, BepiColombo clearly provides us the rare opportunity to pursue MESSENGER achievements by significantly improving our view and understanding of Mercury's exosphere. Moreover, with its two spacecrafts and comprehensive set of instruments, it will provide us a unique capability to observe the complex relations between Mercury's exosphere, surface, magnetosphere and environments.

5 On the Importance to Further Observe, Analyse and Model the Moon and Mercury Na and K Exospheres

Mercury and the Moon Na and K exospheres appear more and more clearly as a complex and intrinsic component of the environments of these objects. Our current view of the Moon and Mercury presents their Na and K exosphere as a product and potential driver of the interaction with their plasma environment, a product and signature of the meteoritic environments and a product, signature but probably also driver of their surface composition.

This complexity is actually what makes so crucial the observations of these two components of Mercury or the Moon exospheres. The exosphere has an obvious role in the formation and variability of the magnetosphere. It contains direct signatures of Mercury and of the Moon interaction with our Sun and the Earth magnetosphere in the case of the Moon. Observing the exosphere was shown to provide an original and highly useful view on how the solar wind interacts with the Moon's crustal magnetic fields or to track the energy and mass exchanges between the solar wind and Mercury's magnetosphere. Mercury and the Moon's ionized environments are essentially formed from their respective exospheres. These newly created ions are then further interacting with the solar wind contributing to shape the electric and magnetic environments of these objects.

These ions are also one of the most easily measurable signatures of Mercury and Moon's surface space weathering. Thanks to the observations of the Moon exosphere, it has been possible to track the effects of the surface bombardment by the meteoroid and by the solar wind but also to use these observations to probe the very uppermost layer of these objects and also to infer the composition of some large size terrains at both the Moon and Mercury. To understand the surface/exosphere cycle is also essential to properly interpret some remote observations of the surface composition, the exosphere readsorption by the surface being potentially one of the explanations for the unusual surface composition on Mercury. Therefore, observing the exosphere helps us to explore the composition of some specific regions of the surface of these objects, the cusps region at Mercury, the locally magnetized regions at the Moon and the permanently shaded regions at Mercury and the Moon.

The exosphere of weakly magnetized objects without thick atmospheres is therefore an intrinsic component of these objects which characteristics contribute to shape the surface, electric and magnetic environments of these objects. Integrating the exosphere as one of the targets of our efforts to understand these objects is therefore essential to properly tackle their complexity.

Acknowledgements The collaboration of the authors was facilitated by support from the International Space Science Institute in the Surface Bounded Exospheres and Interactions in the Solar System Workshop. C. Schmidt, T. Cassidy, R. Vervack and R. Killen acknowledge support from NASA programs 80NSSC19K0790, 80NSSC18K0857 and 80NSSC21K0051. A.A.B. was partially supported by Russian Science Foundation (grant no. 20-12-00105) and by the Kazan Federal University Strategic Academic Leadership Program ("PRIORITY-2030"). R. Killen and M. Horanyi acknowledge support from the LEADER and IMPACT nodes, respectively, of NASA's SSERVI virtual institute. J. M. Jasinski acknowledges support from an appointment to the NASA Postdoctoral Program Fellowship at the Jet Propulsion Laboratory administered by Universities Space Research Association through a contract with the National Aeronautics and Space Administration (NASA). J. M. Jasinski acknowledges support from the Jet Propulsion Laboratory, California Institute of Technology, under a contract with NASA; and NASA's Discovery Data Analysis Program (grant number 80NM0018F0612).

References

S. Aizawa, J.M. Raines, D. Delcourt, N. Terada, N. André, MESSENGER observations of planetary ion characteristics in the vicinity of Kelvin-Helmholtz vortices at Mercury. J. Geophys. Res. Space Phys. **125**, e27871 (2020a). https://doi.org/10.1029/2020JA027871

S. Aizawa et al., Statistical study of non-adiabatic energization and transport in Kelvin-Helmholtz vortices at Mercury. Planet. Space Sci. **193**, 105079 (2020b)

F. Allegrini, M.A. Dayeh, M.I. Desai, H.O. Funsten, S.A. Fuselier, P.H. Janzen et al., Lunar energetic neutral atom (ENA) spectra measured by the interstellar boundary explorer (IBEX). Planet. Space Sci. **85**, 232–242 (2013). https://doi.org/10.1016/j.pss.2013.06.014

B.J. Anderson, C.L. Johnson, H. Korth, M.E. Purucker, R.M. Winslow, J.A. Slavin, S.C. Solomon, R.L. McNutt, J.M. Raines, T.H. Zurbuchen, The global magnetic field of Mercury from messenger orbital observations. Science **333**, 1859–1862 (2011)

G.B. Andrews, T.H. Zurbuchen, B.H. Mauk, H. Malcom, L.A. Fisk, G. Gloeckler, G.C. Ho, J.S. Kelley, P.L. Koehn, T.W. LeFevere, S.S. Livi, R.A. Lundgren, J.M. Raines, The energetic particle and plasma spectrometer instrument on the MESSENGER spacecraft. Space Sci. Rev. **131**(1–4), 523–556 (2007). https://doi.org/10.1007/s11214-007-9272-5

F. Aumayr, H. Winter, Potential sputtering. Philos. Trans. R. Soc. Lond. Ser. A **362**, 77–102 (2003). https://doi.org/10.1098/rsta.2003.1300

S. Barabash, A. Bhardwaj, M. Wieser, R. Sridharan, T. Kurian, S. Varier et al., Investigation of the solar wind-Moon interaction onboard Chandrayaan-1 mission with the SARA experiment. Curr. Sci. **96**(4), 526–532 (2009). http://www.jstor.org/stable/24105464

J. Baumgardner, M. Mendillo, The use of small telescopes for spectral imaging of low light level extended atmospheres in the Solar System. Earth Moon Planets (2009). https://doi.org/10.1007/s11038-009-9314-y

J. Baumgardner, J. Wilson, M. Mendillo, Imaging the sources and full extent of the sodium tail of the planet Mercury. Geophys. Res. Lett. **35**, L03201 (2008)

J. Baumgardner, S. Luettgen, C. Schmidt, M. Mayyasi, S. Smith, C. Martinis, J. Wroten, L. Moore, M. Mendillo, Long-term observations and physical processes in the Moon's extended sodium tail. J. Geophys. Res. **126**(3), e2020JE006671 (2021)

H.L. Bay, J. Bohdansky, W.O. Hofer, J. Roth, Angular distribution and differential sputtering yields for low-energy light-ion irradiation of polycrystalline nickel and tungsten. Appl. Phys. **21**, 327–333 (1980)

R. Behrisch, W. Eckstein (eds.), *Sputtering by Particle Bombardment* (Springer, Berlin, 2007)

M. Benna, D.M. Hurley, T.J. Stubbs, P.R. Mahaffy, R.C. Elphic, Lunar soil hydration constrained by exospheric water liberated by meteoroid impacts. Nat. Geosci. **12**, 333–338 (2019)

A.A. Berezhnoy, Meteoroid bombardment as a source of the lunar exosphere. Adv. Space Res. **45**(1), 70–76 (2010)

A.A. Berezhnoy, Chemistry of impact events on the Moon. Icarus **226**(1), 205–211 (2013)

A.A. Berezhnoy, Chemistry of impact events on Mercury. Icarus **300**, 200–212 (2018)

A.A. Berezhnoy, B.A. Klumov, Impacts as a source of the atmosphere on Mercury. Icarus **195**(2), 511–522 (2008)

A.A. Berezhnoy, K.I. Churyumov, V.V. Kleshchenok, E.A. Kozlova, V. Mangano, V. Pakhomov Yu, V.O. Ponomarenko, V.V. Shevchenko, I. Velikodsky Yu, Properties of the lunar exosphere during the Perseid 2009 meteor shower. Planet. Space Sci. **96**, 90 (2014)

A. Bhardwaj, S. Barabash, Y. Futaana, Y. Kazama, K. Asamura, R. Sridharan et al. Low energy neutral atom imaging on the Moon with the SARA instrument aboard Chandrayaan-1 mission. J. Earth Syst. Sci. **114**(6), 749–760 (2005). https://doi.org/10.1007/BF02715960

A. Bhardwaj, M. Wieser, M.B. Dhanya, S. Barabash, Y. Futaana, M. Holmström et al., The Sub-keV Atom Reflecting Analyzer (SARA) experiment aboard Chandrayaan-1 mission: instrument and observations. Adv. Geosci. **19**, 151–162 (2010)

A. Bhardwaj, M.B. Dhanya, A. Alok et al., A new view on the solar wind interaction with the Moon. Geosci. Lett. **2**, 10 (2015). https://doi.org/10.1186/s40562-015-0027-y

P. Borin, G. Cremonese, F. Marzari, M. Bruno, S. Marchi, Statistical analysis of micrometeoroids flux on Mercury. Astron. Astrophys. **503**(1), 259–264 (2009). https://doi.org/10.1051/0004-6361/200912080

P. Borin, G. Cremonese, F. Marzari, Statistical analysis of the flux of micrometeoroids at Mercury from both cometary and asteroidal components. Astron. Astrophys. **585**, A106 (2016a). https://doi.org/10.1051/0004-6361/201526767

P. Borin, G. Cremonese, F. Marzari, Statistical analysis of the flux of micrometeoroids at Mercury from both cometary and asteroidal components (Corrigendum). Astron. Astrophys. **588**, C3 (2016b). https://doi.org/10.1051/0004-6361/201526767e

P. Borin, G. Cremonese, F. Marzari, A. Lucchetti, Asteroidal and cometary dust flux in the inner solar system. Astron. Astrophys. **605**, A94 (2017). https://doi.org/10.1051/0004-6361/201730617

M.E. Brown, Potassium in Europa's atmosphere. Icarus **151**, 190 (2001)

M.H. Burger, R.M. Killen, R.J. Vervack Jr., E.T. Bradley, W.E. McClintock, M. Sarantos, M. Benna, N. Mouawad, Monte Carlo modeling of sodium in Mercury's exosphere during the first two MESSENGER flybys. Icarus **209**, 63–74 (2010). https://doi.org/10.1016/j.icarus.2010.05.007

M.H. Burger, R.M. Killen, W.E. McClintock, A.W. Merkel, R.J. Vervack Jr., T.A. Cassidy, M. Sarantos, Seasonal variations in Mercury's dayside calcium exosphere. Icarus **238**, 51–58 (2014). https://doi.org/10.1016/j.icarus.2014.04.049

T.A. Cassidy, R.E. Johnson, Monte Carlo model of sputtering and other ejection processes within a regolith. Icarus **176**, 499–507 (2005)

T.A. Cassidy, A.W. Merkel, M.H. Burger, M. Sarantos, R.M. Killen, W.E. McClintock, R.J. Vervack, Mercury's seasonal sodium exosphere: MESSENGER orbital observations. Icarus **248**, 547–559 (2015)

T.A. Cassidy, W.E. McClintock, R.M. Killen, M. Sarantos, A.W. Merkel, R.J. Vervack, M.H. Burger, A cold-pole enhancement in Mercury's sodium exosphere. Geophys. Res. Lett. **43**, 11,121–11,128 (2016)

J.W. Chamberlain, *Physics of the Aurora and Airglow, Physics of the Aurora and Airglow* (Am. Geophys. Union, Washington, 2011). https://agupubs.onlinelibrary.wiley.com/doi/book/10.1029/SP041

J.W. Chamberlain, D.M. Hunten, *Theory of Planetary Atmospheres: An Introduction to Their Physics and Chemistry* (1989)

J-Y. Chaufray, F. Leblanc, Non Maxwellian radiative transfer of the sodium D1 and D2 lines in the exosphere of Mercury. Icarus **223**(2), 975–985 (2013). https://doi.org/10.1016/j.icarus.2013.01.005-2013

A.A. Christou, R.M. Killen, M.H. Burger, The meteoroid stream of comet Encke at Mercury: implications for MErcury Surface, Space ENvironment, GEochemistry, and Ranging observations of the exosphere. Geophys. Res. Lett. **42**(18), 7311 (2015). https://doi.org/10.1002/2015GL065361

M.J. Cintala, Impact-induced thermal ejects in the lunar and Mercurian regoliths. J. Geophys. Res. **97**, 947–973 (1992). https://doi.org/10.1029/91JE02207

A. Colaprete, K. Vargo, M. Shirley, D. Land is, D. Wooden, J. Karcz, B. Hermalyn, A. Cook, An overview of the LADEE ultraviolet-visible spectrometer. Space Sci. Rev. **185**(1–4), 63–91 (2014). https://doi.org/10.1007/s11214-014-0112-0

A. Colaprete, M. Sarantos, D.H. Wooden, T.J. Stubbs, A.M. Cook, M. Shirley, How surface composition and meteoroid impacts mediate sodium and potassium in the lunar exosphere. Science **351**(6270), 249–252 (2016). Eggleton R.A. ed. 2001. The Regolith Glossary

P.J. Coleman, B.R. Lichtenstein, C.T. Russell, L.R. Sharp, G. Schubert, Magnetic fields near the Moon. Geochem. Cosmochem. Acta **36**, 2271–2286 (1972)

S.W.H. Cowley, C.J. Owen, A simple illustrative model of open flux tube motion over the dayside magnetopause. Planet. Space Sci. **37**, 1461–1475 (1989). https://doi.org/10.1016/0032-0633(89)90116-5

D.C. Delcourt, On the supply of heavy planetary material to the magnetotail of Mercury. Ann. Geophys. **31**, 1673–1679 (2013). https://doi.org/10.5194/angeo-31-1673-2013

D.C. Delcourt, T.E. Moore, S. Orsini, A. Milillo, J.-A. Sauvaud, Centrifugal acceleration of ions near Mercury. Geophys. Res. Lett. **29**, 12 (2002). https://doi.org/10.1029/2001GL013829

D.C. Delcourt, S. Grimald, F. Leblanc, J.-J. Bertherlier, A. Millilo, A. Mura, A quantitative model of planetary Na$^+$ contribution to Mercury's magnetosphere. Ann. Geophys. **21**, 1723–1736 (2003)

D.C. Delcourt, K. Seki, N. Terada, T.E. Moore, Centrifugally stimulated exospheric ion escape at Mercury. Geophys. Res. Lett. **39**, L22105 (2012). https://doi.org/10.1029/2012GL054085

R.M. Dewey, J.M. Raines, W. Sun, J.A. Slavin, G. Poh, MESSENGER observations of fast plasma flows in Mercury's magnetotail. Geophys. Res. Lett. **45**, 10,110–10,118 (2018). https://doi.org/10.1029/2018GL079056

R.M. Dewey, J.A. Slavin, J.M. Raines, A.R. Azari, W. Sun, MESSENGER observations of flow braking and flux pileup of dipolarizations in Mercury's magnetotail: evidence for current wedge formation. J. Geophys. Res. Space Phys. **125**, e2020JA028112 (2020). https://doi.org/10.1029/2020JA028112

G.A. DiBraccio, J.A. Slavin, S.A. Boardsen, B.J. Anderson, H. Korth, T.H. Zurbuchen, J.M. Raines, D.N. Baker, R.L. McNutt Jr., S.C. Solomon, MESSENGER observations of magnetopause structure and dynamics at Mercury. J. Geophys. Res. **118**(3), 997–1008 (2013). https://doi.org/10.1002/jgra.50123

G.A. DiBraccio, J.A. Slavin, S.M. Imber, D.J. Gershman, J.M. Raines, C.M. Jackman, S.A. Boardsen, B.J. Anderson, H. Korth, T.H. Zurbuchen, R.L. McNutt, S.C. Solomon, MESSENGER observations of flux ropes in Mercury's magnetotail. Planet. Space Sci. **115**, 77–89 (2015a). https://doi.org/10.1016/j.pss.2014.12.016

G.A. DiBraccio, J.A. Slavin, J.M. Raines, D.J. Gershman, P.J. Tracy, S.A. Boardsen et al. First observations of Mercury's plasma mantle by MESSENGER. Geophys. Res. Lett. **42**, 9666–9675 (2015b). https://doi.org/10.1002/2015GL065805

W. Eckstein, R. Preuss, New fit formulae for the sputtering yield. J. Nucl. Mater. **320**, 209–213 (2003)

J. Egedal, W. Daughton, A. Le, Large-scale electron acceleration by parallel electric fields during magnetic reconnection. Nat. Phys. **8**, 321–324 (2012). https://doi.org/10.1038/nphys2249

R.A. Eggleton (ed.), *The Regolith Glossary* (2001) CRC LEME

R.C. Elphic, G.T. Delory, B.P. Hine, P.R. Mahaffy, M. Horanyi, A. Colaprete, M. Benna, S.K. Noble et al., The lunar atmosphere and dust environment explorer mission. Space Sci. Rev. **185**, 3–25 (2014). https://doi.org/10.1007/s11214-014-0113-z

Y. Futaana, S. Barabash, M. Wieser, M. Holmström, C. Lue, P. Wurz et al., Empirical energy spectra of neutralized solar wind protons from the lunar regolith. J. Geophys. Res. **117**, 05005 (2012)

M.J. Genge, C. Engrand, M. Gounelle, S. Taylor, The classification of micrometeorites. Meteorit. Planet. Sci. **43**(3), 497–515 (2008). https://doi.org/10.1111/j.1945-5100.2008.tb00668. Bibcode:2008M&PS...43..497G

M.V. Gerasimov, B.A. Ivanov, O.I. Yakovlev, Yu.P. Dikov, Physics and chemistry of impacts. Laboratory Astrophysics and Space Research, pp. 279–329 (1998)

D.J. Gershman, J.A. Slavin, J.M. Raines, T.H. Zurbuchen, B.J. Anderson, H. Korth, D.N. Baker, S.C. Solomon, Magnetic flux pileup and plasma depletion in Mercury's subsolar magnetosheath. J. Geophys. Res. Space Phys. **118**, 7181–7199 (2013). https://doi.org/10.1002/2013JA019244

D.J. Gershman, J.A. Slavin, J.M. Raines, T.H. Zurbuchen, B.J. Anderson, H. Korth, D.N. Baker, S.C. Solomon, Ion kinetic properties in Mercury's pre-midnight plasma sheet. Geophys. Res. Lett. **41**, 5740–5747 (2014). https://doi.org/10.1002/2014GL060468

D.J. Gershman, J.M. Raines, J.A. Slavin, T.H. Zurbuchen, T. Sundberg, S.A. Boardsen, B.J. Anderson, H. Korth, S.C. Solomon, MESSENGER observations of multiscale Kelvin- Helmholtz vortices at Mercury. J. Geophys. Res. Space Phys. **120**, 4354–4368 (2015). https://doi.org/10.1002/2014JA020903

C. Grava, R.M. Killen, M. Benna, A.A. Berezhnoy, J.S. Halekas, F. Leblanc, M.N. Nishino, C. Plainaki, J.M. Raines, M. Sarantos, B.D. Teolis, O.J. Tucker, R.J. Vervack, A. Vorburger, Volatiles and refractories in surface-bounded exospheres in the inner Solar System. Space Sci. Rev. **217**, 61 (2021)

J.S. Halekas, Y. Saito, G.T. Delory, W.M. Farrell, New views of the lunar plasma environment. Planet. Space Sci. **59**, 1681–1694 (2011)

J.S. Halekas, D.A. Brain, M. Holmstrom, The Moon's plasma wake, in *Magnetotails in the Solar System*, ed. by A. Keiling, C.M. Jackman, P.A. Delamere. Geophysical Monograph, vol. 207 (American Geophysical, Union, Wiley, USA, 2015)

M. Hapgood, Modelling long-term trends in lunar exposure to the Earth's plasmasheet. Ann. Geophys. **25** (2007). www.ann-geophys.net/25/2037/2007/

L. Hood, G. Schubert, Inhibition of solar wind impingement on Mercury by planetary induction currents. J. Geophys. Res. **84**, 2641–2647 (1979). https://doi.org/10.1029/JA084iA06p02641

L.L. Hood, A. Zakharian, J. Halekas, D.L. Mitchell, R.P. Lin, M.H. Acuña et al., Initial mapping and interpretation of lunar crustal magnetic anomalies using lunar prospector magnetometer data. J. Geophys. Res. **106**, 27825–27840 (2001)

M. Horanyi, Z. Sternovsky, M. Lankton, C. Dumont, S. Gagnard, D. Gathright, E. Grun, D. Hansen, D. James, S. Kempf, B. Lamprecht, R. Srama, J.R. Szalay, G. Wright, The Lunar Dust Experiment (LDEX) onboard the Lunar Atmosphere and Dust Environment Explorer (LADEE) mission. Space Sci. Rev. **185**, 93–113 (2014). https://doi.org/10.1007/s11214-014-0118-7

J.L. Horwitz, Features of ion trajectories in the polar magnetosphere. Geophys. Res. Lett. **11**, 701 (1984)

W.F. Huebner, J.J. Keady, S.P. Lyon, Solar photo rates for planetary atmospheres and atmospheric pollutants. Astrophys. Space Sci. **195**, 1–294 (1992)

D.M. Hunten, A.L. Sprague, Origin and character of the Lunar and Mercurian atmospheres. Adv. Space Res. **19**, 1551 (1997)

D.M. Hunten, G. Cremonese, A.L. Sprague et al., The Leonid meteor shower and the lunar sodium atmosphere. Icarus **136**, 298 (1998)

S.M. Imber, J.A. Slavin, MESSENGER observations of magnetotail loading and unloading: implications for substorms at Mercury. J. Geophys. Res. Space Phys. **122**, 11,402–11,412 (2017). https://doi.org/10.1002/2017JA024332

D. Janches, P. Pokorny, M. Sarantos, J.R. Szalay, M. Horanyi, D. Nesvorny, Constraining the ratio of micrometeoroids from short- and long-period comets at 1 AU from LADEE observations of the lunar dust cloud. Geophys. Res. Lett. **45**, 1713–1722 (2018). https://doi.org/10.1002/2017GL076065

D. Janches, J.S. Bruzzone, P. Pokorny, J.D. Carrillo-Sanchez, M. Sarantos, A comparative study of the seasonal, temporal, and spatial distribution of meteoroids in the upper atmosphere of Venus, Earth and Mars. Planet. Sci. J. **1**, 59 (2020)

D. Janches, C. Apostolos, A.A. Berezhnoy, G. Cremonese, T. Hirai, M. Horany, J.M. Jasinski, M. Sarantos, Meteoroids as one of the sources for exosphere formation on airless bodies in the inner solar system. Space Sci. Rev. **217**, 50 (2021)

J.M. Jasinski, J.A. Slavin, J.M. Raines, G.A. DiBraccio, Mercury's solar wind interaction as characterized by magnetospheric plasma mantle observations with MESSENGER. J. Geophys. Res. Space Phys. **122**, 12,153–12,169 (2017). https://doi.org/10.1002/2017JA024594

J.M. Jasinski, L.H. Regoli, T.A. Cassidy et al., A transient enhancement of Mercury's exosphere at extremely high altitudes inferred from pickup ions. Nat. Commun. **11**, 4350 (2020). https://doi.org/10.1038/s41467-020-18220-2

J.M. Jasinski, T.A. Cassidy, J.M. Raines, A. Milillo, L.H. Regoli, R. Dewey et al., Photoionization loss of Mercury's sodium exosphere: Seasonal observations by MESSENGER and the THEMIS telescope. Geophys. Res. Lett. **48**, e2021GL092980 (2021). https://doi.org/10.1029/2021GL092980

X. Jia, J.A. Slavin, G. Poh, G.A. DiBraccio, G. Toth, Y. Chen et al., MESSENGER observations and global simulations of highly compressed magnetosphere events at Mercury. J. Geophys. Res. **124**, 229–247 (2019). https://doi.org/10.1029/2018JA026166

R.E. Johnson, R. Baragiola, Lunar surface: sputtering and secondary ion mass spectrometry. Geophys. Res. Lett. **18**, 2169–2172 (1991)

R.E. Johnson, F. Leblanc, B.V. Yakshinskiy, T.E. Madey, Energy distribution for desorption of sodium and potassium from ice: the Na/K ratio at Europa. Icarus **156**, 136–142 (2002)

C.L. Johnson, M.E. Purucker, H. Korth, B.J. Anderson, R.M. Winslow, M.M.H. Al Asad, J.A. Slavin, I.I. Alexeev, R.J. Phillips, M.T. Zuber, S.C. Solomon, MESSENGER observations of Mercury's magnetic field structure. J. Geophys. Res. **117**, E00L14 (2012). https://doi.org/10.1029/2012JE004217

M. Kagitani, M. Taguchi, A. Yamazaki et al., Variation in lunar sodium exosphere measured from lunar orbiter SELENE (Kaguya). Planet. Space Sci. **58**, 1660 (2010)

E. Kallio, P. Janhunen, Solar wind and magnetospheric ion impact on Mercury's magnetosphere. Geophys. Res. Lett. **30**, 1877 (2003). https://doi.org/10.1029/2003GL017842

S. Kameda, I. Yoshikawa, M. Kagitani, S. Okano, Interplanetary dust distribution and temporal variability of Mercury's atmospheric Na. Geophys. Res. Lett. **36**(15), L15201 (2009). https://doi.org/10.1029/2009GL039036.

R.M. Killen, Seeing effects on images of Mercury sodium, in *EPSC2020-3, Vol. 14*, European Planetary Science Congress 2020, Granada, Spain (2020) https://doi.org/10.5194/epsc2020-5 (virtual)

R.M. Killen, J.M. Hahn, Impact vaporization as a possible source of Mercury's calcium exosphere. Icarus **250**, 230–237 (2015)

R.M. Killen, W-H. Ip, The surface-bounded atmospheres of Mercury and the Moon. Rev. Geophys. **37**, 361–406 (1999)

R.M. Killen, T.H. Morgan, Diffusion of Na and K in the uppermost regolith of Mercury. J. Geophys. Res. **98**, 23589–23601 (1993)

R.M. Killen, A.E. Potter, T.H. Morgan, Spatial distribution of sodium vapor in the atmosphere of Mercury. Icarus **85**, 145–167 (1990)

R.M. Killen, A. Potter, A. Fitzsimmons, T.H. Morgan, Sodium D2 line profiles: clues to the temperature structure of Mercury's exosphere. Planet. Space Sci. **47**, 1449–1458 (1999)

R.M. Killen, A.E. Potter, P. Reiff, M. Sarantos, B.V. Jackson, P. Hick, B. Giles, Evidence for space weather at Mercury. J. Geophys. Res. **106**, 20509–20525 (2001)

R.M. Killen, M. Sarantos, A.E. Potter, P.H. Reiff, Source rates and ion recycling rates for Na and K in Mercury's atmosphere. Icarus **171**(1), 1–19 (2004)

R. Killen, G. Cremonese, H. Lammer, S. Orsini, A.E. Potter, A.L. Sprague et al., Processes that promote and deplete the exosphere of Mercury. Space Sci. Rev. **132**(2–4), 433–509 (2007)

R.M. Killen, D.E. Shemansky, N. Mouawad, Expected emission from Mercury's exospheric species, and their UV-visible signatures. Astrophys. J. Suppl. Ser. **181**(2), 351–359 (2009)

R.M. Killen, A.E. Potter, D.M. Hurley et al., Observations of the lunar impact plume from the LCROSS event. Geophys. Res. Lett. **37** (2010). https://doi.org/10.1029/2010GL045508

R.M. Killen, T.H. Morgan, A.E. Potter, C. Plymate, R. Tucker, J.D. Johnson, Coronagraphic observations of the lunar sodium exosphere January-June 2017. Icarus **328**, 152–159 (2019)

R.M. Killen, T.H. Morgan, A.E. Potter, G. Bacon, I. Ajang, A.R. Poppe, Coronagraphic observations of the lunar sodium exosphere 2018–2019. Icarus **355** (2021). https://doi.org/10.1016/j.icarus.2020.114155

H. Korth, B.J. Anderson, D.J. Gershman, J.M. Raines, J.A. Slavin, T.H. Zurbuchen, S.C. Solomon, R.L. Mc-Nutt Jr., Plasma distribution in Mercury's magnetosphere derived from MESSENGER Magnetometer and Fast Imaging Plasma Spectrometer observations. J. Geophys. Res. Space Phys. **119**, 2917–2932 (2014). https://doi.org/10.1002/2013JA019567

D.C.P. Kuruppuaratchi, E.J. Mierkiewicz, R.J. Oliversen et al., High-resolution, ground-based observations of the lunar sodium exosphere during the Lunar Atmosphere and Dust Environment Explorer (LADEE) mission. J. Geophys. Res., Planets **123**, 2430 (2018)

H. Lammer, P. Wurz, M.R. Patel, R. Killen, C. Kolb, S. Massetti, S. Orsini, A. Milillo, The variability of Mercury's exosphere by particle and radiation induced surface release processes. Icarus **166**(2), 238–247 (2003). https://doi.org/10.1016/j.icarus.2003.08.006

Y. Langevin, The regolith of Mercury: present knowledge and implications for the Mercury Orbiter mission. Planet. Space Sci. **45**, 31–37 (1997). https://doi.org/10.1016/S0032-0633(96)00098-0

F. Leblanc, R.E. Johnson, Mercury's sodium exosphere. Icarus **164**, 261–281 (2003)

F. Leblanc, R.E. Johnson, Mercury exosphere. I. Global circulation model of its sodium component. Icarus **209**, 280–300 (2010)

F. Leblanc, R.E. Johnson, M.E. Brown, Europa's sodium atmosphere: an ocean source? Icarus **159**, 132–144 (2002)

F. Leblanc, A. Doressoundiram, N. Schneider, V. Mangano, A. Lopez-Ariste, C. Lemen, B. Gelly, C. Barbieri, G. Cremonese, High latitude peaks in Mercury's sodium exosphere: spectral signature using THEMIS Solar Telescope. Geophys. Res. Lett. **35**, L18204 (2008). https://doi.org/10.1029/2008GL035322

F. Leblanc, A. Doressoundiram, N. Schneider, S. Massetti, M. Wedlund, A. López Ariste, C. Barbieri, V. Mangano, G. Cremonese, Short term variations of Mercury's Na exosphere observed with very high spectral resolution. Geophys. Res. Lett. **36**, L07201 (2009). https://doi.org/10.1029/2009GL038089

X. Li, F. Guo, H. Li, G. Li, Particle acceleration during magnetic reconnection in a low-beta plasma. Astrophys. J. **843**, 1 (2017). https://doi.org/10.3847/1538-4357/aa745e

E. Liljeblad, T. Sundberg, T. Karlsson, A. Kullen, Statistical investigation of Kelvin-Helmholtz waves at the magnetopause of Mercury. J. Geophys. Res. **119**(12), 9670–9683 (2014). https://doi.org/10.1002/2014JA020614

S.T. Lindsay, M.K. James, E.J. Bunce, S.M. Imber, H. Korth, A. Martindale, T.K. Yeoman, MESSENGER X-ray observations of magnetosphere-surface interaction on the nightside of Mercury. Planet. Space Sci. **125**, 72–79 (2016)

M.R. Line, E.J. Mierkiewicz, R.J. Oliversen et al., Sodium atoms in the lunar exotail: observed velocity and spatial distributions. Icarus **219**, 609 (2012)

M. Lockwood, M.O. Chandler, J.L. Horwitz, J.H. Waite, T.E. Moore, C.R. Chappell, The cleft ion fountain. J. Geophys. Res. **90**, 9736–9748 (1985)

K. Lodders, B. Fegley, *The Planetary Scientist Companion* (Oxford University Press, London, 1998). 371 pp.

T.E. Madey, B.V. Yakshinskiy, V.N. Ageev, R.E. Johnson, Desorption of alkali atoms and ions from oxide surfaces: relevance to origins of Na and K in atmospheres of Mercury and the Moon. J. Geophys. Res., Planets **103**(E3), 5873–5887 (1998)

P.R. Mahaffy, R. Richard Hodges, M. Benna, T. King, R. Arvey, M. Barciniak, M. Bendt, D. Carigan, T. Errigo, D.N. Harpold, V. Holmes, C.S. Johnson, J. Kellogg, P. Kimvilakani, M. Lefavor, J. Hengemihle, F. Jaeger, E. Lyness, J. Maurer, D. Nguyen, T.J. Nolan, F. Noreiga, M. Noriega, K. Patel, B. Prats, O. Quinones, E. Raaen, F. Tan, E. Weidner, M. Woronowicz, C. Gundersen, S. Battel, B.P. Block, K. Arnett, R. Miller, C. Cooper, C. Edmonson, The Neutral Mass Spectrometer on the Lunar Atmosphere and Dust Environment Explorer mission. Space Sci. Rev. **185**(1–4), 27–61 (2014). https://doi.org/10.1007/s11214-014-0043-9

V. Mangano, A. Milillo, A. Mura, S. Orsini, E. De Angelis, P. Di Lellis A.M. Wurz, The contribution of impulsive meteoritic impact vaporization to the Hermean exosphere. Planet. Space Sci. **55**(11), 1541–1556 (2007). https://doi.org/10.1016/j.pss.2006.10.008

V. Mangano, F. Leblanc, C. Barbieri, S. Massetti, A. Milillo, G. Cremonese, C. Grava, Detection of a southern peak in Mercury's sodium exosphere with the TNG in 2005. Icarus **201**, 424–431 (2009)

V. Mangano, S. Massetti, A. Milillo, A. Mura, S. Orsini, F. Leblanc, Dynamical evolution of sodium anisotropies in the exosphere of Mercury. Planet. Space Sci. **82**, 1–10 (2013)

V. Mangano, S. Massetti, A. Milillo, C. Plainaki, S. Orsini, R. Rispoli, F. Leblanc, THEMIS Na exosphere observations of Mercury and their correlation with in-situ magnetic field measurements by MESSENGER. Planet. Space Sci. **115**, 102–109 (2015)

S. Marchi, A. Morbidelli, G. Cremonese, Flux of meteoroid impacts on Mercury. Astron. Astrophys. **431**(3), 1123–1127 (2005). https://doi.org/10.1051/0004-6361:20041800

S. Massetti, S. Orsini, A. Milillo, A. Mura, E. DeAngelis, H. Lammer, P. Wurz, Mapping of the cusp plasma precipitation on the surface of Mercury. Icarus **166**, 229–237 (2003). https://doi.org/10.1016/j.icarus.2003.08.005

S. Massetti, S. Orsini, A. Milillo, A. Mura, Modelling Mercury's magnetosphere and plasma entry through the dayside magnetopause. Planet. Space Sci. **55**, 1557–1568 (2007). https://doi.org/10.1016/j.pss.2006.12.008

S. Massetti, V. Mangano, A. Milillo, A. Mura, S. Orsini, C. Plainiki, Short-term observations of double peaked Na emission from Mercury's exosphere. Geophys. Res. Lett. (2017). https://doi.org/10.1002/2017GL073090

D.J. McComas, F. Allegrini, P. Bochsler, P. Frisch, H.O. Funsten, M. Gruntman et al., Lunar backscatter and neutralization of the solar wind: first observations of neutral atoms from the Moon. Geophys. Res. Lett. **36**, 12104 (2009)

T.J. McCoy, P.N. Peplowski, F.M. McCubbin, S.Z. Weider, *The Geochemical and Mineralogical Diversity of Mercury* (Cambridge University Press, Cambridge, 2019)

M.A. McGrath, R.E. Johnson, L.J. Lanzerotti, Sputtering of sodium on the planet Mercury. Nature **323**, 694–696 (1986)

M. Mendillo, B. Flynn, J. Baumgardner, Imaging experiments to detect an extended sodium atmosphere on the Moon. Adv. Space Res. **13**(Pergamon), 313 (1993)

M. Mendillo, J. Baumgardner, J. Wilson, Observational test for the solar wind sputtering origin of the Moon's extended sodium atmosphere. Icarus **137**, 13–23 (1999)

A.W. Merkel, R.J. Vervack, R.M. Killen, T.A. Cassidy, W.E. McClintock, L.R. Nittler, M.H. Burger, Evidence connecting Mercury's magnesium exosphere to its magnesium-rich surface terrane. Geophys. Res. Lett. **45**, 6790–6797 (2018)

A. Milillo et al., Investigating Mercury's environment with the two spacecraft BepiColombo mission. Space Sci. Rev. **216**, 93 (2020). https://doi.org/10.1007/s11214-020-00712-8

A. Milillo, V. Mangano, S. Massetti, A. Mura, C. Plainaki, T. Alberti, A. Ippolito, S. Ivanovski, A. Aronica, E. De Angelis, A. Kazakov, R. Noschese, S. Orsini, R. Rispoli, R. Sordini, N. Vertolli, Exospheric Na distributions along the Mercury orbit with the THEMIS telescope. Icarus **355**, 114179 (2021)

L.S. Morrissey, O.J. Tucker, R.M. Killen, D.L. Domingue, S. Nakhla, D.W. Savin, Sputtering of surfaces by ion irradiation: comparing molecular dynamics and binary collision approximation models to laboratory measurements. J. Appl. Phys. **130** (2021). https://doi.org/10.1063/5.0051073

A. Mura, Loss rates and time scales for sodium at Mercury. Planet. Space Sci. **63–64**, 2–7 (2012)

A. Mura, S. Orsini, A. Milillo, D. Delcourt, S. Massetti, Dayside H+ circulation at Mercury and neutral particle emission. Icarus **175**, 305–319 (2005)

A. Mura, A. Millilo, S. Orsini, S. Massetti, Numerical and analytical model of Mercury's exosphere: dependence on surface and external conditions. Planet. Space Sci. **55**, 1569–1583 (2007). https://doi.org/10.1016/j.pss.2006.11.028

A. Mura, P. Wurz, H.I.M. Lichtenegger, H. Schleicher, H. Lammer, D. Delcourt, A. Milillo, S. Orsini, S. Massetti, M.L. Khodachenko, The sodium exosphere of Mercury: comparison between observations during Mercury's transit and model results. Icarus **200**, 1–11 (2009)

D. Nesvorny, P. Jenniskens, H.F. Levison, W.F. Bottke, D. Vokrouhlicky, M. Gounelle, Cometary origin of the zodiacal cloud and carbonaceous micrometeorites. Implications for hot debris disks. Astrophys. J. **713**, 816–836 (2010). https://doi.org/10.1088/0004-637X/713/2/816. arXiv:0909.4322

S. Orsini, V. Mangano, A. Mura, D. Turrini, S. Massetti, A. Milillo, C. Plainaki, The influence of space environment on the evolution of Mercury. Icarus **239**, 281–290 (2013). https://doi.org/10.1016/j.icarus.2014.05.031S

S. Orsini, V. Mangano, A. Milillo, C. Plainaki, A. Mura, J.M. Raines, E. De Angelis, R. Rispoli, F. Lazzarotto, A. Aronica, Mercury sodium exospheric emission as a proxy for solar perturbations transit. Nat. Sci. Rep. **8**, 928 (2018)

P.N. Peplowski, D.J. Lawrence, E.A. Rhodes, A.L. Sprague, T.J. McCoy, B.W. Denevi, L.G. Evans, J.W. Head, L.R. Nittler, S.C. Solomon, K.R. Stockstill-Cahill, S.Z. Weider, Variations in the abundances of potassium and thorium on the surface of Mercury: results from the MESSENGER Gamma-Ray Spectrometer. J. Geophys. Res. **117**, E00L04 (2012)

P.N. Peplowski, L.G. Evans, K.R. Stockstill-Cahill, D.J. Lawrence, J.O. Goldsten, T.J. McCoy, L.R. Nittler, S.C. Solomon, A.L. Sprague, R.D. Starr, S.Z. Weider, Enhanced sodium abundance in Mercury's north polar region revealed by the MESSENGER Gamma-Ray Spectrometer. Icarus **228**, 86–95 (2014)

M. Pezzella, S.N. Yurchenko, J. Tennyson, A method for calculating temperature-dependent photodissociation cross sections and rates. Phys. Chem. Chem. Phys. **23**, 16390–16400 (2021)

G. Poh, J.A. Slavin, X. Jia, J.M. Raines, S.M. Imber, W.-J. Sun, D.J. Gershman, G.A. DiBraccio, K.J. Genestreti, A.W. Smith, Coupling between Mercury and its nightside magnetosphere: cross-tail current sheet asymmetry and substorm current wedge formation. J. Geophys. Res. Space Phys. **122**, 8419–8433 (2017). https://doi.org/10.1002/2017JA024266

P. Pokorny, M. Sarantos, D. Janches, Reconciling the dawn/dusk asymmetry in Mercury's exosphere with the micrometeoroid impact directionality. Astrophys. J. Lett. **842**, L17 (2017). https://doi.org/10.3847/2041-8213/aa775d. arXiv:1706.01461

P. Pokorny, M. Sarantos, D. Janches, A comprehensive model of the meteoroid environment around Mercury. Astrophys. J. **863**, 31 (2018). https://doi.org/10.3847/1538-4357/aad051. arXiv:1807.02749

A.R. Poppe, S. Fatemi, I. Garrick-Bethell, D. Hemingway, M. Holmström, Solar wind interaction with the Reiner Gamma crustal magnetic anomaly: connecting source magnetization to surface weathering. Icarus **266**, 261–266 (2016)

A.R. Poppe, S. Xu, L. Liuzzo, J.S. Halekas, Y. Harada ARTEMIS, Observations of lunar nightside surface potentials in the magnetotail lobes: evidence for micrometeoroid impact charging. Geophys. Res. Let. **48** (2021). https://doi.org/10.1029/2021GL094585

F. Postberg, S. Kempf, J. Schmidt et al., Sodium salts in E-ring ice grains from an ocean below the surface of Enceladus. Nature **459**, 1098 (2009). https://www.nature.com/articles/nature08046

M.J. Poston, G.A. Grieves, A.B. Aleksandrov, C.A. Hibbitts, M.D. Dyar, T.M. Orlando, Temperature programmed desorption studies of water interactions with Apollo lunar samples, 12001 and 72501. Icarus **255**, 24–29 (2015)

A.E. Potter, R.M. Killen, Observations of the sodium tail of Mercury. Icarus **194**(1), 1–12 (2008)

A.E. Potter, T.H. Morgan, Potassium in the atmosphere of Mercury. Icarus **67**, 336–340 (1986)

A.E. Potter, T.H. Morgan, Extended sodium exosphere of the Moon. Geophys. Res. Lett. **15**, 1515 (1988a)

A.E. Potter, T.H. Morgan, Discovery of sodium and potassium vapor in the atmosphere of the Moon. Science **241**, 675–680 (1988b)

A.E. Potter, T.H. Morgan, Evidence for magnetospheric effects on the sodium atmosphere of Mercury. Science **248**, 835 (1990)

A.E. Potter, T.H. Morgan, Coronagraphic observations of the lunar sodium exosphere near the lunar surface. J. Geophys. Res. **103**, 8581–8586 (1998)

A.E. Potter, R.M. Killen, T.H. Morgan, Rapid changes in the sodium exosphere of Mercury. Planet. Space Sci. **47**, 1441–1448 (1999)

A.E. Potter, R.M. Killen, T.H. Morgan, Variation of lunar sodium during passage of the Moon through the Earth's magnetotail. J. Geophys. Res. **105**, 15073–15084 (2000)

A.E. Potter, R.M. Killen, T.H. Morgan, The sodium tail of Mercury. Meteorit. Planet. Sci. **37**, 1165–1172 (2002)

A.E. Potter, R.M. Killen, M. Sarantos, Spatial distribution of sodium on Mercury. Icarus **181**, 1–12 (2006)

A.E. Potter, R.M. Killen, T.H. Morgan, Solar radiation acceleration effects on Mercury sodium emission. Icarus **186**, 571–580 (2007)

A.E. Potter, R.M. Killen, K.P. Reardon, T.A. Bida, Observations of neutral sodium above Mercury during the transit of November 8, 2006. Icarus **226**, 172–185 (2013)

T.H. Prettyman, J.J. Hagerty, R.C. Elphic et al., Elemental composition of the lunar surface: analysis of gamma ray spectroscopy data from Lunar Prospector. J. Geophys. Res., Planets **111**, 12007 (2006). http://doi.wiley.com/10.1029/2005JE002656

J.M. Raines, J.A. Slavin, T.H. Zurbuchen, G. Gloeckler, B.J. Anderson, D.N. Baker, H. Korth, S.M. Krimigis, R.L. McNutt Jr., MESSENGER observations of the plasma environment near Mercury. Planet. Space Sci. **59**(15), 2004–2015 (2011). https://doi.org/10.1016/j.pss.2011.02.004

J.M. Raines, D.J. Gershman, T.H. Zurbuchen, M. Sarantos, J.A. Slavin, J.A. Gilbert et al., Distribution and compositional variations of plasma ions in Mercury's space environment: the first three Mercury years of MESSENGER observations. J. Geophys. Res. Space Phys. **118**(4), 1604–1619 (2013)

J.M. Raines, D.J. Gershman, J.A. Slavin, T.H. Zurbuchen, H. Korth, B.J. Anderson, S.C. Solomon, Structure and dynamics of Mercury's magnetospheric cusp: MESSENGER measurements of protons and planetary ions. J. Geophys. Res. Space Phys. **119**, 6587–6602 (2014). https://doi.org/10.1002/2014JA020120

S.A. Rosborough, R.J. Oliversen, E.J. Mierkiewicz, M. Sarantos, S.D. Robertson, D.C. Kuruppuaratchi et al., High-resolution potassium observations of the lunar exosphere. Geophys. Res. Lett. **46**, 6964–6971 (2019). https://doi.org/10.1029/2019GL083022

M. Sarantos, S. Tsavachidis, The boundary of alkali surface boundary exospheres of Mercury and the Moon. Geophys. Res. Lett. **47**, e2020GL088930 (2020). https://doi.org/10.1029/2020GL088930

M. Sarantos, P.H. Reiff, T.H. Hill, R.M. Killen, A.L. Urquhart, A Bx -interconnected magnetosphere model for Mercury. Planet. Space Sci. **49**, 1629–1635 (2001). https://doi.org/10.1016/S0032-0633(01)00100-3

M. Sarantos, R.M. Killen, A.S. Sharma, J.A. Slavin, Influence of plasma ions on source rates for the lunar exosphere during passage through the Earth's magnetosphere. Geophys. Res. Lett. **35**, L04105 (2008). https://doi.org/10.1029/2007GL032310

M. Sarantos, R.M. Killen, A. Surjalal Sharma, J.A. Slavin, Sources of sodium in the lunar exosphere: modeling using ground based observations of sodium emission and spacecraft data of the plasma. Icarus **205**, 364 (2010)

L. Saul, P. Wurz, A. Vorburger, M. Rodríguez, S.A. Fuselier, D.J. McComas, Solar wind reflection from the lunar surface: the view from far and near. Planet. Space Sci. **84**, 1–4 (2013)

H. Schleicher, G. Wiedemann, H. Wöhl, T. Berkefeld, D. Soltau, Detection of neutral sodium above Mercury during the transit on 2003 May 7. Astron. Astrophys. **425**, 1119–1124 (2004)

C.A. Schmidt, Monte Carlo modeling of North-South asymmetries in Mercury's sodium exosphere. J. Geophys. Res. Space Phys. **118**, 4564–4571 (2013)

C.A. Schmidt, J.K. Wilson, J. Baumgardner, M. Mendillo, Orbital effects on Mercury's escaping sodium exosphere. Icarus **207**, 9–16 (2010)

C.A. Schmidt, J. Baumgardner, M. Mendillo, J.K. Wilson, Escape rates and variability constraints for high-energy sodium sources at Mercury. J. Geophys. Res. Space Phys. **117**, A03301 (2012)

C.A. Schmidt, F. Leblanc, L. Moore, T. Bida, C. Gray, Detection of Mercury's potassium tail, in *DPS Meeting Abstract* (2017). https://dps2017-aas.ipostersessions.com/default.aspx?s=1E-D9-DF-33-AE-10-48-69-02-D9-64-4C-2F-CC-5A-5B

C.A. Schmidt, J. Baumgardner, L. Moore, T.A. Bida, R. Swindle, P. Lierle, The rapid imaging planetary spectrograph: observations of Mercury's sodium exosphere in twilight. Planet. Sci. J. **1**, 1 (2020). https://doi.org/10.3847/PSJ/ab76c9

N.M. Schneider, M.H. Burger, E.L. Schaller et al., No sodium in the vapour plumes of Enceladus. Nature **459**, 1102 (2009)

D.E. Self, J.M.C. Plane, Absolute photolysis cross-sections for $NaHCO_3$, $NaOH$, NaO, NaO_2 and NaO_3: implications for sodium chemistry in the upper mesosphere. Phys. Chem. Chem. Phys. **4**, 16–23 (2002)

E.M. Sieveka, R.E. Johnson, Ejection of atoms and molecules from Io by plasma-ion impact. Astrophys. J. **287**, 418–426 (1984)

P. Sigmund, Sputtering by ion bombardment: theoretical concepts, in *Topics in Applied Physics*. Sputtering by Particle Bombardment I, vol. 47, ed. by R. Behrisch (Springer, Berlin, 1981). Ch. 2

J.A. Slavin, R.E. Holzer, The effect of erosion on the solar wind stand-off distance at Mercury. J. Geophys. Res. **84**, 2076–2082 (1979). https://doi.org/10.1029/JA084iA05p02076

J.A. Slavin, R.E. Holzer, Solar wind flow about the terrestrial planets, 1. Modeling bow shock position and shape. J. Geophys. Res. **86**, 11,401–11,418 (1981). https://doi.org/10.1029/JA086iA13p11401

J.A. Slavin, B.J. Anderson, T.H. Zurbuchen, D.N. Baker, S.M. Krimigis, M.H. Acuña, M. Benna, S.A. Boardsen, G. Gloeckler, R.E. Gold, G.C. Ho, H. Korth, R.L. McNutt Jr., J.M. Raines, M. Sarantos, D. Schriver, S.C. Solomon, P. Trávníček, MESSENGER observations of Mercury's magnetosphere during northward IMF. Geophys. Res. Lett. **36**, L02101 (2009a). https://doi.org/10.1029/2008GL036158

J.A. Slavin, M.H. Acuña, B.J. Anderson, D.N. Baker, M. Benna, S.A. Boardsen, G. Gloeckler, R.E. Gold, G.C. Ho, H. Korth, S.M. Krimigis, R.L. McNutt Jr., J.M. Raines, M. Sarantos, D. Schriver, S.C. Solomon, P. Trávníček, T.H. Zurbuchen, MESSENGER observations of magnetic reconnection in Mercury's magnetosphere. Science **324**, 606–610 (2009b). https://doi.org/10.1126/science.1172011

J.A. Slavin, B.J. Anderson, D.N. Baker, M. Benna, S.A. Boardsen, G. Gloeckler, R.E. Gold, G.C. Ho, H. Korth, S.M. Krimigis, R.L. McNutt Jr., L.R. Nittler, J.M. Raines, M. Sarantos, D. Schriver, S.C. Solomon, R.D. Starr, P.M. Trávníček, T.H. Zurbuchen, MESSENGER observations of extreme loading and unloading of Mercury's magnetic tail. Science **329**, 665–668 (2010). https://doi.org/10.1126/science.1188067

J.A. Slavin, S.M. Imber, S.A. Boardsen, G.A. DiBraccio, T. Sundberg, M. Sarantos, T. Nieves-Chinchilla, A. Szabo, B.J. Anderson, H. Korth, T.H. Zurbuchen, J.M. Raines, C.L. Johnson, R.M. Winslow, R.M. Killen, R.L. McNutt Jr., S.C. Solomon, MESSENGER observations of a flux-transfer-event shower at Mercury. J. Geophys. Res. **117**, A00M06 (2012a). https://doi.org/10.1029/2012JA017926

J.A. Slavin, B.J. Anderson, D.N. Baker, M. Benna, S.A. Boardsen, R.E. Gold, G.C. Ho, S.M. Imber, H. Korth, S.M. Krimigis, R.L. McNutt Jr., J.M. Raines, M. Sarantos, D. Schriver, S.C. Solomon, P. Trávníček, T.H. Zurbuchen, MESSENGER and Mariner 10 flyby observations of magnetotail structure and dynamics at Mercury. J. Geophys. Res. **117**, A01215 (2012b). https://doi.org/10.1029/2011JA016900

J.A. Slavin, H.R. Middleton, J.M. Raines, X. Jia, J. Zhong, W.J. Sun, S. Livi, S.M. Imber, G.K. Poh, M. Akhavan-Tafti, J.Â.M. Jasinski, G.A. DiBraccio, C. Dong, R.M. Dewey, M.L. Mays, MESSENGER observations of disappearing dayside magnetosphere events at Mercury. J. Geophys. Res. Space Phys. **124**, 6613–6635 (2019)

S.M. Smith, J.K. Wilson et al., Discovery of the distant lunar sodium tail and its enhancement following the Leonid meteor shower of 1998. Geophys. Res. Lett. **26**, 1642–1652 (1999)

W.H. Smyth, M.L. Marconi, Theoretical overview and modeling of the sodium and potassium atmospheres of Mercury. Astrophys. J. **441**, 839–864 (1995a)

W.H. Smyth, M.L. Marconi, Theoretical overview and modeling of the sodium and potassium atmospheres of the Moon. Astrophys. J. **443**, 371–392 (1995b). https://ui.adsabs.harvard.edu/abs/1995ApJ...443..371S/abstract

S. Soter, J. Ulrichs, Rotation and heating of the planet Mercury. Nature **214**, 1315–1316 (1967)

T.W. Speiser, Particle trajectory in model current sheets, 1. Analytical solutions. J. Geophys. Res. **70**, 4219 (1965)

A.L. Sprague, Mercury's atmospheric bright spots and potassium variations: a possible cause. J. Geophys. Res. **97**, 18257–18264 (1992)

A.L. Sprague, R.W.H. Kozlowski, D.M. Hunten, Caloris Basin: an enhanced source for potassium in Mercury's atmosphere. Science **249**, 1140–1143 (1990)

A.L. Sprague, R.W.H. Kozlowski, D.M. Hunten, N.M. Schneider, D.L. Domingue, W.K. Wells, W. Schmitt, U. Fink, Distribution and abundance of sodium in Mercury's atmosphere, 1985–1988. Icarus **129**, 506–527 (1997)

A.L. Sprague, D.M. Hunten, R.W.H. Kozlowski, F.A. Grosse, R.E. Hill, R.L. Morris, Observations of sodium in the lunar atmosphere during international lunar atmosphere week, 1995. Icarus **131**, 372–381 (1998).

A.L. Sprague, M. Sarantos, D.M. Hunten et al., The lunar sodium atmosphere: April-May 1998. Can. J. Phys. **90**, 725 (2012). http://www.nrcresearchpress.com/doi/10.1139/p2012-072

R.D. Starr, D. Schriver, L.R. Nittler, S.Z. Weider, P.K. Byrne, G.C. Ho, E.A. Rhodes, C.E. Schlemm II., S.C. Solomon, P.M. Trávníček, MESSENGER detection of electron-induced X-ray fluorescence from Mercury's surface. J. Geophys. Res. **117**, E00L02 (2012). https://doi.org/10.1029/2012JE004118

S.A. Stern, B.C. Flynn, Narrow field imaging of the lunar sodium exosphere. Astron. J. **109**, 35 (1995)

S.T. Suess, B.E. Goldstein, Compression of the Hermaean magnetosphere by the solar wind. J. Geophys. Res. **84**, 3306–3312 (1979). https://doi.org/10.1029/JA084iA07p03306

T. Sundberg, S.A. Boardsen, J.A. Slavin, B.J. Anderson, H. Korth, T.H. Zurbuchen, J.M. Raines, S.C. Solomon, MESSENGER orbital observations of large-amplitude Kelvin-Helmholtz waves at Mercury's magnetopause. J. Geophys. Res. **117**(A4), A04216 (2012). https://doi.org/10.1029/2011JA017268

J.R. Szalay, M. Horanyi, A. Colaprete, M. Sarantos, Meteoritic influence on sodium and potassium abundance in the lunar exosphere measured by LADEE. Geophys. Res. Lett. **43**(12), 6096–6102 (2016). https://doi.org/10.1002/2016GL069541

S.K. Trumbo, M.E. Brown, K.P. Hand, Sodium chloride on the surface of Europa. Sci. Adv. **5**, 7123 (2019)

H. Tsunakawa, F. Takahashi, H. Shimizu, H. Shibuya, M. Matsushima, Surface vector mapping of magnetic anomalies over the Moon using Kaguya and Lunar Prospector observations. J. Geophys. Res. **120**, 1160–1185 (2015). https://doi.org/10.1002/2014JE004785

R.R. Valiev, A.A. Berezhnoy, I.D. Gritsenko, B.S. Merzlikin, V.N. Cherepanov, T. Kurten, C. Wöhler, Photolysis of diatomic molecules as a source of atoms in planetary exospheres. Astron. Astrophys. **633**, A39 (2020)

S. Verani, C. Barbieri, C.R. Benn, G. Cremonese, M. Mendillo, The 1999 quadrantids and the lunar Na atmosphere. Mon. Not. R. Astron. Soc. **327**, 244 (2001)

R. von Steiger, N.A. Schwadron, L.A. Fisk, J. Geiss, G. Gloeckler, S. Hefti, T.H. urbuchen, Composition of quasi-stationary solar wind flows from Ulysses/Solar Wind Ion Composition Spectrometer. J. Geophys. Res. **105**(A12), 27,217–27,238 (2000). https://doi.org/10.1029/1999JA000358

A. Vorburger, P. Wurz, S. Barabash, M. Wieser, Y. Futaana, C. Lue et al., Energetic neutral atom imaging of the lunar surface. J. Geophys. Res. **118**, 3937–3945 (2013)

M. Wieser, S. Barabash, Y. Futaana, M. Holmström, A. Bhardwaj, R. Sridharan, M.B. Dhanya, P. Wurz, A. Schaufelberger, K. Asamura, Extremely high reflection of solar wind protons as neutral hydrogen atoms from regolith in space. Planet. Space Sci. **57**, 2132–2134 (2009a). https://doi.org/10.1016/j.pss.2009.09.012

J.K. Wilson, S.M. Smith, J. Baumgardner, M. Mendillo, Modeling an enhancement of the lunar sodium tail during the Leonid Meteor Shower of 1998. Geophys. Res. Lett. **26**, 1645–1648 (1999)

J.K. Wilson, M. Mendillo, H. Spence, Magnetospheric influence on the Moon's exosphere. J. Geophys. Res. **111**, 107207 (2006)

R.M. Winslow, C.L. Johnson, B.J. Anderson et al., Observations of Mercury's northern cusp region with MESSENGER's magnetometer. Geophys. Res. Lett. **39**, L08112 (2012)

R.M. Winslow, B.J. Anderson, C.L. Johnson, J.A. Slavin, H. Korth, M.E. Purucker, D.N. Baker, S.C. Solomon, Mercury's magnetopause and bow shock from MESSENGER observations. J. Geophys. Res. Space Phys. **118**, 2213–2227 (2013). https://doi.org/10.1002/jgra.50237

R.M. Winslow, C.L. Johnson, B.J. Anderson, D.J. Gershman, J.M. Raines, R.J. Lillis, H. Korth, J.A. Slavin, S.C. Solomon, T.H. Zurbuchen, M.T. Zuber, Mercury's surface magnetic field determined from proton-reflection magnetometry. Geophys. Res. Lett. **41**, 4463–4470 (2014). https://doi.org/10.1002/2014GL060258

R.M. Winslow, N. Lugaz, L.C. Philpott, N.A. Schwadron, C.J. Farrugia, B.J. Anderson, C.W. Smith, Interplanetary coronal mass ejections from MESSENGER orbital observations at Mercury. J. Geophys. Res. Space Phys. **120**, 6101–6118 (2015)

P. Wurz, H. Lammer, Monte-Carlo simulation of Mercury's exosphere. Icarus **164**, 1–13 (2003)

P. Wurz, D. Gamborino, A. Vorburger, J.M. Raines, Heavy ion composition of Mercury's magnetosphere. J. Geophys. Res. **124**, 2603–2612 (2019). https://doi.org/10.1029/2018JA026319

P. Wurz et al., Particles and photons as drivers for particle release from the surfaces of the Moon and Mercury. Space Sci. Rev. (2021), submitted

B.V. Yakshinskiy, T.E. Madey, Photon-stimulated desorption as a substantial source of sodium in the lunar atmosphere. Nature **400**, 642–644 (1999)

B.V. Yakshinskiy, T.E. Madey, Photon-stimulated desorption of Na from a lunar sample: temperature-dependent effects. Icarus **168**, 53–59 (2004). https://doi.org/10.1016/j.icarus.2003.12.007

B.V. Yakshinskiy, T.E. Madey, Temperature-dependent DIET of alkalis from SiO_2 films: comparison with a lunar sample. Surf. Sci. **593**, 202–209 (2005)

I. Yoshikawa, S. Kameda, K. Matsuura, K. Hikosaka, G. Murakami, K. Yoshioka, H. Nozawa, D. Rees, S. Okano, H. Misawa, A. Yamazaki, O. Korablev, Observation of Mercury's sodium exosphere by MSASI in the BepiColombo mission. Planet. Space Sci. **55**(11), 1622–1633 (2007). https://doi.org/10.1016/j.pss.2006.01.01

J. Zhong, W.X. Wan, Y. Wei, J.A. Slavin, J.M. Raines, Z.J. Rong, L.H. Chai, X.H. Han, Compressibility of Mercury's dayside magnetosphere. Geophys. Res. Lett. **42**, 10,135–10,139 (2015). https://doi.org/10.1002/2015GL067063

J. Zhong, L.C. Lee, X.G. Wang, Z.Y. Pu, J.S. He, Y. Wei, W.X. Wan, Multiple X-line reconnection observed in Mercury's magnetotail driven by an interplanetary coronal mass ejection. Astrophys. J. **893**, 1 (2020). https://doi.org/10.3847/2041-8213/ab8380

J. Zhong, Y. Wei, L.C. Lee, J.S. He, J.A. Slavin, Z.Y. Pu, H. Zhang, X.G. Wang, W.X. Wan, Formation of macroscale flux transfer events at Mercury. Astrophys. J. **893**, 1 (2020). https://doi.org/10.3847/2041-8213/ab8566

M.H. Zhu, J. Chang, T. Ma et al., Potassium map from Chang'E-2 constrains the impact of crisium and orientale basin on the Moon. Nat. Sci. Rep. **3**, 1611 (2013)

T.H. Zurbuchen, J.M. Raines, G. Gloeckler, S.M. Krimigis, J.A. Slavin, P.L. Koehn et al., MESSENGER observations of the composition of Mercury's ionized exosphere and plasma environment. Science **321**(5885), 90–99 (2008)

T.H. Zurbuchen, J.M. Raines, J.A. Slavin, D.J. Gershman, J.A. Gilbert, G. Gloeckler et al., MESSENGER observations of the spatial distribution of planetary ions near Mercury. Science **333**(6051), 1862–1865 (2011)

Publisher's Note Springer Nature remains neutral with regard to jurisdictional claims in published maps and institutional affiliations.

Authors and Affiliations

F. Leblanc[1] · C. Schmidt[2] · V. Mangano[3] · A. Mura[3] · G. Cremonese[4] · J.M. Raines[5] · J.M. Jasinski[6] · M. Sarantos[7] · A. Milillo[3] · R.M. Killen[7] · S. Massetti[3] · T. Cassidy[8] · R.J. Vervack Jr.[9] · S. Kameda[10] · M.T. Capria[3] · M. Horanyi[8] · D. Janches[7] · A. Berezhnoy[11,12] · A. Christou[13] · T. Hirai[14] · P. Lierle[2] · J. Morgenthaler[15]

✉ V. Mangano

1   LATMOS/CNRS, Sorbonne Université, UVSQ, IPSL, Paris, France

2   Center for Space Physics, Boston University, Boston, USA

3   INAF/IAPS, Rome, Italy

4   INAF/Osservatorio Astronomico di Padova, Roma, Italy

5   Dept. of Climate and Space Sciences and Engineering, University of Michigan, Ann Arbor, MI, USA

6   NASA Jet Propulsion Laboratory, California Institute of Technology, Pasadena, CA, USA

7   NASA Goddard Space Flight Center, Greenbelt, MD, USA

8   Laboratory of Atmospheric and Space Physics, University of Colorado Boulder, CO, USA

9   Johns Hopkins Applied Physics Laboratory, Laurel, MD, USA

10 Rikkyo University, Tokyo, Japan

11 Sternberg Astronomical Institute, Moscow State University, Moscow, Russia

12 Institute of Physics, Kazan Federal University, Kazan, Russia

13 Armagh Observatory College Hill, Armagh, UK

14 Planetary Exploration Research Center, Chiba Institute of Technology, Chiba, Japan

15 Planetary Science Institute, Tuscon, AZ, USA

Space Science Reviews (2021) 217:74
https://doi.org/10.1007/s11214-021-00846-3

Water Group Exospheres and Surface Interactions on the Moon, Mercury, and Ceres

Norbert Schörghofer[1,2] · Mehdi Benna[3] · Alexey A. Berezhnoy[4,5] ·
Benjamin Greenhagen[6] · Brant M. Jones[7] · Shuai Li[8] · Thomas M. Orlando[7] ·
Parvathy Prem[6] · Orenthal J. Tucker[3] · Christian Wöhler[9]

Received: 31 October 2020 / Accepted: 9 August 2021 / Published online: 1 September 2021
© The Author(s) 2021

Abstract

Water ice, abundant in the outer solar system, is volatile in the inner solar system. On the largest airless bodies of the inner solar system (Mercury, the Moon, Ceres), water can be an exospheric species but also occurs in its condensed form. Mercury hosts water ice deposits in permanently shadowed regions near its poles that act as cold traps. Water ice is also present on the Moon, where these polar deposits are of great interest in the context of future lunar exploration. The lunar surface releases either OH or H_2O during meteoroid showers, and both of these species are generated by reaction of implanted solar wind protons with metal oxides in the regolith. A consequence of the ongoing interaction between the solar wind and the surface is a surficial hydroxyl population that has been observed on the Moon. Dwarf planet Ceres has enough gravity to have a gravitationally-bound water exosphere, and also has permanently shadowed regions near its poles, with bright ice deposits found in the most long-lived of its cold traps. Tantalizing evidence for cold trapped water ice and exospheres of molecular water has emerged, but even basic questions remain open. The relative and absolute magnitudes of sources of water on Mercury and the Moon remain largely unknown. Exospheres can transport water to cold traps, but the efficiency of this process remains uncertain. Here, the status of observations, theory, and laboratory measurements is reviewed.

Keywords Mercury · The Moon · Ceres · Exospheres · Water · Hydroxyl

1 Introduction

The abundance of water ice in permanently shadowed craters on the Moon is key to the future exploration of the lunar surface. The underlying idea was formulated a long time ago (Watson et al. 1961a,b): water from exogenic or endogenic sources is transported laterally through the exosphere and then trapped in permanently shadowed regions (PSRs) near the lunar poles. No part of this process has yet been confirmed. The primary source of water

Surface-Bounded Exospheres and Interactions in the Inner Solar System
Edited by Anna Milillo, Menelaos Sarantos, Benjamin D. Teolis, Go Murakami, Peter Wurz and Rudolf von Steiger

Extended author information available on the last page of the article

on the Moon is unknown. The efficiency of the water exosphere as a transport process is controversial. And studies of the abundance and distribution of ice in lunar PSRs have resulted in divergent answers. What is established is that the cold traps of the planet Mercury are filled with water ice, and cold traps on the Moon contain at least some, but perhaps not much, ice. Research over the last decade has provided a great deal of new information. The MESSENGER spacecraft orbited Mercury from 2012 to 2015. The Lunar Reconnaissance Orbiter (LRO) has continuously provided data since 2009, and the Lunar Atmosphere and Dust Environment Explorer (LADEE) orbited in the equatorial plane of the Moon from 2013 to 2014. The Chandrayaan-1 mission ended in 2009, but some of its measurements have only been analyzed more recently. Ceres, a large asteroid that also has PSRs, was visited by the Dawn spacecraft from 2015 to 2018. These three airless bodies of the inner solar system (Mercury, the Moon, and Ceres) each have (nearly) ice-free surfaces, PSRs, and (at least potentially) gravitationally-bound water exospheres. Here we summarize recent results from orbital measurement campaigns, physical chemistry, and theoretical models about water group exospheres. Some of the measurement techniques do not distinguish between water (H_2O) and hydroxyl (OH), so "water group" (OH or H_2O) is a frequently used term in this review.

The surface-bounded exospheres of Mercury and the Moon have been reviewed by Killen and Ip (1999) and Stern (1999). The exosphere and polar deposits of Mercury after MESSENGER have been reviewed by Killen et al. (2018), McClintock et al. (2018), and Chabot et al. (2018a). Reviews of surface volatiles on the Moon are available by Basilevsky et al. (2012) and Hayne et al. (2014). Lawrence (2017) summarizes our understanding of polar volatiles on Mercury and the Moon. Table 1 lists some of the pertinent observational discoveries.

2 Models of H, H_2, and OH Interactions on Planetary Surfaces

Observations of water products bound to airless bodies like the Moon, Mercury, and Ceres are infrequent and possess sparse temporal and global coverage. Therefore, theoretical models are required to understand the driving dynamics and sources of surficial and exospheric water group molecules. Zeller et al. (1966) demonstrated proton-induced hydroxyl formation experimentally. Extraordinary theoretical studies carried out by Starukhina (2001) of solar wind implantation and H atom diffusion predicted the presence of a widespread OH/H_2O IR 2.8 μm feature well before it was observed in 2009 (Pieters et al. 2009; Sunshine et al. 2009; Clark 2009). Likewise, Johnson (1971), Hodges (1973) and Hartle and Thomas (1974) carried out pioneering studies that predicted the presence of H_2 in the lunar exosphere sourced from the solar wind and primarily lost by thermal escape. In this section we review recent theoretical studies aimed at examining how exogenic hydrogen sources are partitioned within surficial and exospheric inventories of the lunar environment.

2.1 Surface Diffusion Models

Starukhina and Shkuratov (2000) predicted that solar wind protons implanted in oxygen bearing regolith become both physically trapped in defects and chemically trapped as OH, based on experimental evidence (e.g., Zeller et al. 1966; Mattern et al. 1976). The solar wind bombardment simultaneously frees trapped hydrogen, which diffuses to the surface and thermally desorbs into the exosphere as H_2 (Starukhina 2006). This process can be characterized as a diffusive process and the surface concentration of OH molecules and the H_2 desorption flux are proportional to $\sim \exp(-E_a/kT)$, where E_a is the activation energy,

Table 1 Selected observations relevant to water group exospheres and the surface reservoirs they interact with

Planetary Body	Instrument or Method	Spacecraft or Observatory (time period of data collection)	Major discoveries
Mercury	radar	Goldstone, Arecibo, and Very Large Array (1991-1992)	polar ice deposits
Mercury	various	MESSENGER (2011-2015)	relation between ice and permanently shadowed regions
The Moon	neutron spectrometer	Lunar Prospector (1998-1999)	polar regions are enriched in hydrogen
The Moon	near infrared, visible, UV spectrometers	LCROSS (2009)	H_2O and OH detected in impact plume above cold trap
The Moon	M^3 (Moon Mineralogy Mapper)	Chandrayaan-1 (2008-2009)	diurnal and latitude variation in OH-band strength, H_2O exposed in cold traps
The Moon	Diviner	LRO (2009-now)	detailed mapping of cold traps
The Moon	NMS (Neutral Mass Spectrometer)	LADEE (2013-2014)	water group exosphere detected, correlates with meteoroid streams
Ceres	Long-wavelength prime camera	International UV Explorer (1990-1991)	detection of temporary OH exosphere
Ceres	GRaND and Framing Camera	Dawn (2015-2018)	ice-rich subsurface, permanently shadowed regions
Ceres	far infrared	Herschel Space Observatory (2011-2013)	detection of temporary H_2O atmosphere

k is the Boltzmann constant and T is the local surface temperature. This interpretation is consistent with the observations of the lunar OH/H_2O veneer (Pieters et al. 2009; Sunshine et al. 2009; Clark 2009; Li and Milliken 2017; Wöhler et al. 2017b), further described in Sect. 4.1.

Irradiation of regolith grains leads to a dynamic equilibrium between the formation of dangling bonds (immobile atoms with an unsatisfied valence), which are chemical trapping sites, and diffusing hydrogen (Griscom 1984; Fink et al. 1995; Farrell et al. 2015). Farrell et al. (2015, 2017) highlighted that diffusion of hydrogen is best characterized by considering a distribution of activation energies, because exposed grains are known to be defect rich in the top 100 nm due to the space environment (Noble et al. 2005). They found this approach is qualitatively consistent with the latitudinal trends and diurnal modulation of the observed 2.8 μm OH/H_2O feature over the lunar surface (Sunshine et al. 2009; McCord et al. 2011). Farrell et al. (2015, 2017) review the solution of the continuity equation for the steady state hydrogen surface concentration balanced by the solar wind implantation against diffusive losses and retention in the regolith by formation of OH, and the implementation of Monte Carlo models of H retention in the surface using a Gaussian distribution of activation energies to characterize diffusion. These models solely considered the steady state concentration of hydrogen, and all surface loitering H atoms are presumed to be trapped as OH.

Grumpe et al. (2019) adapted the continuity model applied in the Farrell et al. studies to include time dependence, the OH continuity equations in addition to the H continuity equation, and the surface photolysis of OH. A micrometeoroid source of OH and H_2O was also included, but solar wind implantation was found to be the dominant source. The Grumpe et al. continuity model was constrained to the inferred OH surface densities from Moon Mineralogy Mapper M^3 spectra using the thermal correction model applied in Wöhler et al. (2017b), which inferred hydroxyl abundances on order of $1 - 10$ ppm with mild diurnal and latitudinal variations. For comparison, the water concentration in Apollo samples is typically around $40 - 50$ ppm (Li and Milliken 2017), but could potentially have been affected by terrestrial contamination. Grumpe et al. (2019) fit the variation of the 3 μm band depth versus local time of day to extract the H diffusion activation energy, OH photolysis lifetime and region specific OH activation energies.

2.2 H_2 Exospheric Models

Molecular hydrogen was first detected in the lunar exosphere by the Lyman Alpha Mapping Project (LAMP) UV spectrograph on LRO (Stern et al. 2013), and later confirmed by the Chandrayaan-1 Altitudinal Composition Explorer (CHACE) mass spectrometer (Thampi et al. 2015). Hurley et al. (2017) used an exospheric Monte Carlo model to track the density and spatial distribution of H_2 on the Moon produced by micrometeoroid bombardment, solar wind knock-on sputtering and chemical sputtering. Micrometeoroid impacts volatilize implanted hydrogen atoms which may chemically combine as H_2 and thermally desorb from the surface at temperatures > 600 K (Cintala 1992) with a Maxwell Boltzmann flux speed distribution (Brinkmann 1970). Knock-on sputtering occurs via momentum transfer during direct collisions from solar wind ions with regolith atoms and molecules. The sputter ejecta are released from the surface assuming a Sigmund-Thompson energy distribution roughly consistent with sputter theory (Johnson 1990). Hurley et al. (2017) determined that both micrometeoroid vaporization and solar wind sputtering were too energetic to reproduce H_2 exospheric densities of 1200 ± 400 cm^{-3} observed by LAMP, because the solar wind source of protons is insufficient. Solar wind plasma bombardment also releases physically and chemically trapped H atoms which diffuse and thermally desorb from the surface as H_2 at the local surface temperature with a Maxwell Boltzmann flux speed distribution. This process is referred to as chemical sputtering. Hurley et al. estimated H_2 chemical sputtering at rates of ~ 19 g/s consistent with both the LAMP observations and the solar wind source of protons ~ 31.5 g/s.

2.3 Coupled Surface Diffusion – Exospheric Models

The abundance of water group exospheric gases is directly linked to the subsurface diffusion and chemistry of hydrogen atoms with regolith oxides. Tucker et al. (2019) coupled the Farrell et al. Monte Carlo model (using a distribution of activation energies to characterize H atom diffusion) to a time dependent global Monte Carlo model of exospheric degassed H_2. When applying a Gaussian distribution of diffusive activation energies with a peak centered at ~ 0.5 eV and width of ~ 0.1 eV the model simultaneously fit the Moon Mineralogy Mapper M^3 latitudinal and diurnal OH surface concentrations as interpreted from the thermal correction model applied in Li and Milliken (2017), the LAMP H_2 exospheric densities (1200 ± 400 cm^{-3}), and the CHACE exospheric H_2 surface densities as a function of latitude (~ 400 cm^{-3} – 800 cm^{-3}, latitudes $20° - 80°$) obtained while the Moon was in the geomagnetic tail (Tucker et al. 2021). For a peak activation energy of 0.7 eV the model surface was hydrogen retentive and the surface concentration was a couple of orders of magnitude larger, whereas the exospheric H_2 densities were an order of magnitude lower. Likewise, for a peak activation energy of 0.3 eV the surface was hydrogen emissive and significantly underestimated the Li et al. data at mid to high latitudes.

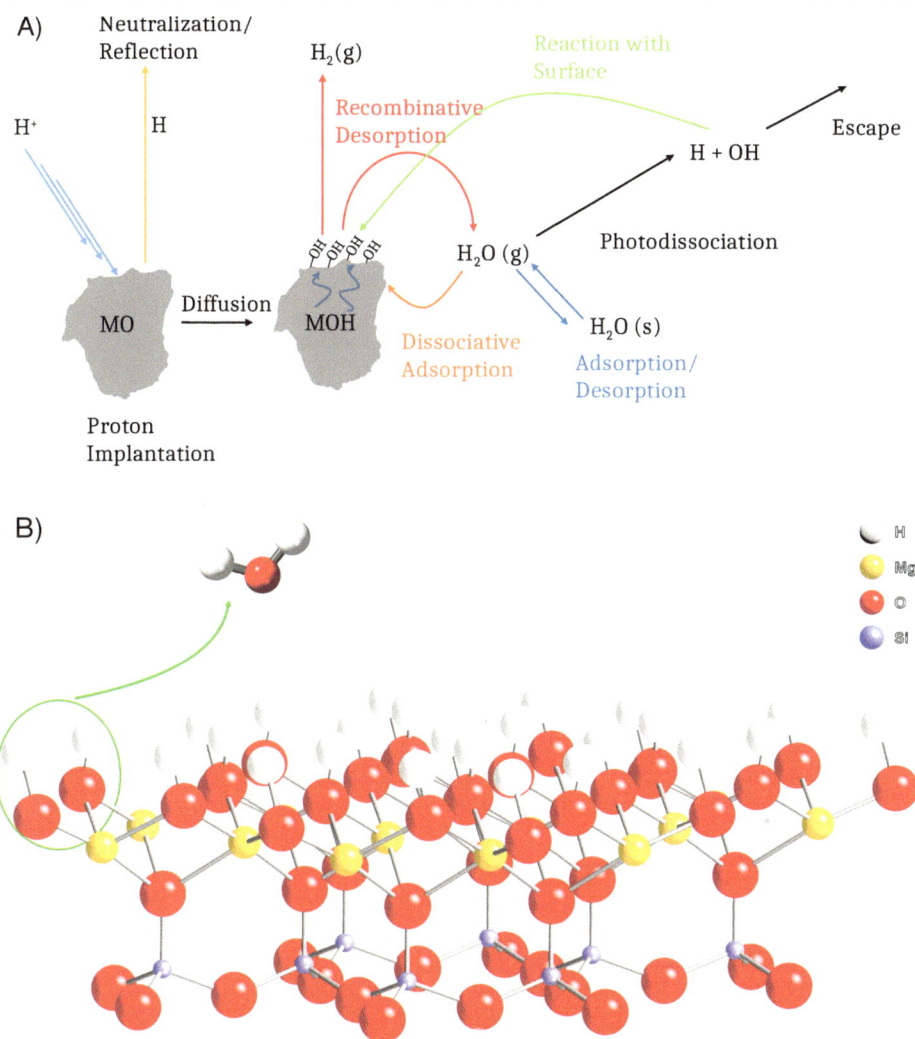

Fig. 1 **A**) Kinetic scheme for the solar-wind-induced water cycle on the surface of Mercury. **B**) Illustration of mineral surface of magnesium silicate with saturated hydroxyls. Upon heating, a reaction takes place between neighboring OH sites (encircled with green) resulting in the formation of gas phase water and an oxygen bridge between cations. Figures adapted from Jones et al. (2020), copyright by AAS

Second order recombinative desorption (RD) of neighboring hydroxyls on surface grains has recently been investigated as an important pathway of gas phase water production on the Moon and other airless bodies (Jones et al. 2018, 2020; Fig. 1), see Sect. 3. Implanted protons react with metal oxides forming bound hydroxyls (M-OH, where M represents a metal atom). Over time, the hydroxyl sites diffuse to the surface where they can recombine with other surficial hydroxyls and desorb as gas phase water. RD is not efficient for silicate grains (50 wt%. of the regolith) at temperatures < 600 K, and the maximum surface temperature on the Moon is about 400 K, but such temperatures can be obtained during micrometeoroid impacts. Jones et al. (2018) derived an upper limit for the dayside H_2O exospheric density of 60 cm^{-3}, consistent with the abundances inferred from LADEE measurements, described in Sect. 5 below.

At Mercury the dayside temperatures at local noon exceed 600 K, and RD may be an important process for H_2O production and transport of water to polar cold traps (Jones et al. 2020). Jones et al. developed a forward model by using the degassed H_2O flux from a surface diffusive-kinetic Monte Carlo model of OH RD in an exospheric transport Monte Carlo model of H_2O to examine its transport to the poles. The model predicts a global H_2O degassing rate of 3×10^{30} molecules per Mercury day (4 g/s). At this source rate, RD is a constant replenishing source of water to the poles and cold traps, and could account for a significant portion of the total water mass trapped as ice over geological time periods, supplementing other potential sources such as comets and meteoroid impacts.

3 Hydroxylation & in-Situ Generation of H_2O

Potential sources of lunar water group species (H_2O and –OH) include: primordial water (Hui et al. 2013; Robinson et al. 2016), water delivered via comets and meteoroids (Arnold 1979; Greenwood et al. 2011) and captured during the impact event (Daly and Schultz 2018; Stopar et al. 2018), and implantation from solar wind (Zeller et al. 1966). A prevalent mechanism for the formation of molecular water on the surfaces of airless bodies may be a thermally activated process known as recombinative desorption (RD) or associative desorption (AD) of hydroxyl (–OH) defects that were made by implantation of solar wind protons.

The formation of chemically bound hydroxyls is a well-known consequence of proton implantation (Zeller et al. 1966; Mattern et al. 1976; Burke et al. 2011; Ichimura et al. 2012; Managadze et al. 2011; Schaible and Baragiola 2014). These solar wind implanted hydroxyls are attributed to the infrared absorption feature centered at 2.8 μm observed on the lunar surface (McCord et al. 2011). Over time, the implanted OH defect sites will diffuse to the surface of the regolith grain in a relatively short amount of time with respect to the planetary and lunar rotation timescales, e.g., within a few lunar hours given the OH diffusion constants (Starukhina and Shkuratov 2000).

Following the buildup of terminal hydroxyl sites on the regolith grain, water can form through RD. Here, RD is a surface-mediated process that describes a chemical reaction between neighboring or interacting termination sites followed by desorption of the molecular product. For example, consider neighboring species representing dangling M-OH bonds (M being a generic metal cation, e.g., Si, Al, Fe, Mg, Ti, etc.). M-OH sites will react to form H_2O while concurrently healing the oxygen bridge defect:

$$\text{M-OH} + \text{M-OH} \rightarrow \text{M-O-M} + H_2O \text{ (g)}.$$

We note that RD on the surface is not completely necessary for the production of molecular water. Recent proton irradiation and thermal annealing studies have shown molecular water to form within the irradiation layer (Zhu et al. 2020; Zeng et al. 2021). The ejection occurs at thermal velocities and is controlled by the local surface temperature. The most probable ejection velocity at 300 K is 610 m s^{-1}. (For comparison, the escape velocity of the Moon is 2.4 km/s.) The porosity and fractal nature of the regolith might affect the ejection angles and velocities.

If a trapped H-atom is present, it can possibly react with a M-OH site forming H_2. The

$$\text{M-OH} + \text{M} \cdots \text{H} \rightarrow \text{M-O-M} + H_2 \text{ (g)}$$

reactions can involve both surface and sub-surface sites. However, previous experimental work has shown that formation of molecular hydrogen is only relevant near the saturation limit ($\sim 10^{17}$ H$^+$ cm^{-2}) (Mattern et al. 1976; Blanford et al. 1985). In addition, molecular hydrogen was suggested to form (Starukhina 2006) via

$$\text{M-OH} + \text{M-OH} \rightarrow \text{M-O-O-M} + H_2 \text{ (g)}.$$

However, experimental work has shown the energy barrier for formation of molecular hydrogen via terminal hydroxyl sites is considerably higher than that of water formation, resulting in nearly 100% yield of water via RD on a highly hydroxylated metal oxide surface (Du et al. 2012). While the experimental conditions are not fully representative of regolith on an airless body, the overall mechanism and the disparity of activation energies/kinetics are validated with proton irradiation studies; Crandall et al. (2019) demonstrated that only 2% of the implanted protons resulted in the formation of H$_2$.

Alternative pathways of in situ water formation through chemical sputtering have also been identified as a possible source term for water on airless bodies (Crider and Vondrak 2000; Gibson 1977; Potter 1995). In particular, a small yield of water ions directly sputtered under keV D$^+$ bombardment from the surface of oligoclase was measured at 10^{-4} D$_2$O$^+$ per incident D$^+$ (Blanford et al. 1985) and 10^{-4} D$_2$O$^+$ per incident D$^+$ from the surface of SiO$_2$ (Managadze et al. 2011). Similarly, water formation was observed from reactions of hydrogen atoms with an oxide layer of stainless steel (Ishibe and Oyama 1979).

In summary, proton implantation into any airless body composed of regolith enriched in minerals of various metal oxides will result in the formation of bound hydroxyls. Upon heating, this results in the formation of molecular hydrogen, and water (Fig. 1). The rate of each is controlled by the local surface temperature, concentration profiles, and the associated activation energies. Since RD and chemical sputtering both require the solar wind, they can be considered solar wind water formation processes.

Typically, the temperature required for complete conversion of the implanted OH via RD is much greater than the highest temperature reached on the lunar surface. For example, the water formation rate via RD from SiO$_2$ peaks at ~ 600 K (Gun'ko et al. 1998). However, other metal oxides have significantly lower peak RD temperatures, e.g., TiO$_2$ is ~ 250 K (Henderson 1994), MgO is 225 K (Stirniman et al. 1996), and Al$_2$O$_3$ is ~ 350 K (Nelson et al. 1998). In addition, time is on the side of water formation on these airless bodies as they often exhibit long diurnal cycles. Consequently, despite the slow reaction rate at 300–400 K, the amount of time the regolith experiences dayside temperatures compensate for the low production rate of water. While the minerals identified above do not exist in pure form in the lunar regolith, the necessary activation energies can be approximated based on the assumption that at the microscopic scale, the lunar regolith will behave in a similar fashion to their respective pure representatives, e.g., the MgO network in pyroxene is similar to pure MgO.

Jones et al. (2018) demonstrated that a minor fraction ($\sim 30\%$) of the total sites produced by the solar wind on an Apollo mare sample (10084) were available for recombinative desorption during typical noontime temperatures on the Moon. Activation will remove some OH sites but not all, resulting in a latitude dependent IR signature (Jones et al. 2018) that has been observed in some interpretations of the M^3 data set (McCord et al. 2011; Li and Milliken 2017; Wöhler et al. 2017b). Overall, RD on the lunar surface results in the polar regions exhibiting the highest concentration of chemically trapped hydrogen in the form of bound hydroxyls, with a latitude dependent OH signal and a relatively (Δ ppm < 1) constant OH signal at all longitudes. In addition, given both the high temperatures on Mercury and the amount of time the regolith experiences those temperatures, recombinative desorption is estimated to account for 10% of the ice on Mercury trapped in the permanently shadowed regions (Jones et al. 2020).

Water formation following a simulated micrometeoroid impact has been demonstrated in the lab as well. Zhu et al. (2019) exposed anhydrous olivine $(Mg,Fe)_2SiO_4$ samples to a D_2^+ ion beam. (Deuterium is used instead of regular hydrogen to distinguish any resulting hydroxyl or molecular water from potential background sources.) After the ion exposure, the samples were exposed to a pulsed infrared laser, followed by temperature-programmed desorption (TPD). The pulsed laser generates intense heating events with temperatures that can reach higher than 1,400 K, similar to temperatures produced by micrometeoroid impacts. Simulations of micrometeoroid impacts were conducted at temperatures of 10 and 300 K. These laboratory simulation experiments demonstrated that water can be generated and released from anhydrous minerals implanted with solar-wind protons through rapid energetic heating, as would occur during micrometeoroid impacts.

In summary, water can be formed *in situ* by thermally induced reactions of solar wind implanted hydroxyls on grain surfaces on airless bodies. A fraction of the water formed from this mechanism is expected to accumulate in cold PSRs (Sect. 8).

4 Observations of Hydroxyl on the Lunar Surface

Lunar rocks are made up of minerals and glasses, with only trace amounts of water (Papike et al. 1991; Greenwood et al. 2011). Some of the hydrogen identified in Apollo samples is indigenous (Saal et al. 2008; Boyce et al. 2010). Hydroxyl found in glasses in Apollo samples has isotopic ratios indicating it is derived from solar wind, rather than from meteoroids or terrestrial contamination (Liu et al. 2012). Native H_2O may be found in the amorphous rims formed on silicate grains through solar wind irradiation, as identified by Bradley et al. (2014) on interplanetary dust particles.

4.1 Spectroscopic Observations of OH/H$_2$O

The Moon Mineralogy Mapper (M^3) was a hyperspectral imaging spectrometer onboard the Indian Chandrayaan-1 spacecraft launched in 2008 (Green et al. 2011). It measured the spectral reflectance on the lunar surface from 0.43 to 3.0 μm which was split into 260 continuous spectral bands ($\sim$10 nm sampling interval) (Green et al. 2011). The M^3 images cover over 95% of the lunar surface. The spectral bands of M^3 data near 3 μm were used to assess the hydration features on the lunar surface. The fundamental stretching of OH and the first overtone of the molecular water vibration both occur near 3 μm. The absorption strength of the reflectance data near 3 μm can be directly linked to the absolute OH/H_2O content.

Lunar surface hydroxyl or water was assessed globally through analyzing the absorption features of OH and H_2O near 3 μm seen by three different missions/instruments, namely Chandrayaan-1 Moon Mineralogy Mapper (M^3) (Pieters et al. 2009), EPOXI near infrared spectrometer (Sunshine et al. 2009), and Cassini Visual and Infrared Mapping Spectrometer (VIMS) (Clark 2009). Both M^3 and EPOXI data (280 m/pixel and 60 km/pixel, respectively) suggest that the 3 μm absorption strength shows strong latitudinal dependence from the mid-latitude to the poles and no pronounced absorptions near 3 μm were seen near the equator (Pieters et al. 2009; Sunshine et al. 2009). It might be due to the much lower spatial resolution of the VIMS data (175 km/pixel) that the OH/H_2O absorptions near 3 μm exhibit only

Fig. 2 Depth of the 3-μm absorption band (OHIBD) inferred from M^3 data and overlaid on M^3 1.579 μm reflectance. In the morning (left), the OHIBD slightly increases from the equator toward the poles. At midday (middle), the OHIBD decreases globally and does not show a dependence on latitude. In the afternoon (right), the OHIBD increases globally and shows a dependence on latitude similar to the morning. At all times of day, the OHIBD is higher in the highlands than in the maria. Image reproduced with permission from Grumpe et al. (2019)

weak latitudinal variation. The estimates of total OH/H$_2$O content from the three datasets at the strongest 3 μm absorption are consistent with each other at ∼1000 ppm at the polar region (Clark 2009; Pieters et al. 2009; Sunshine et al. 2009). Diurnal variation of the 3 μm absorption strength is observed in the M^3 and EPOXI data (Pieters et al. 2009; Sunshine et al. 2009).

Further improved thermal correction of the M^3 data also suggest that the OH/H$_2$O absorptions near 3 μm show strong latitudinal variation and diurnal variation (Li and Milliken 2017; Wöhler et al. 2017b). Li and Milliken (2017) performed laboratory experiments using basaltic glasses and anorthosite that are related to the lunar surface composition to link the 3-μm absorptions strength with the absolute water group content (OH + H$_2$O). The results suggest that the absolute abundances of OH/H$_2$O at the lunar surface varies from < 20 ppm near the equator to ∼1000 ppm near the pole (Li and Milliken 2017). The diurnal variation of OH/H$_2$O near the equator is very weak, which could be due to the low water content at this latitude zone. The strongest diurnal variation of OH/H$_2$O is seen in the mid latitude between ∼25° − 60° (up to 200 ppm), while no obvious diurnal variation is seen at latitudes > ∼60° at both the northern and southern hemispheres (Li and Milliken 2017). The integrated time regolith was exposed to the micrometeoroid and solar wind environment is known as "maturity" (McKay et al. 1991). A strong correlation between the maturity of lunar regolith and mapped OH/H$_2$O content is observed at the mid-high latitudes of the Moon (Li and Milliken 2017). However, different approaches for correcting the thermal effect of the M^3 data result in vastly different spatial and temporal variation of the 3 μm absorption. Bandfield et al. (2018) developed a new method for removing the thermal emission component from the M^3 reflectance spectra and found that there is no latitudinal or diurnal variation in the 3-μm absorption strength.

Wöhler et al. (2017a) processed the M^3 data using thermal and topographic corrections. The 3-μm band depth maps constructed by Wöhler et al. (2017a) for the southern high-latitude area around the crater Boguslawsky indicate that the 3-μm band is stronger at lunar morning than at midday. Wöhler et al. (2017b) constructed nearly global OH integrated band depth (OHIBD) maps (Fig. 2). In the lunar morning (07:00-08:00), the 3-μm band depth in the polar highlands exceeds the value in the equatorial highlands by 20-30%. The lunar maria exhibit OHIBD values lower by 10-15% than the highlands, where high-Ti mare basalt have the lowest OHIBD. At lunar midday (10:00-14:00), the dependence of the OHIBD on latitude vanishes, and the OHIBD in the equatorial highlands is 10-15% lower than in the morning. All maria are associated with low OHIBD values, with the lowest OHIBD occurring in high-Ti maria. In the lunar afternoon (16:00-17:00), the OHIBD is similar to but slightly lower than in the morning. The relative OHIBD decreases by ~30% between morning and midday in the polar highlands, whereas it is nonzero and nearly invariable with time of day in the low-latitude highlands. In high-Ti mare areas, much stronger time-of-day-dependent relative OHIBD variations of ~50-70% can be observed. Wöhler et al. (2017b) explained these observations by the superposed effects of a time-of-day-dependent, weakly bounded OH/H_2O component built up by H adsorption from the solar wind and later removed by thermal evaporation and photolysis, plus a strongly bounded OH/H_2O component that is stable against thermal evaporation and photolysis.

Pieters et al. (2009), McCord et al. (2011), Li and Milliken (2017), Wöhler et al. (2017b), and Bandfield et al. (2018) each found widespread presence of surficial OH/H_2O based on M^3 spectral reflectance data, consistent with observations made through other near-infrared spectrometers (Clark 2009; Sunshine et al. 2009; Honniball et al. 2020). McCord et al. (2011), Starukhina (2012), and Bandfield et al. (2018) concluded the diurnal variation may be an observational artifact, consistent with the hydroxyl dynamics expected from physical chemistry (Sect. 3). The degree of latitude dependence varies among authors. Recent earth-based observations by Honniball et al. (2020) also suggest that the 3 μm band varies with latitude, composition, and local time of day.

The speciation (OH or H_2O or both) on the lunar surface is still unclear from M^3 measurements. The OH vibration band is centered between 2.65 and 2.9 μm and the fundamental stretching vibration band of OH near 2.8 μm, while the 3.0 μm absorption band of H_2O corresponds to the first overtone of its bending vibration. The presence of H_2O makes the whole 3 μm spectra broader and more symmetric, whereas OH exhibits an asymmetric absorption band extending well beyond 3 μm. However, an OH absorption band may also occur at wavelengths of >2.9 μm, depending on the energy characterizing the binding of the hydroxyl with the respective cation; thus, a broad OH absorption band may well be present in the range between 2.65 and 3.5 μm (e.g., Dyar et al. 2010). The wavelength range of M^3 is too limited to accurately determine the full shape and maximum absorption point in the 3 μm region, making it difficult to differentiate between OH and H_2O, particularly if both species are present. The EPOXI and VIMS data cover a wider spectral range, but the possible thermal residual and low spatial resolution complicate the interpretation of the 3 μm band shape and make it difficult to discriminate OH from H_2O in these two datasets. Recent observations in the 6-micron region (Honniball et al. 2021) confirm H_2O is present in a sunlit region (Clavius crater) at abundances of 100-400 μg/g. This water may be stored in glasses, as observed in Apollo samples.

Measurements by LAMP (the Lyman Alpha Mapping Project instrument) in the far-UV also provide evidence for surface hydration on the Moon varying in abundance with both terrain type and local time (Hendrix et al. 2012, 2019). Highlands regions are more hydrated than mare regions, consistent with the results from M^3. Hendrix et al. (2019) interpret the variations in the far-UV spectra as a partial monolayer of water thermally adsorbing and desorbing, but OH is not ruled out. When the Moon crosses the Earth's magnetotail, where the solar wind source of protons is absent, no decrease in surface hydration was observed by LAMP. Recently, Wang et al. (2021) also found that the hydration level in the polar regions remains unchanged during passage through Earth's magnetotail, based on their analysis of M^3 data.

4.2 Local Anomalies

Lunar swirls are bright structures uncorrelated with topography, which are associated with local magnetic fields. A commonly accepted hypothesis is that lunar swirls exist due to shielding of the surface by a preexisting magnetic field, leading to the prevention of soil maturation (e.g., Hood and Schubert 1980; Glotch et al. 2015). An alternative hypothesis is that they were induced by external processes, such as interaction with cometary gas, leading to regolith compaction and/or removal of the uppermost regolith layer (e.g., Schultz and Srnka 1980; Shevchenko 1993; Pinet et al. 2000).

Similar to previous studies (e.g., Kramer et al. 2011; Li and Milliken 2017; Li and Garrick-Bethell 2019), recent results by Hess et al. (2020) indicate a weaker 3-μm band at the swirl locations than on the surrounding surface. This finding supports the assumption that the magnetic field locally reduces the flux of solar wind protons, leading to a reduced rate of OH/H_2O formation in the regolith. However, Hess et al. (2020) also showed that the spectral appearance of the northeastern parts of the Reiner Gamma swirl can be better explained by regolith compaction rather than by immature soil, due to the absence of variations in spectral reddening and in the depth of the mafic absorption bands between on-swirl and off-swirl surfaces. Furthermore, they described a small swirl structure in the western part of Mare Moscoviense that is not associated with a magnetic anomaly and does not show reduced OHIBD values compared to the surrounding surface. These findings provide evidence that external processes might be relevant for the formation of swirls, such as the interaction with cometary gas, which Shevchenko (1993) compared to the effect of a landing rocket jet on the lunar regolith. This would lead to soil compaction and, according to the model by Hapke (2008), increased surface albedo without changing the mafic band depths and spectral slope. Additionally, the cometary gas may also remove the uppermost, highly matured regolith layer and reveal immature material, which would be consistent with the observation by Hess et al. (2020) that spectra indicating a low surface maturity occur at all studied swirls. Using spectra of the lunar surface acquired by the Yutu rover at different distances from the Chang'E-3 lander, Wu and Hapke (2018) demonstrated that a landing rocket jet causes exactly this effect. These considerations suggest that the magnetic field appears to be relevant for extending the lifetime of the swirls by keeping their surfaces immature and bright, rather than for their actual formation.

M^3 data also provided evidence of lunar interior water. An enhanced hydration feature was observed in the central peak of Bullialdus crater and such hydration anomalies were attributed to the lunar interior origin (Klima et al. 2013). Excessive water group signature is observed in almost all large pyroclastic deposits near the equatorial regions, where the M^3 data were acquired under low phase angles and are thus more reliable, which is consistent with an endogenous origin (Milliken and Li 2017). Analysis results from Milliken and Li (2017) also suggest that high-Ti pyroclastic deposits show higher water content than low-Ti ones, indicating either different water contents in magma sources, different degassing history, or both.

5 Observations of the Water Group in the Lunar Exosphere

During the Apollo missions, several exospheric species were identified by the LACE mass spectrometer, such as He, Ar, and CH_4. The Apollo 14 Suprathermal Ion Detector Experiment (SIDE) detected ions of mass 18 amu/q during an event of approximately 14 hours duration (Freeman et al. 1973). Freeman and Hills (1991) concluded that the most probable

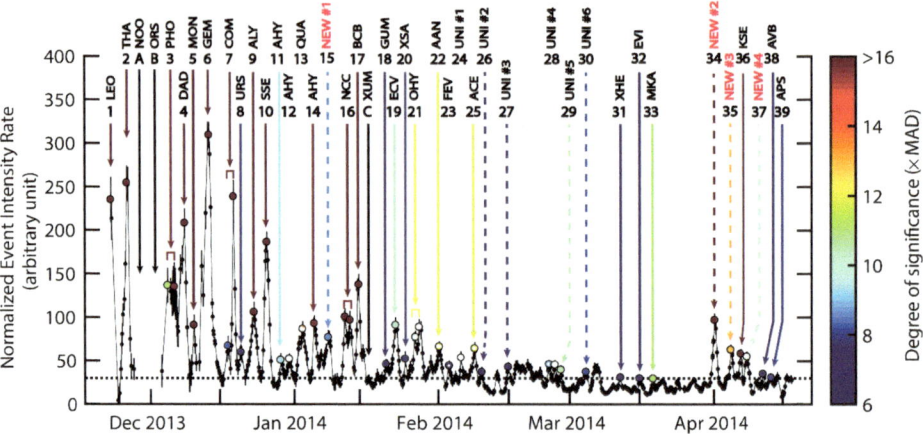

Fig. 3 Identification of sporadic water group events linked to meteoroid streams in the NMS data. The data reveals atypical peaks (data points colored based on their statistical significance) with statistical significance >6 × Median Absolute Deviation (MAD). Solid arrows denote the timing of established streams. Six dashed arrows (labeled "UNI") do not correspond to any established streams. Streams #15, 34, 35, and 37 (labeled "NEW") are believed to correspond to new streams. Stream groups A and B occurred while NMS was not operational. Established stream C was undetected. The dotted line reflects the median observed intensity. The uncertainty error bars (black) reflect 3σ due to counting statistics and data processing. Image reproduced with permission from Benna et al. (2019)

source of these water ions was the Lunar Module (Stern 1999). CHACE (Chandra's Altitudinal Composition Explorer) was a mass spectrometer on the Moon Impact Probe (MIP), which was released from Chandrayaan-1 on November 18, 2008. Based on measurements by CHACE, Sridharan et al. (2010, 2015) inferred H_2O densities on the order of 10^6 cm^{-3}, which far exceeds the upper limit measured later by LADEE, and this interpretation of the data is not widely accepted.

Over the course of its relatively short mission (8 months), the LADEE spacecraft collected measurements of the composition of the lunar exosphere using the Neutral Mass Spectrometer (NMS) instrument. While other key volatiles were rapidly identified (Benna et al. 2015a; Hodges 2016), initial data analyses did not yield direct detection of water group species. An upper limit density for H_2O of 9 cm^{-3} at 4 km altitude above the sunrise terminator was established based on the observed average instrument background (Benna et al. 2015b).

Subsequent analyses revealed that small changes in the instrument's water background captured faint variations in water released into the exosphere, well below what could be detected by direct (instantaneous) observations. These new observations revealed that water fluctuates in the lunar exosphere between a background level of 0.6 cm^{-3} and peak densities of ~40 cm^{-3}. The temporal variation of OH/H_2O is dominated by episodic and short-lived intense events (Fig. 3). These high-intensity events are predominant in the period between November and January during which the Moon encounters a series of well-known, annually occurring, strong meteoroid streams (e.g., Leonids, Geminids, and Quadrantids). Benna et al. (2019) demonstrated that these events captured the signatures of water releases into the lunar exosphere by meteoroid strikes on the lunar surface. The NMS instrument cannot resolve whether the parent exospheric molecule of the measured water is H_2O or OH. Incident hydroxyl radicals would be rapidly converted to H_2O by reaction with atomic hydrogen on the instrument internal surfaces.

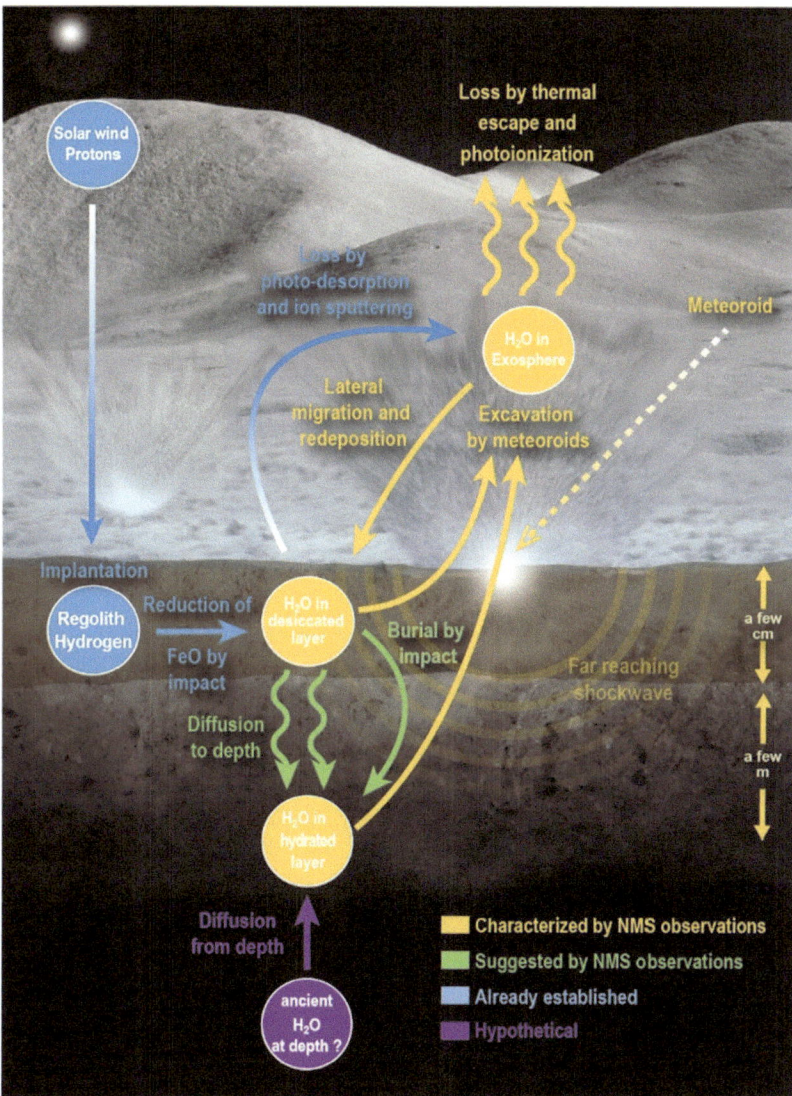

Fig. 4 The lunar water cycle as suggested by the NMS observations. Solar wind-implanted hydrogen is the main exogenous source of water production. Synthesized water is extracted by the far-reaching shock waves generated by large micrometeoroids that strike the surface. The liberated water escapes or is redeposited elsewhere. In order to sustain the water loss from meteoroid impacts, the hydrated soil might require replenishment from a deeper ancient water reservoir. Image reproduced with permission from Benna et al. (2019)

A comparison of the NMS measurements with those of the Lunar Dust EXperiment (LDEX) and an exosphere model shows that the mass flux of released OH/H$_2$O is largely generated by meteoroids in the 0.15 g – 10^6 g mass range, with smaller meteoroids having minimal contribution to the water group exosphere (Benna et al. 2019). The lack of efficiency of smaller impactors was attributed to the presence of a few-centimeter thick desiccated top layer that shielded hydrated regolith below. Only meteoroids larger than 0.15 g are able to pierce this dry stratum and excavate deeper soils. The hydration of the regolith at depth was estimated to be 220 – 520 ppm by weight of H_2O-equivalent based on the intensity of the observed water events.

While the presence of a desiccated top layer is expected based on the spectral observations of the lunar equatorial region (Li and Milliken 2017), the NMS measurements showed that its thickness extends beyond the top millimeter of the soil. The derived thickness of the desiccated regolith layer is commensurate with the depth at which diurnal equatorial temperature swings are 220 – 255 K. This implies that OH/H_2O has a short lifetime within the surface thermally active layer because it is rapidly transported by thermal desorption and diffusion upward to the surface, where it can be lost, or downward deeper in the soil where it can be sequestered.

The NMS observations also reveal that, in order to account for the relatively large intensities of the detected events, impacts must yield an amount of water 5-13 times the impactor mass. Therefore, the released water cannot originate from structural water in hydrous minerals, which requires high shock pressures in order to devolatilize as part of melts and vapors. Instead, it most likely comes from water desorbed from regolith grains by the shock wave that expands across the impact site, well beyond the confines of the volume of the excavated crater (Fig. 4).

If the regolith has a specific surface area of 0.5 $m^2\,g^{-1}$ and a density of 1.7 $g\,cm^{-3}$ at 10 cm (Vasavada et al. 2012), then the maximum amount of adsorbed water would be $\sim$150 ppm using the standard definition of a saturated monolayer (1×10^{19} molecules m^{-2}). However, the temperature of the lunar regolith at 8 cm depth still reaches 260 K over a diurnal cycle (Vasavada et al. 2012). Temperature programmed desorption measurements (Poston et al. 2015; Jones et al. 2020) have shown that regolith does not contain enough high energy binding sites to adsorb water at these temperatures. RD following a meteoroid impact into the solar wind implanted regolith can potentially explain the LADEE observations.

By aggregating all NMS observations, Benna et al. (2019) estimated that the water loss rate due to meteoroid impact is 3.4 – 8.1 $g\,s^{-1}$, which is larger than estimates of water synthesis rate by solar wind (Housley et al. 1973; Arnold 1979). The balance between solar wind production and loss by micrometeoroids would imply that the Moon is in a net water loss regime.

6 Observations of Surface Reservoirs and an Exosphere on Ceres

6.1 Ice Reservoirs

(1) Ceres is the largest of the asteroids, with diameters of $964 \times 964 \times 892$ km and a rotation period of 9.07 h. It is situated in the middle asteroid belt, 2.77 AU from the sun. Ceres has an effective temperature of 166 K and a subsolar temperature of 235 K. Its low density of 2.16 g/cm^3 reveals that water ice is a major constituent, as silicates and iron have a much higher density and the bulk porosity of such a large body could not be large. It also has hydrated minerals, consistent with extensive aqueous alteration during its early history, and spectrally it is a dark, carbonaceous asteroid (McSween et al. 2017). Due to its "roundness", presumably caused by topographic compensation during its early history of radiogenic heating, Ceres is also classified as a dwarf planet.

As an ice-rich body, Ceres has a steady supply of water that can feed an exosphere. For this reason, Ceres must have a water exosphere, the remaining question being how tenuous. With an escape velocity of 0.51 km/s, many species are gravitationally bound at thermal speeds.

Thanks to the Dawn spacecraft, which orbited Ceres 2015–2018, our knowledge of Ceres has expanded dramatically. The Gamma-Ray and Neutron Detector (GRaND, Prettyman et al. 2011) measured the hydrogen concentration within the top ~1 m of the surface (Prettyman et al. 2017). The non-icy portion of Ceres' carbon-bearing regolith contains similar amounts of hydrogen as those present in aqueously altered carbonaceous chondrites. At moderate and high latitudes, the regolith contains high concentrations of hydrogen, consistent with broad expanses of water ice. The latitudinal dependence of the hydrogen concentration is far stronger than the longitudinal dependence, suggesting that the ice content is temperature driven, with no ice within the sensing depth of GRaND at the relatively warm equatorial regions and ice very near the surface in the cold polar regions. Fanale and Salvail (1989) had long predicted an ice table at shallow and latitude-dependent depths. Remarkably, ice has survived at shallow depths over billions of years, despite continuous sublimation and impact bombardment. Based on thermal inertia measurements, Ceres is covered by a dust mantle (Rivkin et al. 2011). Small grain sizes can act as a strong barrier to vapor diffusion, and quantitative estimates place the grain size at one micron to be consistent with the low rate of vapor diffusion (Prettyman et al. 2017; Li et al. 2019). The GRaND measurements, combined with models, suggest that the equatorial regions are ice-free to at least 1 m depth whereas the polar regions may have ice even at cm depth (Schorghofer 2016; Prettyman et al. 2017; Landis et al. 2017, 2019).

The shallow depth of ice suggests that Ceres has lost ice to space only slowly, and on a long-term average the supply to the exosphere is therefore small. If loss is by sublimation only, then the ice table corresponds to a current outgassing rate of only 0.003 kg/s, but excavation of deeper ice by impacts and ice delivered by impactors also contribute to the water exosphere.

In stark contrast to the shallow subsurface, very little ice is exposed on the surface. The large relatively bright spots seen on the surface of Ceres consist mostly of salts. Water ice was detected spectroscopically by Dawn only at a handful of locations (Combe et al. 2016, 2019; Sizemore et al. 2019), and these occurrences are at fresh craters near rim shadows and sometimes associated with landslides. Landis et al. (2019) estimated the number of ice-exposing impacts, and found that impact craters that remain bright (exposed to the surface) could supply 0.08-0.56 kg/s of water vapor on a time average.

With an axis tilt of only 4° relative to the normal of the orbital plane, Ceres has perennially shadowed craters in its polar regions (Hayne and Aharonson 2015, Schorghofer et al. 2016). Within some of these PSRs, bright deposits have been identified that most likely consist of water ice (Platz et al. 2016). This topic will be further reviewed in Sect. 8.

6.2 Observations of an Exosphere (or Lack Thereof)

Numerous attempts have been made to detect an exosphere around Ceres with earth-based and space telescopes (Table 2). Most of these observations resulted in no detection, but the upper bounds that could be obtained were often very high. Two sets of telescopic observations resulted in positive detections. A'Hearn and Feldman (1992) observed escaping OH. If this OH formed by photodissociation of H_2O, the H_2O production rate is roughly 10^{26} s^{-1}. With the Herschel Space Observatory, Küppers et al. (2014) detected H_2O three out of four times. Combined with modeling they estimate the outgassing rate to be 2×10^{26} s^{-1} (6 kg/s). The last observation listed in Table 2 was carried out when Ceres was near perihelion, when any sublimation driven activity can be expected to be enhanced, but no gaseous OH was detected within the sensitivity of the measurement.

The Dawn spacecraft found no direct evidence for an exosphere around Ceres, nor did it carry an instrument designed for that purpose. Reports of a haze (Nathues et al. 2015;

Table 2 List of searches for an atmosphere around Ceres. See text for further description

Authors	Instrument	Species	Abundance (s^{-1})
A'Hearn and Feldman (1992)	Int. UV Explorer	OH	$[H_2O] \approx 10^{26}$
Rousselot et al. (2011)	Very Large Telescope	OH	$[OH] < 7 \times 10^{25}$
Küppers et al. (2014)	Herschel Space Observatory	H_2O	$[H_2O] \approx 10^{26}$
Roth et al. (2016)	Hubble Space Telescope	O	$[H_2O] < 4 \times 10^{26}$
Jia et al. (2017)	Dawn GRaND	(electrons)	$[H_2O] \approx 6 \times 10^{25}$ (?)
McKay et al. (2017)	Apache Point Observatory	O	$[H_2O] < 5 \times 10^{28}$
Roth (2018)	Hubble Space Telescope	H, O, S	$[H_2O] < 4 \times 10^{26}$
Rousselot et al. (2019)	Very Large Telescope	OH	$[H_2O] < 2 \times 10^{26}$

Thangjam et al. 2016) were quickly attributed to measurement uncertainties (Schröder et al. 2017). There were however mysterious electron bursts on three consecutive orbits (Russell et al. 2016). One of several possible explanations for such events would be a bow shock from the interaction of a solar energetic particle (SEP) event with an atmosphere. Jia et al. (2017) modeled such a situation, finding that about 6×10^{25} s^{-1} (2 kg/s) water vapor production would be required to form such a shock.

Villarreal et al. (2017) found a correlation between detections of an exosphere and the inferred presence of solar energetic protons (SEP) at Ceres. Overall, the observations suggest that a transient exosphere, or even atmosphere, may appear sporadically on Ceres, but the intensity, frequency, and cause of such events are uncertain.

6.3 Exosphere Models of Ceres

On Ceres, the thermal speed of water molecules is comparable to the escape speed, and there is significant gravitational fallback. An exosphere or atmosphere around Ceres has a gravitational scale height (a few hundred km) much smaller than the Hill radius. This situation contrasts with comets, where molecules radially stream away from the body. The Cerean exosphere has been modeled with several different types of numerical models, often with disparate results.

Tu et al. (2014) and Schorghofer et al. (2016, 2017a) modelled the exosphere as an ensemble of ballistic trajectories, assuming the molecules thermally accommodate when in contact with the surface. This yields a half-life of the water exosphere of about 7 hr. On average, a water molecule undergoes 3 ballistic hops before it is lost by gravitational escape. These authors also quantified the delivery of water molecules to PSRs, demonstrating that a sizable fraction of outgassed water will end up in cold traps.

In connection with the Herschel Space Observatory observations, Küppers et al. (2014) simulated a collisional atmosphere with a Direct Simulation Monte Carlo (DSMC) model, which resulted in an outgassing rate of 6 kg/s. Formisano et al. (2016) used a comet Single Particle Hydrodynamics model for a vapor-dust mixture, and predicted that an optically thin atmosphere could be maintained for tens of days after an outgassing event.

Fanale and Salvail (1989) and Hayne and Aharonson (2015) explored the possibility that Ceres might have polar water ice caps, and concluded that, with realistic parameters, no polar ice caps are to be expected, consistent with the subsequent observations by the Dawn spacecraft. However, Ceres may have optically thin seasonal caps. Molecular residence times at polar temperatures are long enough to allow the seasonal buildup of water

Fig. 5 Ballistic hops of water molecules according to model calculations. On Mercury and the Moon, water molecules are eventually lost due to photo-destruction; on Ceres most are lost through gravitational escape. Image reproduced with permission from Schorghofer et al. (2017a)

Mercury  The Moon  Ceres

molecules delivered by the exosphere even beyond polar winter (Schorghofer et al. 2017a). This seasonal water reservoir is, at this point, purely a theoretical prediction.

7 Lateral Transport in the Water Exosphere

The lateral transport of water molecules on nominally airless bodies is hypothesized to occur through ballistic hops, punctuated by encounters with the planetary surface (e.g., Watson et al. 1961a,b; Butler 1997).

How water group molecules interact with planetary regoliths remains a subject of active investigation, and a source of significant uncertainty in numerical simulations of water transport. Molecules encountering a planetary surface may be scattered or adsorbed to the surface. Adsorption can take several forms: molecules may be held by van der Waals forces (physisorption), chemically bound (chemisorption) or break apart (dissociative adsorption). The residence time of a molecule on the surface is sensitive to both surface temperature and composition. Laboratory results (e.g., Poston et al. 2015) indicate that surface residence times of water molecules can be characterized by a distribution of desorption activation energies, but in situ activation energies remain to be definitively determined (e.g., Hendrix et al. 2019).

Most volatile transport models assume that molecules fully accommodate to the local surface temperature such that upon desorption, molecular velocities may be drawn from a distribution characteristic of that temperature. Two different velocity distributions for molecules released from a surface commonly appear in the literature: (i) the Maxwell-Boltzmann distribution of velocities for a gas in thermal equilibrium, and (ii) the Maxwell-Boltzmann flux distribution for velocities of molecules crossing a surface (e.g., Brinkmann 1970). From a volatile transport perspective, both approaches appear to yield very similar results (Butler 1997; Schorghofer et al. 2017a). Figure 5 illustrates the repeated ballistic hops of water molecules on Mercury, the Moon, and Ceres according to these models. The average hop distances and hop durations of H_2O molecules launched on the dayside at thermal speeds are about 150 km (4 minutes) on Mercury, 200 km (7 minutes) on the Moon (Schorghofer 2015), and on the order of the body diameter (one hour) on Ceres.

The primary loss mechanisms that act on a migrating water vapor exosphere are photodestruction and gravitational escape. On the Moon and Mercury, photolysis is the dominant loss process. Solar ultraviolet radiation can break apart water molecules into a variety of neutral and charged species, with the primary reaction being dissociation of H_2O into H and OH (Huebner and Mukherjee 2015). The excess energy of dissociation imparts mean velocities of ~ 1 km/s and ~ 18 km/s to OH and H, respectively (Crovisier 1989).

Barring losses to photolysis or surface processes, lateral migration of water molecules continues until a molecule reaches cold traps where surface temperature is sufficiently low that water ice may remain stable over geological timescales, unless otherwise mobilized by impact vaporization or plasma sputtering (Farrell et al. 2019). The fraction of water molecules delivered to a planetary surface that are ultimately cold-trapped depends on the size (gravity) and proximity of the body to the Sun, and the areal coverage of cold traps. On the Moon ∼10% of all water delivered to the surface by various sources is predicted to reach polar cold traps through lateral transport in the exosphere (e.g., Schorghofer 2014). When a comet impact creates a dense atmosphere, between ∼20% to nearly 100% of impact-delivered water that remains gravitationally bound is predicted to reach cold traps at the poles (Berezhnoi and Klumov 1998; Stewart et al. 2011; Prem et al. 2015). In contrast, Ceres' low gravity and Mercury's proximity to the Sun are anticipated to result in less efficient exospheric transport of water to cold traps (Butler 1997; Schorghofer et al. 2017a).

The structure of a water vapor exosphere on the Moon may initially reflect the nature of an episodic source (e.g., Goldstein et al. 2001), but soon approaches a quasi-steady state, characterized by an enhancement of exospheric and surface water at the dawn terminator. On many airless bodies, but particularly the Moon and Mercury, surface temperature rises rapidly at sunrise, causing adsorbed water molecules to desorb. Some of this newly mobilized water falls back to the night side of the dawn terminator while the remainder moves towards the day side, causing the surficial and exospheric enhancements that are seen in model calculations (Schorghofer et al. 2017b; Prem et al. 2018).

It is important to note that steady-state exospheres of molecular water have not yet been observed to exist on the Moon, Mercury, or Ceres, and the behavior described above is currently a model-based prediction rather than an observational fact. (The detection by LADEE could be due to OH or H_2O; those detected on Ceres are sporadic outbursts, see Sects. 5 and 6). Model results also suggest that diurnal variations in surface concentration and exospheric density may be more pronounced at higher latitudes (Sect. 4). Besides diurnal variability, seasonal trends in the lateral transport of water vapor on the Moon have also been recently examined; model results indicate that the cold traps in the northern hemisphere may accumulate more water per unit area than those in the southern hemisphere (Kloos et al. 2019).

As water is photodestroyed and cold-trapped, the mass of a water vapor exosphere decays over time. The photo-dissociation timescale 1 au from the sun is 22 hours at normal solar activity (Crovisier 1989). The decay of the total mass of the exosphere is slower than that due to hiding of water molecules on the night side (or even in topographic shadows of the daytime polar regions). This introduces the length of the solar day as a relevant time scale (Berezhnoy et al. 2003; Prem et al. 2018; Schorghofer et al. 2017b).

Exosphere models have addressed whether collisionless transport can distribute water unevenly between relatively nearby cold traps. Moores (2016) predicted that cold traps farther from the pole accumulate far more water per unit area, while others find the variations to be small (Schorghofer 2014; Prem et al. 2018). An updated version of Moores' model (Kloos et al. 2019) also predicts nearly uniform infall in the polar region, leading to a consensus among models. Cold traps are too small to have any "rain shadow" effect. The average hop distance at lunar dayside temperatures is about 200 km; for comparison the diameter of Shoemaker Crater is 51 km.

Although exospheric models agree in many respects, there are some current points of disagreement. One question that remains to be definitively addressed is the nature of the interaction between water molecules and pristine and radiation damaged grain surfaces. Watson et al. (1961a,b) envisioned that water molecules thermally accommodate when in contact with the surface. Hodges (1991, 2002) argues that water molecules will be chemisorbed instead, preventing them from leaving the surface until released by a more energetic event, such as sputtering or meteoroid impacts. That would effectively shut down lateral transport by the exosphere. Long-term exospheric monitoring by orbital and surface instrumentation could contribute to addressing this question (e.g., Prem et al. 2020). Another partially unresolved question is the significance of isotopic fractionation during the transport of water. Current models for the Moon agree that D/H fractionation during transport is likely to be small, but differ on whether cold traps may be enriched (Crider and Vondrak 2000) or depleted (Schorghofer 2014) in deuterated water. Our understanding of these aspects of volatile transport will likely advance as models continue to leverage experimental and observational data.

Water molecules can also migrate into the porous subsurface, by hopping between grain surfaces. In this sense, the exosphere extends into the subsurface. This process has not been fully explored with models, but under some temperature conditions, subsurface diffusion can lead to the sequestration of ice outside of cold traps (Schorghofer and Taylor 2007; Schorghofer and Aharonson 2014).

8 Cold Traps and Permanently Shadowed Regions

8.1 Temperature and Volatile Stability

Cold traps are special regions on airless bodies where volatiles such as water ice can be stable (Urey 1952; Watson et al. 1961a,b; Arnold 1979). Many cold traps are transient, such as the night side surfaces of slow-rotating airless bodies that are cold enough only for part of the diurnal cycle before temperatures rise and volatiles mobilize. On the other end of the spectrum, some cold traps exist for billions of years in permanent shadow and contain temperatures low enough to trap water ice on these geologic time scales. Here we discuss cold traps capable of sequestering volatiles on Mercury, the Moon, and Ceres.

Low obliquity of the spin axis relative to the ecliptic is a critical component of producing stable cold traps. Without low obliquity, the spin axis will periodically orient in the direction of the sun, warm, and devolatilize cold traps.

Sunlit surfaces on Mercury and the Moon are too warm for water ice to be stable. On these bodies there can be no "ice caps." Therefore, topographic relief, often in the form of impact craters, is the critical second component for stable cold traps. Even shallow craters provide sufficient relief to create regions where the sun does not rise above the local horizon (Vasavada et al. 1999). Scattered visible light and emitted infrared radiation from illuminated surfaces near cold traps contribute to the surface energy balance in shadowed regions. For a bowl-shaped crater the equilibrium temperature of the shadowed region can be calculated analytically (Buhl et al. 1968; Ingersoll et al. 1992); the lower the depth-to-diameter ratio, the colder the shadowed surface. Diviner on LRO has mapped the temperatures on the lunar surface in detail. Figure 6 shows maximum surface temperatures in the south polar region near summer and winter solstice.

Efforts to evaluate the sublimation rates of volatiles under vacuum (e.g., Watson et al. 1961b; Vasavada et al. 1999; Zhang and Paige 2009) differ in detail but agree that at temperatures below about 110 K, water ice will sublimate slowly enough to be stable on geologic timescales. At 110 K, the sublimation rate is about 10 cm/Gyr. Some volatiles, such as Hg and S are less volatile than water ice and thus can be cold trapped at higher temperatures (Watson et al. 1961b). Other volatiles, such as CO_2 and NH_3, are more volatile than water ice ("supervolatiles") and require cold traps with temperatures below 54 K and 66 K respectively (Zhang and Paige 2009; Fig. 7). A loss rate of 1 m/Gyr corresponds to 1 nm (about 3

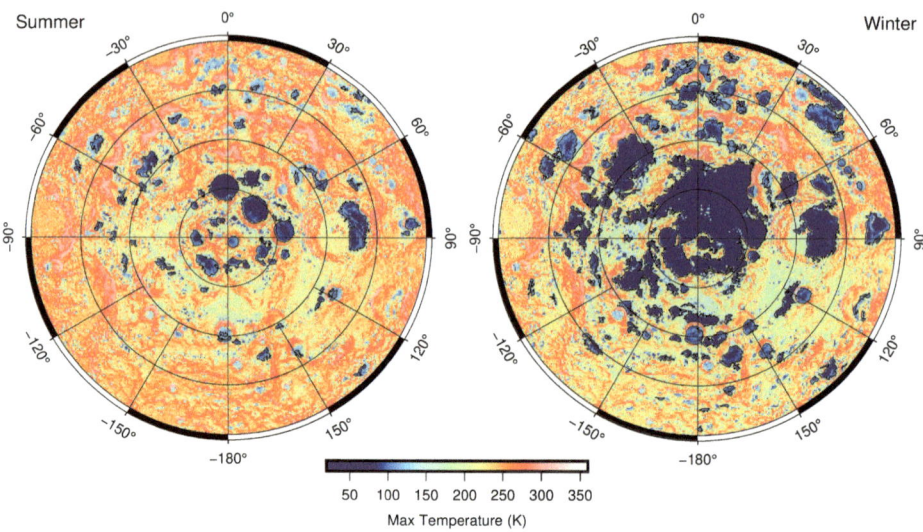

Fig. 6 Summer and winter surface temperatures in the south polar region of the Moon based on Diviner temperature measurements. The maximum is taken with respect to local time, and the data were divided into six seasonal bins for the draconic year, as in Schorghofer and Williams (2020). Shown is the region poleward of 80°S latitude, with black contours for 110 K, the approximate threshold value for cold traps

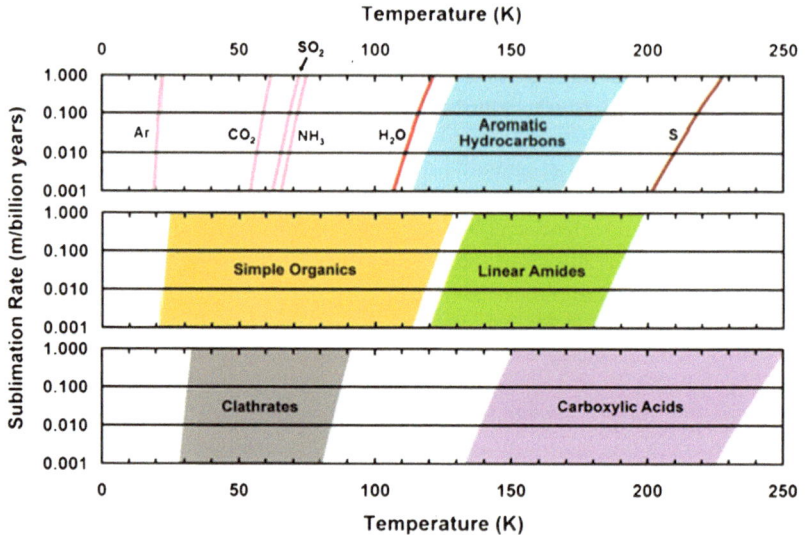

Fig. 7 Vacuum sublimation rates as a function of temperature for various organic and inorganic compounds. Except for sulfur, the volatility of common inorganic compounds is lower than that of water. Image reproduced with permission from Zhang and Paige (2009)

monolayers of H_2O) per year. Hence, water molecules might still move around even though the macroscopic loss rate is negligible.

Ice exposed on the surface is lost not only by sublimation, but also by Lyman-α (UV) radiation from the very local interstellar medium (Morgan and Shemansky 1991) and by sputtering from solar wind directly or from the tail of Earth's magnetosphere (Arnold

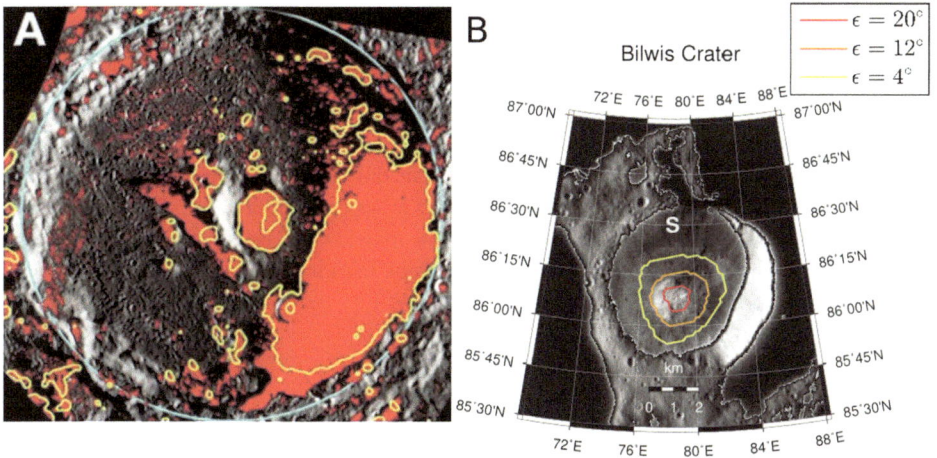

Fig. 8 Ice deposits within PSRs on Mercury and Ceres. **A)** High-reflectance surface within Prokofiev crater on Mercury; the radar-bright region (yellow contour) is located within a PSR (red). Image adapted from Chabot et al. (2014); copyright by GSA. **B)** Bilwis crater on Ceres with a bright crater floor deposit. Colored contours are boundaries of PSRs at various values of the axis tilt ε. No comparable relation between ice deposits and PSRs has yet been identified on the Moon. Image adapted from Ermakov et al. (2017), copyright by AGU

1979; Lanzerotti et al. 1981). Surface ice is further eroded by meteoric impact vaporization and meteoric impact ejection (Farrell et al. 2019), although impact ejecta can also protect volatiles from further sublimation. The surfaces of airless bodies experience a statistical turnover due to meteor impacts (Gault et al. 1974; Arnold 1975; Morris 1978; Crider and Vondrak 2003; Costello et al. 2020).

8.2 Mercury

The axis tilt of Mercury is currently only 2 arc minutes and has remained near zero for at least the past 3.5 Gyr, so it has been anticipated that Mercury has permanently shadowed regions that may have trapped water ice (Thomas 1974). Modeling predicted the surface temperatures in these hypothetical shadowed regions would be low enough for water ice to be stable (Paige et al. 1992; Vasavada et al. 1999). Radar observations revealed highly reflective regions near Mercury's north and south pole with a high circular polarization ratio (Slade et al. 1992; Harmon and Slade 1992; Butler and Muhleman 1993; Harmon et al. 1994). Analysis of these radar data suggest water ice deposits which are very pure and at least several meters thick. Radar observations alone are not a definitive indicator of water ice; for instance, Starukhina (2001, 2012) suggested the dielectric properties of silicates are uncertain at low temperatures and may be an alternative explanation of the radar observations ("cold silicate hypothesis"). However, observations by the MESSENGER spacecraft support the presence of water ice.

The MESSENGER spacecraft found enhanced hydrogen concentrations in the polar region from neutron measurements (Neumann et al. 2013) and optical evidence for water ice on the surface in the PSRs (Chabot et al. 2014, 2018a). The areas of high radar backscatter are strongly correlated with areas that are in permanent shadow (Fig. 8A) and with the predicted abundance of thermally stable water ice (Paige et al. 2013; Chabot et al. 2018b). According to MESSENGER near-infrared reflectance measurements of the surface, the coldest areas are bright, whereas areas with higher predicted temperatures, where water ice is expected to be short-lived on the very surface but long-lived at shallow depth, are dark (Paige et al. 2013). The dark surface deposits may be devolatilized sublimation lag deposits (Neumann et al. 2013; Paige et al. 2013; Hamill et al. 2020) or due to radiation processing of various cold-trapped species (Crites et al. 2013; Delitsky et al. 2017). Mercury's polar deposits have sharp boundaries, which suggests that the deposits are geologically young or that their boundaries are refreshed by an ongoing process (Chabot et al. 2018a). Radar data suggest that the ice deposits must be at least a few meters thick (several radar wavelengths), and they are likely no more than tens of meters thick (Susorney et al. 2019).

8.3 The Moon

PSRs can be mapped by illumination modeling (ray-tracing) with digital elevation maps, and have been identified down to a latitude of $\sim 60°$ (McGovern et al. 2013). PSRs poleward of $80°$ cover 1.6×10^4 km^2 in the south polar region and 1.3×10^4 km^2 in the north polar region (Mazarico et al. 2011; McGovern et al. 2013). Vice versa, some areas are sunlit up to 89% of time (Noda et al. 2008) and known as "peaks of eternal light", although the direct solar illumination is not perfectly continuous.

Cold traps, which lie within PSRs, have been mapped based on surface temperatures measured by the Diviner instrument on LRO (Paige et al. 2010a,b; Hayne et al. 2015; Williams et al. 2019; Schorghofer and Williams 2020; Fig. 6). Near the poles, the Moon's 1.54° obliquity induces significant seasonal changes in surface temperatures, where seasons are defined by the 347 day draconic year. Summer cold traps, which are truly permanently shadowed, cover 1.3×10^4 km^2 poleward of 80°S and 0.53×10^4 km^2 poleward of 80°N when defined by a peak temperature of 110 K (Williams et al. 2019). This amounts to 0.05% of the global surface area. In winter, these areas increase by factors of 2.8 and 4.3, respectively (Fig. 6). The temperatures of near-pole crater Faustini is ~ 30-40 K in winter and ~ 45-70 K in summer. While these temperatures are well below the threshold for water ice stability, they have significant effects on more volatile molecules, such as CO_2. The lowest temperatures measured by Diviner are about 20 K, and may be limited by the regional heat flow from the lunar mantle. On the other extreme, non-polar permanently shadowed regions at mid-latitudes have significant scattered light and can be as warm as ~ 216 K with average temperatures of ~ 134 K.

Radar observations did not reveal signs of pure near-surface ice deposits as they did on Mercury (Nozette et al. 1996; Stacy et al. 1997; Simpson and Tyler 1999; Campbell et al. 2003), but neutron spectroscopy found the hydrogen concentration is enhanced in the polar regions (Feldman et al. 1998, 2000, 2001; Mitrofanov et al. 2010). The LCROSS impact into Cabeus crater released a variety of volatiles (including H_2O, H_2S, NH_3, SO_2, and C_2H_4) and therefore provided direct evidence for these species in lunar cold traps (Colaprete et al. 2010). The concentration of water ice at the impact site was an estimated 5.6±2.9% by mass. There is also evidence for water ice beyond Cabeus crater, based on neutron counts (Feldman et al. 1998, 2000, 2001; Mitrofanov et al. 2010), ultraviolet spectroscopy (Gladstone et al. 2012; Hayne et al. 2015), near infrared spectroscopy (Fisher et al. 2017; Li et al. 2018), and depth-to-diameter ratios of craters (Rubanenko et al. 2019). The locations identified by these individual studies are sometimes inconsistent and contain some identifications of water ice beyond known cold traps in apparently illuminated terrain; ambiguities remain in the interpretation of these datasets.

8.4 Ceres

The axis tilt of Ceres is small, and Hayne and Aharonson (2015) predicted the presence of PSRs before Dawn's arrival. Dawn determined the axis tilt to be 4.03°, and the PSRs of the northern polar region were mapped in two ways. Stereo imaging, more specifically stereo photogrammetry, was used to construct topographic shape models, and these were used to calculate the extent of shadows. In this way, the northern PSRs were mapped and found to cover 1,800 km² or 0.13% of the hemisphere (Schorghofer et al. 2016). Another approach was to stack images from various local times acquired close to the northern summer solstice. This method, of higher spatial resolution, identified hundreds of PSRs totalling 2,200 km² of area. Due to illumination conditions, no comparable analyses could be carried out for the southern polar region, but statistically the topographies of the polar regions are similar.

Some scattered light is available even within shadowed regions, and searches were conducted for ice deposits within PSRs. Bright deposits were discovered in a small number of PSRs (Platz et al. 2016; Ermakov et al. 2017). The largest bright deposit is within Bilwis crater (Fig. 8B). In Zatik crater, bright material extends into sunlight, where it was spectroscopically identified as H_2O ice (Platz et al. 2016). Observations by Dawn also led to an accurate measurement of the moments of inertia of Ceres, which made it possible to reliably backward integrate the obliquity history. The obliquity (axis tilt) of Ceres varies between 2° and 20° with a period of 24.5 kyr (Bills and Scott 2017; Ermakov et al. 2017). These obliquity oscillations may also explain why only a small fraction of PSRs has bright deposits. At maximum axis tilt only a handful of PSRs remain (Ermakov et al. 2017).

8.5 Micro Cold Traps

While orbital measurements have succeeded in predicting and observing cold traps on scales of 100 s of meters to 10 s of kilometers, the prevalence of cold traps on smaller scales has to be inferred statistically. At finer resolution the area of PSRs increases, because the shadow of a large hill is still a contiguous shadow when viewed at finer resolution and shadows of smaller hills are added (Petrov et al. 2003; Gläser et al. 2014). Hayne et al. (2021) found that 10% of the PSR area on the Moon is in patches smaller than 100 m in diameter. The regolith on airless bodies is sufficiently insulating to at least cm-scales, assuming typical roughness values, to maintain temperature gradients of more than 100 K (Bandfield et al. 2015). In theory, this should create heterogeneous terrains of illuminated surfaces and "micro" cold traps extending to vast regions beyond the larger "macro" cold traps (Rubanenko et al. 2018; Hayne et al. 2021). The effects of regolith physical properties on micro cold traps and the subsurface temperatures are areas for future study.

9 Water Cycles

In lieu of a summary, we consider the water cycle from source to sink, or rather semi-cycle, since the water molecules are not recycled.

9.1 Sources

Molecular water can originate from exogenic and endogenic sources. Exogenic sources are comets and hydrated meteoroids. In the long-term, the mass flux of the impactor population is dominated by the largest objects, so there is a probabilistic factor in the amount delivered.

For example, Mercury's massive ice deposits could have resulted from a recent large comet impact (Moses et al. 1999; Ernst et al. 2020; Chabot et al. 2018a), which would explain the apparent difference in the abundance of cold trapped ice compared to the Moon.

In solar-wind generated water, the hydrogen is exogenic and the oxygen endogenic. The solar wind contains almost no deuterium (Stephant and Robert 2014), and an isotopic measurement (D/H) of ice deposits could therefore constrain its origin. Formation and retention of OH, H_2, and H_2O are ongoing surficial processes (Sects. 2-4). Solar-wind implantation of hydrogen induces the formation of hydroxyls. Hydroxyls can then combine to form H_2O. This recombinative desorption requires high temperatures, such as those reached on the dayside of Mercury, or heating by impacts, which occurs on all large airless bodies. Many questions remain. On the Moon, only relatively large meteoroids release water group species, whereas dust-sized impactors do not, suggesting a desiccated layer is present (Sect. 5). The amplitudes of diurnal and latitude variations of the surficial hydroxyl concentration are insufficiently understood, as results of various spectroscopic analyses are inconsistent with one another and with some of the theoretical kinetics derived from laboratory measurements.

Endogenic outgassing of primordial water is more likely in the distant past than in recent history (Needham and Kring 2017; Deutsch et al. 2019; Head et al. 2020). At this point, the major source of molecular water has not been determined on either Mercury or the Moon. In the case of Ceres (Sect. 6), its own ice-rich crust is presumably the major source of its cold-trapped water. Even dry impactors can release water vapor on Ceres and even without impactors, indigenous water molecules slowly diffuse through the overlying ice-free surface layer. Further in the past, cryovolcanoes may have brought ice from greater depths to the surface of Ceres (Ruesch et al. 2016).

9.2 Exospheric Transport and Sinks

On large airless bodies, water molecules undergo ballistic hops rather than escaping to space (Sect. 7). Through a sequence of such hops, water may be transported from any location on the surface to cold traps. The nature of the interaction of returning water molecules upon contact with the pristine surface is uncertain. In one extreme the molecules thermally accommodate and leave the surface again, in the other they are chemically adsorbed. The massive ice deposits in the cold traps of Mercury must have arrived there somehow, but that ice could have resulted from a rare recent comet impact rather than been delivered steadily through exospheric transport. Observations of water group exospheres are rare. Nevertheless, a thermally accommodated water exosphere remains the standard theory awaiting confirmation or disproof.

Water molecules are lost due to gravitational escape, photo-destruction, cold-trapping, and possibly by dissociative chemisorption. On Mercury, photo-destruction is thought to be the major loss process, due to its proximity to the sun and high gravity. On Ceres, gravitational escape dominates. From that perspective, the Moon is expected to transport water more efficiently to the polar cold traps than these two other planetary bodies, yet the Moon appears to have a lower concentration of cold-trapped water ice than the other two bodies.

At low temperature, volatiles can accumulate over time even when exposed to vacuum. Permanently shadowed regions (PSRs) are closely related to cold traps, although they never perfectly coincide (Sect. 8). Thick ice deposits are found in the Hermean cold traps, often covered by a dark lag deposit. On the Moon, very little ice has been detected on the very surface of cold traps, but the LCROSS impact experiment and two sets of neutron spectroscopy measurements (one by the Lunar Prospector and the other by LRO) have provided clear evidence for a bulk-enrichment of hydrogen in the polar regions. On Ceres, the spin axis tilt varies periodically, so the present-day PSRs and cold traps are larger than the truly permanent PSRs and cold traps. Bright crater floor deposits are seen in several of the truly permanent PSRs. Cold-trapped water may subsequently be destroyed by space weathering (Farrell et al. 2019).

9.3 Outlook

The study of the Hermean, lunar, and Cerean water cycles has only just begun. Theories have been in place for decades, but they are far from established and some might not survive at all. Whereas interactions of solar wind protons with the surface have long been discussed, the quantitative aspects (such as mineral-dependent diffusivities of hydrogenated molecules and the corresponding activation energies, and the efficiency of H_2 and H_2O formation) are only slowly becoming clear. There were also observational surprises, such as the existence of a latitude-dependent hydroxyl population on the Moon and the fact that water group species are released during meteoroid showers, in excess of the water contained in the impacting meteoroids. Lateral transport in the gravitationally-bound surface-bounded water exospheres of the large airless bodies remains a theoretical concept, as there is insufficient observational data to put these models to a test. Observations that distinguish between surficial OH and H_2O and exospheric water group neutrals and ions would provide important constraints to decipher the driving physics. Cold trapping of water ice is expected to occur on all three bodies, but the question is confounded by significant discrepancies among proxy measurements of the water content in the lunar polar regions.

This last topic will surely benefit from in-situ measurements by upcoming landed missions to the polar regions of the Moon. The Polar Resources Ice Mining Experiment-1 (PRIME-1), MoonRanger and NASA's Volatiles Investigating Polar Exploration Rover (VIPER) will explore the south polar region of the Moon in 2022 and 2023. Understanding the distribution of water ice at the lunar poles is of major international interest; further planned missions to the south pole include the Lunar Polar Exploration Mission (LUPEX, a collaboration between the Japanese and Indian space agencies), Chang'e 6 and 7 (by the Chinese National Space Administration), Luna 25 (by Russia's Roscosmos), and Luna 27 (a collaboration between Roscosmos and the European Space Agency). Commercially operated lunar landers currently under development also plan to carry science payloads that will characterize the lunar exosphere and surface thermal environment at and beyond the poles. Several innovative orbital instruments will also search for volatiles. Observations of the lunar surface (over an extended infrared spectra range and new radar wavelengths) and exosphere by the Chandrayaan 2 orbiter are underway since 2019. A group of thirteen CubeSats, including Lunar Flashlight, Lunar IceCube, and LunaH-Map, is expected to be launched in 2021. ShadowCam on the Korea Pathfinder Lunar Orbiter, launching in 2022, is designed to image shadowed regions using scattered light. Lunar Trailblazer will carry a short-wave infrared imaging spectrometer that is an improvement over M^3. With or without these new missions, laboratory experiments using simulants and returned samples along with comprehensive and capable theoretical models are important tools for developing an understanding of the processes that govern the presence of water on the surfaces and in the exospheres of the large airless bodies of the inner solar system.

Acknowledgements The authors are grateful for the hospitality of the International Space Science Institute (ISSI) during the Workshop on "Surface Bounded Exospheres and Interactions in the Solar System" that took place January 20-24, 2020 in Bern, Switzerland. OT gratefully acknowledges the scientific contributions of NASA Heliophysicist Richard Hartle who passed away on 19 February 2019 at the age of 82. AB was partially supported by Russian Science Foundation grant no. 20-12-00105. BTG appreciated SSERVI support through NNA14AB02A (VORTICES). BJ and TO acknowledge support through SSERVI REVEALS. NS was supported by SSERVI cooperative agreement NNH16ZDA001N (TREX). PP acknowledges support from SSERVI through the LEADER and ICE Five-O teams. OT was supported in part by SSERVI LEADER. CW was partially supported by RFBR-DFG grant no. WO 1800/7-1.

References

M.F. A'Hearn, P.D. Feldman, Water vaporization on Ceres. Icarus **98**, 54–60 (1992). https://doi.org/10.1016/0019-1035(92)90206-M

J.R. Arnold, Monte Carlo simulation of turnover processes in the lunar regolith. Proc. Lunar Planet. Sci. Conf. **6**, 2375–2395 (1975)

J.R. Arnold, Ice in the lunar polar regions. J. Geophys. Res. **84**(B10), 5659 (1979). https://doi.org/10.1029/JB084iB10p05659

J.L. Bandfield, P.O. Hayne, J.-P. Williams, B.T. Greenhagen, D.A. Paige, Lunar surface roughness derived from LRO Diviner Radiometer observations. Icarus **248**, 357–372 (2015). https://doi.org/10.1016/j.icarus.2014.11.009

J.L. Bandfield, M.J. Poston, R.L. Klima, C.S. Edwards, Widespread distribution of OH/H_2O on the lunar surface inferred from spectral data. Nat. Geosci. **11**, 173–177 (2018). https://doi.org/10.1038/s41561-018-0065-0

A.T. Basilevsky, A.M. Abdrakhimov, V.A. Dorofeeva, Water and other volatiles on the Moon: a review. Sol. Syst. Res. **46**(2), 89–107 (2012). https://doi.org/10.1134/S0038094612010017

M. Benna, P.R. Mahaffy, J.S. Halekas, R.C. Elphic, G.T. Delory, Variability of helium, neon, and argon in the lunar exosphere as observed by the LADEE. Geophys. Res. Lett. **42**, 3723–3729 (2015a). https://doi.org/10.1002/2015GL064120

M. Benna, D.M. Hurley, T.J. Stubbs, P.R. Mahaffy, R.C. Elphic, Observations of meteoroidal water in the lunar exosphere by the LADEE NMS instrument, in *Annual Meeting of the Lunar Exploration Analysis Group*, Columbia, MD (2015b)

M. Benna, D.M. Hurley, T.J. Stubbs, P.R. Mahaffy, R.C. Elphic, Lunar soil hydration constrained by exospheric water liberated by meteoroid impacts. Nat. Geosci. **12**, 333–338 (2019). https://doi.org/10.1038/s41561-019-0345-3

A.A. Berezhnoi, B.A. Klumov, Lunar ice: can its origin be determined? JETP Lett. **68**, 163–167 (1998). https://doi.org/10.1134/1.567840

A.A. Berezhnoy, N. Hasebe, T. Hiramoto, B.A. Klumov, Possibility of the presence of S, SO_2, and CO_2 at the poles of the Moon. Publ. Astron. Soc. Jpn. **55**, 859–870 (2003). https://doi.org/10.1093/pasj/55.4.859

B.G. Bills, B.R. Scott, Secular obliquity variations of Ceres and Pallas. Icarus **284**, 59–69 (2017). https://doi.org/10.1016/j.icarus.2016.10.024

G. Blanford, P. Borgesen, M. Maurette, W. Moller, B. Monart, Hydrogen and water desorption on the Moon: approximate on-line simulations, in *Lunar Bases and Space Activities of the 21st Century*, ed. by W.W. Mendell (Lunar and Planetary Institute, Houston, 1985), pp. 603–610

J.W. Boyce, Y. Liu, G.R. Rossman, Y. Guan, J.M. Eiler, E.M. Stolper, L.A. Taylor, Lunar apatite with terrestrial volatile abundances. Nature **466**(7305), 466–469 (2010). https://doi.org/10.1038/nature09274

J.P. Bradley, H.A. Ishii, J.J. Gillis-Davis et al., Detection of solar wind-produced water in irradiated rims on silicate minerals. Proc. Natl. Acad. Sci. **111**(5), 1732–1735 (2014). https://doi.org/10.1073/pnas.1320115111

R.T. Brinkmann, Departures from Jeans' escape rate for H and He in the Earth's atmosphere. Planet. Space Sci. **18**, 449–478 (1970). https://doi.org/10.1016/0032-0633(70)90124-8

D. Buhl, W.J. Welch, D.G. Rea, Reradiation and thermal emission from illuminated craters on the lunar surface. J. Geophys. Res. **73**(16), 5281–5295 (1968). https://doi.org/10.1029/JB073i016p05281

D. Burke, C. Dukes, J.-H. Kim, J. Shi, M. Famá, R. Baragiola, Solar wind contribution to surficial lunar water: laboratory investigations. Icarus **211**, 1082 (2011). https://doi.org/10.1016/j.icarus.2010.11.007

B.J. Butler, D.O. Muhleman, Mercury: full-disk radar images and the detection and stability of ice at the north pole. J. Geophys. Res. **98**(E8), 15003–15023 (1993). https://doi.org/10.1029/93JE01581

B.J. Butler, The migration of volatiles on the surfaces of Mercury and the Moon. J. Geophys. Res., Planets **102**, 19283–19291 (1997). https://doi.org/10.1029/97JE01347

B.A. Campbell, D.B. Campbell, J.F. Chandler, A.A. Hine, M.C. Nolan, P.J. Perillat, Radar imaging of the lunar poles. Nature **426**(6963), 137–138 (2003). https://doi.org/10.1038/426137a

N.L. Chabot, C.M. Ernst, B.W. Denevi et al., Images of surface volatiles in Mercury's polar craters acquired by the MESSENGER spacecraft. Geology **42**(12), 1051–1054 (2014). https://doi.org/10.1130/G35916.1

N.L. Chabot, D.J. Lawrence, G.A. Neumann et al., Mercury's polar deposits, in *Mercury: The View After MESSENGER* (Cambridge University Press, Cambridge, 2018a). https://doi.org/10.1017/9781316650684.014

N.L. Chabot, E.E. Shread, J.K. Harmon, Investigating Mercury's South polar deposits: Arecibo radar observations and high-resolution determination of illumination conditions. J. Geophys. Res., Planets **123**(2), 666–681 (2018b). https://doi.org/10.1002/2017JE005500

M.J. Cintala, Impact-induced thermal effects in the lunar and mercurian regoliths. J. Geophys. Res., Planets **97**(E1), 947–973 (1992). https://doi.org/10.1029/91JE02207

R.N. Clark, Detection of adsorbed water and hydroxyl on the Moon. Science **326**(5952), 562–564 (2009). https://doi.org/10.1126/science.1178105

A. Colaprete, P. Schultz, J. Heldmann et al., Detection of water in the LCROSS ejecta plume. Science **330**(6003), 463–468 (2010). https://doi.org/10.1126/science.1186986

J.-P. Combe, T.B. McCord, F. Tosi et al., Detection of local H_2O exposed at the surface of Ceres. Science **353**, aaf3010 (2016). https://doi.org/10.1126/science.aaf3010

J.-P. Combe, A. Raponi, F. Tosi et al., Exposed H_2O-rich areas detected on Ceres with the dawn visible and InfraRed mapping spectrometer. Icarus **318**, 22–41 (2019). https://doi.org/10.1016/j.icarus.2017.12.008

E.S. Costello, R.R. Ghent, P.G. Lucey, The mixing of lunar regolith: vital updates to a canonical model. Icarus **314**, 327–344 (2020). https://doi.org/10.1016/j.icarus.2018.05.023

P.B. Crandall, J.J. Gillis-Davis, R.I. Kaiser, Untangling the origin of molecular hydrogen in the lunar exosphere. Astrophys. J. **887**, 27 (2019). https://doi.org/10.3847/1538-4357/ab4e1f

D.H. Crider, R.R. Vondrak, The solar wind as a possible source of lunar polar hydrogen deposits. J. Geophys. Res., Planets **105**, 26773–26782 (2000). https://doi.org/10.1029/2000JE001277

D.H. Crider, R.R. Vondrak, Space weathering effects on lunar cold trap deposits. J. Geophys. Res., Planets **108**(E7), 5079 (2003). https://doi.org/10.1029/2002JE002030

S.T. Crites, P.G. Lucey, D.J. Lawrence, Proton flux and radiation dose from galactic cosmic rays in the lunar regolith and implications for organic synthesis at the poles of the Moon and Mercury. Icarus **226**(2), 1192–1200 (2013). https://doi.org/10.1016/j.icarus.2013.08.003

J. Crovisier, The photodissociation of water in cometary atmospheres. Astron. Astrophys. **213**, 459–464 (1989)

R.T. Daly, P.H. Schultz, The delivery of water by impacts from planetary accretion to present. Sci. Adv. **4**, eaar2632 (2018). https://doi.org/10.1126/sciadv.aar2632

M.L. Delitsky, D.A. Paige, M.A. Siegler et al., Ices on Mercury: chemistry of volatiles in permanently cold areas of Mercury's north polar region. Icarus **281**, 19–31 (2017). https://doi.org/10.1016/j.icarus.2016.08.006

A.N. Deutsch, J.W. Head III., G.A. Neumann, Age constraints of Mercury's polar deposits suggest recent delivery of ice. Earth Planet. Sci. Lett. **520**, 26–33 (2019). https://doi.org/10.1016/j.epsl.2019.05.027

Y. Du, N.G. Petrik, N.A. Deskins et al., Hydrogen reactivity on highly-hydroxylated TiO_2(110) surfaces prepared via carboxylic acid adsorption and photolysis. Phys. Chem. Chem. Phys. **14**, 3066 (2012). https://doi.org/10.1039/C1CP22515D

M.D. Dyar, C.A. Hibbitts, T.M. Orlando, Mechanisms for incorporation of hydrogen in and on terrestrial planetary surfaces. Icarus **208**(1), 425–437 (2010). https://doi.org/10.1016/j.icarus.2010.02.014

A.I. Ermakov, E. Mazarico, S.E. Schroder et al., Ceres's obliquity history and implications for the permanently shadowed regions. Geophys. Res. Lett. **44**, 2652–2661 (2017). https://doi.org/10.1002/2016GL072250

C.M. Ernst, N.L. Chabot, O.S. Barnouin, Examining the potential contribution of the Hokusai impact to water ice on Mercury. J. Geophys. Res., Planets **123**, 2628–2646 (2020). https://doi.org/10.1029/2018JE005552

F.P. Fanale, J.R. Salvail, The water regime of asteroid (1) Ceres. Icarus **82**(1), 97–110 (1989). https://doi.org/10.1016/0019-1035(89)90026-2

W.M. Farrell, D.M. Hurley, M.I. Zimmerman, Solar wind implantation into lunar regolith: hydrogen retention in a surface with defects. Icarus **255**, 116–126 (2015). https://doi.org/10.1016/j.icarus.2014.09.014

W.M. Farrell, D.M. Hurley, V.J. Esposito, J.L. McLain, M.I. Zimmerman, The statistical mechanics of solar wind hydroxylation at the Moon, within lunar magnetic anomalies, and at Phobos. J. Geophys. Res., Planets **122**(1), 269–289 (2017). https://doi.org/10.1002/2016JE005168

W.M. Farrell, D.M. Hurley, M.J. Poston et al., The young age of the LAMP-observed frost in lunar polar cold traps. Geophys. Res. Lett. **46**, 8680–8688 (2019). https://doi.org/10.1029/2019GL083158

W.C. Feldman, S. Maurice, A.B. Binder, B.L. Barraclough, R.C. Elphic, D.J. Lawrence, Fluxes of fast and epithermal neutrons from lunar prospector: evidence for water ice at the lunar poles. Science **281**(5382), 1496–1500 (1998). https://doi.org/10.1126/science.281.5382.1496

W.C. Feldman, D.J. Lawrence, R.C. Elphic, B.L. Barraclough, S. Maurice, I. Genetay, A.B. Binder, Polar hydrogen deposits on the Moon. J. Geophys. Res., Planets **105**(E2), 4175–4195 (2000). https://doi.org/10.1029/1999JE001129

W.C. Feldman, S. Maurice, D.J. Lawrence et al., Evidence for water ice near the lunar poles. J. Geophys. Res., Planets **106**(E10), 23231–23251 (2001). https://doi.org/10.1029/2000JE001444

D. Fink, J. Krauser, D. Nagengast et al., Hydrogen implantation and diffusion in silicon and silicon dioxide. Appl. Phys. A **61**(4), 381–388 (1995). https://doi.org/10.1007/BF01540112

E.A. Fisher, P.G. Lucey, M. Lemelin et al., Evidence for surface water ice in the lunar polar regions using reflectance measurements from the Lunar Orbiter Laser Altimeter and temperature measurements from the Diviner Lunar Radiometer Experiment. Icarus **292**, 74–85 (2017). https://doi.org/10.1016/j.icarus.2017.03.023

M. Formisano, M.C. De Sanctis, G. Magni, C. Federico, M.T. Capria, Ceres water regime: surface temperature, water sublimation and transient exo(atmo)sphere. Mon. Not. R. Astron. Soc. **455**, 1892–1904 (2016). https://doi.org/10.1093/mnras/stv2344

J.W. Freeman Jr., H.K. Hills, The Apollo lunar surface water vapor event revisited. Geophys. Res. Lett. **18**(11), 2109–2112 (1991). https://doi.org/10.1029/91GL02625

J.W. Freeman, H.K. Hills, R.A. Lindeman, R.R. Vondrak, Observations of water vapor ions at the lunar surface. Moon **8**(1–2), 115–128 (1973). https://doi.org/10.1007/BF00562753

D.E. Gault, F. Hörz, D.E. Brownlee, J.B. Hartung, Mixing of the lunar regolith. Proc. Lunar Planet. Sci. Conf. **5**, 2365–2386 (1974)

E.K. Gibson, Production of simple molecules on the surface of Mercury. Phys. Earth Planet. Inter. **15**, 303 (1977). https://doi.org/10.1016/0031-9201(77)90038-3

G.R. Gladstone, K.D. Retherford, A.F. Egan et al., Far-ultraviolet reflectance properties of the Moon's permanently shadowed regions. J. Geophys. Res., Planets **117**(E12), E00H04 (2012). https://doi.org/10.1029/2011JE003913

P. Gläser, F. Scholten, D. De Rosa et al., Illumination conditions at the lunar south pole using high resolution Digital Terrain Models from LOLA. Icarus **243**, 78–90 (2014). https://doi.org/10.1016/j.icarus.2014.08.013

T.D. Glotch, J.L. Bandfield, P.G. Lucey et al., Formation of lunar swirls by magnetic field standoff of the solar wind. Nat. Commun. **6**, 6189 (2015). https://doi.org/10.1038/ncomms7189

D.B. Goldstein, J.V. Austin, E.S. Barker, R.S. Nerem, Short-time exosphere evolution following an impulsive vapor release on the Moon. J. Geophys. Res., Planets **106**, 32841–32845 (2001). https://doi.org/10.1029/2000JE001326

R.O. Green, C. Pieters, P. Mouroulis et al., The Moon Mineralogy Mapper (M3) imaging spectrometer for lunar science: Instrument description, calibration, on-orbit measurements, science data calibration and on-orbit validation. J. Geophys. Res., Planets **116**(E10), E00G19 (2011). https://doi.org/10.1029/2011JE003797

J.P. Greenwood, S. Itoh, N. Sakamoto, P. Warren, L. Taylor, H. Yurimoto, Hydrogen isotope ratios in lunar rocks indicate delivery of cometary water to the Moon. Nat. Geosci. **4**, 79 (2011). https://doi.org/10.1038/ngeo1050

D.L. Griscom, Thermal bleaching of X-ray induced defect centers in high purity fused silica by diffusion of radiolytic molecular hydrogen. J. Non-Cryst. Solids **19**(68), 301–325 (1984)

A. Grumpe, C. Wöhler, A.A. Berezhnoy, V.V. Shevchenko, Time-of-day-dependent behavior of surficial lunar hydroxyl/water: observations and modeling. Icarus **321**, 486–507 (2019). https://doi.org/10.1016/j.icarus.2018.11.025

V. Gun'ko, V.I. Zarko, B.A. Chuikov et al., Temperature-programmed desorption of water from fumed silica, silica/titania, and silica/alumina. Int. J. Mass Spectrom. Ion Process. **172**, 161 (1998). https://doi.org/10.1016/S0168-1176(97)00269-3

C.D. Hamill, N.L. Chabot, E. Mazarico et al., New illumination and temperature constraints of Mercury's volatile polar deposits. Planet. Sci. J. **1**, 57 (2020). https://doi.org/10.3847/PSJ/abb1c2

B. Hapke, Bidirectional reflectance spectroscopy. 6. Effects of porosity. Icarus **195**(2), 918–926 (2008). https://doi.org/10.1016/j.icarus.2008.01.003

J.K. Harmon, M.A. Slade, Radar mapping of Mercury: full-disk images and polar anomalies. Science **258**(5082), 640–643 (1992). https://doi.org/10.1126/science.258.5082.640

J.K. Harmon, M.A. Slade, R.A. Velez, A. Crespo, M.J. Dryer, J.M. Johnson, Radar mapping of Mercury's polar anomalies. Nature **369**(6477), 213–215 (1994). https://doi.org/10.1038/369213a0

R.E. Hartle, G.E. Thomas, Neutral and ion exosphere models for lunar hydrogen and helium. J. Geophys. Res. **79**(10), 1519–1526 (1974). https://doi.org/10.1029/JA079i010p01519

P.O. Hayne, O. Aharonson, Thermal stability of ice on Ceres with rough topography. J. Geophys. Res. **120**, 1567–1584 (2015). https://doi.org/10.1002/2015JE004887

P.O. Hayne, A.P. Ingersoll, D.A. Paige et al., New approaches to lunar ice detection and mapping, Keck Institute for Space Studies Report (2014). https://resolver.caltech.edu/CaltechAUTHORS:20190213-134457058

P.O. Hayne, A. Hendrix, E. Sefton-Nash, Evidence for exposed water ice in the Moon's south polar regions from Lunar Reconnaissance Orbiter ultraviolet albedo and temperature measurements. Icarus **255**, 58–69 (2015). https://doi.org/10.1016/j.icarus.2015.03.032

P.O. Hayne, O. Aharonson, N. Schorghofer, Micro cold traps on the Moon. Nat. Astron. **5**, 169–175 (2021). https://doi.org/10.1038/s41550-020-1198-9

J.W. Head, L. Wilson, A.N. Deutsch, M.J. Rutherford, A.E. Saal, Volcanically-induced transient atmospheres on the Moon: assessment of duration, significance and contributions to polar volatile traps. Geophys. Res. Lett. **47**, e2020GL089509 (2020). https://doi.org/10.1029/2020GL089509

M.A. Henderson, The interaction of water with solid surfaces: fundamental aspects revisited. Surf. Sci. **46**, 1–308 (1994). https://doi.org/10.1016/S0167-5729(01)00020-6

A.R. Hendrix, K.D. Retherford, G.R. Gladstone et al., The lunar far-UV albedo: indicator of hydration and weathering. J. Geophys. Res., Planets **117**, E12001 (2012). https://doi.org/10.1029/2012JE004252

A.R. Hendrix, D. Hurley, K.D. Retherford et al., Diurnally migrating lunar water: evidence from ultraviolet data. Geophys. Res. Lett. **46**, 2417–2424 (2019). https://doi.org/10.1029/2018GL081821

M. Hess, C. Wöhler, M. Bhatt et al., Processes governing the VIS/NIR spectral reflectance behavior of lunar swirls. Astron. Astrophys. **639**, A12 (2020). https://doi.org/10.1051/0004-6361/201937299

R.R. Hodges Jr., Helium and hydrogen in the lunar atmosphere. J. Geophys. Res. **78**(34), 8055–8064 (1973). https://doi.org/10.1029/JA078i034p08055

R.R. Hodges Jr., Exospheric transport restrictions on water ice in lunar polar traps. Geophys. Res. Lett. **18**(11), 2113–2116 (1991). https://doi.org/10.1029/91GL02533

R.R. Hodges Jr., Ice in the lunar polar regions revisited. J. Geophys. Res., Planets **107**(E2), 5011 (2002). https://doi.org/10.1029/2000JE001491

R.R. Hodges Jr., Methane in the lunar exosphere: implications for solar wind carbon escape. Geophys. Res. Lett. **43**, 6742–6748 (2016). https://doi.org/10.1002/2016GL068994

C.I. Honniball, P.G. Lucey, C.M. Ferrari-Wong et al., Telescopic observations of lunar hydration: variations and abundance. J. Geophys. Res., Planets **125**, e2020JE006484 (2020). https://doi.org/10.1029/2020JE006484

C.I. Honniball, P.G. Lucey, S. Li et al., Molecular water detected on the sunlit Moon by SOFIA. Nat. Astron. **5**, 121–127 (2021). https://doi.org/10.1038/s41550-020-01222-x

L.L. Hood, G. Schubert, Lunar magnetic anomalies and surface optical properties. Science **208**(4439), 49–51 (1980). https://doi.org/10.1126/science.208.4439.49

R.M. Housley, R.W. Grant, N.E. Paton, Origin and characteristics of excess Fe metal in lunar glass welded aggregates. Geochim. Cosmochim. Acta, Suppl. 4 **3**, 2737–2749 (1973)

W.F. Huebner, J. Mukherjee, Photoionization and photodissociation rates in solar and blackbody radiation fields. Planet. Space Sci. **106**, 11–45 (2015). https://doi.org/10.1016/j.pss.2014.11.022

H. Hui, A.H. Peslier, Y. Zhang, C.R. Neal, Water in lunar anorthosites and evidence for a wet early Moon. Nat. Geosci. **6**, 177 (2013). https://doi.org/10.1038/ngeo1735

D.M. Hurley, J.C. Cook, K.D. Retherford et al., Contributions of solar wind and micrometeoroids to molecular hydrogen in the lunar exosphere. Icarus **283**, 31–37 (2017). https://doi.org/10.1016/j.icarus.2016.04.019

A.S. Ichimura, A.P. Zent, R.C. Quinn, M.R. Sanchez, L.A. Taylor, Hydroxyl (OH) production on airless planetary bodies: evidence from H^+/D^+ ion-beam experiments. Earth Planet. Sci. Lett. **345–348**, 90–94 (2012). https://doi.org/10.1016/j.epsl.2012.06.027

A.P. Ingersoll, T. Svitek, B.C. Murray, Stability of polar frosts in spherical bowl-shaped craters on the Moon, Mercury, and Mars. Icarus **100**(1), 40–47 (1992). https://doi.org/10.1016/0019-1035(92)90016-Z

Y. Ishibe, H. Oyama, Reduction of metal oxide layer with hydrogen atoms dissociated on a hot rhenium filament. J. Nucl. Mater. **85**, 1191 (1979). https://doi.org/10.1016/0022-3115(79)90423-9

Y.D. Jia, M.N. Villarreal, C.T. Russell, Possible Ceres bow shock surfaces based on fluid models. J. Geophys. Res. Space Phys. **122**, 4976–4987 (2017). https://doi.org/10.1002/2016JA023712

F.S. Johnson, Lunar atmosphere. Rev. Geophys. **9**(3), 813–823 (1971). https://doi.org/10.1029/RG009i003p00813

B.M. Jones, A. Aleksandrov, K. Hibbitts, M.D. Dyar, T.M. Orlando, Solar wind-induced water cycle on the Moon. Geophys. Res. Lett. **45**(20), 10,959–10,967 (2018). https://doi.org/10.1029/2018GL080008

B.M. Jones, M. Sarantos, T.M. Orlando, A new in situ quasi-continuous solar-wind source of molecular water on Mercury. Astrophys. J. **891**(2), L43 (2020). https://doi.org/10.3847/2041-8213/ab6bda

R. Killen, W.-H. Ip, The surface-bounded atmospheres of Mercury and the Moon. Rev. Geophys. **37**, 361–406 (1999). https://doi.org/10.1029/1999RG900001

R. Killen, M.H. Burger, R.J. Vervack Jr., T.A. Cassidy, Understanding Mercury's exosphere: models derived from MESSENGER observations, in *Mercury: The View After MESSENGER* (Cambridge University Press, Cambridge, 2018). https://doi.org/10.1017/9781316650684.016

R. Klima, J. Cahill, J. Hagerty, D. Lawrence, Remote detection of magmatic water in Bullialdus crater on the Moon. Nat. Geosci. **6**(9), 737–741 (2013). https://doi.org/10.1038/ngeo1909

J.L. Kloos, J.E. Moores, J. Sangha et al., The temporal and geographic extent of seasonal cold trapping on the Moon. J. Geophys. Res., Planets **124**, 1935–1944 (2019). https://doi.org/10.1029/2019JE006003

G. Kramer et al., M3 spectral analysis of lunar swirls and the link between optical maturation and surface hydroxyl formation at magnetic anomalies. J. Geophys. Res. **116**, E00G18 (2011)

M. Küppers, L. O'Rourke, D. Bockelée-Morvan et al., Localized sources of water vapour on the dwarf planet (1) Ceres. Nature **505**, 525–527 (2014). https://doi.org/10.1038/nature12918

M.E. Landis, S. Byrne, N. Schorghofer et al., Conditions for sublimating water ice to supply Ceres' exosphere. J. Geophys. Res. **122**, 1984–1995 (2017). https://doi.org/10.1002/2017JE005335

M.E. Landis, S. Byrne, J.-Ph. Combe et al., Water vapor contribution to Ceres' exosphere from observed surface ice and postulated ice-exposing impacts. J. Geophys. Res., Planets **124**, 61–75 (2019). https://doi.org/10.1029/2018JE005780

L.J. Lanzerotti, W.L. Brown, R.E. Johnson, Ice in the polar regions of the Moon. J. Geophys. Res. **86**, 3949 (1981). https://doi.org/10.1029/JB086iB05p03949

D.J. Lawrence, A tale of two poles: toward understanding the presence, distribution, and origin of volatiles at the polar regions of the Moon and Mercury. J. Geophys. Res., Planets **122**(1), 21–52 (2017). https://doi.org/10.1002/2016JE005167

S. Li, I. Garrick-Bethell, Surface water at lunar magnetic anomalies. Geophys. Res. Lett. **46**(24), 14318–14327 (2019). https://doi.org/10.1029/2019GL084890

S. Li, R.E. Milliken, Water on the surface of the Moon as seen by the Moon Mineralogy Mapper: distribution, abundance, and origins. Sci. Adv. **3**, e1701471 (2017). https://doi.org/10.1126/sciadv.1701471

S. Li, P.G. Lucey, R.E. Milliken et al., Direct evidence of surface exposed water ice in the lunar polar regions. Proc. Natl. Acad. Sci. **115**(36), 8907–8912 (2018). https://doi.org/10.1073/pnas.1802345115

J.Y. Li, S.E. Schröder, S. Mottola et al., Spectrophotometric modeling and mapping of Ceres. Icarus **322**, 144–167 (2019). https://doi.org/10.1016/j.icarus.2018.12.038

Y. Liu, Y. Guan, Y. Zhang, G.R. Rossman, J.M. Eiler, L.A. Taylor, Direct measurement of hydroxyl in the lunar regolith and the origin of lunar surface water. Nat. Geosci. **5**(11), 779–782 (2012). https://doi.org/10.1038/ngeo1601

G.G. Managadze, V.T. Cherepin, Y.G. Shkuratov, V.N. Kolesnik, A.E. Chumikov, Simulating OH/H_2O formation by solar wind at the lunar surface. Icarus **215**, 449 (2011). https://doi.org/10.1016/j.icarus.2011.06.025

P.L. Mattern, G.J. Thomas, W. Bauer, Hydrogen and helium implantation in vitreous silica. J. Vac. Sci. Technol. **13**, 430–436 (1976). https://doi.org/10.1116/1.568938

E. Mazarico, G.A. Neumann, D.E. Smith, M.T. Zuber, M.H. Torrence, Illumination conditions of the lunar polar regions using LOLA topography. Icarus **211**(2), 1066–1081 (2011). https://doi.org/10.1016/j.icarus.2010.10.030

W.E. McClintock, T.A. Cassidy, A.W. Merkel et al., Observations of Mercury's exosphere: composition and structure, in *Mercury: The View After MESSENGER* (Cambridge University Press, Cambridge, 2018). https://doi.org/10.1017/9781316650684.015

T.B. McCord, L.A. Taylor, J.P. Combe, G. Kramer, C.M. Pieters, J.M. Sunshine, R.N. Clark, Sources and physical processes responsible for OH/H_2O in the lunar soil as revealed by the Moon Mineralogy Mapper (M3). J. Geophys. Res., Planets **116**(4), 1–22 (2011). https://doi.org/10.1029/2010JE003711

J.A. McGovern, D.B. Bussey, B.T. Greenhagen, D.A. Paige, J.T. Cahill, P.D. Spudis, Mapping and characterization of non-polar permanent shadows on the lunar surface. Icarus **223**(1), 566–581 (2013). https://doi.org/10.1016/j.icarus.2012.10.018

D.S. McKay, G. Heiken, A. Basu et al., The lunar regolith, in *Lunar Sourcebook*, ed. by G.H. Heiken et al. (Cambridge University Press, New York, 1991), pp. 285–356, Chap. 7

A.J. McKay, D. Bodewits, J.-Y. Li, Observational constraints on water sublimation from 24 Themis and 1 Ceres. Icarus **286**, 308–313 (2017). https://doi.org/10.1016/j.icarus.2016.09.032

H.Y. McSween Jr., J.P. Emery, A.S. Rivkin et al., Carbonaceous chondrites as analogs for the composition and alteration of Ceres. Meteorit. Planet. Sci. **53**(9), 1793–1804 (2017). https://doi.org/10.1111/maps.12947

R.E. Milliken, S. Li, Remote detection of widespread indigenous water in lunar pyroclastic deposits. Nat. Geosci. **10**(8), 561–565 (2017). https://doi.org/10.1038/ngeo2993

I.G. Mitrofanov, A.B. Sanin, W.V. Boynton et al., Hydrogen mapping of the lunar south pole using the LRO neutron detector experiment LEND. Science **330**(6003), 483–486 (2010). https://doi.org/10.1126/science.1185696

J.E. Moores, Lunar water migration in the interval between large impacts: heterogeneous delivery to permanently shadowed regions, fractionation, and diffusive barriers. J. Geophys. Res., Planets **121**, 46–60 (2016). https://doi.org/10.1002/2015JE004929

T.H. Morgan, D.E. Shemansky, Limits to the lunar atmosphere. J. Geophys. Res. **96**(A2), 1351–1367 (1991). https://doi.org/10.1029/90JA02127

R.V. Morris, In situ reworking/gardening/of the lunar surface-Evidence from the Apollo cores, in *Lunar and Planetary Science Conference Proceedings*, vol. 9 (1978), pp. 1801–1811

J.I. Moses, K. Rawlins, L. Zahnle, L. Dones, External sources of water for Mercury's putative ice deposits. Icarus **137**(2), 197–221 (1999). https://doi.org/10.1006/icar.1998.6036

A. Nathues, M. Hoffmann, M. Schaefer et al., Sublimation in bright spots on (1) Ceres. Nature **528**, 237–240 (2015). https://doi.org/10.1038/nature15754

D.H. Needham, D.A. Kring, Lunar volcanism produced a transient atmosphere around the ancient Moon. Earth Planet. Sci. Lett. **478**, 175–178 (2017). https://doi.org/10.1016/j.epsl.2017.09.002

C.E. Nelson, J.W. Elam, M.A. Cameron, M.A. Tolbert, S.M. George, Desorption of H_2O from a hydroxylated single-crystal α-Al_2O_3 (0001) surface. Surf. Sci. **416**, 341 (1998). https://doi.org/10.1016/S0039-6028(98)00439-7

G.A. Neumann, J.F. Cavanaugh, X. Sun et al., Bright and dark polar deposits on Mercury: evidence for surface volatiles. Science **339**(6117), 296–300 (2013). https://doi.org/10.1126/science.1229764

S.K. Noble, L.P. Keller, C.M. Pieters, Evidence of space weathering in regolith breccias I: lunar regolith breccias. Meteorit. Planet. Sci. **40**(3), 397–408 (2005). https://doi.org/10.1111/j.1945-5100.2005.tb00390.x

H. Noda, H. Araki, S. Goossens et al., Illumination conditions at the lunar polar regions by KAGUYA (SELENE) laser altimeter. Geophys. Res. Lett. **35**, L24203 (2008). https://doi.org/10.1029/2008GL035692

S. Nozette, C.L. Lichtenberg, P. Spudis et al., The Clementine bistatic radar experiment. Science **274**(5292), 1495–1498 (1996). https://doi.org/10.1126/science.274.5292.1495

D.A. Paige, S.E. Wood, A.R. Vasavada, The thermal stability of water ice at the poles of Mercury. Science **258**(5082), 643–646 (1992). https://doi.org/10.1126/science.258.5082.643

D.A. Paige, M.C. Foote, B.T. Greenhagen et al., The lunar reconnaissance orbiter diviner lunar radiometer experiment. Space Sci. Rev. **150**(1–4), 125–160 (2010a). https://doi.org/10.1007/s11214-009-9529-2

D.A. Paige, M.A. Siegler, J.A. Zhang et al., Diviner lunar radiometer observations of cold traps in the Moon's south polar region. Science **330**(6003), 479–482 (2010b)

D.A. Paige, M.A. Siegler, J.K. Harmon et al., Thermal stability of volatiles in the north polar region of Mercury. Science **339**(6117), 300–303 (2013). https://doi.org/10.1126/science.1231106

J. Papike, L. Taylor, S. Simon, Lunar minerals, in *Lunar Sourcebook*, vol. 5 (Cambridge University Press, Cambridge, 1991), pp. 121–182, Chap. 5

D.V. Petrov, Y.G. Shkuratov, D.G. Stankevich, V.V. Shevchenko, E.A. Kozlova, The area of cold traps on the lunar surface. Sol. Syst. Res. **37**(4), 260–265 (2003). https://doi.org/10.1023/A:1025022130047

C.M. Pieters, J.N. Goswami, R.N. Clark et al., Character and spatial distribution of OH/H_2O on the surface of the Moon seen by M3 on Chandrayaan-1. Science **326**, 468–572 (2009). https://doi.org/10.1126/science.1178658

P.C. Pinet, V.V. Shevchenko, S.D. Chevrel, Y. Daydou, C. Rosemberg, Local and regional lunar regolith characteristics at Reiner Gamma formation: optical and spectroscopic properties from Clementine and Earth-based data. J. Geophys. Res. **105**(E4), 9457–9475 (2000). https://doi.org/10.1029/1999JE001086

T. Platz, A. Nathues, N. Schorghofer et al., Surface water-ice deposits in the northern shadowed regions of Ceres. Nat. Astron. **1**, 0007 (2016). https://doi.org/10.1038/s41550-016-0007

M.J. Poston, G.A. Grieves, A.B. Aleksandrov et al., Temperature programmed desorption studies of water interactions with Apollo lunar samples 12001 and 72501. Icarus **255**, 24–29 (2015). https://doi.org/10.1016/j.icarus.2014.09.049

A.E. Potter, Chemical sputtering could produce sodium vapor and ice on Mercury. Geophys. Res. Lett. **22**, 3289 (1995). https://doi.org/10.1029/95GL03181

P. Prem, N.A. Artemieva, D.B. Goldstein, P.L. Varghese, L.M. Trafton, Transport of water in a transient impact-generated lunar atmosphere. Icarus **255**, 148–158 (2015). https://doi.org/10.1016/j.icarus.2014.10.017

P. Prem, D.B. Goldstein, P.L. Varghese, L.M. Trafton, The influence of surface roughness on volatile transport on the Moon. Icarus **299**, 31–45 (2018). https://doi.org/10.1016/j.icarus.2017.07.010

P. Prem, D.M. Hurley, D.B. Goldstein, P.L. Varghese, The evolution of a spacecraft-generated lunar exosphere. J. Geophys. Res., Planets **125**, 8 (2020). https://doi.org/10.1029/2020JE006464

T.H. Prettyman, W.C. Feldman, H.Y. McSween et al., Dawn's gamma ray and neutron detector. Space Sci. Rev. **163**, 371–459 (2011). https://doi.org/10.1007/s11214-011-9862-0

T.H. Prettyman, Y. Yamashita, M.J. Toplis et al., Extensive water ice within Ceres' aqueously altered regolith: evidence from nuclear spectroscopy. Science **355**, 55–59 (2017). https://doi.org/10.1126/science.aah6765

A.S. Rivkin, J.Y. Li, R.E. Milliken et al., The surface composition of Ceres. Space Sci. Rev. **163**, 95–116 (2011). https://doi.org/10.1007/s11214-010-9677-4

K.L. Robinson, J.J. Barnes, K. Nagashima et al., Water in evolved lunar rocks: evidence for multiple reservoirs. Geochim. Cosmochim. Acta **188**, 244 (2016). https://doi.org/10.1016/j.gca.2016.05.030

L. Roth, Constraints on water vapor and sulfur dioxide at Ceres: exploiting the sensitivity of the Hubble Space Telescope. Icarus **305**, 149–159 (2018). https://doi.org/10.1016/j.icarus.2018.01.011

L. Roth, N. Ivchenko, K.D. Retherford et al., Constraints on an exosphere at Ceres from Hubble Space Telescope observations. Geophys. Res. Lett. **43**, 2465–2472 (2016). https://doi.org/10.1002/2015GL067451

P. Rousselot, E. Jehin, J. Manfroid et al., A search for water vaporization on Ceres. Astron. J. **142**, 125 (2011). https://doi.org/10.1088/0004-6256/142/4/125

P. Rousselot, C. Opitom, E. Jehin et al., Search for water outgassing of (1) Ceres near perihelion. Astron. Astrophys. **628**, A22 (2019). https://doi.org/10.1051/0004-6361/201935738

L. Rubanenko, E. Mazarico, G.A. Neumann, D.A. Paige, Ice in micro cold traps on Mercury: implications for age and origin. J. Geophys. Res., Planets **123**, 2178–2191 (2018). https://doi.org/10.1029/2018JE005644

L. Rubanenko, J. Venkatraman, D.A. Paige, Thick ice deposits in shallow simple craters on the Moon and Mercury. Nat. Geosci. **12**(8), 597–601 (2019). https://doi.org/10.1038/s41561-019-0405-8

O. Ruesch, T. Platz, P. Schenk et al., Cryovolcanism on Ceres. Science **353**, aaf4286 (2016). https://doi.org/10.1126/science.aaf4286

C.T. Russell, C.A. Raymond, E. Ammannito et al., Dawn arrives at Ceres: exploration of a small, volatile-rich world. Science **353**, 1008–1010 (2016). https://doi.org/10.1126/science.aaf4219

A.E. Saal, E.H. Hauri, M.L. Cascio et al., Volatile content of lunar volcanic glasses and the presence of water in the Moon's interior. Nature **454**, 192–195 (2008). https://doi.org/10.1038/nature07047

M.J. Schaible, R.A. Baragiola, Hydrogen implantation in silicates: the role of solar wind in SiOH bond formation on the surfaces of airless bodies in space. J. Geophys. Res., Planets **119**, 2017 (2014). https://doi.org/10.1002/2014JE004650

N. Schorghofer, Migration calculations for water in the exosphere of the Moon: dusk-dawn asymmetry, heterogeneous trapping, and D/H fractionation. Geophys. Res. Lett. **41**, 4888–4893 (2014). https://doi.org/10.1002/2014GL060820

N. Schorghofer, Two-dimensional description of surface-bounded exospheres with application to the migration of water molecules on the Moon. Phys. Rev. E **91**(5), 052154 (2015). https://doi.org/10.1103/PhysRevE.91.052154

N. Schorghofer, Predictions of depth-to-ice on asteroids based on an asynchronous model of temperature, impact stirring, and ice loss. Icarus **276**, 88–95 (2016). https://doi.org/10.1016/j.icarus.2016.04.037

N. Schorghofer, O. Aharonson, The lunar thermal ice pump. Astrophys. J. **788**(2), 169 (2014). https://doi.org/10.1088/0004-637X/788/2/169

N. Schorghofer, G.J. Taylor, Subsurface migration of H_2O at lunar cold traps. J. Geophys. Res., Planets **112**(E2), E02010 (2007). https://doi.org/10.1029/2006JE002779

N. Schorghofer, J.P. Williams, Mapping of ice storage processes on the Moon with time-dependent temperatures. Planet. Sci. J. **1**, 54 (2020). https://doi.org/10.3847/PSJ/abb6ff

N. Schorghofer, E. Mazarico, T. Platz et al., The permanently shadowed regions of dwarf planet Ceres. Geophys. Res. Lett. **43**, 6783–6789 (2016). https://doi.org/10.1002/2016GL069368

N. Schorghofer, S. Byrne, M.E. Landis et al., The putative cerean exosphere. Astrophys. J. **850**, 85 (2017a). https://doi.org/10.3847/1538-4357/aa932f

N. Schorghofer, P. Lucey, J.P. Williams, Theoretical time variability of mobile water on the Moon and its geographic pattern. Icarus **298**, 111–116 (2017b). https://doi.org/10.1016/j.icarus.2017.01.029

S.E. Schröder, S. Mottola, U. Carsenty et al., Resolved spectrophotometric properties of the Ceres surface from Dawn Framing Camera images. Icarus **288**, 201–225 (2017). https://doi.org/10.1016/j.icarus.2017.01.026

P.H. Schultz, L.J. Srnka, Cometary collisions on the moon and Mercury. Nature **284**, 22–26 (1980). https://doi.org/10.1038/284022a0

V.V. Shevchenko, Observable evidence for cometary impacts on the Moon and their age. Astron. Rep. **37**(3), 314–319 (1993)

R.A. Simpson, G.L. Tyler, Reanalysis of Clementine bistatic radar data from the lunar south pole. J. Geophys. Res., Planets **104**(E2), 3845–3862 (1999). https://doi.org/10.1029/1998JE900038

H. Sizemore, B.E. Schmidt, D.A. Buczkowski et al., A global inventory of ice-related morphological features on dwarf planet Ceres: implications for the evolution and current state of the cryosphere. J. Geophys. Res., Planets **124**, 1650–1689 (2019). https://doi.org/10.1029/2018JE005699

M.A. Slade, B.J. Butler, D.O. Muhleman, Mercury radar imaging: evidence for polar ice. Science **258**, 635–639 (1992). https://doi.org/10.1126/science.258.5082.635

R. Sridharan, S.M. Ahmed, T.P. Das, P. Sreelatha, P. Pradeepkumar, N. Naik, G. Supriya, The sunlit lunar atmosphere: A comprehensive study by CHACE on the Moon Impact Probe of Chandrayaan-1. Planet. Space Sci. **58**(12), 1567–1577 (2010). https://doi.org/10.1016/j.pss.2010.07.027

R. Sridharan, S.M. Ahmed, T.P. Das, P. Sreelatha, P. Pradeepkumar, N. Naik, G. Supriya, Corrigendum to "The sunlit lunar atmosphere: a comprehensive study by CHACE on the Moon Impact Probe of Chandrayaan-1". Planet. Space Sci. **111**, 167–168 (2015). https://doi.org/10.1016/j.pss.2014.12.021. [Planet. Space Sci. **58**, 1567-1577 (2010)]

N.J.S. Stacy, D.B. Campbell, P.G. Ford, Arecibo radar mapping of the lunar poles: a search for ice deposits. Science **276**(5318), 1527–1530 (1997). https://doi.org/10.1126/science.276.5318.1527

L.V. Starukhina, Water detection on atmosphereless celestial bodies: alternative explanations of the observations. J. Geophys. Res., Planets **106**(E7), 14701–14710 (2001). https://doi.org/10.1029/2000JE001307

L.V. Starukhina, Polar regions of the Moon as a potential repository of solar-wind-implanted gases. Adv. Space Res. **37**(1), 50–58 (2006). https://doi.org/10.1016/j.asr.2005.04.033

L.V. Starukhina, Water on the Moon: what is derived from the observations? in *The Moon: Prospective Energy and Material Resources*, ed. by V. Badescu (Springer, Berlin, 2012), pp. 57–85. https://doi.org/10.1007/978-3-642-27969-0_3

L.V. Starukhina, Y.G. Shkuratov, The lunar poles: water ice or chemically trapped hydrogen? Icarus **147**(2), 585–587 (2000). https://doi.org/10.1006/icar.2000.6476

A. Stephant, F. Robert, The negligible chondritic contribution in the lunar soils water. Proc. Natl. Acad. Sci. **111**(42), 15007–15012 (2014). https://doi.org/10.1073/pnas.1408118111

S.A. Stern, The lunar atmosphere: history, status, current problems, and context. Rev. Geophys. **37**(4), 453–491 (1999). https://doi.org/10.1029/1999RG900005

S.A. Stern, J.C. Cook, J.Y. Chaufray et al., Lunar atmospheric H_2 detections by the LAMP UV spectrograph on the Lunar Reconnaissance Orbiter. Icarus **226**(2), 1210–1213 (2013). https://doi.org/10.1016/j.icarus.2013.07.011

B.D. Stewart, E. Pierazzo, D.B. Goldstein, P.L. Varghese, L.M. Trafton, Simulations of a comet impact on the Moon and associated ice deposition in polar cold traps. Icarus **215**(1), 1–16 (2011). https://doi.org/10.1016/j.icarus.2011.03.014

M. Stirniman, C. Huang, R. Scott Smith, S. Joyce, B.D. Kay, The adsorption and desorption of water on single crystal MgO (100): the role of surface defects. J. Chem. Phys. **105**, 1295 (1996). https://doi.org/10.1063/1.471993

J.D. Stopar, B.L. Jolliff, E.J. Speyerer, E.I. Asphaug, M.S. Robinson, Potential impact-induced water-solid reactions on the Moon. Planet. Space Sci. **162**, 157–169 (2018). https://doi.org/10.1016/j.pss.2017.05.010

J.M. Sunshine, T.L. Farnham, L.M. Feaga, O. Groussin, F. Merlin, R.E. Milliken, M.F. A'Hearn, Temporal and spatial variability of lunar hydration as observed by the Deep Impact Spacecraft. Science **326**(5952), 565–568 (2009). https://doi.org/10.1126/science.1179788

H.C. Susorney, P.B. James, C.L. Johnson et al., The thickness of radar-bright deposits in Mercury's northern hemisphere from individual Mercury Laser Altimeter tracks. Icarus **323**, 40–45 (2019). https://doi.org/10.1016/j.icarus.2019.01.016

S.V. Thampi, R. Sridharan, T.P. Das, S.M. Ahmed, J.A. Kamalakar, A. Bhardwaj, The spatial distribution of molecular hydrogen in the lunar atmosphere—new results. Planet. Space Sci. **106**, 142–147 (2015). https://doi.org/10.1016/j.pss.2014.12.018

G. Thangjam, M. Hoffmann, A. Nathues et al., Haze at Occator crater on dwarf planet Ceres. Astrophys. J. Lett. **833**, 2 (2016). https://doi.org/10.3847/2041-8213/833/2/L25

G.E. Thomas, Mercury: does its atmosphere contain water? Science **183**(4130), 1197–1198 (1974). https://doi.org/10.1126/science.183.4130.1197

L. Tu, W.-H. Ip, Y.-C. Wang, A sublimation-driven exospheric model of Ceres. Planet. Space Sci. **104**, 157–162 (2014). https://doi.org/10.1016/j.pss.2014.09.002

O.J. Tucker, W.M. Farrell, R.M. Killen, D.M. Hurley, Solar wind implantation into the lunar regolith: Monte Carlo simulations of H retention in a surface with defects and the H_2 exosphere. J. Geophys. Res., Planets **124**(2), 278–293 (2019). https://doi.org/10.1029/2018JE005805

O.J. Tucker, W.M. Farrell, A.R. Poppe, On the effect of magnetospheric shielding on the lunar hydrogen cycle. J. Geophys. Res., Planets **126**, e2020JE006552 (2021). https://doi.org/10.1029/2020JE006552

H. Urey, *The Planets, Their Origin and Development* (Yale University Press, New Haven, 1952)

A.R. Vasavada, D.A. Paige, S.E. Wood, Near-surface temperatures on Mercury and the Moon and the stability of polar ice deposits. Icarus **141**(2), 179–193 (1999). https://doi.org/10.1006/icar.1999.6175

A.R. Vasavada, J.L. Bandfield, B.T. Greenhagen, Lunar equatorial surface temperatures and regolith properties from the diviner lunar radiometer experiment. J. Geophys. Res., Planets **117**(E12), E00H18 (2012). https://doi.org/10.1029/2011JE003987

M.N. Villarreal, C.T. Russell, J.G. Luhmann et al., The dependence of the cerean exosphere on solar energetic particle events. Astrophys. J. Lett. **838**, L8 (2017). https://doi.org/10.3847/2041-8213/aa66cd

H. Wang, H.Z. Zhang, J. Shi et al., Earth wind as a possible exogenous source of lunar surface hydration. Astrophys. J. Lett. **907**(2), L32 (2021). https://doi.org/10.3847/2041-8213/abd559

K. Watson, B. Murray, H. Brown, On the possible presence of ice on the Moon. J. Geophys. Res. **66**, 1598–1600 (1961a). https://doi.org/10.1029/JZ066i005p01598

K. Watson, B. Murray, H. Brown, The behavior of volatiles on the lunar surface. J. Geophys. Res. **66**, 3033–3045 (1961b). https://doi.org/10.1029/JZ066i009p03033

J.P. Williams, B.T. Greenhagen, D.A. Paige et al., Seasonal polar temperatures on the Moon. J. Geophys. Res., Planets **124**(10), 2505–2521 (2019). https://doi.org/10.1029/2019JE006028

C. Wöhler, A. Grumpe, A.A. Berezhnoy, E.A. Feoktistova, N.A. Evdokimova, K. Kapoor, V.V. Shevchenko, Temperature regime and water/hydroxyl behavior in the crater Boguslawsky on the Moon. Icarus **285**, 118–136 (2017a). https://doi.org/10.1016/j.icarus.2016.12.026

C. Wöhler, A. Grumpe, A.A. Berezhnoy, V.V. Shevchenko, Time-of-day-dependent global distribution of lunar surficial water/hydroxyl. Sci. Adv. **3**(9), 1–11 (2017b). https://doi.org/10.1126/sciadv.1701286

Y. Wu, B. Hapke, Spectroscopic observations of the Moon at the lunar surface. Earth Planet. Sci. Lett. **484**, 145–153 (2018). https://doi.org/10.1016/j.epsl.2017.12.003

E.J. Zeller, L.B. Ronca, P.W. Levy, Proton-induced hydroxyl formation on the lunar surface. J. Geophys. Res. **71**(20), 4855–4860 (1966). https://doi.org/10.1029/JZ071i020p04855

X. Zeng, H. Tang, X.Y. Li, X. Zeng, Y. Wen, J. Liu, Y. Zou, Experimental investigation of OH/H_2O in H^+-irradiated plagioclase: implications for the thermal stability of water on the lunar surface. Earth Planet. Sci. Lett. **560**, 116806 (2021). https://doi.org/10.1016/j.epsl.2021.116806

J.A. Zhang, D.A. Paige, Cold-trapped organic compounds at the poles of the Moon and Mercury: Implications for origins. Geophys. Res. Lett. **36**(16), L16203 (2009). https://doi.org/10.1029/2009GL038614

C. Zhu, P.B. Crandall, J.J. Gillis-Davis, H.A. Ishii, J.P. Bradley, L.M. Corley, R.I. Kaiser, Untangling the formation and liberation of water in the lunar regolith. Proc. Natl. Acad. Sci. **116**, 11165–11170 (2019). https://doi.org/10.1073/pnas.1819600116

C. Zhu, S. Góbi, M.J. Abplanalp et al., Regenerative water sources on surfaces of airless bodies. Nat. Astron. **4**(1), 45–52 (2020). https://doi.org/10.1038/s41550-019-0900-2

Publisher's Note Springer Nature remains neutral with regard to jurisdictional claims in published maps and institutional affiliations.

Authors and Affiliations

Norbert Schörghofer[1,2] [iD] · Mehdi Benna[3] [iD] · Alexey A. Berezhnoy[4,5] [iD] · Benjamin Greenhagen[6] [iD] · Brant M. Jones[7] [iD] · Shuai Li[8] [iD] · Thomas M. Orlando[7] [iD] · Parvathy Prem[6] [iD] · Orenthal J. Tucker[3] [iD] · Christian Wöhler[9] [iD]

✉ N. Schörghofer
  norbert@psi.edu

1   Planetary Science Institute, Tucson, AZ, USA

2   Planetary Science Institute, Honolulu, HI, USA

3   NASA Goddard Space Flight Center, Greenbelt, MD, USA

4   Moscow State University, Moscow, Russia

5   Kazan Federal University, Kazan, Russia

 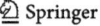

[6] Johns Hopkins Applied Physics Laboratory, Laurel, MD, USA

[7] School of Chemistry and Biochemistry, Georgia Institute of Technology, Atlanta, GA, USA

[8] Hawaii Institute of Geophysics and Planetology, University of Hawaii, Honolulu, HI, USA

[9] Image Analysis Group, Technical University of Dortmund, Dortmund, Germany

Space Science Reviews (2022) 218:15
https://doi.org/10.1007/s11214-022-00876-5

The Exosphere as a Boundary: Origin and Evolution of Airless Bodies in the Inner Solar System and Beyond Including Planets with Silicate Atmospheres

Helmut Lammer[1] · Manuel Scherf[1,2,3] · Yuichi Ito[4,5] · Alessandro Mura[6] ·
Audrey Vorburger[7] · Eike Guenther[8] · Peter Wurz[7] · Nikolai V. Erkaev[9,10,11] ·
Petra Odert[2]

Received: 19 May 2021 / Accepted: 16 February 2022 / Published online: 5 April 2022
© The Author(s) 2022

Abstract

In this review we discuss all the relevant solar/stellar radiation and plasma parameters and processes that act together in the formation and modification of atmospheres and exospheres that consist of surface-related minerals. Magma ocean degassed silicate atmospheres or thin gaseous envelopes from planetary building blocks, airless bodies in the inner Solar System, and close-in magmatic rocky exoplanets such as CoRot-7b, HD 219134 b and 55 Cnc e are addressed. The depletion and fractionation of elements from planetary embryos, which act as the building blocks for proto-planets are also discussed. In this context the formation processes of the Moon and Mercury are briefly reviewed. The Lunar surface modification since its origin by micrometeoroids, plasma sputtering, plasma impingement as well as chemical surface alteration and the search of particles from the early Earth's atmosphere that were collected by the Moon on its surface are also discussed. Finally, we address important questions on what can be learned from the study of Mercury's environment and its solar wind interaction by MESSENGER and BepiColombo in comparison with the expected observations at exo-Mercurys by future space-observatories such as the JWST or ARIEL and ground-based telescopes and instruments like SPHERE and ESPRESSO on the VLT, and vice versa.

Keywords Mercury · Moon · Dust · Planetary embryos · Young Sun · Rocky close-in exoplanets

1 Introduction

The exosphere is a thin gaseous atmospheric layer that surrounds a planet or natural satellite where atoms and/or molecules are gravitationally bound, but where the gas density is so low that the particles are collisionless. In ordinary planets with dense atmospheres, the exosphere is the uppermost atmospheric layer, where the gas density thins out and becomes one with outer space (e.g., Chamberlain 1963). The lower boundary of an exosphere is called exobase or critical level, which corresponds to the altitude where barometric conditions no

---

Surface-Bounded Exospheres and Interactions in the Inner Solar System
Edited by Anna Milillo, Menelaos Sarantos, Benjamin D. Teolis, Go Murakami, Peter Wurz and Rudolf von Steiger

Extended author information available on the last page of the article

longer apply, and the atmosphere temperature becomes nearly a constant. At this altitude, the mean free path of upward traveling atmospheric species is equal to the scale height (e.g., Chamberlain 1963; Bauer and Lammer 2004). Under these conditions, the pressure scale height is equal to the density scale height of the main constituent. Since the dimensionless Knudsen number Kn is defined as the ratio of the mean free path length to a representative physical length scale, it means that the exobase lies in the atmospheric level where $Kn \approx 1$.

However, there are several planetary bodies in inner planetary systems that have no dense atmospheres with the Earth-centric structures like troposphere, stratospheres, mesospheres, thermospheres below their exospheres. Rocky bodies from planetesimals to magmatic planetary embryos, asteroids, natural satellites like the Moon and low gravity close-in planets like Mercury or even some "rocky" exoplanets that orbit at close-in distances around their host stars possess only an exosphere or a gravitationally bounded atmosphere/exosphere environment that consist of outgassed and surface released (moderately) volatile elements.

At planetary bodies where an exosphere acts as a boundary between the surface layer and the surrounding environment, the exobase corresponds to the surface. Such an exosphere is a thin gaseous envelope where atoms and molecules are released from the surface by various processes such as thermal release, photon- or electron-stimulated desorption, particle sputtering and micrometeorite impact vaporization (e.g., Domingue et al. 2014; Wurz et al. 2021, this journal).

Atoms and molecules that are emitted from the surface are ejected on ballistic trajectories until they collide again with the surface, where they can alter the chemistry of the surface material, modifying optical surface properties, etc. together with external plasma (e.g., Killen et al. 2007; Wurz et al. 2021, this journal). In case the particles are released into the exosphere with energies that are larger than the escape energy, they are lost from the body since they move through a collisionless environment. Smaller planetary bodies (i.e., planetesimals, planetary embryos, asteroids, small moons, etc.), where the surface material experiences escape velocity are not considered to have exospheres.

Recent studies related to the outgassing of noble gases (Ar, Ne, Kr, Xe) and the before mentioned moderately volatile elements and their losses from planetesimals and growing magmatic low mass planetary embryos indicate that evaporation processes should have depleted the planetary building blocks from their initial abundance in non-chondritic and chondritic rocky materials (Hin et al. 2017; Young et al. 2019; Sossi et al. 2019; Benedikt et al. 2020; Lammer et al. 2020a,b). These findings further indicate that accretional vapor loss from magmatic planetary building blocks shapes planetary compositions. A steady-state rock vapor or an atmosphere/exosphere envelope that consists of the body's minerals forms above magma oceans within minutes to hours and results from a balance between rates of magma evaporation and atmospheric escape (Young et al. 2019; Benedikt et al. 2020).

A main evidence for evaporation-related losses during planet formation is heavy isotope enrichment in several rock-forming elements relative to chondrites that are found in various differentiated bodies in the Solar System (Young et al. 2019; Sossi et al. 2019; Benedikt et al. 2020). It was also shown by Young et al. (2019) and Benedikt et al. (2020) that magmatic planetary embryos with masses that are lower than that of the Moon, the gravity is too weak for the build-up of a dense silicate atmosphere. Because the low gravity and hot surface temperatures act together, all outgassed elements will escape immediately to space and the planetary building block will be depleted in noble gases and moderately volatile elements (Lammer et al. 2020a).

One can expect that terrestrial planets that formed close to their star, or in case of Mercury to the Sun, might have accreted significantly volatile depleted material after the gas disk dissipated and during the so-called giant impact phase. Lower mass planets that formed further out and migrated inward to close-orbital distances would have lost their primordial H_2-He-dominated atmospheres due to EUV-driven hydrodynamic escape (Owen and Wu 2017; van Eylen et al. 2018; Armstrong et al. 2019). That there is a sub-Neptune-desert or a photoevaporation valley in close-orbital distances is also confirmed by exoplanet observations with the Kepler space telescope (McDonald et al. 2019). This sub-Neptune desert or so-called Fulton gap (Fulton et al. 2017; Fulton and Petigura 2018) is an observed scarity of planets with radii between $\approx$ 1.4–2 Earth radii (R_{Earth}). It is expected that thermal escape of sub-Neptunes in close orbital distances would lead to a population of "hot" rocky cores with smaller radii at small separations from their parent stars, and planets with thick hydrogen- and helium-dominated envelopes with larger radii at larger distances. The bimodality in the distribution was confirmed with higher-precision data in the California-Kepler Survey in 2017 (Fulton et al. 2017), which was shown to match the predictions of the mass-loss hypothesis.

If these bodies also accreted volatile-rich materials after these periods and after they have grown to masses too high for the delivered volatiles to efficiently escape, they may result in planets with a high metal to silicate ratio, while the crust remains volatile-rich such as expected for Mercury (e.g., Peplowski et al. 2012; Nittler et al. 2018). At more massive higher metal/silicate ratio exoplanets at close-in orbits around their host stars, dayside surface temperatures above 1500 K can originate so that magma oceans or magma lakes remain over the planet's lifetime (Schaefer and Fegley 2009; Valencia et al. 2010; Ito et al. 2015; Miguel et al. 2019; Venot et al. 2020). In such cases, rock vapor atmospheres can originate above the hot surface and stellar wind plasma interactions with the mineralogical atmosphere/exosphere environment will occur (Mura et al. 2011; Guenther et al. 2011; Vidotto et al. 2018).

During the planet formation process large planetesimals, planetary embryos and the growing protoplanets develop magma oceans due to heating of the decay of radioactive elements, particularly due to the short-lived ^{26}Al, ^{60}Fe (Lichtenberg et al. 2016; O'Neill et al. 2020), gravitational energy released upon accretion (Albarède and Blichert-Toft 2007; Elkins-Tanton 2012) and collisions (e.g., Morbidelli et al. 2012; Brasser 2013; Johansen et al. 2015; Lammer et al. 2021). Analysis of some rocks on the Moon, Mars, and Vesta indicate such an early widespread silicate melting and fractional crystallization afterwards. The crystallization ages of these rocks agree with the age range of primary planetary formation until $\leq$ 4.4 Gyr (Elkins-Tanton 2012; and references therein), indicating that accretionary and radiogenic heat produces mantle melting. The lifetime of a magma ocean depends on a number of parameters such as:

- the size of the planetary body;
- the amount of accreted radioactive elements;
- the temperature of the accreting material;
- the time between collisions with large planetary embryos;
- the existence of a conductive boundary layer;
- and the existence of an atmosphere.

The before mentioned magmatic bodies can be separated in two categories where transient magma oceans are present, first: magmatic planetesimals and planetary embryos that belong to the building blocks of planets with too low masses for their gravity to bind outgassed constituents (Hin et al. 2017; Young et al. 2019; Benedikt et al. 2020). On such bodies, outgassed silicates and moderately volatile elements escape to space or form only tiny atmospheres near the surface, which are in balance between outgassing and escape rates but are lost when the outgassing process decreases, and the magma ocean solidifies.

The second kind of bodies at which transient magma oceans are present are the early planets after they finished their accretion. When the magma ocean solidifies, depending on the oxidation stage of the magmatic mantle, either predominantly H_2O, and CO_2 for oxidized, and H_2, and CO for reduced conditions, respectively, will be outgassed and steam atmospheres build up (Lebrun et al. 2013; Salvador et al. 2017; Nikolaou et al. 2019; Bower et al. 2019; Herbort et al. 2020). The further evolution of these dense atmospheres depends on the stellar XUV flux evolution, orbit location, water inventory, and the volcanic activity of the respective planet.

Until the photometrical discovery of the first detected "rocky" exoplanet Corot-7b by the French-led CoRoT space telescope in February 2009 (Léger et al. 2009), only transient magma oceans as discussed above were know. With Corot-7b's radius of 1.58 ± 0.1 R_{Earth} (Léger et al. 2009) and a mass of 7.42 ± 1.21 Earth masses (M_{Earth}; Hatzes et al. 2011) the planet's bulk density lies close to the density-radius relationship of Mercury. Due to its close orbital distance d of 0.0172 ± 0.00029 AU the planet has an equilibrium temperature of ≤ 1800 K at its dayside which is hot enough to expect a magma ocean or magma ponds. Since this discovery, more of such close-in "rocky" exoplanets with expected permanent magma oceans/ponds on their dayside were found. One can also expect that this type of planets will outgas volatile and moderately volatile elements from such magmatic hot regions so that detectable silicate atmospheres may build up until the elemental reservoir in the magma ocean depletes. If the gravity of such budies is high enough, however, to keep atmospheres that consist of the planet's silicates, they experience extreme stellar wind plasma interactions that shape their exospheres to cometary-like structures (see Sect. 5.3 for a detailed discussion). This can be observed in the future by large ground and space-based telescopes.

In this review, we are focusing on the origin and evolution of airless bodies in the inner Solar System and on rocky exoplanets in close-in orbits. We, therefore, do not discuss planets with magma ocean outgassed primary atmospheres.

In Sect. 2 we discuss the latest knowledge on the radiation and plasma environment of young stars and the Sun. This is important if one is interested in radiation and particle related release processes of minerals from the surfaces of planetary building blocks as well as the historical exposure of the Hermean and Lunar surfaces. In Sect. 3 we discuss the depletion and fractionation of rock-forming elements from planetesimals to magmatic planetary embryos. Section 4 investigates the origin of the Moon and its exosphere evolution including fingerprints from the Earth's ancient atmosphere, in Sect. 5 we address the characteristics of Mercury and its formation hypotheses, and compare the planet with more massive close-in rocky exo-planets where the stars luminosity form magma oceans and related silicate atmospheres with extended exospheres. Before we conclude the review, we discuss in Sect. 6 the possibilities for observations of silicate-like atmospheres/exospheres from close-in hot higher metal/silicate ratio type exoplanets.

2 Radiation and Plasma Environment of Young Stars and the Sun

2.1 X-Ray and EUV Evolution

To understand the evolution of airless bodies in the inner Solar System and rocky close-in exoplanets, it is important to reconstruct the evolution of the solar and stellar radiation and plasma environments over time. The X-ray and EUV flux evolution of young stars, together often subsumed as XUV (≤ 91.2 nm), is particularly important since short wavelength radiation drives loss processes, not only on planetary bodies with extended atmospheres but also within the exospheres of airless bodies (Wurz et al. 2021, this journal). The XUV flux

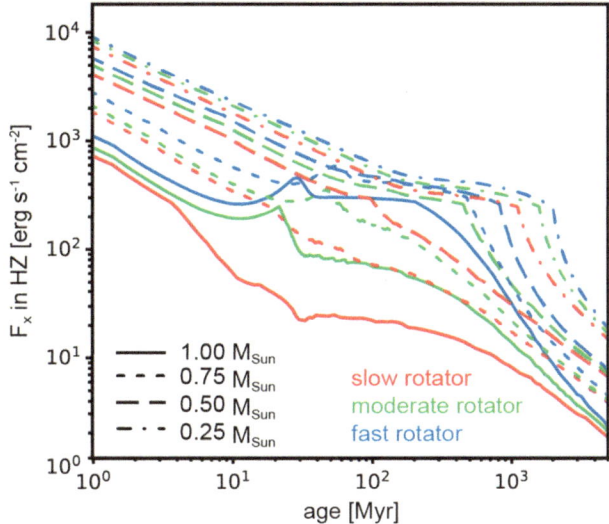

Fig. 1 Evolution of the X-ray surface flux for slow, moderate, and fast rotators scaled to the respective habitable zones of stars with masses of 0.25, 0.5, 0.75, and 1.0 M_{Sun} according to Johnstone et al. (2021a)

from the young Sun, for instance, leads to photoionization of particles in the exospheres of the Moon and Mercury. A higher number of particles gets ionized for higher XUV fluxes, which reduces the return flux onto the surface of these bodies, thereby increasing escape. The radiation from the young Sun also heats up the thermospheres of magma ocean degassed atmospheres (see Sect. 3) which leads to atmospheric expansion and strong thermal escape rates (e.g., Benedikt et al. 2020). Since stellar radiation scales with $1/d^2$, close-in rocky exoplanets experience a far more extreme radiation environment than any Solar System objects. CoRoT-7b, for instance, at an orbit of 0.0172 AU is irradiated by ~ 3400 times higher XUV fluxes than present-day Earth at 1 AU. Since such high radiation significantly affects exospheres of such bodies, it is crucial to understand the early radiation environments of young stars and the Sun.

The XUV flux evolution of a star is dependent on its initial rotation rate with faster rotating stars showing higher initial fluxes than moderate or slow rotators (e.g., Tu et al. 2015; Johnstone et al. 2015a, 2015b, 2021a). For solar-like stars, all rotators, however, show an initial saturation phase with XUV fluxes being as high as 400 to 500 times the present-day solar value that can last for about 5 to 150 million years (Myr), depending on whether the star was a slow or a fast rotator or something in between (Tu et al. 2015). The torque from the stellar mass loss, however, slows down the initial rotation rate until the different rotators converge towards one single track which, for solar-like stars, happens after about 1 billion years (Gyr) (Johnstone et al. 2015b). The convergence of the different tracks happens later for lower-mass stars, but the difference between the various rotational tracks gets less and less pronounced for decreasing stellar masses (Johnstone et al. 2021a). Moreover, Johnstone et al. (2021a) found that the total emitted XUV flux from M- or K-type stars is generally lower than for G- or even F-type stars. On the other hand, any body that receives the same amount of bolometric luminosity as around a G star would, therefore, experience much higher and longer lasting XUV flux exposure corresponding to its orbit.

Figure 1 shows the evolution of the X-ray surface flux for slow, moderate, and fast rotators with masses of 1, 0.75, 0.5, and 0.25 solar masses (M_{Sun}) as an example scaled to the corresponding habitable zones of these stars according to Johnstone et al. (2021a). Here, the different rotators are defined as the 5th (slow), 50th (moderate), and 95th (fast) percentiles of the rotational distribution of the investigated stellar sample of Johnstone et al. (2021a). To show the difference of the short-wavelength radiation received by a body for the same bolometric luminosity at different host stars, it is common to illustrate this effect within the habitable zone. As mentioned above, close-in airless bodies will likewise receive an even higher radiation than depicted within the exemplary Fig. 1.

This also illustrates that the determination of the early evolution of the Sun, and hence also of the planets, is not straight forward since it is not possible to infer its initial rotation rate from its present-day value. However, there are several different studies suggesting that the Sun likely was a slow, or at most, a slow to moderate rotating young G-type star (e.g., Saxena et al. 2019; Lammer et al. 2020a,b; Johnstone et al. 2021b). An N_2-dominated atmosphere would not have been stable under the strong XUV flux of a fast rotator during the Earth's Archean eon (Johnstone et al. 2021b), while the present-day noble gas ratios of Ar, Ne, and Ar/Ne in the atmospheres of Earth and Venus can only be reproduced in case that the young Sun was a slow, or a slow to moderate rotator. Likewise, the present-day moderately volatile surface composition of the Moon can also be reproduced through sputtering if the Sun was a slow rotator (Saxena et al. 2019).

Besides the XUV flux evolution, one also must take into account the related evolution of solar and stellar flares since their increased burst of radiation can significantly affect escape processes at the airless bodies in the Solar System and beyond (see Wurz et al. 2021, this journal). Again, fast rotators happen to flare more often than moderate or slow rotators with more massive stars, having more energetic flares than lower-mass stars, and flare rate generally decreasing for older stars (e.g. Davenport et al. 2019; Johnstone et al. 2021a). While at present-day, the strongest flares observed for the Sun are in the range of $\sim 10^{32}$ erg (e.g., Emslie et al. 2005; Emslie and Massone 2012), this value can reach up to 10^{37} erg for solar-like stars and $\sim 10^{33}$ erg for low-mass stars in the range of 0.1–0.2 M_{Sun} (e.g., Wu et al. 2015; Yang and Liu 2019). Also, here, it has to be pointed out that even though the flare energies are lower for lower-mass stars, their input into their respective HZs is likely higher than in the case of more massive stars (Johnstone et al. 2021a).

The far-ultraviolet (FUV, 91.2–200 nm) and the ultraviolet flux (UV, 200–400 nm) are also increasing towards the past, but less significant (e.g., Ribas et al. 2005; Claire et al. 2012). UV related processes like photo-stimulated desorption (PSD; Wurz et al. 2021, this journal) were, therefore, likely also more efficient than at present-day.

2.2 Bolometric Luminosity Evolution

The bolometric luminosity L_{bol} of a star is the integrated flux over all wavelength ranges and is predominantly shaped by the optical. On the contrary to the XUV evolution, for a G star the bolometric luminosity generally increases over time after reaching the zero-age main sequence (ZAMS; Fig. 2). After the arrival of the Sun at ZAMS, L_{bol} was about 30% lower than at present-day (Baraffe et al. 2015; Gough 1981; e.g., Newman and Rood 1977). The reason for the subsequent increase in L_{bol} is due to nuclear fusion of hydrogen to helium in the core of the Sun (e.g., Feulner 2012; Gough 1981). While He is accumulating, the molecular weight of the core is also increasing, which leads to a contraction of the core and a therewith connected increase in heat to keep the star stable, with the latter resulting in a higher luminosity output.

For the Sun, $L_{\text{bol}}(t)$ can be estimated as a function of time t through the following approximation by Gough (1981), i.e.,

$$L_{\text{bol}}(t) = \left[1 + \frac{2}{5} \left(1 - \frac{t}{t_\odot} \right) \right]^{-1} L_{\text{bol},\odot},$$

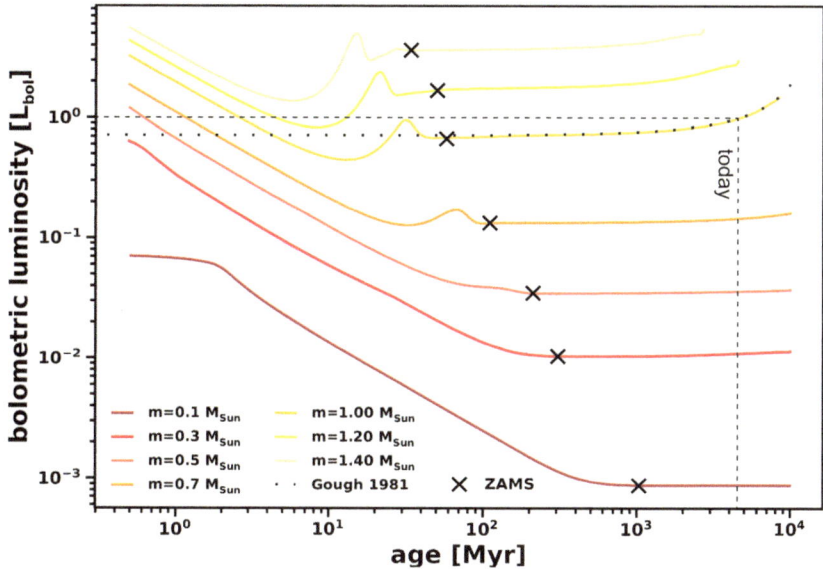

Fig. 2 The evolution of L_{bol} over time for different stellar masses according to the stellar evolution model of Baraffe et al. (2015). The black crosses indicate the respective ages at which these stellar masses reach the zero age main sequence (ZAMS)

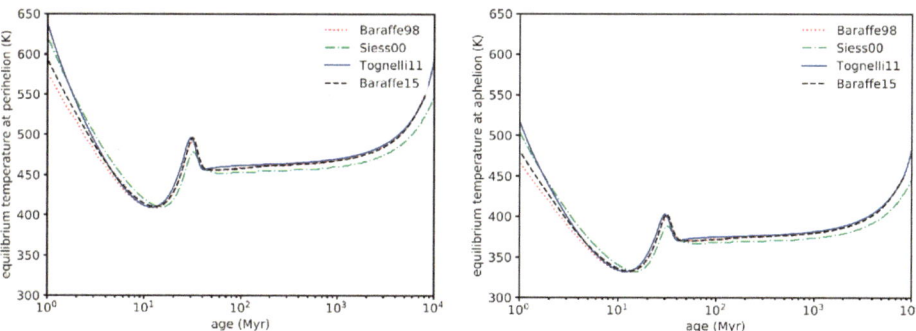

Fig. 3 The evolution of the equilibrium temperature for Mercury at perihelion (0.307 AU; left), and aphelion (0.466 AU; right) for different bolometric luminosity evolution models by Baraffe et al. (1998), Siess et al. (2000), Tognelli et al. (2011), and Baraffe et al. (2015)

where $L_{bol,\odot} = 3.85 \times 10^{26}$ W is the present-day bolometric luminosity, and $t_{\odot} = 4.57$ Gyr is the age of the Sun. This equation correlates very well with the evolution of the Sun's bolometric luminosity except for the first ~ 0.1 Gyr. This can be seen in Fig. 3, which shows the evolution of L_{bol} for different solar masses as calculated with the stellar evolution model of Baraffe et al. (2015). One can also see the Sun's settling onto the main sequence which is accompanied by radial shrinking and the conversion of gravitational energy into heat. When the proton-proton nuclear reaction chain sets in (e.g., Bethe 1939), L_{bol} suddenly increases before settling again 30% below the present-day value. Compared to this steady increase of $\sim 1\%$ per 100 Myr, fluctuations over a whole solar cycle are rather low, being in the range of 0.1% (Solanki et al. 2013).

While solar-like and more massive stars show an increase in bolometric luminosity already after the first few 10s of Myr, low mass stars need much longer to settle onto the main sequence. They show a significant decrease of L_{bol} for the first few 100 Myr (e.g., Baraffe et al. 2015; Spada et al. 2013); as can be seen in Fig. 2, late M stars with a mass of 0.1 $M_\odot$ decrease in L_{bol} by about two orders of magnitude within the first ~ 500 Myr. Such a strong decrease over a relatively long timeframe does not only affect the potential early habitability of terrestrial planets orbiting such stars but might also affect the early evolution of airless bodies' exospheres significantly.

The bolometric luminosity can be used for an estimate of the equilibrium temperature T_{eq} of Mercury and other more or less airless rocky close-in exoplanets. As an example, Fig. 3 shows estimates of the evolution of T_{eq} for Mercury's orbital distance at aphelion and perihelion over the lifetime of the Solar System by using various stellar evolutionary tracks for a solar-mass star. The models and selected parameters are: [M/H] $= 0$, Y $= 0.282$, $\alpha = 1.9$ (Baraffe et al. 1998); Y $= 0.28$, Z $= 0.02$ (Siess et al. 2000); Y $= 0.288$, Z $= 0.02$, $\alpha = 1.68$ (Tognelli et al. 2011); [M/H] $= 0$, Y $= 0.28$, $\alpha = 1.6$ (Baraffe et al. 2015). Here, [M/H] is the metallicity, Y the He content, Z the metal content, and α the mixing length parameter. Particularly for the first ~ 30 Myr, T_{eq} is significantly varying which is important if one wants to simulate magma ocean degassed atmospheres at Mercury, but also at close-in exoplanets of other stars. Due to these hotter T_{eq}-phases and hence environments, magma oceans of planetary embryos solidify slower so that the outgassing and related depletion of rock-forming elements takes longer as well. Moreover, thermal release of surface-elements is more efficient on airless bodies in inner systems. Depending on the gravity of close-in planets or planetary embryos, higher luminosities can also enhance the thermal escape of primordial atmospheric gas so that bodies in close-in orbital distances loose their gaseous envelopes via boil-off (Owen and Wu 2016; Lammer et al. 2016, 2018).

2.3 Plasma Environment in Time

As for the XUV flux evolution, solar and stellar mass loss is dependent on the rotational evolution of the respective star. Even though the solar wind did not remove a significant amount of mass from the Sun, it removes angular momentum through magnetic field stresses thereby leading to a rotational spin down and, thus, consequently, to a decrease in mass loss and wind over time (e.g., Kraft 1967; Skumanich 1972; Wood 2004; Johnstone et al. 2015a,b; Vidotto 2021). That the solar mass loss and the therewith connected solar wind might have been higher in the past is also supported by observational studies of stellar astrospheres (e.g., Wood et al. 2002, 2005) for solar-like stars older than ~ 700 Myr. For younger stars, however, these studies might even indicate a significantly lower mass loss, as can be seen in Fig. 4a (ξ Boo compared). Wood et al. (2005) only extrapolated the mass loss evolution of the Sun (black solid and dash-dotted lines) back to 700 Myr since the young binary star ξ Boo (with spectral types G8V and K4V; black dot in Fig. 4a) was found to have a significantly lower mass loss rate.

More recent simulations by Airapetian and Usmanov (2016) with a three-dimensional magnetohydrodynamic Alfvén wave driven solar wind model, retrieved mass loss rates for solar-like stars that are in the same range as found by Wood et al. (2005), as can also be seen in Fig. 4a. Further observations of the mass loss rates of solar-like stars, particularly of those of a young age, might be needed to retrieve a clearer picture on the evolution of the solar and stellar plasma environments. Another important factor when considering the plasma environment and its impact on the evolution of airless bodies' exospheres are Coronal Mass Ejections (CMEs) which often show a significant increase in ambient particle

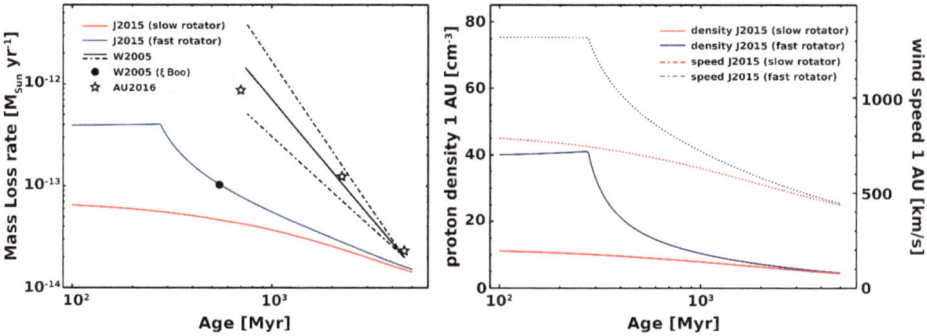

Fig. 4 Left: The mass loss evolution of solar like stars. While observations of stellar astrospheres (black solid and dashed lines as mean and upper and lower limits, respectively) by Wood et al. (2005) indicate a significant increase in solar mass loss for younger stars, a contradicting observation of the binary star ξ Boo (black circle) by Wood et al. (2005) suggest clearly lower loss rates for stars younger than ~ 700 Myr (however, further observations might be necessary to support smaller loss rates for young ages). The three displayed stars show simulations by Airapetian and Usmanov (2016) which lie well within the mass loss evolution estimate by Wood et al. (2005). The red and blue lines show the simulated evolution of the mass loss for a slow and fast rotator by Johnstone et al. (2015a, 2015b). Right: The evolution of proton density (solid lines) and wind speed (dotted lines) of the slow wind for slow and fast rotating solar-like stars in the Model A by Johnstone et al. (2015a, 2015b). Left figure adopted from Scherf and Lammer (2021)

density and velocities up to > 2000 km/s (e.g., Gopalswamy 2004; Chen 2011; Webb and Howard 2012). While a specific planetary body in the present Solar System is only hit by about ~ 6–16% of all CMEs, this increases to more than 31% for solar-like stars with an age of about 700 Myr (Kay et al. 2019). As Kay et al. (2019) found for these ages via studying the solar twin k^1 Ceti, CMEs are more frequently focused onto the Ecliptic plain due to the coronal magnetic field reflecting it towards the astrospheric current sheet. In addition, CMEs might have been significantly more frequent in the past than at present-day (e.g., Odert et al. 2017), and likely even stronger (e.g., Airapetian et al. 2016). Extreme space weather events might, therefore, have played a crucial role in the evolution of airless bodies.

3 Outgassing from Transient Magma Oceans and Depletion of Elements from Low-Mass Embryos

Planets accrete mass by numerous collisions between small objects, which accumulate to planetesimals, and planetary embryos. During these collisions transient magma oceans originate in many planetary bodies in the early solar and extrasolar systems, determining the initial conditions for diverse evolutionary paths of terrestrial planets (Deng et al. 2020). Within the first ≈ 3 Myr after the origin of the Solar System, planetesimals and larger planetary embryos develop magmatic pools, oceans, and some can perhaps completely melt by the heating of short-lived radioactive elements (e.g., Urey 1955; Fish et al. 1960; Elkins-Tanton 2012; Lichtenberg et al. 2016, 2018, 2019) such as ^{26}Al, and ^{60}Fe, frequent collisions (e.g., Safronov and Zvjagina 1969; Wetherill 1980; Tonks and Melosh 1993; Schlichting et al. 2015), and gravitational energy (Albarède and Blichert-Toft 2007). It was found by Lichtenberg et al. (2016, 2018) and Neumann et al. (2020) that the heating by the before mentioned short-lived radioisotopes, followed by internal differentiation and fast volatile outgassing determined to a large extent the thermal history and interior structure of these planetary building blocks and, hence, their final composition during the earliest stages of planetary formation. Neumann et al. (2020) studied the energy balance in small bodies that are heated by decay of radioactive elements and compaction-driven water-rock separation in a dust-water/ice-empty pores mixture. Additionally, these authors considered also second-order processes, such as accretional heating, hydrothermal circulation, and ocean or ice convection and found that precursors of bodies like Ceres in the inner Solar System could have been wet and/or dry.

Collisional erosion (O'Neill and Palme 2008; Carter et al. 2015; Bonsor et al. 2015; Boujibar et al. 2015; Carter et al. 2018; Allibert et al. 2021) fractionates elements such as Si, Fe and Mg (i.e., Fe/Mg, Si/Fe) according to their incompatibility with mantle minerals during melting, while losses of outgassed elements preferentially remove volatiles. Furthermore, it was shown by Sossi et al. (2019) that moderately volatile species can be fractionated from each other through their loss from large planetesimals or planetary embryos in dependence of their equilibrium pressure.

Depending on the body's oxygen fugacity (e.g., Sossi et al. 2019), its temperature, bulk composition and solidification path of the magma ocean (Elkins-Tanton 2008, 2012), thermodynamic studies indicate that moderately volatile rock-forming elements such as Na, K, Mg, Ca, Si, etc. are outgassed from the magmatic surface (Schaefer and Fegley 2007; Fegley et al. 2016; Odert et al. 2018; Young et al. 2019; Sossi et al. 2019). Certainly, losses of these outgassed species caused by thermal escape and collisional erosion modified the bulk composition not only of the planetary building blocks but also the composition of the terrestrial planets where they have been incorporated during accretion (Lammer et al. 2020a). The elevated Mn/Na ratio of smaller rocky bodies relative to chondrites most likely reflects the oxygenation of the magma ocean stage, because Na is more volatile under oxidized conditions than Mn (e.g., O'Neill and Palme 2008; Siebert et al. 2018). Furthermore, Earth's Si/Mg ratio indicated that proto-Earth most likely evolved through escape from the accreting building blocks (Fegley et al. 2016).

The fast escape of the outgassed radioactive isotope ^{40}K, from magmatic planetesimals or larger building blocks may further alter their composition and structure and hence the accreting protoplanets. The amount of the ^{40}K isotope is very important since its radioactive decay contributes to the thermal evolution of the interiors of young telluric planetary bodies (i.e., planetary embryos, terrestrial planets, satellites, see O'Neill et al. 2020), and processes such as the development of long-lived magnetic dynamos (e.g., Turcotte and Schubert 2002; Murthy et al. 2003; Nimmo et al. 2004; Nimmo and Kleine 2015).

Hin et al. (2017), Young et al. (2019) and, more recently, Benedikt et al. (2020) studied the losses of magma-related outgassed rock-forming elements from large planetesimals and planetary embryos. Benedikt et al. (2020) showed that for Moon-mass and smaller bodies with magma layers, no dense silicate atmosphere or even an outgassed steam atmosphere can build up. It was found by these studies that, if these bodies have magma oceans and a sufficiently high surface temperature (i.e., 1500–3000 K), then escape should be immediate. Because of the high surface temperature and the low gravity of these bodies, the so-called unitless escape or Jeans parameter $\lambda = (GM_{pl} \, m_i / kT_{surf} r_{pl})$, which compares the gravitational energy with the thermal energy, of the outgassed rock-forming elements indicate an immediate hydrodynamic loss (Young et al. 2019; Benedikt et al. 2020), so that a dense rock-vapor atmosphere cannot build up. G is the gravitational constant, M_{pl} and m_i the masses of the protoplanetary body and a particular element, k is the Boltzmann constant; T_{surf} is the temperature at the surface of the body with radius r_{pl}.

The thermal escape regime changes over a narrow range of λ where the escape is purely hydrodynamic for values that are ≤ 2 to 3, whereas for $\lambda \geq 6$ it is not (Volkov et al. 2011; Erkaev et al. 2015). Benedikt et al. (2020) found that for a Moon-mass-embryo that is surrounded by a 2500 K magma ocean all elements such as Na, Mg, Si, K, Ar, Ne, O, H_2O

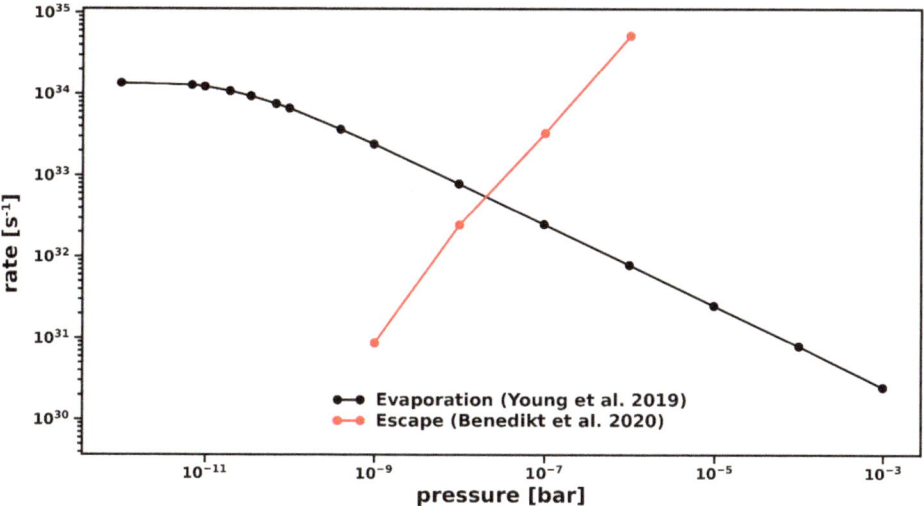

Fig. 5 Outgassing of a silicate atmosphere with a mean weight of 34 unified atomic mass units (amu) from the magma ocean of a planetary embryo with $m = 0.1$ moon masses (M_{Moon}) and a surface temperature of $T_{\text{surf}} = 1500$ K as calculated by Young et al. (2019), and the corresponding escape flux as simulated by Benedikt et al. (2020). Figure adopted from Benedikt et al. (2020)

would have a λ that is < 6, i.e., they are immediately lost to space. Heavier elements and molecules such as Fe, FeO or SiO, however, have escape parameters that are slightly above the critical value. For these elements, one can expect that they will also experience high thermal loss rates; they, furthermore, will be ionized and picked-up by the solar wind.

Young et al. (2019) and Benedikt et al. (2020) modeled the outgassing and loss of a rock-vapor atmosphere with an average mass of $m_{\text{av}} \approx 34$ g/mol from a planetary embryo with the mass of $\approx 0.01\ M_{\text{Moon}}$ by using a melt composition of an Enstatite chondrite without Fe. Figure 5 shows the outgassing rates as a function of pressure for an escaping silicate atmosphere of such a body. One can see that the equilibrium between outgassing and escape is reached at a pressure of $\approx 2 \times 10^{-8}$ bar and a hydrodynamic escape rate of $\approx 5.5 \times 10^{32}\ \text{s}^{-1}$. It should also be noted that for such high escape rates such bodies can also lose noble gases like Ar and Ne, but as it was shown by Benedikt et al. (2020) and in Lammer et al. (2020b), the initial fractionation of ^{36}Ar/^{38}Ar and ^{20}Ne/^{22}Ne does not change significantly. From the results of these studies one can expect that the building blocks of proto-Mercury, Venus, and Earth, etc., were highly depleted in volatile and moderately volatile elements. This would indicate that growing protoplanets accreted significantly volatile depleted material, which is also in agreement with Sossi et al. (2019), Lammer et al. (2020a, 2021) and Herbort et al. (2020) who showed that Earth's volatiles are the result of the accretion of smaller building blocks which experienced various levels of volatile losses.

Elemental data that have been collected from Venus (e.g., Morgan and Anders 1980; Basilevsky 1997), Earth (e.g., Lyubetskaya and Korenaga 2007; Arevalo et al. 2009), and Mars (e.g., Taylor 2013; Yoshizaki and McDonoug 2020) are consistent with the before mentioned hypothesis. However, it would be expected that Mercury is less volatile compared to Venus, Earth and Mars due to its closer orbit around the Sun (e.g., Cameron et al. 1988) and if a giant impact (see Sect. 5.2), as suggested by some researchers, was involved (e.g., Smith 1979; Benz et al. 1988), the volatile depletion would be even enhanced (see Sect. 5.1.2). For example, Earth's Moon that originated by a giant impact–high temperature

Table 1 Lunar parameters (from D. Williams, NASA planetary fact sheet)

Mass (10^{24} kg)	0.073	Rotation Period (hours)	656	Orbital Period (days)	88	Mean Temperature (C)	−20
Diameter (km)	3475	Length of Day (hours)	709	Orbital Velocity (km/s)	47	Surface Pressure (bars)	0
Density (kg/m^3)	3340	Dist. from Sun (10^6 km)	57.9	Orbital Inclination (°)	5.1	Global Magnetic Field	No
Gravity (m/s^2)	1.6	Perihelion (10^6 km)	46	Orbital Eccentricity	0.055	Bond albedo	0.11
Escape Velocity (km/s)	2.4	Aphelion (10^6 km)	69.8	Obliquity to Orbit (°)	6.7	Visual geom. albedo	0.12

event (see Sect. 4.1) resulted in the volatile poorest body analyzed so far (e.g., Hartmann and Davis 1975). However, from the MESSENGER X-ray and γ-ray spectrometer data and Earth-based observations of the planet's Na and K exosphere, it is now known that this is not the case and Mercury's crust/mantle is volatile-rich (e.g., Peplowski et al. 2011). As discussed in more details in Sect. 5.1.2, if one compares Mercury's K/Th, K/U or Cl/K surface ratios with the before mentioned terrestrial planets, one finds that these ratios are close to that of Mars but slightly higher (Peplowski et al. 2012; Evans et al. 2015; Nittler et al. 2018). Compared to Mercury, Venus' and Earth's K/Th ratios are ≈ 2.5–3.5 times lower, respectively (Nittler et al. 2018).

Mercury is the most reduced terrestrial planet. From experiments it is known that at low oxygen fugacities, elements that are typically considered lithophile can become more siderophile (e.g., Chabot and Drake 1999; Bouhifd et al. 2007; Mills et al. 2007; McCubbin et al. 2012). Because of this it is possible that the surface elements, as measured by MESSENGER's instruments, were modified due to Mercury's strongly reduced oxygen fugacity, which could have affected these elements through metal/silicate partitioning during the planet's core formation, as suggested by McCubbin et al. (2012). On the other hand, a later accretion of un-depleted chondritic material, at the same time when Earth obtained its volatiles, may be another explanation for the puzzling innermost planet in the Solar System.

To summarize, the studies reviewed within this section indicate that the building blocks of terrestrial planets are depleted in volatile and moderately volatile elements as soon as they formed magma oceans or magma pools. These findings agree with Marty (2012) and Lammer et al. (2020b) who found that Earth accreted $\approx 0.95\ M_{\text{Earth}}$ from extremely depleted building blocks.

4 Origin and Exosphere Evolution over the Moon's History

4.1 Moon Forming Hypotheses

The Moon (see Table 1 for a parameter list) formed most likely between 50 to 200 Myr after the origin of the Solar System, with earlier ages being more likely as suggested by several different isotopic dating systems (see, e.g., Lock et al. 2020, for an extensive discussion). The Moon forming event itself created an Earth-Moon system with a few significant characteristics that have to be explained by any Lunar formation hypothesis. Besides the Pluto-Charon system, the Earth-Moon size ratio is the lowest in the whole Solar System, with the Moon having roughly 27% of the Earth's radius. On the other hand, the Lunar metal core only comprises $\sim$ 1–2% of its total mass while the Earth's core makes up about 30% of the whole planet. Besides this difference in composition, the Moon is also significantly more depleted in volatile and moderately volatile elements, while its major elements and isotopic abundances show a striking similarity to the composition of the Earth (e.g., Canup et al. 2021). Other characteristics worth noting are the high total angular momentum of the Earth-Moon system with an initially rapidly spinning Earth, and evidence that the Moon formed hot and held a deep magma ocean.

Several different theories on the Lunar origin were published over the last decades that try to solve these characteristics. Early explanations included co-accretion, capture of a planetary embryo as well as disintegrating capture in which a planetesimal passed through the Roche-lobe of the Earth and re-accreted later, and fission from a rapidly spinning Earth (see, e.g., Wood 1986, for a discussion of these earlier theories). However, these were not able to account for most of the specific characteristics listed above. The co-accretion hypothesis, for instance, was not able to explain the high angular momentum of the system and the small Lunar iron core, while the capture from a different orbital position seems to be difficult to reconcile with the similar isotopic composition between the Earth and the Moon. Another model, however, that emerged in the 1970s was the giant impact model (e.g., Hartmann and Davis 1975; Cameron and Ward 1976) which seemed to be able to account for the small Lunar iron core, the high angular momentum, the similar oxygen isotopic composition, and the deep magma ocean of the Moon (Wood 1986). While older impact models (e.g., Benz et al. 1986, 1987, 1989; Cameron and Benz 1991) were able to reproduce the iron-poor core of the Moon, it was later found that only quite a narrow range of grazing impact scenarios with a Mars-mass embryo colliding approximately with mutual escape velocity was able to additionally reproduce the angular momentum of the system, as well as the Lunar size (Canup and Asphaug 2001; Canup 2004a,b, 2008).

The canonical (Fig. 6, left panel) and other giant impact scenarios, however, predicted that the Moon should have primarily been formed by material from the impactor by typically 70–80% (Canup et al. 2021), thereby leading to a distinct Lunar isotopic composition, even though different studies showed a significant isotopical similarity between the Earth and the Moon (e.g., Zhang et al. 2012; Kruijer et al. 2015; Touboul et al. 2015). If the impacting embryo was similar to Mars with regard to its oxygen isotopic composition, less than 5% of this body should have been re-accreted to the Moon while > 95% should have been delivered from the Earth's mantle (e.g., Canup 2012; Canup et al. 2021). However, it is yet unclear how the composition of the impactor could have looked like. While some argue that it might have been comparable to Mars (e.g., Pahlevan and Stevenson 2007), other studies point towards an Earth-like composition (e.g, Dauphas et al. 2014; Mastrobuono-Battisti et al. 2015; Dauphas 2017; see also Canup et al. 2021 for a review of this topic). In case that the impactor differed isotopically from the Earth, equilibration (e.g., Pahlevan and Stevenson 2007; Lock et al. 2018) must have taken place, a process through which the isotopic signature of the impactor gets erased by vapor mixing between the post-impact planet and the debris disk.

In more recent years, it was shown that the canonical impact scenarios can be extended by a broader range of impact parameter space that can explain most of the above Earth-Moon system characteristics. A hit-and-run collision, for instance, with an increased impact velocity and steeper impact angle was proposed by Reufer et al. (2012). These simulations result in disks that are composed only by about 40–60% from the impactor but angular momentums that are about 30–40% too high, meaning that some mechanisms for losing angular momentum has to act afterwards. Multiple giant impacts (Rufu et al. 2017) and an

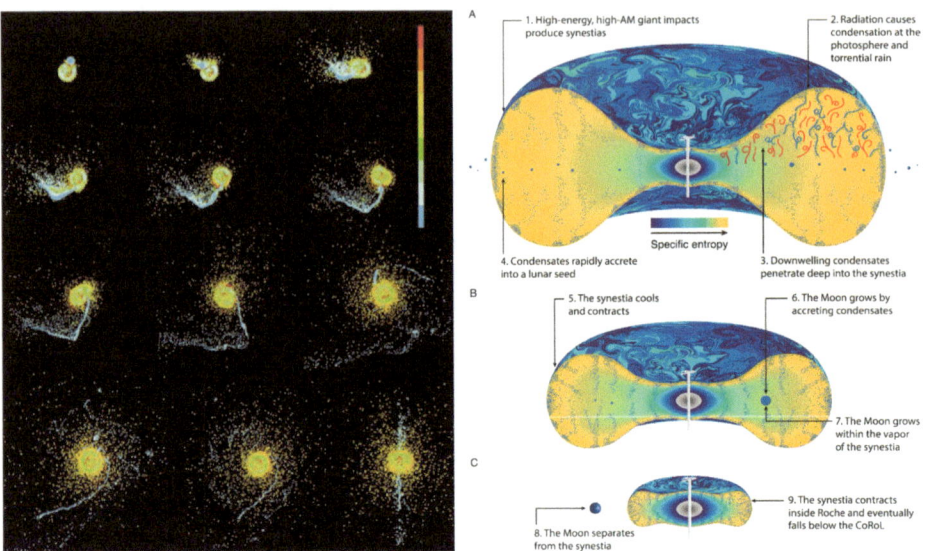

Fig. 6 The canonical giant Moon forming impact (left) by Canup and Asphaug (2001) vs the new concept of a synestia (right) by Lock and Stewart (2017); see text for further information. The left figure covers the first 23 hours from the impact, the color scale illustrates the thermal state of the matter with blue and dark green being condensed matter. AM in the right figure stands for angular momentum, CoRoL for corotation limit. Left figure from Canup and Asphaug (2001); right figure from Lock et al. (2018)

impact onto a magma ocean dominated proto-Earth (Hosono et al. 2019) have also been shown to potentially be able to create the Moon. In addition, it was demonstrated that the angular momentum could have also been significantly altered after the Moon forming impact through lunar tidal evolution, offering more energetic events than previously thought (e.g., Canup 2012; Cuk and Stewart 2012; Wisdom and Tian 2015; Cuk et al. 2016; Tian et al. 2017). Kokubo and Genda (2010) even found that low velocity impacts, as in the canonical Moon forming scenario, are relatively rare, so that high energetic events might be favorable.

Such high energy events might commonly form so-called synestias, as was found by Lock and Stewart (2017) and Lock et al. (2018). Synestias (Fig. 6, right panel) are partially vaporized and rapidly rotating large biconcave disk-shaped objects with an angular momentum that exceeds the corotation limit. Lock and Stewart (2017) and Lock et al. (2018) proposed that equilibration within the high entropy regions of the synestia occurs due to turbulent mixing, which could ultimately lead to a Moon with Earth-like composition. However, the synestia is differentiated into iron and silicate layers and about 75% of the silicate mass is comprised by a low-entropy inner layer followed by a 25% high-entropy outer layer. While the outer layer is mixed well, there is only little mixing between the different layers (Lock et al. 2018). The Moon itself forms in the outer region of the synestia, where droplets condense onto a Lunar seed that was accreted from large debris of the impact. Lock et al. (2018) proposed that radial transport of silicate rain droplets due to gas drag might be able to mix and homogenize the different regions, but only up to about 50% of the emerging body. Intra-impact mixing will, thus, still be needed, in case that the different bodies have significant diverging isotopic compositions (Lock et al. 2020). However, synestias generally allow a greater mixing than the canonical impact theory (see, e.g., Fig. 9 in Lock et al. 2020, where synestia simulations are compared with canonical impact scenarios for various impactors with different isotopic compositions). While synestia simulations (Lock and Stewart 2017;

Table 2 Summary of Moon-forming hypotheses and their physical plausibility

	Earth-Moon size ratio	Small metal core	Depletion of volatile elements	Isotopic similarity to Earth	Angular momentum of Earth-Moon system	hot formation of the Moon	Physical plausibility
Co-accretion[1]	maybe	no	maybe	maybe	no	maybe	maybe
Capture of planetary embryo[1]	maybe	no	maybe	likely no	maybe	maybe	likely no
Disintegrative capture[1]	maybe	maybe	maybe	no	maybe	maybe	likely no
Fission from rapidly rotating Earth[1]	no	yes	maybe	maybe	no	yes	no
Canonical impact scenario[2,3,4,5,6]	yes	yes	yes	likely no	only narrow of scenarios	yes	yes
Hit-and-run collision[7]	yes	yes	yes	likely no	likely no	yes	yes
Multiple giant impacts[8]	yes	yes	yes	likely yes	yes	yes	yes
Impact onto magma ocean dominated proto-Earth[9]	yes	yes	yes	likely yes	yes	yes	yes
High energetic impact events, synestias[10,11,12,13,14,15]	yes	yes	yes	likely yes	yes	yes	yes

[1]Wood (1986), [2]Hartmann and Davis (1975), [3]Cameron and Ward (1976), [4]Benz et al. (1986) [5]Cameron and Benz (1991), [6]Canup et al. (2021), [7]Reufer et al. (2012), [8]Rufu et al. (2017), [9]Hosono et al. (2019), [10]Canup (2012), [11]Cuk et al. (2016), [12]Kokubo and Genda (2010), [13]Lock and Stewart (2017), [14]Lock et al. (2018), [15]Nielsen et al. (2021)

Lock et al. 2018) can successfully mix bodies with a difference in ^{17}O of up to 0.3‰, canonical impacts (Canup 2004a,b, 2008) were only successful for $\Delta^{17}O \sim 0.01$‰ (Lock et al. 2020).

It finally has to be noted that, as suggested by Lock and Stewart (2017), previously investigated high energy impacts (e.g., Canup 2012; Cuk and Stewart 2012) are also likely to produce synestias, but this was not yet known at earlier times. This is also exemplified by a similarly successful mixing of planetary bodies with up to $\Delta^{17}O \sim 0.3$‰ within the simulations of Cuk and Stewart (2012) than for synestias (Lock et al. 2020). Lock and Stewart (2017) further suggest that almost all planets transition through a synestia at least once during their accretion. However, as stated above, the impactor might nevertheless have been isotopically similar to proto-Earth to explain the similarities between the Earth and the Moon. Such a similarity was recently also suggested by Nielsen et al. (2021) who found that the vanadium isotopic composition of the Moon is offset from the bulk silicate Earth's value by 0.18 ± 0.04 parts per thousand towards the chondritic value. These authors propose that this isotopic fractionation resulted from terrestrial core formation prior to the giant impact which further suggests that no post-giant impact equilibration through a synestia or other alternative impact geometries could have taken place. According to Nielsen et al. (2021), this result also implies evidence for the canonical giant impact scenario and for a common isotopic reservoir in the inner Solar System out of which the impactor and proto-Earth must have accreted. Table 2 summarizes the various Moon-forming hypotheses and their physical plausibility, discussed before.

One can see in Table 2, that, by our current understanding, multiple giant impacts, an impact onto a magma ocean proto-Earth or high energetic impact events might best reproduce the formation of the Moon. After the Moon might have formed by one of these catastrophes (or a potential combination of these) and after the magma ocean solidified, its surface was exposed to frequent impacts, a decreasing XUV flux over time and the solar wind over the whole history of the Solar System. Different processes, as described within Sect. 4.3, weathered the upper surface layers of the Moon which also modified its composition over time.

4.2 The Lunar Water Inventory

From various Apollo missions, pyroclastic glass beads indicate that their water content, which can be traced back to different eruptive events, is ≤ 100 ppm and in many cases approach the detection limits of the Secondary-Ion Mass Spectrometry (SIMS) and Fourier-Transform-Infrared spectroscopy (FTIR) detection limits (Hauri et al. 2017). Saal et al. (2008) estimated that lunar magma has lost $\geq 90\%$ of their pre-eruptive H_2O-budget via degassing.

This is in agreement with Hauri et al. (2011), who measured up to 1,200 ppm H_2O in melt inclusions contained within olivine crystals from Apollo 17 orange glass samples and showed that this lunar magma contained F, Cl, and S in abundances similar as discovered in Earth's mid-ocean ridge basalts. The findings of magmatic H_2O in volcanic lunar samples indicate that the origin of the Moon and its evolution must have had processes that allow for the accretion and retention of most volatiles that were present in the Solar System (Saal et al. 2008; Hauri et al. 2011, 2017; Füri et al. 2014; Chen et al. 2015).

Additional studies regarding water in lunar apatite that found analogous abundances to Earth's apatite, indicate further evidence that the Moon's interior contains significant amounts of magmatic water (e.g., Boyce et al. 2010; McCubbin et al. 2010; Anand et al. 2014; Barnes et al. 2014). From these findings, it can be expected that lunar magma likely contained much more water and other volatiles prior to eruption than we currently measure in the degassed glassy melt droplets (e.g., Hauri et al. 2017).

It is expected from Elkins-Tanton et al. (2011) that the lunar magma ocean had a depth of ≈ 500–1000 km and crystallized within ≈ 10 Myr if a stable crust existed throughout its crystallization phase. According to Solomon and Longhi (1977) and Meyer et al. (2010), the timescale for the solidification of the lunar magma ocean could be as long as 100–200 Myr if the loss of heat was greatly reduced. The difference of a factor of 10, demonstrates that the timescale for the lunar magma ocean solidification is highly sensitive to the details of heat loss through its surface. As discussed in Sect. 3, the mass of the Moon is too low to retain an atmosphere (Hauri et al. 2017; Benedikt et al. 2020), so that the Moon's magmatic surface will release volatiles into the vacuum of space. According to Elkins-Tanton and Grove (2011), the degassing efficiency could have been mitigated or modified by the presence of a surface crust. However, this magma ocean phase is most likely the time period during which the Moon could have gained its H_2O between the formation event and full solidification from impacting hydrous meteorites (Elkins-Tanton et al. 2011; Hauri et al. 2015, 2017).

Besides the detection of water in Apollo samples, hydroxyl and/or H_2O bearing minerals were recently also detected through combined observations of the Indian space mission Chandrayaan-1, and the flybys of NASA's Deep Impact and Cassini missions at the surface of the Moon within permanently shaded regions (PSR) at the poles (Clark 2009; Pieters et al. 2009; Sunshine et al. 2009; see also Fig. 7a). This water may be produced due to solar wind interaction with the surface (Tucker et al. 2019; Jones et al. 2020; see also Fig. 7b) and is subsequently trapped within PSR resulting in an abundance of 10 to 1000 ppm and locally even higher (Clark 2009). This was confirmed by Li et al. (2018), who found direct evidence for

Fig. 7 **a)** Distribution of surface ice (cyan) within permanently shaded regions at the Moon's south (left) and north pole (right) as detected by NASA's Moon Mineralogy Mapper instrument (Image credit: NASA). **b)** Illustration of the kinetic scheme that describes the H^+ induced H_2O cycle that is related to recombinative desorption, dissociative adsorption, adsorption, photodissociation, kinetic escape OH/H radical reaction, photo-stimulated desorption and desorption (after Jones et al. 2020)

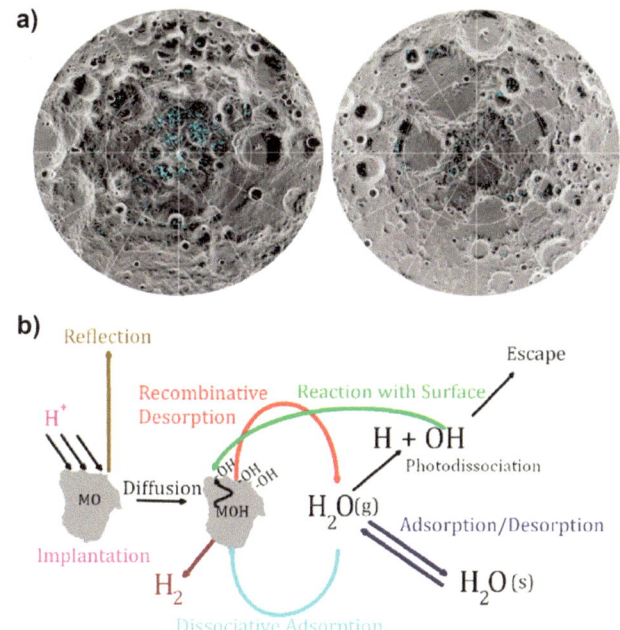

surface-exposed water ice in the polar regions of the Moon with the Moon Mineralogy Mapper instrument, again on Chandrayaan-1. Observations with the NASA/DLR Stratospheric Observatory for Infrared Astronomy (SOFIA) detected molecular water to be present at the Lunar surface with an abundance of 100 to 400 ppm (Honniball et al. 2021). These authors propose that distribution of water might be the result of local geology and may be restricted to a small latitude range. Hayne et al. (2021) further investigated the surface distribution of potential cold traps and found a total area of $\sim 40,000$ km^2 out of which 60% is located in the south with the majority at latitudes $> 80°$. This opens up a wider distribution of water ice at the Moon which might be accessible and an important resource for future human exploration (e.g., Hayne et al. 2021).

4.3 Large Impacts, Micrometeorites, and Sputtering Effects on the Lunar Surface Composition over Time

The lunar surface is covered with large and small craters, which originated by impacts. The size of these craters ranging from microscopic sizes up to several hundreds of kilometers in diameters. The larger the impactor, the larger is the excavation depth and the volume of the melt. For instance, Pieters (1993) suggest that the unusual surface composition at the lunar South Pole Aitken (SPA) basin is likely due to the exposure of mantle or lower crustal material. In addition, the ejecta of large impactors may also cover large areas on the Lunar surface. As an example, and contrary to Pieters (1993), it was also suggested that the anomalous chemical composition of SPA could also be due to the accumulation of impact ejecta from the nearside Serenitatis basin, as suggested by Wieczorek et al. (2001), Wieczorek and Zuber (2001). The SPA basin with its diameter of 2900 km has a depth between 6.2 and 8.2 km and it is one of the largest known impact craters in the Solar System, and the largest, oldest, and deepest basin recognized on the Moon (Petro and Pieters 2004). Airless bodies such as the Moon lack generally erosional processes, with the possible exception of volcanism (Spudis 1997), and as a result, impact debris accumulate at the object's surface.

According to Wieczorek and Zuber (2001) the ejecta of large impacts like the one that formed the South Pole Aitken basin could cover a large fraction of the lunar surface or even the whole body. Additionally, more frequent but smaller impactors such as micrometeorites, stir and mix the upper surface layer until today. It has been estimated from the analysis of Lunar Reconnaissance Orbiter (LRO) satellite data that the top centimeter of the lunar surface is overturned every $\geq 80,000$ years (Speyerer et al. 2016). However, one should keep in mind that the thickness of the ejecta of the early large impactors such as the one that formed the South Pole Aitken basin may exceed the thickness of the layer where micrometeorites affect.

Micro-meteoroid bombardment, for example, has shattered, fragmented, churned, and homogenised the Lunar surface (creating a loose layer of fine-grained regolith in the process), while solar wind plasma induces radiolysis, the chemical alteration of the topmost 1–3 atomic surface layers (Behrisch and Eckstein 2007). In addition, both interaction processes result in particle release, through micro-meteoroid impact vaporization on one hand and through sputtering on the other hand. During their ballistic flight, these particles become part of the Moon's exosphere before they either escape or return to the surface. Since some species have a higher probability to return to the Lunar surface than others, these processes result in a chemical alteration (fractionation) of the Lunar surface over time. In this section we will review how the micro-meteoroid flux has changed in the past ~ 4 Gyr, how much this process and sputtering have contributed to the Lunar exosphere, and to what degree they have chemically altered the Lunar surface as a function of time.

Micro-meteoroid impacts were much more frequent in the early days than they are today, though it is difficult to tell by how much exactly. Unlike large meteoroids that leave impact craters withstanding time, micro-meteoroids unfortunately leave no lasting observational evidence. The early meteoroid impact rates thus have to be estimated from the better-known flux of larger, crater and impact basin producing impactors. Several Lunar bombardment models have been proposed to explain the Lunar crater chronology. These models all fit the derived Lunar cratering chronologies well, but differ notably in shape, especially during the first Gyr, where, according to some models, the late heavy bombardment was said to have taken place.

Four models are mostly discussed by the scientific community: the 'smooth decline' model with half-lives of 50 Myr to 100 Myr (Hiesinger et al. 2012; Neukum et al. 2001; Wilhelms et al. 1987; Zahnle et al. 2007), the 'single cataclysm' model with a late heavy bombardment around 3.9 Ga ago (Ryder 2002, 2003), the 'multiple cataclysm' models with several spikes in the first Gyr (Tera et al. 1974), and the sawtooth cataclysm (Morbidelli et al. 2012). Figure 8 shows representative samples of all of these models for the first Gyr. Whereas the 'single cataclysm' model also known as Lunar cataclysm or late heavy bombardment model was most commonly favoured in the past (Ryder 2002, 2003), a paradigm-shift has led to the scientific community nowadays favouring the 'steady decrease' model (see e.g., Fernandes et al. 2013; Fritz et al. 2014; Hopkins and Mojzsis 2015; Boehnke and Harrison 2016; Mojzsis et al. 2019). According to the 'steady decrease' model, the micro-meteoroid impact flux was roughly in the order of four magnitudes higher at the beginning of the Solar System than at present-day.

Today, micro-meteoroids and solar-wind ions chemically alter the Lunar surface to about equal measures (see, e.g., Killen et al. 2012). This was not the case at the beginning of the Moon's life, though. Monte-Carlo modelling reveals that, in fact, the contribution to the exosphere due to micro-meteoroid bombardment was more than 1'000 times more substantial than the contribution by sputtering in the beginning (Vorburger et al. 2020).

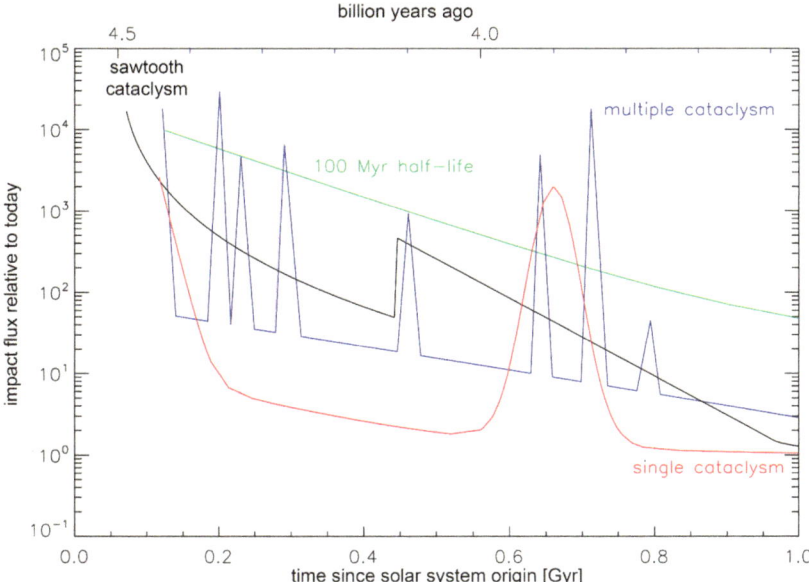

Fig. 8 Different models for the Lunar impact history for the first billion years (see text for further information)

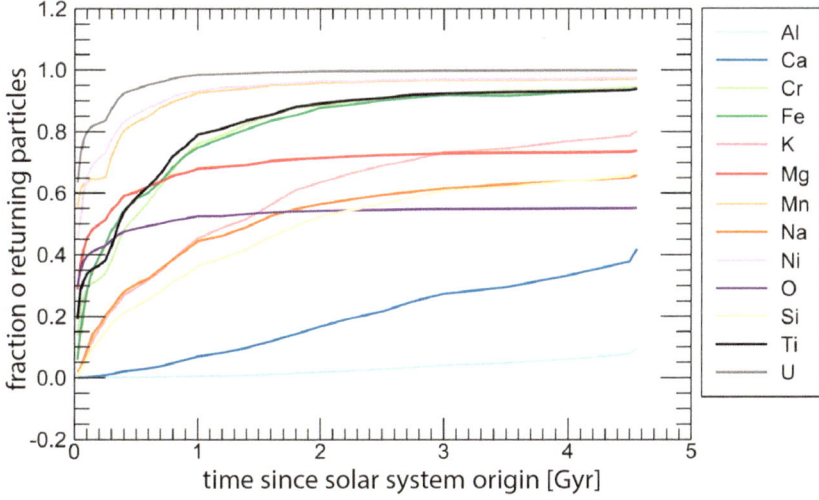

Fig. 9 The 12 most abundant species and uran as the fraction of particles that return to the Lunar surface as a function of time

In addition, as mentioned above, different return rates for different species lead to a chemical fractionation of the Lunar surface with time. Figure 9 shows for the 12 most abundant species and uran the fraction of particles that return to the Lunar surface as a function of time. Whereas at the beginning the return rates were rather low (due to the high solar UV flux resulting in high ionisation losses), the return rates today are persistently higher, and range from $\sim 10\%$ (aluminium) to almost 100% (uranium). Out of the 13 species, uranium continuously exhibits the highest return rate (due to its high mass and low ionization rate), whereas aluminum continuously exhibits the lowest return rate (due to its very high ionization rate). Over time, this leads to an enrichment of uranium and a depletion of aluminum in the Lunar surface material. Another species of interest poses the moderately volatile element potassium, because the Moon's surface material is found to be anomalously low in K/U when compared to the terrestrial planets (see e.g., Taylor and Jakes 1974; Anderson 2005; Lucey 2006; Peplowski et al. 2011; Taylor and Wieczorek 2014).

As one can see from Fig. 9, the return rate for potassium ranges from less than 1% to $\sim 80\%$ of the return rate of uranium. The persistently lower return rate of potassium when compared to uranium has thus led to a decrease in the Lunar K/U ratio with time and might explain the low K/U ratio in the Lunar surface material observed today.

Due to the species' different and varying return rates, the chemical composition of the Lunar surface as we observe it today does thus not reflect the Lunar surface composition right after the Moon's formation. These changes thus have to be factored in when trying to deduce the original Lunar surface composition from today's measurements and observations.

Saxena et al. (2019) modeled the effect of paleo space weather onto the Lunar surface and tried to reproduce the present-day potassium and sodium abundances in the regolith via sputtering of frequent CME impacts and argue that the present abundances can best be reproduced in case that the early Sun was a slow rotator (see also Sect. 2.1). Although space weather as modelled by Saxena et al. (2019), however, will modify the chemical composition of the upper surface layers, it may not be the most important process. Apollo drill cores of the Lunar surface indicate non-monotonic variations in the composition of the upper meter of regolith. This indicates that the Lunar surface has been overlain many times by impact **ejecta**. As pointed out above, one should note that the modification of the upper layers over evolutionary time scales is very complex. Interplanetary dust, cometary dust and larger meteorites permanently bombard the surfaces of airless bodies like the Moon or Mercury. This processes not only add external material to the upper regolith layers, but churns and vitrifies it. Some of these debris will escape but some will also be retained. Therefore, the results of Saxena et al. (2019) should be taken with care.

4.4 Fingerprints of Early Earth's Atmosphere on the Lunar Surface

Early Earth's atmosphere might have been susceptible to strong atmospheric escape, particularly of hydrogen from dissociated H_2O and CH_4 (e.g., Zahnle et al. 2019) and nitrogen (e.g., Tian et al. 2008; Lammer et al. 2018, 2019; Gebauer et al. 2020; Johnstone et al. 2021b; Sproß et al. 2021) due to the increased XUV flux and plasma environment from the young Sun (see Sect. 2). It was already suggested some time ago by Marty et al. (2003) and Ozima et al. (2005) that nitrogen from such an "Earth Wind" together with light noble gases could have been implanted onto the Moon. This idea was backed-up by strong variations of N, He, Ne and Ar isotopic extra-Lunar implantations into the Lunar regolith by as much as 30% (Ozima et al. 2005). These authors argued, this cannot be explained due to solar wind implantation alone, since no fractionation process at the Sun or within the solar wind could explain such strong variation.

Ozima et al. (2005) calculated the escape from a terrestrial atmosphere that was directly exposed to the solar wind due to the intrinsic geomagnetic field not yet being present and found loss rates similar to present-day Venus for an atmosphere that was not significantly expanded compared to present-day. These authors further argued that the Earth wind could have only implanted terrestrial ions into the Lunar regolith until Earth did not possess a

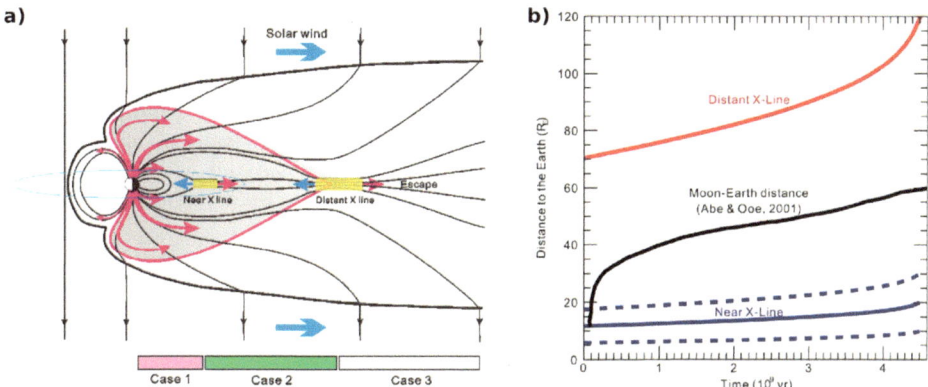

Fig. 10 The transport of Earth's atmospheric ions through the magnetotail onto the Lunar surface according to Wei et al. (2020). Panel **a**) shows the terrestrial magnetosphere, for Case 1 (pink), the Moon is located within the near X-line, for Case 2 (green), between the distant X-line and the near X-line, and for Case 3 (white) beyond the distant X-line Wei et al. (2020) point out that at present-day mainly Case 2 applies, and Case 1 only for 1% of the time. Panel **b**) illustrates the distances of the near and distant X-lines and the position of the Moon relative to the Earth over time. Figures from Wei et al. (2020)

magnetosphere, since a magnetic field would have prevented strong atmospheric escape; such an effect could have, therefore, also been used as a tracer for the onset of the geomagnetic field. By mixing their escape rates with solar wind implantation, they found scenarios for which the present nitrogen and light noble gas isotopic ratios at the Lunar surface could have been explained. Ozima et al. (2005) further pointed out that this hypothesis could be tested by measurements on the far side of the Moon since the Earth wind could have been only implanted onto the near side.

However, as more recent studies have shown, Earth's atmosphere could have been significantly expanded in the past due to the strong XUV flux from the Sun (Tian et al. 2008; Lammer et al. 2018; Johnstone et al. 2021b). Atmospheric escape might have, therefore, also been significant even though an intrinsic geomagnetic field was already present which is also supported by remnant magnetizations in zircons that date back even until 4.1 Ga (e.g., Tarduno et al. 2014, 2015). Polar outflow, for instance, might have even been much more significant in the past than at present-day (e.g., Kislyakova et al. 2020). This is also backed by the "missing xenon paradox" (e.g., Hébrard and Marty 2014; Zahnle et al. 2019), i.e., the isotopic fractionation of heavy xenon isotopes which might best be explained by strong polar outflow of hydrogen ions that dragged away Xe^{+} ions during the Archean eon (Zahnle et al. 2019).

Part of this outflow will be transported along the field lines into the tail of the terrestrial magnetosphere. Some of them will then also be transported back towards the Earth through magnetic reconnection at the distant and near x-lines (see Fig. 10), of which the distant one is always beyond the Moon's orbit and the near one depending on the space weather conditions. Since the Moon crosses the Earth's magnetotail, Wei et al. (2020) hypothesized that the outflowing charged particles from the Earth's atmosphere might not only have been implanted into the nearside of the Moon but that also the farside regolith should show a recognizable enrichment of, e.g., [15]N. These authors, in addition, suggest that measuring these implants on both sides provide a way to study the evolution of the Earth's atmosphere and magnetosphere. They are concluding that within the last 3.5 Gyr after the Lunar volcanism and magnetic field shut down, the nearside of the Moon has been impacted by $\sim$ 400 km/s atmospheric ions for most of this time, as well as the farside by $\sim$ 100 km/s atmospheric ions that were deflected back from the distant X-line (see Fig. 10). In addition, the farside is presently also impacted by about 1% of its time by 400 km/s atmospheric ions originating from the near X-line, even though this might change depending on the Earth's dipole field evolution (see Fig. 10b). The farside, however, was not impacted by atmospheric ions at times when the terrestrial dipole field was too weak or at periods of geomagnetic reversals. Before 3.5 Ga, at the time when the Moon potentially yet had its own intrinsic magnetic field, Wei et al. (2020) suggest that atmospheric ions could have been implanted at the Lunar magnetic poles.

Besides nitrogen, oxygen could have also been transported to the Lunar surface. Several different studies (e.g., Hashizume and Chaussidon 2005, 2009; Ireland et al. 2006) found anomalous oxygen isotopic components in the Lunar soil, particularly the provenance of a ^{16}O-poor component which remained weakly understood. However, already Ozima et al. (2007) proposed that oxygen ejected from the upper atmosphere and transported through the magnetotail onto the Lunar surface might explain this ^{16}O-poor component. A more recent study by Terada et al. (2017) analyzed measurements of the Japanese Kaguya spacecraft and found that it observed a significant amount of 1–10 keV O^+ ions only when the Moon crossed the Earth's plasma sheet. The high energy that would allow penetration depths that would fit to the previous measurements within the Lunar regolith, together with ^{16}O-poor mass-independent fractionation within the Earth's upper atmosphere (Thiemens 2006) let these authors conclude that this component was indeed implanted into the Lunar surface by the Earth wind with at least 2.6×10^4 ions c m^{-2} s^{-1}. Terada et al. (2017) further concluded, in agreement with Ozima et al. (2005) and Wei et al. (2020), that the Lunar surface might indeed be a window into the last few Gyr for the Earth's atmosphere.

Finally, Wang et al. (2021) suggest that the Earth wind partially contributes to Lunar surface hydration, specifically to the OH/H$_2$O abundance at the Lunar poles. When the Moon crosses the terrestrial magnetosphere, OH/H$_2$O production, which is normally triggered by the proton flux from the solar wind, was observed to not decrease, even though the proton flux at the relevant energy range of 1 keV is by two orders of magnitude lower within the magnetotail than in the solar wind. However, Wang et al. (2021) found that other energy ranges (below 325 eV, and above 4 keV) that are more efficient within the magnetotail than outside together with heavy ions from the Earth wind such as N^+ and O^+ are contributing to the production of OH/H$_2$O, thereby compensating the lower proton flux at 1 keV. O^+ ions, for instance, were measured with the Kaguya spacecraft to show an ion flux at the lunar surface when encountering the terrestrial plasma sheet of $\sim 7 \times 10^{-19}$ g cm^{-2} s^{-1} (Terada et al. 2017). The present-day micrometeoroite flux, for comparison, is by about 3 orders of magnitude higher, i.e., $\sim 7 \times 10^{-16}$ g cm^{-2} s^{-1}.

As one can see from this discussion, future Lunar missions that sample the far- and the near side but also spectroscopic missions might be able to reveal not only information about the Lunar surface and exosphere but also about the history of the Earth's magnetosphere and atmosphere.

4.5 Noble Gas Isotopes on the Lunar Surface: Archive of the Solar Wind

No indigenous noble gas component has been unambiguously identified in Lunar rocks (Marty et al. 2003). The composition of noble gases in the Lunar exosphere (largely inferred from studies of gas trapped in Lunar regolith samples) indicates that some of them might be dominated by a solar wind source (e.g. Hodges and Hoffman 1975; Wieler et al. 1996), some probably arise from the interior of the Moon or other external sources, like comets. Internal sources are supported by observations of episodic outgassing of radon (Gorenstein et al. 1974a, 1974b; Hodges 1973), and have been reviewed by Lawson et al. (2005).

Solar wind impinging on the Lunar surface might be a direct contributor of volatile species in the Lunar exosphere. Because the solar wind impinges on the Lunar surface with energies of about 1 keV/nuc H, He and other solar wind species are absorbed in the surface material (the regolith grains, rocks, ...), are trapped, and accumulate in the regolith grains. Since the noble gases do not chemical bind within the regolith grains, a fraction of the noble gases is subsequently released via diffusion to the surface to become part of the Lunar exosphere (e.g., Hinton and Taeusch 1964; Johnson 1971; Hodges 1973). Very efficient retention of implanted H and He has been shown long ago, however with prolonged ion irradiation saturation of the implantation occurs (Lord 1968).

Let us consider the noble gases that are implanted into the Lunar soil first. Assuming that the Lunar soils are saturated with noble gas atoms (Schultz et al. 1978) one obtains an equilibrium between the flux of implanted solar wind noble gas ions $f_{\mathrm{SW,i}}$ and the flux of released noble gas atoms $f_{\mathrm{rel,i}}$ of species i from the soil by diffusion, i.e. $f_{\mathrm{rel,i}} = f_{\mathrm{SW,i}}$. Once the noble gases are released into the exosphere they stay there because they do not chemically bind to the surface, and they become permanent gases of the exosphere. The loss fluxes from the exosphere are Jeans escape f_{esc} and photo-ionisation f_{ion}. These loss fluxes and loss fractions from the exosphere can be calculated, e.g., with a Monte Carlo code (Wurz et al. 2012). For a stable population of the exosphere with noble gases there has to be an equilibrium between the input by the solar wind and the loss from the exosphere:

$$f_{\mathrm{SW,i}} = f_{\mathrm{esc,i}} + f_{\mathrm{ion,i}}. \tag{1}$$

Dividing by the released flux $f_{\mathrm{rel,i}}$, we get

$$\frac{f_{\mathrm{SW,i}}}{f_{\mathrm{rel,i}}} = \frac{f_{\mathrm{esc,i}}}{f_{\mathrm{rel,i}}} + \frac{f_{\mathrm{ion,i}}}{f_{\mathrm{rel,i}}} = r_{\mathrm{esc,i}} + r_{\mathrm{ion,i}}, \tag{2}$$

with r being the loss fraction for the respective process and species. Thus, in equilibrium we get for the apparent flux of noble gases released from the surface

$$f_{\mathrm{rel,i}} = \frac{f_{\mathrm{SW,i}}}{r_{\mathrm{esc,i}} + r_{\mathrm{ion,i}}}, \tag{3}$$

which is higher than the flux by diffusion from the regolith into the exosphere because of the accumulation of noble gases in the exosphere. The density of species i in the exosphere at the surface and the column density are then

$$n_i(0) = f_{\mathrm{rel,i}} \sqrt{\frac{8 k_{\mathrm{B}} T}{\pi m_i}} \quad \text{and} \quad N_{\mathrm{C,i}} = \int_0^\infty n_i(r) dr. \tag{4}$$

A possible additional loss process might be cold trapping of some noble gases in the permanently shadowed craters near the poles (e.g. Hodges 1980), which is not included in this calculation. Since the exospheric loss rates are small the noble gases are enriched in the Lunar exosphere until the flux of escaping particles matches the influx by the solar wind. Table shows a calculation of the amount of noble gases in the Lunar exosphere based on a solar wind ion flux of

$$f_{\mathrm{SW}_i} = v_{\mathrm{SW}} n_{\mathrm{SW}_i} \approx 400 \cdot 10^3 \, \frac{\mathrm{m}}{\mathrm{s}} \times 8 \cdot 10^6 \, \mathrm{m}^{-3} = 3.2 \cdot 10^{12} \, \mathrm{m}^{-2} \mathrm{s}^{-1}.$$

Table also shows the literature data for the abundance of noble gases in the Lunar exosphere and Fig. 11 shows the resulting density profiles for the noble gases resulting from the solar wind implantation into the soil.

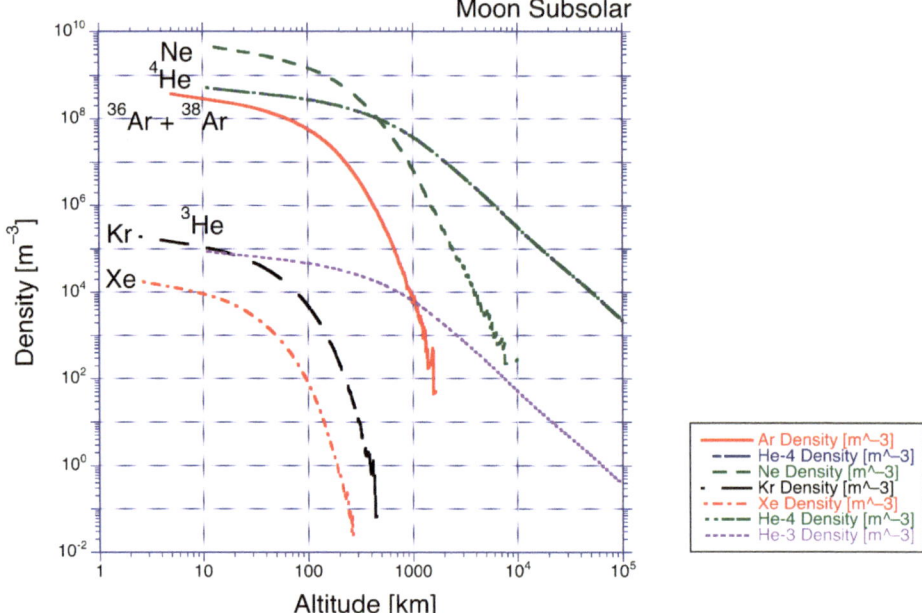

Fig. 11 Density profiles for noble gases resulting from solar wind implantation assuming equilibrium between solar wind input and exospheric escape, based on Monte Carlo calculations (Wurz et al. 2007, 2012)

For ^{4}He the model results in about a factor 5 less than the observed value, but origin of the Lunar ^{4}He is mostly from radioactive decay. The contribution of solar wind helium to the Lunar He exosphere is small, with a calculated surface density of $n_0 = 540$ cm^{-3} at the sub-solar point compared to the Apollo measurement of 2'000 cm^{-3} on the dayside (Heiken et al. 1991). The range for the He exosphere density at the surface given in Table is because it is known already from the Apollo missions that there is a diurnal variation in the densities of ^{40}Ar and ^{4}He, and likely by the other volatile species (Stern 1999; Benna et al. 2015). There is a similar situation for argon as is for helium, the ^{40}Ar is from radioactive decay of ^{40}K. The ^{40}Ar density from the Apollo measurements at the surface is $n_0 = 40'000$ cm^{-3} at sunrise indicating some condensation on the night side (Stern 1999). From recent LADEE measurements about $n_0 = 8'000$ cm^{-3} at the subsolar meridian (Benna et al. 2015). Table 3 and Fig. 11 show the combined ^{36}Ar and ^{38}Ar density of solar wind origin as $n_0 = 374$ cm^{-3}. The Apollo measurements gave a ^{40}Ar:^{36}Ar ratio of approximately 10:1, implying a ^{36}Ar surface density of about 800 cm^{-3}, which means that the solar wind contribution of argon to the Lunar exospheric argon inventory is at most half. For Ne we get reasonable agreement between the model and the observations, we calculate a surface density of 4900 cm^{-3} that compares favourably with the measurements $(4$–$10) \cdot 10^3$ cm^{-3} (Heiken et al. 1991). For Kr and Xe the model predictions are much lower than the existing upper limits from observations.

A similar calculation has been made for solar wind protons being implanted in the Lunar soil recently (Wurz et al. 2012), and assuming that the final released species is H$_2$. A density of $n_0 = 2100$ cm^{-3} was predicted from this model, compared to the Apollo measurement of $(2.5$–$9.9) \cdot 10^3$ cm^{-3} on the dayside (Heiken et al. 1991). The recent measurements from the LAMP UV spectrograph on the Lunar Reconnaissance Orbiter gave a value of $n_0 = 1200 \pm 400$ cm^{-3} (Stern et al. 2013).

Table 3 Calculation of the amount of noble gases in the Lunar exosphere assuming equilibrium between solar wind input and exospheric escape, based on Monte Carlo calculations (Wurz et al. 2007, 2012)

Species	SW flux	Loss fractions	Model results, sub-solar point	Literature values
He	$f_{SW}(H) \times 0.03$	$r_{esc,He} = 0.15$	$n_0 = 5.44 \cdot 10^2$ cm^{-3}	$n_0 = (2\text{--}40) \cdot 10^3$ cm^{-3}, day–night (Stern 1999)
		$r_{ion,He} = 1.3 \cdot 10^{-3}$	$N_C = 1.72 \cdot 10^{10}$ cm^{-2}	$N_C = 1 \cdot 10^{11}$ cm^{-2} (Killen and Ip 1999)
^{36}Ar, ^{38}Ar	$f_{SW}(H) \times 3.55 \cdot 10^{-6}$	$r_{esc,Ar} = 0$	$n_0 = 3.74 \cdot 10^2$ cm^{-3}	$n_0 = (0.3\text{--}30) \cdot 10^2$ cm^{-3}, day–night (Stern 1999)
		$r_{ion,Ar} = 7.8 \cdot 10^{-5}$	$N_C = 1.90 \cdot 10^9$ cm^{-2}	
Ne	$f_{SW}(H) \times 7.41 \cdot 10^{-5}$	$r_{esc,Ne} = 5 \cdot 10^{-6}$	$n_0 = 4.22 \cdot 10^3$ cm^{-3}	$n_0 = 2 \cdot 10^3$ cm^{-3}, $N_C = (4\text{--}20) \cdot 10^{10}$ cm^{-2} (Killen and Ip 1999)
		$r_{ion,Ne} = 8.8 \cdot 10^{-5}$	$N_C = 4.54 \cdot 10^{10}$ cm^{-2}	$n_0 = 3 \cdot 10^3$ cm^{-3} (Das et al. 2016)
Kr	$f_{SW}(H) \times 1.91 \cdot 10^{-9}$	$r_{esc,Kr} = 0$	$n_0 = 1.90 \cdot 10^{-1}$ cm^{-3}	$n_0 < 2 \cdot 10^4$ cm^{-3} (Stern 1999)
		$r_{ion,Kr} = 1.26 \cdot 10^{-4}$	$N_C = 4.25 \cdot 10^5$ cm^{-2}	
Xe	$f_{SW}(H) \times 1.86 \cdot 10^{-10}$	$r_{esc,Xe} = 0$	$n_0 = 1.82 \cdot 10^{-2}$ cm^{-3}	$n_0 < 3 \cdot 10^3$ cm^{-3} (Stern 1999)
		$r_{ion,Xe} = 1.60 \cdot 10^{-4}$	$N_C = 2.82 \cdot 10^4$ cm^{-2}	

It finally has to be noted that the solar wind might also provide information on the isotopic composition of the solar atmosphere, the bulk Sun and hence the protoplanetary nebula, since it was found that early solar wind was trapped in solar gas rich soils and breccias on the Lunar surface (Anders and Grevesse 1989; Pepin et al. 1999; Palma et al. 2002). ^{36}Ar/^{38}Ar ratio from such samples that did not originate from the lunar mantle but originated from Ar isotopes of the early solar wind show a divergence compared to the modern solar wind ratio, which indicates that the solar wind at some time in the past had a ^{36}Ar/^{38}Ar ratio that was above today's values (Becker et al. 2003).

5 Mercury vs. "Exo-Mercurys"

5.1 Mercury: The Innermost Planet in the Solar System

5.1.1 The Parameter Space of Mercury

Every planet in the Solar System is unique in some respect, making all eight of them precious as archetypes for the studies of exoplanets. Among them, Mercury (see Table 4 for a list of Mercury's parameters) is probably the one with the most uniqueness (Solomon 2003); it is

Table 4 Mercury's parameter list (from D. Williams, NASA planetary fact sheet)

Mass (10^{24} kg)	0.33	Rotation Period (hours)	1407	Orbital Period (days)	88	Mean Temperature (C)	167
Diameter (km)	4879	Length of Day (hours)	4222	Orbital Velocity (km/s)	47	Surface Pressure (bars)	0
Density (kg/m^3)	5427	Dist. from Sun (10^6 km)	57.9	Orbital Inclination (°)	7	Global Magnetic Field	Yes
Gravity (m/s^2)	3.7	Perihelion (10^6 km)	46	Orbital Eccentricity	0.205	Bond albedo	0.088
Escape Velocity (km/s)	4.3	Aphelion (10^6 km)	69.8	Obliquity to Orbit (°)	0.034	Visual geom. albedo	0.142

the closest to the Sun, the smallest in size and lightest in mass, but these are only the most obvious peculiarities.

In view of possible future direct observations of exoplanets, it is worth noting the way Mercury reflects light. The bond albedo and the geometric albedo are 0.09 and 0.14, respectively, which are very similar to the Moon values, and over certain terrains the albedo is greater than that of similar terrains on the Moon (see, e.g., Mallama 2017). Mercury's reflection depends on its uppermost layer, which is made of regolith, consisting of fragmental material derived from the impact of meteoroids over billions of years that covers more coherent bedrock, formed by processing of the older material (Eggleton 2001). With respect to the Moon, that of Mercury is more mature, with smaller grain sizes and larger proportions of glassy particles (Langevin 1997), probably because of the differences in the respective size and energy distribution of impactors. Obvious differences in the solar wind, cosmic ray and UV fluxes also play a role when comparing Mercury to the Moon and other objects.

In terms of internal structure, and its importance as a paradigm of planetary evolution, Mercury occupies a very solitary position in the distribution of Solar System objects as well. Two facts are of uttermost importance in this respect: Mercury is, together with the Earth, the only inner planet with an **intrinsic** magnetic field (Ness et al. 1975), and it also shares with the Earth the highest mean density in the Solar System.

The dipole moment of Mercury is about 330 nT. It is much smaller than the Earth's one, and it has barely the required strength to efficiently interact with the solar wind in a similar way (i.e., resulting in a bow shock, a magnetosheath, and a magnetosphere). Hence the magnetosphere of Mercury is a miniature version of the Earth's one, and, in comparison, Mercury occupies a much larger fraction of it. The observation of such a magnetic field or a magnetosphere is the indirect evidence (although not conclusive) that the planet has an electrically conducting fluid shell surrounding a solid inner core.

Mercury's mean density is almost equal to that of the Earth (5.44 g/cm^3 and 5.52 g/cm^3, respectively), which is the highest among all planets. However, the Earth's has a much larger internal pressure, which compresses the core and, once this effect is considered, it results that Mercury probably contains a much larger fraction of iron than the Earth and hence of all other planets. Therefore, the ways the masses and volumes are subdivided in core, mantle and crust are quite different. The core, made of liquid iron and other metals accounts for $\sim 42\%$ of Mercury's volume (i.e., $> 60\%$ of its mass), and only 17% of the Earth's volume (Harder and Schubert 2001; for the inner core to be molted: Margot et al. 2007). The silicates that form the mantle fill a shell of only ≈ 400–450 km thickness (Rivoldini and van Hoolst 2013; Wardinski et al. 2019), while at the Earth this is the most of the volume. Finally, the crust at Mercury, differently from the Earth, has no plate tectonics as the presence of impact craters suggests (Spudis et al. 1998), so that the signs of the impacts on the surface of Mercury can virtually last forever, even though space weathering and surface processes surely play a role (Orsini et al. 2014).

In terms of being the easiest paradigm for close-in exoplanets, we must note that it is not only the closer to its parent star, but also it has the most eccentric (0.205) and inclined (7°) orbit of any other planet in the Solar System. This implies also that the ratio between solar radiation at perihelion and aphelion is the largest (2.3). Even if not the hottest, the surface of Mercury experiences the greatest thermal excursion among all the planets (from ~ 100 K at night to ~ 700 K at the subsolar point during part of its orbit). Such extreme thermal gradient should be accounted when considering the net mass loss from the planet (Kang et al. 2021).

Close-in extrasolar planets are easily locked to synchronous rotation with their host star because of the strong tidal interaction (Gladman et al. 1996). However, in the Solar System we don't have an example of a tidally-locked planet. Only moons experience complete tidal locking, and the closest, precious example among planets is the 3:2 spin-orbit resonance of Mercury (~ 59 vs ~ 88 days, respectively), which has likely been achieved through similar dissipative processes. At perihelion, however, Mercury experience (very slow) retrograde motion for a couple of weeks, so that apparently it behaves almost like a tidally locked planet. Since the obliquity of Mercury is close to 0 it does not experience seasons, but thanks to the 3:2 resonance the same hemisphere always faces the Sun at alternate perihelion passages, producing the so-called hot poles regions at 0° and 180° longitudes (in opposition to the cold poles at 90° and 270°), which may help in understanding the physics of "eyeball" exoplanets, which are tidally locked planets, for which tidal locking induces spatial features in the geography or composition of the planet resembling an eyeball (Angerhausen et al. 2013).

5.1.2 Formation Hypotheses of Mercury

As already outlined in the previous section, Mercury has some peculiar physical and chemical characteristics that must be considered before constructing a valid formation hypothesis. Besides its high density and metal/silicate ratio, which was already known since Mariner 10 (1974–1975), the more recent MESSENGER mission (2008–2015) revealed and confirmed two additional and slightly surprising characteristics. In contrast to the high metal/silicate ratio that would suggest a strong depletion of volatile and moderately volatile elements if one assumes solar composition, MESSENGER showed that Mercury's surface is indeed quite volatile rich (e.g., Peplowski et al. 2012; Weider et al. 2015; Nittler et al. 2018). The planet has a very high ratio of K/Th (see Fig. 12a) and a surface abundance of ~ 1300 ppm K (Peplowski et al. 2012) which is well above Earth's value but yet below chondritic, while the moderately volatile lithophile elements Si, Ca, Al, and Mg are approximately chondritic (Weider et al. 2015). In addition, the volatile elements Na (2.6–5 wt%) and Cl (1200–2500 ppm) are highly abundant and near-chondritic as well (Peplowski et al. 2015; Evans et al. 2015; see also Fig. 12b), while the Cl/K ratio is comparable to Mars (Evans et al. 2015). S, another moderately volatile element, is also highly abundant on its surface with ~ 4 wt% (Nittler et al. 2011; Evans et al. 2012), a value that is about an order of magnitude higher than at Earth. Even substantial water-ice deposits were discovered within the permanently shadowed craters at the poles (Chabot et al. 2018).

Besides the high amounts of volatiles, MESSENGER also confirmed that Mercury is extremely reduced (McCubbin et al. 2012) and its surface highly depleted in Fe of any form (Murchie et al. 2015). The oxygen fugacity $f(O_2)$ of Mercury's interior was estimated

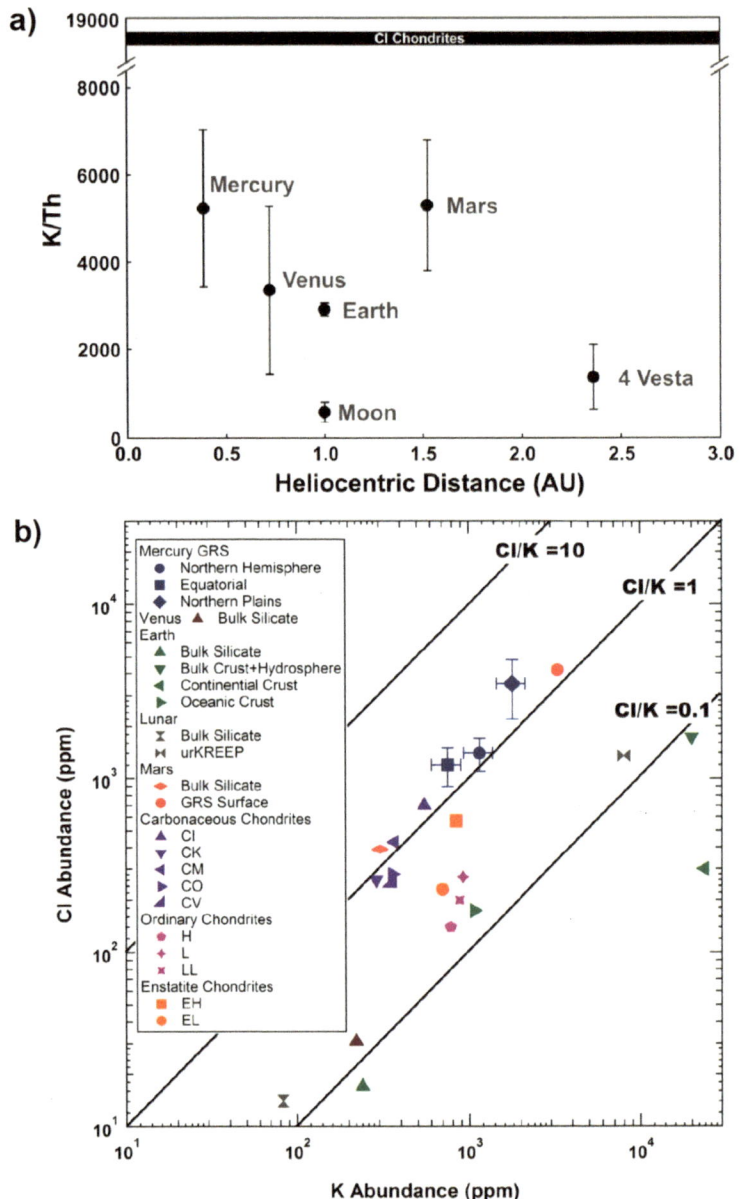

Fig. 12 **a**) K/Th for different Solar System bodies and for CI Chondrites (figure from McCubbin et al. 2012). **b**) Cl vs K for different Solar System bodies and chondrites (figure from Evans et al. 2015)

by McCubbin et al. (2012) to be between IW-6.3 and IW-2.6, with the upper limit being relatively unlikely (Zolotov et al. 2013). Here, IW stands for the $f(O_2)$ of the standard equilibrium reaction buffer between iron metal and wüstite. Therefore, IW-6.3 describes an $f(O_2)$ that is $10^{-6.3}$ below IW. Such exceptionally low $f(O_2)$ suggests extremely reduced conditions during the accretion period of the Hermean protoplanet. Earth, for comparison,

⌂ Springer

witnessed significantly less reduced conditions during formation (e.g., Frost et al. 2008); its modern mid-ocean ridge basalts, a tracer for early conditions, show an upper mantle $f(O_2)$ of IW+2 (Cottrell and Kelley 2011). The protocrust of Mars was estimated to have an $f(O_2)$ between IW-1 and IW+1 (e.g., Hirschmann and Withers 2008). Mercury is, there-fore, one of the most reduced objects in the Solar System, and only enstatite chondrites and aubrites, as well as some Calcium-Aluminum-rich inclusions (CAIs) potentially show similarly low oxygen fugacities (e.g., Ebel and Stewart 2017). Any formation hypothesis of Mercury should, therefore, be able to explain, i) the high iron to silicate ratio, ii) the surpris-ingly high abundance of volatiles, and iii) the extremely reduced nature of the planet. By now, however, none of the theories below either addresses all these characteristics or is even able to explain all of them sufficiently.

One of the oldest theories proposed to explain the high iron/silicate ratio of Mercury is the giant impact hypothesis (Benz et al. 1988). In this proposal, a smaller body impacts with high energy onto a bigger and differentiated proto-Mercury with a similar metal/silicate ratio as the Earth to strip off a substantial part of the silicate mantle while the two iron cores merge into one. Even though part of the stripped-away mantle will be reaccreted back onto the remaining protoplanet, Benz et al. (2007) found that high energy impacts might indeed be able to account for the anomalously high metal to silicate ratio at Mercury, and, therefore, also for its mean density. This is in contrast with a more recent numerical study by Carter et al. (2015) which suggests that reaccreting debris will be problematic and that giant impacts cannot explain such a high variability in the metal/silicate ratio. Furthermore, Stewart et al. (2013, 2016) found that such high energy impact events might even vaporize the entire mantle, and would likely recondense back onto the core, simulations out of which the concept of a synestia evolved (Lock and Stewart 2017 and Lock et al. 2018), one of the theories for the Lunar Moon forming impact (see Sect. 4.1). The giant impact model can, moreover, not account for the volatile rich mantle of Mercury. Whether such an origin might be reconcilable with the low oxidation state of the planet remains by now unknown.

Another theory related to the giant impact theory was developed by Asphaug and Reufer (2014), i.e., the theory of inefficient accretion. Their numerical hydrocode simulations show that Mercury could have been stripped off its mantle by one or several high-speed hit-and-run collisions (with less energy than in the original giant impact model) with a larger target planet (see Fig. 13). While proto-Mercury loses most of its mantle, the silicates get then reaccreted onto the larger embryo which might have ended up accreting onto Venus or Earth. Asphaug and Reufer (2014) also illustrated that if Mercury and Mars are relics out of 20 original planetary embryos that mainly formed the other terrestrial planets, then it is sta-tistically likely that one of the remaining ends up with a stripped-off mantle. While these less energetic hit-and-run collisions might allow the survival of more volatiles than in the giant impact scenario, it is yet neither clear whether this would be sufficient, nor how likely it is that the remaining proto-Mercury will not accrete onto the larger body (e.g., Ebel and Stewart 2017). Helffrich et al. (2019), however, argue that a stripping of Mercury's mantle through impacts might at least be compatible with the high sulfur and low iron composition of its surface.

A more recent study by Chau et al. (2018) simulated different impact scenarios, i.e., the giant impact, hit-and-run, and a multiple collision scenario. They found that a single giant impact as well as the hit-and-run scenario require highly tuned impact parameters to achieve Mercury's mass and Fe/Si ratio, while a multiple-collision impact scenario escapes fine-tuning, and allows a volatile rich Hermean surface due to the relatively low impact energies. However, this scenario is constrained by timing several collisions within a relatively short timeframe, and by the volatile-rich composition of the planet's surface. Chau et al. (2018)

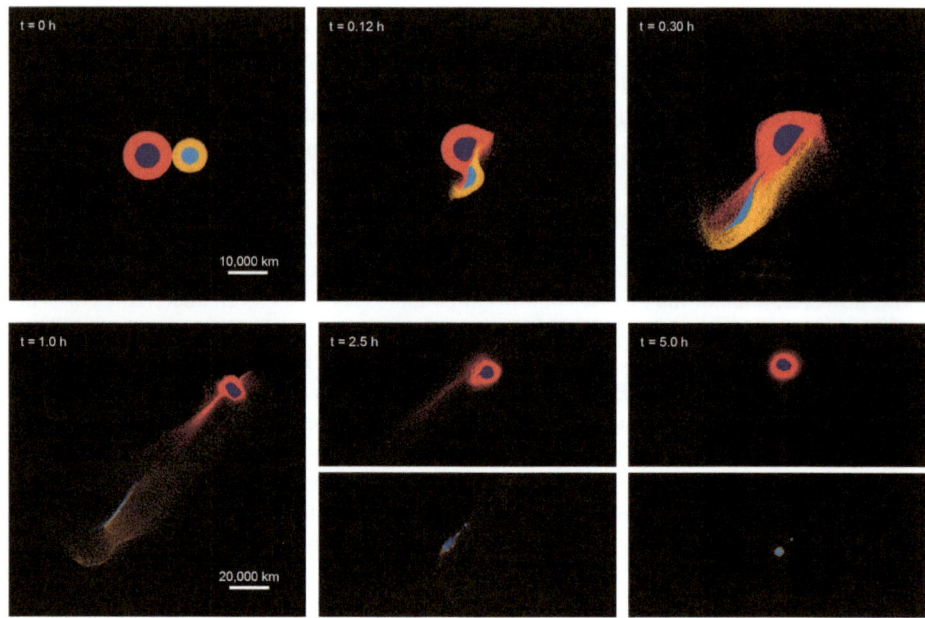

Fig. 13 Forming Mercury ($M_1 = 4.52\ M_{\mathrm{Mercury}}$) through a single hit and run collision with a more massive object ($M_2 = 0.85\ M_{\mathrm{Earth}}$) according to Asphaug and Reufer (2014). Tens of hours after the collision, the mantle of M_2 is dispersed and a metallic remnant remains. The bigger body might further accrete onto Earth and Venus. Figure from Asphaug and Reufer (2014)

finally conclude that it might be possible to form Mercury through collisions, but that it is difficult. The latter is in agreement with a study by Clement et al. (2019) who found that it is highly unlikely with a probability of less than 1% to form Mercury, and the Mercury-Venus dynamical spacing together with the correct terrestrial planets' orbital excitations through a set of collisions. In a subsequent work, however, Clement et al. (2021) tested another hypothesis based on Volk and Gladman (2015) who suggested that Mercury might have been the lone surviving relic out of a cataclysm of several large planets inside the orbit of Venus. While Volk and Gladman (2015) assumed the cataclysmic planets to have masses similar or higher than Earth, Clement et al. (2021) performed numerical simulations with a multiplanet system inside of Venus of Mars-sized mass each. They found that perturbations and collisions could have indeed resulted in a higher metal/silicate ratio inner planet. In a follow-up study, Clement and Chambers (2021) were even able to recreate Mercury and its orbital spacing with Venus through collisions and a mass-depleted disk of ~ 0.1–$0.25\ M_{\mathrm{Earth}}$ inside 0.5 AU. However, these authors neither address the potentially volatile rich mantle of Mercury nor its reduced nature.

Finally, O'Neill and Palme (2008) suggested that collisional erosion, i.e., the stripping of mantle and crust material through smaller impacts, could have significantly altered the iron/silicate ratio of planetary bodies. Svetsov (2011) subsequently simulated whether such impacts could have also stripped away most of the mantle of Mercury and found that a proto-Mercury with an iron/silicate ratio comparable to the Earth could have been altered to the present Hermean composition through small impactor with an impact velocity of ~ 30 km/s that together added up to a mass bigger than present Mercury. However, as already written

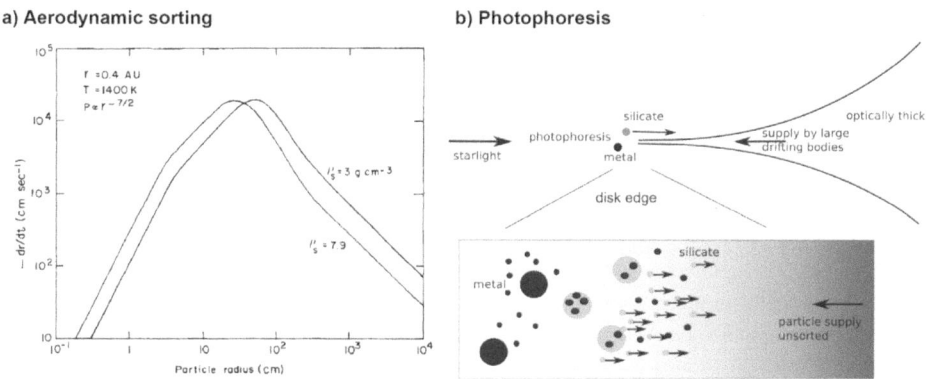

Fig. 14 While aerodynamic sorting (**a**), as proposed by Weidenschilling (1978), preferentially drags silicate boulders into the Sun (illustrated by the drag-induced radial velocity difference between the less dense "silicate" and denser "iron" in (**a**) as calculated by Weidenschilling 1978), photophoresis, as suggested by Wurm et al. (2013), pushes silicate grains outwards, away from the Sun. For the latter, the nebula gas already has to be dissipated, while aerodynamic sorting only works during the nebula phase. Both processes are potential contributors to Mercury's anomalous Fe/Si ratio. Left figure from Weidenschilling (1978); right from Wurm et al. (2013)

above, the more recent study by Clement et al. (2019) found that such impacts will likely not significantly alter the Fe/Si ratios of the proto-planets due to reaccretion.

Besides a collisional origin, there are two further main theories that were developed to explain Mercury's extraordinary characteristics, i.e., the post-accretion evaporation models and metal-silicate fractionation within the solar nebula. The post-accretion evaporation model was first suggested by Cameron (1985) who proposed that very high temperatures within the solar nebula led to the volatilization of the Hermean mantle which was later carried away by the strong solar wind of the young Sun. Fegley and Cameron (1987) calculated that about 70–80% from an initially chondritic mantle should have been evaporated to lead to the present metal/silicate ratio of Mercury. However, this model is neither in agreement with the volatile-rich remaining mantle as measured by MESSENGER, nor with the present knowledge about atmospheric escape, since the initial protoplanet with a mass more than twice as big as the present Mercury, would likely have been too massive to allow for such strong thermal or non-thermal escape processes at Mercury's orbit around the young Sun to evaporate a total amount greater than its present-day mass.

To form Mercury through metal-silicate fractionation within the solar nebula, several different processes were by now suggested. The first one was already put forward by Weidenschilling (1978) who suggested that aerodynamic sorting prior to accretion might result in a metal-rich inner nebula (Fig. 14a). While the gas drag decays the orbit of lighter silicate-dominated boulders with sizes > 1 m faster than denser metal-rich bodies, the latter might get enriched within the feeding zone of proto-Mercury, thereby leading to an elevated Fe/Si ratio within the accreting proto-planet. Another fractionation process was explored by Wurm et al. (2013) who suggested that photophoresis (Fig. 14b) might contribute to the anomalous metal to silicate ratio of Mercury. Photophoresis could fractionate millimeter-sized high thermal conductivity materials such as iron from lower thermal conductivity solids such as silicates since the latter would be preferentially pushed outwards into the colder and optically thin disk due to the thermal gradient that would build up within the silicate grains. This process would deplete the innermost part of the solar nebula from silicates, as pointed out by Wurm et al. (2013). For this process to work the feeding zone of Mercury should have been optically thin, which might have been the case as observations of cleared inner disks within extrasolar nebulae might suggest (e.g. Espaillat et al. 2014). Both processes – aerodynamic fractionation and photophoresis, however, could have counteracted each other, and, as pointed out by Ebel and Stewart (2017), a model addressing different chemical and physical factors influencing small solids within the inner disk might be needed to better justify the influence of these processes onto the Hermean iron/silicate ratio.

Another fractionation process within the solar nebula was put forward by Ebel and Alexander (2011) who found that the equilibrium condensation in systems enriched in C-enriched and O-depleted dust can form condensates with Fe/Si ratios reaching up to 50% of bulk Mercury at a temperature of 1650 K and a total pressure of 10 Pa, since Si remains in the vapor at such conditions. Ebel and Alexander (2011) further point out that such conditions can explain the formation of enstatite chondrite and aubrite parent body compositions and might also explain Mercury's anomalous composition since the planet could have formed from enstatite chondrite parent planetesimals. That Mercury's feeding zone might have indeed an enstatite-rich environment was also proposed by Pignatale et al. (2016) through simulating the vertical settling and radial drift of dust grains using a thermodynamic equilibrium model. However, origin mechanism proposed by Ebel and Alexander (2011) relies on the assumption that C-rich dust reached the inner Solar System, which is yet unknown (e.g., Peplowski et al. 2016; Vander Kaaden and McCubbin 2015).

A final fractionation process suggested so far was first put forward by Hubbard (2014) who proposed magnetic fractionation. They argue that dust grains rich in metallic iron can attract each other magnetically, and that magnetically induced collision speeds might be high enough to knock-off loosely bound silicates from the grains, thereby enriching and growing metal grains. Hubbard (2014) further argues that the magnetic field requirements for "magnetic erosion" are only fulfilled within the inner disk. This work motivated Kruss and Wurm (2018) to perform experiments on how magnetic fields up to 7 mT influence the aggregation and size of dust clusterings. They found that the cluster size depends on the strength of the magnetic field and the ratio between iron and quartz. If planetesimal formation is sensitive to the largest aggregates, Kruss and Wurm (2018) conclude, then planetesimals will preferentially grow iron-rich in the inner region of protoplanetary disks. In a follow-up study, Kruss and Wurm (2020) extended their experiments by adding pure quartz aggregates to the iron-rich aggregates. They found that their mechanism still works, but a certain fraction of iron-rich material has already to be present to trigger the magnetic enrichment. In case that there are more than 80% nonmagnetic aggregates, the mechanism will halt. A sufficient iron fraction, however, should have been present in the inner disk, as Kruss and Wurm (2020) argue. While these formation hypotheses can address the high iron to silicate ratio of Mercury and might partially also address its volatile rich mantle, none of these models can address the extremely reduced nature of Mercury.

Furthermore, it also has to be pointed out that there might be an additional way to at least reproduce Mercury's high Fe/Si ratio and its volatile rich mantle, which we will call "accretional evaporation", but a comprehensive study on the following idea is yet missing. As discussed in Sect. 3 it has been shown by Young et al. (2019) and Benedikt et al. (2020), small planetary embryos are significantly affected by the loss of volatile and moderately volatile elements. At early stages of planetary accretion and after these bodies are set free from the dissipating solar nebula, such elements will outgas from the magma ocean of the embryos and immediately be lost to space through hydrodynamic escape. This evaporation is strongest for smaller bodies close to the Sun and might lead to a significant loss of the silicate mantle in case that the magma ocean can be protracted over several Myr through frequent impacts, which should have been likely as was at least already shown for Mars

Table 5 Summary of different Mercury-formation hypotheses

	High metal/silicate ratio	Volatile rich surface	Extremely reduced nature of Mercury
Classical giant impact hypothesis[1, 2, 3, 4, 5]	only fine-tuned parameters	no	unknown
Inefficient accretion[6, 7, 8]	only fine-tuned parameters	only fine-tuned parameters	unknown
Multiple collision scenario[9, 10]	unlikely	unlikely	unknown
Cataclysmic relic[11, 12, 13]	likely yes	unknown	unknown
Erosion through smaller impacts[14, 15]	maybe	unknown	unknown
Post-accretion evaporation model[16, 17]	likely no	no	unknown
Aerodynamic sorting[7, 18]	maybe counteracting photophoresis	no	unknown
Photophoresis[7, 19]	maybe counteracting aerodynamic sorting	no	unknown
Equilibrium condensation[20]	maybe – depending on whether C-rich dust reached inner Solar System	unknown	likely yes
Magnetic fractionation[21]	maybe	maybe	unknown
Accretional evaporation[22]	likely yes	yes	unknown

[1]Benz et al. (1988), [2]Benz et al. (2007), [3]Carter et al. (2015), [4]Stewart et al. (2013), [5]Stewart et al. (2016), [6]Asphaug and Reufer (2014), [7]Ebel and Stewart (2017), [8]Helffrich et al. (2019), [9]Chau et al. (2018), [10]Clement et al. (2019), [11]Clement et al. (2021), [12]Volk and Gladman (2015), [13]Clement and Chambers (2021), [14]O'Neill and Palme (2008), [15]Svetsov (2011), [16]Cameron (1985), [17]Cameron et al. (1987), [18]Weidenschilling (1978), [19]Wurm et al. (2013), [20]Pignatale et al. (2016), [21]Hubbard (2014), [22]hypothesis discussed in this review article

(Maindl et al. 2015). When these planetary embryos grow, either through collisions with each other or through accretion of chondritic material, the escape will be significantly diminished after the escape regime changes from hydrodynamic to Jeans escape, a shift that might happen after the embryo reaches several Moon masses (Mercury's core holds about 4 Moon masses). After this change in the escape regime, the proto-planet could proceed accreting volatile rich material, for instance from carbonaceous chondrites, thereby building up a small volatile-rich mantle until accretion stops. However, no one modeled this potential formation theory in detail, and it also remains unclear so far, how such a process might have affected the redox state of Mercury.

Table 5 summarizes the before discussed formation hypotheses of Mercury. As one can see, there is at present no hypothesis that either sufficiently explains the planet's high metal/silicate ratio, its volatile rich surface and the extremely reduced nature within just one theory, or that is sufficiently well studied to already allow answering all of these three characteristics. It is, furthermore, certainly also likely that several of the mentioned hypotheses and processes acted together in the formation of the Solar System's innermost planet.

Further studies are, therefore, needed to resolve the puzzle of Mercury's formation. It might, however, be likely that several of the above-mentioned processes might have played out together to form the innermost planet of the Solar System.

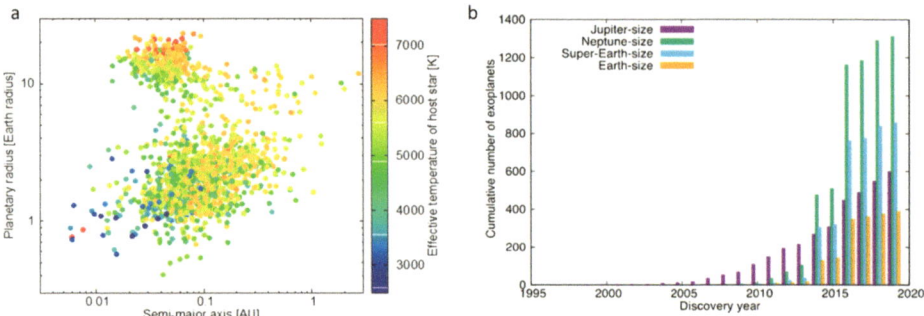

Fig. 15 Number and property of exoplanets detected until 2020. Panel (**a**) Two-dimensional distribution of discovered exoplanets in sizes and orbits smaller than 3 AU. The colors show the different effective temperatures of their host stars. Panel (**b**) Cumulative number of discovered exoplanets with different sizes; Jupiter-size (violet, $r_{pl} > 6\ R_{Earth}$), Neptune-size (green, $6\ R_{Earth} > r_{pl} > 2\ R_{Earth}$), super-Earth-size (light blue, $2\ R_{Earth} > r_{pl} > 1.25\ R_{Earth}$) and Earth-size (orange, $1.25\ R_{Earth} > r_{pl}$). The data has been taken from an open exoplanet catalog database (https://exoplanetarchive.ipac.caltech.edu, 27/7/2020)

5.2 Close-in Rocky Exoplanets with High Metal/Silicate Ratios

Exoplanets are unique objects in planetary science, because they have a wide variety of characters regarding masses, sizes, orbital elements, and host star types; some are similar to the Solar System planets. The size of discovered exoplanets ranges from sub-Earth- to super-Jupiter-size, and they orbit around the various type of their host stars such as Sun-like stars and M dwarfs with semimajor axis ranging from one hundredths of Earth's orbit, as shown in Fig. 15a.

After the initial discovery phase of extrasolar planets in which gas-giants were discovered, the focus has now shifted to planets with less than ten Earth-masses, corresponding to objects that are smaller than two, or three times the Earth. Was the mere discovery of objects the main achievement of the past years, we are now approaching the realm of studying their nature. It has now become clear that there are at least two different species of low-mass planets. One kind has extended atmospheres that are presumably hydrogen-dominated, others at first glance seem to be bare rocks.

Most of them are close to their host stars and discovered by transit measurement (more exactly, by the Kepler transiting exoplanet survey) because the measurement has a strong bias in favor of close-in planets. Also, their masses distribute over a range from Mars-mass to a few Jupiter-masses (see also Winn and Fabrycky 2015, for a review). In the last decade, the number of detected small exoplanets rapidly have increased, as shown in Fig. 15b.

Up to now 96.8% of the known planets smaller than 3 R_{Earth} orbit closer than Mercury in our Solar System (http://exoplanet.eu). The measurements of the radii of the planets alone, however, gives us only an incomplete picture; what we need are measurements of the radii and the masses, which would give us the density of the planets. Fridlund et al. (2020) argued that an accuracy of better than 15% for the masses and better than 5% are required to find out what the nature of the planets are (Fig. 15a). Then, over 1000 exoplanets whose radii are less than 2 R_{Earth} have been discovered until 2020. Such exoplanets are called super-Earths or sometimes called Earth-like planets, although they could be super-Mercurys or sub-Neptunes, terrestrial planets that could not get rid of their primordial atmospheres after the disk dispersed (Lammer et al. 2020b).

These low mass planets are one of the most common group of exoplanets while the first discovered exoplanet 51 Pegasi b was a Jupiter-sized planet (Mayor and Queloz 1995).

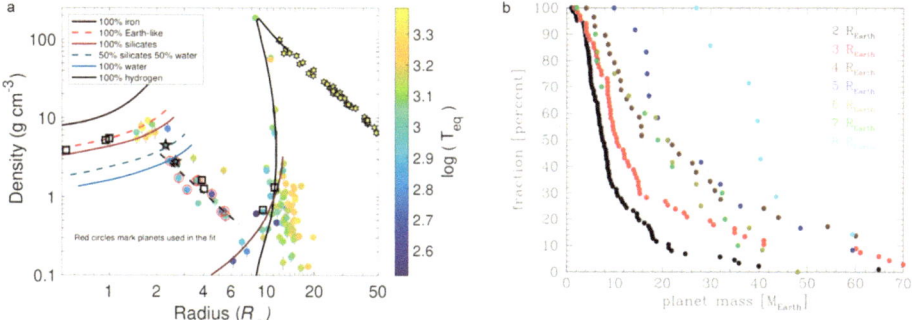

Fig. 16 Panel (**a**) Density–radius diagram of planets orbiting G-type host stars with masses determined within at least 15% and radii with an accuracy better than 5%. The two-star symbols represent TOI-763 b and c. The black squares are the Solar System planets, and the bright yellow star symbols at radii between 12 and 50 R_{Earth} are red dwarf stars from Persson et al. (2018). The theoretical mass-radius curves are from Zeng et al. (2016) except the H-He model taken from Baraffe et al. (2003, 2008). The black dashed line represents a linear fit to the ice planets marked with red circles (figure and description taken from Fridlund et al. 2020). The colors correspond to the different equilibrium temperatures T_{eq} of the planets. Panel (**b**) Cumulative frequency of planet masses for different planet radii between 2 and 8 R_{Earth}. For example, 50% of the planet with 2 R_{Earth} have masses larger than 6 M_{Earth}. The dashed lines show the detection limits of 30 and 100 RV measurements with an instrument that delivers an accuracy of only 10 ms^{-1} for planets with orbital periods of 1 and 10 days. From left to right: 100 RV-measurements and orbital period of one day, 30-RV measurements and orbital period of one day, 100 RV-measurements and orbital period of ten days and 30 RV-measurements and orbital period of ten days

Fressin et al. (2013) estimated the occurrence rates of planets in different classes of planetary radii such as giant planets (6–22 R_{Earth}), large Neptunes (4–6 R_{Earth}), small Neptunes (2–4 R_{Earth}), super-Earths (1.25–2 R_{Earth}), and Earth-sized planets (0.8–1.25 R_{Earth}) using the data of planets and planet candidates measured by Kepler telescope. They showed that the occurrence rates of planets with orbital periods $P \leq 85$ days per one star are $2 \pm 0.22\%$ for giant planets, $1.97 \pm 0.23\%$ for large Neptunes, $23.5 \pm 1.6\%$ for small Neptunes, $23.0 \pm 2.4\%$ for super-Earths and $18.4 \pm 3.7\%$ for Earth-sized planets.

Also, Fulton et al. (2017), Fulton and Petigura (2018) showed that the close-in small planet distribution splits into two classes consisting of planets with radii of $< 1.5\ R$ and planets with radii of 2–3 R_{Earth} in periods of < 100 days based on an occurrence rate analysis using the precise radius measurements from the California-Kepler Survey. Based on planetary formation and evolution models (e.g., Owen and Wu 2017; Jin and Mordasini 2018) which have reproduced the bimodal distribution of close-in exoplanets; most of the close-in small exoplanets which are included in the former class are likely bare rocky planets that have lost their primordial hydrogen-rich atmosphere due to photo-evaporation. Figure 17 shows the statistical mass-estimate of the known low mass exoplanets.

5.3 Stellar Wind Plasma Interaction from Mercury to Close-in Rocky Exoplanets

If a rocky exoplanet orbits very close to its host star it looses all its volatiles and may form like airless bodies such as Mercury an exosphere or even atmosphere that consists of its minerals. The interaction of stellar wind plasma with the exospheres of such planets and possibly with their magnetospheric environment can be studied with numerical models that have been developed for Mercury or the Moon. Virtually any source of energy can give surface-bounded particles the necessary momentum to overcome the surface binding energy. At Mercury, and in general on any planetary surface, the energy source can be the

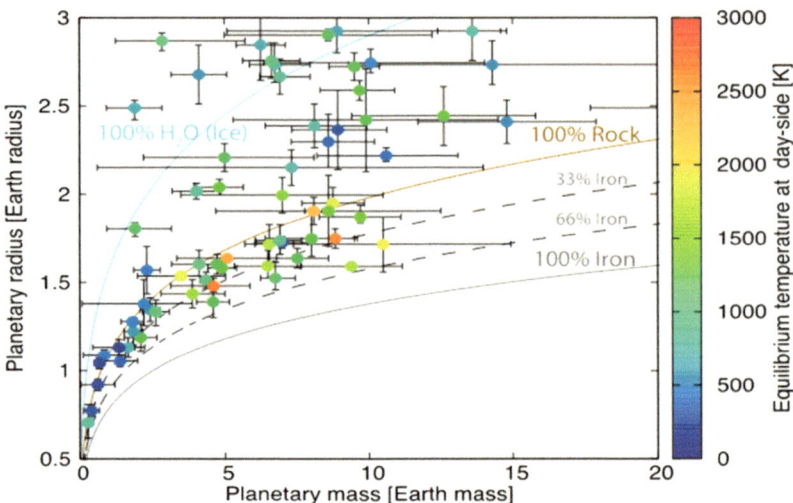

Fig. 17 The mass-radius diagram of exoplanets with radii of less than 3 R_{Earth} and masses constrained not as only the upper or lower limit values. The bars show the error bar of measured values and the colors shows the different dayside equilibrium temperatures with zero planetary albedo. Solid lines show the mass and radius relationships of planets composed of only H_2O (light blue), rock (brown) and iron (grey). The dashed lines also show the mass and radius relationships of rocky planets with iron fractions of 33% and 66%. The mass and radius relationships are calculated using the formula theoretically derived in Fortney et al. (2007). The data has been taken from an open exoplanet catalogue database (https://exoplanetarchive.ipac.caltech.edu, 27/7/2020)

planetary heat (TD, thermal desorption), the impact of solar wind or other plasma particles (IS, ion sputtering), the solar UV flux (PSD, photon-stimulated desorption), and micrometeorite impact vaporization (MMIV). Depending on the species (volatiles or refractories), local time, orbital position, and solar wind conditions, these different sources can have different relative importance (see Wurz et al. 2021, this journal). Because Na is the easiest element to be observed from Earth, and because of its intrinsic importance in the exosphere of Mercury, it is by far the most studied element but still the mechanisms of release are not completely understood and recent MESSENGER observations (Cassidy et al. 2016) seem to add complexity to previous established models (Mura et al. 2009; Leblanc and Johnson 2010). Once neutrals are released, they fill the exosphere travelling in ballistic orbits, and may be pushed in the anti-stellar direction by the radiation pressure of the star, forming a cometary-like tail or coma (Potter et al. 2002; see also Fig. 18, Na population).

Mercury's interaction with the solar wind is in some aspect similar to that of the Earth: it's magnetic field responses to the plasma flow resulting in a magnetosphere with reconnection events (Slavin et al. 2009); the solar wind can partially penetrate and charge-exchange with exospheric neutrals (Mura et al. 2003); exospheric particles are photoionized and can become pick-up ions in the solar wind (Sarantos et al. 2009). Other phenomena are more peculiar: Mercury's exosphere is partially generated by solar wind precipitating onto the polar cusps via ion sputtering (Sarantos et al. 2007); the solar wind also causes weathering of the surface (Domingue et al. 2014). Numerical models of the interaction of Mercury with the solar winds can be traced back as early as 2000 (Kabin et al. 2000; Ip and Kopp 2002; Kallio and Janhunen 2003a, 2003b, 2004; Massetti et al. 2003; Mura et al. 2003, 2005, 2006; Wurz et al. 2010, etc.).

Fig. 18 Simulated Na and Ca$^+$ tails for a close-in, super-Mercury exoplanet. The arrows indicate the velocity of the planet/reference frame (V_p), the velocity of the stellar wind in the inertial reference frame (V_{sw}) and in the non-inertial planetary reference frame (V_{sw}^*). The high Ca$^+$ tail inclination is exaggerated to match a case of high stellar wind aberration. The sodium tail is the feature on the left. (Adapted from Mura et al. 2011)

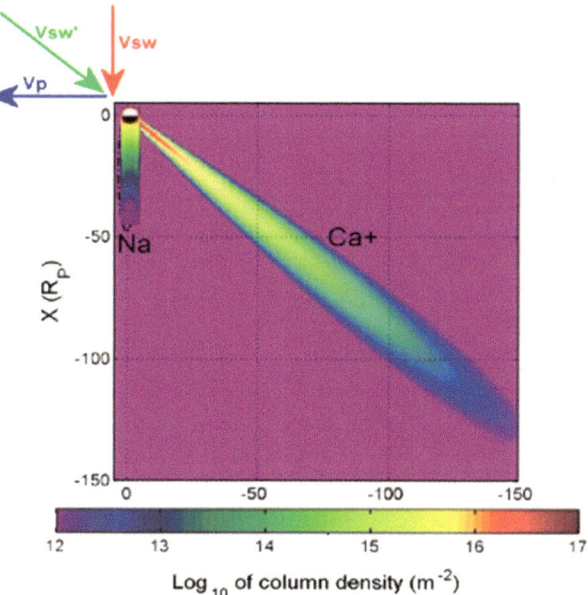

Charge-exchange results in the formation of Energetic Neutral Atoms (ENA) that escape the planetary environment. While they are usually considered a diagnostic signal (Roelof et al. 1985; Orsini and Mililllo 1999) for orbiting instrumentation, in extreme conditions they may be detected over interplanetary distances and result in a substantial mass loss (Holmström et al. 2008; Kislyakova et al. 2014). In all respects, Mercury is the natural case study or paradigm for the investigation of the interaction of a close-in exoplanet with the stellar wind. We know that the interaction of the solar wind with a planet falls into 4 basic cases: planet with or without an atmosphere, and with or without an intrinsic magnetic field. The Earth, Mercury, Venus, and Mars can fill the 2×2 grid of all possibilities. In all 4 cases, the solar wind is known to impact to or to induce the formation of peculiar structures: a bow shock (a "fast shock" surface where the solar wind starts to be perturbed), a magnetopause (the tangential discontinuity surface where the passage of matter is not permitted, and the stellar wind flows around it), a magnetosheath (the region between these two surfaces), and a magnetosphere (the cavity inside the magnetopause). In the non-magnetized case, the conductivity of the ionosphere causes the formation of induced structures, even in extreme cases such as Mars where there is almost no atmosphere; the residual crustal magnetic field is capable of generating mini magnetospheres (Breus et al. 2005, see also Ness et al. 2000). The temporal variability of the solar wind parameters (pressure and IMF) is known to play a substantial role in changing the shape and size of the magnetosphere. In the simplest approach, one can calculate the stand-off point, where the pressures of the solar wind (left-hand side) and the magnetic field pressure (right-hand side) are balanced:

$$\rho \upsilon^2 = \frac{4}{2\mu_0} \left(\frac{M}{r_{so}^3} \right)^2,$$

where ρ and υ are the stellar wind number density and bulk velocity, r_{so} is the stand-off distance and M is the planetary magnetic moment (Walker and Russell 1985; Grießmeier et al. 2005). The main problem is that, for almost all known exoplanets, none of these values

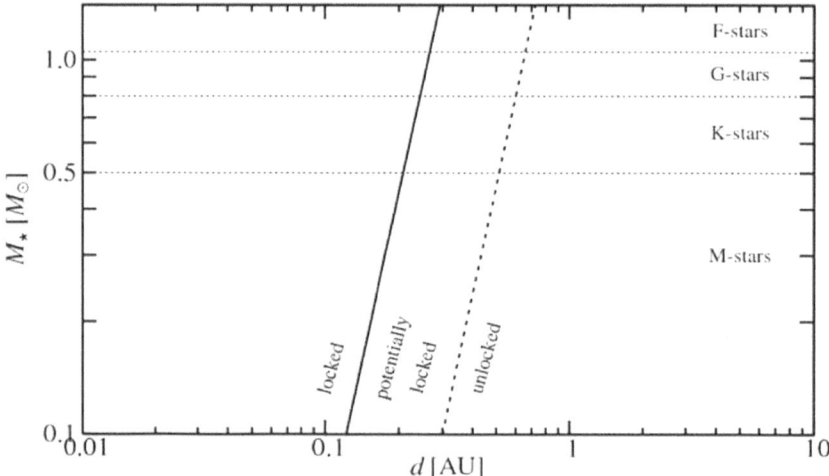

Fig. 19 Tidally locked (left) versus freely rotating (right) regime for "super-Earth" planets as a function of orbital distance d and mass M^* of the host star (figure from Mura et al. 2011)

are known, both on the stellar wind side and on the planetary side of the equation. The absence of an intrinsic magnetic field as that of Earth or stronger is sometimes associated with a tidally-locked planet (this has been recently questioned, see Reiners and Christensen 2010), so that, from the definition of the Love number k2 in Gladman et al. (1996) and their formula for the time of tidal locking, one can infer the preferred regime for exoplanets as a function of orbital distance and mass of the host star (Fig. 19) which is, however, just an educated guess.

Mercury's magnetic field was measured for the first time by Mariner 10 (Connerney and Ness 1988). The updated value of the magnetic moment is 195 ± 10 nT R_M^3, which is sufficient to keep the magnetopause stagnation point at about 1.5 R_M from planetary center (at the Earth, this is $\sim 10\ R_E$). Because of the polarity of the intrinsic magnetic field, reconnection and plasma precipitation occur at the cusps, especially when the interplanetary magnetic field component B_z is negative, while the B_x component plays a role in determining north-south asymmetries (Sarantos et al. 2001). More recent MESSENGER observations (Anderson et al. 2011) refined our knowledge of the magnetic field and revealed an offset of the dipole of almost 500 kilometers northward of the geographic equator. Such shift could in principle lead to a larger cusp region in the south dayside hemisphere, in principle, even if the excess of induced sodium release has not been observed according to recent surveys (Milillo et al. 2021).

In summary, since Mercury has a quite weak intrinsic dipole, it may be used, in some way, as an example for both exoplanetary cases (with/without an intrinsic magnetic field). In fact, in some extreme cases, the stand-off distance at Mercury can be so close to the planet that a large part of the daily surface is exposed to solar wind precipitation (Leblanc et al. 2003; Kallio and Janhunen 2003a). It is worth noting that, in case of a close-in exoplanet with an atmosphere, the situation is not dramatically different; protons are likely to impact the uppermost atmosphere layer. Such protons can produce energization of neutral particles, known as atmospheric sputtering (e.g., Johnson 1990; Lammer and Bauer 1993) in a similar way as ion sputtering (Fig. 20).

Part of the neutral particles of the exosphere are eventually photoionized by the solar radiation, leading to the formation of an 'exo-ionosphere'. At Mercury, the exo-ionosphere

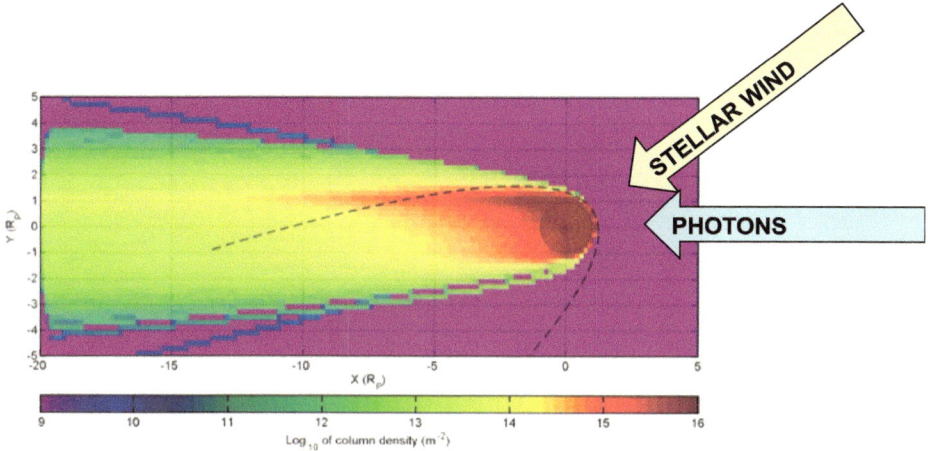

Fig. 20 Simulation of a Na population, with the formation of a tail, due to ion sputtering (or atmospheric sputtering), for a super-Mercury exoplanet with significant stellar wind aberration

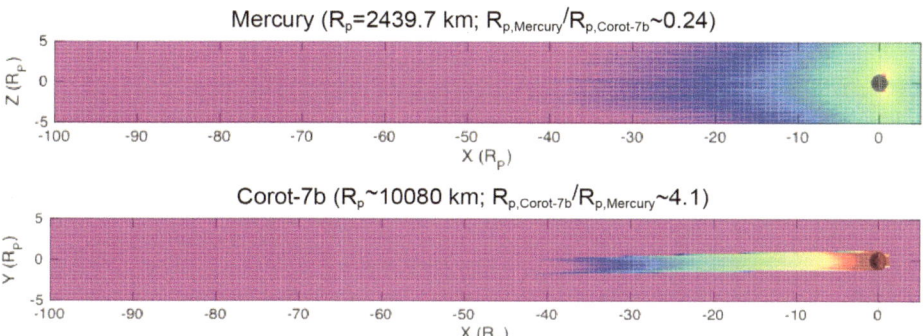

Fig. 21 Typical simulated case of Mercury's Na tail (top) in comparison with CoRoT-7b (bottom) with the stellar wind parameters as given in Mura et al. (2011). Please note that the radius of CoRoT-7b is about 4 times larger than Mercury's

is confined inside the magnetosphere, but if the internal magnetic field is not strong enough the exo-ionosphere would be dragged by the stellar wind flow and form an ionized tail that is much larger than the Na tail (Fig. 18). In this respect, Mercury is surely the most appropriate paradigm in the Solar System, as the long-studied sodium tail from both ground based observatories (e.g. Potter et al. 2003) and from space (McClintock et al. 2008, 2009) have revealed much on the evolution of this planet (Orsini et al. 2014). Exosphere transit observations (Schleicher et al. 2004; Mura et al. 2009) had greatly boosted the knowledge of the interaction between the solar wind and Mercury. Hence, it is possible to have an indication on the atmospheric stability by looking at the exospheric tails of exoplanets (Mura et al. 2011; Guenther et al. 2011), formed by those species that are subject to radiation pressure acceleration. This is a fashion similar to the well-studied Mercury case (Killen et al. 2007).

Such ion tails are much larger on close-in exo-Mercury's like CoRoT-7b. Figure 21 shows the colour-coded density of Ca^+ ions in the xy plane, integrated along the z direction, under the assumption that the close-in exoplanet has a negligible magnetic field (Mura et al. 2011). The Ca^+ tail has a scale length of $\approx 10\, r_{pl}$, when the Ca^+ ions become

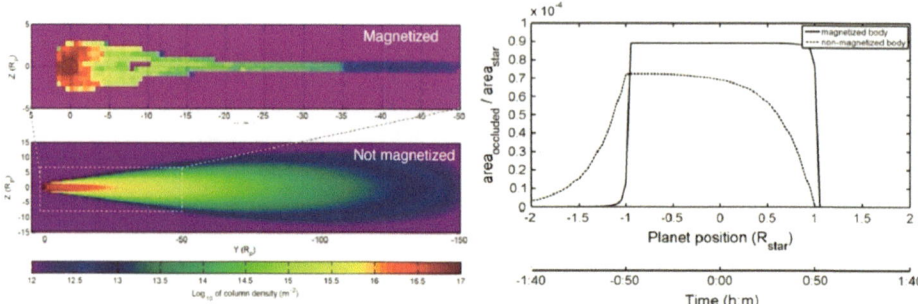

Fig. 22 Left: preliminary simulation of the Ca^+ tail, as column densities integrated along the line of sight, for a close in exoplanet, for two cases: with an intrinsic magnetic field (top-left) and without (bottom-left). Note that the scaling is different: in the case of a magnetized body, the Ca^+ population is confined in the magnetosphere, in the case of a non magnetized body, Ca^+ has a cometary shape. Because of the planet relative velocity w.r.t. the stellar wind, the "magnetosphere" is not in the direction of the observer; for this reason, a rapidly orbiting close-in exoplanet has a favorable geometry condition, but, in general, the integrated column densities look different from almost any point of view. Right: simulation of an occultation experiment for the two cases. Fraction of occluded area, due to Ca^+ absorption, function of the planet position with respect to the host star ($x = 0$ is the center of the star, from Mura et al. 2011)

Ca^{++}. One should note that under such extreme conditions the Ca^+ density falls off exponentially but is still noticeable at a distance of $\approx 100\, r_{pl}$. The Na tail with its different inclination is also shown in Fig. 18 as a reference. In this example, neutral particles in the exosphere of a close-in exoplanet are very rapidly ionized by stellar UV. The resulting ions are picked up by the stellar wind, dragged in the anti-star direction, and form a tail, in this case an ion tail.

As shown in Fig. 22, depending on the presence of a significant dipole magnetic field of the planet, such an ion tail is either confined inside a magnetosphere (if the body has a magnetic field able to form one) or is dragged away by the stellar wind itself (Mura et al. 2011). While the direction of the neutral atom tail is regulated by the radiation pressure and hence is in the anti-stellar direction, the ionized tail is in the direction of the stellar wind that, in some case, may have a significant aberration.

Recently Vidotto et al. (2018) presented a 3D study of the formation of refractory-rich exospheres around the rocky planets HD 219134 b (see Table 6), a close-in exo-Mercury with a mass of $\sim 4.7\, M_{Earth}$ and a radius of $\sim 1.6\, R_{Earth}$. The planet orbits around an 11 Gyr old star of spectral class K3V at 0.0388 AU, which is about ten times closer than Mercury. Since the temperature (~ 1000 K) of this planet is too cool to host a magma ocean, it may have an exosphere formed by sputtering of surface particles due to the exposure to a dense stellar wind. If so, the sputtering will release refractory elements from the entire dayside of the planet. As seen Fig. 23, sputtered elements such as O and Mg will create an extended neutral exosphere with densities $\geq 10\, cm^{-3}$ over several planetary radii. Vidotto et al. (2018) also investigated the detectability of such an exosphere with the current UV instruments and found that it is currently very unlikely to be observed.

In the next section we will discuss the different atmospheres that might evolve on exo-Mercurys.

5.3.1 What Kind of Atmospheres Do We Expect for Close-in Rocky Exoplanets

The measured density (i.e., mass and radius) allows us to infer the bulk composition of the planet. Although one can expect a diverse iron fractionation of rocky exoplanets, the den-

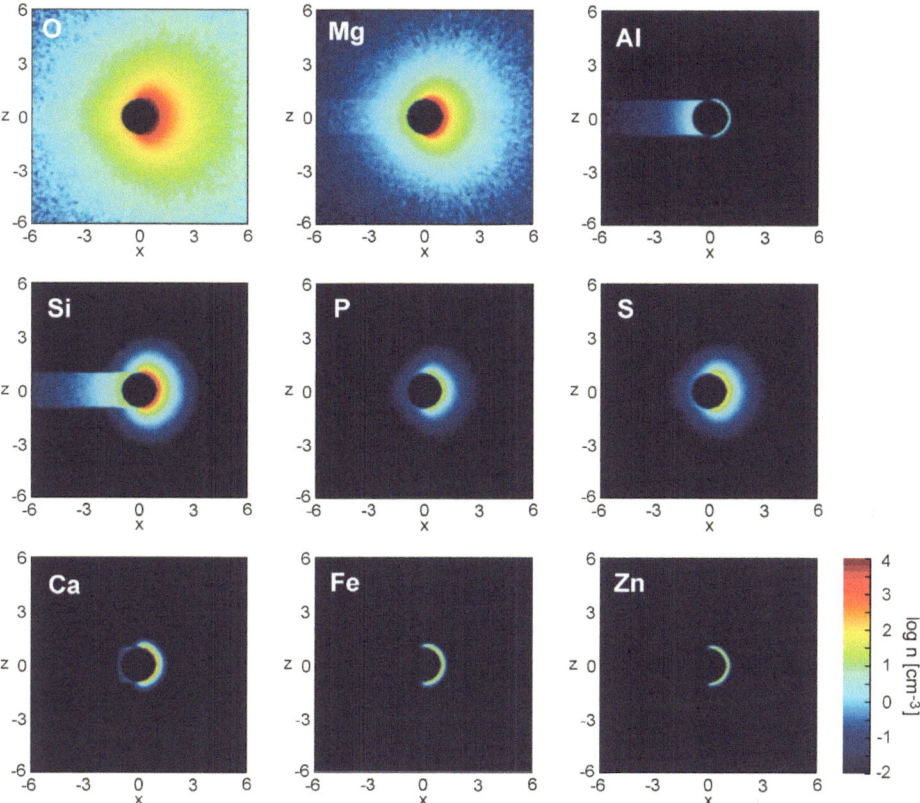

Fig. 23 The sputtering induced exosphere for different atomic species of HD 219134 b as simulated by Vidotto et al. (2018). Figure from Vidotto et al. (2018)

sity may give an information about the fraction of the iron core for some of the close-in low mass exoplanets, although most of the masses of the small exoplanets have not been estimated yet, i.e., from 732 exoplanets with radii $< 1.5\ R_{Earth}$, only 74 have mass estimates (data from NASA Exoplanet Archive as of 30 August 2021)[1]. As shown before, Fig. 17 shows the measured mass and radius of exoplanets with radii of $< 3\ R_{Earth}$. Most of exoplanets with radii of $< 2\ R_{Earth}$ distribute along the mass and radius relationships for rocky planets while the others should consist of large fractions of H_2-dominant, H_2O-vapor (see also, Zeng et al. 2019) or even He-dominant envelopes. If one considers rock and iron as the component of planetary bulk composition, there is a variety of rocky exoplanets' compositions such as pure rocky composition, Earth-like composition (Fe $\sim 30\%$) and higher metal/silicate ratio composition (Fe $\sim 70\%$). In addition, the radiative equilibrium temperatures of rocky exoplanets are ranging from that of Earth to high temperatures up to a few thousand K, as shown in Fig. 16. Most of the rocky exoplanets with measured masses and radii have orbits less than 0.1 AU and, thus, they are likely tidally locked. Then, the dayside equilibrium temperatures of rocky exoplanets around G-type or K-type stars are high enough to melt and vaporize rock (1500 K is a typical melting point of rock) due to strong stellar irradiation while the planets with equilibrium temperatures of hundreds K are orbiting around M dwarfs.

Table 6 List of parameters for HD 219134 b (values from Motalebi et al. 2015; Gillon et al. 2017)

Mass (10^{24} kg)	≈ 28.3	Rotation Period (hours)	74.16	Orbital Period (days)	3.09	Mean Temperature (C)	1015
Diameter (km)	≈ 20435	Length of Day (hours)	tidally locked	Orbital Velocity (km/s)	≈ 136	Surface Pressure (bars)	?
Density (kg/m^3)	≈ 6360	Dist. from star (10^6 km)	5.8	Orbital Inclination (°)	85.06	Global Magnetic Field	?
Gravity (m/s^2)	≈ 18	Perihelion (10^6 km)	5.8	Orbital Eccentricity	0	Bond albedo	?
Escape Velocity (km/s)	≈ 19.2	Aphelion (10^6 km)	5.8	Obliquity to Orbit (°)	?	Visual geom. albedo	?

Motivated by detections of exoplanets with higher metal/silicate ratio compositions, the evolution and formation scenarios explaining such a metal-rich composition have been also argued in an exoplanetary science field. The detection of the high-density rocky exoplanets, K2-229b and Kepler-107c, was reported by recent observations (Santerne et al. 2018; Bonomo et al. 2019). K2-229b has a radius of 1.16 (+0.06, −0.05) R_{Earth}, a mass of 2.59 ± 0.43 M_{Earth} and a semi-major axis of 0.012 AU. Its bulk density and equilibrium temperature with zero planetary albedo at the dayside are 8.9 ± 2.1 g/cm^3 and 2300 K, respectively. Also, Kepler-107c has a radius of 1.60 ± 0.026 R_{Earth}, a mass of 9.39 ± 1.77 M_{Earth} and a semi-major axis of 0.06 AU. Its bulk density and day-side equilibrium temperature are 12.65 ± 1.77 g/cm^3 and 1600 K, respectively. Their masses and radii are consistent with a metal-rich composition like Mercury (i.e., about 70% metallic core and 30% rocky mantle). As with the evolution and formation scenarios for Mercury, there are some possible origins for such a high-density exoplanet; The photoevaporation of a rocky mantle, giant impacts and formation in the metallic iron-rich region of a proto-planetary disk (e.g., Cameron 1985; Benz et al. 1988; Lewis 1972; see also Sect. 5.2). Especially, a giant impact is the likely origin of Kepler-107c (Bonomo et al. 2019). This is because the innermost planet, Kepler-107b, has similar density with pure rocky composition and, thus, is less dense than Kepler-107c. While the de-trend of the two planet's densities in orbits is inconsistent with the photoevaporation scenario and the iron-rich composition of Kepler-107c would not be primordial considering the Fe/Si and Mg/Si ratios derived from the host star abundance as a proxy of the protoplanetary disk composition, the mass and radius of Kepler-107c matches theoretical predictions from collisional mantle stripping through a giant impact (Marcus et al. 2010).

When comparing the mass and density measurements with models, we have to make an assumption about the relative abundance of the elements. It is a common habit to use simply the solar abundance. However, the abundances reflect the history of the material from which the star and planets have formed. That is, how many supernovae of type I and II, there were, whether the material was enriched by a Wolf-Rayet star, and so on. For example, the anomalous abundance in ^{26}Al in the early solar-system was due the wind of a Wolf-Rayet star that had more than 20 M_{Sun} (Portegies Zwart 2019). Thus, the relative abundances of all stars are different and so are their planets.

Furthermore, the abundances of elements within the proto-planetary disk differs from inside out. Metals are in the inner part, ices in the other part and silicates in the middle. As outlined, in Sect. 5.2, the high iron abundance of Mercury could also be due to the location where it formed, as a result of photophoresis which separated iron and silicates in the disk (Wurm et al. 2013). The temperature of the disk was 1300 and 1450 K at its current location (Gail 1998) which is too high for silicates. The fact that chemical abundances change with

Table 7 List of parameters for 55 Cnc e (values from Fischer et al. 2008; Dawson and Fabrycky 2010; Bourrier et al. 2018)

Mass (10^{24} kg)	≈ 48.3	Rotation Period (hours)	17.7	Orbital Period (days)	0.737	Mean Temperature (C)	≈ 2570
Diameter (km)	≈ 24340	Length of Day (hours)	tidally locked	Orbital Velocity (km/s)	≈ 234	Surface Pressure (bars)	?
Density (kg/m^3)	≈ 6400	Dist. from star (10^6 km)	2.32	Orbital Inclination (°)	83.59	Global Magnetic Field	?
Gravity (m/s^2)	≈ 22	Perihelion (10^6 km)	2.43	Orbital Eccentricity	0.05	Bond albedo	?
Escape Velocity (km/s)	≈ 23	Aphelion (10^6 km)	2.20	Obliquity to Orbit (°)	?	Visual geom. albedo	?

the distance from the host star also means that the compositions of rocky planets constrain where it formed (Kane et al. 2020). According to Plotnykov and Valencia (2020), iron enrichment and perhaps depletion must have happened before gas dispersal, if the chemical diversity of highly irradiated planets is the result of atmospheric evaporation. Studies of the composition of the atmospheres, or the material released from the surface, thus, promise to give us important insights in planet formation.

Unfortunately, at least observationally, there is a third class of low-mass planets. These are planets with intermediate densities. The prototype of this class is GJ1214b. The mass and radius and density of GJ 1214b are 6.16 ± 0.91 M_{Earth} and 2.71 ± 0.24 R_{Earth}, and 1.6 ± 0.6 g/cm^3, respectively (Anglada-Escudé et al. 2013). A model with a rocky core and a hazy atmosphere as well as a model with a mixture of rock and ice fits the data equally well (Rogers and Seager 2010). Observations of the atmospheres of such planets are possibly the best way to lift the ambiguity. Thus, atmospheric studies are of key importance for our understanding of these planets.

So far, the best-studied close-in rocky exoplanet that is permanently molten on its tidally locked side is 55 Cancri e (55 Cnc e; see Table 7. Infrared-phase curve allows to determine the location of the hottest part on the surface of the planet and the phase-curve in the optical is a crude albedo map of the surface. Eclipse observations and phase curves have already been obtained for the rocky planet 55 Cnc e which has only 8.59 M_{Earth}-masses. The thermal emission phase curve shows that the hottest spot is not at the sub-stellar point but 41 ± 12 degrees east of it (Demory et al. 2016a). This implies that the planet has an atmosphere that must even have a pressure of 1.4 bar (Angelo and Hu 2017). Repeated infrared observations have shown that the depth of the eclipse varies with time, which is further evidence for an atmosphere (Deming et al. 2015; Demory et al. 2016b; Tamburo et al. 2018). The best explanation for these results is that the substellar hemisphere is covered by material that has a high reflectivity, which varies in size. This material could be some kind of bright clouds, or haze in an atmosphere. Although 55 Cnc e is considered to be a rocky planet, the eclipse methods showed that it has an atmosphere. Combining the data from five transits, Ridden-Harper et al. (2016) presented hints of Na and the Ca$^+$ lines originating in the atmosphere of the planet. If real, this signal would correspond to an optically thick Ca$^+$ atmosphere that is five times larger than the Roche lobe. However, Jindal et al. (2020) recently showed that 55 Cnc e either has an atmosphere with a high mean molecular weight and/or clouds, or no atmosphere at all.

The atmospheres of close-in rocky planets are likely very different from the atmospheres of the Earth, or from that of the gas-giants that we have studied so far. Shortly after the

Table 8 List of parameters for CoRoT-7b (values from Léger et al. 2009, 2011; Hatzes et al. 2011)

Mass (10^{24} kg)	≈ 44.4	Rotation Period (hours)	1407	Orbital Period (days)	0.85	Mean Temperature (C)	≈ 1000–1500
Diameter (km)	≈ 20155	Length of Day (hours)	tidally locked	Orbital Velocity (km/s)	≈ 220	Surface Pressure (bars)	?
Density (kg/m^3)	≈ 10400	Dist. from star (10^6 km)	2.57	Orbital Inclination (°)	80.1	Global Magnetic Field	?
Gravity (m/s^2)	≈ 29	Perihelion (10^6 km)	2.57	Orbital Eccentricity	0	Bond albedo	?
Escape Velocity (km/s)	≈ 24	Aphelion (10^6 km)	2.57	Obliquity to Orbit (°)	?	Visual geom. albedo	?

discovery of CoRoT-7b (Léger et al. 2009, ; see Table 8), several authors came up with the idea of a lava or magma ocean (see also next section) at the substellar point (Briot and Schneider 2010; Barnes et al. 2010, and Rouan et al. 2011). The reason that such a planet should have a magma ocean is not only the intense radiation by the host star. If the host star is strongly magnetic and the planet along its close-in orbit crosses the stellar magnetic field lines, the sub-surface regions of ultra-short period planets can be subject to induction heating (Kislyakova et al. 2017, 2018), which can also produce magmatic layers at the upper mantle or below the surface or can increase volcanic activity.

Mura et al. (2011) developed a model for a higher metal/silicate ratio atmosphere of CoRoT-7b. This model shows that the atmospheric loss rates of CoRoT-7b could be 2–3 orders of magnitude higher than that of Mercury, forming an ionized tail of escaping particles (see Fig. 18 and Fig. 21). This tail should be observable in the Na and Ca resonance lines. A first attempt to detect the atmosphere of CoRoT-7b using high-resolution spectroscopy, however, resulted only in upper limits (Guenther et al. 2011).

Luckily, we now know a number of ultra-short period planets orbiting stars that are closer and brighter than CoRoT-7b. We mentioned already 55 Cnc e above. Another interesting target is HD 3167, which is 13 times brighter than CoRoT-7. Guenther and Kislyakova (2020) searched for the signatures of a higher metal/silicate ratio exosphere and volcanic activity at HD3167b and obtained upper limits of $I_p/I^* = 10^{-4}$ to 10^{-3} for the most important tracers, where I_p is the planetary, and I^* the stellar flux. These upper limits are lower than claimed detections by Ridden-Harper et al. (2016) for 55 Cnc e. Another interesting object is Kepler-1520b, because it has a dusty tail containing material that is escaping from the planet (Schlawin et al. 2018). More observations of more objects are certainly required. Perhaps these phenomena are episodic and repeated observations will lead to a detection.

While on Earth-like planets inside the habitable zone H_2O is the most abundant volcanic gas (Schmincke 2004), volcanically active exoplanets in close orbital distances are expected to eject mostly SO_2, which would then dissociate into oxygen and sulphur atoms (Kislyakova et al. 2018). While planets with a high surface pressure are expected to have carbon-rich and dry volcanic gases, low pressure atmospheres will contain mostly sulphur-rich gases (Gaillard and Scaillet 2014). Possible tracers for such an atmosphere are the same lines that have been detected in the plasma torus of Jupiter's Moon Io. These are [S III] 3722, [O II] 3726, [O II] 3729, [S II] 4069, [S II] 4076, [O III] 5008, [O I] 6300, [S III] 6312, [S II] 6716, [S II] 6731 Å lines, as well as the Na D-lines (Brown 1974; Brown et al. 1975; Brown and Yung 1976; Kupo et al. 1976; Brown and Shemansky 1982; Morgan and Pilcher 1982; Thomas 1993, 1996; Kueppers and Jockers 1995, 1997; Guenther and Kislyakova 2020).

Fig. 24 Artist impression of a magma ocean planet, specifically of CoRoT-7b. Impression by ESO/L. Calcada

In the following Sect. 5.4 we will discuss in more detail magma ocean-related silicate atmospheres that should originate around very hot rocky close-in exoplanets.

5.4 Magma Ocean Related Silicate Atmospheres

Most of the potential rocky exoplanets are hot enough to melt and vaporize rock due to strong stellar irradiation, as shown in Fig. 24. Also, about 500 discovered super-Earths or Earth-sized planets which are potentially rocky planets have substellar-point equilibrium temperatures high enough (> 1500 K) for rock to melt and vaporize itself (https://exoplanetarchive.ipac.caltech.edu, 27/7/2020). These include a first discovered high-density exoplanet, CoRoT-7 b (Léger et al. 2009; Queloz et al. 2009), whose substellar-point temperature are estimated to be about 2500 K with zero planetary albedo. If such planets are indeed rocky, they are thought to be entirely or partially covered with magma oceans and possibly have gases vaporized from the magma oceans as their secondary atmospheres composed of mainly sodium like Mercury (e.g., Schaefer and Fegley 2009; Léger et al. 2011). Such close-in rocky exoplanets are sometimes called lava planets or magma-ocean planets. One may think the category of lava planets or magma-ocean planets includes a young terrestrial planet just after its formation or a large collisional event. Thus, we refer to the rocky exoplanet as a close-in molten exoplanet just to clarify it.

5.4.1 Close-in Molten Exoplanets

Close-in molten exoplanets would have lost their primordial hydrogen-rich atmospheres due to hydrodynamic escape caused by X-ray and UV irradiation from their host stars (see Owen 2019, for a review). Escape of highly-irradiated hydrogen-rich atmospheres is known

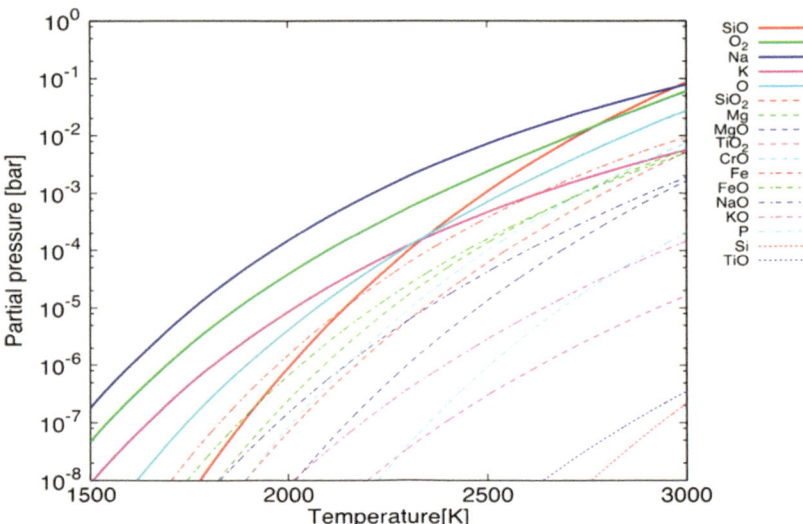

Fig. 25 Composition of gas in chemical equilibrium with molten silicate. Partial pressures of the gas species are shown as functions of magma temperature. These partial pressures are calculated by a chemical equilibrium model of Ito et al. (2015) assuming volatile-free bulk silicate Earth composition (see Ito et al. 2015, for the details)

to occur in an energy-limited fashion (Sekiya et al. 1981; Watson et al. 1981) and the hydrodynamic escape process has been intensively studied (e.g., Lammer et al. 2003; Yelle 2004; García Muñoz 2007; Kubyshkina et al. 2018; Kubyshkina and Vidotto 2021). For example, in the case of CoRoT-7 b, Valencia et al. (2010) demonstrated that the cumulative escaped mass of a hydrogen-rich atmosphere could be a few Earth masses with an analysis based on energy-limited formula and thus, would have lost the primordial atmosphere. Therefore, they are expected to have secondary atmospheres. The compositions of the secondary atmospheres on rocky exoplanets are predicted to be diverse by theorists because the atmospheric components could be outgassed from their interiors and/or got material brought by impacts of smaller bodies (e.g., Elkins-Tanton and Seager 2008).

The secondary atmospheres of close-in molten exoplanets are likely composed of materials directly vaporized from their magma ocean. Then, impacts of smaller bodies would not affect the main composition of the atmospheres because of the rapid vaporization/condensation (i.e., gas-melt equilibrium condition) unless the impacts are massive enough to change the magma composition. Some theoretical models have predicted the atmospheric composition based on gas-melt chemical equilibrium calculations. If close-in molten planets are dry, they likely have atmospheres composed of rocky materials such as Na, K, O_2 and SiO (Schaefer and Fegley 2009; Miguel et al. 2011; Ito et al. 2015; Ito and Ikoma 2021). On the other hand, if close-in molten planets have volatile elements such as H, C, N, S and Cl, they likely have atmospheres composed mainly of H_2O and/or CO_2 with rocky vapors such as Na and SiO (Schaefer et al. 2012; Herbort et al. 2020). These volatile species should be efficiently outgassed from their magma ocean (e.g., Schaefer and Fegley 2007) but they could selectively disappear through massive escape due to strong stellar irradiation (Valencia et al. 2010; Léger et al. 2011), Then, the planets become dry having only low-volatility elements (i.e., refractory elements) and silicate atmospheres.

The total pressure and composition of a silicate atmosphere highly depend on the temperature of the magma ocean (Schaefer and Fegley 2009; Miguel et al. 2011; Ito et al. 2015). Figure 25 shows the partial pressures of gas species in a silicate atmosphere vaporized from magma ocean with volatile-free bulk silicate Earth composition. Na is the most abundant species in most of the temperature range, while SiO increases with temperature and becomes the most abundant one for $T > 2800$ K. Also, the total vapor pressure is as small as about 10^{-7} bars at $T = 1500$ K and 0.1 bar at $T = 3000$ K. These properties depend on magma composition, but they are almost same in typical rocky compositions of Earth; Earth's MORB, bulk crust and upper crust (Ito et al. 2015).

The strong dependence of the total pressure on temperature would cause strong winds from the planetary sub-stellar point to the terminator. If the magma composition are well mixed the atmospheric pressure remains close to their local saturation values because of the rapid vaporization/condensation of gases (Castan and Menou 2011) but, if not, the atmosphere and magma are compositionally variegated in latitudes/longitudes due to materials transport via the wind (Kite et al. 2016). In vertical, silicate atmospheres have thermal inversion structures for the substellar-point equilibrium temperature higher than 2300 K but have isothermal profiles for lower temperatures (Ito et al. 2015). The thermal inversion is caused by Far-UV absorption of SiO and visible absorption of Na and K.

Ito and Ikoma (2021) developed a 1-D hydrodynamic model for the study of UV-irradiated silicate atmospheres that contain Na, Mg, O, Si, their ions and electrons. Their model also includes the thermal and photochemistry, molecular diffusion, thermal conduction, X-ray and UV heating, and radiative line cooling. It was found that most of the host stars energy of these short wave lengths is lost by the radiative emission of Na, Mg, Mg^+, Si^+_2, Na^+_3 and Si^+_3. Ito and Ikoma (2021) found that a magmatic Earth-sized planet at an orbit of 0.02 AU around a young solar-type star, with the above given silicate atmosphere develops a low X-ray and UV heating efficiency which is in the order of 1×10^{-3}, which corresponds to a total thermal mass loss rate of ≈ 0.3 M_{Earth}/Gyr. It was found that efficient cooling in such a silicate atmosphere of a 1 M_{Earth}-mass planet yields photo-evaporation rates that are not large enough for modifying the planetary mass and bulk composition largely. On the other hand, one should expect that no dense silicate atmosphere might accumulate because as discussed in Sect. 5.3 the dense stellar wind at 0.02 AU will most likely erode the atmosphere via ion pick up and other nonthermal loss processes.

Recent transit observations through photometric and spectroscopic methods have reported the signal of the atmospheric components or temperature of close-in exoplanets whose densities are consistent with rocky ones. Such characterization is important to know not only what exoplanets are like but also how they were evolved and formed. One of the most famous close-in super-Earths is 55 Cnc e (see Table 7 and Sect. 5.3.1). The exoplanet orbits a bright G8V ($V = 5.95$) star, which allows for measurements with a higher signal-to-noise ratio than that for other super-Earths. The 1 σ upper limit on the measured 55 Cnc e's density reaches the pure rocky regime without iron core. Thus, the planet is possibly a dense rocky planet with a relatively large atmosphere, or a planet made of lighter materials as water and carbon but with a small atmosphere.

The thermal phase curve and transmission spectra of 55 Cnc e are observed. Demory et al. (2016b) found the day night temperature difference on 55 Cancri e are large by analyzing the thermal phase curve. The temperature averaged over the dayside is 2700 ± 270 K that is about twice higher than the night-side temperature, 1380 ± 400 K. The significant eastward hot-spot shift was also measured. These features seem imbalanced from the viewpoint of climate theory (Hammond and Pierrehumbert 2017). This is because, as the mean molecular weight of an atmosphere increases, the atmosphere tends to have a larger day–night temperature contrast but a smaller eastward phase shift (Zhang and Showman 2017).

Also, the transmission spectra of 55 Cnc e have suggested the atmosphere contains a significant amount of light gases such as hydrogen (Tsiaras et al. 2016) but water vapor was not found (Esteves et al. 2017; Jindal et al. 2020). These might suggest that 55 Cnc e has a volatile-rich atmosphere. On the other hand, Bourrier et al. (2019) reported the detection of variability in 55 Cnc's FUV emission lines induced by O, C^+, $C2^+$, N_4^+, Si^+, Si_2^+ and Si_3^+ before/during/after the transit of 55 Cnc e. They concluded the variations are unlikely to originate from purely the host star and purely the planet. Ridden-Harper et al. (2016) reported the possible detection of a sodium and calcium exosphere escaping from 55 Cnc e during transit while escaping hydrogen from 55 Cnc e was not detected (Ehrenreich et al. 2012). This might suggest that 55 Cnc e have a silicate atmosphere but, taken as a whole, these observations seem hard to reach a consensus with each other and atmospheric models for now.

5.4.2 Evaporating Close-in Rocky Exoplanets

Rocky planets that may be currently evaporating have been detected around the three stars; KIC 12557548, K2-22 and KOI-2700 (e.g., Rappaport et al. 2012, 2014; Sanchis-Ojeda et al. 2015). Such planets are called evaporating planets. While planet transits are usually symmetric and periodic without any significant variations in their shape or depth over time, the variation in transit depth and the ingress-egress asymmetry of the transit light curve of KIC 12557548 were found (Rappaport et al. 2012). The variation and asymmetry of transit light curves have been interpreted as a piece of the evidence of catastrophic evaporation of small rocky planets with an evaporated dust tail (e.g., Rappaport et al. 2012; Perez-Becker and Chiang 2013; Kawahara et al. 2013; Croll et al. 2014). Also, since the variation period of transits for KIC 12257548 is consistent within 1-σ of the rotation period of its host star, the evaporation of the planet KIC 12257548 b may be correlated with stellar activity such as UV emission and magnetic field (Kawahara et al. 2013). The estimated mass loss rate for KIC 12257548 b is about one Earth mass per Gyr while an upper limit of the detected planet's radius is only about one Earth radius (e.g., Rappaport et al. 2012; Brogi et al. 2012; Kawahara et al. 2013; van Lieshout et al. 2014). As the equilibrium temperature of an evaporating planet is high (e.g., about 2000 K for KIC 12257548 b), a thermal ("Parker-type") wind from low-mass rocky planets is one of possible interspersions for such massive evaporation (Perez-Becker and Chiang 2013).

5.5 What Can We Learn from Future Research on Close-in Rocky Exoplanets?

Low-mass rocky planets orbiting close to their host stars are very interesting to study, because they give us many new insights how such planets form and evolve. Precise measurements of the mass and radius indicate that some of them have not lost their primordial atmospheres that presumably contains only a few percent of the mass of the planet. These atmospheres are likely to be H_2/He-dominated but it is possible that they contain material released from the rocky cores underneath. The next generation of instruments will allow us to study these atmospheres. For the first time we might be able to find out what these planets are made of which puts strong constrains on how and where they formed.

ARIEL (Atmospheric Remote-sensing Infrared Exoplanet Large-survey) (Puig et al. 2018; Tinetti et al. 2016) is a satellite that is especial designed for the study of exoplanets atmospheres. The telescope has an elliptical primary of 1.1×0.7 m, and the launch is scheduled for 2028. The VIS-channel has three photometric bands: 0.5–0.6 μm, 0.6–0.81 μm, and 0.81–0.1. μm. The NIR-channel obtains low-resolution spectra in the wavelength ranges

1.1–1.95 μm (with a resolving power R of $R = \lambda/\Delta\lambda = 20$ with $\Delta\lambda$ as the smallest difference that can be distinguished at wavelength λ), 1.95–3.9 μm ($R = 100$), and 3.9–7.8 μm ($R = 30$). Simulations show that a signal-to-nose ratio of 10 is already sufficient to detect the most prominent chemical species. The simulation of the 2.7 R_{Earth} planet GJ1214b furthermore shows that it will be possible to detect features in the spectrum even if the ratio of the planetary radius r_{pl} to the stellar radius r_{star}, i.e., $(r_{\text{pl}}/r_{\text{star}})2$ is only 1% larger in the lines than in the continuum if 100 transits are averaged (ARIEL mission proposal). This means that with ARIEL it is possible to study planets with less than 10 M_{Earth} even if they have only slightly extended atmospheres.

Another important mission for exoplanet research will be JWST with its 6.5 m aperture telescope. JWST offers the following observing modes: 0.7–5 μm ($R = 100$; NIRspec prism), 0.7–2.5 μm ($R = 700$; NIRISS grism+ prism), 2.5–5 μm ($R = 1700$; NIRcam grisms), 5–12 μm ($R = 70$; MIRI LRS prisms). Simulations by Greene (2011) show that a single eclipse observation of GJ1214b in the 2.5–5 μm range will already constrain the most prominent molecular species in the atmosphere like H_2O, CH_4 and NH_3.

Studies of magmatic rocky planets that lost their primordial atmospheres or never had some are also important. A big leap forward will be the high-resolution spectrographs on the next generation telescopes. One of the new instruments is the Visible Echelle Spectrograph – G-CLEF (Chicago-Large Earth Finder) of the Giant Magellan Telescope (GMT) is vacuum-enclosed and fiber-fed spectrograph that covers the wavelength region from 0.35 μm to 0.95 μm. The Precision Radial Velocity (PRV) mode provides a resolution $R = 108,000$ (Szentgyorgyi et al. 2014). High-resolution spectroscopy at infrared wavelength benefit from the higher brightness ratio between the planet and the star and they have the advantage that most molecular lines are at infrared wavelengths. METIS ('Mid-infrared Extremely Large Telescope (ELT) Imager and Spectrograph') is an L/M and N-band instrument that covers the wavelength region from 3–14 μm. In the L/M band (3–5.3 μm) it will provide a spectral resolution of $R = 100,000$ (Brandl et al. 2010; Lenzen et al. 2010). High Resolution Echelle Spectrometer (HiRES) is an Initiative to realize a high-resolution spectrograph for the European-Extremely Large Telescope (E-ELT). The wavelength coverage will be unusually large, 0.36 to 1.8 μm, possibly even up to 2.48 μm. The spectral resolution will be of the order of $R = 100,000$. HiRISE is the proposed fiber coupling between the direct imager SPHERE and the spectrograph CRIRES+ at the Very Large Telescope. Figure 26 shows the detection limits of 5 σ for HiRISE derived for a bright nearby young star (H = 3:5, 19 pc, 20 Myr), compared to the 20% best SPHERE/SHINE detection limits. We overplot state-of-the-art population synthesis models based on the core accretion formation scenario (Mordasini 2018; Emsenhuber et al. 2020a, 2020b).

The ELT instruments will not only outperform the existing ones because they are more modern, but also because this method is basically only limited by the signal-to-noise ratio that can be achieved in a given amount of time. Even high resolution than $R = 100,000$ could have its advantages (Lopez-Morales et al. 2019).

The combination of high-contrast imaging and high-spectral-resolution can yield higher quality spectra than either of the two methods alone (Otten et al. 2021). This is even true if the planet is not seen directly in the high-contrast image, because the adaptive optics system still helps improving the light-ratio between the star and the planet. Combining the VLT instruments SPHERE and ESPRESSO might even allow to detect Earth-mass planet Proxima b despite the planet-to-star contrast ratio of 107. However, 20 to 40 observing nights on the VLT are needed for that (Lovis et al. 2017).

By studying such silicate atmospheres, we will obtain a deeper understanding of the mama ocean phase through which all rocky planets evolve. The detection of the outgassed silicate atmosphere would be a revelation. Perhaps many of these planets also have

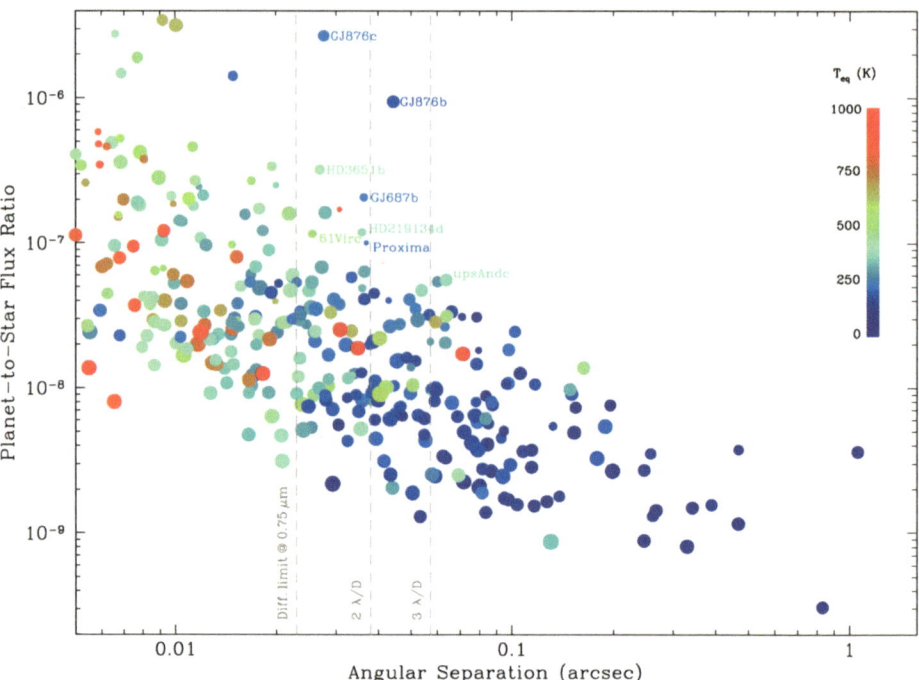

Fig. 26 Estimated planet-to-star contrast in reflected light for known exoplanets as a function of angular separation from their host star. Dot size is proportional to the logarithm of planet mass, while the color scale represents equilibrium temperature (assuming a Bond albedo of 0.3). Vertical dashed lines indicate the di fraction limit, $2\lambda/D$ and $3\lambda/D$ thresholds for the 8.2-m VLT at 750 nm (corresponding to the O_2 A-band). The combination of a 10^3–10^4 contrast enhancement from SPHERE to the high spectral resolution of ESPRESSO can reveal the planetary spectral features and disentangle them from the stellar ones. The estimated planet-to-star contrast for Proxima b is 10^{-7} in reflected light. This is challenging, but a 5-σ detection would be possible by observing the object with SPHERE+ESPRESSO for 20–40 nights

cometary-like tail structures, similar but more massive than those of Mercury. If we can detect them, we would also be able to put constrains on the composition of these planets. Perhaps the next generation of instruments, or just continued observations of known planets, or the discovery of such planets orbiting nearby stars may make such observations possible. After a long phase where we knew nothing about small exoplanets that they exist, we are now getting pieces of evidence what they really are and that they are not like any of the planets of our Solar System.

6 Conclusions

The origin and evolution of various airless planetary bodies from large planetesimals to planetary embryos to Earth's Moon and Mercury as well as recently discovered hot, rocky and most likely magmatic exoplanets were discussed. It is pointed out that by studying the origin and evolution of the Moon, Mercury and other airless planetary bodies, one has to consider that these bodies and hence their surfaces experienced a far more extreme bombardment by solar/stellar radiation and plasma during the first hundreds of million years compared to that of today. These extreme environments will have affected and modified together with frequent meteoritic impacts the uppermost surface layers of these bodies. We used the Moon as an example for that and addressed also the indications that the Lunar surface may contain "fingerprints" of escaped particles from early Earth's upper atmosphere that was heated and expanded due to the higher XUV-flux of the young Sun. We compared Mercury's solar wind interaction with that expected at known higher metal/silicate ratio planets that orbit very close to their host stars and speculate which kind of mineral-like atmospheric/exospheric environments these planets will have. In the near future one can expect that space-observatories such as the JWST or ARIEL and large telescopes like the VLT with its SPHERE and ESPRESSO instruments may allow us to detect atmospheres/exospheres that consist of rock-vapor or minerals by high-contrast-imaging and high-spectral-resolution. The detection of such atmospheres/exospheres around higher metal/silicate ratio planets may reveal their composition so that one can compare them with formation theories of Mercury in the Solar System.

Acknowledgements The authors thank ISSI for supporting the fruitful ISSI-workshop "Surface Bounded Exospheres and Interactions in the Solar System". E. Guenther acknowledges that this work was generously supported by the Thüringer Ministerium für Wirtschaft, Wissenschaft und Digitale Gesellschaft. N.V. Erkaev acknowledges support by the Ministry of Science and Higher Education project No. 075-15-2020-780.

Funding Open access funding provided by Österreichische Akademie der Wissenschaften.

References

V.S. Airapetian, A.V. Usmanov, Reconstructing the solar wind from its early history to current epoch. Astrophys. J. **817**, L24 (2016). https://doi.org/10.3847/2041-8205/817/2/L24

V.S. Airapetian, A. Glocer, G. Gronoff, E. Hébrard, W. Danchi, Prebiotic chemistry and atmospheric warming of early Earth by an active young Sun. Nat. Geosci. **9**, 452–455 (2016). https://doi.org/10.1038/ngeo2719

F. Albarède, J. Blichert-Toft, The split fate of the early Earth, Mars, Venus, and Moon. C. R. Géosci. **339**, 917–927 (2007). https://doi.org/10.1016/j.crte.2007.09.006

L. Allibert, S. Charnoz, J. Siebert, S.A. Jacobson, S.N. Raymond, Quantitative estimates of impact induced crustal erosion during accretion and its influence on the Sm/Nd ratio of the Earth. Icarus **363**, 114412 (2021). https://doi.org/10.1016/j.icarus.2021.114412

M. Anand, R. Tartèse, J. Barnes, Understanding the origin and evolution of water in the Moon through lunar sample studies. Philos. Trans. R. Soc. A **372**, 20130254 (2014)

E. Anders, N. Grevesse, Abundances of the elements: meteoritic and solar. Geochim. Cosmochim. Acta **53**, 197–214 (1989). https://doi.org/10.1016/0016-7037(89)90286-X

D.L. Anderson, *New Theory of the Earth* (Cambridge University Press, Cambridge, 2005)

B.J. Anderson et al., The global magnetic field of Mercury from MESSENGER orbital observations. Science **333**, 1859 (2011). https://doi.org/10.1126/science.1211001

I. Angelo, R. Hu, A case for an atmosphere on super-Earth 55 Cancri e. Astron. J. **154**, 232 (2017). https://doi.org/10.3847/1538-3881/aa9278

D. Angerhausen, H. Sapers, A.C. Vieira Araujo, HABEBEE: habitability of Eyeball-Exo-Earths. Astrobiology **13**, 309–314 (2013). https://doi.org/10.1089/ast.2012.0846

G. Anglada-Escudé, B. Rojas-Ayala, A.P. Boss, A.J. Weinberger, J.P. Lloyd, GJ 1214 reviewed. Trigonometric parallax, stellar parameters, new orbital solution, and bulk properties for the super-Earth GJ 1214b. Astron. Astrophys. **551**, A48 (2013). https://doi.org/10.1051/0004-6361/201219250

R. Arevalo, W.F. McDonough, M. Luong, The K/U ratio of the silicate Earth: insights into mantle composition, structure and thermal evolution. Earth Planet. Sci. Lett. **278**, 361–369 (2009). https://doi.org/10.1016/j.epsl.2008.12.023

D.J. Armstrong, F. Meru, D. Bayliss, G.M. Kennedy, D. Veras, A gap in the mass distribution for warm Neptune and terrestrial planets. Astrophys. J. **880**(1), L1 (2019). https://doi.org/10.3847/2041-8213/ab2ba2. ISSN 2041-8213

E. Asphaug, Impact origin of the Moon. Annu. Rev. Earth Planet. Sci. **42**, 551–578 (2014)

E. Asphaug, A. Reufer, Mercury and other iron-rich planetary bodies as relics of inefficient accretion. Nat. Geosci. **7**, 564–568 (2014). https://doi.org/10.1038/ngeo2189

I. Baraffe, G. Chabrier, F. Allard, P.H. Hauschildt, Evolutionary models for solar metallicity low-mass stars: mass-magnitude relationships and color-magnitude diagrams. Astron. Astrophys. **337**, 403–412 (1998)

I. Baraffe, G. Chabrier, T.S. Barman, F. Allard, P.H. Hauschildt, Evolutionary models for cool brown dwarfs and extrasolar giant planets. The case of HD 209458. Astron. Astrophys. **402**, 701–712 (2003). https://doi.org/10.1051/0004-6361:20030252

I. Baraffe, G. Chabrier, T. Barman, Structure and evolution of super-Earth to super-Jupiter exoplanets. I. Heavy element enrichment in the interior. Astron. Astrophys. **482**, 315–332 (2008). https://doi.org/10.1051/0004-6361:20079321

I. Baraffe, D. Homeier, F. Allard, G. Chabrier, New evolutionary models for pre-main sequence and main sequence low-mass stars down to the hydrogen-burning limit. Astron. Astrophys. **577**, A42 (2015). https://doi.org/10.1051/0004-6361/201425481

R. Barnes, S.N. Raymond, R. Greenberg, B. Jackson, N.A. Kaib, CoRoT-7b: super-Earth or super-Io? Astrophys. J. **709**, L95–L98 (2010). https://doi.org/10.1088/2041-8205/709/2/L95

J.J. Barnes, R. Tartèse, M. Anand, F.M. McCubbin, I.A. Franchi, N.A. Starkey, S.S. Russel, The origin of water in the primitive Moon as revealed by the lunar highlands samples. Earth Planet. Sci. Lett. **390**, 244–252 (2014)

A.T. Basilevsky, Venera 8 landing site geology revisited. J. Geophys. Res. **102**, 9257–9262 (1997). https://doi.org/10.1029/97JE00413

S. Bauer, H. Lammer, *Planetary Aeronomy: Atmosphere Environments in Planetary Systems* (Springer, Berlin, 2004)

R.H. Becker, R.N. Clayton, E.M. Galimov, H. Lammer, R.O. Pepin, R. Wieler, Isotopic signatures of volatiles in terrestrial planets – working group report. Space Sci. Rev. **106**, 377–410 (2003). https://doi.org/10.1023/A:1024610325914

R. Behrisch, W. Eckstein, *Sputtering by Particle Bombardment*, vol. 110 (2007)

M.R. Benedikt, M. Scherf, H. Lammer, E. Marcq, P. Odert, M. Leitzinger, N.V. Erkaev, Escape of rock-forming volatile elements and noble gases from planetary embryos. Icarus **347**, 113772 (2020)

M. Benna, P.R. Mahaffy, J.S. Halekas, R.C. Elphic, G.T. Delory, Variability of helium, neon, and argon in the lunar exosphere as observed by the LADEE NMS instrument. Geophys. Res. Lett. **42**, 3723–3729 (2015). https://doi.org/10.1002/2015GL064120

W. Benz, W.L. Slattery, A.G.W. Cameron, The origin of the Moon and the single-impact hypothesis I. Icarus **66**, 515–535 (1986). https://doi.org/10.1016/0019-1035(86)90088-6

W. Benz, W.L. Slattery, A.G.W. Cameron, The origin of the Moon and the single-impact hypothesis, II. Icarus **71**, 30–45 (1987). https://doi.org/10.1016/0019-1035(87)90160-6

W. Benz, W.L. Slattery, A.G.W. Cameron, Collisional stripping of Mercury's mantle. Icarus **74**, 516–528 (1988)

W. Benz, A.G.W. Cameron, H.J. Melosh, The origin of the Moon and the single-impact hypothesis III. Icarus **81**, 113–131 (1989). https://doi.org/10.1016/0019-1035(89)90129-2

W. Benz, A. Anic, J. Horner, J.A. Whitby, The origin of Mercury. Space Sci. Rev. **132**, 189–202 (2007). https://doi.org/10.1007/s11214-007-9284-1

H.A. Bethe, Energy production in stars. Phys. Rev. **55**, 434–456 (1939). https://doi.org/10.1103/PhysRev.55.434

P. Boehnke, T.M. Harrison, Illusory late heavy bombardments. Proc. Natl. Acad. Sci. **113**, 10802–10806 (2016). https://doi.org/10.1073/pnas.1611535113

A.S. Bonomo et al., A giant impact as the likely origin of different twins in the Kepler-107 exoplanet system. Nat. Astron. **3**, 416–423 (2019)

A. Bonsor, Z.M. Leinhardt, P.J. Carter, T. Elliott, M.J. Walter, S.T. Stewart, A collisional origin to Earth's non-chondritic composition? Icarus **247**, 291–300 (2015). https://doi.org/10.1016/j.icarus.2014.10.019

M.A. Bouhifd et al., Potassium partitioning into molten iron alloys at high-pressure: implications for Earth's core. Phys. Earth Planet. Inter. **160**, 22–33 (2007). https://doi.org/10.1016/j.pepi.2006.08.005

356

A. Boujibar, D. Andrault, N. Bolfan-Casanova, M.A. Bouhifd, J. Monteux, Cosmochemical fractionation by collisional erosion during the Earth's accretion. Nat. Commun. **6**, 8295 (2015). https://doi.org/10.1038/ncomms9295

V. Bourrier et al., The 55 Cancri system reassessed. Astron. Astrophys. **619**, A1 (2018)

V. Bourrier et al., High-energy environment of super-Earth 55 Cancri e. I. Far-UV chromospheric variability as a possible tracer of planet-induced coronal rain. Astron. Astrophys. **615**, A117 (2019)

D. Bower, D. Kitzmann, A.S. Wolf, P. Sanan, C. Dorn, A.V. Oza, Linking the evolution of terrestrial interiors and an early outgassed atmosphere to astrophysical observations. Astron. Astrophys. **631**, A103 (2019). https://doi.org/10.1051/0004-6361/201935710

J.W. Boyce, Y. Liu, G.R. Rossman, Y. Guan, J.M. Eiler, E.M. Stolper, L.A. Taylor, Lunar apatite with terrestrial volatile abundances. Nature **466**, 466–469 (2010)

B.R. Brandl et al., Instrument concept and science case for the mid-IR E-ELT imager and spectrograph METIS, in *Ground-Based and Airborne Instrumentation for Astronomy III*, vol. 7735 (2010). https://doi.org/10.1117/12.857346

R. Brasser, The formation of Mars: building blocks and accretion time scale. Space Sci. Rev. **174**, 11–25 (2013). https://doi.org/10.1007/s11214-012-9904-2

T.K. Breus et al., The effects of crustal magnetic fields and the pressure balance in the high latitude ionosphere/atmosphere at Mars. Adv. Space Res. **36**, 2043–2048 (2005). https://doi.org/10.1016/j.asr.2005.01.100

D. Briot, J. Schneider, Occurrence, physical conditions, and observations of super-Ios and hyper-Ios, in *Pathways Towards Habitable Planets*, vol. 430 (2010), p. 409

M. Brogi et al., Evidence for the disintegration of KIC 12557548 b. Astron. Astrophys. **545**, L5 (2012)

R.A. Brown, Optical line emission from Io, in *Exploration of the Planetary System*, vol. 65 (1974), pp. 527–531

R.A. Brown, D.E. Shemansky, On the nature of S II emission from Jupiter's hot plasma torus. Astrophys. J. **263**, 433–442 (1982). https://doi.org/10.1086/160515

R.A. Brown, Y.L. Yung, Io, its atmosphere and optical emissions, in *IAU Colloq. 30: Jupiter: Studies of the Interior, Atmosphere, Magnetosphere and Satellites* (1976), pp. 1102–1145

R.A. Brown, R.M. Goody, F.J. Murcray, F.H. Chaffee, Further studies of line emission from Io. Astrophys. J. **200**, L49–L53 (1975). https://doi.org/10.1086/181894

A.G.W. Cameron, The partial volatilization of Mercury. Icarus **64**, 285–294 (1985)

A.G.W. Cameron, W. Benz, The origin of the moon and the single impact hypothesis IV. Icarus **92**, 204–216 (1991). https://doi.org/10.1016/0019-1035(91)90046-V

A.G.W. Cameron, W.R. Ward, The origin of the Moon, in *Lunar and Planetary Science Conference* (1976)

A.G.W. Cameron, W. Benz, W.L. Slattery, Planetary collision calculations: origin of Mercury. Lunar Planet. Sci. Conf. **18**, 151 (1987)

A.G.W. Cameron, W. Benz, B. Fegley, W.L. Slattery, The strange density of Mercury – theoretical considerations, in *Mercury* (University of Arizona Press, Tucson, 1988), pp. 692–708

R.M. Canup, Dynamics of lunar formation. Annu. Rev. Astron. Astrophys. **42**, 441–475 (2004a). https://doi.org/10.1146/annurev.astro.41.082201.113457

R.M. Canup, Simulations of a late lunar-forming impact. Icarus **168**, 433–456 (2004b). https://doi.org/10.1016/j.icarus.2003.09.028

R.M. Canup, Lunar-forming collisions with pre-impact rotation. Icarus **196**, 518–538 (2008). https://doi.org/10.1016/j.icarus.2008.03.011

R.M. Canup, Forming a Moon with an Earth-like composition via a giant impact. Science **338**, 1052 (2012). https://doi.org/10.1126/science.1226073

R.M. Canup et al. (2021). Origin of the Moon. arXiv e-prints

R.M. Canup, E. Asphaug, Origin of the Moon in a giant impact near the end of the Earth's formation. Nature **412**, 708–712 (2001)

P.J. Carter, Z.M. Leinhardt, T. Elliott, M.J. Walter, S.T. Stewart, Compositional evolution during rocky protoplanet accretion. Astrophys. J. **813**, 72 (2015). https://doi.org/10.1088/0004-637X/813/1/72

P.J. Carter, Z.M. Leinhardt, T. Elliott, S.T. Stewart, M.J. Walter, Collisional stripping of planetary crusts. Earth Planet. Sci. Lett. **484**, 276–286 (2018). https://doi.org/10.1016/j.epsl.2017.12.012

T.A. Cassidy et al., A cold-pole enhancement in Mercury's sodium exosphere. Geophys. Res. Lett. **43**, 11,121–11,128 (2016). https://doi.org/10.1002/2016GL071071

T. Castan, K. Menou, Atmospheres of hot super-earths. Astrophys. J. **743**, L36 (2011)

N.L. Chabot, M.J. Drake, Potassium solubility in metal: the effects of composition at 15 kbar and 1900 °C on partitioning between iron alloys and silicate melts. Earth Planet. Sci. Lett. **172**, 323–335 (1999). https://doi.org/10.1016/S0012-821X(99)00208-3

N.L. Chabot, E.E. Shread, J.K. Harmon, Investigating Mercury's south polar deposits: Arecibo radar observations and high-resolution determination of illumination conditions. J. Geophys. Res., Planets **123**, 666–681 (2018). https://doi.org/10.1002/2017JE005500

J.W. Chamberlain, Planetary coronae and atmospheric evaporation. Planet. Space Sci. **11**, 901–960 (1963). https://doi.org/10.1016/0032-0633(63)90122-3

A. Chau, C. Reinhardt, R. Helled, J. Stadel, Forming Mercury by giant impacts. Astrophys. J. **865**, 35 (2018). https://doi.org/10.3847/1538-4357/aad8b0

P.F. Chen, Coronal mass ejections: models and their observational basis. Living Rev. Sol. Phys. **8**, 1 (2011). https://doi.org/10.12942/lrsp-2011-1

Y. Chen, Y. Zhang, Y. Liu, Y. Guan, J. Eiler, E.M. Stolper, Water, fluorine, and sulfur concentrations in the lunar mantle. Earth Planet. Sci. Lett. **427**, 37–46 (2015)

M.W. Claire, J. Sheets, M. Cohen, I. Ribas, V.S. Meadows, D.C. Catling, The evolution of solar flux from 0.1 nm to 160 μm: quantitative estimates for planetary studies. Astrophys. J. **757**, 95 (2012). https://doi.org/10.1088/0004-637X/757/1/95

R.N. Clark, Detection of adsorbed water and hydroxyl on the Moon. Science **326**(5952), 562 (2009). https://doi.org/10.1126/science.1178105

M.S. Clement, J.E. Chambers, Dynamical avenues for Mercury's origin II: in-situ formation in the inner terrestrial disk. arXiv e-prints (2021)

M.S. Clement, N.A. Kaib, J.E. Chambers, Dynamical constraints on Mercury's collisional origin. Astron. J. **157**, 208 (2019). https://doi.org/10.3847/1538-3881/ab164f

M.S. Clement, J.E. Chambers, A.P. Jackson, Dynamical avenues for Mercury's origin. I. The lone survivor of a primordial generation of short-period protoplanets. Astron. J. **161**, 240 (2021). https://doi.org/10.3847/1538-3881/abf09f

J.E.P. Connerney, N.F. Ness, Mercury's magnetic field and interior, in *Mercury* (University of Arizona Press, Tucson, 1988), pp. 494–513

E. Cottrell, K.A. Kelley, The oxidation state of Fe in MORB glasses and the oxygen fugacity of the upper mantle. Earth Planet. Sci. Lett. **305**, 270–282 (2011). https://doi.org/10.1016/j.epsl.2011.03.014

B. Croll et al., Multiwavelength observations of the candidate disintegrating sub-Mercury KIC 12557548b. Astrophys. J. **786**, 100 (2014)

M. Cuk, S.T. Stewart, Making the Moon from a fast-spinning Earth: a giant impact followed by resonant despinning. Science **338**, 1047 (2012). https://doi.org/10.1126/science.1225542

M. Cuk, D.P. Hamilton, S.J. Lock, S.T. Stewart, Tidal evolution of the Moon from a high-obliquity, high-angular-momentum Earth. Nature **539**, 402–406 (2016). https://doi.org/10.1038/nature19846

T.P. Das, S.V. Thampi, A. Bhardwaj, S.M. Ahmed, R. Sridharan, Observation of Neon at mid and high latitudes in the sunlit lunar exosphere: results from CHACE aboard MIP/Chandrayaan-1. Icarus **272**, 206–211 (2016)

N. Dauphas, The isotopic nature of the Earth's accreting material through time. Nature **541**, 521–524 (2017)

N. Dauphas, C. Burkhardt, P. Warren, F.-Z. Teng, Geochemical arguments for an Earth-like Moon-forming impactor. Philos. Trans. R. Soc. Lond. Ser. A **372**, P20130244 (2014). https://doi.org/10.1098/rsta.2013.0244

J.R.A. Davenport, K.R. Covey, R.W. Clarke, A.C. Boeck, J. Cornet, S.L. Hawley, The evolution of flare activity with stellar age. Astrophys. J. **871**, 241 (2019). https://doi.org/10.3847/1538-4357/aafb76

R.I. Dawson, D.C. Fabrycky, Radial velocity planets de-aliased: a new, short period for super-Earth 55 Cnc e. Astrophys. J. **722**, 937–953 (2010). https://doi.org/10.1088/0004-637X/722/1/937

D. Deming et al., Spitzer secondary eclipses of the dense, modestly-irradiated, giant exoplanet HAT-P-20b using pixel-level decorrelation. Astrophys. J. **805**, 132 (2015). https://doi.org/10.1088/0004-637X/805/2/132

B.-O. Demory et al., A map of the large day-night temperature gradient of a super-Earth exoplanet. Nature **532**, 207–209 (2016a)

B.-O. Demory, M. Gillon, N. Madhusudhan, D. Queloz, Variability in the super-Earth 55 Cnc e. Mon. Not. R. Astron. Soc. **455**, 2018–2027 (2016b). https://doi.org/10.1093/mnras/stv2239

J. Deng, Z. Du, B.B. Karki, D.B. Ghosh, K.K.M. Lee, A magma ocean origin to divergent redox evolutions of rocky planetary bodies and early atmospheres. Nat. Commun. **11**, 1–7 (2020). https://doi.org/10.1038/s41467-020-157

D.L. Domingue et al., Mercury's weather-beaten surface: understanding Mercury in the context of lunar and asteroidal space weathering studies. Space Sci. Rev. **181**, 121–214 (2014). https://doi.org/10.1007/s11214-014-0039-5

D.S. Ebel, C.M.O. Alexander, Equilibrium condensation from chondritic porous IDP enriched vapor: implications for Mercury and enstatite chondrite origins. Planet. Space Sci. **59**, 1888–1894 (2011). https://doi.org/10.1016/j.pss.2011.07.017

D.S. Ebel, S.T. Stewart, The elusive origin of Mercury. arXiv e-prints (2017)

R.A. Eggleton (ed.), The regolith glossary. Cooperative Research Centre for Landscape Evolution and Mineral Exploration, Canberra (2001)

D. Ehrenreich et al., Hint of a transiting extended atmosphere on 55 Cancri b. Astron. Astrophys. **547**, A18 (2012)

L.T. Elkins-Tanton, Linked magma ocean solidification and atmospheric growth for Earth and Mars. Earth Planet. Sci. Lett. **271**, 181–191 (2008). https://doi.org/10.1016/j.epsl.2008.03.062

L.T. Elkins-Tanton, Magma oceans in the inner solar system. Annu. Rev. Earth Planet. Sci. **40**, 113–139 (2012). https://doi.org/10.1146/annurev-earth-042711-105503

L.T. Elkins-Tanton, T.L. Grove, Water (hydrogen) in the lunar mantle: results from petrology and magma ocean modeling. Earth Planet. Sci. Lett. **307**, 173–179 (2011)

L.T. Elkins-Tanton, S. Seager, Ranges of atmospheric mass and composition of super-Earth exoplanets. Astrophys. J. **685**, 1237–1246 (2008)

L.T. Elkins-Tanton, S. Burgess, Q.Z. Yin, The lunar magma ocean: reconciling the solidification process with lunar petrology and geochronology. Earth Planet. Sci. Lett. **304**, 326–336 (2011)

A. Emsenhuber, C. Mordasini, M. Mayor, M. Maxime, S. Udry, L. Mishra, Y. Alibert, W. Benz, E. Asphaug, The New Generation Planetary Population Synthesis (NGPPS): comparison with the HARPS GTO survey, in *Europlanet Science Congress Abstracts EPSC2020-339* (2020a). https://doi.org/10.5194/epsc2020-339

A. Emsenhuber, C. Saverio, E. Asphaug, T.S.J. Gabriel, S.R. Schwartz, F. Roberto, Realistic on-the-fly outcomes of planetary collisions. II. Bringing machine learning to N-body simulation. Astrophys. J. **891**, 6 (2020b). https://doi.org/10.3847/1538-4357/ab6de5

A.G. Emslie, A.M. Massone, Bayesian confidence limits of electron spectra obtained through regularized inversion of solar hard X-ray spectra. Astrophys. J. **759**, 122 (2012). https://doi.org/10.1088/0004-637X/759/2/122

A.G. Emslie, B.R. Dennis, G.D. Holman, H.S. Hudson, Refinements to flare energy estimates: a followup to "Energy partition in two solar flare/CME events" by A.G. Emslie et al. J. Geophys. Res. Space Phys. **110**, A11103 (2005). https://doi.org/10.1029/2005JA011305

N.V. Erkaev, H. Lammer, P. Odert, Y.N. Kulikov, K.G. Kislyakova, Extreme hydrodynamic atmospheric loss near the critical thermal escape regime. Mon. Not. R. Astron. Soc. **448**, 1916–1921 (2015). https://doi.org/10.1093/mnras/stv130

C. Espaillat et al., An observational perspective of transitional disks, in *Protostars and Planets VI* (2014), p. 497. https://doi.org/10.2458/azu_uapress_9780816531240-ch022

L.J. Esteves, E.J.W. de Mooij, R. Jayawardhana, C. Watson, R. de Kok, A search for water in a super-Earth atmosphere: high-resolution optical spectroscopy of 55Cancri e. Astron. J. **153**, 268 (2017)

L.G. Evans et al., Major-element abundances on the surface of Mercury: results from the MESSENGER Gamma-Ray Spectrometer. J. Geophys. Res., Planets **117**, E00L07 (2012). https://doi.org/10.1029/2012JE004178

L.G. Evans et al., Chlorine on the surface of Mercury: MESSENGER gamma-ray measurements and implications for the planet's formation and evolution. Icarus **257**, 417–427 (2015). https://doi.org/10.1016/j.icarus.2015.04.039

B. Fegley, A.G.W. Cameron, A vaporization model for iron/silicate fractionation in the Mercury protoplanet. Earth Planet. Sci. Lett. **82**, 207–222 (1987). https://doi.org/10.1016/0012-821X(87)90196-8

B. Fegley, N.S. Jacobson, K.B. Williams, J.M.C. Plane, L. Schaefer, K. Lodders, Solubility of rock in steam atmospheres of planets. Astrophys. J. **824**, 103 (2016). https://doi.org/10.3847/0004-637X/824/2/103

V.A. Fernandes, J. Fritz, B.P. Weiss, I. Garrick-Bethell, D.L. Shuster, The bombardment history of the Moon as recorded by ^{40}Ar-^{39}Ar chronology. Meteorit. Planet. Sci. **48**, 241–269 (2013)

G. Feulner, The faint young Sun problem. Rev. Geophys. **50**, RG2006 (2012). https://doi.org/10.1029/2011RG000375

D.A. Fischer et al., Five planets orbiting 55 Cancri. Astrophys. J. **675**, 790–801 (2008). https://doi.org/10.1086/525512

R.A. Fish, G.G. Goles, E. Anders, The record in the meteorites. III. On the development of meteorites in asteroidal bodies. Astrophys. J. **132**, 243 (1960). https://doi.org/10.1086/146918

J.J. Fortney, M.S. Marley, J.W. Barnes, Planetary radii across five orders of magnitude in mass and stellar insolation: application to transits. Astrophys. J. **659**, 1661–1672 (2007)

F. Fressin et al., The false positive rate of Kepler and the occurrence of planets. Astrophys. J. **766**, 81 (2013)

M. Fridlund et al., The TOI-763 system: sub-Neptunes orbiting a Sun-like star. Mon. Not. R. Astron. Soc. **498**, 4503–4517 (2020). https://doi.org/10.1093/mnras/staa2502

J. Fritz, B. Bitsch, E. Kührt, A. Morbidelli, C. Tornow, K. Wünnemann, V.A. Fernandes, J.L. Grenfell, H. Rauer, R. Wagner, S.C. Werner, Earth-like habitats in planetary systems. Planet. Space Sci. **98**, 254–267 (2014)

D.J. Frost, U. Mann, Y. Asahara, D.C. Rubie, The redox state of the mantle during and just after core formation. Philos. Trans. R. Soc. Lond. Ser. A **366**, 4315–4337 (2008). https://doi.org/10.1098/rsta.2008.0147

B.J. Fulton, E.A. Petigura, The California-Kepler survey. VII. Precise planet radii leveraging Gaia DR2 reveal the stellar mass dependence of the planet radius gap. Astron. J. **156**, 264 (2018)

B.J. Fulton, E.A. Petigura, A.W. Howard, H. Isaacson, G.W. Marcy, P.A. Cargile, L. Hebb, L.M. Weiss, J.A. Johnson, T.D. Morton, E. Sinukoff, I.J.M. Crossfield, L.A. Hirsch, The California-Kepler survey. III. A gap in the radius distribution of small planets. Astrophys. J. **154**(3), 109 (2017). https://doi.org/10.3847/1538-3881

E. Füri, E. Deloule, A. Gourenko, B. Marty, New evidence for chondritic lunar water from combined D/H and noble gas analyses of single Apollo 17 volcanic glasses. Icarus **214**, 109–120 (2014). https://doi.org/10.1016/j.icarus.2013.10.029

H.-P. Gail, Chemical reactions in protoplanetary accretion disks. IV. Multicomponent dust mixture. Astron. Astrophys. **332**, 1099–1122 (1998)

F. Gaillard, B. Scaillet, A theoretical framework for volcanic degassing chemistry in a comparative planetology perspective and implications for planetary atmospheres. Earth Planet. Sci. Lett. **403**, 307–316 (2014). https://doi.org/10.1016/j.epsl.2014.07.009

A. García Muñoz, Physical and chemical aeronomy of HD 209458b. Planet. Space Sci. **55**, 1426–1455 (2007). https://doi.org/10.1016/j.pss.2007.03.007

S. Gebauer et al., Atmospheric nitrogen when life evolved on Earth. Astrobiology **20**, 1413–1426 (2020). https://doi.org/10.1089/ast.2019.2212

M. Gillon et al., Two massive rocky planets transiting a K-dwarf 6.5 parsecs away. Nat. Astron. **1**, 0056 (2017). https://doi.org/10.1038/s41550-017-0056

B. Gladman, D.D. Quinn, P. Nicholson, R. Rand, Synchronous locking of tidally evolving satellites. Icarus **122**, 166–192 (1996). https://doi.org/10.1006/icar.1996.0117

N. Gopalswamy, A global picture of CMEs in the inner heliosphere, in *The Sun and the Heliosphere as an Integrated System* (2004), p. 201. https://doi.org/10.1007/978-1-4020-2831-9_8

P. Gorenstein, L. Golub, P. Bjorkholm, Detection of radon at the edges of lunar maria with the Apollo alpha-particle spectrometer. Science **183**, 411–413 (1974a)

P. Gorenstein, L. Golub, P. Bjorkholm, Radon emanations from the Moon, spatial and temporal variability. Earth Moon Planets **9**, 129 (1974b)

D.O. Gough, Solar interior structure and luminosity variations. Sol. Phys. **74**, 21–34 (1981). https://doi.org/10.1007/BF00151270

G. Greene, Optimizing architectures for multi mission archives, in *Astronomical Data Analysis Software and Systems XX*, vol. 442 (2011), p. 3

J.-M. Grießmeier, U. Motschmann, G. Mann, H.O. Rucker, The influence of stellar wind conditions on the detectability of planetary radio emissions. Astron. Astrophys. **437**, 717–726 (2005). https://doi.org/10.1051/0004-6361:20041976

E.W. Guenther, K.G. Kislyakova, Searching for volcanic activity and a higher metal/silicate ratio exosphere of the super-Earth HD3167 b. Mon. Not. R. Astron. Soc. **491**, 3974–3982 (2020). https://doi.org/10.1093/mnras/stz3288

E.W. Guenther et al., Constraints on the exosphere of CoRoT-7b. Astron. Astrophys. **525**, A24 (2011). https://doi.org/10.1051/0004-6361/201014868

M. Hammond, R.T. Pierrehumbert, Linking the climate and thermal phase curve of 55 Cancri e. Astrophys. J. **849**, 152 (2017)

H. Harder, G. Schubert, Sulfur in Mercury's core? Icarus **151**, 118–122 (2001). https://doi.org/10.1006/icar.2001.6586

W.K. Hartmann, D.R. Davis, Satellite-sized planetesimals and lunar origin. Icarus **24**, 504–515 (1975). https://doi.org/10.1016/0019-1035(75)90070-6

K. Hashizume, M. Chaussidon, A non-terrestrial ^{16}O-rich isotopic composition for the protosolar nebula. Nature **434**, 619–622 (2005). https://doi.org/10.1038/nature03432

K. Hashizume, M. Chaussidon, Two oxygen isotopic components with extra-selenial origins observed among lunar metallic grains – in search for the solar wind component. Geochim. Cosmochim. Acta **73**, 3038–3054 (2009). https://doi.org/10.1016/j.gca.2009.02.024

A.P. Hatzes et al., The mass of CoRoT-7b. Astrophys. J. **743**, 75 (2011). https://doi.org/10.1088/0004-637X/743/1/75

E.H. Hauri, T. Weinreich, A.E. Saal, M.C. Rutherford, J.A. Van Orman, High pre-eruptive water contents preserved in lunar melt inclusions. Science **333**, 213–215 (2011)

E.H. Hauri, A.E. Saal, M.J. Rutherford, J.A. Van Orman, Water in the Moon's interior: truth and consequences. Earth Planet. Sci. Lett. **409**, 252–264 (2015)

E.H. Hauri, A.E. Saal, M. Nakajima, M. Anand, M.J. Rutherford, J.A. Van Orman, M. Le Voyer, Origin and evolution of water in the Moon's interior. Annu. Rev. Earth Planet. Sci. **45**, 89–111 (2017)

P.O. Hayne, O. Aharonson, N. Schörghofer, Micro cold traps on the Moon. Nat. Astron. **5**, 169 (2021). https://doi.org/10.1038/s41550-020-1198-9

E. Hébrard, B. Marty, Coupled noble gas-hydrocarbon evolution of the early Earth atmosphere upon solar UV irradiation. Earth Planet. Sci. Lett. **385**, 40–48 (2014). https://doi.org/10.1016/j.epsl.2013.10.022

G. Heiken, D. Vaniman, B.M. French, *Lunar Sourcebook* (Cambridge University Press, New York, 1991)

G. Helffrich, R. Brasser, A. Shahar, The chemical case for Mercury mantle stripping. Prog. Earth Planet. Sci. **6**, 66 (2019). https://doi.org/10.1186/s40645-019-0312-z

O. Herbort, P. Woitke, Ch. Helling, A. Zerkle, The atmospheres of rocky exoplanets. I. Outgassing of common rock and the stability of liquid water. Astron. Astrophys. **636**, A71 (2020). https://doi.org/10.1051/0004-6361/201936614

H. Hiesinger, C.H. van der Bogert, J.H. Pasckert, L. Funcke, L. Giacomini, L.R. Ostrach, M.S. Robinson, How old are young Lunar craters? J. Geophys. Res., Planets **117**(E12), E00H10 (2012)

R.C. Hin et al., Magnesium isotope evidence that accretional vapour loss shapes planetary compositions. Nature **549**, 511–515 (2017). https://doi.org/10.1038/nature23899

F.L. Hinton, D.R. Taeusch, Variation of the lunar atmosphere with the strength of the solar wind. J. Geophys. Res. **69**, 1341–1347 (1964)

M.M. Hirschmann, A.C. Withers, Ventilation of CO_2 from a reduced mantle and consequences for the early Martian greenhouse. Earth Planet. Sci. Lett. **270**, 147–155 (2008). https://doi.org/10.1016/j.epsl.2008.03.034

R.R. Hodges Jr., Helium and hydrogen in the lunar atmosphere. J. Geophys. Res. **78**, 8055–8064 (1973)

R.R. Hodges Jr., Lunar cold traps and their influence on argon-40, in *Proceedings of the 11th Lunar and Planetary Science Conference*, Houston, TX, USA, 17–21 March 1980, vol. 3 (Pergamon Press, New York, 1980), pp. 2463–2477

R.R. Hodges Jr., J.H. Hoffman, Implications of atmospheric ^{40}Ar escape on the interior structure of the Moon, in *Proceedings of the 6th Lunar Science Conference* (1975), pp. 3039–3047

M. Holmström, A. Ekenbäck, F. Selsis, T. Penz, H. Lammer, P. Wurz, Energetic neutral atoms as the explanation for the high-velocity hydrogen around HD 209458b. Nature **451**, 970–972 (2008). https://doi.org/10.1038/nature06600

C.I. Honniball, P.G. Lucey, S. Li, S. Shenoy, T.M. Orlando, C.A. Hibbitts, D.M. Hurley et al., Molecular water detected on the sunlit Moon by SOFIA. Nat. Astron. **5**, 121 (2021). https://doi.org/10.1038/s41550-020-01222-x

M.D. Hopkins, S.J. Mojzsis, A protracted timeline for Lunar bombardment from mineral chemistry, Ti thermometry and U-Pb geochronology of Apollo 14 melt breccia zircons. Icarus **24**, 504–514 (2015)

N. Hosono, S-i. Karato, J. Makino, T.R. Saitoh, Terrestrial magma ocean origin of the Moon. Nat. Geosci. **12**, 418–423 (2019). https://doi.org/10.1038/s41561-019-0354-2

A. Hubbard, Explaining Mercury's density through magnetic erosion. Icarus **241**, 329–335 (2014). https://doi.org/10.1016/j.icarus.2014.06.032

W.-H. Ip, A. Kopp, MHD simulations of the solar wind interaction with Mercury. J. Geophys. Res. Space Phys. **107**, 1348 (2002). https://doi.org/10.1029/2001JA009171

T.R. Ireland, P. Holden, M.D. Norman, J. Clarke, Isotopic enhancements of ^{17}O and ^{18}O from solar wind particles in the lunar regolith. Nature **440**, 776–778 (2006). https://doi.org/10.1038/nature04611

Y. Ito, M. Ikoma, Hydrodynamic escape of mineral atmosphere from hot rocky exoplanet. I. Model description. Mon. Not. R. Astron. Soc. **502**, 750–771 (2021). https://doi.org/10.1093/mnras/staa3962

Y. Ito, M. Ikoma, H. Kawahara, H. Nagahara, Y. Kawashima, T. Nakamoto, Theoretical emission spectra of atmospheres of hot rocky super-Earths. Astrophys. J. **801**, 144 (2015)

S. Jin, C. Mordasini, Compositional imprints in density-distance-time: a rocky composition for close-in low-mass exoplanets from the location of the valley of evaporation. Astrophys. J. **853**, 163 (2018)

A. Jindal et al., Arid or Cloudy: Characterizing the Atmosphere of the super-Earth 55 Cancri e using High-Resolution Spectroscopy. arXiv e-prints (2020)

A. Johansen, M.-M. Mac Low, P. Lacerda, M. Bizzarro, Growth of asteroids, planetary embryos, and Kuiper belt objects by chondrule accretion. Sci. Adv. **1**, 1500109 (2015). https://doi.org/10.1126/sciadv.1500109

T.-V. Johnson, Galilean satellites: narrowband photometry 0.30 to 1.10 microns. Icarus **14**, 94–111 (1971). https://doi.org/10.1016/0019-1035(71)90104-7

R.E. Johnson, *Energetic Charged-Particle Interactions with Atmospheres and Surfaces*. Physics and Chemistry in Space, vol. 69 (Springer, Berlin, 1990)

C.P. Johnstone, M. Güdel, T. Lüftinger, G. Toth, I. Brott, Stellar winds on the main-sequence. I. Wind model. Astron. Astrophys. **577**, A27 (2015a). https://doi.org/10.1051/0004-6361/201425300

C.P. Johnstone, M. Güdel, I. Brott, T. Lüftinger, Stellar winds on the main-sequence. II. The evolution of rotation and winds. Astron. Astrophys. **577**, A28 (2015b). https://doi.org/10.1051/0004-6361/201425301

C.P. Johnstone, M. Bartel, M. Güdel, The active lives of stars: a complete description of the rotation and XUV evolution of F, G, K, and M dwarfs. Astrophys. J. **649**, A96 (2021a). https://doi.org/10.1051/0004-6361/202038407

C.P. Johnstone, H. Lammer, K. Kislyakova, M. Scherf, M. Güdel, The young Sun's XUV-activity as a constraint for lower CO_2-limits in the Earth's Archean atmosphere. Earth Planet. Sci. Lett. **576**, 117197 (2021b). https://doi.org/10.1016/j.epsl.2021.117197

B.M. Jones, M. Sarantos, T.M. Orlando, A new in situ quasi-continuous solar-wind source of molecular water on Mercury. Astrophys. J. **891**, L43 (2020). https://doi.org/10.3847/2041-8213/ab6bda

K. Kabin, T.I. Gombosi, D.L. DeZeeuw, K.G. Powell, Interaction of Mercury with the solar wind. Icarus **143**, 397–406 (2000). https://doi.org/10.1006/icar.1999.6252

E. Kallio, P. Janhunen, Modelling the solar wind interaction with Mercury by a quasi-neutral hybrid model. Ann. Geophys. **21**, 2133–2145 (2003a). https://doi.org/10.5194/angeo-21-2133-2003

E. Kallio, P. Janhunen, Solar wind and magnetospheric ion impact on Mercury's surface. Geophys. Res. Lett. **30**, 1877 (2003b). https://doi.org/10.1029/2003GL017842

E. Kallio, P. Janhunen, The response of the Hermean magnetosphere to the interplanetary magnetic field. Adv. Space Res. **33**, 2176–2181 (2004). https://doi.org/10.1016/S0273-1177(03)00447-2

S.R. Kane, R.M. Roettenbacher, C.T. Unterborn, B.J. Foley, M.L. Hill, A volatile-poor formation of LHS 3844b based on its lack of significant atmosphere. Planet. Sci. J. **1**, 36 (2020). https://doi.org/10.3847/PSJ/abaab5

W. Kang, F. Ding, R. Wordsworth, S. Seager, Escaping outflows from disintegrating exoplanets: day-side versus night-side escape. Astrophys. J. **906**, 67 (2021)

H. Kawahara, T. Hirano, K. Kurosaki, Y. Ito, M. Ikoma, Starspots-transit depth relation of the evaporating planet candidate KIC 12557548b. Astrophys. J. **776**, L6 (2013)

C. Kay, V.S. Airapetian, T. Lüftinger, O. Kochukhov, Frequency of coronal mass ejection impacts with early terrestrial planets and exoplanets around active solar-like stars. Astrophys. J. **886**, L37 (2019). https://doi.org/10.3847/2041-8213/ab551f

R.M. Killen, W.-H. Ip, The surface-bounded atmospheres of Mercury and the Moon. Rev. Geophys. **37**(3), 361–406 (1999)

R. Killen et al., Processes that promote and deplete the exosphere of Mercury. Space Sci. Rev. **132**, 433–509 (2007). https://doi.org/10.1007/s11214-007-9232-0

R.M. Killen, D.M. Hurley, W.M. Farrell, The effect on the Lunar exosphere of a coronal mass ejection passage. J. Geophys. Res. **117**, E00K02 (2012)

K.G. Kislyakova, M. Holmström, H. Lammer, P. Odert, M.L. Khodachenko, Magnetic moment and plasma environment of HD 209458b as determined from Ly-alpha observations. Science **346**, 981–984 (2014). https://doi.org/10.1126/science.1257829

K.G. Kislyakova et al., Magma oceans and enhanced volcanism on TRAPPIST-1 planets due to induction heating. Nat. Astron. **1**, 878–885 (2017). https://doi.org/10.1038/s41550-017-0284-0

K.G. Kislyakova et al., Effective induction heating around strongly magnetized stars. Astrophys. J. **858**, 105 (2018). https://doi.org/10.3847/1538-4357/aabae4

K.G. Kislyakova et al., Evolution of the Earth's polar outflow from mid-Archean to present. J. Geophys. Res. Space Phys. **125**, e2020JA027837 (2020). https://doi.org/10.1029/2020JA027837

E.S. Kite, B. Fegley, L. Schaefer, E. Gaidos, Atmosphere-interior exchange on hot, rocky exoplanets. Astrophys. J. **828**, 80 (2016)

E. Kokubo, H. Genda, Formation of terrestrial planets from protoplanets under a realistic accretion condition. Astrophys. J. **714**, L21–L25 (2010). https://doi.org/10.1088/2041-8205/714/1/L21

R.P. Kraft, Studies of stellar rotation. V. The dependence of rotation on age among solar-type stars. Astrophys. J. **150**, 551 (1967). https://doi.org/10.1086/149359

T.S. Kruijer, T. Kleine, M. Fischer-Gödde, P. Sprung, Lunar tungsten isotopic evidence for the late veneer. Nature **520**, 534–537 (2015). https://doi.org/10.1038/nature14360

M. Kruss, G. Wurm, Seeding the formation of Mercurys: an iron-sensitive bouncing barrier in disk magnetic fields. Astrophys. J. **869**, 45 (2018). https://doi.org/10.3847/1538-4357/aaec78

M. Kruss, G. Wurm, Composition and size dependent sorting in preplanetary growth: seeding the formation of Mercury-like planets. Planet. Sci. J. **1**, 23 (2020). https://doi.org/10.3847/PSJ/ab93c4

M. Kruss, G. Wurm, Composition and size dependent sorting in preplanetary growth: seeding the formation of higher metal/silicate ratio planets. Planet. Sci. J. **1**, 23 (2020). https://doi.org/10.3847/PSJ/ab93c4

D. Kubyshkina, A.A. Vidotto, How does the mass and activity history of the host star affect the population of low-mass planets? Mon. Not. R. Astron. Soc. (2021). https://doi.org/10.1093/mnras/stab897

D. Kubyshkina et al., Grid of upper atmosphere models for 1–40 M_{Earth} planets: application to CoRoT-7 b and HD 219134 b,c. Astron. Astrophys. **619**, A151 (2018). https://doi.org/10.1051/0004-6361/201833737

M. Kueppers, K. Jockers, Fabry-Perot imaging of O III lambda = 5007 Angstrom emission in the Io plasma torus. Astrophys. J. **441**, L101 (1995). https://doi.org/10.1086/187800

M. Kueppers, K. Jockers, A multi-emission imaging study of the Io plasma torus. Icarus **129**, 48–71 (1997). https://doi.org/10.1006/icar.1997.5760

I. Kupo, Y. Mekler, Y. Mekler, A. Eviatar, Detection of ionized sulfur in the Jovian magnetosphere. Astrophys. J. **205**, L51–L53 (1976). https://doi.org/10.1086/182088

H. Lammer, S.J. Bauer, Atmospheric mass loss from Titan by sputtering. Planet. Space Sci. **41**, 657–663 (1993). https://doi.org/10.1016/0032-0633(93)90049-8

H. Lammer, F. Selsis, I. Ribas, E.F. Guinan, S.J. Bauer, W.W. Weiss, Atmospheric loss of exoplanets resulting from stellar X-ray and extreme-ultraviolet heating. Astrophys. J. **598**, L121–L124 (2003)

H. Lammer, N.V. Erkaev, L. Fossati, I. Juvan, P. Odert, P.E. Cubillos, E. Guenther, K.G. Kislyakova, C.P. Johnstone, T. Lüftinger, M. Güdel, Identifying the 'true' radius of the hot sub-Neptune CoRoT-24b by mass-loss modelling. Mon. Not. R. Astron. Soc. **461**, L62–L66 (2016)

H. Lammer, A.L. Zerkle, S. Gebauer, N. Tosi, L. Noack, M. Scherf, E. Pilat-Lohinger, M. Güdel, J.L. Grenfell, M. Godolt, A. Nikolaou, Origin and evolution of the atmospheres of early Venus, Earth and Mars. Astron. Astrophys. Rev. **26**, 2 (2018)

H. Lammer, L. Spross, L. Grenfell, M. Scherf, L. Fossati, M. Lendl, P. Cubillos, The role of N_2 as a geo-biosignature for the detection and characterization of Earth-like habitats. Astrobiology **19**, 927–950 (2019). https://doi.org/10.1089/ast.2018.1914

H. Lammer, M. Scherf, H. Kurokawa, Y. Ueno, C. Burger, Z. Leinhard, T. Maindl, C.P. Johnstone, M. Leizinger, M. Benedikt, L. Fossati, B. Marty, B. Fegley, P. Odert, K.G. Kislyakova, Loss and fractionation of noble gas isotopes and moderate volatile elements from planetary embryos and early Venus, Earth and Mars. Space Sci. Rev. **216**, 74 (2020a). https://doi.org/10.1007/s11214-020-00701-x

H. Lammer, M. Leitzinger, M. Scherf, P. Odert, C. Burger, D. Kubyshkina, C.P. Johnstone, T. Maindl, C.M. Schäfer, M. Güdel, N. Tosi, A. Nikolaou, E. Marcq, N.V. Erkaev, L. Noak, K.G. Kislyakova, L. Fossati, E. Pilat-Lohinger, F. Ragossnig, E.A. Dorfi, Measured atmospheric $^{36}Ar/^{38}Ar$, $^{20}Ne/^{22}Ne$, $^{36}Ar/^{22}Ne$ noble gas isotope and bulk K/U ratios constrain the early evolution of Venus and Earth. Icarus **339**, 11351 (2020b)

H. Lammer, R. Brasser, A. Johansen, M. Scherf, M. Leitzinger, Formation of Venus, Earth and Mars: constrained by isotopes. Space Sci. Rev. **217**, 7 (2021). https://doi.org/10.1007/s11214-020-00778-4

Y. Langevin, The regolith of Mercury: present knowledge and implications for the Mercury Orbiter mission. Planet. Space Sci. **45**, 31–37 (1997). https://doi.org/10.1016/S0032-0633(96)00098-0

S.L. Lawson, W.C. Feldman, D.J. Lawrence, K.R. Moore, R.C. Elphic, R.D. Belian, Recent outgassing from the lunar surface: the Lunar Prospector alpha particle spectrometer. J. Geophys. Res. **110**, E09009 (2005). https://doi.org/10.1029/2005JE002433

F. Leblanc, R.E. Johnson, Mercury exosphere I. Global circulation model of its sodium component. Icarus **209**, 280–300 (2010). https://doi.org/10.1016/j.icarus.2010.04.020

F. Leblanc, J.G. Luhmann, R.E. Johnson, M. Liu, Solar energetic particle event at Mercury. Planet. Space Sci. **51**, 339–352 (2003). https://doi.org/10.1016/S0032-0633(02)00207-6

T. Lebrun, H. Massol, E. Chassefière, A. Davaille, E. Marcq, P. Sarda, F. Leblanc, G. Brandeis, Thermal evolution of an early magma ocean in interaction with the atmosphere. J. Geophys. Res. **118**, 1155–1176 (2013)

A. Léger et al., Transiting exoplanets from the CoRoT space mission. VIII. CoRoT-7b: the first super-Earth with measured radius. Astron. Astrophys. **506**, 287–302 (2009)

A. Léger et al., The extreme physical properties of the CoRoT-7b super-Earth. Icarus **213**, 1–11 (2011)

R. Lenzen et al., METIS: system engineering and optical design of the mid-infrared E-ELT instrument, in *Ground-Based and Airborne Instrumentation for Astronomy III*, vol. 7735 (2010). https://doi.org/10.1117/12.856242

J.S. Lewis, Metal/silicate fractionation in the Solar System. Earth Planet. Sci. Lett. **15**, 286–290 (1972)

S. Li, P.G. Lucey, R.E. Milliken, P.O. Hayne, E. Fisher, J.-P. Williams, D.M. Hurley et al., Direct evidence of surface exposed water ice in the lunar polar regions. Proc. Natl. Acad. Sci. USA **115**, 8907 (2018). https://doi.org/10.1073/pnas.1802345115

T. Lichtenberg, G.J. Golabek, T.V. Gerya, M.R. Meyer, The effects of short-lived radionuclides and porosity on the early thermo-mechanical evolution of planetesimals. Icarus **274**, 350–365 (2016)

T. Lichtenberg, G.J. Golabek, C.P. Dullemond, M. Schönbächler, T.V. Gerya, M.R. Meyer, Impact splash chondrule formation during planetesimal recycling. Icarus **302**, 27–43 (2018)

T. Lichtenberg, T. Keller, R.F. Katz, G.J. Golabek, T.V. Gerya, Magma ascent in planetesimals: control by grain size. Earth Planet. Sci. Lett. **507**, 154–165 (2019). https://doi.org/10.1016/j.epsl.2018.11.034

S.J. Lock, S.T. Stewart, The structure of terrestrial bodies: impact heating, corotation limits, and synestias. J. Geophys. Res., Planets **122**, 950–982 (2017). https://doi.org/10.1002/2016JE005239

S.J. Lock et al., The origin of the Moon within a terrestrial synestia. J. Geophys. Res., Planets **123**, 910–951 (2018). https://doi.org/10.1002/2017JE005333

S.J. Lock, K.R. Bermingham, R. Parai, M. Boyet, Geochemical constraints on the origin of the Moon and preservation of ancient terrestrial heterogeneities. Space Sci. Rev. **216**, 109 (2020)

M. Lopez-Morales et al., Detecting Earth-like biosignatures on rocky exoplanets around nearby stars with ground-based extremely large telescopes. Bull. Am. Astron. Soc. **51**, 162 (2019)

H.C. Lord, Hydrogen and helium ion implantation into olivine and enstatite: retention coefficients, saturation concentrations, and temperature-release profiles. J. Geophys. Res. **73**, 5271–5280 (1968). https://doi.org/10.1029/JB073i016p05271

C. Lovis et al., Atmospheric characterization of Proxima b by coupling the SPHERE high-contrast imager to the ESPRESSO spectrograph. Astron. Astrophys. **599**, A16 (2017). https://doi.org/10.1051/0004-6361/201629682

P. Lucey, Understanding the lunar surface and space-moon interactions. Rev. Mineral. Geochem. **60**(1), 83–219 (2006)

T. Lyubetskaya, J. Korenaga, Chemical composition of Earth's primitive mantle and its variance: 1. Method and results. J. Geophys. Res., Solid Earth **112**, B03211 (2007). https://doi.org/10.1029/2005JB004223

T.I. Maindl et al., Impact induced surface heating by planetesimals on early Mars. Astron. Astrophys. **574**, A22 (2015). https://doi.org/10.1051/0004-6361/201424256

A. Mallama, The spherical bolometric albedo of planet Mercury. arXiv e-prints (2017)

R.A. Marcus, D. Sasselov, L. Hernquist, S.T. Stewart, Minimum radii of super-Earths: constraints from giant impacts. Astrophys. J. **712**, L73–L76 (2010)

J.L. Margot, S.J. Peale, R.F. Jurgens, M.A. Slade, I.V. Holin, Large longitude libration of Mercury reveals a molten core. Science **316**, 710 (2007). https://doi.org/10.1126/science.1140514

B. Marty, The origins and concentrations of water, carbon, nitrogen and noble gases on Earth. Earth Planet. Sci. Lett. **313**, 56–66 (2012)

B. Marty, K. Hashizume, M. Chaussidon, R. Wieler, Nitrogen isotopes on the Moon: archives of the solar and planetary contributions to the inner solar system. Space Sci. Rev. **106**, 175–196 (2003). https://doi.org/10.1023/A:1024689721371

S. Massetti, S. Orsini, A. Millilo, A. Mura, E. De Angelis, H. Lammer, P. Wurz, Mapping of the cusp plasma precipitation on the surface of Mercury. Icarus **166**, 229–237 (2003). https://doi.org/10.1016/j.icarus.2003.08.005

A. Mastrobuono-Battisti, H.B. Perets, S.N. Raymond, A primordial origin for the compositional similarity between the Earth and the Moon. Nature **520**, 212–215 (2015). https://doi.org/10.1038/nature14333

M. Mayor, D. Queloz, A Jupiter-mass companion to a solar-type star. Nature **378**, 355–359 (1995)

W.E. McClintock et al., Mercury's exosphere: observations during MESSENGER's first Mercury flyby. Science **321**, 92 (2008). https://doi.org/10.1126/science.1159467

W.E. McClintock et al., MESSENGER observations of Mercury's exosphere: detection of magnesium and distribution of constituents. Science **324**, 610 (2009). https://doi.org/10.1126/science.1172525

F.M. McCubbin, A. Steele, E.H. Hauri, H. Nekvasil, S. Yamashita, R.J. Hemley, Nominally hydrous magmatism on the Moon. Proc. Natl. Acad. Sci. USA **107**, 11223–11228 (2010)

F.M. McCubbin, M.A. Riner, K.E. Vander Kaaden, L.K. Burkemper, Is Mercury a volatile-rich planet? Geophys. Res. Lett. **39**, L09202 (2012). https://doi.org/10.1029/2012GL051711

G.D. McDonald, L. Kreidberg, E. Lopez, The sub-Neptune desert and its dependence on stellar type: controlled by lifetime X-ray irradiation. Astrophys. J. **876**(1), 22 (2019). https://doi.org/10.3847/1538-4357/ab1095

J. Meyer, L. Elkins-Tanton, J. Wisdom, Coupled thermal-orbital evolution of the early Moon. Icarus **208**, 1–10 (2010)

Y. Miguel, L. Kaltenegger, B. Fegley, L. Schaefer, Compositions of hot super-Earth atmospheres: exploring Kepler candidates. Astrophys. J. **742**, L19 (2011)

Y. Miguel, A. Cridland, C.W. Ormel, J.J. Fortney, S. Ida, Diverse outcomes of planet formation and composition around low-mass stars and brown dwarfs. Mon. Not. R. Astron. Soc. **491**(2), 1998–2009 (2019)

A. Milillo et al., Exospheric Na distributions along the Mercury orbit with the THEMIS telescope. Icarus **355**, 114179 (2021). https://doi.org/10.1016/j.icarus.2020.114179

N.M. Mills, C.B. Agee, D.S. Draper, Metal silicate partitioning of cesium: implications for core formation. Geochim. Cosmochim. Acta **71**, 4066–4081 (2007). https://doi.org/10.1016/j.gca.2007.05.024

S.J. Mojzsis, R. Brasser, N.M. Kelly, O. Abramov, S.C. Werner, Onset of giant planet migration before 4480 million years ago. Astrophys. J. **881**, 44 (2019). https://doi.org/10.3847/1538-4357/ab2c03

A. Morbidelli, S. Marchi, W.F. Bottke, D.A. Kring, A sawtooth-like timeline for the first billion years of lunar bombardment. Earth Planet. Sci. Lett. **355**, 144–151 (2012). https://doi.org/10.1016/j.epsl.2012.07.037

C. Mordassini, Planetary population synthesis. in *Handbook of Exoplanets* ed. by H.J. Deeg, J.A. Belmonte. Cambridge Planetary Science (Springer, Berlin, 2018), p. 2425. https://doi.org/10.1007/978-3-319-55333-7_143

J.W. Morgan, E. Anders, Chemical composition of Earth, Venus, and Mercury (planets/solar nebula/element abundances/mantle/core). Proc. Natl. Acad. Sci. USA **77**(12), 6,973–6,977 (1980)

J.S. Morgan, C.B. Pilcher, Plasma characteristics of the Io torus. Astrophys. J. **253**, 406–421 (1982). https://doi.org/10.1086/159645

F. Motalebi et al., The HARPS-N rocky planet search. I. HD 219134 b: a transiting rocky planet in a multi-planet system at 6.5 pc from the Sun. Astron. Astrophys. **584**, A72 (2015). https://doi.org/10.1051/0004-6361/201526822

A. Mura, A. Milillo, S. Orsini, D. Delcourt, R. D'Amicis, Proton circulation around Mercury and ENA emission. Earth Planet. Sci. Lett. **433**, 249–256 (2003). EGS-AGU-EUG Joint Assembly. The neon isotopes in chondrites and on Earth

A. Mura, S. Orsini, A. Milillo, D. Delcourt, S. Massetti, E. De Angelis, Dayside H^+ circulation at Mercury and neutral particle emission. Icarus **175**, 305–319 (2005). https://doi.org/10.1016/j.icarus.2004.12.010

A. Mura, S. Orsini, A. Milillo, A.M. Di Lellis, E. De Angelis, Neutral atom imaging at Mercury. Planet. Space Sci. **54**, 144–152 (2006). https://doi.org/10.1016/j.pss.2005.02.009

A. Mura et al., The sodium exosphere of Mercury: comparison between observations during Mercury's transit and model results. Icarus **200**, 1–11 (2009). https://doi.org/10.1016/j.icarus.2008.11.014

A. Mura et al., Comet-like tail-formation of exospheres of hot rocky exoplanets: possible implications for CoRoT-7b. Icarus **211**, 1–9 (2011). https://doi.org/10.1016/j.icarus.2010.08.015

S.L. Murchie et al., Orbital multispectral mapping of Mercury with the MESSENGER Mercury dual imaging system: evidence for the origins of plains units and low-reflectance material. Icarus **254**, 287–305 (2015). https://doi.org/10.1016/j.icarus.2015.03.027

V.R. Murthy, W. van Westrenen, Y. Fei, Experimental evidence that potassium is a substantial radioactive heat source in planetary cores. Nature **423**, 163–165 (2003). https://doi.org/10.1038/nature01560

N.F. Ness, K.W. Behannon, R.P. Lepping, Y.C. Whang, The magnetic field of Mercury, 1. J. Geophys. Res. **80**, 2708 (1975). https://doi.org/10.1029/JA080i019p02708

N.F. Ness et al., Effects of magnetic anomalies discovered at Mars on the structure of the Martian ionosphere and solar wind interaction as follows from radio occultation experiments. J. Geophys. Res. **105**, 15991–16004 (2000). https://doi.org/10.1029/1999JA000212

G. Neukum, B.A. Ivanov, W.K. Hartmann, Cratering records in the inner solar system in relation to the lunar reference system. Space Sci. Rev. **96**, 55–86 (2001)

W. Neumann, R. Jaumann, J. Castillo-Rogez, C.A. Raymond, C.T. Russel, Ceres' partial differentiation: undifferentiated crust mixing with a water-rich mantle. Astron. Astrophys. **633**, A117 (2020). https://doi.org/10.1051/0004-6361/201936607

M.J. Newman, R.T. Rood, Implications of solar evolution for the Earth's early atmosphere. Science **198**, 1035–1037 (1977). https://doi.org/10.1126/science.198.4321.1035

S.G. Nielsen, D.V. Bekaert, M. Auro, Isotopic evidence for the formation of the Moon in a canonical giant impact. Nat. Commun. **12**, 1817 (2021). https://doi.org/10.1038/s41467-021-22155-7

A. Nikolaou, N. Katyal, N. Tosi, M. Godolt, J.L. Grenfell, H. Rauer, What factors affect the duration and outgassing of the terrestrial magma ocean? Astrophys. J. **875**, 11 (2019). https://doi.org/10.3847/1538-4357/ab08ed

F. Nimmo, T. Kleine, *Early Differentiation and Core Formation*. Washington DC American Geophysical Union Geophysical Monograph Series, vol. 212 (2015), pp. 83–102. https://doi.org/10.1002/9781118860359.ch5

F. Nimmo, G.D. Price, J. Brodholt, D. Gubbins, The influence of potassium on core and geodynamo evolution. Geophys. J. Int. **156**, 363–376 (2004). https://doi.org/10.1111/j.1365-246X.2003.02157.x

L.R. Nittler et al., The major-element composition of Mercury's surface from MESSENGER X-ray spectrometry. Science **333**, 1847 (2011). https://doi.org/10.1126/science.1211567

L.R. Nittler, N.L. Chabot, T.L. Grove, P.N. Peplowski, The chemical composition of Mercury, in *Mercury: The View After MESSENGER*, ed. by S. Solomon, L. Nittler, B. Anderson. Cambridge Planetary Science (Cambridge University Press, Cambridge, 2018), pp. 30–51. https://doi.org/10.1017/9781316650684.003

P. Odert, M. Leitzinger, A. Hanslmeier, H. Lammer, Stellar coronal mass ejections – I. Estimating occurrence frequencies and mass-loss rates. Mon. Not. R. Astron. Soc. **472**, 876–890 (2017). https://doi.org/10.1093/mnras/stx1969

P. Odert, H. Lammer, N.V. Erkaev, A. Nikolaou, H.I.M. Lichtenegger, C.P. Johnstone, K.G. Kislyakova, M. Leitzinger, N. Tosi, Escape and fractionation of volatiles and noble gases from Mars-sized planetary embryos and growing protoplanets. Icarus **307**, 327–346 (2018)

H.S.C. O'Neill, H. Palme, Collisional erosion and the non-chondritic composition of the terrestrial planets. Philos. Trans. R. Soc. Lond. Ser. A **366**, 4205–4238 (2008). https://doi.org/10.1098/rsta.2008.0111

C. O'Neill, H.S.C. O'Neill, M. Jellinek, On the distribution and variation of radioactive heat producing elements within meteorites, the Earth, and planets. Space Sci. Rev. **216**, 37 (2020). https://doi.org/10.1007/s11214-020-00656-z

S. Orsini, A. Milillo, Magnetospheric plasma loss processes in the Earth's ring current and energetic neutral atoms. Nuovo Cimento C **22C**, 633–648 (1999)

S. Orsini et al., The influence of space environment on the evolution of Mercury. Icarus **239**, 281–290 (2014). https://doi.org/10.1016/j.icarus.2014.05.031

G.P.P.L. Otten et al., Direct characterization of young giant exoplanets at high spectral resolution by coupling SPHERE and CRIRES+. Astron. Astrophys. **646**, A150 (2021). https://doi.org/10.1051/0004-6361/202038517

J.E. Owen, Atmospheric escape and the evolution of close-in exoplanets. Annu. Rev. Earth Planet. Sci. **47**, 67–90 (2019)

J.E. Owen, Y. Wu, Atmospheres of low-mass planets: the "boil-off". Astron. J. **817**(2), 107 (2016)

J.E. Owen, Y. Wu, The evaporation valley in the Kepler planets. Astrophys. J. **847**, 29 (2017)

M. Ozima, K. Seki, N. Terada, Y.N. Miura, F.A. Podosek, H. Shinagawa, Terrestrial nitrogen and noble gases in lunar soils. Nature **436**, 655–659 (2005). https://doi.org/10.1038/nature03929

M. Ozima, Q.-Z. Yin, H. Seki, F. Podosek, K. Zahnle, Biotic Earth wind as the origin of oxygen isotope anomalies in contemporary lunar regolith, in *Lunar and Planetary Science Conference* (2007)

K. Pahlevan, D.J. Stevenson, Equilibration in the aftermath of the lunar-forming giant impact. Earth Planet. Sci. Lett. **262**, 438–449 (2007). https://doi.org/10.1016/j.epsl.2007.07.055

R.L. Palma, R.H. Becker, R.O. Pepin, D.J. Schlutter, Irradiation records in regolith materials, II: solar wind and solar energetic particle components in helium, neon, and argon extracted from single lunar mineral grains and from the Kapoeta howardite by stepwise pulse heating. Geochim. Cosmochim. Acta **66**, 2929–2958 (2002). https://doi.org/10.1016/S0016-7037(99)00002-2

R.O. Pepin, R.H. Becker, D.J. Schlutter, Irradiation records in regolith materials. I: isotopic compositions of solar-wind neon and argon in single lunar mineral grains. Geochim. Cosmochim. Acta **63**, 2145–2162 (1999). https://doi.org/10.1016/S0016-7037(99)00002-2

P.N. Peplowski, L.G. Evans, S.A. Hauck, T.J. McCoy, W.V. Boynton, J.J. Gillis-Davis, D.S. Ebel, J.O. Goldsten, D.K. Hamara, D.J. Lawrence, R.L. McNutt, L.R. Nittler, S.C. Solomon, E.A. Rhodes, A.L. Sprague, R.D. Starr, K.R. Stockstill-Cahill, Radioactive elements on Mercury's surface from MESSENGER: implications for the planet's formation and evolution. Science **333**(6051), 1850 (2011)

P.N. Peplowski et al., Variations in the abundances of potassium and thorium on the surface of Mercury: results from the MESSENGER Gamma-Ray Spectrometer. J. Geophys. Res., Planets **117**, E00L04 (2012). https://doi.org/10.1029/2012JE004141

P.N. Peplowski et al., Geochemical terranes of Mercury's northern hemisphere as revealed by MESSENGER neutron measurements. Icarus **253**, 346–363 (2015). https://doi.org/10.1016/j.icarus.2015.02.002

P.N. Peplowski et al., Remote sensing evidence for an ancient carbon-bearing crust on Mercury. Nat. Geosci. **9**, 273–276 (2016). https://doi.org/10.1038/ngeo2669

D. Perez-Becker, E. Chiang, Catastrophic evaporation of rocky planets. Mon. Not. R. Astron. Soc. **433**, 2294–2309 (2013)

C.M. Persson et al., Super-Earth of 8 $M_\oplus$ in a 2.2-day orbit around the K5V star K2-216. Astron. Astrophys. **618**, A33 (2018). https://doi.org/10.1051/0004-6361/201832867

N.E. Petro, C.M. Pieters, Surviving the heavy bombardment: ancient material at the surface of South Pole-Aitken basin. J. Geophys. Res. **109**(E6), E06004 (2004). https://doi.org/10.1029/2003je002182

C.M. Pieters, Crustal diversity of the Moon: compositional analyses of Galileo solid state imaging data. J. Geophys. Res. **98**(E9), 17127–17148 (1993). https://doi.org/10.1029/93JE01221

C.M. Pieters et al., Character and spatial distribution of OH/H_2O on the surface of the Moon seen by M^3 on Chandrayaan-1. Science **326**(5952), 568 (2009). https://doi.org/10.1126/science.1178658

F.C. Pignatale, K. Liffman, S.T. Maddison, G. Brooks, 2D condensation model for the inner Solar Nebula: an enstatite-rich environment. Mon. Not. R. Astron. Soc. **457**, 1359–1370 (2016). https://doi.org/10.1093/mnras/stv3003

M. Plotnykov, D. Valencia, Chemical fingerprints of formation in rocky super-Earths' data. Mon. Not. R. Astron. Soc. **499**, 932–947 (2020). https://doi.org/10.1093/mnras/staa2615

S. Portegies Zwart, The formation of solar-system analogs in young star clusters. Astron. Astrophys. **622**, A69 (2019). https://doi.org/10.1051/0004-6361/201833974

A.E. Potter, R.M. Killen, T.H. Morgan, The sodium tail of Mercury. Meteorit. Planet. Sci. **37**, 1165–1172 (2002). https://doi.org/10.1111/j.1945-5100.2002.tb00886.x

A.E. Potter, R.M. Killen, T.H. Morgan, Long-term observations of sodium on Mercury, in *AAS/Division for Planetary Sciences Meeting Abstracts #35* (2003)

L. Puig et al., The phase a study of the ESA M4 mission candidate ARIEL. Exp. Astron. **46**, 211–239 (2018). https://doi.org/10.1007/s10686-018-9604-3

D. Queloz et al., The CoRoT-7 planetary system: two orbiting super-Earths. Astron. Astrophys. **506**, 303–319 (2009)

366

S. Rappaport et al., Possible disintegrating short-period super-Mercury orbiting KIC 12557548. Astrophys. J. **752**, 1 (2012)

S. Rappaport, T. Barclay, J. DeVore, J. Rowe, R. Sanchis-Ojeda, M. Still, KOI-2700b a planet candidate with dusty effluents on a 22 hr orbit. Astrophys. J. **784**, 40 (2014)

A. Reiners, U.R. Christensen, A magnetic field evolution scenario for brown dwarfs and giant planets. Astron. Astrophys. **522**, A13 (2010). https://doi.org/10.1051/0004-6361/201014251

A. Reufer, M.M.M. Meier, W. Benz, R. Wieler, A hit-and-run giant impact scenario. Icarus **221**, 296–299 (2012). https://doi.org/10.1016/j.icarus.2012.07.021

I. Ribas, E.F. Guinan, M. Güdel, M. Audard, Evolution of the solar activity over time and effects on planetary atmospheres. I. High-energy irradiances (1–1700 Å). Astrophys. J. **622**, 680–694 (2005). https://doi.org/10.1086/427977

A.R. Ridden-Harper et al., Search for an exosphere in sodium and calcium in the transmission spectrum of exoplanet 55 Cancri e. Astron. Astrophys. **593**, A129 (2016)

A. Rivoldini, T. van Hoolst, The interior structure of Mercury constrained by the low-degree gravity field and the rotation of Mercury. Earth Planet. Sci. Lett. **377–378**, 62–72 (2013). https://doi.org/10.1016/j.epsl.2013.07.021

E.C. Roelof, D.G. Mitchell, D.J. Williams, Energetic neutral atoms (-0.5–50 keV) from the ring current: IMP 7/8 and ISSE 1. J. Geophys. Res. **90**, 10991–11008 (1985). https://doi.org/10.1029/JA090iA11p10991

L.A. Rogers, S. Seager, Three possible origins for the gas layer on GJ 1214b. Astrophys. J. **716**, 1208–1216 (2010). https://doi.org/10.1088/0004-637X/716/2/1208

D. Rouan et al., The orbital phases and secondary transits of Kepler-10b. A physical interpretation based on the lava-ocean planet model. Astrophys. J. **741**, L30 (2011). https://doi.org/10.1088/2041-8205/741/2/L30

R. Rufu, O. Aharonson, H.B. Perets, A multiple-impact origin for the Moon. Nat. Geosci. **10**, 89–94 (2017). https://doi.org/10.1038/ngeo2866

G. Ryder, Mass flux in the ancient Earth-Moon system and benign implications for the origin of life on Earth. J. Geophys. Res., Planets **107**(E4), 6-1–6-13 (2002)

G. Ryder, Bombardment of the Hadean Earth: wholesome or deleterious? Astrobiology **3**, 3–6 (2003)

A.E. Saal, E.H. Hauri, M.L. Cascio, J.A. Van Orman, M.C. Rutherford, R.F. Cooper, Volatile content of lunar volcanic glasses and the presence of water in the Moon's interior. Nature **454**, 192–195 (2008)

V.S. Safronov, E.V. Zvjagina, Relative sizes of the largest bodies during the accumulation of planets. Icarus **10**, 109–115 (1969). https://doi.org/10.1016/0019-1035(69)90013-X

A. Salvador, H. Massol, A. Davaille, E. Marcq, P. Sarda, E. Chassefière, The relative influence of H_2O and CO_2 on the primitive surface conditions and evolution of rocky planets. J. Geophys. Res. **122**, 1458–1486 (2017). https://doi.org/10.1002/2017JE005286

R. Sanchis-Ojeda et al., The K2-ESPRINT project I: discovery of the disintegrating rocky planet K2-22b with a cometary head and leading tail. Astrophys. J. **812**, 112 (2015)

A. Santerne et al., An Earth-sized exoplanet with a higher metal/silicate ratio composition. Nat. Astron. **2**, 393–400 (2018)

M. Sarantos, P.H. Reiff, T.W. Hill, R.M. Killen, A.L. Urquhart, A B_x-interconnected magnetosphere model for Mercury. Planet. Space Sci. **49**, 1629–1635 (2001). https://doi.org/10.1016/S0032-0633(01)00100-3

M. Sarantos, R.M. Killen, D. Kim, Predicting the long-term solar wind ion-sputtering source at Mercury. Planet. Space Sci. **55**, 1584–1595 (2007). https://doi.org/10.1016/j.pss.2006.10.011

M. Sarantos et al., Sodium-ion pickup observed above the magnetopause during MESSENGER's first Mercury flyby: constraints on neutral exospheric models. Geophys. Res. Lett. **36**, L04106 (2009). https://doi.org/10.1029/2008GL036207

P. Saxena, R.M. Killen, V. Airapetian, N.E. Petro, N.M. Curran, A.M. Mandell, Was the Sun a slow rotator? Sodium and potassium constraints from the lunar regolith. Astrophys. J. **876**, L16 (2019). https://doi.org/10.3847/2041-8213/ab18fb

L. Schaefer, B. Fegley, Outgassing of ordinary chondritic material and some of its implications for the chemistry of asteroids, planets, and satellites. Icarus **186**, 462–483 (2007)

L. Schaefer, B. Fegley, Chemistry of silicate atmospheres of evaporating super-Earths. Astrophys. J. **703**, L113–L117 (2009)

L. Schaefer, K. Lodders, B. Fegley, Vaporization of the Earth: application to exoplanet atmospheres. Astrophys. J. **755**, 41 (2012)

M. Scherf, H. Lammer, Did Mars possess a dense atmosphere during the first 400 million years? Space Sci. Rev. **217**, 2 (2021). https://doi.org/10.1007/s11214-020-00779-3

E. Schlawin, T.P. Greene, M. Line, J.J. Fortney, M. Rieke, Clear and cloudy exoplanet forecasts for JWST: maps, retrieved composition, and constraints on formation with MIRI and NIRCam. Astron. J. **156**, 40 (2018). https://doi.org/10.3847/1538-3881/aac774

H. Schleicher, G. Wiedemann, H. Wöhl, T. Berkefeld, D. Soltau, Detection of neutral sodium above Mercury during the transit on 2003 May 7. Astron. Astrophys. **425**, 1119–1124 (2004). https://doi.org/10.1051/0004-6361:20040477

H.E. Schlichting, R. Sari, A. Yalinewich, Atmospheric mass loss during planet formation: the importance of planetesimal impacts. Icarus **247**, 81–94 (2015). https://doi.org/10.1016/j.icarus.2014.09.053

H.-U. Schmincke, *Volcanismus* (Springer, Berlin, 2004), p. 324. https://doi.org/10.1007/978-3-642-18952-4

L. Schultz, H.W. Weber, B. Spettel, H. Hintenberger, H. Waenke, Noble gas and element distribution in agglutinate grain size separates of different density, in *Proceedings of the 9th Lunar and Planetary Science Conference (A79-39176 16-91)*, vol. 2, Houston, TX, USA, 13–17 March 1978 (Pergamon Press, New York, 1978), pp. 2221–2232

M. Sekiya, C. Hayashi, K. Nakazawa, Dissipation of the primordial terrestrial atmosphere due to irradiation of the solar EUV. Prog. Theor. Phys. **64**, 1968–1985 (1981). https://doi.org/10.1143/PTP.64.1968

J. Siebert, P.A. Sossi, I. Blanchard, B. Mahan, J. Badro, F. Moynier, Chondritic Mn/Na ratio and limited post-nebular volatile loss of the Earth. Earth Planet. Sci. Lett. **485**, 130–139 (2018). https://doi.org/10.1016/j.epsl.2017.12.042

L. Siess, E. Dufour, M. Forestini, An Internet server for pre-main sequence tracks of low- and intermediate-mass stars. Astron. Astrophys. **358**, 593–599 (2000)

A. Skumanich, Time scales for Ca II emission decay, rotational braking, and lithium depletion. Astrophys. J. **171**, 565 (1972). https://doi.org/10.1086/151310

J.A. Slavin et al., MESSENGER observations of magnetic reconnection in Mercury's magnetosphere. Science **324**, 606 (2009). https://doi.org/10.1126/science.1172011

J.V. Smith, Mineralogy of the planets: a voyage in space and time. Mineral. Mag. **43**, 1–89 (1979). https://doi.org/10.1180/minmag.1979.043.325.01

S.K. Solanki, N.A. Krivova, J.D. Haigh, Solar irradiance variability and climate. Annu. Rev. Astron. Astrophys. **51**, 311–351 (2013). https://doi.org/10.1146/annurev-astro-082812-141007

S.C. Solomon, Mercury: the enigmatic innermost planet. Earth Planet. Sci. Lett. **216**, 441–455 (2003). https://doi.org/10.1016/S0012-821X(03)00546-6

S.C. Solomon, J. Longhi, Magma oceanography. 1. Thermal evolution. Lunar Planet. Sci. Conf. Abstr. **8**, 884 (1977)

P.A. Sossi, S. Klemme, H.S.C. O'Neill, J. Berndt, F. Moynier, Evaporation of moderately volatile elements from silicate melts: experiments and theory. Geochim. Cosmochim. Acta **260**, 204–231 (2019)

F. Spada, P. Demarque, Y.-C. Kim, A. Sills, The radius discrepancy in low-mass stars: single versus binaries. Astrophys. J. **776**, 87 (2013). https://doi.org/10.1088/0004-637X/776/2/87

E.J. Speyerer, R.Z. Povilaitis, M.S. Robinson, P.C. Thomas, R.V. Wagner, Quantifying crater production and regolith overturn on the Moon with temporal imaging. Nature **538**, 215–218 (2016). https://doi.org/10.1038/nature19829. PMID 27734864

L. Sproß, M. Scherf, V.I. Shematovich, D. Bisikalo, H. Lammer, Life as the only reason for the existence of N_2-O_2-dominated atmospheres. Astron. Rep. **65**, 275–296 (2021)

P.D. Spudis, Mercury: geology, in *Encyclopedia of Planetary Science*. Encyclopedia of Earth Science Series (Springer, Dordrecht, 1997). https://doi.org/10.1007/1-4020-4520-4_252

P.D. Spudis, S. Nozete, C. Lichtenberg, R. Bonner, W. Ort, R. Malaret, M. Robinson, E. Shoemaker, The clementine bistatic radar experiment: evidence for ice on the Moon. Sol. Syst. Res. **32**, 17–22 (1998)

S.A. Stern, The lunar atmosphere: history, status, current problems, and context. Rev. Geophys. **37**(4), 453–491 (1999)

S.A. Stern, J.C. Cook, J.-Y. Chaufray, P.D. Feldman, G.R. Gladstone, K.D. Retherford, Lunar atmospheric H_2 detections by the LAMP UV spectrograph on the Lunar Reconnaissance Orbiter. Icarus **226**, 1210–1213 (2013)

S.T. Stewart, Z.M. Leinhardt, M. Humayun, Giant impacts, volatile loss, and the K/Th ratios on the Moon, Earth, and Mercury, in *Lunar and Planetary Science Conference* (2013)

S.T. Stewart et al., Mercury impact origin hypothesis survives the volatile crisis: implications for terrestrial planet formation, in *Lunar and Planetary Science Conference* (2016)

J.M. Sunshine, T.L. Farnham, L.M. Feaga, O. Groussin, F. Merlin, R.E. Milliken, M.F. A'Hearn, Temporal and spatial variability of lunar hydration as observed by the Deep Impact Spacecraft. Science **326**, 565 (2009). https://doi.org/10.1126/science.1179788

V. Svetsov, Cratering erosion of planetary embryos. Icarus **214**, 316–326 (2011). https://doi.org/10.1016/j.icarus.2011.04.026

A. Szentgyorgyi et al., A preliminary design for the GMT-Consortium Large Earth Finder (G-CLEF), in *Ground-Based and Airborne Instrumentation for Astronomy*, vol. 9147 (2014). https://doi.org/10.1117/12.2056741

P. Tamburo, A. Mandell, D. Deming, E. Garhart, Confirming variability in the secondary eclipse depth of the super-Earth 55 Cancri e. Astron. J. **155**, 221 (2018). https://doi.org/10.3847/1538-3881/aabd84

368

J.A. Tarduno, E.G. Blackman, E.E. Mamajek, Detecting the oldest geodynamo and attendant shielding from the solar wind: implications for habitability. Phys. Earth Planet. Inter. **233**, 68–87 (2014). https://doi.org/10.1016/j.pepi.2014.05.007

J.A. Tarduno, R.D. Cottrell, W.J. Davis, F. Nimmo, R.K. Bono, A Hadean to Paleoarchean geodynamo recorded by single zircon crystals. Science **349**, 521–524 (2015). https://doi.org/10.1126/science.aaa9114

G.J. Taylor, The bulk composition of Mars. Chem. Erde **73**, 401–420 (2013)

S.R. Taylor, P. Jakes, The geochemical evolution of the Moon, in *Lunar and Planetary Science Conference Proceedings*, vol. 5 (1974), pp. 1287–1305

G.J. Taylor, M.A. Wieczorek, Lunar bulk chemical composition: a post-gravity recovery and interior laboratory reassessment. Philos. Trans. R. Soc. Lond. Ser. A **372**, 20130242 (2014)

F. Tera, D.A. Papanastassiou, G.J. Wasserburg, Isotopic evidence for a terminal Lunar cataclysm. Earth Planet. Sci. Lett. **22**, 1–21 (1974)

K. Terada, S. Yokota, Y. Saito, N. Kitamura, K. Asamura, M.N. Nishino, Biogenic oxygen from Earth transported to the Moon by a wind of magnetospheric ions. Nat. Astron. **1**, 0026 (2017). https://doi.org/10.1038/s41550-016-0026

M.H. Thiemens, History and applications of mass-independent isotope effects. Annu. Rev. Earth Planet. Sci. **34**, 217–262 (2006). https://doi.org/10.1146/annurev.earth.34.031405.125026

N. Thomas, Detection of [O iii] lambda 5007 emission from the Io plasma torus. Astrophys. J. **414**, L41 (1993). https://doi.org/10.1086/186991

N. Thomas, High resolution spectra of Io's neutral potassium and oxygen clouds. Astron. Astrophys. **313**, 306–314 (1996)

F. Tian, J.F. Kasting, H.-L. Liu, R.G. Roble, Hydrodynamic planetary thermosphere model: 1. Response of the Earth's thermosphere to extreme solar EUV conditions and the significance of adiabatic cooling. J. Geophys. Res., Planets **113**, E05008 (2008). https://doi.org/10.1029/2007JE002946

Z. Tian, J. Wisdom, L. Elkins-Tanton, Coupled orbital-thermal evolution of the early Earth-Moon system with a fast-spinning Earth. Icarus **281**, 90–102 (2017). https://doi.org/10.1016/j.icarus.2016.08.030

G. Tinetti et al., The science of ARIEL (Atmospheric Remote-sensing Infrared Exoplanet Large-survey), in *Space Telescopes and Instrumentation 2016: Optical, Infrared, and Millimeter Wave*, vol. 9904 (2016). https://doi.org/10.1117/12.2232370

E. Tognelli, P.G. Prada Moroni, S. Degl'Innocenti, The Pisa pre-main sequence tracks and isochrones. A database covering a wide range of Z, Y, mass, and age values. Astron. Astrophys. **533**, A109 (2011). https://doi.org/10.1051/0004-6361/200913913

W.B. Tonks, H.J. Melosh, Magma ocean formation due to giant impacts. J. Geophys. Res. **98**, 5319–5333 (1993). https://doi.org/10.1029/92JE02726

M. Touboul, I.S. Puchtel, R.J. Walker, Tungsten isotopic evidence for disproportional late accretion to the Earth and Moon. Nature **520**, 530–533 (2015). https://doi.org/10.1038/nature14355

A. Tsiaras et al., Detection of an atmosphere around the super-Earth 55 Cancri e. Astrophys. J. **820**, 99 (2016)

L. Tu, C.P. Johnstone, M. Güdel, H. Lammer, The extreme ultraviolet and X-ray Sun in time: high-energy evolutionary tracks of a solar-like star. Astron. Astrophys. **577**, L3 (2015)

O.J. Tucker, W.M. Farrell, R.M. Killen, D.M. Hurley, Solar wind implantation into the Lunar regolith: Monte Carlo simulations of H retention in a surface with defects and the H_2 exosphere. J. Geophys. Res. **124**, 278–293 (2019). https://doi.org/10.1029/2018JE005805

D.L. Turcotte, G. Schubert, *Geodynamics* (Cambridge University Press, Cambridge, 2002), p. 472. ISBN 0521661862

H.C. Urey, The cosmic abundances of potassium, uranium, and thorium and the heat balances of the Earth, the Moon, and Mars. Proc. Natl. Acad. Sci. **41**, 127–144 (1955). https://doi.org/10.1073/pnas.41.3.127

D. Valencia, M. Ikoma, T. Guillot, N. Nettelmann, Composition and fate of short-period super-Earths. The case of CoRoT-7b. Astron. Astrophys. **516**, A20 (2010)

V. van Eylen, C. Agentoft, M.S. Lundkvist, H. Kjeldsen, J.E. Owen, B.J. Fulton, E. Petigura, I. Snellen, An asteroseismic view of the radius valley: stripped cores, not born rocky. Mon. Not. R. Astron. Soc. **479**(4), 4786–4795 (2018)

R. van Lieshout, M. Min, C. Dominik, Dusty tails of evaporating exoplanets. I. Constraints on the dust composition. Astron. Astrophys. **572**, A76 (2014)

K.E. Vander Kaaden, F.M. McCubbin, Exotic crust formation on Mercury: consequences of a shallow, FeO-poor mantle. J. Geophys. Res., Planets **120**, 195–209 (2015). https://doi.org/10.1002/2014JE004733

O. Venot, T. Cavalié, R. Bounaceur, P. Tremblin, L. Brouillard, R. Lhoussaine Ben Brahim, New chemical scheme for giant planet thermochemistry. Update of the methanol chemistry and new reduced chemical scheme. Astron. Astrophys. **634**, A78 (2020). https://doi.org/10.1051/0004-6361/201936697

A. Vidotto, The evolution of the solar wind. Living Rev. Sol. Phys. **18**, 1 (2021). https://doi.org/10.1007/s41116-021-00029-w

A.A. Vidotto, M. Opher, V. Jatenco-Pereira, T.I. Gombosi, Simulations of winds of weak-lined T Tauri stars. II. The effects of a tilted magnetosphere and planetary interactions. Astrophys. J. **720**, 1262–1280 (2010). https://doi.org/10.1088/0004-637X/720/2/1262

A.A. Vidotto et al., Characterization of the HD 219134 multi-planet system II. Stellar-wind sputtered exospheres in rocky planets b & c. Mon. Not. R. Astron. Soc. **481**, 5296–5306 (2018). https://doi.org/10.1093/mnras/sty2130

K. Volk, B. Gladman, Consolidating and crushing exoplanets: did it happen here? Astrophys. J. **806**, L26 (2015). https://doi.org/10.1088/2041-8205/806/2/L26

A.N. Volkov, R.E. Johnson, O.J. Tucker, J.T. Erwin, Thermally driven atmospheric escape: transition from hydrodynamic to Jeans escape. Astrophys. J. **729**, L24 (2011). https://doi.org/10.1088/2041-8205/729/2/L24

A. Vorburger, P. Wurz, M. Scherf, H. Lammer, A. Galli, V. Assis Fernandes, Chemical composition of the Moon's 'primary' crust – a clue at a terrestrial origin, in *EGU General Assembly Conference Abstracts* (2020)

R.J. Walker, C.T. Russell, Flux transfer events at the Jovian magnetopause. J. Geophys. Res. **90**, 7397–7404 (1985). https://doi.org/10.1029/JA090iA08p07397

H.Z. Wang et al., Earth wind as a possible exogenous source of lunar surface hydration. Astrophys. J. **907**, L32 (2021). https://doi.org/10.3847/2041-8213/abd559

I. Wardinski, B. Langlais, E. Thébault, Correlated time-varying magnetic fields and the core size of Mercury. J. Geophys. Res. **124**, 2178–2197 (2019). https://doi.org/10.1029/2018JE005835

A.J. Watson, T.M. Donahue, J.C.G. Walker, The dynamics of a rapidly escaping atmosphere: applications to the evolution of Earth and Venus. Icarus **48**, 150–166 (1981)

D.F. Webb, T.A. Howard, Coronal mass ejections: observations. Living Rev. Sol. Phys. **9**, 3 (2012). https://doi.org/10.12942/lrsp-2012-3

Y. Wei et al., Implantation of Earth's atmospheric ions into the nearside and farside lunar soil: implications to geodynamo evolution. Geophys. Res. Lett. **47**, e2019GL086208 (2020). https://doi.org/10.1029/2019GL086208

S.J. Weidenschilling, Iron silicate fractionation and the origin of Mercury. Icarus **35**, 99–111 (1978). https://doi.org/10.1016/0019-1035(78)90064-7

S.Z. Weider et al., Evidence for geochemical terranes on Mercury: global mapping of major elements with MESSENGER's X-Ray Spectrometer. Earth Planet. Sci. Lett. **416**, 109–120 (2015). https://doi.org/10.1016/j.epsl.2015.01.023

G.W. Wetherill, Formation of the terrestrial planets. Annu. Rev. Astron. Astrophys. **18**, 77–113 (1980)

M.A. Wieczorek, M.T. Zuber, A Serenitatis origin for the Imbrian grooves and South Pole-Aitken thorium anomaly. J. Geophys. Res. **106**, 27,853–27,864 (2001)

M. Wieczorek, M. Zuber, R. Phillips, The role of magma buoyancy on the eruption of lunar basalts. Earth Planet. Sci. Lett. **185**, 71–83 (2001). https://doi.org/10.1016/S0012-821X(00)00355-1

R. Wieler, K. Kehm, A. Meshik et al., Secular changes in the xenon and krypton abundances in the solar wind recorded in single lunar grains. Nature **384**, 46–49 (1996). https://doi.org/10.1038/384046a0

D.E. Wilhelms, J.F. McCauley, N.J. Trask, *The Geologic History of the Moon* (1987)

J.N. Winn, D.C. Fabrycky, The occurrence and architecture of exoplanetary systems. Annu. Rev. Astron. Astrophys. **53**, 409–447 (2015)

J. Wisdom, Z.L. Tian, Early evolution of the Earth-Moon system with a fast-spinning Earth. Icarus **256**, 138–146 (2015). https://doi.org/10.1016/j.icarus.2015.02.025

J.A. Wood, Moon over Mauna Loa: a review of hypotheses of formation of Earth's Moon, in *Origin of the Moon* (1986), pp. 17–55

B.E. Wood, Astrospheres and solar-like stellar winds. Living Rev. Sol. Phys. **1**, 2 (2004). https://doi.org/10.12942/lrsp-2004-2

B.E. Wood, H.-R. Müller, G.P. Zank, J.L. Linsky, Measured mass-loss rates of solar-like stars as a function of age and activity. Astrophys. J. **574**, 412–425 (2002). https://doi.org/10.1086/340797

B.E. Wood, H.-R. Müller, G.P. Zank, J.L. Linsky, S. Redfield, New mass-loss measurements from astrospheric Ly-alpha absorption. Astrophys. J. **628**, L143–L146 (2005). https://doi.org/10.1086/432716

C.-J. Wu, W.-H. Ip, L.-C. Huang, A study of variability in the frequency distributions of the superflares of G-type stars observed by the Kepler mission. Astrophys. J. **798**, 92 (2015). https://doi.org/10.1088/0004-637X/798/2/92

G. Wurm, M. Trieloff, H. Rauer, Photophoretic separation of metals and silicates: the formation of higher metal/silicate ratio planets and metal depletion in chondrites. Astrophys. J. **769**, 78 (2013). https://doi.org/10.1088/0004-637X/769/1/78

P. Wurz, U. Rohner, J.A. Whitby, C. Kolb, H. Lammer, P. Dobnikar, J.A. Martín-Fernández, The lunar exosphere: the sputtering contribution. Icarus **191**, 486–496 (2007). https://doi.org/10.1016/j.icarus.2007.04.034

P. Wurz, J.A. Whitby, U. Rohner, J.A. Martìn-Fernàndez, H. Lammer, C. Kolb, Self-consistent modelling of Mercury's exosphere by sputtering, micro-meteorite impact and photon-stimulated desorption. Planet. Space Sci. **58**, 1599–1616 (2010). https://doi.org/10.1016/j.pss.2010.08.003

P. Wurz, D. Abplanalp, M. Tulej, H. Lammer, A neutral gas mass spectrometer for the investigation of lunar volatiles. Planet. Space Sci. **74**, 264–269 (2012)

P. Wurz et al., Particles and photons as drivers for particle release from the surfaces of the Moon and Mercury. Space Sci. Rev. (2021, this journal). https://doi.org/10.1007/s11214-022-00875-6

H. Yang, J. Liu, The flare catalog and the flare activity in the Kepler mission. Astrophys. J. Suppl. Ser. **241**, 29 (2019). https://doi.org/10.3847/1538-4365/ab0d28

R.V. Yelle, Aeronomy of extra-solar giant planets at small orbital distances. Icarus **170**, 167–179 (2004)

T. Yoshizaki, W.F. McDonoug, The composition of Mars. Geochim. Cosmochim. Acta **273**, 137–162 (2020)

E.D. Young, A. Shahar, F. Nimmo, H.E. Schlichting, E.A. Schauble, H. Tang, J. Labidi, Near-equilibrium isotope fractionation during planetesimal evaporation. Icarus **323**, 1–15 (2019)

K. Zahnle, N. Arndt, C. Cockell, A. Halliday, E. Nisbet, F. Selsis, N.H. Sleep, Emergence of a habitable planet. Space Sci. Rev. **129**, 35–78 (2007)

K.J. Zahnle, D.C. Catling, Strange messenger: a new history of hydrogen on Earth, as told by Xenon. Geochim. Cosmochim. Acta **244**, 56–85 (2019). https://doi.org/10.1016/j.gca.2018.09.017

L. Zeng, D.D. Sasselov, S.B. Jacobsen, Mass-radius relation for rocky planets based on PREM. Astrophys. J. **819**, 127 (2016). https://doi.org/10.3847/0004-637X/819/2/127

L. Zeng et al., Growth model interpretation of planet size distribution. Proc. Natl. Acad. Sci. **116**, 9723–9728 (2019)

X. Zhang, A.P. Showman, Effects of bulk composition on the atmospheric dynamics on close-in exoplanets. Astrophys. J. **836**, 73 (2017)

J. Zhang, N. Dauphas, A.M. Davis, I. Leya, A. Fedkin, The proto-Earth as a significant source of lunar material. Nat. Geosci. **5**, 251–255 (2012). https://doi.org/10.1038/ngeo1429

M.Y. Zolotov, A.L. Sprague, S.A. Hauck, L.R. Nittler, S.C. Solomon, S.Z. Weider, The redox state, FeO content, and origin of sulfur-rich magmas on Mercury. J. Geophys. Res., Planets **118**, 138–146 (2013). https://doi.org/10.1029/2012JE004274

Publisher's Note Springer Nature remains neutral with regard to jurisdictional claims in published maps and institutional affiliations.

Authors and Affiliations

Helmut Lammer[1] · Manuel Scherf[1,2,3] · Yuichi Ito[4,5] · Alessandro Mura[6] · Audrey Vorburger[7] · Eike Guenther[8] · Peter Wurz[7] · Nikolai V. Erkaev[9,10,11] · Petra Odert[2]

✉ H. Lammer
helmut.lammer@oeaw.ac.at

[1] Space Research Institute, Austrian Academy of Sciences, Schmiedlstr. 6, 8042 Graz, Austria

[2] Institute of Physics/IGAM, University of Graz, Graz, Austria

[3] Institute of Geodesy, Technical University of Graz, Graz, Austria

[4] Dept. of Physics and Astronomy, Faculty of Maths and Physical Sciences, University College London, Gower Street, WC1E 6BT, London, UK

[5] National Astronomical Observatory of Japan, Osawa 2-21-1, Mitaka, Tokyo 181-8588, Japan

[6] Istituto de Fisica dello Spazio Interplanetario-CNR, Rome, Italy

[7] Physikalisches Institut, University of Bern, Bern, Switzerland

[8] Thüringer Landessternwarte Tautenburg – Karl-Schwarzschild-Observatorium, Sternwarte 5, 07778 Tautenburg, Germany

9   Institute of Computational Modelling, Siberian Branch of the Russian Academy of Sciences, 660036 Krasnoyarsk, Russian Federation

10   The Applied Mechanics Department, Siberian Federal University, 660074 Krasnoyarsk, Russian Federation

11   Institute of Laser Physics, Siberian Branch of the Russian Academy of Sciences, 630090, Novosibirsk, Russian Federation

Space Science Reviews (2023) 219:49
https://doi.org/10.1007/s11214-023-00994-8

Future Directions for the Investigation of Surface-Bounded Exospheres in the Inner Solar System

Anna Milillo[1] · Menelaos Sarantos[2] · Cesare Grava[3] · Diego Janches[2] ·
Helmut Lammer[4] · Francois Leblanc[5] · Norbert Schorghofer[6] · Peter Wurz[7] ·
Benjamin D. Teolis[3] · Go Murakami[8]

Received: 9 December 2022 / Accepted: 10 August 2023 / Published online: 14 September 2023
© The Author(s) 2023

Abstract

Surface-bounded exospheres result from complex interactions between the planetary environment and the rocky body's surface. Different drivers including photons, ion, electrons, and the meteoroid populations impacting the surfaces of different bodies must be considered when investigating the generation of such an exosphere. Exospheric observations of different kinds of species, i.e., volatiles or refractories, alkali metals, or water group species, provide clues to the processes at work, to the drivers, to the surface properties, and to the release efficiencies. This information allows the investigation on how the bodies evolved and will evolve; moreover, it allows us to infer which processes are dominating in different environments. In this review we focus on unanswered questions and measurements needed to gain insights into surface release processes, drivers, and exosphere characterizations. Future opportunities offered by upcoming space missions, ground-based observations, and new directions for modelling are also discussed.

Keywords Surface-bounded exosphere · Mercury · Moon · Surface release processes

1 Introduction

Exospheres of airless bodies result from the complex interactions between the external agents and the surface, and the surface properties are crucial for determining the efficiencies of the various sources (Teolis et al. 2023). Different drivers in the environment must be considered when the exosphere generation mechanism is investigated. Particularly important are the ion (Wurz et al. 2022) and the meteoroid (Janches et al. 2021) populations impacting the surfaces of different bodies. In Fig. 1, a comparative scheme of the drivers and the released surface materials at Mercury and at the Moon is shown. In fact, exosphere observations of different class of species, i.e., volatiles or refractories (Grava et al. 2021a), alkali metals (Leblanc et al. 2022) or water group species (Schörghofer et al. 2021), can shed light on the release processes at work, not only on current solar system airless bodies,

Surface-Bounded Exospheres and Interactions in the Inner Solar System
Edited by Anna Milillo, Menelaos Sarantos, Benjamin D. Teolis, Go Murakami, Peter Wurz and Rudolf von Steiger

Extended author information available on the last page of the article

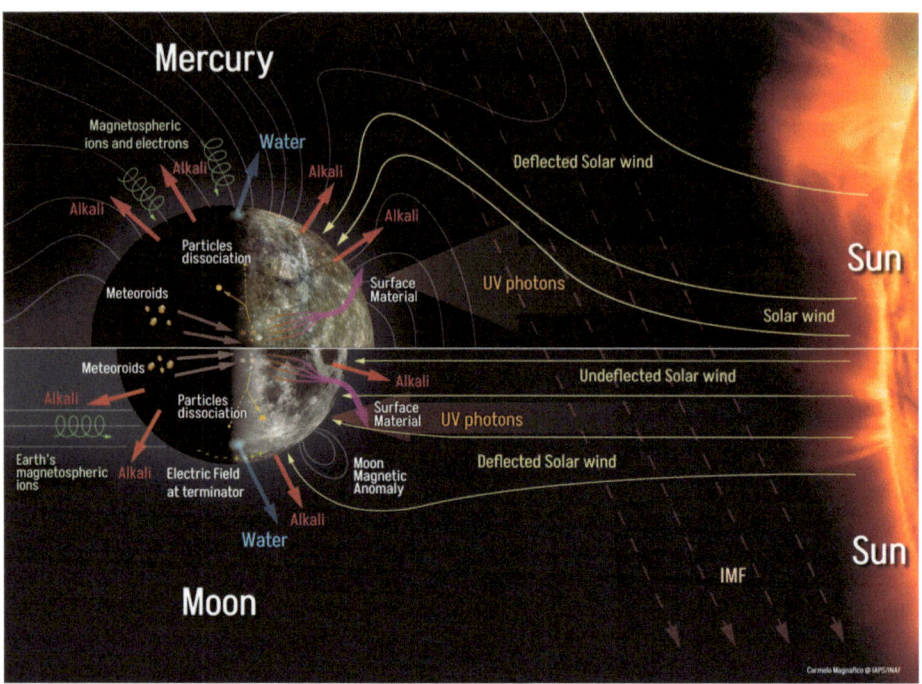

Fig. 1 Schematics of the processes at the two main airless bodies, i.e. Mercury and the Moon

but also on early solar system bodies such as planetary embryos and even close-in rocky exoplanets with magmatic surfaces (Lammer et al. 2022).

The most suitable and best studied examples of airless Solar system rocky bodies are the Moon and Mercury, but also asteroids and the two satellites of Mars are subject to similar interactions, with the big difference being the role of gravity (Schläppi et al. 2008; Plainaki et al. 2009; Nénon et al. 2019).

Already before the 1970s it was known that the Moon does not have a dense atmosphere (Hinton and Taeusch 1964), but the first detection of Ar and He in the lunar exosphere was obtained during the Apollo missions (Hoffman et al. 1973; Hodges 1973a) (see Killen and Ip 1999 and references therein). In the following decade, Na and K have been observed using ground-based techniques (Potter and Morgan 1988). Lunar pickup have been observed near the Moon originally with the Suprathermal Ion Spectrometer (STICS) on the WIND spacecraft (Mall et al. 1998) and more recently measurements of Kaguya (Tanaka et al. 2009; Yokota et al. 2009, 2020), Chang'E−1 (Wang et al. 2011), the ARTEMIS mission (e.g.: Halekas et al. 2012; Poppe et al. 2012, Zhou et al. 2013), and the Lunar Atmospheric and Dust Environment Explorer (LADEE) mission (Halekas et al. 2015; Poppe 2016) confirmed the detections of H_2^+, He^+, C^+, O^+, Ne^+, Na^+, Al^+, Si^+/CO^+, K^+, and $40Ar^+$ (Poppe et al. 2022). At Mercury, only an upper limit for neutral oxygen has been obtained (Shemansky 1988). In both cases, it is possible that the oxygen is in molecular form in the exosphere. In the meanwhile other species have been identified in the exosphere: CH_4, Ne, Rn, and OH/H_2O (Stern 1999; Benna et al. 2015; Hodges 2016).

Lunar noble gases (helium, argon, and neon) were amongst the first exospheric species measured by mass spectrometry during the Apollo 17 mission (Hoffman et al. 1973). Those detections have been confirmed by the Lunar Atmosphere and Dust Environment Explorer

(Benna et al. 2015; Hodges and Mahaffy 2016) and Chandrayaan-1 (e.g., Dhanya et al. 2021). These gases are exemplary species for studying the gas-regolith interaction because they indicate various degrees of gas-surface bonding. The neon density at the surface increases from dusk to dawn as the regolith surface cools off, exhibiting the behaviour expected for a non-condensable species. In fact, for such species, where the gas atoms do not lose sufficient energy upon impact with the surface, or even when they do, the bond is so weak that they immediately (within micro to nanoseconds) are emitted again, the exospheric density at the surface, n, is expected to be inversely related to the surface temperature, T: $n \sim T^{-5/2}$ (Hodges and Johnson 1968). Helium is also a gas that does not freeze out at the coldest night-time temperatures, but exhibits a pre-dawn peak and a decrease towards dawn, deviating from the predicted behavior for non-condensable species because of the larger ballistic range for these light atoms. And last, argon density decreases from dusk to dawn and then increases at dawn, which is the behavior consistent with a condensable species.

The measurements of these and other volatile species, like methane (Hodges 2016) and molecular hydrogen (Stern et al. 2013), raised questions about precisely how gases lose energy when in contact with regolith. For example, there has been a long-standing debate on whether helium atoms in the lunar exosphere should be thermalized with the surface. This issue has been resolved only recently, with spectroscopic observations confirming the thermalization of lunar exospheric helium (Grava et al. 2021b). As another example, the adsorbable nature of argon from measurements of the LACE (Lunar Atmosphere Composition Experiment) mass spectrometer deployed on the lunar surface by the Apollo 17 mission was not expected for a noble gas. Exosphere models require the adoption of unexpectedly high values of the activation energy for desorption of argon-40 (e.g., Bernatowicz and Podosek 1991; Hodges and Mahaffy 2016; Hodges 2016) to match the measurements. This was initially attributed to the cleanliness of the pristine lunar regolith. More recently, Kegerreis et al. (2017) and Sarantos and Tsavachidis (2020, 2021) noted the role of regolith microstructure in prolonging the ability to temporarily retain argon and other gases by Knudsen diffusion within the first mm of regolith. This should also apply to deeper and long-term sequestration of water (Schörghofer 2022) and species of similar volatility. About 30% of the ejected argon atoms are ultimately adsorbed in the Permanently Shadowed Regions (PSR) near the lunar poles (Grava et al. 2015), and another fraction ends up in seasonal polar shadows from which they can be re-emitted every half year, creating seasons (Kegerreis et al. 2017; Hodges 2018). In fact, the polar regions of the Moon are expected to contain a vast repository of volatile gases trapped in the regolith. These atoms and molecules were either transported to the poles via exospheric ballistic hopping, and/or they were synthesized from simpler atomic constituents which landed on grains that act as catalysts.

The source of volatile gases is the lunar interior as well as implanted solar wind. At least some of the helium, ranging by different estimates between 10% (Hodges 1975) to 15–20% (Benna et al. 2015; Grava et al. 2021b) and even up to 40% (Hurley et al. 2016), appears to be effusing from the interior the surface originating from the decay of radiogenic elements. LADEE measurements discovered an enhancement in exosphere density above western maria and Oceanus Procellarum (Benna et al. 2015) that is likely related to the enhanced abundance of ^{40}K at those locations (Kegerreis et al. 2017). Selenographic variations have been demonstrated in other gases whose distribution on the ground is non-uniform such as potassium (Colaprete et al. 2016; Rosborough et al. 2019).

The first observation of Mercury's exosphere has been obtained by Mariner 10, during its fly-bys of Mercury. The UVS spectrometer revealed H, He and an upper limit of atomic oxygen as constituents in its exosphere (Broadfoot et al. 1974, 1976).

Mariner 10 fly-bys revealed the existence of a weak internal global magnetic field (Ness et al. 1975, 1976) with the dipole axis approximately aligned with its spin axis. This dipole field is strong enough to maintain a small magnetosphere populated with plasma originating from the solar wind and from the planet's exosphere. This observation implies a more complex interaction of the Mercury with the solar wind and Interplanetary Magnetic Field (IMF) through its small magnetosphere.

Since the second half of 1980s up to present day, Earth-based observations revealed important features about the exosphere of Mercury (see reviews by Killen et al. 2007; Leblanc et al. 2022), but they are limited to species observable through Earth's atmosphere, which are Na, K, and Ca) (e.g.: Potter and Morgan 1985, 1986; Bida et al. 2000). It is quite challenging to observe Mercury with telescopes operated during night, because of its vicinity to the Sun; Mercury is visible only for a few hours before sunrise or after sunset, so it is difficult to investigate the variability of the exosphere. Since the 1990s advanced technologies allowed much-prolonged day-time observations with Solar telescopes, thus allowing studies of short time variabilities (Mangano et al. 2015; Massetti et al. 2017; Leblanc et al. 2022). The main observables in the ground-based observations are the D1 and D2 emission lines of the Na exosphere with its variable distribution sometime spread to the whole sunlit hemisphere, sometime with two peaks in Northward and Southward hemispheres (e.g.: Potter and Morgan 1990; Potter et al. 1999). The ground-based observations of the D2 line emission of K show similar distributions (Potter and Morgan 1986, 1997). This variability and increase in emission intensity just below regions (cusps) where the solar wind is expected to enter and impact to the surface, made it clear that there is a correlation between alkali distribution and ion impact onto the surface (Killen et al. 2001). The strict relation of Na exosphere variability with plasma impact onto the surface is supported by the lucky observation of Na ground-based observation during an interplanetary Coronal Mass Ejection (iCME) arrival at Mercury registered by MESSENGER on 20 September 2012 (Orsini et al. 2018). In fact, during this observation the Na exosphere showed a distribution that changed when the dense iCME plasma likely compressed the dayside magnetosphere.

Despite this evidence, there are still many unexplained features that require further observations, laboratory measurements and modelling that will be described in the next sections. In fact, the ion sputtering process, as studied in laboratory experiments, has a low efficiency for solar wind ions on rocky regolith, and cannot justify the observed high column densities of Na (up to 10^{11} cm^2) with an altitude profile consistent with temperatures of about 1200 K typical of Photon Stimulated Desorption (PSD) release process (Cassidy et al. 2015; Wurz et al. 2022).

Another specific feature of the Na exosphere is the formation of an anti-sunward tail that is strongly variable along Mercury's orbit (e.g: Potter et al. 2002a,b; Potter and Killen 2008), clearly proportional to the solar radiation pressure. This Na tail is a visual display of the Na loss rate from the planet (Schmidt et al. 2012) comparable to other exospheric species losses (Wurz et al. 2019). Given this loss, the investigation of Na source-sink balance is still an open question. In fact, it is still not clear how the Na content at Mercury's surface evolves in time.

The observations of MESSENGER during flybys and during the whole mission confirmed the presence of an internal magnetic dipole moment of 190 nT R$_M^3$ and found that the dipole is offset northward by about 0.2 R$_M$. (Anderson et al. 2011). Furthermore, MESSENGER MAG and FIPS observations depicted a really dynamic magnetosphere, with a strong coupling with the external solar wind conditions and a high reconnection rate that produces a frequent and efficient solar wind entry and circulation inside the magnetosphere (Slavin et al. 2021). For the first time MESSENGER observed planetary heavy ions in the magnetosphere: Ca$^+$ by Mercury Atmospheric and Surface Composition Spectrometer (MASCS) and He$^+$, water groups and Na$^+$-Mg$^+$-Si$^+$ groups by FIPS (Zurbuchen et al. 2011; Raines et al. 2013). The heavy ion populations in the magnetosphere implies surface release mechanisms directly in ionized state or an efficient photoionization of the exospheric components (Wurz et al. 2019). On the other hand, if these heavy ions impact onto the surface at day and night sides, they may contribute to produce a second-generation exosphere (e.g.: Delcourt et al. 2003; Milillo et al. 2005, 2020).

MASCS UV and Vis spectrometer regularly observed Na (mainly in the low latitudes regions), Ca and, for the first time, Mg atoms in the exosphere (McClintock et al. 2008; Killen et al. 2007). And later Bida and Killen (2017) detected from HIRES ground-based observations traces of the minor species Al and Fe.

The MESSENGER Na observations analysed together with the ground-based observations, showed that superimposed on the short-term variability there is a long-term variability of Na exosphere along the orbit. The global Na density as well as asymmetries in local time and latitudes distributions have a clear recurrence along Mercury's year (Potter et al. 2006; Cassidy et al. 2015; Milillo et al. 2021; Leblanc et al. 2022). The science community is still debating on a global scenario able to fully explain these behaviours.

Finally, MESSENGER detected Mn and Al in Mercury's exosphere (Vervack et al. 2016), the latter was previously observed with low signal/noise ratio by ground-based observations by Bida and Killen (2016). Unexpectedly for a highly oxidated surface (Wurz et al. 2010), the presence of atomic oxygen in the exosphere was still not confirmed, thus a debate on where it is and in which form is still ongoing.

The MESSENGER observations showed a clear dichotomy in the exospheres of different kind of species; in fact, while moderately volatile component, like alkali, i.e. Na and K, distributions have a long term variability along Mercury's orbit and a short term variability especially at high latitudes, the refractory component like Ca and Mg have predominantly a source in the ram hemisphere (in direction of the planet velocity) (Burger et al. 2014), where most of the meteoroids impacts the surface due to the relative velocity of Mercury and the dust disk particles (Pokorný et al. 2018; Janches et al. 2021). The correlation with the expected meteoroid impacts is proved also by the Ca enhancement in the exosphere when Mercury crosses the trajectory of the comet 2P/Encke dust stream (Killen and Hahn 2015). Thus, for these species a Micrometeoroid Impact Vaporization (MIV) is the expected primary generation process. Nevertheless, the high scale height of these components, firstly spotted from the challenging Ca ground-based observation by Bida et al. (2000) and confirmed and quantified (corresponding to a characteristic temperature above 20,000 K, inferred from the scale height) by MESSENGER/MASCS measurements is not easily explainable by the even more energetic surface release processes, like ion sputtering or MIV. This mismatch between the expected and observed distribution of refractory species, opens debate on the possible multiple complex processes able to energize the particles after the release from the surface. The release of molecules (e.g. oxides), which are subsequently dissociated into atoms, could be the pathway for explaining the observed high characteristic energy of some refractory species (Killen 2016; Grava et al. 2021a).

It is likely that a complex mechanism acting in the solar wind - magnetosphere - surface interaction should be invoked to fully explain the Na distributions at Mercury. Given that the whole surface material is released via MIV, the role of this process in the release of different species is also an open point. Any progress on this subject will be dependent on a better understanding on what controls different types of diffusion. The development of models of volatile diffusion coupled with exospheric 3D models and dedicated laboratory experiments are of great importance for the interpretations of the observations.

Despite the crucial role of the water in astrobiology studies, the relative and absolute magnitudes of sources of members of the water group (H_2O, OH, ...) on Mercury and

Fig. 2 Near surface ice stability is modelled to be possible over vastly larger areas than strict PSR (white areas). The coloured regions represent different depths to reach ice stability in the top 1 meter of regolith. (From Paige et al. 2010)

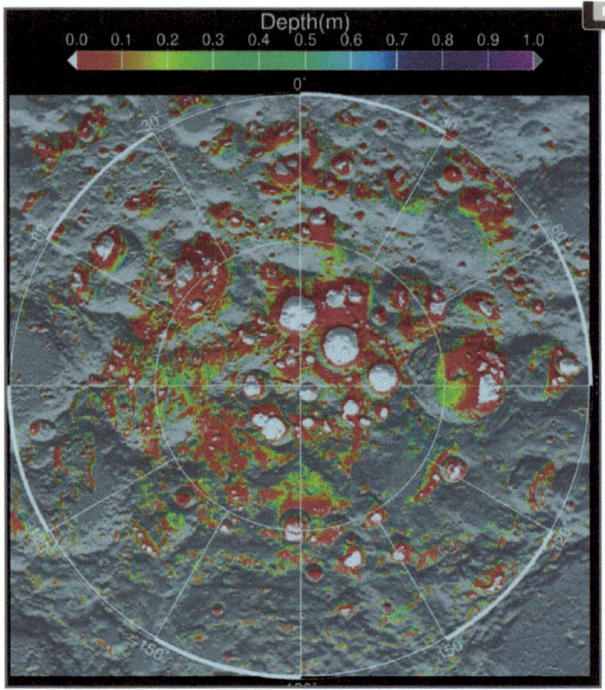

on the Moon remain uncertain. Water released into the exosphere can potentially migrate globally, and become trapped at the cold traps at PSR (Watson et al. 1961; Paige et al. 1992; Schörghofer et al. 2021) (Fig. 2). Slade et al. (1992) observed radar-bright spots at the northern pole at Mercury with Arecibo and Goldstone radar telescopes, later confirmed by MESSENGER observations to be water ice (Chabot et al. 2018). Evidence for cold trapped ice on the Moon comes from neutron spectroscopy (Feldman et al. 1998; Lawrence 2017), the LCROSS impact experiment (Colaprete et al. 2010), UV albedo ratios (Hayne et al. 2015), and near-infrared spectroscopy (Li et al. 2018); see Lucey et al. (2022) for a more detailed review. For analogy, bright spots inside cold traps have been identified also on Ceres by images from the Dawn spacecraft and are likely also made up of ice (Platz et al. 2017). At this point it remains unclear whether this cold-trapped ice was delivered by an exosphere or deposited during rare events that sporadically create temporary atmospheres. It is also possible that the dominant formation and accumulation mechanism may vary from body to body and with distance from the Sun.

Complementary to the exospheric observations are laboratory experiments for investigating the planetary analogues interacting with different drivers, like ions, electrons and UV photons. In Sect. 2, some open points in the surface release processes and the way the next studies will try to answer are described.

For a better understanding of the interaction between the surface and the external environment, we need to characterize the drivers, i.e., impacting plasma, dust distribution, solar UV irradiation of at the surface, as well as the generated exosphere. In Sect. 3, some observations that will help to characterize the drivers thanks to the near-future missions around the Moon and Mercury are described and suggested. In Sect. 4, we describe the open questions in the investigation of the composition, dynamics, sources and loss rates of the exosphere.

Coordinated space and ground-based observations are suggested and expected outcomes are depicted.

Finally, for a deep and global understanding of the exospheric environment we need models for the generation and circulation of the exosphere. Comparison of simulated exospheres with observations of different species will be an essential tool for interpreting the results, for validating theories of exospheric generation and eventually for showing where there are knowledge gaps. In Sect. 5, a short review of more advanced models for airless bodies is given and the direction of further model improvements are suggested. Summary of the expected results of the study of airless bodies' interaction with their parent star and next mission outcomes are described in the final Sect. 6.

2 Open Points on Surface Release and Space Weathering Processes

The observations of the Hermean and lunar exospheres are on spatial scales commensurate with the dimensions of the object, from ~ 100 km size features (like the sputter contribution at the magnetospheric cusps), to global exospheres (e.g., the He exosphere), to extended tails of Na of several, even 1000 planetary radii (Potter et al. 2002a,b; Schmidt et al. 2012). The external drivers causing the population of these exospheres are varied in their spatial and temporal extent (Wurz et al. 2022).

However, for the quantitative understanding of the release processes also the microphysics has to be understood very well. Particle release processes act on very small spatial scales, all the way down to the atomic scale on the surface. It is important to consider in the analysis the actual material, which is fine grained regolith with highly structured grains resulting from the micro-meteorite gardening over millions of years. Moreover, the very surface of these grains is chemically and physically altered by processes of space weathering (ion impact, solar irradiation, ...).

Laboratory measurements of thermal desorption rates and residence time on grains are needed to better understand the surface-exosphere relationship and thus explain the dependence of volatiles' exospheres with local time and solar zenith angle. Experiments that derive yields, cross sections, and threshold energy for PSD and electron-stimulated desorption, which are major source processes for some volatiles, would ultimately need improved simulations of surface-bounded exospheres. Desired simulations are the study of diffusion of volatiles on soil grains, the effect of topography (both at the micro-scale, such as shadow from grains, and at the macro-scale, such as mountains and craters) on the exosphere, and the destruction of deposits of frozen volatiles in PSRs from micrometeoroid bombardment and photolysis from Lyman-alpha photons and cosmic rays.

Despite many research studies are available on processes of interaction of ions, electrons, and photons with the surfaces and their effects on surface modification and particle release, the grains of rocky regolith on planetary surfaces often do not have the same properties as surface analogues investigated in laboratory studies. Generally, the theoretical modelling of surface physics processes has to rely on simple surfaces, which are far away from realistic surfaces to be encountered on planetology. Therefore, we should also explore more complicated surfaces to provide suitable information for studies of particle release in planetary systems. A study accounting for the structured regolith surface was done by Szabo et al. (2022a, 2022b).

In the following sub-sections, a brief overview of surface release processes is provided. More detailed descriptions can be found in Wurz et al. (2022).

2.1 Thermal Desorption

Thermal desorption of solids is well understood for pure species, e.g., the sublimation of water (e.g. Fray and Schmitt 2009). The situation becomes more complicated for mixtures, e.g., CO_2 in water ice, where the sublimation fluxes for H_2O and CO_2 as a function of temperature will depend on the mixing ratio. The most complicated case are adsorbed layers, at the level of monolayer or fractions of it, on surfaces. Here, the examples are Na atoms on the mineral surface. Clearly, the sublimation flux of pure Na would give too high release fluxes. The binding energy of Na to the mineral substrate (e.g. the regolith grains) will be a function of the mineral itself, the structure of the surface (ideally flat, but in reality highly structured), the association of the Na atoms (i.e., localized Na islands or individual Na atoms scattered of the surface), and the exact location of the Na atoms on surface features. For example, the activation energy for release from the surfaces is lowest when the Na atom is on a piece of flat surface, higher when located at a step, and even higher in the corner of a structure. Thus, the microscopic structure of the surface is important. Activation energies for thermal desorption for structured surfaces have to be studied in the laboratory. Recent laboratory experiments, in fact, will focus on sticking coefficients and residence times of Na atoms on the surface, which is important information for the circulation of Na between the exosphere and the surface (Sarantos and Tsavachidis 2021).

2.2 Sputtering by Ion Impact

Sputtering by ion impact is a well-studied process in surface science for many decades since sputtering is used for a range of industrial and analytical applications. However, the sample studies are often not representative of actual surfaces encountered in planetary science. Studies of sputtering from mineral grains, considering fractured surfaces, with significant porosity are mostly missing. Anyway, the theoretical background for understanding the effect of porosity on sputtering is currently developed (Szabo et al. 2022a, 2022b). Moreover, the top surface of 50–100 nm of the regolith grains are space weathered (Pieters and Noble 2016), resulting in different composition and crystal structure, which strongly will affect the sputter yields. Moving from the microscopic to the macroscopic scale, there is a range of minerals present in the regolith, given by the mix of grains in the surface, which will have different sputter yields for the species. Combining these varieties into macroscopic sputter yields for provinces on the lunar or Hermean surface still has not been done. For example, several lunar magnetic anomalies are spatially correlated with surface regions of high albedo of the lunar regolith, known as "lunar swirls" (e.g. Denevi et al. 2014). A possible explanation is that the deflection of solar wind protons in highly magnetized regions prevents or reduces space weathering and darkening of soils (Hood and Williams 1989). This idea is also supported by the OH depletion that is the signature of reduced proton implanting onto the surface (Schaible and Baragiola 2014). The ion impact at the edge of the mini-magnetosphere above the magnetic anomalies, on the contrary, will cause enhanced diffusion inside the regolith grains, ion sputtering, back-scattering of ions and neutral atoms (see also Sect. 3)

The generation of hydroxyl and then molecular water by the interaction of solar wind protons with silicate grains is crucial to our understanding of water formation on large airless bodies (Schörghofer et al. 2021). The efficiency of this process needs to be quantified in laboratory experiments for a variety of compositions and surface properties (ideally with lunar samples).

2.3 Photon Stimulated Desorption

PSD is a process that operates on the atomic scale. A UV photon is absorbed by atoms localized on the surface which may result in the electronic excitation of an atom residing on the surface, e.g. Na, rendering into an anti-binding state causing release of this atom. Although Desorption Induced by Electronic Transitions (DIET), by photons and electrons, has been studied in the surface science community for many decades (see Wurz et al. 2022 and references therein), the particular surfaces relevant for planetology have been studied only for very few cases. Studies on the dependence of the release of Na, K, and others from photon wavelength and the corresponding cross-sections of the interaction are mostly missing for regolith grains. Also, the energetics, i.e., the energy distributions of the released particles are not well studied, and often thermal distributions are assumed in the modelling and analysis of observations even though this process is a DIET process and not a thermal process.

2.4 Electron Stimulated Desorption

Electron stimulated desorption (ESD) is a surface release process comparable to the PSD at microscales (see Wurz et al. 2022 and references therein) even if the driver has a different nature and distribution. Since both electrons (in the energy range 15-hundreds eV) and UV photons produce the same electronic excitation on surface atoms, the velocity distributions of the released particles are the same. The cross sections of the processes are quite similar, but since the electron fluxes are generally much lower than the photon fluxes this process is generally masked by PSD in the dayside of the bodies. The efficiency of this process at Mercury and Moon could be investigated in the nightside surface.

2.5 Micrometeoroid Impact Vaporization

Most of the dust particles hitting the surface of Mercury and the Moon are meteoroids, with sizes in the range of typically 1 to 100 μm. Thus, the affected volume at the impact size is of similar dimension, a microscopic process on the surface. The high speed of the impacting dust particles gives rise to an impact plume, releasing material from the surface. This material is broken up pieces from the surface, atoms, molecules, ions. The temperature of the shock-induced cloud just after impact (the first 100 ns) was estimated to be between 15,000 K and 27,000 K. The temperature and pressure quickly decrease during adiabatic expansion of the cloud reaching a quenching temperature, typically 2000–5000 K. During the rapid expansion of the plume, non-equilibrium processes take place resulting also in the formation of molecules (Berezhnoy 2013, 2018). Thus, the input to the exosphere is not just atoms, but also molecules. For the latter their internal temperature (rotation and vibration) influences how long they will survive in the exosphere before they separate into their atomic constituents. Little information is available about the thermodynamic conditions in such plumes, and laboratory studies with dust impacts at the relevant impact speeds could be performed for a combination of relevant impactor and surface materials (Cintala 1992).

Another area which requires further work for MIV process description concerns laboratory experiments. Experimental results from Koschny and Grün (2001) predict lunar dust cloud density values higher by four orders of magnitude than those inferred from LDEX measurements (Pokorný et al. 2019). This discrepancy could be partially due to the very low velocity impacts (1–12 km/s) experimented on ice-rich surfaces for estimating the yield. Clearly experiments better matched experimental conditions are needed to advance in this area.

3 Open Points on the Characterization of the Drivers (Meteoroids, Ions and Electrons)

As mentioned in the previous section, the external drivers causing the exosphere generation are varied in their spatial and temporal extent (Wurz et al. 2022).

The Sun illuminates the entire dayside, but because of the irradiation dependence on the solar zenith angle there is a longitudinal and latitudinal variation of its effect on the thermal release and release via photon stimulated desorption. The ion sputtering is driven by the solar wind ions and magnetospheric ions hitting the surface. For Mercury, the ion bombardment displacement at the surface depends on the configuration of the magnetosphere, which adjusts itself to the solar wind plasma and magnetic field parameters on short time scales. The global situation is simpler for the Moon, spending most of its time in the solar wind, but there is also the passage through the terrestrial magnetosphere which sends different plasma populations to the lunar surface (Kallio et al. 2019). At some locations on the lunar surface there are magnetic anomalies that could affect ion bombardment distribution. Lastly, meteoroid impacts release material into the exosphere. There are different sources of meteoroids in the inner solar system (e.g., Pokorný et al. 2018; Janches et al. 2021). Most of the interplanetary dust is distributed in a disk in the ecliptic plane. In addition to the steady flux of meteoroids, there are occasional crossings with dust streams from comets (e.g. meteoroid showers; i.e.: Killen and Hahn 2015). Understanding the spatial and temporal variability of these external drivers is important to interpret the observation on these spatial and temporal scales, both ground-based and with spacecraft.

3.1 Ions and Electrons

MESSENGER magnetic field and ion measurements provided a huge amount of information about the interaction between the Sun and the hermean magnetosphere. Moreover, the observation and analysis of Flux Transfer Events (FTE) showed that this is the most intense and common way to transfer particles from the solar wind to the magnetosphere (Raines et al. 2015). In the cusps, the solar wind is channelled downward by FTE in form of filaments, FTE showers are frequently observed and are associated to Na^+ - group enhancement probably released after direct ion sputtering of solar wind onto the surface (Fig. 3, Sun et al. 2022).

The iCME impact onto the Mercury's magnetosphere could produce a strong magnetopause compression, in extreme cases the magnetopause is so close to the planet that almost the whole dayside surface is exposed to ion impacts (Slavin et al. 2019). The Na ground-based observations obtained by THEMIS telescope during an iCME encounter with Mercury seems to validate this interpretation showing a variable Na exosphere distributed at mid latitudes in nominal conditions that expands to the whole dayside at iCME passage (Orsini et al. 2018), but much more observations of IMF and local solar wind simultaneously to exosphere imaging at high-time resolution are required for a confirmation of this scenario.

Anyway, the increase in the statistics of Na exosphere measurements at Mercury during different solar activities could not be enough to explain the complex behaviour of the Na exosphere and its relationship with the impacting plasma. For a final proof of the process that generates these observed planetary ion populations and causes the exosphere variability, a full set of simultaneous observations would be needed, from driver to the resulting release. We need measurements of

- the upstream solar wind and IMF for evaluating how the external conditions affect the plasma precipitation,

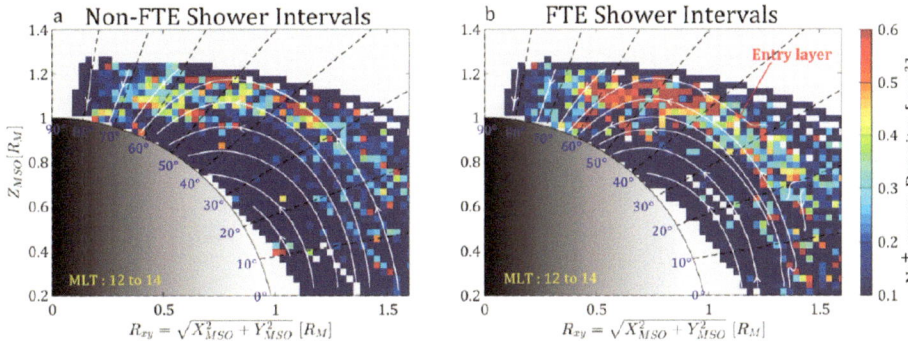

Fig. 3 The magnetic field topology as well as MESSENGER's spatial distribution measurements of the sodium-group (Na^+-group) ions during (a) intervals without flux transfer event (FTE) showers and (b) intervals with FTE showers, shown in the R-Z plane. Colors indicate the observed density of the Na^+-group ions. The white lines represent the magnetic field lines obtained through the average magnetic fields measured by MESSENGER during the intervals without FTE showers and with FTE showers, respectively. (Sun et al. 2022)

- plasma and magnetic field in situ measurements to characterize the plasma directed toward the surface,
- mapping of the area where particles are precipitating, obtainable by the detection of backscattered and neutralized impacting ions,
- characterization (temporal and spatial variation in densities, mass components and vertical profiles) of the exosphere released after ion precipitation events,
- measurements of planetary ions directed outward from the planet.

While the single-spacecraft mission, MESSENGER, could not observe all these targets simultaneously, this ambitious goal will be addressed at Mercury thanks to the ESA-JAXA BepiColombo mission, which was launched in 2018. In fact, two spacecraft (Mio and Mercury Planetary Orbit – MPO) will be placed in orbit around the planet in 2025. In this way, the external conditions will be monitored together with the close-to-surface populations (Fig. 4, Milillo et al. 2020).

At Mercury the interaction scenario, i.e. the precipitation path, is even more complicated by the induction currents generated by the solar wind plasma flux in the large metallic core close to the surface. In fact, during fast events of magnetic compression, the induction currents act against the compression of the day-side magnetosphere (Jia et al. 2015, 2019; Dong et al. 2019). Eventually, the global current system within the magnetosphere and the surface is still an open question that can be solved only by multi-vantage point observations that allow discrimination between the inner and outer magnetic components.

Transport of plasma and magnetic flux from the dayside to the nightside loads the magnetotail with plasma, until it is released by reconnection in the tail (Slavin et al. 2021; Imber and Slavin 2017). This process accelerates plasma toward the nightside of the planet, where a fraction of it may impact the surface near the open/closed field line boundary in the plasma sheet horns, that are located at mid latitudes in the northern hemisphere and at lower latitudes in the southern hemisphere where the boundary is shifted northward together with the internal magnetic dipole (Raines et al. 2015). Different planetary ions, directly released from the surface or generated in the exosphere after photoionization, circulate into the magnetosphere, part of these populations, after experiencing acceleration processes, can impact the surface thus generating a second generation of exosphere (Milillo et al. 2020). The effect on

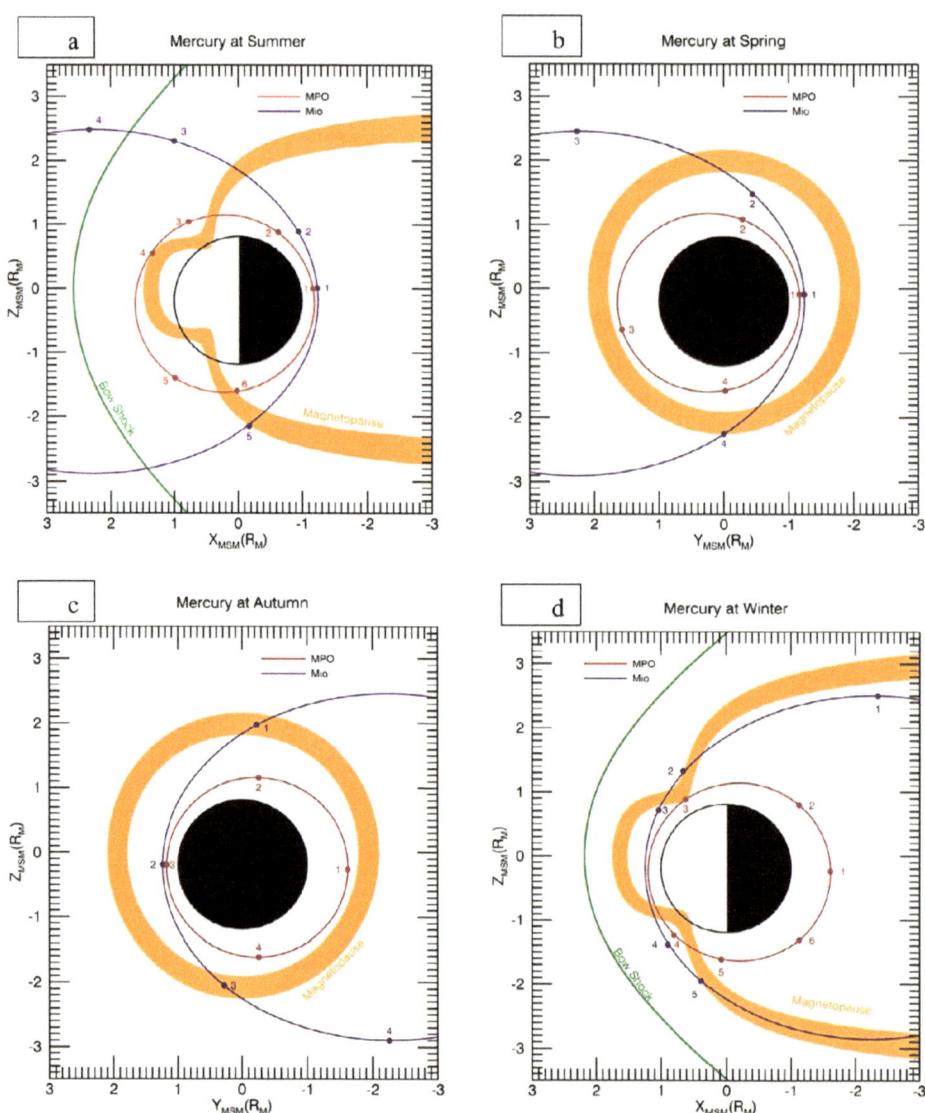

Fig. 4 Schematic view of perihelion/Summer (a), Autumn (c), aphelion/Winter (d) and Spring (b) Bepi-Colombo orbits configurations (from Milillo et al. 2020). The planet Mercury is represented by the black circle (filled in the nightside). The red and blue lines show the MPO and Mio orbits after insertion. The orange area represents the variability (1σ) of the magnetopause according to the 3D-model of Zhong et al. (2015) which includes indentions for the cusp regions. The green line represents the approximate position of the bow shock (Winslow et al. 2013)

the exosphere of the nightside ion precipitation has still not been investigated since nightside exosphere measurements were not allowed with the MESSENGER UV-Vis spectrometer MASCS. The in-situ measurements by BepiColombo/MPO with the SERENA-STROFIO mass spectrometer performed simultaneously with ion measurements will allow the investigation of the nightside exosphere generation processes for the first time (Milillo et al. 2020; Orsini et al. 2021).

Signature of electron impact onto Mercury's nightside surface has been identified by the X-Rays observations of MESSENGER (Lindsay et al. 2016) in agreement with analysis of depolarization signatures by Dewey et al. (2020). The electron impact mapping has been obtained mainly for the northern hemisphere since MESSENGER orbit was highly eccentric with periherm close to the north pole. The ESD produced after the impact of the electrons onto the surface has never been observed. The simultaneous observations of electrons and volatile components of the exosphere are required to obtain hints on the efficiency of this process. The BepiColombo mission will provide a more accurate electron precipitation mapping at both hemispheres thanks to the optimal orbit of the MPO spacecraft; the observations of the signatures of reconnection and depolarization from Mio spacecraft in the magnetotail, will help to identify the origin or trajectory of the impacting electron population; furthermore, the measurements of the nightside volatile component of the exosphere obtained with a mass spectrometer will allow to evaluate the effect of the electron stimulated desorption where most of other surface release processes are not active (Milillo et al. 2020).

Generally, the interpretation of single- or double-point observations in the magnetosphere requires advanced simulations of ion circulation in Mercury's environment allowing to follow the particle trajectories and to connect the measurements (see Sect. 6).

In contrast to Mercury, ions precipitate more freely onto the lunar surface. The Moon is exposed to plasma environments with different plasma characteristics as it orbits around the Earth. It spends nearly a quarter of its orbit in the terrestrial magnetosphere, when we see it as full Moon, and the rest of the time in Earth's magnetosheath or in the solar wind (Kallio et al. 2019). Outside the terrestrial magnetosphere there are highly variable conditions (from low to fast streams, from low densities solar wind to dense iCME events) (Wurz et al. 2022). When the Moon is in the solar wind, the general distribution of solar wind impact onto the surface is only driven by the geometry (a cosine law from the subsolar point) and it is negligible at the night side. The lunar surface is charged positively on the dayside due to the emission of photoelectrons from the dayside and negatively on the nightside due to the difference between the electron and proton fluxes (Halekas et al. 2011). This makes it possible to have ion precipitation also in the regions close to the terminator by deflected solar wind (Halekas et al. 2011; Vorburger et al. 2016). Sputtering by electrons is also thought to be the dominant erosion process for potential surface frosts in lunar cold traps (Farrell et al. 2019).

The situation is different in the localized regions where there are magnetic anomalies. Here the solar wind is deflected and flows along the mini magnetospheric-cavities, thus impacting the surface at the boundary of the region and leaving the magnetized region shielded (Futaana et al. 2013), the estimated difference being about 50% (Vorburger et al. 2013). This has been demonstrated by the IBEX and Chandrayaan-1 observations of the back-scattered neutralized solar wind (McComas et al. 2009; Wieser et al. 2010; Vorburger et al. 2012, 2013) and of the reflected electrons and ions observed from and Chandrayaan-1 and SELENE (Anderson et al. 1975; Lue et al. 2011; Saito et al. 2010, 2012). The velocity distributions of downward-travelling particles are altered from those of the pristine ambient plasma also after the interaction with plasma waves in the near-Moon space (Halekas et al. 2012; Harada et al. 2014a,b; Fatemi et al. 2015; Wurz et al. 2022). The detailed precipitation map of the solar wind at the magnetic anomaly regions together with local plasma simulations will allow us to thoroughly analyse the interaction in these complex regions. This could aid in the characterization of lunar sites for human colonization.

When the Moon passes through the terrestrial magnetotail both sunward and antisunward flows commonly exist (Troshichev et al. 1999; Øieroset et al. 2002). Heavy ions could be present especially during geomagnetic activities (Seki et al. 1996; Poppe et al. 2016). The intensity of back-scattered particles is higher in these conditions, and it seems less sensitive to the magnetic anomalies, hence the shielding is less efficient. This is probably due to higher plasma temperatures and energies, and low Mach number (Allegrini et al. 2013).

When the Moon is located in the foreshock region, the high-energy ions back-streaming from the bow shock can directly access the lunar surface (Benson et al. 1975; Nishino et al. 2017) and produce ion sputtering. We still do not exactly know which fraction of these ions impact the nightside lunar surface.

The plasma monitoring upstream of the Moon could be obtained in the future by the international space station Deep Space Gateway, that will be located between the Earth and the Moon orbit and that will include a full plasma package (Dandouras et al. 2023). This continuous monitor coupled with exospheric measurements performed by dedicated missions in orbit close to the Moon surface (ISRO Chandrayaan, NASA-ESA Artemis and NASA CLPS, Chinese Lunar Exploration programs, Korea Pathfinder Lunar Orbiter – KPLO) will allow to perform statistical studies on plasma and exosphere variations.

In spite of the extensive modelling of the sputtered component (e.g., Wurz et al. 2007; Sarantos et al. 2012), the detection of the exosphere generated by ion sputtering is difficult because of low energy range and low intensity of the products. Little evidence of an energetic exosphere potentially produced by this sputtering has been obtained (Wang et al. 2021). Vorburger et al. (2013) reported that, when the solar wind was at high-helium content, the Chandryaan-1/CENA sensor measured a slightly higher heavier/light mass ratio than at nominal solar wind conditions. Most of the measurements were due to backscattered solar wind; nevertheless, this observation could be the signature of higher sputtering efficiency due to He^{++} impacts (supposed to be about 20% more than the yield of H^+). The sputtered density is expected to be much reduced over large magnetic anomalies as the deceleration of solar wind protons reduces the sputtering yield (Poppe et al. 2014); in fact, several lunar magnetic anomalies are also correlated with lunar swirls (e.g., Denevi et al. 2014), this seems to confirm local intense space weathering activity (see also Sect. 2).

3.2 Micrometeoroid

While micrometeoroids refill the surface with gardening, their impacts onto the surface of airless bodies are surely one of the major drivers of surface release, especially for refractories and molecules that are tightly bound to the minerals (Janches et al. 2021). The release material is proportional to the incoming flux and to the velocity of the meteoroids (Cintala 1992). Different origins have been identified for the micrometeoroid populations at Earth's and at Mercury's orbits. The main sources for the inner solar system meteoroid populations are particles originating from Main Belt Asteroids (MBAs), Jupiter Family Comets (JFCs), Halley Type and Oort Cloud Comets (HTCs and OCCs) (Janches et al. 2021). Each family of meteoroid has a characteristic trajectory and velocity distribution, so that anisotropic distribution of meteoroids in arrival direction may produce seasonal, diurnal and planetographic variability of incoming meteoroids (Fentzke and Janches 2008; Janches et al. 2018; Pokorný et al. 2017; Janches et al. 2021).

In particular, measurement of Ca and Mg in the exosphere of Mercury showed that the micrometeoroid impact vaporization is the main source mechanisms for these refractory species, producing a clear asymmetric distribution toward dawn (Mercury's ram direction) and a clear seasonal modulation proportional to the expected dust distribution (Burger et al. 2014; Pokorný et al. 2018) (Fig. 5). In fact, the eccentric and inclined (7° with respect to

Fig. 5 Above: Total vaporization flux as a function of True Anomaly Angle (x-axis), and the impact velocity (y-axis) at Mercury. The units are g cm^{-2} s^{-1} per 2 km s^{-1} bin. From Pokorný et al. (2018). Below: Ca source rate determined using a source with T = 70,000 K (black curve) compared to sources derived from MESSENGER/MASCS data (red curve with error bars) along Mercury's year (Burger et al. 2014)

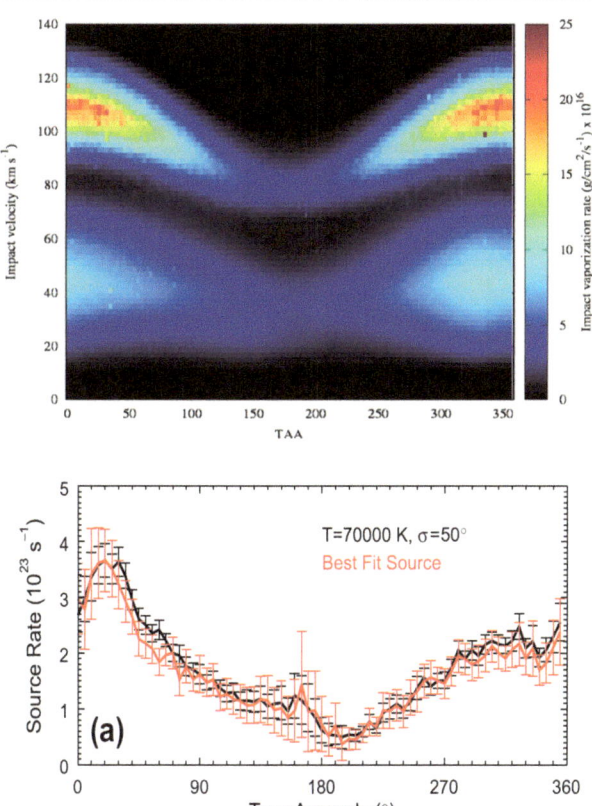

the ecliptic plane orbit where the dust disk is distributed) orbit of Mercury passes through highly variable micrometeoroid intensities. Regions closer to the Sun have higher micrometeoroid densities and the relative velocities with the planet are higher, as well. Kameda et al. (2009) reported a relation of the exospheric Na global intensities with respect to the distance from the expected dust disk, but Milillo et al. (2021) reported that the Na distributions generally peak at the hemisphere farther from the dust disk (North above the disk and South below it). The increase of global Ca content in the exosphere at the comet 2P/Encke dust stream crossing was observed the by MESSENGER/MASCS UV spectrometer (Fig. 5 bottom panel) (Killen and Hahn 2015), but the intensity of this increase is still to be quantitatively explained (Killen 2016; Plainaki et al. 2017).

For the first time, BepiColombo mission will be able to measure the dust distribution from the Mio spacecraft and the exospheric distributions of different species and molecules (Milillo et al. 2020).

Regarding the Moon, the LDEX measurements provided compelling evidence that our understanding of how meteoroids influence the lunar surface must be revisited. A detailed description of these findings is presented in Janches et al. (2021). In summary, the measured fluxes showed that the Moon is engulfed in a permanently present, but highly variable dust exosphere that is most dense at 5–8 hrs of lunar local time, with a peak density tilted somewhat sunward of the dawn terminator (Fig. 6). Several authors have shown that Long Period Comets (LPC) produced meteoroids (i.e., HTC and OCC; Janches et al. 2021) should play a major role in the production of the observed ejecta cloud in the Moon's equatorial plane.

Fig. 6 The modelled annually averaged lunar dust density distribution for particles with a ≥ 0.3 μm. The Sun is on the left and the apex motion of the Moon about the Sun is towards the top of the page (Szalay and Horányi 2016)

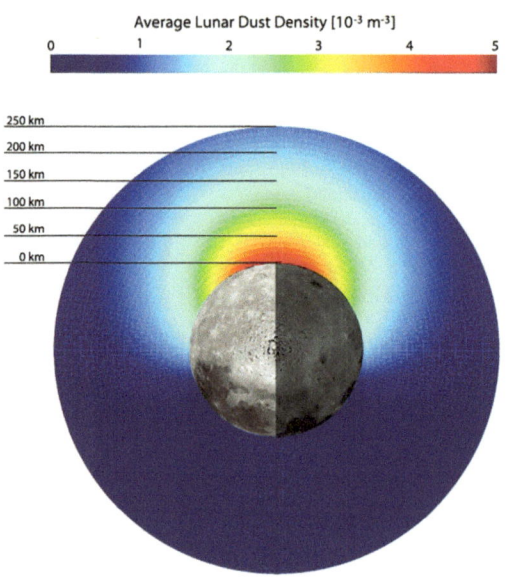

Furthermore, the cloud density is modulated by both the Moon's orbital motion about the Earth and about the Sun. The tilting of the ejecta cloud toward the Sun seems to be more pronounced earlier during the LADEE mission (November 2013), while the LDEX signal became more centred around the dawn terminator toward the end of the mission (April 2014) showing a clear seasonal variability.

Efforts of modelling the influence of meteoroids on the lunar surface parallel those at Mercury and differ again on the meteoroid populations included in the different treatments. The effect of gravitational focusing plays a significant role in shaping the lunar and terrestrial meteoroid environment and the night-side to day-side asymmetry, although reproduced by the models, still have unknown physical effects that require further investigation.

The absolute mass flux of meteoroids onto the Moon is also a critical quantity that cannot be fully constrained with LDEX observations. For example, the total flux of MBA meteoroids cannot be constrained by modelling LDEX observations because they produce a negligible contribution to the total ejecta mass production rate due to their very low velocity. Furthermore, to stay consistent with Earth-based estimates of the mass flux ratio of short-to-long period comets (Carrillo-Sánchez et al. 2016), Pokorný et al. (2019) finally concluded that the total mass accreted at the Moon is approximately 1.4 t/day assuming 43.3 t/day at Earth, where the individual contribution of meteoroid populations are: JFCs $\sim$72.6%, HTCs $\sim$12.8% and OCCs $\sim$10.0%. An important note is that these results represent one of many possible fits to the available LDEX measurements and that the solution space to provide a similar or better fit is wide due to the limited selenographic coverage of LADEE.

JFCs meteoroids are concentrated close to the ecliptic plane, arriving from direction towards and away from the Sun (helion and anti-helion sources). HTC and OCC meteoroids impact the Moon mainly towards the apex direction while MBA meteoroids have radiants ranging from all directions and are hence able to populate the anti-apex source. Like at Earth, the apex source has average impact velocities exceeding 55 km/s, while the toroidal and helion/anti-helion sources are in general populated by meteoroids a factor of two slower. Due to the smaller gravitational focusing on the Moon, JFC and MBA meteoroids contribute 2.5 and 5 times less in terms of the mass flux to the lunar meteoroid environment, respectively,

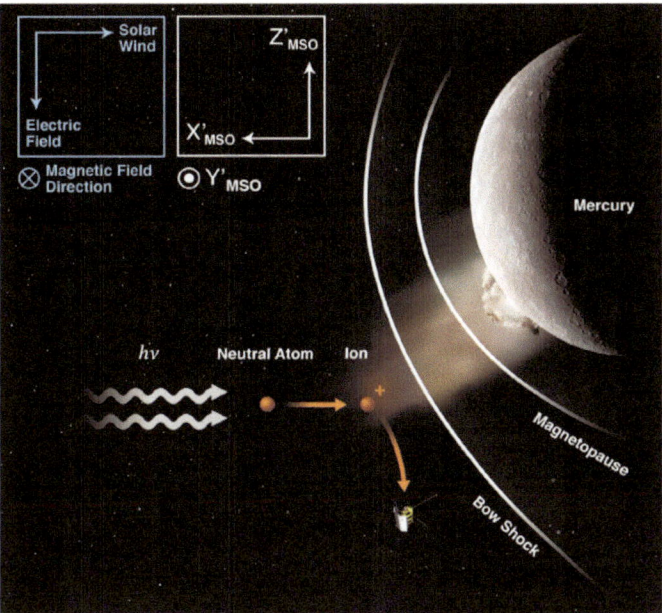

Fig. 7 A schematic from Jasinski et al. (2020) showing the photoionization of neutral particles released from the surface of Mercury due to a large impactor. The newly photoionized particles were observed as pickup ions by the MESSENGER spacecraft in the solar wind upstream of the bow shock

than at Earth. As a result of the broad latitude distribution of cometary impactors, the entire lunar surface can be exposed to impacts with velocities as high as 30 km/s, where the near ecliptic directions can produce impacts with velocities up to 72 km/s.

Finally, Pokorný et al. (2019) showed that the meteoroid mass flux and, consequently, the impact vaporization flux and ejecta mass production rate experience yearly and monthly variations that can be well represented by a sum of two sine functions with periods of one year and 29.5 days (synodic period of the Moon). The mass flux variations amount to 3.3% of the yearly average mass flux, while monthly variations amount to only 0.2%. For the case of the impact vaporization flux accounts for 6–8%, while monthly variations are around 4-5%. When the full spectrum of impact velocities is taken into account, the apex/dawn terminator source is dominating both the impact vaporization flux and the ejecta mass production rate for any day of the year. This expected total vapor rate is higher than considered in lunar exosphere models (Sarantos et al. 2012), meaning that the role of impact vaporization in supplying the lunar exosphere with metals may have been previously underestimated, especially for species like Na and K which do not condense.

In addition to the quasi-continuous flux of micrometeoroid, which contribute to the exosphere generation, sporadic impacts of meteoroid bigger than 1 cm must be considered. Their frequency is much lower than the impact rate of the micrometeoroid populations, but their contribution to the instantaneous exosphere density could be dominant (Mangano et al. 2007). In fact, MESSENGER revealed the signature of a major meteoroid impact by detecting the pickup planetary ions in the solar wind upstream of the bow shock probably originating from atoms released after a meteoroid impact and subsequently photoionized (Jasinski et al. 2020) (Fig. 7).

We can expect to record much more detections of these events during the BepiColombo mission lifetime including mission extensions (Mangano et al. 2007). It will be possible to search for the impact crater after these events, thanks to the possibility to obtain high spatial resolution imaging of the surface with the camera suite SimbioSys on board of MPO (Cremonese et al. 2020).

In the incoming decade, dust detectors on board DESTINY (Krüger et al. 2019) and IMAP (McComas et al. 2018) missions will allow monitoring of dust at 1 AU, thus providing an important tool for constraining the meteoroid contribution to the lunar exosphere formation.

3.3 Solar UV Variability

Together with ions, electrons, and micrometeoroids, solar photons in the UV range arriving at the surface are an important factor in the release of some species softly bound to the surface, such as volatiles and alkali atoms. In fact, the PSD process releases atoms or molecules adsorbed on the surface, i.e., species that are not chemically bonded within a mineral (Wurz et al. 2022).

At Mercury and the Moon, the PSD is considered the most efficient surface release process for Na when coupled with the action of ion impact (Mura et al. 2009). In fact, the vertical profile of the Na densities at Mercury is consistent with a characteristic temperature of 1200 K, compatible with the PSD energy distribution (Cassidy et al. 2015).

The solar photon fluxes exhibit short time variability due to the emission of solar flares and long-time variability due to the orbital eccentricity especially on Mercury. The intense photon flux varies by more than a factor 2 along the orbit of Mercury, due to the inverse square distance dependence). At the Moon the UV flux is less variable along the orbit that is less eccentric.

Variabilities of some orders of magnitudes in EUV emission are expected in less than few hours during solar flares (Werner et al. 2022); in the Lyα line an increase about 20% has been observed in major flares (Milligan 2015).

Finally, along the 11-year solar cycle the Lyα flux varies in the range $(3.5–6) \cdot 10^{11}$ ph/(cm^2 s) at 1 AU and at solar maximum the flare index (describing the number, size and brightness of the flaring areas (Ozguc et al. 2021)) reaches a value up to 20 times the solar minimum periods (Bruevich and Yakunina 2017).

It is estimated that in the early phases of the Sun 10- or 100-times higher UV fluxes were emitted (Ribas et al. 2005), so that we can expect that PSD was even more relevant in the first stages of the Mercury's history (Orsini et al. 2014).

To better constrain the effect of UV variability in the volatile components of the exospheres we would need simultaneous short- and long-term observations of the Sun.

For Mercury, the BepiColombo mission will provide systematic observations during its nominal lifetime (2 years) and possible extensions (one or two more years) (Benkhoff et al. 2021).

4 Open Points on Exosphere Investigations

Being collisionless, a surface-bounded exosphere can be considered as the sum of different single-species exospheres. In fact, past exospheric observations of airless bodies revealed quite different distributions and dynamics of the different species around the body, mostly depending on chemical properties. The main families can be grouped in: highly volatiles and water groups, refractories and molecules and moderately volatiles alkali (Grava et al. 2021a; Schörghofer et al. 2021; Leblanc et al. 2022).

The volatiles are those elements are weakly bound to the surface; they are easily released and have a low sticking efficiency. Examples of this group are hydrogen, water groups, helium and methane. The refractories are elements with a very high melting point and strongly bound to other atoms, and above all they are easily oxidizable. They are on the opposite scale of surface-exosphere interaction compared to volatiles. The alkali metals, like Na and K, have their outermost electron in an s-orbital and this shared electron configuration results in a high reactivity.

4.1 Volatiles and Water Group Species

In general, exosphere densities of solar-wind-derived volatiles are expected to scale with the solar wind flux: the higher the flux of solar wind ions of a given species, the higher the exospheric content of the corresponding neutral species. This is the case for example with helium: on the Moon, the lunar exospheric helium density decreases when the Moon enters in Earth's magnetotail (Feldman et al. 2012), and the lunar surface is shielded from the solar wind bombardment (in this case from alpha particles). When the Moon exits the magnetotail a few days later, the exospheric density of ^{4}He quickly recovers to nominal levels (Grava et al. 2021b). Observations from LADEE showed also that exospheric He responds to specific solar wind streams (Benna et al. 2015). But for other solar-wind derived species, this relationship is not as straightforward. Neon, for example, is also a solar-wind-derived species, but its photo-ionization lifetime is 3 months. Therefore, it does not show short-term variations due to solar wind fluctuations. It does show long-term fluctuations: the pre-dawn exospheric density measured by LACE during nominal solar wind conditions was about one order of magnitude lower than the exospheric density measured by LADEE from orbit, during a iCME. However, modelling efforts by Killen et al. (2019) revealed that the surface-exosphere interaction of neon is not as simple as previously thought: the neon lifetime required to match LACE data is 4.5 days, 20 times shorter than the photo-ionization lifetime for nominal solar wind conditions (100 days), and comparable to helium, a gas that is lost mainly through thermal escape.

At Mercury, close to subsolar point Mariner-10 detected a H double temperature vertical density profile that is justified for highly volatile species that mix population thermalized with the dayside surface (420 K) together with low temperatures population thermalized with the night side (110 K) and circulating in the dayside (Hunten et al. 1988; Grava et al. 2021a). Exospheric hydrogen was detected also at the Moon in molecular form H_2 (Stern et al. 2013). The implantation of solar-wind protons into the lunar soil generates an OH-veneer and, in small amounts, molecular water, a subject reviewed in Schörghofer et al. (2021) (Fig. 8). These interactions of the solar wind with the amorphized grain surface layers need to be understood in far more detail, so we can quantify the production rate of H_2 versus H_2O and the degree of latitudinal and possibly diurnal variation of the OH surface population. Another central question is to what degree water molecules can repeatedly hop on the lunar surface, which defines the ability of an exosphere to transport water to polar cold traps.

The Moon offers an ideal laboratory to study the fate of solar wind ions recycled as neutral to form the exosphere of an airless body. One example is the neutralization of solar wind alpha particles to create helium. The recent finding that observations of lunar helium are consistent with full thermal accommodation with the surface (Grava et al. 2021b) suggests that, even if it is expected that a good portion is back-scattered and should not interact with grains

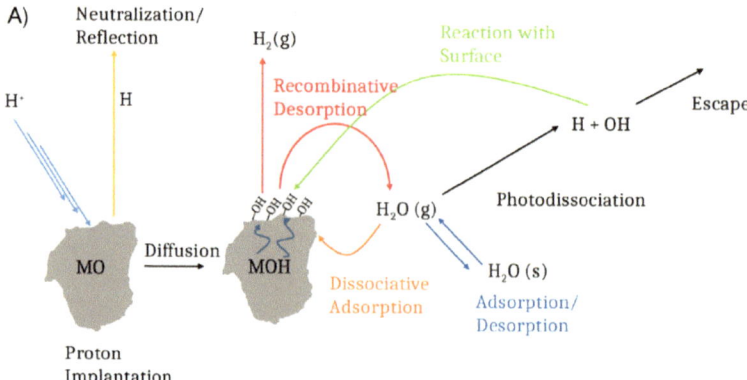

Fig. 8 Kinetic scheme for the solar-wind-induced water cycle on the surface of Mercury. Upon heating, a reaction takes place between neighboring OH sites (encircled with green) resulting in the formation of gas phase water and an oxygen bridge between cations (Schörghofer et al. 2021)

more than once, some incident helium atoms are expected to experience multiple collisions with different grains of the regolith prior to reflection from the lunar surface, thus losing energy. Methane (CH_4) is another example. It has been detected by LADEE/NMS. Hodges (2016) showed that the recombination of solar wind carbon ions with hydrogen atoms in the soil is a pathway for production of methane in the lunar exosphere (methane can offer insights on the fate of H as well, as pointed out by Tucker et al. 2019). However, carbon-bearing species can also come from micrometeoroids so the conclusion obtained should be taken with some caveats. Hodges (2016) predicted that CO should also be present in significant amounts in the lunar exosphere, owing to its photoionization lifetime, which is 9 times longer than methane's. Unfortunately, artifacts in the mass channel 28 prevented the LADEE NMS from detecting CO, but LACE detected a post-sunrise peak in exospheric density at the same mass channel, which were attributed to either N_2 or CO (Hoffman et al. 1973). Despite the absence of artifacts, LADEE NMS did not detect CO_2. More measurements are clearly desired to understand the solar wind ions recycling at the lunar surface.

A central role of the exospheres of condensable species is their ability to transport molecules from any location on the surface to cold traps, where they accumulate in condensed form as ices. At the current state of knowledge, these processes are expected and plausible, but they have never been proved. The definite detection of an exosphere of molecular water on a large airless body with a silicate-rich surface would be a crucial observation. The few existing observations are limited by ambiguities. CHACE-1 made a one-time measurement on the Moon (Sridharan et al. 2010) that exceeds the upper limit from the Apollo era (Stern 1999). LADEE did not distinguish between OH and H_2O (Benna et al. 2019). On Mercury, no instrument has yet allowed such a measurement to be performed. And also on Ceres, for example, most attempts failed to detect an H_2O exosphere, with two exceptions involving instrumentation that is no longer available (Schörghofer et al. 2021). In total, no measurement reveals whether such an exosphere is global or localized in the shape of a plume. Definitive observations of a water exosphere and its properties would greatly advance our framework of understanding exospheres and ices on these planetary bodies.

Another key task is to determine the ability of exospheres of condensable species for lateral transport. In other words, whether molecules undergo repeated ballistic hops or are chemisorbed indefinitely on the surface after the first hop. So far, observations have provided only indirect evidence for or against this hypothesis, and, likewise, theory has been used to argue for or against repeated ballistic hops. The answer to this question may to a large degree determine the abundance of water on the Moon, an essential resource for sustained robotic and human presence on the Moon.

The interaction of H and H_2O with radiation-damaged silicate-rich grains needs to be understood far more comprehensively. One aspect, already mentioned, is the ability of H_2O molecules to repeatedly desorb. The other is the rate at which solar wind interactions generate molecular water by interaction of solar wind protons with metal oxides. The activation energies and processes on these complex amorphized surfaces can be investigated with further laboratory measurements and solid state modelling.

Volatiles are also useful to study outgassing from the interior. Evidence from outgassing from the interior of the Moon has been provided by several instruments. The Apollo 17 LACE mass spectrometer deployed on the lunar surface detected ^{40}Ar, which is the radiogenic product of the decay of ^{40}K within the crust. A small fraction of the lunar exospheric helium, also detected by LACE, is also endogenic, coming from the radioactive decay of thorium and uranium (Hodges 1977). The Alpha Particle Spectrometers (APS) onboard the Apollo 15 and 16 command modules and Lunar Prospector detected alpha particles produced by the decay of radon and polonium (Gorenstein and Bjorkholm 1973; Bjorkholm et al. 1973; Lawson et al. 2005). Data collected from these two types of instruments reveal that the outgassing of these radiogenic elements is variable, both spatially and temporally. Some of the regions that were actively outgassing radon during the Apollo era were not active during the Lunar Prospector survey, 30 years later. Radiogenic gases concentration on the Moon appears to peak in pyroclastic deposits and prominent young craters such as Aristarchus and Alphonsus, and at the mare-highlands boundaries. These are all regions with either a thin crust (such as in pyroclastic deposits) or a fractured terrain (the Maria edges, such as the landing site of Apollo 17), which both facilitate the outgassing from the lunar interior into the exosphere (Fig. 9). Some radiogenic gases detected so far (radon, polonium) condense on the cold lunar nightside surface and on PSRs. Therefore, long-term monitoring of the exospheric density of these gases can constrain the outgassing rate and thus the amount of incompatible elements thorium and uranium, benefiting the study of the origin of the Moon (and Mercury, assuming these elements will be detected by BepiColombo).

4.2 Refractories

Being very "sticky", i.e., with a very high activation energy for desorption, refractories are released only by energetic processes, such as MIV and sputtering by energetic ions, mostly from the solar wind. As such, they are important species to study the response of surface-bounded exospheres to changes in the external environment (solar wind and micrometeoroid flux).

For certain species a double mechanism has been proposed. For example, for calcium, detected at Mercury, it has been suggested that first it is released from the surface by an energetic process (MIV or ion sputtering) in the form of calcium oxide (CaOH, Ca(OH)$_2$, or CaO). After the first plume expansion, it is subsequently dissociated by photons or electrons, or via unimolecular decay, with an excess of energy, resulting in different molecules or atoms with additional energy imparted to Ca products (Berezhnoy 2018). The final products after the impact depend on the quenching temperature of the expanding cloud that is expected to be in the range 3000-4000 K. At temperatures $\leq$3750 K in the impact-produced cloud Ca(OH)$_2$ dominates over both atomic Ca, CaO and CaOH, while at higher temperatures it is considered that the predominant form of the initial calcium ejecta is CaO (Berezhnoy 2018). Recently, Moroni et al. (2023) showed that observed Ca density at Mercury can be quantitatively explained only if the quenching temperature is below 3750 K.

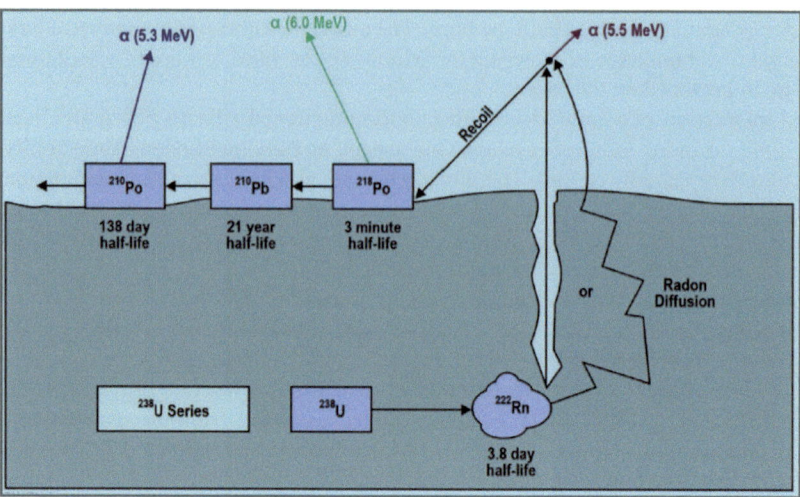

Fig. 9 Scheme of radon decay, with alpha particle energies pertaining to each product. The short half-life of radon makes it a useful species to constrain regions of active outgassing (Grava et al. 2021a or Grava et al. 2021b, adapted from Lawson et al. 2005, JGR, 110)

The release near dawn, necessary to explain an excess of Ca detected there (Burger et al. 2014), is consistent with micrometeoroid bombardment, which peaks at dawn due to the motion of the planet. Moreover, a regular excess concentration of Ca in the Hermean exosphere at certain true anomaly angles has been explained by the intersection of Mercury's orbit with that of comet 2P/Encke (Killen and Hahn 2015). Nevertheless, as mentioned in Sect. 3.2, a quantitative estimate of the excess Ca produced has not yet been obtained (Killen 2016; Plainaki et al. 2017). Other features in the yearly Ca distributions, like a decrease while approaching the perihelion or a secondary maximum while approaching the aphelion, still need to be explained, as well (Fig. 5 bottom panel).

This suggests that other mechanisms are likely at play, waiting to be uncovered by future observations from the ground or from BepiColombo's suite of instruments. It is also possible that Ca is released in another molecular form, such as CaS. Mg is another species of interest. Detected at Mercury by MESSENGER, as for Ca, the temperature (energy) of Mg atoms in the exosphere is twice that expected from MIV. Mg is important because of the link to Mg-rich regions at the surface. Merkel et al. (2018) showed that the exospheric Mg abundance peaks when Mg-rich regions are exposed at dawn at perihelion.

More observations, laboratory measurements, and simulations are needed to study the exospheric distribution and dynamics. In particular, much more atomic and molecular species are expected to be released into Mercury's and Moon's exospheres. The nightside exospheres are still almost not constrained by measurements since most of the observations, especially for Mercury, have been performed by remote sensing instruments that collects emission lines of photon-excited atoms.

4.3 Alkali Metals

Alkali more than other species, like refractories and highly volatile components are released by complex mechanisms. Due to their high reactivity but relatively low binding energy with the surface makes their release is affected by thermal radiation, UV photons, ion and electron

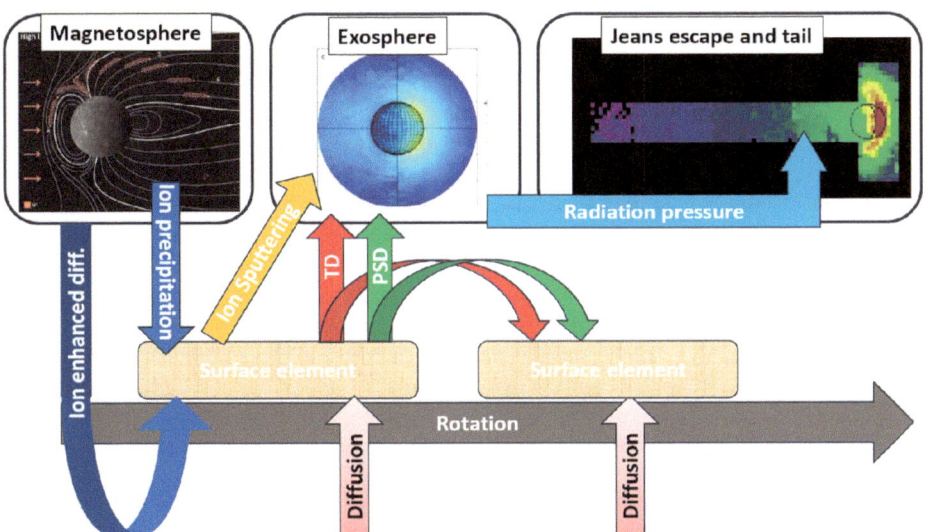

Fig. 10 Scheme of Sodium circulation between Mercury' surface and exosphere. It can be applied to the lunar Na exosphere by considering a different plasma precipitation map (revised from Leblanc et al. 2022)

impact and micrometeoroid impact, radiative diffusion inside the regolith, and triggered by energetic particle radiation and chemical modification of the surface (Wurz et al. 2022; Leblanc et al. 2022). On the other hand, they (especially Na) are the species most observed in the environment of Mercury and the Moon since Na and K have two resonance emission lines in the visible range, close and near the solar emission maximum, so they are easily observable by ground telescopes.

Beside the set of observations and modelling that could help to progress in our understanding of the Na and K exospheres around the Moon and Mercury, the relations between magnetosphere, surface and exosphere can be specifically addressed by combining various observations and new theoretical and experimental developments.

The very first external driver that controls the alkali exospheres of the Moon and Mercury is the solar photon flux. This flux has three different effects on the sodium and potassium exospheres. First, it contributes to the ejection of these volatiles from the surface by either thermal desorption or by photon stimulated desorption. A detailed description of these two mechanisms can be found in Wurz et al. (2022), as well as in Leblanc et al. (2022) (Fig. 10). The EUV/UV range of the solar photon flux also leads to the photo-ionization of these atoms, which is very probably the most efficient process for creating sodium and potassium exospheric ions. But, the solar photon absorption and emission also induces an anti-sunward force, the solar radiation pressure, at the origin of the very extended exospheric tail at Mercury (Schmidt 2013) but also at the Moon (Baumgardner et al. 2021).

To observe how efficient might be the solar flux might be in controlling the exosphere of Mercury and the Moon, one approach is to image the diurnal and annual variabilities of the exosphere. It needs, however, a large set of observations and the help of models taking into account all possible mechanisms at the origin of the alkali atoms in the exospheres (see Leblanc et al. 2022). For example, waiting for the BepiColombo nominal mission, when we will have in situ detection and global imaging of the Na distributions (Milillo et al. 2020), we need more ground-based observations to obtain a more complete dataset for the investigation of the dawn-dusk Mercury's seasonal variability (Milillo et al. 2021). In fact, ground-based observations with dawn/dusk view at all True Anomaly Angle (TAA, i.e., Mercury's angular distance from perihelion) are still not available. Furthermore, we need more statistics for investigating the northern/southern peak occurrence along the orbit.

The two opposite effects in terms of exosphere production and loss, photo-ionization and photon-desorption, are controlled by the same EUV/UV spectral range also during variable photon flux in the solar events. Solar flares are relatively short (few tens of minutes) with respect to the typical time scale of the global exosphere (few hours) which makes any signature in the exosphere difficult to image directly in its neutral component.

Another key driver of the alkali exospheres, that has been highly debated along decades, is the relation between Mercury's magnetosphere and exosphere. It is known since the very first ground-based observations of Mercury's sodium exosphere (Potter and Morgan 1990), that sporadic peaks in sodium emissions at high latitudes exist in Mercury's exosphere. It has been postulated that what controls these maxima was the effect of the precipitating solar wind particles onto the surface which sputtered volatiles in Mercury's exosphere. No evidence for a similar process has ever been observed around the Moon. Actually, the relatively long time scale of this structure (Leblanc et al. 2008) rather suggests that the sodium atoms were not ejected directly from the surface but rather brought to the upper surface by radiative induced diffusion inside the regolith as suggested at the Moon when passing through the Earth magnetospheric tail (Wilson et al. 2006; Sarantos et al. 2008) and for Mercury sodium exosphere (Mura et al. 2009).

But without a simultaneous knowledge of the solar wind conditions, it is extremely difficult to infer the real efficiency of solar wind particles in populating the exosphere. Ground-based observations during iCME (Orsini et al. 2018) were used to infer such a relation. However, this remains very speculative because ground-based observations are also strongly dependent on the Earth atmospheric condition. On the other hand, recent studies (Sun et al. 2022) show that during event of ion precipitation enhancements (Flux Transfer Events) the Na-group ions population above the cusp of Mercury's magnetosphere (interpreted as ionized component of the sputtered material) increases, as well. It implies that any interpretation of the apparent positions and structures of the reconstructed exosphere needs to be done considering the intrinsic limited spatial and temporal resolutions. This is particularly true when the images of Mercury's exospheric emission are obtained on hour time scale (Mangano et al. 2009), a time scale significantly longer than the typical solar wind perturbation duration. Fortunately, ground based observatories able to target Mercury are gradually incorporating adaptive optics which should significantly improve their spatial resolution and allows a much better tracking of Mercury spatial variability induced on short time scales. Moreover, BepiColombo forthcoming insertion around Mercury will provide a new opportunity (after MESSENGER) to correlate in situ solar wind observation with ground based and in situ observations of Mercury's exosphere (Milillo et al. 2020). In the Moon case, high spatial and temporal resolution measurements of Na could allow the investigation of the surface response to solar wind variability especially in proximity of the magnetic anomalies (see Sect. 3).

Some directions of investigations of the alkali exospheres can be summarized as follows (Leblanc et al. 2022):

- In-situ measurements are essential to obtain simultaneous observations of drivers and the resulting exosphere. Thus, detailed study of the effect of impact of iCME or meteoroid showers can be performed by in-situ measurements.

- Ground based observations have the capability to monitor and globally image the Na and K exospheres.

- Modelling developments remain essential to further understand ground based, MESSENGER, and future observations. In particular, these new models should provide a better description of the interaction between exosphere and surface by introducing a detailed description of the regolith and of the fate of the exosphere through it. The effect of space weathering on the surface release should be parametrized in the models.

- Laboratory experiments have a key role in addressing the weathering of the surface and the effects on the release efficiency.

4.4 Open Points on Exosphere Dynamics, Sources, and Loss

The exosphere composition of the airless bodies is related to the surface composition, weighted by the surface-release process efficiencies and particle dynamics. Information on exosphere composition can be obtained indirectly via heavy ion detection. Planetary ions are mostly generated via photoionization of the exospheric components. The circulation of charged particles in the magnetosphere (in the case of Mercury) or as pick up ions in the solar wind (at the Moon) allows the detection of planetary species at higher distances.

For the Moon the exosphere generation processes are similar to the Mercury case but with different weight and geometry and at lower gravity. Over extended time scales, the different gravity field of Mercury compared to the Moon is expected to have produced differences in exospheric loss according to the efficiency of different surface release processes for different species (Wurz et al. 2010). This would result in an alteration of the surface composition at Mercury, in contrast to the Moon, where the exospheric loss is similar for all refractory elements (Wurz et al. 2007). For example, the Na/K ratio in Mercury's exosphere cannot be simply associated with the primary abundances but different transport and loss must be considered. Both Na and K could be accelerated anti-sunward by radiation pressure, photoionized and captured by the solar wind. However, they can also re-impact the surface and ionized (followed by neutralization) or neutral state. The higher mass of K relative to Na results in a smaller scale height and a larger gyroradius, which may result in more rapid net loss of K (Potter et al. 2002a,b). Taking into account different loss processes, the observed Na/K exospheric ratio may be consistent with an initial abundance close to solar or meteoritic ones (Leblanc and Doressoundiram 2011). On the other hand, in the case of asteroids, gravity is negligible so the exosphere radiates from the body at a net loss, thus a survey of the exosphere gives direct information of the pristine material of the body (Plainaki et al. 2009).

Unfortunately, up to now only measurements at low mass resolution are available in Mercury's environment from MESSENGER. The ion spectrometers on board both Mercury's orbiters of BepiColombo will allow the identification of new species, also the heaviest ones which can be energized and circulated in the magnetosphere (Milillo et al. 2020).

For the Moon, multiple missions with instruments relevant to exospheres and exosphere-surface interactions are expected to launch soon. Already at the Moon are CHACE-2 (a neutral mass spectrometer) (Das et al. 2020) and IIRS the infrared spectrometer on Chandrayaan-2 (already in orbit) (Chowdhury et al. 2020).

MESSENGER UVVS/MASCS observations have clearly shown that the surface composition might have a direct impact on the exosphere composition in the case of its Mg component (Merkel et al. 2018). No similar evidence has been obtained in the case of the Alkali exospheres, even if the high latitude peak in surface density has been tentatively explained by the preferential exospheric sodium reabsorption of the surface in these cold regions (Peplowski et al. 2014). Here again, BepiColombo forthcoming detailed observation of the surface composition combined with in situ and remote observations of the exosphere might help us to improve our understanding of this relation (Hiesinger et al. 2020).

A prevalence of Na emission in the equatorial dawn hemisphere has been first spotted by ground-based observations (Sprague et al. 1997; Potter et al. 2002a,b; Schleicher et al. 2004) and then confirmed by the MESSENGER/MASCS and by other statistical studies of ground-based images (Fig. 11 panel b) (Milillo et al. 2021). This local time asymmetry has been explained considering a global Na circulation in the exosphere and preferred condensation of the alkali atoms in the cold regions of the nightside surface (Mura et al. 2009). The MESSENGER observations showed a dawn enhancement in the outbound leg of Mercury's orbit but also a prevalence in the equatorial dusk regions between TAA 180° and 270° where the surface exposed at lower average temperatures is sunlit (Cassidy et al. 2016) (Fig. 11 panel a). Such intriguing correlation has been interpreted by the long-term effects of the thermal forcing on the surface capability to trap sodium atoms and to release it when these longitudes face the Sun. How the thermal forcing acts on the sodium surface reservoir is not well understood but could be due to the migration of this volatile species through the first hundred μm of the regolith (Sarantos and Tsavachidis 2020), a migration which would be controlled by the thermal gradient with depth and the radiative induced diffusion of the sodium atoms through the grains and regolith. This peak is not consistent with models of thermal desorption (Leblanc and Johnson 2003; Mura et al. 2009), which have all the Na reservoir, which was built up during night, quickly desorbed – and thus depleted – in the early morning hours. The simulation cannot explain this behaviour without considering extra internal sources of Na (Leblanc et al. 2022).

Porosity also affects mobility on the surface of volatile elements. For alkali atoms (Na and K), similarly to what described in Killen et al. 2004, surface diffusion reduces the net desorption of adsorbates (Sarantos and Tsavachidis 2020). This reduction in desorption rate is temperature-dependent, as it depends on porosity (and thus mobility), PSD yield, and sticking coefficient. Therefore, only when temperature reaches 500 K does thermal desorption start contributing to mobility. On the Moon, the difference in surface grains mobility (diffusion) between K and Na may explain the difference in longevity of Na and K measured from LADEE: K is heavier, so it has reduced mobility, and hence it is more easily photo-destroyed, while Na is more long-lived.

The seasonal variation of the total content of Na in the exosphere shows a distribution with a maximum at aphelion and a secondary maximum at perihelion (Cassidy et al. 2015; Milillo et al. 2021). The drivers able to release Na are maximal at perihelion, so that the observed trend is not straightforward. The suggested explanation of this trend includes the variable contribution of the radiation pressure that pushes Na atoms into the tail and the variable photoionization rate along Mercury's eccentric orbit and thermal inertia of the surface (Rognini et al. 2022). The quantitative justification of an exospheric maximum at aphelion could be due to the combined effect of the 3:2 orbital resonance, and the plasma or micrometeoroid precipitation in the nightside along the orbit (Mura et al. 2023).

Added to the equatorial dawn/dusk asymmetry of the Na distributions, a northward or southward peak prevalence along the TAA has also been discovered by analysing THEMIS ground-based images (Milillo et al. 2021). Specifically, it seems that between TAA 0° and 180° in the outbound leg of the orbit (from periherm to apoherm) coinciding with Mercury's orbit above the dust disk, the Na peak is observed more frequently at Northern latitudes, on the contrary, the peak is observed more frequently Southward in the inbound leg (TAA from 180° to 360°) coinciding with Mercury below the dusk disk (Fig. 11 panel c). This was not expected; in fact, if the Na surface release was related to ion impact, a predominance of the southward peak, where the cusp and loss cone are expected to be wider because of the magnetic dipole northward shift, should be seen. Furthermore, if the Na surface release was related to ion impact onto the surface, we would not expect a north-south asymmetry linked to TAA. The ion precipitation which in turn is linked to the IMF conditions varies on short time scales and should not depend on season. A link of the north-south asymmetry to TAA

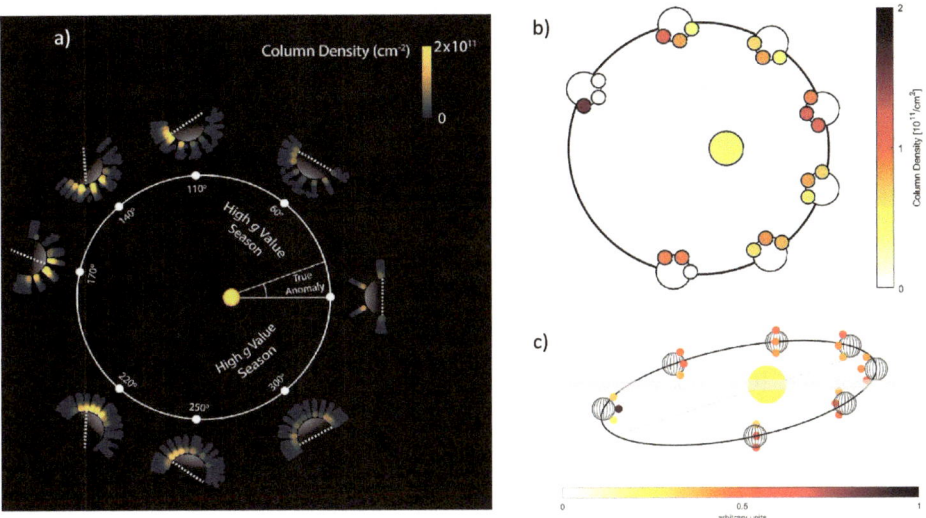

Fig. 11 Averaged Na column density measurements along Mercury orbit. a) equatorial distributions show-ing dawn/dusk asymmetries derived from MESSENGER/MASCS observations (Cassidy et al. 2016): dawn prevalence between TAA 90° and 180°, and dusk prevalence between 180° and 270°. b) equatorial distri-butions from THEMIS ground-based observations (Milillo et al. 2021) that confirms the dawn prevalence, but cannot confirm the dusk prevalence observed by MESSENGER. Grey bullets represent no data are avail-able. c) Latitudinal distribution showing that statistically the Northern/Southern peaks are more frequent in the outbound/inbound leg when Mercury is above/below the expected interplanetary dusk disk (Milillo et al. 2021)

opens new hypotheses on the Na generation mechanisms. On the other side it also seems to contradict the role of micrometeoroids impacts that should be higher closer to the dust disk, so in the opposite hemispheres.

Eventually, the importance of the meteoroid bombardment in supplying part of the ob-served exospheric species has been demonstrated by MESSENGER (as an example in the case of the exospheric Ca and Mg, see Janches et al. 2021 and Grava et al. 2021a). The role of this release process on the alkali population in the exosphere is still not clear. While different studies suggest that it is generally small (e.g.: Wurz and Lammer 2003; Wurz et al. 2010) Kameda et al. (2009) found a correlation between the total Na exospheric intensity and the distance from the interplanetary dust disk, and Gamborino et al. (2019) suggest that the MMIV process is the predominant mechanism populating equatorial high altitudes dur-ing the MESSENGER observations. BepiColombo offers a unique perspective thanks to the possibility to quantify the dust flux, its temporal and spatial distributions by Mercury Dust Monitor (Kobayashi et al. 2020). The combination of measurements of the incident dust flux and in situ/remote sensing observations of the various components of the exosphere will be a unique opportunity to fully address investigation the exospheric signatures of the perma-nent bombardment of the surface by this dust. In fact, this is a key mechanism in the space weathering of the surface but also a major driver of Mercury and the Moon exospheres.

5 Requirements for Advanced Exosphere Modelling

Different MonteCarlo (MC) models have been proposed for describing the exospheres of Moon, Mercury, and asteroids. Although the numerical implementation of how to track millions of test particle trajectories to estimate macroscopically measurable parameters is straightforward, advances are required when it comes to several input parameters that are required or assumed in these models. These include the speed distributions of ejected atoms and molecules, the initial directions of emission, the specification of what occurs during surface-atom interactions (e.g., degree of energy and momentum accommodation), the differing amounts of a constituent at different geographic locations, the directionality of drivers such as incident ions and meteoroids, and other factors (e.g., porosity). Many of these improvements necessitate that new laboratory experiments be performed.

5.1 Modelling Status

Foundational work in the use of Monte Carlo sampling of particle trajectories to estimate exospheric measurables was described by Hodges (1980). Early work focused on explaining measurements of helium and hydrogen-bearing gases at the Moon (Hodges 1973b; Crider and Vondrak 2000), with those papers explaining the algorithms used for the sampling of initial locations and velocities for the few test particles that stand for a much larger ensemble of gas atoms and molecules. The same methods were later applied to simulate argon (e.g., Grava et al. 2015; Hodges 2018), methane (Hodges 1975, 2016), and the water group gas constituents of the Moon (e.g., Crider and Vondrak 2000, 2003; Smolka et al. 2023). The more recent studies include an increasingly sophisticated treatment of the gas-regolith interactions, including the chemical conversions of species on the regolith surface.

Sodium has received much attention in modelling due to the availability of interesting lunar and Hermean observations since the mid 1980s. A first model describing the Na and K exosphere along Mercury's orbit was developed by Smyth (1986) and updated by Smyth and Marconi (1995). This model included a Na population released at 2.6 km/s plus an ambient (thermally accommodated) population generated after the first bouncing onto the surface. It considered the photoionization and the important role of radiation pressure for the escape along the orbit showing the anticorrelation between radiation pressure and Na exosphere observations. This model considered that the escape is replaced by a general source of Na due to micrometeoroid gardening.

The longevity of the exospheric reservoir was since studied with more sophisticated time-dependent sodium adsorbate models. Mura et al. (2009) (IAPS exosphere MC model), Leblanc and Johnson (2003), and Leblanc and Johnson (2010) developed Na exosphere models where the non-uniform planet rotation and the sticking efficiency are taken into account, so that the seasonality of Na release is better described. Additionally, Mura et al. (2009) included the effect of the enhanced efficiency for Na release by PSD after ion impact on the dayside as well as in the night side. By only considering the PSD as the main surface release process for Na and a simple surface thermal model, the observed yearly distribution maximum at aphelion cannot be reproduced. Only by adding a more realistic thermal model, which accounts for the regolith's thermal inertia, can a secondary peak at aphelion be reproduced (Rognini et al. 2022), but still far different than the observed seasonally varying densities (Cassidy et al. 2015; Milillo et al. 2021). Recently Mura et al. (2023) presented an analytical model that explains the peaks at aphelion and perihelion with the combined effect of the 3:2 orbital resonance, and the modulation of the magnitude of sources and losses along the orbit. The processes due to plasma and micrometeoroid impacts produce an accumulation of un-bound Na atoms in the nightside. These atoms are released when they are exposed to Sun light at dawn following the specific 3:2 spin to orbit resonance. Finally, the enhancement of Na column density at aphelion, and the dawn-dusk asymmetry in the outbound leg are due to the orbit resonance and are stable features of Mercury's exosphere.

As mentioned in Sect. 4, other observables that cannot be reproduced by the state-of-the-art models are a dusk prevalence in Na equatorial exosphere at TAA 180°-270° (Cassidy et al. 2016); the Na increase northward/southward in the outbound/inbound orbit leg, corresponding also to above/below ecliptic plane (Milillo et al. 2021); and the short- term variation of Na and its relationship with the plasma precipitation. Thus, no model yet succeeds in describing the seasonal sodium circulation. One area that merits further study is the incorporation into the boundary conditions of global models of the effects that regolith imposes on gas release because of micro-shadows and Knudsen diffusion (Sarantos and Tsavachidis 2020, 2021). Besides, the relative importance of thermal or photon desorption is still under discussion, since, according to Gamborino et al. (2019), close to the subsolar point, the Na atoms are mostly released by TD, and no more reservoir is left for PSD release. The actual temperature at which the surface transitions from retentive to emitting is a function of the binding energy of adsorbates to the surface, and if the surface is heterogeneous (e.g., consisting of different minerals), the high energy sites are filled first due to surface diffusion, leading to non-linear release with temperature (Sarantos and Tsavachidis 2021) and thus suppressing thermal desorption. Existing experiments may not have resolved these distributions as they were conducted at larger spatial scales. Thus, sensitive measurements of alkali adsorption and desorption on lunar samples at low coverages are required for improved models.

More reactive species are easier to simulate because the boundary conditions (reactions, no thermal desorption) are simpler to implement in a numerical program. Predictions for other Hermean species were presented by Wurz and Lammer (2003) using a bidimensional MC model that expanded in altitude the expected or observed exospheric densities of different species at the surface according to the energy distribution of each involved process. The Wurz et al. (2010) model update considered a more realistic surface mineralogical composition. Their results pointed out that globally the ion sputtering and MIV contributions in populating the exosphere are minor, but they could be relevant for specific refractory elements.

As MESSENGER and ground-based observations became available, some of the models of refractory species were refined, mainly in the assumptions of speed distributions of ejected atoms. Burger et al. (2014) developed a model for deriving the Ca distribution after micrometeoroid impact vaporization to be compared to the MESSENGER Ca observations, while Sarantos et al. (2011) presented an exospheric model for Mg and compared to the MESSENGER observations. They found that both these exospheres present a highly energetic component (equivalent temperature > 20,000 K) that requires a multiple step process for its generation. Plainaki et al. (2017) presented a MC model derived from the IAPS exosphere MC model for comparing the observed Ca distributions due to a possible source of micrometeoroid shower at the 2P/Encke comet trajectory crossing. This model considers for the first time after MIV the Ca release in molecular form (CaO) and shock induced photodissociation and neutralization process, as suggested by Killen (2016). The expected mapping of the impact stream onto the surface, as derived by Christou et al. (2015), was considered as input. Their results showed a well reproduced exospheric distribution, but the expected energetic Ca intensities were about 2 orders of magnitude less than observed. The study by Berezhnoy (2018) on the MIV expanded cloud shows that the resulting exospheric components from Ca bearing rocks, have higher fraction of atomic Ca than considered before (CaO 3 times more than energetic Ca).

Recently the updated IAPS MC model (Mura et al. 2007) has been used to reproduce the yearly Ca exospheric distribution considering the micrometeoroid input from Pokorný

et al. (2018) and the Berezhnoy (2018) results (Moroni et al. 2023). They found that the total exospheric Ca content is consistent with the observed one if the plume quenching temperature is less than 3750 K and not about 4000 K as considered in previous studies. In their study, relevant features are still not reproduced and they did not include the comet dust streams so that to fully describe the short and long time variabilities and the local and global features in the distributions still different assumptions have to be considered.

Predictions for a variety of refractory species (Wurz et al. 2007; Sarantos et al. 2012) cannot be tested due to the unavailability of measurements, but improved models should incorporate the non-uniform distribution of species on the surface (Prettyman et al. 2006) following those measurements of potassium which show correlation of the exosphere and surface distributions (Colaprete et al. 2016). Also, multispecies models that incorporate all existing constraints on the hydrogen and water cycle are required.

5.2 Open Issues on Exospheric Modelling

As we move forward, an improved description of gas-surface interaction, ranging from the nanoscale to the planetary scale, is required for higher fidelity models.

On the nanoscale, inputs from Molecular Dynamics (MD) simulations will provide information from first principles on the energy exchange and angular distribution of gases from smooth and rough surfaces. At the moment, such MD calculations are performed for solar wind hydrogen (Leblanc et al. 2023) and should be extended to other species discussed here (e.g., sodium). New models should include release in molecular form and consider chemical processes in the exosphere. Another effect to be considered at the atomic scale is surface diffusion between adjacent sites. Present modelling practices assume that the binding energy of an adsorbate is assigned at the time of impact and is never updated. When instead surface diffusion is permitted, the release of adsorbates at low coverages, such as experienced in most lunar regions, could be suppressed at some temperatures and enhanced at higher temperatures due to the population of sites of the highest binding energy through diffusive motion along grain boundaries (Sarantos and Tsavachidis 2021).

On the microscale, emphasis should be placed on accounting for the effects of microroughness consistent due to the granular nature of regolith. The effect of small-scale shadows (<10 cm) in global models could be acknowledged through a trapping probability (Hayne et al. 2021). An additional effect to be included is that shadows at the scale of several grains can be a substantial sink for photosensitive species like sodium (Sarantos and Tsavachidis 2020). The sputtering yields for different plasma populations as well as the angular distributions for sputtered atoms should incorporate the effect of surface microroughness (e.g., Cupak et al. 2021). For species of intermediate volatility and at low coverages, new models must account for the fact that release from regolith is gradual due to Knudsen diffusion into the subsurface, which decelerates desorption (e.g., Sarantos and Tsavachidis 2021). The effect of Knudsen diffusion has been included in local simulations to estimate volatile retention at specific locations of the Moon (Schörghofer 2022; Reiss et al. 2021). However, the only global exosphere model to account for explicit diffusion into the subsurface is that of Teolis et al. (2023), and toy models approximating the effect of subsurface diffusion were proposed by Kegerreis et al. (2017). Although the effect of diffusion can be approximated in a one-layer model by adopting an effective distribution of binding sites (Sarantos and Tsavachidis 2020), or, equivalently, by waiting times for desorption that are not exponentially distributed (e.g., log-normally distributed), solving Fick's law (Teolis et al. 2023) is the preferred solution. In the absence of experimental data, calculations with sphere packings like those of Sarantos and Tsavachidis (2021) can be used to estimate

Knudsen and surface diffusion coefficients for global models. And finally, temperature gradients in the first few centimeters of the regolith have been adopted in simulations of volatile pumping and retention at the local scale, but not in global exosphere simulations. This is a necessary step. Calculations of these gradients based on LRO radiometer measurements exist for the Moon, and radiometer data from BepiColombo can be directly used to constrain Mercury models in the near future.

On the planetary scale, improvements to the directionality of the drivers and the effects of topography should be considered. Better description of the drivers as new inputs become available is an obvious step. Such information may include the meteoroid (Pokorný et al. 2017, 2018, 2019) and plasma directionality, the dust distribution in the interplanetary space, surface inhomogeneity and realistic compositions. New models must include large-scale topography ($\sim$1 km scale or more), which is especially important near the poles and terminators.

6 Expected Results from the Next Future Observations

In the coming years, space exploration of the inner solar system will see a golden age. In addition to the BepiColombo mission (Benkhoff et al. 2021) that will systematically observe Mercury with two spacecraft, other missions in the inner heliosphere are devoted to Sun and IMF observations, like Solar Orbiter (Müller et al. 2020) and the Parker Solar Probe (Fox et al. 2016). Moreover, the lunar exploration program, boosted by the human colonization aim, includes tens of orbital and landed missions from multiple nations and, for the first time, also by private entities. Multiple lunar landings are planned during 2023-2027 under NASA's Commercial Lunar Payload Services (CLPS) program. At the moment, flights to 11 delivery sites have been selected (Lacus Mortis, Mare Crisium, Shackleton, Reiner Gamma, Nobile, Schrödinger, Gruithuisen Domes, South Polar region, and other sites yet to be defined). The selected payloads contain neutral gas mass spectrometers, plasma spectrometers, magnetometers, and drills and sample collection for in situ analyses of volatile content. Of particular interest within the CLPS program is the Volatiles Investigating Polar Exploration Rover (VIPER) landing at Nobile Crater. VIPER will drill at several locations over a period of 100 days to determine the volatile content of sites in the south polar region as a function of depth. It carries an infrared spectrometer, a mass spectrometer, and a neutron spectrometer.

Deployment of surface instruments is also planned by human astronauts in about 2026 during the Artemis program. The ARTEMIS III Science Definition Team (SDT) Report contains several priority measurements that are relevant to exospheric processes and drivers, making it likely that some instruments with such objectives will be deployed. The Priority Goals of the SDT Report that are relevant are: 7 L-1) Understand the plasma properties near the lunar surface and how they respond to external drivers, particularly across the terminator; 7 L-2) Understand the origin of lunar surface potentials, how they evolve between sunlit and shadowed regions, and under what circumstances they pose a threat to exploration; 2E) Learn how water vapor and other volatiles are released from the surface and migrate to the poles, 5B) Heliospheric investigations using the Moon; and 7K) Understand lunar dust behaviour.

Relevant measurements from lunar orbit are expected to be conducted from the Lunar Gateway. The planned scientific payload for this station includes electron and ion spectrometers, as well as magnetometers and radiation dosimeters. Given its elliptic orbit around the Moon, the payload on the Lunar Gateway can be expected to detect lunar ions from the exosphere and surface.

Missions from other agencies include Chandrayaan 2 (ISRO), which recently refined the composition of the surface, making first distributions of global sodium on the surface available (Narendranath et al. 2022), KPLO launched in August 2022 for characterization of the moon topography and magnetism. SELENE (KAGUYA), JAXA mission, obtained important results for the lunar environment characterization (i.e. Nagaoka et al. 2021; Yokota et al. 2020). Some major results are the identification of the reflected ions after solar wind impact onto the Moon surface (Saito et al. 2010) and the study of the interaction of plasma sheet electrons with lunar surface identified as an empty region in the electron distributions function, that was consistent with the presence of a relatively strong electric field (~ 10 mV/m) around the Moon when it is in the Earth's plasma sheet (Harada et al. 2010). Ion acceleration by the spacecraft potential and the electron beam accelerated by the potential difference between lunar surface and spacecraft were observed simultaneously when SELENE was in the Earth's magnetotail (Saito et al. 2014). These observation enable a possible way to derive the night side lunar surface potential and spacecraft potential from the observed data (Saito et al. 2014).

After the successful SELENE mission, JAXA and ISRO plan an international collaborative mission, Lunar Polar Exploration Mission (LUPEX) (https://www.exploration.jaxa.jp/e/program/lunarpolar/), to obtain the data on the quantity and forms of the water resources present on the Moon, to determine the feasibility of utilizing such resources for sustainable space exploration activities in the future. An Exospheric Mass Spectrometer is included in the payload.

Finally, also the Chinese space administration (CNSA) has an extensive lunar exploration program (https://www.cnsa.gov.cn/english/index.html). Chang'e-5 mission successfully returned lunar near-side samples. Future program includes the Chang'e-6, Chang'e-7, and Chang'e-8 missions. Chang'e-6 is poised to collect samples from the far side of the moon, marking a mission that will be the first of its kind in human history. Chang'e-7 mission will land on the lunar south pole and search for water. The final goal is to establish a scientific research station at the moon's south polar region.

This conjunction of so many important missions offers the opportunity to have unprecedented measurements of the exospheres of airless bodies, well represented by Moon and Mercury, together with the observation of the external conditions, so that, detailed investigations of the active mechanisms responsible for planet interactions with the Sun or generally with the interplanetary medium will be possible.

Improvements in ground-based observation technologies will provide a new global view of the planets at high spatial and temporal resolution to be coupled with the space observations.

The development of laboratory experiments and theoretical models and simulations related to the exosphere generation mechanisms is required for the interpretation of the expected new space data. From this point of view, fundamental new results are being achieved also thanks to the availability of powerful new computing resources. The expected outstanding results from the near future will play a crucial role in the investigation of the solar system evolution and of other planetary systems conditions.

For a better understanding of the evolution of the Sun-Earth-Moon system, a detailed investigation of the lunar exosphere, in addition to samples from the far- and near-side of the lunar surface, are of great importance. The evolutionary record of the Earth's atmosphere (e.g., Ozima et al. 2005; Terada et al. 2017; Wei et al. 2020; Lammer et al. 2022), along with traces of the young solar wind (e.g., Hodges and Hoffman 1975; Wieler et al. 1996; Marty et al. 2003) might be stored to some extent within the lunar regolith through implanted particles. During the Hadean and Archean eons, the Earth's upper atmosphere is

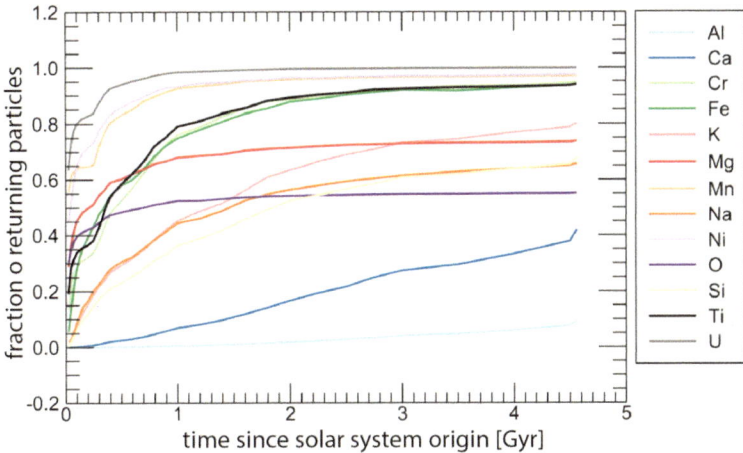

Fig. 12 The fraction of returning exospheric particles of the twelve most abundant species together with uranium that return to the Lunar surface as a function of time (Lammer et al. 2022)

thought to have been hotter and more extended than at present due to the absorption of the higher XUV flux of the young Sun (e.g., Lammer et al. 2018; Johnstone et al. 2021) indicating that atmospheric losses, i.e., the so-called 'Earth-wind', might have been significantly higher (Kislyakova et al. 2020). These escaping ions from Earth's upper atmosphere were implanted into the regolith and their isotopic signatures can be measured through sampling the surface of the Moon. One can expect that these implanted particles/isotopes (N, He, Ne, Xe and Ar, etc.) will not only cause strong variations of up to 30% in the lunar regolith (Ozima et al. 2005) but these enhancements of exogenic regolith implantations should also be observed in the exosphere when they are released from the surface by various processes, as discussed in Wurz et al. (2022). As expected by Lammer et al. (2018) and Wei et al. (2020), the escape of charged particles from the Earth's atmosphere should show a recognizable enrichment of, e.g., ^{15}N isotopes at affected areas at the nearside and farside of the lunar regolith and hence in the exosphere above these areas.

Moreover, the contribution to the lunar exosphere by micro-meteoroid bombardment was over 1000 times more relevant during early times than the contribution by sputtering (Lammer et al. 2022) while today micro-meteoroids and solar-wind ions chemically alter the lunar surface more or less in similar ways (see, e.g., Killen et al. 2012).

Additionally, to the enhanced micro-meteoroid bombardment, one can see in Fig. 12 that different return rates of various exospheric species lead to a chemical fractionation of the lunar surface over time. Due to the differing return rates of various exospheric elements, one can expect that the chemical composition of the present lunar surface does thus not reflect the surface composition after the Moons' origin. Due to a much higher solar XUV flux, the return rates were lower billions of years ago as compared with today. From the 13 species shown in Fig. 12, uranium continuously exhibits the highest return rate due to its high mass and low ionization rate, while Al continuously exhibits the lowest return rate due to its high ionization rate and lower mass. Over time, this leads to an enrichment of uranium and a depletion of aluminium in the Lunar surface material. In a similar way, the lower return rate of K compared to U has thus led to a decrease in the lunar K/U ratio over time and might also contribute to the explanations of the low K/U ratio in the lunar surface material observed today. Thus, a detailed study of the lunar exosphere and the relevant sources and sinks of its elements together with surface samples in the future that are collected from a large variety of different areas, can shed further light onto the evolution of the Earth-Moon system including early Earth's atmosphere and magnetosphere, the early solar wind and even the impact history of the inner Solar System. Finally, more and better data on the isotope abundances of the Moon itself of its volatile content and its related exosphere can help to separate the different lunar origin hypotheses that are discussed in Lammer et al. (2022).

Better understanding the origin of the innermost airless body in the Solar System, Mercury, is one of the greatest riddles in planetary sciences. This is specifically true since Messenger discovered that the mantle of Mercury is surprisingly volatile rich (e.g., Peplowski et al. 2011), a scientific finding that contradicted most of its previously established formation hypothesis.

The BepiColombo mission will provide unprecedented comprehensive measurements of the exosphere of Mercury. Its payload includes a mass spectrometer, an UV-Vis spectrometer, and a Na imager. The in-situ measurements of atomic and molecular exospheric densities on the day and night side, the column densities of different species and the spatial distribution of Na disk as well as the tail will be available at the same time (Milillo et al. 2020). The detailed investigation of its surface composition and magnetic field by Bepi-Colombo will help to form a clearer picture on Mercury's origin, evolution, and internal structure. These expected findings together with the discovery and investigation of so-called "Exo-Mercuries" and their statistical properties in other stellar systems by present and future ground- and space-based instrumentation such as the European Extremely Large Telescope (E-ELT), the Giant Magellan Telescope (GMT), JWST, and Ariel will tell us about their frequency and to distinguish the different formation pathways between ultra-close-in orbiting planets with silicate atmospheres and/or a stripped mantle, and planets on Mercury-like orbital distances.

Acknowledgements The authors thank ISSI for supporting the fruitful ISSI-workshop "Surface Bounded Exospheres and Interactions in the Solar System". AM is grateful to Carmelo Magnafico for the artwork of Fig. 1.

Funding Open access funding provided by Istituto Nazionale di Astrofisica (INAF) within the CRUI-CARE Agreement.

Declarations

Competing Interests The authors declare no competing interests.

References

Allegrini F, Dayeh MA, Desai MI, Funsten HO, Fuselier SA, Janzen PH, McComas DJ, Möbius E, Reisenfeld DB, Rodríguez M DF, Schwadron N, Wurz P (2013) Lunar energetic neutral atom (ENA) spectra measured by the interstellar boundary explorer (IBEX). Planet Space Sci 85:232–242. https://doi.org/10.1016/j.pss.2013.06.014

Anderson KA, Lin RP, McGuire RE, McCoy JE (1975) Measurement of lunar and planetary magnetic fields by reflection of low energy electrons. Space Sci Instrum 1:439–470

Anderson BJ et al (2011) The global magnetic field of Mercury from MESSENGER orbital observations. Science 333:1859. https://doi.org/10.1126/science.1211001

Baumgardner J, Luettgen S, Schmidt C, Mayyasi M, Smith S, Martinis C, Wroten J, Moore L, Mendillo M (2021) Long-term observations and physical processes in the Moon's extended sodium tail. J Geophys Res 126(3):e2020JE006671. https://doi.org/10.1029/2020JE006671

Benkhoff J et al (2021) BepiColombo – mission overview and science goals. Space Sci Rev 217:90. https://doi.org/10.1007/s11214-021-00861-4

Benna M, Mahaffy PR, Halekas JS, Elphic RC, Delory GT (2015) Variability of helium, neon, and argon in the lunar exosphere as observed by the LADEE NMS instrument. Geophys Res Lett 42(10):3723–3729. https://doi.org/10.1002/2015GL064120

Benna M, Hurley DM, Stubbs TJ, Mahaffy PR, Elphic RC (2019) Lunar soil hydration constrained by exospheric water liberated by meteoroid impacts. Nat Geosci 12:333–338. https://doi.org/10.1038/s41561-019-0345-3

Benson J, Freeman JW, Hills HK (1975) The lunar terminator ionosphere. In: Proc of the 6th Lunar and planetary science conference proceedings, vol 6, pp 3013–3021

Berezhnoy AA (2013) Chemistry of impact events on the Moon. Icarus 226(1):205–211. https://doi.org/10.1016/j.icarus.2013.05.030

Berezhnoy AA (2018) Chemistry of impact events on Mercury. Icarus 300:210–222. https://doi.org/10.1016/j.icarus.2017.08.034

Bernatowicz TJ, Podosek FA (1991) Argon adsorption and the lunar atmoshphere. In: Proc. Lunar Planet. Sci. Conf., vol 21, pp 307–313

Bida TA, Killen RM (2017) Observations of the minor species Al and Fe in Mercury's exosphere. Icarus 289:227–238. https://doi.org/10.1016/j.icarus.2016.10.019

Bida T, Killen RM, Morgan TH (2000) Discovery of Ca in the atmosphere of Mercury. Nature 404:159–161. https://doi.org/10.1038/35004521

Bjorkholm P, Golub L, Gorenstein P (1973) Detection of a nonuniform distribution of polonium-210 on the Moon with the Apollo 16 alpha particle spectrometer. Science 180(4089):957–959

Broadfoot AL, Kumar S, Belton MJS, McElroy MB (1974) Mercury's atmosphere from Mariner 10: preliminary results. Science 185(4146):166–169

Broadfoot AL, Shemansky DE, Kumar S (1976) Mariner 10: Mercury atmosphere. Geophys Res Lett 3(10):577–580

Bruevich EA, Yakunina GV (2017) Flare activity of the sun and variations in its UV emission during cycle 24. Astrophysics 60(3). https://doi.org/10.1007/s10511-017-9492-7

Burger MH, Killen RM, McClintock WE, Merkel AW, Vervack RJ, Cassidy TA, Sarantos M (2014) Seasonal variability in Mercury's dayside calcium exosphere. Icarus 238:51–58. https://doi.org/10.1016/j.icarus.2014.04.049

Carrillo-Sánchez JD, Nesvorný D, Pokorný P, Janches D, Plane JMC (2016) Sources of cosmic dust in the Earth's atmosphere. Geophys Res Lett 43(23):11979. https://doi.org/10.1002/2016GL071697

Cassidy TA, Merkel AW, Burger MH, Sarantos M, Killen RM, McClintock WE, Vervack RJ (2015) Mercury's seasonal sodium exosphere: MESSENGER orbital observations. Icarus 248:547–559. https://doi.org/10.1016/j.icarus.2014.10.037

Cassidy TA, McClintock WE, Killen RM, Sarantos M, Merkel AW, Vervack RJ, Burger MH (2016) A cold-pole enhancement in Mercury's sodium exosphere. Geophys Res Lett 43:11,121–11,128. https://doi.org/10.1002/2016GL071071

Chabot NL, Shread EE, Harmon JK (2018) Investigating Mercury's south polar deposits: Arecibo radar observations and high-resolution determination of illumination conditions. J Geophys Res, Planets 123:666–681. https://doi.org/10.1002/2017JE005500

Chowdhury AR et al (2020) Imaging infrared spectrometer onboard Chandrayaan-2 orbiter. Curr Sci 118(3):368–375. https://doi.org/10.18520/cs/v118/i3/368-375

Christou AA, Killen RM, Burger MH (2015) The meteoroid stream of comet encke at Mercury: implications for Mercury surface, space ENvironment, GEochemistry, and ranging observations of the exosphere. Geophys Res Lett 42:7311–7318. https://doi.org/10.1002/2015GL065361

Cintala MJ (1992) Impact-induced thermal effects in the lunar and Mercurian regoliths. J Geophys Res 97:947–973. https://doi.org/10.1029/91JE02207

Colaprete A et al (2010) Detection of water in the LCROSS ejecta plume. Science 330:463. https://doi.org/10.1126/science.1186986

Colaprete A, Sarantos M, Wooden DH, Stubbs TJ, Cook AM, Shirley M (2016) How surface composition and meteoroid impacts mediate sodium and potassium in the lunar exosphere. Science 351(6270):249–252. https://doi.org/10.1126/science.aad2380

Cremonese G et al (2020) SIMBIO-SYS: cameras and spectrometer for the BepiColombo mission. Space Sci Rev 216(5):1–78. https://doi.org/10.1007/s11214-020-00704-8

Crider DH, Vondrak RR (2000) The solar wind as a possible source of lunar polar hydrogen deposits. J Geophys Res, Planets 105(E11):26773–26782

Crider DH, Vondrak RR (2003) Space weathering effects on lunar cold trap deposits. J Geophys Res, Planets 108(E7)

Cupak C, Szabo PS, Biber H, Stadlmayr R, Grave C, Fellinger M, Brötzner J, Wilhelm RA, Möller W, Mutzke A, Moro MV, Aumayr F (2021) Sputter yields of rough surfaces: importance of the mean surface inclination angle from nano- to microscopic rough regimes. Appl Surf Sci 570:151204. https://doi.org/10.1016/j.apsusc.2021.151204

Dandouras I et al (2023) Space plasma physics science opportunities for the lunar orbital platform - gateway. Front. Astron. Space Sci., vol 10. https://doi.org/10.3389/fspas.2023.1120302

Das TP, Thampi SV, Dhanya MB, Naik N, Sreelatha P, Pradeepkumar P et al (2020) Chandra's atmospheric composition explorer-2 onboard Chandrayaan-2 to study the lunar neutral exosphere. Curr Sci 118:202–209. https://doi.org/10.18520/cs/v118/i2/202-209

Dhanya MB et al (2021) Argon-40 in lunar exosphere: observations from Chace-2 on Chandrayaan-2 orbiter. Geophys Res Lett 48(20):e2021GL094970. https://doi.org/10.1029/2021GL094970

Delcourt DC, Grimald S, Leblanc F, Bertherlier J-J, Millilo A, Mura A (2003) A quantitative model of planetary Na+ contribution to Mercury's magnetosphere. Ann Geophys 21:1723–1736. https://doi.org/10.5194/angeo-21-1723-2003

Denevi BW, Robinson MS, Boyd AK, Sato H, Hapke BW, Hawke BR (2014) Characterization of space weathering from lunar reconnaissance orbiter camera ultraviolet observations of the Moon. J Geophys Res, Planets 119(5):976–997. https://doi.org/10.1002/2013JE004527

Dewey RM, Slavin JA, Raines JM, Azari AR, Sun W (2020) MESSENGER observations of flow braking and flux pileup of dipolarizations in Mercury's magnetotail: evidence for current wedge formation. J Geophys Res 125:e2020JA028112. https://doi.org/10.1029/2020JA028112

Dong C, Wang L, Hakim A, Bhattacharjee A, Slavin JA, DiBraccio GA, Germaschewski K (2019) Global ten-moment multifluid simulations of the solar wind interaction with Mercury: from the planetary conducting core to the dynamic magnetosphere. Geophys Res Lett 46:11588–11596. https://doi.org/10.1029/2019GL083180

Farrell WM, Hurley DM, Poston MJ, Hayne PO, Szalay JR, McLain JL (2019) The young age of the LAMP-observed frost in lunar polar cold traps. Geophys Res Lett 46(15):8680–8688. https://doi.org/10.1029/2019GL083158

Fatemi S, Lue C, Holmstrom M, Poppe AR, Wieser M, Barabash S, Delory GT (2015) Solar wind plasma interaction with Gerasimovich lunar magnetic anomaly. J Geophys Res Space Phys 120:4719–4735. https://doi.org/10.1002/2015JA021027

Feldman WC, Maurice S, Binder AB, Barraclough BL, Elphic RC, Lawrence DJ (1998) Fluxes of fast and epithermal neutrons from lunar prospector: evidence for water ice at the lunar poles. Science 281:1496–1500. https://doi.org/10.1126/science.281.5382.1496

Feldman PD, Hurley DM, Retherford KD, Gladstone GR, Stern SA, Pryor W, Parker JW, Kaufmann DE, Davis MW, Versteeg MH (2012) Temporal variability of lunar exospheric helium during January 2012 from LRO/LAMP. Icarus 221(2):854–858. https://doi.org/10.1016/j.icarus.2012.09.015

Fentzke JT, Janches D (2008) A semi-empirical model of the contribution from sporadic meteoroid sources on the meteor input function observed at arecibo. J Geophys Res Space Phys 113:A03304. https://doi.org/10.1029/2007JA012531

Fox NJ, Velli MC, Bale SD et al (2016) The solar probe plus mission: humanity's first visit to our star. Space Sci Rev 204:7. https://doi.org/10.1007/s11214-015-0211-6

Fray N, Schmitt B (2009) Sublimation of ices of astrophysical interest: a bibliographic review. Planet Space Sci 57:2053–2080. https://doi.org/10.1016/j.pss.2009.09.011

Futaana Y, Barabash S, Wieser M, Lue C, Wurz P, Vorburger A, Bhardwaj A, Asamura K (2013) Remote energetic neutral atom imaging of electric potential over a lunar magnetic anomaly. Geophys Res Lett 40:262–266. https://doi.org/10.1002/grl.50135

Gamborino D, Vorburger A, Wurz P (2019) Mercury's sodium exosphere: an ab initio calculation to interpret MESSENGER observations. Ann Geophys 37:455–470. https://doi.org/10.5194/angeo-2018-109

Gorenstein P, Bjorkholm P (1973) Detection of Radon emanation from the crater Aristarchus by the Apollo 15 alpha particle spectrometer. Science 179(4075):792–794

Grava C, Chaufray J-Y, Retherford KD, Gladstone GR, Greathouse TK, Hurley DM, Hodges RR, Bayless AJ, Cook JC, Stern SA (2015) Lunar exospheric argon modeling. Icarus 255:135–147. https://doi.org/10.1016/j.icarus.2014.09.029

Grava C, Killen RM, Benna M, Berezhnoy AA, Halekas JS, Leblanc F, Nishino MN, Plainaki C, Raines JM, Sarantos M, Teolis BD, Tucker OJ, Vervack RJ, Vorburger A (2021a) Volatiles and refractories in surface-bounded exospheres in the inner Solar System. Space Sci Rev 217:61. https://doi.org/10.1007/s11214-021-00833-8

Grava C, Hurley DM, Feldman PD, Retherford KD, Greathouse TK, Pryor WR, Gladstone GR, Halekas JS, Mandt K, Wyrick DY, Davis MW, Egan AF, Kaufman DE, Versteeg M, Stern SA (2021b) LRO/LAMP observations of the lunar helium exosphere: constraints on thermal accommodation and outgassing rate. Mon Not R Astron Soc 501(3):4438–4451. https://doi.org/10.1093/mnras/staa3884

Halekas JA, Angelopoulos V, Sibeck DG, Khurana KK, Russell CT, Delory GT et al (2011) First results from ARTEMIS, a new two-spacecraft lunar mission: counter-streaming plasma populations in the lunar wake. Space Sci Rev 165(1–4):93–107. https://doi.org/10.1007/s11214-010-9738-8

Halekas JA, Poppe AR, Delory GT, Sarantos M, Farrell WM, Angelopoulos V, McFadden JP (2012) Lunar pickup ions observed by ARTEMIS: spatial and temporal distribution and constraints on species and source locations. J Geophys Res, Planets 117:E06006. https://doi.org/10.1029/2012JE004107

Halekas JS, Benna M, Mahaffy PR, Elphic RC, Poppe AR, Delory GT (2015) Detections of lunar exospheric ions by the LADEE neutral mass spectrometer. Geophys Res Lett 42(13):5162–5169. https://doi.org/10.1002/2015gl064746

Harada Y et al (2010) Interaction between terrestrial plasma sheet electrons and the lunar surface: SELENE (Kaguya) observations. Geophys Res Lett 37(19):L10202. https://doi.org/10.1029/2010GL044574

Harada Y, Futaana Y, Barabash S, Wieser M, Wurz P, Bhardwaj A, Asamura K, Saito Y, Yokota S, Tsunakawa H, Machida S (2014a) Backscattered energetic neutral atoms from the Moon in the Earth's plasma sheet observed by Chandarayaan-1/subkeV atom reflecting analyser instrument. J Geophys Res 119(5):3573–3584. https://doi.org/10.1002/2013JA019682

Harada Y, Halekas JS, Poppe AR, Kurita S, McFadden JP (2014b) Extended lunar precursor regions: electron wave interaction. J Geophys Res 119:9160–9173. https://doi.org/10.1002/2014JA020618

Hayne PO et al (2015) Evidence for exposed water in the Moon's south polar regions from lunar reconnaissance orbiter ultraviolet albedo and temperature measurements. Icarus 255:58–69

Hayne PO, Aharonson O, Schörghofer N (2021) Micro cold traps on the Moon. Nat Astron 5(2):169–175

Hiesinger H et al (2020) Studying the composition and mineralogy of the Hermean surface with the Mercury Radiometer and Thermal Infrared Spectrometer (MERTIS) for the BepiColombo mission: an update. Space Sci Rev 216. https://doi.org/10.1007/s11214-020-00732-4

Hinton FL, Taeusch DR (1964) Variation of the lunar atmosphere with the strength of the solar wind. J Geophys Res 69(7):1341–1347

Hodges RR Jr (1973a) Differential equation for exospheric lateral transportation and its application to terrestrial hydrogen. J Geophys Res 78(31):7340–7346

Hodges RR Jr (1973b) Helium and hydrogen in the lunar atmosphere. J Geophys Res 78(34):8055–8064

Hodges RR (1975) Formation of the lunar atmosphere. Moon 14(1):139

Hodges RR Jr (1977) Release of radiogenic gases from the moon. Phys Earth Planet Inter 14(3):282–288. https://doi.org/10.1016/0031-9201(77)90178-9

Hodges RR Jr (1980) Methods for Monte Carlo simulation of the exospheres of the Moon and Mercury. J Geophys Res Space Phys 85(A1):164–170

Hodges RR Jr (2016) Methane in the lunar exosphere: implications for solar wind carbon escape. Geophys Res Lett 43(13):6742–6748. https://doi.org/10.1002/2016GL068994

Hodges RR Jr (2018) Semiannual oscillation of the lunar exosphere: implications for water and polar ice. Geophys Res Lett 45(15):7409–7416. https://doi.org/10.1029/2018GL077745

Hodges RR Jr, Johnson FS (1968) Lateral transport in planetary exospheres. J Geophys Res 73(23):7307–7317

Hodges RR Jr, Hoffman JH (1975) Implications of atmospheric [40]Ar escape on the interior structure of the Moon. In: Proceedings of the 6th lunar science conference, pp 3039–3047

Hodges RR Jr, Mahaffy PR (2016) Synodic and semiannual oscillations of argon-40 in the lunar exosphere. Geophys Res Lett 43(1):22–27. https://doi.org/10.1002/2015GL067293

Hoffman JH, Hodges RR Jr, Johnson FS, Evans DE (1973) Lunar atmospheric 1414 composition results from Apollo 17. In: Proc of the 4th Lunar and Planetary Science Conference

Hood LL, Williams CR (1989) The lunar swirls: distribution and possible origins. In: Proc 19th Lunar and Planetary Science Conference (A89-36486 15-91). LPI, Houston, pp 99–113

Hunten DM, Morgan TM, Shemansky DM (1988) The Mercury atmosphere. In: Mercury (A89-43751 19-91). University of Arizona Press, Tucson, pp 562–612

Hurley DM, Cook JC, Benna M, Halekas JS, Feldman PD, Retherford KD, Hodges RR, Grava C, Mahaffy P, Gladstone GR, Greathouse T, Kaufmann DE, Elphic RC, Stern AS (2016) Understanding temporal and spatial variability of the lunar helium atmosphere using simultaneous observations from LRO, LADEE, and ARTEMIS. Icarus 273:45–52. https://doi.org/10.1016/j.icarus.2015.09.011

Imber SM, Slavin JA (2017) MESSENGER observations of magnetotail loading and unloading: implications for substorms at Mercury. J Geophys Res Space Phys 122:11,402–11,412. https://doi.org/10.1002/2017JA024332

Janches D, Pokorny P, Sarantos M, Szalay JR, Horanyi M, Nesvorny D (2018) Constraining the ratio of micrometeoroids from short- and long-period comets at 1 AU from LADEE observations of the lunar dust cloud. Geophys Res Lett 45:1713–1722. https://doi.org/10.1002/2017GL076065

Janches D, Apostolos C, Berezhnoy AA, Cremonese G, Hirai T, Horany M, Jasinski JM, Sarantos M (2021) Meteoroids as one of the sources for exosphere formation on airless bodies in the inner solar system. Space Sci Rev 217:50. https://doi.org/10.1007/s11214-021-00827-6

Jasinski JM, Regoli LH, Cassidy TA et al (2020) A transient enhancement of Mercury's exosphere at extremely high altitudes inferred from pickup ions. Nat Commun 11:4350. https://doi.org/10.1038/s41467-020-18220-2

Jia X, Slavin JA, Gombosi TI, Daldorff LKS, Toth G, van der Holst B (2015) Global MHD simulations of Mercury's magnetosphere with coupled planetary interior: induction effect of the planetary conducting core on the global interaction. J Geophys Res Space Phys 120:4763–4775. https://doi.org/10.1002/2015JA021143

Jia X, Slavin JA, Poh G, DiBraccio GA, Toth G, Chen Y et al (2019) MESSENGER observations and global simulations of highly compressed magnetosphere events at Mercury. J Geophys Res 124:229–247. https://doi.org/10.1029/2018JA026166

Johnstone CP, Lammer H, Kislyakova K, Scherf M, Güdel M (2021) The young Sun's XUV-activity as a constraint for lower CO2-limits in the Earth's Archean atmosphere. Earth Planet Sci Lett 576:117197. https://doi.org/10.1016/j.epsl.2021.117197

Kallio E, Dyadechkin S, Wurz P, Khodachenko M (2019) Space weathering on the Moon: farside-nearside solar wind precipitation asymmetry. Planet Space Sci 166:9–22. https://doi.org/10.1016/j.pss.2018.07.013

Kameda S, Yoshikawa I, Kagitani M, Okano S (2009) Interplanetary dust distribution and temporal variability of Mercury's atmospheric Na. Geophys Res Lett 36(15):L15201. https://doi.org/10.1029/2009GL039036

Kegerreis JA, Eke VR, Massey RJ, Beaumont SK, Elphic RC, Teodoro LF (2017) Evidence for a localized source of the argon in the lunar exosphere. J Geophys Res, Planets 122(10):2163–2181

Killen RM (2016) Pathways for energization of Ca in Mercury's exosphere. Icarus 268:32–36. https://doi.org/10.1016/j.icarus.2015.12.035

Killen RM, Ip W-H (1999) The surface-bounded atmospheres of Mercury and the Moon. Rev Geophys 37:361–406. https://doi.org/10.1029/1999RG900001

Killen RM, Hahn JM (2015) Impact vaporization as a possible source of Mercury's calcium exosphere. Icarus 250:230–237. https://doi.org/10.1016/j.icarus.2014.11.035

Killen RM, Potter AE, Reiff P, Sarantos M, Jackson BV, Hick P, Giles B (2001) Evidence for space weather at Mercury. J Geophys Res 106:20509–20525. https://doi.org/10.1029/2000JE001401

Killen RM, Sarantos M, Potter AE, Reiff P (2004) Source rates and ion recycling rates for Na and K in Mercury's atmosphere. Icarus 171:1–19. https://doi.org/10.1016/j.icarus.2004.04.007

Killen RM, Cremonese G, Lammer H, Orsini S, Potter AE, Sprague AL, Wurz P, Khodachenko M, Lichtenegger HIM, Milillo A, Mura A (2007) Processes that promote and deplete the exosphere of Mercury. Space Sci Rev 132:433–509. https://doi.org/10.1007/s11214-007-9232-0

Killen RM, Hurley DM, Farrell WM (2012) The effect on the lunar exosphere of a coronal mass ejection passage. J Geophys Res 117:e27837. https://doi.org/10.1029/2011JE004011

Killen RM, Williams DR, Park J, Tucker OJ, Kim S-J (2019) The lunar neon exosphere seen in LACE data. Icarus 329:246–250. https://doi.org/10.1016/j.icarus.2019.04.018

Kislyakova KG, Johnstone CP, Scherf M, Holmström M, Alexeev II, Lammer H, Khodachenko ML, Güdel M (2020) Evolution of the Earth's polar outflow from mid-Archean to present. J Geophys Res 125(8):e27837. https://doi.org/10.1029/2020JA027837

Kobayashi M et al (2020) Mercury Dust Monitor (MDM) onboard the Mio Orbiter of the BepiColombo mission. Space Sci Rev 216. https://doi.org/10.1007/s11214-020-00775-7

Koschny D, Grün E (2001) Impacts into ice-silicate mixtures: ejecta mass and size distributions. Icarus 154:402–411. https://doi.org/10.1006/icar.2001.6708

Krüger H, Strub P, Srama R, Kobayashi M, Arai T, Kimura H, Hirai T, Moragas-Klostermeyer G, Altobelli N, Sterken VJ, Agarwal J, Sommer M, Grün E (2019) Modelling DESTINY+ interplanetary and interstellar dust measurements en route to the active asteroid (3200) Phaethon. Planet Space Sci 172:22–42. https://doi.org/10.1016/j.pss.2019.04.005

Lammer H, Zerkle AL, Gebauer S, Tosi N, Noack L, Scherf M, Pilat-Lohinger E, Güdel M, Grenfell JL, Godolt M, Nikolaou A (2018) Origin and evolution of the atmospheres of early Venus, Earth and Mars. Astron Astrophys Rev 26:2. https://doi.org/10.1007/s00159-018-0108-y

Lammer H, Scherf M, Ito Y, Mura A, Vorburger A, Guenther E, Wurz P, Erkaev NV, Odert P (2022) The exosphere as a boundary: origin and evolution of airless bodies in the inner solar system and beyond including planets with silicate atmospheres. Space Sci Rev 218:15. https://doi.org/10.1007/s11214-022-00876-5

410

Lawrence DJ (2017) A tale of two poles: toward understanding the presence, distribution, and origin of volatiles at the polar regions of the Moon and Mercury. J Geophys Res, Planets 122(1):21–52. https://doi.org/10.1002/2016JE005167

Lawson SL, Feldman WC, Lawrence DJ, Moore KR, Elphic RC, Belian RD, Maurice S (2005) Recent outgassing from the lunar surface: the lunar prospector alpha particle spectrometer. J Geophys Res, Planets 110(E9):E09009. https://doi.org/10.1029/2005JE002433

Leblanc F, Johnson RE (2003) Mercury's sodium exosphere. Icarus 164:261–281. https://doi.org/10.1016/S0019-1035(03)00147-7

Leblanc F, Johnson RE (2010) Mercury exosphere. I. Global circulation model of its sodium component. Icarus 209:280–300. https://doi.org/10.1016/j.icarus.2010.04.020

Leblanc F, Doressoundiram A (2011) Mercury exosphere: II. The sodium/potassium ratio. Icarus 211:10–20. https://doi.org/10.1016/j.icarus.2010.09.004

Leblanc F, Doressoundiram A, Schneider N, Mangano V, Lopez-Ariste A, Lemen C, Gelly B, Barbieri C, Cremonese G (2008) High latitude peaks in Mercury's sodium exosphere: spectral signature using THEMIS solar telescope. Geophys Res Lett 35:L18204. https://doi.org/10.1029/2008GL035322

Leblanc F, Schmidt C, Mangano V, Mura A, Cremonese G, Raines JM, Jasinski JM, Sarantos M, Milillo A, Killen RM, Massetti S, Cassidy T, Vervack RJ Jr, Kameda S, Capria MT, Horanyi M, Janches D, Berezhnoy A, Christou A, Hirai T, Lierle P, Morgenthaler J (2022) Comparative Na and K Mercury and Moon exospheres. Space Sci Rev 218:2. https://doi.org/10.1007/s11214-022-00871

Leblanc F, Deborde R, Tramontina D, Bringa E, Chaufray JY, Aizawa S, Modolo R, Morrissey L, Woodson A, Verkercke S, Dukes C (2023) On the origins of backscattered solar wind energetic neutral hydrogen from the Moon and Mercury. Planet Space Sci 229:105660. https://doi.org/10.1016/j.pss.2023.105660

Li S et al (2018) Direct evidence of surface exposed water ice in the lunar polar regions. Proc Natl Acad Sci 115:8907–8912

Lindsay ST, James MK, Bunce EJ, Imbera SM, Korth H, Martindale A, Yeoman TK (2016) MESSENGER X-ray observations of magnetosphere–surface interaction on the nightside of Mercury. Planet Space Sci 125:72–79. https://doi.org/10.1016/j.pss.2016.03.005

Lucey PG et al (2022) Volatile interactions with the lunar surface. Geochem 82:125858

Lue C, Futaana Y, Barabash S, Wieser M, Holmström M, Bhardwaj A, Dhanya MB, Wurz P (2011) Strong influence of lunar crustal fields on the solar wind flow. Geophys Res Lett 38:L03202. https://doi.org/10.1029/2010GL046215

Mall U, Kirsch E, Cierpka K, Wilken B, Söding A, Neubauer F et al (1998) Direct observation of lunar pick-up ions near the Moon. Geophys Res Lett 25(20):3799–3802. https://doi.org/10.1029/1998GL900003

Mangano V, Milillo A, Mura A, Orsini S, De Angelis E, Di Lellis P, Wurz AM (2007) The contribution of impulsive meteoritic impact vaporization to the Hermean exosphere. Planet Space Sci 55(11):1541–1556. https://doi.org/10.1016/j.pss.2006.10.008

Mangano V, Leblanc F, Barbieri C, Massetti S, Milillo A, Cremonese G, Grava C (2009) Detection of a southern peak in Mercury's sodium exosphere with the TNG in 2005. Icarus 201:424–431. https://doi.org/10.1016/j.icarus.2009.01.016

Mangano V, Massetti S, Milillo A, Plainaki C, Orsini S, Rispoli R, Leblanc F (2015) THEMIS Na exosphere observations of Mercury and their correlation with in-situ magnetic field measurements by MESSENGER. Planet Space Sci 115:102–109. https://doi.org/10.1016/j.pss.2015.04.001

Marty B, Hashizume K, Chaussidon M, Wieler R (2003) Nitrogen isotopes on the Moon: archives of the solar and planetary contributions to the inner solar system. Space Sci Rev 106:175–196. https://doi.org/10.1023/A:1024689721371

Massetti S, Mangano V, Milillo A, Mura A, Orsini S, Plainiki C (2017) Short-term observations of double peaked Na emission from Mercury's exosphere. Geophys Res Lett. https://doi.org/10.1002/2017GL073090

McClintock WE, Vervack RJ Jr, Todd Bradley E, Killen RM, Sprague AL, Izenberg NR (2008) Mercury's exosphere: observations MESSENGER's first Mercury flyby. Science 321:92–94. https://doi.org/10.1126/science.1159467

McComas DJ, Allegrini F, Bochsler P, Frisch P, Funsten HO, Gruntman M et al (2009) Lunar backscatter and neutralization of the solar wind: first observations of neutral atoms from the Moon. Geophys Res Lett 36(12):L12104. https://doi.org/10.1029/2009GL038794

McComas DJ et al (2018) Interstellar mapping and acceleration probe (IMAP): a new NASA mission. Space Sci Rev 214(8):116. https://doi.org/10.1007/s11214-018-0550-1

Merkel AW, Vervack RJ Jr, Cassidy TA, Killen RM, McClintock WE, Nittler LR, Burger MH (2018) Evidence connecting Mercury's Mg exosphere to its Magnesium-rich Surface Terrane. Geophys Rev Lett 45(14). https://doi.org/10.1029/2018GL078407

Milillo A, Wurz P, Orsini S, Delcourt D, Kallio E, Killen RM et al (2005) Surface-exosphere-magnetosphere system of Mercury. Space Sci Rev 117(3–4):397–443. https://doi.org/10.1007/s11214-005-3593-z

Milillo A, Fujimoto M, Murakami G, Benkhoff J, Zender J, Aizawa S et al (2020) Investigating Mercury's environment with the two-spacecraft BepiColombo mission. Space Sci Rev 216(5):1–78. https://doi.org/10.1007/s11214-020-00712-8

Milillo A, Mangano V, Massetti S, Mura A, Plainaki C, Alberti T, Ippolito A, Ivanovski S, Aronica A, De Angelis E, Kazakov A, Noschese R, Orsini S, Rispoli R, Sordini R, Vertolli N (2021) Exospheric Na distributions along the Mercury orbit with the THEMIS telescope. Icarus 355:114179. https://doi.org/10.1016/j.icarus.2020.114179

Milligan RO (2015) Extreme ultra-violet spectroscopy of the lower solar atmosphere during solar flares. Sol Phys 290:3399–3423. https://doi.org/10.1007/s11207-015-0748-2

Moroni M, Mura A, Milillo A, Plainaki C, Mangano V, Alberti T, Andre N, Aronica A, De Angelis E, Del Moro D, Kazakov A, Massetti S, Orsini S, Rispoli R, Sordini R (2023) Micro-meteoroids impact vaporization as source for Ca and CaO exosphere along Mercury's orbit. Icarus 401. https://doi.org/10.1016/j.icarus.2023.115616

Müller D, St. Cyr OC, Zouganelis I et al (2020) The solar orbiter mission. Science overview. Astron Astrophys 642:A1. https://doi.org/10.1051/0004-6361/202038467

Mura A, Milillo A, Orsini S, Massetti S (2007) Numerical and analytical model of Mercury's exosphere: dependence on surface and external conditions. Planet Space Sci 55:1569–1583. https://doi.org/10.1016/j.pss.2006.11.028

Mura A, Wurz P, Lichtenegger HIM, Schleicher H, Lammer H, Delcourt D, Milillo A, Orsini S, Massetti S, Khodachenko ML (2009) The sodium exosphere of Mercury: comparison between observations during Mercury's transit and model results. Icarus 200:1–11. https://doi.org/10.1016/j.icarus.2008.11.014

Mura A Plainaki C Milillo A Mangano V Alberti T Massetti S Orsini S Moroni M De Angelis S Rispoli R Sordini R (2023) The yearly variability of the sodium exosphere of Mercury: a toy model. Icarus 394:115441. https://doi.org/10.1016/j.icarus.2023.115441

Nagaoka H, Ohtake M, Shirai N, Karouji Y, Kayama M, Daket Y, Hasebe N, Ebihara M (2021) Investigation of the source region of the lunar-meteorite group with the remote sensing datasets: implication for the origin of mare volcanism in mare imbrium. Icarus 371:114690. https://doi.org/10.1016/j.icarus.2021.114690

Narendranath S, Pillai NS, Tadepalli SP, Sarantos M, Vadodariya K, Sarwade A, Radhakrishna V, Tyagi A (2022) Sodium distribution on the Moon. Astrophys J Lett 937:L23. https://doi.org/10.3847/2041-8213/ac905a

Nénon Q et al (2019) Phobos surface sputtering as inferred from MAVEN ion observations. J Geophys Res 124:3385–3401. https://doi.org/10.1029/2019JE006197

Ness NF, Behannon KW, Lepping RP, Whang YC (1975) The magnetic field of Mercury, 1. J Geophys Res 80:2708. https://doi.org/10.1029/JA080i019p02708

Ness NF Behannon KW Lepping RP Whang YC 1976) Observations of Mercury's magnetic field. Icarus 28(4):479–488. https://doi.org/10.1016/0019-1035(76)90121-4

Nishino MN, Harada Y, Saito Y, Tsunakawa H, Takahashi F, Yokota S, Matsushima M, Shibuya H, Shimizu H (2017) Kaguya observations of the lunar wake in the terrestrial foreshock: surface potential change by bow-shock reflected ions. Icarus 293:45–51. https://doi.org/10.1016/j.icarus.2017.04.005

Orsini S, Mangano V, Mura A, Turrini D, Massetti S, Milillo A, Plainaki C (2014) The influence of space environment on the evolution of Mercury. Icarus 239:281–290. https://doi.org/10.1016/j.icarus.2014.05.031

Orsini S, Mangano V, Milillo A, Plainaki C, Mura A, Raines JM, De Angelis E, Rispoli R, Lazzarotto F, Aronica A (2018) Mercury sodium exospheric emission as a proxy for solar perturbations transit. Sci Rep 8:928. https://doi.org/10.1038/s41598-018-19163-x

Øieroset M Lin RP Phan TD Larson DE Bale SD (2002) Evidence for Electron Acceleration up to 300 k eV in the Magnetic Reconnection Diffusion Region of Earth's Magnetotail. Phys Rev Lett 89:195001. https://doi.org/10.1103/PhysRevLett.89.195001

Orsini S et al (2021) SERENA: particle instrument suite for Sun-Mercury interaction insights on-board Bepi-Colombo. Space Sci Rev 217(11):1–107. https://doi.org/10.1007/s11214-020-00787-3

Ozguc A, Kilcik AK, Sarp V, Yeşilyaprak H, Pektaş R (2021) Periodic variation of solar flare index for the last solar cycle (Cycle 24), SI predictions of solar activity cycle and its association with geomagnetic activity. Adv Astron 5391091. https://doi.org/10.1155/2021/5391091

Ozima M, Seki K, Terada N, Miura YN, Podosek FA, Shinagawa H (2005) Terrestrial nitrogen and noble gases in lunar soils. Nature 436:655–659. https://doi.org/10.1038/nature03929

Paige DA, Wood SE, Vasavada AR (1992) The thermal stability of water ice at the poles of Mercury. Science 258(5082):643–646. https://doi.org/10.1126/science.258.5082.643

Paige DA, Siegler MA, Zhang JA et al (2010) Diviner lunar radiometer observations of cold traps in the Moon's south polar region. Science 330(6003):479–482. https://doi.org/10.1126/science.1187726

Peplowski PN, Evans LG, Hauck SA, McCoy TJ, Boynton WV, Gillis-Davis JJ, Ebel DS, Goldsten JO, Hamara DK, Lawrence DJ, McNutt RL, Nittler LR, Solomon SC, Rhodes EA, Sprague AL, Starr RD, Stockstill-Cahill KR (2011) Radioactive elements on Mercury's surface from MESSENGER: implications for the planet's formation and evolution. Science 333(6051):1850. https://doi.org/10.1126/science.1211576

Peplowski PN, Evans LG, Stockstill-Cahill KR, Lawrence DJ, Goldsten JO, McCoy TJ, Nittler LR, Solomon SC, Sprague AL, Starr RD, Weider SZ (2014) Enhanced sodium abundance in Mercury's North polar region revealed by the MESSENGER gamma-ray spectrometer. Icarus 228:86–95. https://doi.org/10.1016/j.icarus.2013.09.007

Pieters CM, Noble SK (2016) Space weathering on airless bodies. J Geophys Res 121:1865–1884. https://doi.org/10.1002/2016JE005128

Plainaki C, Milillo A, Orsini S, Mura A, DeAngelis E, DiLellis AM, Dotto E, Livi S, Mangano V, Massetti S, Palumbo ME (2009) Space weathering on near-Earth objects investigated by neutral-particle detection. Planet Space Sci 57:384–392. https://doi.org/10.1016/j.pss.2008.12.002

Plainaki C, Mura A, Milillo A, Orsini S, Livi S, Mangano V, Massetti S, Rispoli R, DeAngelis E (2017) Investigation of the possible effects of comet Encke's meteoroid stream on the Ca exosphere of Mercury. J Geophys Res Planet 122:1217–1226. https://doi.org/10.1002/2017JE005304

Platz T, Nathues A, Schorghofer N et al (2017) Surface water-ice deposits in the northern shadowed regions of Ceres. Nat Astron 1:0007. https://doi.org/10.1038/s41550-016-0007

Pokorný P, Sarantos M, Janches D (2017) Reconciling the dawn/dusk asymmetry in Mercury's exosphere with the micrometeoroid impact directionality. Astrophys J Lett 842:L17. https://doi.org/10.3847/2041-8213/aa775d

Pokorný P, Sarantos M, Janches D (2018) A comprehensive model of the meteoroid environment around Mercury. Astrophys J 863(1):31. https://doi.org/10.3847/1538-4357/aad051

Pokorný P, Janches D, Sarantos M, Szalay JR, Horányi M, Nesvorný D, Kuchner MJ (2019) Meteoroids at the Moon: orbital properties, surface vaporization, and impact ejecta production. J Geophys Res, Planets 124:752–778. https://doi.org/10.1029/2018JE005912

Poppe AR (2016) An improved model for interplanetary dust fluxes in the outer Solar System. Icarus 264:369–386. https://doi.org/10.1016/j.icarus.2015.10.001

Poppe AR, Sarantos M, Halekas JS, Delory GT, Saito Y, Nishino M (2014) Anisotropic solar wind sputtering of the lunar surface induced by crustal magnetic anomalies. Geophys Res Lett 41(14):4865–4872

Poppe AR, Samad R, Halekas JS, Sarantos M, Delory GT, Farrell WM et al (2012) ARTEMIS observations of lunar pick-up ions in the terrestrial magnetotail. Geophys Res Lett 39(L17104). https://doi.org/10.1029/2012gl052909

Poppe AR, Halekas JS, Szalay JR, Horányi M, Levin Z, Kempf S (2016) LADEE/LDEX observations of lunar pickup ion distribution and variability. Geophys Res Lett 43(7):3069–3077. https://doi.org/10.1002/2016gl068393

Poppe AR, Halekas JS, Harada Y (2022) A comprehensive model for pickup ion formation at the Moon. J Geophys Res, Planets 127(10):e2022JE007422. https://doi.org/10.1029/2022JE007422

Potter AE, Killen RM (2008) Observations of the sodium tail of Mercury. Icarus 194(1):1–12. https://doi.org/10.1016/j.icarus.2007.09.023

Potter AE, Morgan TH (1985) Discovery of sodium in the atmosphere of Mercury. Science 229:651–653

Potter AE, Morgan TH (1986) Potassium in the atmosphere of Mercury. Icarus 67:336–340

Potter AE, Morgan TH (1988) Discovery of sodium and potassium vapor in the atmosphere of the Moon. Science 241:675–680. https://doi.org/10.1126/science.229.4714.651

Potter AE, Morgan TH (1990) Evidence for magnetospheric effects on the sodium atmosphere of Mercury. Science 248:835. https://doi.org/10.1126/science.248.4957.835

Potter AE, Morgan TH (1997) Sodium and potassium atmospheres of Mercury. Planet Space Sci 45(1):95–100. https://doi.org/10.1016/S0032-0633(96)00100-6

Potter AE, Killen RM, Morgan TH (1999) Rapid changes in the sodium exosphere of Mercury. Planet Space Sci 47:1441–1448. https://doi.org/10.1016/S0032-0633(99)00070-7

Potter AE, Anderson CM, Killen RM, Morgan TH (2002a) Ratio of sodium to potassium in the Mercury exosphere. J Geophys Res 107(E6). https://doi.org/10.1029/2000JE001493

Potter AE, Killen RM, Morgan TH (2002b) The sodium tail of Mercury. Meteorit Planet Sci 37:1165–1172

Potter AE, Killen RM, Sarantos M (2006) Spatial distribution of sodium on Mercury. Icarus 181:1–12. https://doi.org/10.1016/j.icarus.2005.10.026

Prettyman TH, Hagerty JJ, Elphic RC, Feldman WC, Lawrence DJ, McKinney GW, Vaniman DT (2006) Elemental composition of the lunar surface: Analysis of gamma ray spectroscopy data from Lunar Prospector. J Geophys Res, Planets 111(E12). https://doi.org/10.1029/2005JE002656

Raines JM, Gershman DJ, Zurbuchen TH, Sarantos M, Slavin JA, Gilbert JA et al (2013) Distribution and compositional variations of plasma ions in Mercury's space environment: the first three Mercury years of MESSENGER observations. J Geophys Res Space Phys 118(4):1604–1619. https://doi.org/10.1029/2012JA018073

Raines JM, DiBraccio GA, Cassidy TA, Delcourt DC, Fujimoto M, Jia X, Mangano V, Milillo A, Sarantos M, Slavin JA, Wurz P (2015) Plasma sources in planetary magnetospheres: Mercury. Space Sci Rev 192(1):1–54. https://doi.org/10.1007/s11214-015-0193-4

Reiss P, Warren T, Sefton-Nash E, Trautner R (2021) Dynamics of subsurface migration of water on the Moon. J Geophys Res, Planets 126(5):e2020JE006742

Ribas I, Guinan EF, Güdel M, Audard M (2005) Evolution of the solar activity over time and effects on planetary atmospheres. I. High-energy irradiances (1-1700 Å). Astrophys J 622:680. https://doi.org/10.1086/427977

Rognini E, Mura A, Capria MT, Milillo A, Zinzi A, Galluzzi V (2022) Effects of Mercury surface temperature on the sodium abundance in its exosphere. Planet Space Sci 212:105397. https://doi.org/10.1016/j.pss.2021.105397

Rosborough SA, Oliversen RJ, Mierkiewicz EJ, Sarantos M, Robertson SD, Kuruppuaratchi DCP, Derr NJ, Gallant MA, Roesler FL (2019) High-resolution potassium observations of the lunar exosphere. Geophys Res Lett 46(12):6964–6971. https://doi.org/10.1029/2019GL083022

Saito Y, Yokota S, Asamura K, Tanaka T, Nishino MN, Yamamoto T, Terakawa Y, Fujimoto M, Hasegawa H, Hayakawa H, Hirahara M, Hoshino M, Machida S, Mukai T, Nagai T, Nagatsuma T, Nakagawa T, Nakamura M, Oyama K, Sagawa E, Sasaki S, Seki K, Shinohara I, Terasawa T, Tsunakawa H, Shibuya H, Matsushima M, Shimizu H, Takahashi F (2010) In-flight performance and initial results of plasma energy angle and composition experiment (PACE) on SELENE (Kaguya). Space Sci Rev 154:265–303. https://doi.org/10.1007/s11214-010-9647-x

Saito Y, Nishino MN, Fujimoto M, Yamamoto T, Yokota S, Tsunakawa H, Shibuya H, Matsushima M, Shimizu H, Takahashi F (2012) Simultaneous observation of the electron acceleration and ion deceleration over lunar magnetic anomalies. Earth Planets Space 64:83–92. https://doi.org/10.5047/eps.2011.07.011

Saito Y, Nishino MN, Yokota S, Tsunakawa H, Matsushima M, Takahashi F, Shibuya H, Shimizu H (2014) Night side lunar surface potential in the Earth's magnetosphere. Adv Space Res 54(10):1985–1992. https://doi.org/10.1016/j.asr.2013.05.011

Sarantos M, Killen RM, McClintock WE, Bradley ET, Vervack RJ Jr, Benna M, Slavin JA (2011) Limits to Mercury's magnesium exosphere from MESSENGER second flyby observations. Planet Space Sci 59(15):1992–2003. https://doi.org/10.1016/j.pss.2011.05.002

Sarantos M, Killen RM, Sharma AS, Slavin JA (2008) Influence of plasma ions on source rates for the lunar exosphere during passage through the Earth's magnetosphere. Geophys Res Lett 35:L04105. https://doi.org/10.1029/2007GL032310

Sarantos M, Killen RM, Glenar DA, Benna M, Stubbs TJ (2012) Metallic species, oxygen and silicon in the lunar exosphere: Upper limits and prospects for LADEE measurements. J Geophys Res Space Phys 117(A3). https://doi.org/10.1029/2011JA017044

Sarantos M, Tsavachidis S (2020) The boundary of alkali surface boundary exospheres of Mercury and the Moon. Geophys Res Lett 47:e2020GL088930. https://doi.org/10.1029/2020GL088930

Sarantos M, Tsavachidis S (2021) Lags in desorption of lunar volatiles. Astrophys J Lett 919(2):L14. https://doi.org/10.3847/2041-8213/ac205b

Schaible MJ, Baragiola RA (2014) Hydrogen implantation in silicates: the role of solar wind in SiOH bond formation on the surfaces of airless bodies in space. J Geophys Res, Planets 119:2017. https://doi.org/10.1002/2014JE004650

Schläppi B, Altwegg K, Wurz P (2008) Asteroid exosphere: a simulation for the Rosetta flyby targets (2867) Steins and (21) Lutetia. Icarus 195:674–685. https://doi.org/10.1016/j.icarus.2007.12.021

Schleicher H, Wiedemann G, Wöhl H, Berkefeld T, Soltau D (2004) Detection of neutral sodium above Mercury during the transit on 2003 May 7. Astron Astrophys 425:1119–1124. https://doi.org/10.1051/0004-6361:20040477

Schmidt CA (2013) Monte Carlo modeling of north-south asymmetries in Mercury's sodium exosphere. J Geophys Res Space Phys 118:4564. https://doi.org/10.1002/jgra.50396

Schmidt CA Baumgardner J Mendillo M Wilson JK (2012) Escape rates and variability constraints for high-energy sodium sources at Mercury. J Geophys Res Space Phys 117:A03301. https://doi.org/10.1029/2011JA017217

Schörghofer N, Benna M, Berezhnoy AA, Greenhagen B, Jones BM, Li S, Orlando TM, Prem P, Tucker OJ, Wöhler C (2021) Water group exospheres and surface interactions on the Moon, Mercury, and Ceres. Space Sci Rev 217:74. https://doi.org/10.1007/s11214-021-00846-3

Schörghofer N (2022) Gradual sequestration of water at lunar polar conditions due to temperature cycles. Astrophys J Lett 927(2):L34. https://doi.org/10.3847/2041-8213/ac5a48

Seki K, Hirahara M, Terasawa T, Shinohara I, Mukai T, Saito Y, Machida S, Yamamoto T, Kokubun S (1996) Coexistence of Earth-origin O+ and solar wind-origin H+/He++ in the distant magnetotail. Geophys Res Lett 23(9):985–988. https://doi.org/10.1029/96GL00768

Shemansky DE (1988) Revised atmospheric species abundances at Mercury: the debacle of bad g values. The Mercury Messenger, LPI Contribution No. 2712

Slade MA, Butler BJ, Muhleman DO (1992) Mercury radar imaging: evidence for polar ice. Science 258:635–639. https://doi.org/10.1126/science.258.5082.635

Slavin JA, Middleton HR, Raines JM, Jia X, Zhong J, Sun W-J, Livi S, Imber SM, Poh G-K, Akhavan-Tafti M, Jasinski JM, DiBraccio GA, Dong C, Dewey RM, Mays ML (2019) MESSENGER observations of disappearing dayside magnetosphere events at Mercury. J Geophys Res Space Phys 124:6613–6635. https://doi.org/10.1029/2019JA026892

Slavin JA, Imber SM, Raines JM (2021) A dungey cycle in the life of Mercury's magnetosphere, Chap. 34. In: Maggiolo R et al (eds) Magnetospheres in the solar system. Geophysical monograph, vol 259. AGU, pp 535–556. https://doi.org/10.1002/9781119815624.ch34

Smolka A, Nikolić D, Gscheidle C, Reiss P (2023) Coupled H, H2, OH, and H2O lunar exosphere simulation framework and impacts of conversion reactions. Icarus 397:115508

Smyth WH (1986) Nature and variability of Mercury's sodium atmosphere. Nature 323:696–699

Smyth WH, Marconi ML (1995) Theoretical overview and modeling of the sodium and potassium atmospheres of Mercury. Astrophys J 441:839–864. https://doi.org/10.1086/175407

Sprague AL, Kozlowski RWH, Hunten DM, Schneider NM, Domingue DL, Wells WK, Schmitt W, Fink U (1997) Distribution and abundance of sodium in Mercury's atmosphere, 1985–1988. Icarus 129:506–527. https://doi.org/10.1006/icar.1997.5784

Sridharan R, Ahmed SM, Das TP, Sreelatha P, Padeepkumar P, Naik N, Supriya G (2010) The sunlit lunar atmosphere: a comprehensive study by Chace on the moon impact probe of Chandrayaan-1. Planet Space Sci 58:1567–1577. https://doi.org/10.1016/j.pss.2010.07.027

Stern SA (1999) The lunar atmosphere: history, status, current problems, and context. Rev Geophys 37(4):453–491. https://doi.org/10.1029/1999RG900005

Stern SA, Cook JC, Chaufray JY, Feldman PD, Gladstone GR, Retherford KD (2013) Lunar atmospheric H2 detections by the LAMP UV spectrograph on the lunar reconnaissance orbiter. Icarus 226(2):1210–1213. https://doi.org/10.1016/j.icarus.2013.07.011

Sun W, Slavin JA, Milillo A, Orsini S, Jia X, Raines JM, Livi S, Jasinski JM, Dewey RM, Fu S, Zhao J, Zong Q-G, Saito Y, Li C (2022) MESSENGER observations of planetary ion enhancements at Mercury's northern magnetospheric cusp during flux transfer event showers. J Geophys Res Space Phys 127:e2022JA030280. https://doi.org/10.1029/2022JA030280

Szabo S, Poppe AR, Biber H, Mutzke A, Pichler J, Jäggi N, Galli A, Wurz P, Aumayr F (2022a) Deducing lunar regolith porosity from energetic neutral atom emission. Geophys Res Lett 49:e2022GL101232. https://doi.org/10.1029/2022GL101232

Szabo PS, Cupak C, Biber H, Jäggi N, Galli A, Wurz P, Aumayr F (2022b) A theoretical model for the sputtering of rough surfaces. Surfaces and Interfaces 30. https://doi.org/10.1016/j.surfin.2022.101924

Szalay JR, Horányi M (2016) Lunar meteoritic gardening rate derived from in situ LADEE/LDEX measurements. Geophys Res Lett 43(10):4893. https://doi.org/10.1002/2016GL069148

Tanaka T, Saito Y, Yokota S, Asamura K, Nishino MN, Tsunakawa H et al (2009) First in situ observation of the moon-originating ions in the Earth's magnetosphere by MAP-PACE on SELENE (KAGUYA). Geophys Res Lett 36(22):L22106. https://doi.org/10.1029/2009gl040682

Teolis B, Sarantos M, Schorghofer N, Jones B, Grava C, Mura A, Prem P, Greenhagen B, Capria MT, Cremonese G, Lucchetti A, Galluzzi V (2023) Surface exospheric interactions. Space Sci Rev 219:4. https://doi.org/10.1007/s11214-023-00951-5

Terada K, Yokota S, Saito Y, Kitamura N, Asamura K, Nishino MN (2017) Biogenic oxygen from Earth transported to the Moon by a wind of magnetospheric ions. Nat Astron 1:0026. https://doi.org/10.1038/s41550-016-0026

Troshichev O, Kokubun S, Kamide Y, Nishida A, Mukai T, Yamamoto T (1999) Convection in the distant magnetotail under extremely quiet and weakly disturbed conditions. J Geophys Res 104(A5):10249–10264. https://doi.org/10.1029/1998JA900141

Tucker OJ, Farrell WM, Killen RM, Hurley DM (2019) Solar wind implantation into the lunar regolith: Monte Carlo simulations of H retention in a surface with defects and the H2 exosphere. J Geophys Res, Planets 124:278–293. https://doi.org/10.1029/2018JE005805

Vervack RJ Jr, Killen RM, McClintock WE, Merkel AW, Burger MH, Cassidy TA, Sarantos M, Cassidy TA (2016) New discoveries from MESSENGER and insights into Mercury's exosphere. Geophys Res Lett 43:11,545–11,551. https://doi.org/10.1002/2016GL071284

Vorburger A, Wurz P, Barabash S, Wieser M, Futaana Y, Holmström M, Bhardwaj A, Asamura K (2012) Energetic neutral atom observations of magnetic anomalies on the lunar surface. J Geophys Res 117:A07208. https://doi.org/10.1029/2012JA017553

Vorburger A, Wurz P, Barabash S, Wieser M, Futaana Y, Lue C, Holmström M, Bhardwaj A, Dhanya MB, Asamura K (2013) Energetic neutral atom imaging of the lunar surface. J Geophys Res 118(7):3937–3945. https://doi.org/10.1002/jgra.50337

Vorburger A, Wurz P, Barabash S, Futaana Y, Wieser M, Bhardwaj A, Dhanya MB, Asamura K (2016) Transport of solar wind plasma onto the lunar nightside surface. Geophys Res Lett 43:10586–10594. https://doi.org/10.1002/2016GL071094

Yokota S, Saito Y, Asamura K, Tanaka T, Nishino MN, Tsunakawa H et al (2009) First direct detection of ions originating from the Moon by MAP-PACE IMA onboard SELENE (KAGUYA). Geophys Res Lett 36:L11201. https://doi.org/10.1029/2009gl038185

Yokota S et al (2020) KAGUYA observation of global emissions of indigenous carbon ions from the Moon. Sci Adv 6:19. https://doi.org/10.1126/sciadv.aba1050

Wang X-D, Zong QG, Wang JS, Cui J, Reme H, Dandouras I et al (2011) Detection of m/q = 2 pickup ions in the plasma environment of the Moon: the trace of exospheric. Geophys Res Lett 38(L14204). https://doi.org/10.1029/2011gl047488

Wang HZ et al (2021) Energetic neutral atom distribution on the lunar surface and its relationship with solar wind conditions. Astrophys J Lett 922:L41. https://doi.org/10.3847/2041-8213/ac34f3

Watson K, Murray BC, Brown H (1961) The behavior of volatiles on the lunar surface. J Geophys Res 66:3033–3045

Wei Y et al (2020) Implantation of Earth's atmospheric ions into the nearside and farside lunar soil: implications to geodynamo evolution. Geophys Res Lett 47:e2019GL086208. https://doi.org/10.1029/2019GL086208

Werner E, Leblanc F, Chaufray JY, Modolo R, Aizawa S, Hadid L, Baskevitch C (2022) Modeling the impact of a strong X-class solar flare on the ion composition in Mercury's magnetosphere. Geophys Res Lett 49:e2021GL096614. https://doi.org/10.1029/2021GL096614

Wieler R, Kehm K, Meshik A et al (1996) Secular changes in the xenon and krypton abundances in the solar wind recorded in single lunar grains. Nature 384:46–49. https://doi.org/10.1038/384046a0

Wieser M, Barabash S, Futaana Y, Holmström M, Bhardwaj A, Sridharan R, Dhanya MB, Schaufelberger A, Wurz P, Asamura K (2010) First observation of a mini-magnetosphere above a lunar magnetic anomaly using energetic neutral atoms. Geophys Res Lett 37:L05103. https://doi.org/10.1029/2009GL041721

Winslow RM Anderson BJ, Johnson CL, Slavin JA Korth H et al (2013) Mercury's magnetopause and bow shock from MESSENGER magnetometer observations. J Geophys Res Space Phys 1185:2213–2227 https://doi.org/10.1002/jgra.50237

Wilson JK, Mendillo M, Spence H (2006) Magnetospheric influence on the Moon's exosphere. J Geophys Res 111:107207. https://doi.org/10.1029/2005JA011364

Wurz P, Lammer H (2003) Monte-Carlo simulation of Mercury's exosphere. Icarus 164(1):1–13. https://doi.org/10.1016/S0019-1035(03)00123-4

Wurz P, Rohner U, Whitby JA, Kolb C, Lammer H, Dobnikar P, Martín-Fernández JA (2007) The lunar exosphere: the sputtering contribution. Icarus 191(2):486–496. https://doi.org/10.1016/j.icarus.2007.04.034

Wurz P, Whitby JA, Rohner U, Martín-Fernández JA, Lammer H, Kolb C (2010) Self-consistent modelling of Mercury's exosphere by sputtering, micro-meteorite impact and photon-stimulated desorption. Planet Space Sci 58:1599–1616. https://doi.org/10.1016/j.pss.2010.08.003

Wurz P, Fatemi S, Galli A et al (2022) Particles and photons as drivers for particle release from the surfaces of the Moon and Mercury. Space Sci Rev 218:10. https://doi.org/10.1007/s11214-022-00875-6

Wurz P, Gamborino D, Vorburger A, Raines JM (2019) Heavy ion composition of Mercury's magnetosphere. J Geophys Res 124:2603–2612. https://doi.org/10.1029/2018JA026319

Zhou X-Z, Angelopoulos V, Poppe AR, Halekas JS (2013) ARTEMI observations of lunar pickup ions: mass constraints on ion species. J Geophys Res 118(9):1766–1774. https://doi.org/10.1002/jgre.20125

Zurbuchen TH, Raines JM, Slavin JA, Gershman DJ, Gilbert JA, Gloeckler G et al (2011) MESSENGER observations of the spatial distribution of planetary ions near Mercury. Science 333(6051):1862–1865. https://doi.org/10.1126/science.1211302

Publisher's Note Springer Nature remains neutral with regard to jurisdictional claims in published maps and institutional affiliations.

Authors and Affiliations

Anna Milillo[1] [iD] **· Menelaos Sarantos**[2] **· Cesare Grava**[3] **· Diego Janches**[2] **· Helmut Lammer**[4] **· Francois Leblanc**[5] **· Norbert Schorghofer**[6] **· Peter Wurz**[7] **· Benjamin D. Teolis**[3] **· Go Murakami**[8]

✉ A. Milillo
anna.milillo@inaf.it

M. Sarantos
menelaos.sarantos-1@nasa.gov

C. Grava
cesare.grava@swri.org

D. Janches
diego.janches@nasa.gov

H. Lammer
helmut.lammer@oeaw.ac.at

P. Wurz
peter.wurz@unibe.ch

B.D. Teolis
bteolis@swri.edu

G. Murakami
go@stp.isas.jaxa.jp

1   Institute of Space Astrophysics and Planetology, INAF via del Fosso del Cavaliere 100, 00133, Rome, Italy

2   NASA-Goddard Space Flight Center, Greenbelt, MD 20771, USA

3   SwRI, San Antonio, TX, 78230, USA

4   Austrian Academy of Sciences, Space Research Institute, Schmiedlstraße 6, 8042 Graz, Austria

5   LATMOS/CNRS, Sorbonne Université, UVSQ, IPSL, Paris, France

6   Planetary Science Institute, Tucson, AZ, USA

7   Physics Institute, University of Bern, Bern, Switzerland

8   Institute of Space and Astronautical Science, Japan Aerospace Exploration Agency, Sagamihara, Kanagawa, Japan